THE DATA BOOK OF ASTRONOMY

Also available from Institute of Physics Publishing

The Wandering Astronomer
Patrick Moore

The Photographic Atlas of the Stars
H. J. P. Arnold, Paul Doherty and Patrick Moore

THE DATA BOOK OF ASTRONOMY

PATRICK MOORE

INSTITUTE OF PHYSICS PUBLISHING
BRISTOL AND PHILADELPHIA

British Library Cataloguing-in-Publication Data

A catalogue record for this book is available from the British Library.

ISBN 0 7503 0620 3

Library of Congress Cataloging-in-Publication Data are available

Publisher: Nicki Dennis
Production Editor: Simon Laurenson
Production Control: Sarah Plenty
Cover Design: Kevin Lowry
Marketing Executive: Colin Fenton

Published by Institute of Physics Publishing, wholly owned by The Institute of Physics, London

Institute of Physics Publishing, Dirac House, Temple Back, Bristol BS1 6BE, UK

US Office: Institute of Physics Publishing, The Public Ledger Building, Suite 1035, 150 South Independence Mall West, Philadelphia, PA 19106, USA

Printed in the UK by Bookcraft, Midsomer Norton, Somerset

CONTENTS

FOREWORD

This book may be regarded as the descendant of the *Guinness Book of Astronomy*, which was originally published in 1979 and ran to seven editions. However, the present book is different; it is far more comprehensive, and sets out to provide a quick reference for those who are anxious to check on astronomical facts.

Obviously much has been left out, and not everyone will agree with my selection, but I hope that the result will be of use. It is up to date as of May 2000; no doubt it will need revision even before it appears in print!

ACKNOWLEDGMENTS

Many people have helped me in the production of this book. Remaining errors or omissions are entirely my responsibility. I am most grateful to:

Dr. Peter Cattermole
Dr. Allan Chapman
Dr. Gilbert Fuelder
David Hawksett
Dr. Eleanor Helin
Michael Hendrie
Professor Garry Hunt
John Isles
Chris Lintott
Dr. John Mason
Brian May
Dr. Paul Murdin
Iain Nicolson
Dr. John Rogers
Professor F. Richard Stephenson
Professor Martin Ward
Dr. David Whitehouse
Professor Iwan Williams
Professor Sir Arnold Wolfendale

and on the production side to Robin Rees, and to Nicki Dennis and Simon Laurenson of the Institute of Physics Publishing.

To all these – thank you.

Patrick Moore
Selsey
May 2000

AUTHOR'S NOTE

In this book, I have retained references to the USSR with respect to past results. Now that the USSR has broken up, future developments come under the heading of the *Commonwealth of Independent States*.

METRIC CONVERSION

The current practice of giving lengths in metric units rather than imperial ones has been followed. To help in avoiding confusion, the following table may be found useful.

Centimetres		Inches	Kilometres		Miles
2.54	1	0.39	1.61	1	0.62
5.08	2	0.79	3.22	2	1.24
7.62	3	1.18	4.83	3	1.86
10.16	4	1.58	6.44	4	2.49
12.70	5	1.97	8.05	5	3.11
15.24	6	2.36	9.66	6	3.73
17.78	7	2.76	11.27	7	4.35
20.32	8	3.15	12.88	8	4.97
22.86	9	3.54	14.48	9	5.59
25.40	10	3.94	16.09	10	6.21
50.80	20	7.87	32.19	20	12.43
76.20	30	11.81	48.28	30	18.64
101.6	40	15.75	64.37	40	24.86
127.0	50	19.69	80.47	50	31.07
152.4	60	23.62	96.56	60	37.28
177.8	70	27.56	112.7	70	43.50
203.2	80	31.50	128.7	80	49.71
228.6	90	35.43	144.8	90	55.92
254.0	100	39.37	160.9	100	62.14

WEBSITES

Readers may find the following websites of interest.

http://www.nasa.gov/search/index.html
http://nssdc.gsfc.nasa.gov/
http://ecf.hq.eso.org/astroweb/yp_astro_resources.html
http://www.ast.cam.ac.uk/indext.html
http://wwwflag.wr.usgs.gov/USGSFlag/Space/nomen/
http://oposite.stsci.edu/pubinfo/subject.html
http://www.mtwilson.edu/Science/index.html

1 THE SOLAR SYSTEM

The Solar System consists of one star (the Sun), the nine principal planets, their satellites and lesser bodies such as asteroids, comets and meteoroids, plus a vast amount of thinly-spread interplanetary matter. The Sun contains more than 99% of the mass of the system, and Jupiter is more massive than all the other planets combined. The centre of gravity of the Solar System lies just outside the surface of the Sun, due mainly to the mass of Jupiter.

The Solar System is divided into two parts. There are four relatively small, rocky planets (Mercury, Venus, the Earth and Mars), beyond which come the asteroids, of which only one (Ceres) is over 900 km in diameter. Next come the four giants (Jupiter, Saturn, Uranus and Neptune), plus Pluto, which is smaller than our Moon and has an unusual orbit which brings it at times closer in than Neptune. Pluto may not be worthy of true planetary status, and may be only the largest member of the 'Kuiper Belt' swarm of asteroidal-sized bodies moving in the far reaches of the Solar System. However, Pluto does seem to be in a class of its own, and in size is intermediate between the smallest principal planet (Mercury) and the largest asteroid (Ceres). Planetary data are given in Table 1.1.

It now seems that the distinctions between the various classes of bodies in the Solar System are less clear-cut than has been previously thought. For example, it is quite probable that some 'near-Earth asteroids', which swing away from the main swarm, are ex-comets which have lost their volatiles; and some of the smaller satellites of the giant planets are almost certainly ex-members of the asteroid belt which were captured long ago.

All planets and asteroids move round the Sun in the same sense, and so do the larger satellites in orbit round their primary planets, although some of the small asteroidal-sized satellites have retrograde motion (for example, the four outer members of Jupiter's family and Phœbe in Saturn's). The orbits of the main planets are not greatly inclined to that of the Earth, apart from Pluto (17°), so that to draw a plan of the planetary system on a flat piece of paper is not grossly inaccurate. However, some asteroids have highly-inclined orbits, and so do many comets. It is now thought that short-period comets, all of which have direct motion, come from the Kuiper Belt, while long-period comets, many of which move in a retrograde sense, come from the more distant Oort Cloud.

Most of the planets rotate in the same sense as the Earth, but Venus and Pluto have retrograde rotation, while Uranus is unique in having an axial inclination which is greater than a right angle. The cause of these anomalies is unclear.

ORIGIN OF THE SOLAR SYSTEM

In investigating the origin of the planetary system we do have one important piece of information: the age of the Earth is certainly of the order of 4.6 thousand million[1] years and the Sun, in some form, must obviously be rather older than this. Meteorites are, in general, found to be of about the same age, while the oldest lunar rocks are only slightly younger.

Many theories have been proposed. In 1796 the French astronomer Pierre Simon de Laplace put forward the Nebular Hypothesis, which was in some ways not unlike earlier ideas due to Thomas Wright in England and Immanuel Kant in Germany, but was much more credible. Laplace started with a vast gas cloud, disk-shaped and in slow rotation, which shrank steadily and threw off rings, each of which condensed into a planet, while the central part of the cloud became the Sun. However, it was found that a ring of this sort would not condense into a planet. Moreover, according to the Nebular Hypothesis, most of the angular momentum of the Solar System would reside in the Sun, which would be in quick rotation; actually, most of the angular momentum is due to the giant planets.

---

[1] I avoid using 'billion', because the American billion (now generally accepted) is equal to a thousand million, while the old English billion was equal to a million million.

Table 1.1. Basic data for the planetary system. The orbital data for the planets change slightly from one revolution to another.

Name	Mean distance from Sun (km)	Orbital period	Orbital eccentricity	Orbital inclination	Equatorial diameter (km)	Equatorial rotation period	Number of satellites
Mercury	57 900 000	87.97 days	0.206	7°0′15″.5	4878	58.6 days	0
Venus	108 200 000	224.7 days	0.007	178°	12 104	243.2 days	0
Earth	149 598 000	23h 56m 4s	0.017	0	12 756	23h 56m 4s	1
Mars	227 940 000	687.0 days	0.093	1°51′	6794	24h 37m 23s	2
Jupiter	778 340 000	11.86 years	0.048	1°18′16″	143 884	9h 50m 30s	16
Saturn	1427 000 000	29.5 years	0.056	2°29′21″	120 536	10h 14m	18
Uranus	2869 600 000	84.0 years	0.047	0°46′23″	51 118	17h 14m	20
Neptune	4496 700 000	164.8 years	0.009	1°34′20″	50 538	16h 6m	8
Pluto	5900 000 000	247.7 years	0.248	17°9′	2324	6d 9h 17m	1

In 1901 T. C. Chamberlin and F. R. Moulton, in America, worked out a theory according to which the planets were pulled off the Sun by the action of a passing star; a cigar-shaped tongue of material would be pulled out and this would break up into planets, with the largest planets (Jupiter and Saturn) in the middle part of the system, where the thickest part of the 'cigar' would have been. Again there were insuperable mathematical objections, and a modification of the theory by A. W. Bickerton (New Zealand), involving a 'partial impact', was no better. The original theory was popularized by Sir James Jeans during the first half of the 20th century, but it has now been abandoned. If it had been valid, planetary systems would have been very rare in the Galaxy; close encounters between two stars seldom occur.

Later, G. P. Kuiper proposed that the Sun had a binary companion which never condensed into a proper star, but was spread around to produce planet-forming material; but again there were mathematical objections, and the theory never met with wide support.

Modern theories are much more akin to Laplace's than to later proposals. It is thought that the Solar System began in a huge gas-and-dust cloud, part of which started to collapse and to rotate – possibly triggered off by the effects of a distant supernova. A 'solar nebula' was produced, and in a relatively short period, perhaps 100 000 years, the core turned into what may be called a protostar, the effects of which forced the solar nebula into a flattened, rotating disk. The temperature rose at the centre, and the proto-Sun became a true star; for a while it went through what is known as the T Tauri stage, sending out a strong 'stellar wind' which forced outward the lightest gases, notably hydrogen and helium. The planets built up by accretion. The inner, rocky planets lacked the lightest materials, while in the more distant regions, where the temperature was much lower, the giant planets could form. Jupiter and Saturn grew rapidly enough to draw in material from the solar nebula; Uranus and Neptune, slower to form, could not do so in the same way, because by the time they had become sufficiently massive the nebula had more or less dispersed. This is why Uranus and Neptune contain lesser amounts of hydrogen and helium and more 'ices'. Nuclear processes began in the Sun, and the Solar System began to assume its present form, although at first the Sun was less luminous than it is now.

In its early stages there was a great deal of material which did not condense into planetary form, and the planets were subjected to heavy bombardment, resulting in impact cratering. The main bombardment ended around 4000 million years ago, but the effects of it are still very obvious, as can be seen from the structures visible on the surfaces of the rocky bodies (see Table 1.2).

At present the Solar System is essentially stable, and will remain so until the Sun leaves the Main Sequence and becomes a giant star. This will certainly result in the destruction of the inner planets, so that the Solar System as we know it does have a limited life-span.

Table 1.2. Descriptive terms for surface features.

Catena (catenæ)	Chain of craters
Cavus (cavi)	Hollows or irregular depressions
Chaos	Area of broken terrain
Chasma (chasmata)	Canyon
Colles	Small hills
Corona (coronæ)	Ovoid-shaped feature
Crater	Bowl-shaped depression, either volcanic or impact
Dorsum (dorsa)	Ridge
Facula (faculæ)	Bright spot
Farrum (fara)	Pancake-like structure
Flexus (flexus)	Linear feature
Fluctus (fluctus)	Flow terrain
Fossa (fossæ)	'Ditch'; long, narrow, shallow depression
Labes (labes)	Landslide
Labyrinthus	Complex of intersecting valleys or canyons
Lacus	'Lake'; small plain (only used for the Moon)
Linea (lineæ)	Elongated marking
Macula (maculæ)	Dark spot
Mare	'Sea'; large darkish plain
Mensa (mensæ)	Mesa; flat-topped elevation
Mons (montes)	Mountain
Oceanus	Very large Mare (used only for the Moon, and only once!)
Palus	'Swamp'; small plain (used only for the Moon)
Patera (pateræ)	Shallow crater with scalloped edge
Planitia	Low-lying plain
Planum	Plateau or elevated plain
Promontorium	'Cape' or headline (used only for the Moon)
Regio	Region
Reticulum (reticula)	Reticular pattern of features (Latin *reticulum*, a net)
Rima (rimæ)	Fissure
Rupes (rupes)	Scarp
Scopulus	Lobate or irregular scarp
Sinus	'Bay'; small plain
Sulcus (sulci)	Sub-parallel ridges and furrows
Terra	Extensive 'land' mass (not now used for the Moon)
Tessera (tesseræ)	Terrain with polygonal pattern (once termed 'parquet')
Tholus (tholi)	Small hill or mountain, dome-shaped
Undæ	Dunes
Vallis (valles)	Valley
Vastitas	Widespread lowland plain

2 THE SUN

The Sun, the controlling body of the Solar System, is the only star close enough to be studied in detail. It is 270 000 times closer than the nearest stars beyond the Solar System, those of the Alpha Centauri group. Data are given in Table 2.1.

Table 2.1. The Sun: data.

Distance from Earth:
  mean 149 597 893 km (1 astronomical unit (a.u.))
  max. 152 103 000 km
  min. 147 104 000 km
Mean parallax: 8″.794
Distance from centre of the Galaxy: ∼28 000 light-years
Velocity round centre of Galaxy: ∼220 km s^{-1}
Period of revolution round centre of Galaxy: ∼225 000 000 years
(1 'cosmic year')
Velocity toward solar apex: 19.5 km s^{-1}
Apparent diameter: mean 32′01″
              max. 32′25″
              min. 31′31″
Equatorial diameter: 1 391 980 km
Density, water = 1: mean 1.409
Volume, Earth = 1: 1 303 600
Mass, Earth = 1: 332 946
Mass: 2×10^{27} tonnes (>99% of the mass of the entire Solar System)
Surface gravity, Earth = 1: 27.90
Escape velocity: 617.7 km s^{-1}
Luminosity: 3.85×10^{23} kW
Solar constant (solar radiation per second vertically incident at unit area at 1 a.u. from the Sun); 1368 W m^{-2}
Mean apparent visual magnitude: −26.78 (600 000 times as bright as the full moon)
Absolute magnitude: +4.82
Spectrum: G2
Temperature: surface 5500 °C
          core ∼15 000 000 °C
Rotation period: sidereal, mean: 25.380 days
              synodic, mean: 27.275 days
Time taken for light to reach the Earth, at mean distance: 499.012 s (8.3 min)
Age: ∼4.6 thousand million years

DISTANCE

The first known estimate of the distance of the Sun was made by the Greek philosopher Anaxagoras (500–428 BC). He assumed the Earth to be flat, and gave the Sun's distance as 6500 km (using modern units), with a diameter of over 50 km. A much better estimate was made by Aristarchus of Samos, around 270 BC. His value, derived from observations of the angle between the Sun and the exact half-moon, was approximately 4 800 000 km; his method was perfectly sound in theory, but the necessary measurements could not be made with sufficient accuracy. (Aristarchus also held the belief that the Sun, not the Earth, is the centre of the planetary system.) Ptolemy (c AD 150) increased the distance to 8 000 000 km, but in his book published in AD 1543 Copernicus reverted to only 3 200 000 km. Kepler, in 1618, gave a value of 22 500 000 km.

The first reasonably accurate estimate of the Earth–Sun distance (the astronomical unit) was made in 1672 by G. D. Cassini, from observations of the parallax of Mars. Some later determinations are given in Table 2.2.

One early method involved transits of Venus across the face of the Sun, as suggested by J. Gregory in 1663 and extended by Edmond Halley in 1678; Halley rightly concluded that transits of Mercury could not give accurate results because of the smallness of the planet's disk. In fact, the transit of Venus method was affected by the 'Black Drop' –the apparent effect of Venus drawing a strip of blackness after it during ingress on to the solar disk, thus making precise timings difficult. (Captain Cook's famous voyage, during which he discovered Australia, was made in order to take the astronomer C. Green to a suitable site (Tahiti) in order to observe the transit of 1769.)

Results from the transits of Venus in 1874 and 1882 were still unsatisfactory, and better estimates came from the parallax measurements of planets and (particularly) asteroids. However, Spencer Jones' value as derived from the close approach of the asteroid Eros in 1931 was too high.

Table 2.2. Selected estimates of the length of the astronomical unit.

Year	Authority	Method	Parallax (arcsec)	Distance (km)
1672	G. D. Cassini	Parallax of Mars	9.5	138 370 000
1672	J. Flamsteed	Parallax of Mars	10	130 000 000
1770	L. Euler	1769 transit of Venus	8.82	151 225 000
1771	J. de Lalande	1769 transit of Venus	8.5	154 198 000
1814	J. Delambre	1769 transit of Venus	8.6	153 841 000
1823	J. F. Encke	1761 and 1769 transits of Venus	8.5776	153 375 000
1867	S. Newcomb	Parallax of Mars	8.855	145 570 000
1877	G. Airy	1874 transit of Venus	8.754	150 280 000
1877	E. T. Stone	1874 transit of Venus	8.884	148 080 000
1878	J. Galle	Parallax of asteroids Phocæa and Flora	8.87	148 290 000
1884	M. Houzeau	1882 transit of Venus	8.907	147 700 000
1896	D. Gill	Parallax of asteroid Victoria	8.801	149 480 000
1911	J. Hinks	Parallax of asteroid Eros	8.807	149 380 000
1925	H. Spencer Jones	Parallax of Mars	8.809	149 350 000
1939	H. Spencer Jones	Parallax of asteroid Eros	8.790	149 670 000
1950	E. Rabe	Motion of asteroid Eros	8.798	149 526 000
1962	G. Pettengill	Radar to Venus	8.794 0976	149 598 728
1992	Various	Radar to Venus	8.794 148	149 597 871

The modern method – radar to Venus – was introduced in the early 1960s by astronomers in the United States. The present accepted value of the astronomical unit is accurate to a tiny fraction of 1%.

ROTATION

The first comments about the Sun's rotation were made by Galileo, following his observations of sunspots from 1610. He gave a value of rather less than one month.

The discovery that the Sun shows differential rotation – i.e. that it does not rotate as a solid body would do – was made by the English amateur Richard Carrington in 1863; the rotational period at the equator is much shorter than that at the poles. Synodic rotation periods for features at various heliographic latitudes are given in Table 2.3. Spots are never seen either at the poles or exactly on the equator, but from 1871 H. C. Vogel introduced the method of measuring the solar rotation by observing the Doppler shifts at opposite limbs of the Sun.

Table 2.3. Synodic rotation period for features at various heliographic latitudes.

Latitude (°)	Period (days)
0	24.6
10	24.9
20	25.2
30	25.8
40	27.5
50	29.2
60	30.9
70	32.4
80	33.7
90	34.0

THE SOLAR CONSTANT

This may be defined as being the amount of energy in the form of solar radiation which is normally received on unit area at the top of the Earth's atmosphere; it is roughly equal to the amount of energy reaching ground level on a clear day. The first measurements were made by Sir John Herschel in 1837–8, using an actinometer (basically a bowl of water; the estimate was made by the rate at which the bowl was heated). He gave a value which is about half the actual figure. The modern value is 1.95 cal cm^2 min^{-1} (1368 W m^2).

SOLAR PHOTOGRAPHY

The first photograph of the Sun – a Daguerreotype – seems to have been taken by Lerebours, in France, in 1842. However, the first good Daguerreotype was taken by Fizeau and Foucault, also in France, on 2 April 1845, at the request of F. Arago. In 1854 B. Reade used a dry collodion plate to show mottling on the disk.

The first systematic series of solar photographs was taken from Kew (outer London) from 1858 to 1872, using equipment designed by the English amateur Warren de la Rue. Nowadays the Sun is photographed daily from observatories all over the world, and there are many solar telescopes designed specially for this work. Many solar telescopes are of the 'tower' type, but the largest solar telescope now in operation, the McMath Telescope at Kitt Peak in Arizona, looks like a large, white inclined tunnel. At the top is the upper mirror (the heliostat), 203 cm in diameter; it can be rotated, and sends the sunlight down the tunnel in a fixed direction. At the bottom of the 183 m tunnel is a 152 cm mirror, which reflects the rays back up the tunnel on to the half-way stage where a flat mirror sends the rays down through a hole into the solar laboratory, where the analyses are carried out. This means that the heavy equipment in the solar laboratory does not have to be moved at all.

SUNSPOTS

The bright surface of the Sun is known as the photosphere, and it is here that we see the dark patches which are always called sunspots. Really large spot-groups may be visible with the naked eye, and a Chinese record dating back to 28 BC describes a patch which was 'a black vapour as large as a coin'. There is a Chinese record of an 'obscuration' in the Sun, which may well have been a spot, as early as 800 BC.

The first observer to publish telescope drawings of sunspots was J. Fabricius, from Holland, in 1611, and although his drawings are undated he probably saw the spots toward the end of 1610. C. Scheiner, at Ingoldtädt, recorded spots in March 1611, with his pupil C. B. Cysat. Scheiner wrote a tract which came to the notice of Galileo, who claimed to have been observing sunspots since November 1610. No doubt all these observers recorded spots telescopically at about the same time (the date was close to solar maximum) but their interpretations differed. Galileo's explanation was basically correct. Scheiner regarded the spots as dark bodies moving round the Sun close to the solar surface; Cassini, later, regarded them as mountains protruding through the bright surface. Today we know that they are due to the effects of bipolar magnetic field lines below the visible surface.

Direct telescopic observation of the Sun through any telescope is highly dangerous, unless special filters or special equipment is used. The first observer to describe the projection method of studying sunspots may have been Galileo's pupil B. Castelli. Galileo himself certainly used the method, and said (correctly) that it is 'the method that any sensible person will use'. This seems to dispose of the legend that he ruined his eyesight by looking straight at the Sun through one of his primitive telescopes.

A major spot consists of a darker central portion (umbra) surrounded by a lighter portion (penumbra); with a complex spot there may be many umbræ contained in one penumbral mass. Some 'spots' at least are depressions, as can be seen from what is termed the Wilson effect, announced in 1774 by A. Wilson of Glasgow. He found that with a regular spot, the penumbra toward the limbward side is broadened, compared with the opposite side, as the spot is carried toward the solar limb by virtue of the Sun's rotation. From these observations, dating from 1769, Wilson deduced that the spots must be hollows. The Wilson effect can be striking, although not all spots and spot-groups show it.

Some spot-groups may grow to immense size. The largest group on record is that of April 1947; it covered an area of 18 130 000 000 km^2, reaching its maximum on 8 April. To be visible with the naked eye, a spot-group must cover 500 millionths of the visible hemisphere. (One millionth of the hemisphere is equal to 3 000 000 km^2.)

A large spot-group may persist for several rotations. The present record for longevity is held by a group which

Table 2.4. Zürich sunspot classification.

A	Small single unipolar spot, or a very small group of spots without penumbra.
B	Bipolar sunspot group with no penumbra.
C	Elongated bipolar sunspot group. One spot must have penumbra.
D	Elongated bipolar sunspot group with penumbra on both ends of the group.
E	Elongated bipolar sunspot group with penumbra on both ends. Longitudinal extent of penumbra exceeds 10° but not 15°.
F	Elongated bipolar sunspot group with penumbra on both ends. Longitudinal extent of penumbra exceeds 15°.
H	Unipolar sunspot group with penumbra.

lasted for 200 days, between June and December 1943. On the other hand, very small spots, known as pores, may have lifetimes of less than an hour. A pore is usually regarded as a feature no more than 2500 km in diameter.

The darkest parts of spots – the umbræ – still have temperatures of around 4000 °C, while the surrounding photosphere is at well over 5000 °C. This means that a spot is by no means black, and if it could be seen shining on its own the surface brightness would be greater than that of an arc-lamp.

The accepted Zürich classification of sunspots is given in Table 2.4.

Sunspots are essentially magnetic phenomena, and are linked with the solar cycle. Every 11 years or so the Sun is active, with many spot-groups and associated phenomena; activity then dies down to a protracted minimum, after which activity builds up once more toward the next maximum. A typical group has two main spots, a leader and a follower, which are of opposite magnetic polarity.

The magnetic fields associated with sunspots were discovered by G. E. Hale, from the United States, in 1908. This resulted from the Zeeman effect (discovered in 1896 by the Dutch physicist P. Zeeman), according to which the spectral lines of a light source are split into two or three components if the source is associated with a magnetic field. It was Hale who found that the leader and the follower of a two-spot group are of opposite polarity – and that the conditions are the same over a complete hemisphere of the Sun, although reversed in the opposite hemisphere. At the end of each cycle the whole situation is reversed, so that it is fair to say that the true cycle (the 'Hale cycle') is 22 years in length rather than 11.

The magnetic fields of spots are very strong, and may exceed 4000 G. With one group, seen in 1967, the field reached 5000 G. The preceding and following spots of a two-spot group are joined by loops of magnetic field lines which rise high into the solar atmosphere above. The highly magnetized area in, around and above a bipolar sunspot group is known as an *active region*.

The modern theory of sunspots is based upon pioneer work carried out by H. Babcock in 1961. The spots are produced by bipolar magnetic regions (i.e. adjacent areas of opposite polarity) formed where a bunch of concentrated field lines emerges through the photosphere to form a region of outward-directed or positive field; the flux tube then curves round in a loop, and re-enters to form a region of inward-directed or negative field. This, of course, explains why the leader and the follower are of opposite polarity.

Babcock's original model assumed that the solar magnetic lines of force run from one magnetic pole to the other below the bright surface. An initial polar magnetic field is located just below the photosphere in the convective zone. The Sun's differential rotation means that the field is 'stretched' more at the equator than at the poles. After many rotations, the field has become concentrated as toroids to either side of the equator, and spot-groups are produced. At the end of the cycle, the toroid fields have diffused poleward and formed a polar field with reversed polarity, and this explains the Hale 22-year cycle.

Each spot-group has its own characteristics, but in general the average two-spot group begins as two tiny specks at the limit of visibility. These develop into proper spots, growing and also separating in longitude at a rate of around 0.5 km s^{-1}. Within two weeks the group has reached its maximum length, with a fairly regular leader together with a less regular follower. There are also various minor spots and clusters; the axis of the main pair has rotated until it is roughly parallel with the solar equator. After the group has reached its peak, a decline sets in; the leader is usually the last survivor. Around 75% of groups fit into this pattern, but others do not conform, and single spots are also common.

ASSOCIATED PHENOMENA

Plages are bright, active regions in the Sun's atmosphere, usually seen around sunspot groups. The brightest features of this type seen in integrated light are the *faculæ*.

The discovery of faculæ was made by C. Scheiner, probably about 1611. Faculæ (Latin, 'torches') are clouds of incandescent gases lying above the brilliant surface; they are composed largely of hydrogen, and are best seen near the limb, where the photosphere is less bright than at the centre of the disk (in fact, the limb has only two-thirds the brilliance of the centre, because at the centre we are looking down more directly into the hotter material). Faculæ may last for over two months, although their average lifetime is about 15 days. They often appear in areas where a spot-group is about to appear, and persist after the group has disappeared.

Polar faculæ are different from those of the more central regions, and are much less easy to observe; they are most common near the minimum of the sunspot cycle, and have latitudes higher than 65° north or south, with lifetimes ranging from a few days to no more than 12 min. They may well be associated with coronal plumes.

Even in non-spot zones, the solar surface is not calm. The photosphere is covered with granules, which are bright, irregular polygonal structures; each is around 1000 km in diameter, and may last from 3 to 10 min (8 min is about the average). They are vast convective cells of hot gases, rising and falling at average speeds of about 0.5 km s^{-1}; the gases rise at the centre of the granule and descend at the edges, so that the general situation has been likened to a boiling liquid, although the photosphere is of course entirely gaseous. They cover the whole photosphere, except at sunspots, and it has been estimated that at any one moment the whole surface contains about 4 000 000 granules. At the centre of the disk the average distance between granules is of the order of 1400 km. The granular structure is easy to observe; the first really good pictures of it were obtained from a balloon, Stratoscope II, in 1957.

Supergranulation involves large organized cells, usually polygonal, measuring around 30 000 km in diameter; each contains several hundreds of individual granules. They last from 20 h to several days, and extend up into the chromosphere (the layer of the Sun's atmosphere immediately above the photosphere). Material wells up at

Table 2.5. Classification of solar flares.

Area (square degrees)	Classification
Over 24.7	4
12.5–24.7	3
5.2–12.4	2
2.0–5.1	1
Less than 2	s

F = faint, N = normal, B = bright. Thus the most important flares are classified as 4B.

the centre of the cell, spreading out to the edges before sinking again.

Spicules are needle-shaped structures rising from the photosphere, generally along the borders of the supergranules, at speeds of from 10 to 30 km s^{-1}. About half of them fade out at peak altitude, while the remainder fall back into the photosphere. Their origin is not yet completely understood.

Flares are violent, short-lived outbursts, usually occurring above active spot-groups. They emit charged particles as well as radiations ranging from very short gamma-rays up to long-wavelength radio waves; they are most energetic in the X-ray and EUV (extreme ultra-violet) regions of the electromagnetic spectrum. They produce shock waves in the corona and chromosphere, and may last for around 20 min, although some have persisted for 2 h and one, on 16 August 1989, persisted for 13 h. They are most common between 1 and 2 years after the peak of a sunspot cycle. They are seldom seen visually. The first flare to be observed in 'white' light was observed by R. Carrington on 1 September 1859, but generally flares have to be studied with spectroscopic equipment or the equivalent. Observed in hydrogen light, they are classified according to area. The classification is given in Table 2.5.

It seems that flares are explosive releases of energy stored in complex magnetic fields above active areas. They are powered by magnetic reconnection events, when oppositely-directed magnetic fields meet up and reconnect to form new magnetic structures. As the field lines snap into their new shapes, the temperature rises to tens of millions of degrees in a few minutes, and clouds of plasma are sent outward through the solar atmosphere into space;

the situation has been likened to the sudden snapping of a tightly-wound elastic band. These huge 'bubbles' of plasma, containing thousands of millions of tons of material, are known as Coronal Mass Ejections (CMEs). The particles emitted by the flare travel at a slower speed than the radiations and reach Earth a day or two later, striking the ionosphere and causing 'magnetic storms' – one of which, on 13 March 1977, disrupted the entire communications network in Quebec. Auroræ are also produced. Cosmic rays and energetic particles sent out by flares may well pose dangers to astronauts moving above the protective screen of the Earth's atmosphere, and, to a much lesser extent, passengers in very high-flying aircraft.

Flares are, in fact, amazingly powerful and a major flare may release as much energy as 10 000 million one-megaton nuclear bombs. Some of the ejected particles are accelerated to almost half the speed of light.

THE SOLAR CYCLE

The first suggestion of a solar cycle seems to have come from the Danish astronomer P. N. Horrebow in 1775–6, but his work was not published until 1859, by which time the cycle had been definitely identified. In fact the 11-year cycle was discovered by H. Schwabe, a Dessau pharmacist, who began observing the Sun regularly in 1826 – mainly to see whether he could observe the transit of an intra-Mercurian planet. In 1851 his findings were popularized by W. Humboldt. A connection between solar activity and terrestrial phenomena was found by E. Sabine in 1852, and in 1870 E. Loomis, at Yale, established the link between the solar cycle and the frequency of auroræ.

The cycle is by no means perfectly regular. The mean value of its length since 1715 has been 11.04 years, but there are marked fluctuations; the longest interval between successive maxima has been 17.1 years (1788 to 1805) and the shortest has been 7.3 years (1829.9 to 1837). Since 1715, when reasonably accurate records began, the most energetic maximum has been that of 1957.9; the least energetic maximum was that of 1816. (See Table 2.6)

There are, moreover, spells when the cycle seems to be suspended, and there are few or no spots. Four of these spells have been identified with fair certainty: the Oort Minimum (1010–1050), the Wolf Minimum

Table 2.6. Sunspot maxima and minima, 1718–1999.

Maxima	Minima
1718.2	1723.5
1727.5	1734.0
1738.7	1745.0
1750.5	1755.2
1761.5	1766.5
1769.7	1777.5
1778.4	1784.7
1805.2	1798.3
1816.4	1810.6
1829.9	1823.3
1837.2	1833.9
1848.1	1843.5
1860.1	1856.0
1870.6	1867.2
1883.9	1878.9
1894.1	1899.6
1907.0	1901.7
1917.6	1913.6
1928.4	1923.6
1937.4	1933.8
1947.5	1944.2
1957.8	1954.3
1968.9	1964.7
1979.9	1976.5
1990.8	1986.8
	1996.8

(1280–1340), the Spörer Minimum (1420–1530) and the Maunder Minimum (1645–1715). Of these the best authenticated is the last. Attention was drawn to it in 1894 by the British astronomer E. W. Maunder, based on earlier work by F. G. W. Spörer in Germany.

Maunder found, from examining old records, that between 1645 and 1715 there were virtually no spots at all. It may well be significant that this coincided with a very cold spell in Europe; during the 1680s, for example, the Thames froze every winter, and frost fairs were held on it. Auroræ too were lacking; Edmond Halley recorded that he saw his first aurora only in 1716, after forty years of watching.

Records of the earlier prolonged minima are fragmentary, but some evidence comes from the science of tree rings, dendrochronology, founded by an astronomer, A. E. Douglass. High-energy sonic rays which pervade the

Galaxy transmute a small amount of atmospheric nitrogen to an isotope of carbon, C-14, which is radioactive. When trees assimilate carbon dioxide, each growth ring contains a small percentage of carbon-14, which decays exponentially with a half-life of 5730 years. At sunspot maximum, the magnetic field ejected by the Sun deflects some of the cosmic rays away from the Earth, and reduces the level of carbon-14 in the atmosphere, so that the tree rings formed at sunspot maximum have a lower amount of the carbon-14 isotope. Careful studies were carried out by F. Vercelli, who examined a tree which lived between 275 BC and AD 1914. Then, in 1976, J. Eddy compared the carbon-14 record of solar activity with records of sunspots, auroræ and climatic data, and confirmed Maunder's suggestion of a dearth of spots between 1645 and 1715. Yet strangely, although there were virtually no records of telescopic sunspots during this period, naked-eye spots were recorded in China in 1647, 1650, 1655, 1656, 1665 and 1694; whether or not these observations are reliable must be a matter for debate. There is strong evidence for a longer cycle superimposed on the 11-year one.

The law relating to the latitudes of sunspots (Spörer's Law) was discovered by the German amateur F. G. W. Spörer in 1861. At the start of a new cycle after minimum, the first spots appear at latitudes between 30° and 45° north or south. As the cycle progresses, spots appear closer to the equator, until at maximum the average latitude of the groups is only about 15° north or south. The spots of the old cycle then die out (before reaching the equator), but even before they have completely disappeared the first spots of the new cycle are seen at the higher latitudes. This was demonstrated by the famous 'Butterfly Diagram', first drawn by Maunder in 1904.

The Wolf or Zürich sunspot number for any given day, indicating the state of the Sun at that time, was worked out by R. Wolf of Zürich in 1852. The formula is $R = k(10g + f)$, where R is the Zürich number, g is the number of groups seen, f is the total number of individual spots seen and k is a constant depending on the equipment and site of the observer (k is usually not far from unity). The Zürich number may range from zero for a clear disk up to over 200. A spot less than about 2500 km in diameter is officially classed as a pore.

Rather surprisingly, the Sun is actually brightest at spot maximum. The greater numbers of sunspots do not compensate for the greater numbers of brilliant plages.

SPECTRUM AND COMPOSITION OF THE SUN

The first intentional solar spectrum was obtained by Isaac Newton in 1666, but he never took these investigations much further, although he did of course demonstrate the complex nature of sunlight. The sunlight entered the prism by way of a hole in the screen, rather than a slit.

In 1802 W. H. Wollaston, in England, used a slit to obtain a spectrum and discovered the dark lines, but he merely took them to be the boundaries between different colours of the rainbow spectrum. The first really systematic studies of the dark lines were carried out in Germany by J. von Fraunhofer, from 1814. Fraunhofer realized that the lines were permanent; he recorded 5740 of them and mapped 324. They are still often referred to as the Fraunhofer lines.

The explanation was found by G. Kirchhoff, in 1859 (initially working with R. Bunsen). Kirchhoff found that the photosphere yields a rainbow or continuous spectrum; the overlying gases produce a line spectrum, but since these lines are seen against the rainbow background they are reversed, and appear dark instead of bright. Since their positions and intensities are not affected, each line may be tracked down to a particular element or group of elements. In 1861–2 Kirchhoff produced the first detailed map of the solar spectrum. (His eyesight was affected, and the work was actually finished by his assistant, K. Hofmann.) In 1869 Anders Ångström, of Sweden, studied the solar spectrum by using a grating instead of a prism, and in 1889 H. Rowland produced a detailed photographic map of the solar spectrum. The most prominent Fraunhofer lines in the visible spectrum are given in Table 2.7.

By now many of the chemical elements have been identified in the Sun. The list of elements which have and have not been identified is given in Table 2.8. The fact that the remaining elements have not been detected does not necessarily mean that they are completely absent; they may be present, although no doubt in very small amounts.

So far as relative mass is concerned, the most abundant element by far is hydrogen (71%). Next comes helium

Table 2.7. The most prominent Fraunhofer lines in the visible spectrum of the Sun.

Letter	Wavelength (Å)	Identification	Letter	Wavelength (Å)	Identification
A	7593	O_2			
a	7183	H_2O			
B	6867	O_2			

(These three are telluric lines – due to the Earth's intervening atmosphere.)

Letter	Wavelength (Å)	Identification	Letter	Wavelength (Å)	Identification
C($H\alpha$)	6563	H	b_4	5167	Mg
D_1	5896		F($H\beta$)	4861	H
D_2	5890	Na	f($H\gamma$)	4340	H
E	5270	Ca, Fe	G	4308	Fe, Ti
	5269	Fe	g	4227	Ca
b_1	5183	Mg	h($H\delta$)	4102	H
b_2	5173	Mg	H	3968	Ca^{11}
b_3	5169	Fe	K	3933	

(One Ångström (Å) is equal to one hundred-millionth part of a centimetre; it is named in honour of Anders Ångström. The diameter of a human hair is roughly 500 000 Å.

To convert Ångströms into nanometres, divide all wavelengths by 10, so that, for instance, $H\alpha$ becomes 656.3 nm.)

(27%). All the other elements combined make up only 2%. The numbers of atoms in the Sun relative to one million atoms of hydrogen are given in Table 2.9.

Helium was identified in the Sun (by Norman Lockyer, in 1868) before being found on Earth. Lockyer named it after the Greek $\eta\lambda\iota o\varsigma$, the Sun. It was detected on Earth in 1894 by Sir William Ramsay, as a gas occluded in cleveite.

For a time it was believed that the corona contained another element unknown on Earth, and it was even given a name – coronium – but the lines, described initially by Harkness and Young at the eclipse of 1869, proved to be due to elements already known. In 1940 B. Edlén, of Sweden, showed that the coronium lines were produced by highly ionized iron and calcium.

SOLAR ENERGY

Most of the radiation emitted by the Sun comes from the photosphere, which is no more than about 500 km deep. It is easy to see that the disk is at its brightest near the centre; there is appreciable limb darkening – because when we look at the centre of the disk we are seeing into deeper and hotter layers. It is rather curious to recall that there were once

suggestions that the interior of the Sun might be cool. This was the view of Sir William Herschel, who believed that below the bright surface there was a temperature region which might well be inhabited – and he never changed his view (he died in 1822). Few of his contemporaries agreed with him, but at least his reputation ensured that the idea of a habitable Sun would be taken seriously. And as recently as 1869 William Herschel's son, Sir John, was still maintaining that a sunspot was produced when the luminous clouds rolled back, bringing the dark, solid body of the Sun itself into view[1].

Spectroscopic work eventually put paid to theories of this kind. The spectroheliograph, enabling the Sun to be photographed in the light of one element only, was invented by G. E. Hale in 1892; its visual equivalent, the

[1] It may be worth recalling that in 1952 a German lawyer, Godfried Büren, stated that the Sun had a vegetation-covered inner globe, and offered a prize of 25 000 marks to anyone who could prove him wrong. The leading German astronomical society took up the challenge, and won a court case, although whether the prize was actually paid does not seem to be on record! So far as I know, the last serious protagonist of theories of this sort was an English clergyman, the Reverend P. H. Francis, who held a degree in mathematics from Cambridge University. His 1970 book, *The Temperate Sun*, is indeed a remarkable work.

Table 2.8. The chemical elements and their occurrence in the Sun. The following is a list of elements 1 to 92. ∗ = detected in the Sun. R = included in H. A. Rowland's list published in 1891. For elements 43, 61, 85–89 and 91 the mass number is that of the most stable isotope.

Atomic number	Name	Atomic weight	Occurrence in the Sun
1 H	Hydrogen	1.008	∗ R
2 He	Helium	4.003	∗
3 Li	Lithium	6.939	∗ (in sunspots)
4 Be	Beryllium	9.013	∗ R
5 B	Boron	10.812	∗ (in compound)
6 C	Carbon	12.012	∗ R
7 N	Nitrogen	14.007	∗
8 O	Oxygen	16.000	∗
9 F	Fluorine	18.999	∗ (in compound)
10 Ne	Neon	20.184	∗
11 Na	Sodium	22.991	∗ R
12 Mg	Magnesium	24.313	∗ R
13 Al	Aluminium	26.982	∗ R
14 Si	Silicon	28.090	∗ R
15 P	Phosphorus	30.975	∗
16 S	Sulphur	32.066	∗
17 Cl	Chlorine	35.434	
18 A	Argon	39.949	∗ (in corona)
19 K	Potassium	39.103	∗ R
20 Ca	Calcium	40.080	∗ R
21 Sc	Scandium	44.958	∗ R
22 Ti	Titanium	47.900	∗ R
23 V	Vanadium	50.944	∗ R
24 Cr	Chromium	52.00	∗ R
25 Mn	Manganese	52.94	∗ R
26 Fe	Iron	55.85	∗ R
27 Co	Cobalt	58.94	∗ R
28 Ni	Nickel	58.71	∗ R
29 Cu	Copper	63.55	∗ R
30 Zn	Zinc	65.37	∗ R
31 Ga	Gallium	69.72	∗
32 Ge	Germanium	72.60	∗ R
33 As	Arsenic	74.92	
34 Se	Selenium	78.96	
35 Br	Bromine	79.91	
36 Kr	Krypton	83.80	
37 Rb	Rubidium	85.48	∗ (in spots)
38 Sr	Strontium	87.63	∗ R
39 Y	Yttrium	88.91	∗ R

Table 2.8. (Continued)

Atomic number	Name	Atomic weight	Occurrence in the Sun
40 Zr	Zirconium	91.22	∗ R
41 Nb	Niobium	92.91	∗ R
42 Mo	Molybdenum	95.95	∗ R
43 Tc	Technetium	99	
44 Ru	Ruthenium	101.07	∗
45 Rh	Rhodium	102.91	∗ R
46 Pd	Palladium	106.5	∗ R
47 Ag	Silver	107.87	∗ R
48 Cd	Cadmium	112.41	∗ R
49 In	Indium	114.82	∗ (in spots)
50 Sn	Tin	118.70	∗ R
51 Sb	Antimony	121.76	∗
52 Te	Tellurium	127.61	
53 I	Iodine	126.91	
54 Xe	Xenon	131.30	
55 Cs	Cæsium	132.91	
56 Ba	Barium	137.35	∗ R
57 La	Lanthanum	138.92	∗ R
58 Ce	Cerium	140.13	∗ R
59 Pr	Praseodymium	140.91	∗
60 Nd	Neodymium	144.25	∗ R
61 Pm	Promethium	147	
62 Sm	Samarium	150.36	∗
63 Eu	Europium	151.96	∗
64 Gd	Gadolinium	157.25	∗
65 Tb	Terbium	158.93	∗
66 Dy	Dysoprosium	162.50	∗
67 Ho	Holmium	164.94	
68 Er	Erbium	167.27	∗ R
69 Tm	Thulium	168.94	∗
70 Yb	Ytterbium	173.04	∗
71 Lu	Lutecium	174.98	∗
72 Hf	Hafnium	178.50	∗
73 Ta	Tantalum	180.96	∗
74 W	Tungsten	183.86	∗
75 Re	Rhenium	186.3	
76 Os	Osmium	190.2	∗
77 Ir	Iridium	192.2	∗
78 Pt	Platinum	195.1	∗
79 Au	Gold	197.0	∗
80 Hg	Mercury	200.6	
81 Tl	Thallium	204.4	
82 Pb	Lead	207.2	∗ R
83 Bi	Bismuth	209.0	
84 Po	Polonium	210	

Table 2.8. (Continued)

Atomic number	Name	Atomic weight	Occurrence in the Sun
85 At	Astatine	211	
86 Rn	Radon	222	
87 Fr	Francium	223	
88 Ra	Radium	226	
89 Ac	Actinium	227	
90 Th	Thorium	232	*
91 Pa	Protoactinium	231	
92 U	Uranium	238	

The remaining elements are 'transuranic' and radioactive, and have not been detected in the Sun. They are:

93 Np	Neptunium	237
94 Pu	Plutonium	239
95 Am	Americium	241
96 Cm	Curium	242
97 Bk	Berkelium	243
98 Cf	Californium	244
99 Es	Einsteinium	253
100 Fm	Fermium	254
101 Md	Mendelevium	254
102 No	Nobelium	254
103 Lw	Lawrencium	257
104 Rf	Rutherfordium	—
105 Ha	Hahnium	—
106 Sg	Seaborgium	—
107 Ns	Neilsborium	—
108 Hs	Hassium	—
109 Mt	Meitnerium	—

Table 2.9. Relative frequency of numbers of atoms in the Sun.

Hydrogen	1 000 000
Helium	85 000
Oxygen	600
Carbon	420
Nitrogen	87
Silicon	45
Magnesium	40
Neon	37
Iron	32
Sulphur	16
Aluminium	3
Calcium	2
Sodium	2
Nickel	2
Argon	1

spectrohelioscope, was invented in 1923, also by Hale. In 1933 B. Lyot, in France, developed the Lyot filter, which is less versatile but more convenient, and also allows the Sun to be studied in the light of one element only.

But how did the Sun produce its energy?

One theory, proposed by J. Waterson and, in 1848, by J. R. Mayer, involved meteoritic infall. Mayer found that a globe of hot gas the size of the Sun would cool down in 5000 years or so if there were no other energy source, while a Sun made up of coal, and burning furiously enough to produce as much heat as the real Sun actually does, would be turned into ashes after a mere 4600 years. Mayer therefore assumed that the energy was produced by meteorites striking the Sun's surface.

Rather better was the contraction theory, proposed in 1854 by H. von Helmholtz. He calculated that if the Sun contracted by 60 m per year, the energy produced would suffice to maintain the output for 15 000 000 years. This theory was supported later by the great British physicist Lord Kelvin. However, it had to be abandoned when it was shown that the Earth itself is around 4600 million years old – and the Sun could hardly be younger than that. In 1920 Sir Arthur Eddington stated that atomic energy was necessary, adding 'Only the inertia of tradition keeps the contraction hypothesis alive – or, rather, not alive, but an unburied corpse'.

The nuclear transformation theory was worked out by H. Bethe in 1938, during a train journey from Washington to Cornell University. Hydrogen is being converted into helium, so that energy is released and mass is lost; the decrease in mass amounts to 4 000 000 tons per second. Bethe assumed that carbon and nitrogen were used as catalysts, but C. Critchfield, also in America, subsequently showed that in solar-type stars the proton–proton reaction is dominant.

Slight variations in output occur, and it is often claimed that it is these minor changes which have led to the Ice Ages which have affected the Earth now and then throughout its history, but for the moment at least the Sun is a stable, well-behaved Main Sequence star.

The core temperature is believed to be around 15 000 000 °C, and the density to be about 10 times as dense as solid lead. The core extends one-quarter of the way from the centre of the globe to the outer surface; about

37% of the original hydrogen has been converted to helium. Outside the core comes the radiative zone, extending out to 70% between the centre and the surface; here, energy is transported by radiative diffusion. In the outer layers it is convection which is the transporting agency.

It takes radiation about 170 000 years to work its way from the core to the bottom of the convective zone, where the temperature is over 2 000 000 °C. This seems definite enough, but we have to admit that our knowledge of the Sun is far from complete. In particular, there is the problem of the neutrinos – or lack of them.

Neutrinos are particles with no 'rest' mass and no electrical charge, so that they are extremely difficult to detect. Theoretical considerations indicate that the Sun should emit vast quantities of them, and in 1966 efforts to detect them were begun by a team from the Brookhaven National Laboratory in the USA, led by R. Davis. The 'telescope' is located in the Homestake Gold Mine in South Dakota, inside a deep mineshaft, and consists of a tank of 454 600 l of cleaning fluid (tetrachloroethylene). Only neutrinos can penetrate so far below ground level (cosmic rays, which would otherwise confuse the experiment, cannot do so). The cleaning fluid is rich in chlorine, and if a chlorine atom is struck by a neutrino it will be changed into a form of radioactive argon – which can be detected. The number of 'strikes' would therefore provide a key to the numbers of solar neutrinos.

In fact, the observed flux was much smaller than had been expected, and the detector recorded only about one-third the anticipated numbers of neutrinos. The same result was obtained by a team in Russia, using 100 tons of liquid scintillator and 144 photodetectors in a mine in the Donetsk Basin. Further confirmation came from Kamiokande in Japan, using light-sensitive detectors on the walls of a tank holding 3000 tons of water. When a neutrino hits an electron it produces a spark of light, and the direction of this, as the electron moves, tells the direction from which the neutrino has come – something which the Homestake detector cannot do. Another sort of detector, in Russia, uses gallium-71; if hit by a neutrino, this gallium will be converted to germanium-71. Another gallium experiment has been set up in Gran Sasso, deep in the Apennines, and yet another detector is in the Caucasus Mountains. In each case the neutrino flux in unexpectedly low. There are also indications that the neutrinos are least plentiful around the time of sunspot maximum, although as yet the evidence is not conclusive.

Various theories have been proposed to explain the paucity of solar neutrinos. It is known that neutrinos are of several different kinds, and the Homestake detector can trap only those with relatively low energies; even so, the number of events recorded each month should have been around 25, whereas actually it was on average no more than 9. If the Sun's core temperature is no more than around 14 000 000 °C, as against the usually assumed 15 600 000 °C, the neutrino flux would be easier to explain, but this raises other difficulties. Another suggestion is that the core temperature is reduced by the presence of WIMPs (Weakly Interacting Massive Particles). A WIMP is quite different from ordinary matter, and is said to have a mass from 5 to 10 times that of a proton, but the existence of WIMPs has not been proved, and many authorities are decidedly sceptical about them. At the moment it is fair to say that the solar neutrino problem remains unsolved.

Predictably, the Sun sends out emissions along the whole range of the electromagnetic spectrum. Infra-red radiation was detected in 1800 by William Herschel, during an examination of the solar spectrum; he noted that there were effects beyond the limits of red light. In 1801 J. Ritter detected ultra-violet radiation, by using a prism to produce a solar spectrum and noting that paper soaked in NaCl was darkened if held in a region beyond the violet end of the visible spectrum. Cosmic rays from the Sun were identified by Scott Forbush in 1942, and in 1954 he established that cosmic-ray intensity decreases when solar activity increases (Forbush effect).

The discovery of radio emission from the Sun was made by J. S. Hey and his team, on 27–28 February 1942. Initially, the effect was thought to be due to German jamming of radar transmitters. The first radar contact with the Sun was made in 1959, by Eshleman and his colleagues at the Stanford Research Institute in the United States.

Solar X-rays are blocked by the Earth's atmosphere, so that all work in this field has to be undertaken by space research methods. The first X-ray observations of the Sun were made in 1949 by investigators at the United States Naval Research Laboratory.

SOLAR PROBES

The first attempt at carrying out solar observations from high altitude was made in 1914 by Charles Abbott, using an automated pyrheliometer launched from Omaha by hydrogen-filled rubber balloons. The altitude reached was 24.4 km, and in 1935 a balloon, Explorer II, took a two-man crew to the same height. The initial attempt at solar research using a modern-type rocket was made in 1946, when a captured and converted German V.2 was launched from White Sands, New Mexico; it reached 55 km and recorded the solar spectrum down to 2400 Å. The first X-ray solar flares were recorded in 1956, from balloon-launched rockets, although solar X-rays had been identified as early as 1948.

Many solar probes have now been launched. (In 1976 one of them, the German-built Helios 2, approached the Sun to within 45 000 000 km.) The first vehicle devoted entirely to studies of the Sun was OSO 1 (Orbiting Solar Observatory 1) of 1962; it carried 13 experiments, obtaining data at ultra-violet, X-ray and gamma-ray wavelengths.

Extensive solar observations were made by the three successive crews of the first US space-station, Skylab, in 1973–4. The equipment was able to monitor the Sun at wavelengths from visible light through to X-rays. The last of the crews left Skylab on 8 February 1974, although the station itself did not decay in the atmosphere until 1979. Solar work was also undertaken by many of the astronauts on the Russian space-station Mir, from 1986.

One vehicle of special note was the Solar Maximum Mission (SMM), launched on 14 February 1980 into a circular, 574 km orbit. It was designed to study the Sun during the peak of a cycle. Following a fault, the vehicle was repaired in April 1984 by a crew from the Space Shuttle, and finally decayed on 2 December 1989. The Ulysses probe (1990) was designed to study the poles of the Sun, which can never be seen well from Earth. The Japanese probe Yohkoh ('Sunbeam') has been an outstanding success, as has SOHO (the Solar and Heliospheric Observatory) from 1995. SOHO has, indeed, played a major rôle in the new science of helioseismology.

A selected list of solar probes is given in Table 2.10.

HELIOSEISMOLOGY

The first indications of solar oscillation were detected as long ago as 1960; the period was found to be 5 minutes, and it was thought that the effects were due to a surface 'ripple' in the outer 10 000 km of the Sun's globe. More detailed results were obtained in 1973 by R. H. Dicke, who was attempting to make measurements of the polar and equatorial diameters of the Sun to see whether there were any appreciable polar flattening. Dicke found that the Sun was 'quivering like a jelly', so that the equator bulges as the poles are flattened, but the maximum amplitude is only 5 km, and the velocities do not exceed 10 m s^{-1}.

This was the real start of the science of helioseismology. Seismology involves studies of earthquake waves in the terrestrial globe, and it is these methods which have told us most of what we know about the Earth's interior. Helioseismology is based on the same principle. Pressure waves – in effect, sound waves – echo and resonate through the Sun's interior. Any such wave moving inside the Sun is bent or refracted up to the surface, because of the increase in the speed of sound with increased depth. When the wave reaches the surface it will rebound back downward, and this makes the photosphere move up and down. The amplitude is a mere 25 m, with a temperature change of 0.005 °C, but these tiny differences can be measured by the familiar Doppler principle involving tiny shifts in the positions of well-defined spectral lines. Waves of different frequencies descent to different depths before being refracted up toward the surface – and the solar sound waves are very low-pitched; the loudest lies about $12\frac{1}{2}$ octaves below the lowest note audible to human beings. There are, of course, a great many frequencies involved, so that the whole situation is very complex indeed.

Various ground-based programmes are in use – such as GONG, the Global Oscillation Network Group, made up of six stations spread out round the Earth so that at least one of them can always be in sunlight. However, more spectacular results have come from spacecraft, of which one of the most important is Soho – the Solar and Heliospheric Observatory.

Soho was put into an unusual orbit. It remains 1 500 000 km from the Earth, exactly on a line joining the Earth to the Sun; its period is therefore 365.2 days – the same as ours – and it remains in sunlight, and in contact with Earth, all the time. It lies in a stable point, known as a Lagrangian point, so that as seen from Earth it is effectively motionless. It was launched on 2 December 1995, and after a series of manœuvres arrived at its Lagrangian point in

Table 2.10. Solar missions.

Name	Launch data	Nationality	Experiments
Pioneer 4	3 Mar 1959	American	Lunar probe, but in solar orbit: solar flares.
Vanguard 3	18 Sept 1959	American	Solar X-rays.
Pioneer 5	11 Mar 1960	American	Solar orbit, 0.806 × 0.995 a.u. Flares, solar wind. Transmitted until 26 June 1960, at 37 000 000 km Earth.
OSO 1	7 Mar 1962	American	Orbiting Solar Observatory 1. Earth orbit, 553 × 595 km. Lost on 6 Aug 1963.
Cosmos 3	24 Apr 1962	Russian	Earth orbit, 228 × 719 km; decayed after 176 days. Solar and cosmic radiation.
Cosmos 7	28 July 1962	Russian	Earth orbit, 209 × 368 km. Monitoring solar flares during Vostok 3 and 4 missions. Decayed after 4 days.
Explorer 18—IMP	26 Nov 1963	American	Interplanetary Monitoring Platform 1. Earth orbit, 125 000 × 202 000 km. Provision for flare warn manned missions.
OGO 1	4 Sept 1964	American	Orbiting Geophysical Observatory 1. Earth–Sun relationships.
OSO 2	3 Feb 1965	American	Earth–Sun relationships. Flares.
Explorer 30—Solrad	18 Nov 1965	American	Solar radiation and X-rays; part of the IQSY programme (International Year of the Quiet Sun).
Pioneer 6	16 Dec 1965	American	Solar orbit, 0.814 × 0.985 a.u. First detailed space analysis of solar atmosphere.
Pioneer 7	17 Aug 1966	American	Solar orbit, 1.010 × 1.125 a.u. Solar atmosphere. Flares.
OSO 3	8 Mar 1967	American	Earth–Sun relationships. Flares.
Cosmos 166	16 June 1967	Russian	Earth orbit, 260 × 577 km. Solar radiation. Decayed after 130 days.
OSO 4	18 Oct 1967	American	Earth–Sun relationships. Flares.
Pioneer 8	13 Dec 1967	American	Solar orbit, 1.00 × 1.10 a.u. Solar wind; programme with Pioneers 6 and 7.
Cosmos 215	19 Apr 1968	Russian	Solar orbit, 260 × 577 km. Solar radiation. Decayed after 72 days.
Pioneer 9	8 Nov 1968	American	Solar orbit, 0.75 × 1.0 a.u. Solar wind, flares etc.
HEOS 1	5 Dec 1968	American	High-Energy Orbiting Satellite. Earth orbit, 418 × 112 400 km. With HEOS 2, monitored 7 years of the 11-year solar cycle.
Cosmos 262	26 Dec 1968	Russian	Earth orbit, 262 × 965 km. Solar X-rays and ultra-violet.
OSO 5	22 Jan 1969	American	Earth orbit, 550 km, inclination 32°.8. General solar studies.
OSO 6	9 Aug 1969	American	Earth orbit, 550 km, inclination 32°.8. General solar studies, as with OSO 5.
Shinsei SS1	28 Sept 1971	Japanese	Earth orbit, 870 × 1870 km. Operated for 4 months.
OSO 7	29 Sept 1971	American	Earth orbit, 329 × 575 km. General studies: solar X-ray, ultra-violet, EUV. Operated until 9 July despite having been put into the wrong orbit.
HEOS 2	31 Jan 1972	American	Earth orbit. High-energy particles, in conjunction with HEOS 1.
Prognoz 1	14 Apr 1972	Russian	First of a series of Russian solar wind and X-ray satellites. (Prognoz = Forecast.) Earth orbit, 965 × 200 000 km.
Prognoz 2	29 June 1972	Russian	Earth orbit, 550 × 200 000 km. Solar wind and X-ray studies.
Prognoz 3	15 Feb 1973	Russian	Earth orbit, 590 × 200 000 km. General solar studies, including X-ray and gamma-rays.
Intercosmos 9	19 Apr 1973	Russian–Polish	Earth orbit, 202 × 1551 km, inclination 48°. Solar radio emissions.
Skylab	14 May 1973	American	Manned missions. Three successive crews. Decayed 11 July 1979.
Intercosmos 11	17 May 1974	Russian	Earth orbit, 484 × 526 km. Solar ultra-violet and X-rays.
Explorer 52—Injun	3 June 1974	American	Solar wind.
Helios 1	1 Dec 1974	German	American-launched. Close-range studies; went to 48 000 000 km from the Sun.

Table 2.10. (Continued)

Name	Launch data	Nationality	Experiments
Aryabhāta	19 Apr 1975	Indian	Russian-launched. Solar neutrons and gamma-radiation.
OSO 8	18 Jun 1975	American	General studies, including solar X-rays.
Prognoz 4	22 Dec 1975	Russian	Earth orbit, 634 × 199 000 km. Continuation of Prognoz programmes.
Helios 2	15 June 1976	German	American-launched. Close-range studies: went to 45 000 000 km from the Sun.
Prognoz 5	25 Nov 1976	Russian	Earth orbit, 510 × 199 000 km. Carried Czech and French experiments.
Prognoz 6	22 Sep 1977	Russian	Earth-orbit. Effects of solar X-rays and gamma-rays on Earth's magnetic field.
Solar Maximum Mission	14 Feb 1980	American	Long-term vehicle. Repaired in orbit April 1984; decayed December 1989.
Ulysses	6 Oct 1990	European	American-launched. Solar Polar probe.
Yohkoh	30 Aug 1991	Japanese	X-ray studies of the Sun.
Koronos-1	3 May 1994	Russian	Long-term studies; carries coronagraph and X-ray telescope.
Wind	1 Nov 1994	American	Solar–terrestrial relationships.
SOHO	2 Dec 1995	European	Wide range of studies.
Polar	24 Feb 1996	American	Solar–terrestrial relationships. Polar orbit.
Cluster	4 June 1996	American	Failed to orbit.
TRACE	1 Apr 1998	American	Studies of solar transition region.

In addition, some satellites (such as the SPARTAN probes) have been released from the Space Shuttles and retrieved a few days later.

February 1996. There was an alarm on 25 June 1996, when contact was lost, and it was feared that the whole mission had come to an end; but Soho was reacquired on 27 July, from the Arecibo radio telescope in Puerto Rico, and was in full operation again by 20 September.

Soho has been immensely informative. For example, it has detected vast solar tornadoes whipping across the Sun's surface, with gusts up to 500 000 km h^{-1}. There are jet streams below the visible surface, and definite 'belts' in which material moves more quickly than the gases to either side. There has been a major surprise, too, with regard to the Sun's general rotation. On the surface, the rotation period is 25 days at the equator, rising to 27.5 days at latitudes 40° north or south and as much as 34 days at the poles. This differential rotation persists to the base of the convection zone, but here the whole situation changes; the equatorial rotation slows down and the higher-latitude rotation speeds up. The two rates become equal at a distance about half-way between the surface and the centre of the globe; deeper down, the Sun rotates in the manner of a solid body. As yet it must be admitted that the reasons for this bizarre behaviour are unknown.

It has also been found that the entire outer layer of the Sun, down to about 24 000 km, is slowly but steadily flowing from the equator to the poles. The polar flow rate is no more than 80 km h^{-1}, as against the rotation speed of 6400 km h^{-1}, but this is enough to transport an object from the equator to the pole in little over a year.

THE SOLAR ATMOSPHERE

With the naked eye, the outer surroundings of the Sun – the solar atmosphere – can be seen only during a total solar eclipse. With modern-type equipment, or from space, they can however be studied at any time, although the outer corona is more or less inaccessible except by using space research methods. The structure of the Sun is summarized in Table 2.11.

Above the photosphere, rising to 5000 km, is the chromosphere ('colour-sphere'), so named because its hydrogen content gives it a strong red colour as seen during a total eclipse. The temperature rises quickly with altitude (remembering that the scientific definition of temperature depends upon the speeds at which the atomic particles

Table 2.11. Structure of the Sun.

Core	The region where energy is being generated. The outer edge lies about 175 000 km from the Sun's centre. The core temperature about 15 000 000 °C; the density 150 g cm^{-3} (10 times the density of lead). The temperature at the outer edge is about half the central value.
Radiative zone	Extends from the outer edge of the core to the interface layer, i.e. from 25 pc to 70 pc of the distance from the centre to the surface. Temperatures range from 7 000 000 °C at the base to 2 000 000 °C at the top; the density decreases from 20 g cm^{-3} (about the density of lead) to 0.2 g cm^{-3} (less than the density of water).
Interface layer	Separates the radiative zone from the convective zone. The solar magnetic field is generated by a magnetic dynamo in this layer.
Convective zone	Extends from 200 000 km to the visible surface. At the bottom of the zone the temperature is low enough for heavier ions to retain electrons; the material then inhibits the flow of radiation, and the trapped heat leads to 'boiling' at the surface. Convective motions are seen as granules and supergranules.
Photosphere	The visible surface; temperature 5700 °C, density 0.000 0002 g cm^{-3} (1/10 000 of that of the Earth's air at sea level). Sunspots are seen here. Faculæ lie on and a few hundred km above the bright surface.
Chromosphere	The layer above the photosphere, extending to 5000 km above the bright surface. The temperature increases rapidly with altitude, until the chromosphere merges with the transition layer. The Fraunhofer lines in the solar spectrum are produced in the chromosphere, which acts as a 'reversing layer'. During a total eclipse the chromosphere appears as a red ring round the lunar disk (hence the name: colour-sphere).
Transition region	A narrow layer separating the chromosphere from the higher-temperature corona.
Corona	The outer atmosphere; temperature up to 2000 000 °C, density on average about 10^{-15} g cm^{-3}. The solar wind originates here.
Heliosphere	A 'bubble' in space produced by the solar wind and inside which the Sun's influence is dominant.
Heliopause	The outer edge of the heliosphere, where the solar wind merges with the interstellar medium and loses its identity; the distance from the Sun is probably about 150 a.u.

move around, and is by no means the same as the ordinary definition of 'heat'; the chromosphere is so rarefied that it certainly is not 'hot'). The dark Fraunhofer lines in the solar spectrum are produced in the chromosphere.

Rising from the chromosphere are the prominences, structures with chromospheric temperatures embedded in the corona. They were first described in detail by the Swedish observer Vassenius at the total eclipse of 1733, although he believed them to belong to the Moon rather than to the Sun. (They may have been recorded earlier, in 1706, by Stannyan at Berne.) It was only after the eclipse of 1842 that astronomers became certain that they are solar rather than lunar.

Prominences (once, misleadingly, known as Red Flames) are composed of hydrogen. Quiescent prominences may persist for weeks or even months, but eruptive prominences show violent motions, and may attain heights of several hundreds of thousands of kilometres. Following the eclipse of 19 August 1868, J. Janssen (France) and Norman Lockyer (England) developed the method of observing them spectroscopically at any time. By observing at hydrogen wavelengths, prominences may be seen against the bright disk of the Sun as dark filaments, sometimes termed flocculi. (Bright flocculi are due to calcium.)

Above the chromosphere, and the thin transition region, comes the corona, the 'pearly mist' which extends outward from the Sun in all directions. It has no definite boundary; it simply thins out until its density is no greater than that of the interplanetary medium. The density is in fact very low – less than one million millionth of that of the Earth's air at sea-level, so that its 'heat' is negligible even though the temperature reaches around 2000 000 °C. Because of its high temperature, it is brilliant at X-ray wavelengths.

Seen during a total eclipse, the corona is truly magnificent. The first mention of it may have been due to the Roman writer Plutarch, who lived from about AD 46 to 120. Plutarch's book 'On the Face in the Orb of the Moon' contains a reference to 'a certain splendour' around the eclipsed Sun which could well have been the corona. The corona was definitely recorded from Corfu during the eclipse of 22 December 968. The astronomer Clavius saw it at the eclipse of 9 April 1567, but regarded it as merely the uncovered edge of the Sun; Kepler showed that this could not be so, and attributed it to a lunar atmosphere. After observing the eclipse of 16 June 1806 from Kindehook, New York, the Spanish astronomer Don José Joaquin de Ferrer pointed out that if the corona were due to a lunar atmosphere, then the height of this atmosphere would have to be 50 times greater than that of the Earth, which was clearly unreasonable. However, it was only after careful studies of the eclipses of 1842 and 1851 that the corona and the prominences were shown unmistakably to belong to the Sun rather than to the Moon.

There is some evidence that during eclipses which occurred during the Maunder Minimum (1645–1715) the corona was virtually absent, although the records make it impossible to be sure. Certainly the shape of the corona at spot-maximum is more symmetrical than at spot-minimum, when there are streamers and 'wings' – as was first recognized after studies of the eclipses of 1871 and 1872.

The high temperature of the corona was for many years a puzzle. It now seems that the cause is to be found in what is termed 'magnetic reconnection'. This occurs when magnetic fields interact to produce what may be termed short circuits; the fields 'snap' to a new, lower-energy state, rather reminiscent of the snapping of a twisted rubber band. Vast amounts of energy are released, and can produce flares and other violent phenomena as well as causing the unexpectedly high coronal temperature. A reconnection event was actually recorded, on 8 May 1998, from a spacecraft, TRACE (the Transition Region and Coronal Explorer), which had been launched on 1 April 1998 specifically to study the Sun at a time when solar activity was starting to rise toward the peak of a new cycle.

THE SOLAR WIND

The corona is the source of what is termed the solar wind – a stream of particles being sent out from the Sun all the time. The first suggestion of such a phenomenon was made in the early 1950s, when it was realized that the Sun's gravitational pull is not strong enough to retain the very high-temperature coronal gas, so that presumably the corona was expanding and was being replenished from below. L. Biermann also drew attention to the fact that the tails of comets always point away from the Sun, and he concluded that the ion or gas tails are being 'pushed outward' by particles from the Sun. In this he was correct. (The dust tails are repelled by the slight but definite pressure of solar radiation.) In 1958 E. N. Parker developed the theory of the expanding corona, and his conclusions were subsequently verified by results from space-craft. One of these was Mariner 2, sent to Venus in 1962. En route, Mariner not only detected a continuously flowing solar wind, but also observed fast and slow streams which repeated at 27 day intervals, suggesting that the source of the wind rotated with the Sun.

The solar wind consists of roughly equal numbers of protons and electrons, with a few heavier ions. It leads to a loss of mass of about 10^{12} tons per year (which may sound a great deal, but is negligible by solar standards). As the wind flows past the Earth its density is of the order of 5 atoms cm^{-3}; the speed usually ranges between 200 and 700 km s^{-1}, with an average value of 400 km s^{-1}, although the initial speed away from the Sun may be as high as 900 km s^{-1}.

The fast component of the wind comes from low solar latitudes; the average velocity is of the order of 800 km s^{-1}. The slow component comes from coronal holes, where the density is below average; coronal holes are often found near the poles, and here the magnetic field lines are open, making it easier for wind particles to escape. The wind is 'gusty', and when at its most violent the particles bombard the Earth's magnetosphere, producing magnetic storms and displays of auroræ.

From Earth it is difficult to study the polar regions of the Sun, because our view is always more or less broadside-on; the same is true of most space-craft. The only way to obtain a good view of the solar poles is to send a probe out of the ecliptic, and this was done with Ulysses, launched from Cape Canaveral on 6 October 1990. It was first sent out to Jupiter, and on 8 February 1992 it flew past the Giant Planet, using Jupiter's strong gravitational pull to send it into the required orbit. It flew over the Sun's south pole on 26 June 1994, and over the north pole on 31 July 1995. Some of the findings were unexpected; the magnetic conditions in the polar regions were not quite what had been anticipated.

Note that Ulysses will never fly close to the Sun, and in fact it will always remain outside the orbit of the Earth. Its own orbital period is six years.

How far does the solar wind extend? Probably out to a distance of about 150 a.u., where it will merge with the interstellar medium and cease to be identifiable. This 'heliopause' marks the outer edge of the heliosphere, the area of space inside which the Sun's influence is dominant.

ECLIPSES OF THE SUN

A solar eclipse occurs when the Moon passes in front of the Sun; strictly speaking, the phenomenon is an occultation of the Sun by the Moon. Eclipses may be total (when the whole of the photosphere is hidden), partial, or annular (when the Moon's apparent diameter is less than that of the Sun, so that a ring of the photosphere is left showing round the lunar disk: Latin *annulus*, a ring). Recent and future solar eclipses are listed in Tables 2.12, 2.13, 2.14 and 2.15.

The solar corona can be well seen from Earth only during a total eclipse. In 1930 B. Lyot built and tested

Table 2.12. Solar eclipses 1923–1999. T = total, P = partial, A = annular.

Date	Type	Area	Date	Type	Area
1923 Mar 16	A	S Africa	1941 Mar 27	A	S Pacific, S America
1923 Sept 10	T	California, Mexico	1941 Sept 21	T	China, Pacific
1924 Mar 5	P	S Africa	1942 Mar 16/17	P	S Pacific, Antarctic
1924 July 31	P	Antarctic	1942 Aug 12	P	Invisible in Britain
1924 Aug 26	P	Iceland, N Russia, Japan	1942 Sept 10	P	Britain
1925 Jan 24	T	North-eastern USA	1943 Feb 4/5	T	Japan, Alaska
1925 July 20/1	A	New Zealand, Australia	1943 Aug 1	A	Pacific
1926 Jan 14	T	E Africa, Indian Ocean, Borneo	1944 Jan 25	T	Brazil, Atlantic, Sudan
1926 July 9/10	A	Pacific	1944 July 20	A	India, New Guinea
1927 Jan 3	A	New Zealand, S America	1945 Jan 14	A	Australia, New Zealand
1927 June 29	T	England, Scandinavia	1945 July 9	T	Canada, Greenland, N Europe
1927 Dec 24	P	Polar zone	1946 Jan 3	P	Invisible in Britain
1928 May 19	T	S Atlantic	1946 May 30	P	S Pacific
1928 June 17	P	N Siberia	1946 June 29	P	Arctic
1928 Nov 12	P	England to India	1946 Nov 23	P	N America
1929 May 9	T	Indian Ocean, Philippines	1947 May 20	T	Pacific, Equatorial Africa, Kenya
1929 Nov 1	A	Newfoundland, C Africa, Indian Ocean	1947 Nov 12	A	Pacific
			1948 May 8/9	A	E Asia
1930 Apr 28	T	Pacific	1948 Nov 1	T	Kenya, Pacific
1930 Oct 21/2	T	S Pacific to S America	1949 Apr 28	P	Britain
1931 Apr 17/18	P	Arctic	1949 Oct 21	P	New Zealand, Australia
1931 Sept 12	P	Alaska, N Pacific	1950 Mar 18	A	S Atlantic
1931 Oct 11	A	S America, S Pacific, Antarctic	1950 Sept 12	T	Siberia, N Pacific
1932 Mar 7	A	Antarctic	1951 Mar 7	A	Pacific
1932 Aug 31	T	USA	1951 Sept 1	A	Eastern USA, C and S Africa
1933 Feb 24	A	S America, C Africa	1952 Feb 25	T	Africa, Arabia, Russia
1933 Aug 21	A	Iran, India, N Australia	1952 Aug 20	A	S America
1934 Feb 13/14	T	Pacific	1953 Feb 13/14	P	E Asia
1934 Aug 10	A	S Africa	1953 July 11	P	Arctic
1935 Jan 5	P	No land surface	1953 Aug 9	P	Pacific
1935 Feb 3	P	N America	1954 Jan 5	A	Antarctic
1935 June 30	P	Britain	1954 June 30	T	Iceland, Norway, Sweden, Russia, India
1935 July 30	P	No land surface			
1935 Dec 25	A	New Zealand, south S America	1954 Dec 25	A	S Africa, S Indian Ocean
1936 June 19	T	Greece, Turkey, Siberia, Japan	1955 June 20	T	S Asia, Pacific, Philippines
1936 Dec 13/14	A	Australia, New Zealand	1955 Dec 14	A	Sudan, Indian Ocean, China
1937 June 8	T	Pacific, Chile	1956 June 8	T	S Pacific
1937 Dec 2/3	A	Pacific	1956 Dec 2	P	Europe, Asia
1938 May 29	T	S Atlantic	1957 Apr 29/30	A	Arctic
1938 Nov 21/2	P	E Asia, Pacific coast of N America	1957 Oct 23	T	Antarctica
1938 Apr 19	A	Alaska, Arctic	1958 Apr 19	A	Indian Ocean, Pacific
1939 Oct 12	T	Antarctic	1958 Oct 12	T	Pacific
1940 Apr 7	A	USA, Pacific	1959 Apr 8	A	S Indian Ocean, Pacific
1940 Oct 1	T	Brazil, S Atlantic, S Africa	1959 Oct 2	T	N Atlantic, N Africa

Table 2.12. (Continued)

Date	Type	Area	Date	Type	Area
1960 Mar 27	P	Australia, Antarctica	1977 Oct 12	T	Pacific, Peru, Brazil
1960 Sept 20/1	P	N America, E Siberia	1978 Apr 7	P	Antarctic
1961 Feb 15	T	France, Italy, Greece,	1978 Oct 2	P	Arctic
		Yugoslavia, Russia	1979 Feb 26	T	Pacific, USA, Canada, Greenland
1961 Aug 11	A	S Atlantic, Antarctica	1980 Aug 10	A	S Pacific, Brazil
1962 Feb 4/5	T	Pacific	1981 Feb 4	A	Pacific, S Australia, New Zealand
1962 July 31	A	S America, C Africa	1981 July 31	T	Russia, N Pacific
1963 Jan 25	A	Pacific, S Africa	1982 Jan 25	P	Antarctic
1963 July 20	T	Japan, north N America, Pacific	1982 June 21	P	Antarctic
1964 Jan 14	P	Tasmania, Antarctica	1982 July 20	P	Arctic
1964 July 9	P	N Canada, Arctic	1982 Dec 15	P	Arctic
1964 Dec 3/4	P	NE Asia, Alaska, Pacific	1983 June 11	T	Indian Ocean, E Indies, Pacific
1965 May 30	T	Pacific, New Zealand, Peru coast	1983 Dec 4	A	Atlantic, Equatorial Africa
1965 Nov 23	A	Russia, Tibet, E Indies	1984 May 30	A	Pacific, Mexico, USA, N Africa
1966 May 20	A	Greece, Russia	1984 Nov 22/3	T	E Indies, S Pacific
1966 Nov 12	T	S America, Atlantic	1985 May 19	P	Arctic
1967 May 9	P	N America, Iceland,	1985 Nov 12	T	S Pacific, Antarctica
		Scandinavia	1986 Apr 9	P	Antarctic
1967 Nov 2	T	S Atlantic	1986 Oct 3	T	N Atlantic
1968 Mar 28/9	P	Pacific, Antarctica	1987 Mar 29	T	Argentina, C Africa, Indian Ocean
1968 Sept 22	T	Arctic, Mongolia, Siberia	1987 Sept 23	A	Russia, China, Pacific
1969 Mar 18	A	Indian Ocean, Pacific	1988 Mar 18	T	Indian Ocean, E Indies, Pacific
1969 Sept 11	A	Peru, Bolivia	1989 Mar 7	P	Arctic
1970 Mar 7	T	Mexico, USA, Canada	1989 Aug 31	P	Antarctic
1970 Aug 31/	T	East Indies, Pacific	1990 Jan 26	A	Antarctic
Sept 1			1990 July 22	T	Finland, Russia, Pacific
1971 Feb 25	P	Europe, NW Africa	1991 Jan 15	A	Pacific, New Zealand, SW Australia
1971 July 22	P	Alaska, Arctic	1991 July 11	T	Pacific, Mexico, Hawaii
1971 Aug 20	P	Australasia, S Pacific	1992 Jan 4	A	Pacific
1972 Jan 16	A	Antarctica	1992 June 30	T	Atlantic
1972 July 10	T	Alaska, Canada	1992 Dec 24	P	Arctic
1973 Jan 4	A	Pacific, S Atlantic	1992 May 21	P	Arctic
1973 June 30	T	Atlantic, N Africa, Kenya,	1993 Nov 13	P	Antarctic
		Indian Ocean	1994 May 10	A	Pacific, Mexico, USA, Canada
1973 Dec 24	A	Brazil, Atlantic, N Africa	1994 Nov 3	T	Peru, Brazil, S Atlantic
1974 June 20	T	Indian Ocean	1995 Apr 29	A	S Pacific, Peru, S Atlantic
1974 Dec 13	P	N and C America	1995 Oct 24	T	Iran, India, Borneo, Pacific
1975 May 11	P	Europe, N Asia, Arctic	1996 Apr 17	P	Antarctic
1975 Nov 3	P	Antarctic	1997 Mar 9	T	Siberia, Arctic
1976 Apr 29	A	NW Africa, Turkey, China	1997 Sept 2	P	Antarctic
1976 Oct 23	T	Tanzania, Indian Ocean,	1998 Feb 26	T	Pacific, Venezuela, Atlantic
		Australia	1998 Aug 22	A	Indian Ocean, E Indies, Pacific
1977 Apr 18	A	Atlantic, SW Africa, Indian	1999 Feb 16	A	Indian Ocean, Australia, Pacific
		Ocean	1999 Aug 11	T	Atlantic, England, Turkey, India

Table 2.13. Solar eclipses 2000–2010. T = total, P = partial, A = annular.

Date	Mid-eclipse (GMT)	Type	Maximum length of totality/annularity m	s	Obscuration (percent)	Area
2000 Feb 5	13	P	—		56	Antarctic
2000 July 31	02	P	—		60	Arctic
2000 Dec 25	18	P	—		72	Arctic
2001 June 21	12	T	4	56	—	Atlantic, South Africa
2001 Dec 14	21	A	3	54	—	Central America, Pacific
2002 June 10	24	A	1	13	—	Pacific
2002 Dec 4	08	T	2	04	—	S Africa, Indian Ocean, Australia
2003 May 31	04	A	3	37	—	N Scotland, Iceland
2003 Nov 23	23	T	1	57	—	Antarctic
2004 Apr 19	14	P	—		74	Antarctic
2004 Oct 14	03	P	—		93	Arctic
2005 Apr 8	21	T	0	42	—	Pacific, N of S America
2005 Oct 3	11	A	4	32	—	Atlantic, Spain, Africa, Indian Ocean
2006 Mar 29	10	T	4	07	—	Atlantic, N Africa, Turkey, Russia
2006 Sept 22	12	A	7	09	—	NE of S America, Atlantic, S Indian Ocean
2007 Mar 19	03	P	—	—	88	N America, Japan
2007 Sep 11	13	P	—	—	75	S America, Antarctic
2008 Feb 7	04	A	2	14	—	S Pacific, Antarctica
2008 Aug 1	10	T	2	27	—	N Canada, Greenland, Siberia, China
2009 Jan 26	08	A	7	56	—	S Atlantic, Indian Ocean, Sri Lanka, Borneo
2009 July 22	03	T	6	40	—	India, China, Pacific
2010 Jan 15	07	A	11	11	—	Africa, Indian Ocean
2010 July 11	20	T	5	20	—	Pacific

There will be total eclipses on 2012 Nov 13, 2013 Nov 3, 2015 Mar 20, 2016 Mar 9, 2017 Aug 21, 2019 July 2, 2020 Dec 14, 2021 Dec 4, 2023 Apr 20, 2024 Apr 8, 2026 Aug 12, 2027 Aug 2, 2028 July 22, 2030 Nov 25, 2031 Nov 14, 2033 Mar 30, 2034 Mar 20, 2035 Sept 2, 2037 July 13, 2038 Dec 26 and 2039 Dec 15

a coronagraph, located at the Pic du Midi Observatory (altitude 2870 m); this instrument produces an 'artificial eclipse' inside the telescope. With it Lyot was able to examine the inner corona and its spectrum, but the outer corona remained inaccessible.

The greatest number of eclipses possible in one year is seven; thus in 1935 there were five solar and two lunar eclipses, and in 1982 there were four solar and three lunar. The least number possible in one year is two, both of which must be solar, as in 1984.

The length of the Moon's shadow varies between 381 000 km and 365 000 km, with a mean of 372 000 km. As the mean distance of the Moon from the Earth is 384 000 km, the shadow is on average too short to reach the

Table 2.14. British annular eclipses, 1800–2200.

Date	Location
1820 Sept 7	Shetland
1836 May 15	N Ireland, S Scotland
1847 Oct 9	S Ireland, Cornwall
1858 Mar 15	Dorset to the Wash
1921 Apr 8	NW Scotland, Orkney, Shetland
2003 May 21	Scotland
2173 Apr 12	Hebrides

Earth's surface, so that annular eclipses are more frequent than total eclipses in the ratio of five to four. On average there are 238 total eclipses per century. During the 21st century there will be 224 solar eclipses; 68 total, 72 annular,

Table 2.15. British total eclipses, 1–2200[a].

Date		Location	Date		Location
21	June 19	Cornwall, Sussex	1185	May 1	Scotland
28	July 10	S Ireland, Cornwall	1230	May 14	Almost all England
118	Sept 3	Cornwall, Sussex	1330	July 16	N Scotland
122	June 21	Faroe Islands; between Shetland and Orkney	1339	July 7	Between Shetland and Orkney
129	Feb 6	Wales to Humberside	1424	June 28	Orkney, Shetland
143	May 2	Annular/total; total in S Ireland, annular in Wales	1433	June 17	Scotland
			1440	Feb 3	Near miss of Outer Hebrides
158	July 13	London	1598	Feb 25	Wales, S Scotland
183	Mar 11	N Ireland, N England, S Scotland	1630	June 10	Cork, Scilly Isles
228	Mar 23	Almost all Ireland, England, Wales	1652	Apr 8	Anglesey, Scotland
303	Sept 27	Scotland	1654	Aug 12	Grampian, Aberdeen
319	May 6	London	1679	Apr 10	W Ireland
364	June 16	N Scotland, Orkney	1699	Sept 23	SE tip of Scotland
393	Nov 20	London	1715	May 3	Cornwall, London, Norfolk
413	Apr 16	S Ireland, N Wales, W Midlands	1724	May 22	S Wales, Hampshire, London
458	May 28	Wales to Lincolnshire	1925	Jan 24	Near miss of Outer Hebrides
565	Feb 16	Channel Islands	1927	June 29	Wales, Preston, Giggleswick
594	July 23	Ireland, N England, S Scotland	1954	June 30	Northernmost Scotland (Unst)
639	Sept 3	Wales, Midlands	1999	Aug 11	Cornwall, Devon, Alderney
664	May 1	N Ireland, N England, S Scotland	2015	Mar 20	Faroes; misses Scotland
758	Apr 12	Kent	2081	Sept 3	Channel Islands
849	May 25	Shetland Islands	2090	Sept 23	S Ireland, Cornwall
865	Jan 1	Central Ireland, Cumberland	2133	June 3	Hebrides, Scotland
878	Oct 29	London	2135	Oct 7	S Scotland, N England, N Wales
885	June 16	N Ireland, Scotland	2142	May 25	Channel Islands
968	Dec 22	Scilly, Cornwall, Jersey	2151	June 14	Scotland, N London, Kent
1023	Jan 24	Cornwall, Wales, S Scotland	2160	June 4	Cork, Land's End
1133	Aug 2	Scotland, NE England	2189	Nov 8	Cork, Cornwall
1140	Mar 20	Wales to Norfolk	2200	Apr 14	N Ireland, Isle of Man, Lake District

[a] Calculations by Sheridan Williams, whom I thank for allowing me to quote them.

seven annular/total (that is to say, annular along most of the track) and 77 partial[2].

The track of totality across the Earth's surface can never be more than 272 km wide, and in most cases the width is much less than this. A partial eclipse is seen to either side of the track of totality, although some partial eclipses are not total or annular anywhere on Earth.

The longest possible duration of totality is 7 min 31 s. This has never been observed, but at the eclipse of 20 June 1955 totality over the Philippines lasted for 7 min 8 s.

The longest totality during the 21st century will be on 22 July 2009 (6 min 30 s). The shortest possible duration of totality can be less than 1 s. This happened at the eclipse of 3 October 1986, which was annular along most of the central track, but total for about 1/10 s over a restricted area in the North Atlantic Ocean. (So far as I know, it was not observed.) The shortest total eclipse of the 21st century will be that of 6 December 2067: a mere 8 s.

Annularity can last for longer; the maximum is as much as 12 min 24 s. The annular eclipse of 15 January 2010 will last for 11 min 8 s – that of 16 December 2085 for only 19 s. The largest partial eclipse of the 21st century will be that of 11 April 2051, when the Sun will be 98.5%

[2] The calculations were made by Fred Espenak of NASA. I thank him for allowing me to quote them.

obscured. On 24 October 2098 the obscuration amounts to no more than 0.004%.

The longest totality ever observed was during the eclipse of 30 June 1973. A Concorde aircraft, specially equipped for the purpose, flew underneath the Moon's shadow, keeping pace with it so that the scientists on board (including the British astronomer John Beckman) saw a totality lasting for 72 min. They were carrying out observations at millimetre wavelengths, and at their height of 55 000 feet were above most of the water vapour in our atmosphere which normally hampers such observations. They were also able to see definite changes in the corona and prominences during their flight. The Moon's shadow moves over the Earth at up to 3000 km h^{-1}, so that only Concorde can easily match it.

The first recorded solar eclipse seems to have been that of 2136 BC, seen in China during the reign of the Emperor Chung K'ang. A famous story is attached to it. The Chinese believed that during an eclipse the Sun was being attacked by a hungry dragon, and the only remedy was to beat drums, bang gongs, shout and wail, and in general make as much noise as possible in order to scare the dragon away. Not surprisingly this procedure always worked. It was the duty of the Court astronomers to give warning of a forthcoming eclipse, and it has been said that on this occasion the astronomers, who rejoiced in the names of Hsi and Ho, forgot – with the result that they were executed for negligence. Alas, there can be no doubt that this story is apocryphal.... The next eclipse which may be dated with any certainty is that of 1375 BC, described on a clay tablet found at Ugarit in Syria.

Predictions were originally made by studies of the Saros period. This is the period after which the Sun, Moon and node arrive back at almost the same relative positions. It amounts to 6585.321 solar days, or approximately 18 years 11 days. Therefore, an eclipse tends to be followed by another eclipse in the same Saros series 18 years 11 days later, although conditions are not identical, and the Saros is at best a reasonable guide. (For example, the eclipse of June 1927 was total over parts of England, but the 'return', in July 1945, was not.) One Saros series lasts for 1150 years; it includes 64 eclipses, of which 43 or 44 are total, while the rest are partial eclipses seen from the polar zones of the Earth.

The first known predictions about which we have reasonably reliable information were made by the Greeks. There does seem good evidence that the eclipse of 25 May 585 BC was predicted by Thales of Miletus, the first of the great Greek philosophers. It occurred near sunset in the Mediterranean area, and is said to have put an end to a battle between the forces of King Alyattes of the Lydians and King Cyraxes of the Medes; the combatants were so alarmed by the sudden darkness that they concluded a hasty peace.

Eclipse stories and legends are plentiful. Apparently the Emperor Louis of Bavaria was so frightened by the eclipse of 840 that he collapsed and died, after which his three sons engaged in a ruinous war over the succession. There was also the curious case of General William Harrison (later President of the United States) when he was Governor of Indiana Territory, and was having trouble with the Shawnee prophet Tenskwatawa. He decided to ridicule him by claiming that he could make the Sun stand still and the Moon to 'alter its course'. Unluckily for him, the prophet knew more astronomy than the General, and he was aware that an eclipse was due on 16 July 1806. He therefore said that he would demonstrate his own power by blotting out the Sun. A crowd gathered at the camp, and the prophet timed his announcement at just the right moment, so that Harrison was nonplussed (although in 1811 he did destroy the Shawnee forces at the Battle of Tippecanoe).

The first total solar eclipse recorded in the United States was that of 24 June 1778, when the track passed from Lower California to New England. Two years later, on 21 October 1780, a party went to Penobscot, Maine, to observe an eclipse; it was led by S. Williams of Harvard and had been given 'free passage' by the British forces. Unfortunately, a mistake in the calculations meant that the astronomers went to the wrong place, and remained outside the track of totality. The first American expedition to Europe was more successful; on 28 July 1851 G. P. Bond took a party to Scandinavia, and obtained good results.

Astronomers have always been ready to run personal risks to study eclipses, and one man who demonstrated this in 1870 was Jules Janssen, a leading French expert concerning all matters relating to the Sun. The eclipse was due on 22 December. Janssen was in Paris, but the city was surrounded by the German forces, and there was no obvious escape route. Janssen's solution was to fly out in a hot-air balloon. He arrived safely at Oran – only to be met with an overcast sky. He could certainly count himself unlucky.

In Britain, eclipse records go back a long way. The first account comes from the Anglo-Saxon Chronicle; the eclipse took place on 15 February 538, four years after the death of Cerdic, the first King of the West Saxons. The Sun was two-thirds eclipsed from London.

The celebrated chronicler William of Malmesbury gave a graphic description of the eclipse of 1133: the Sun 'shrouded his glorious face, as the poets say, in hideous darkness, agitating the hearts of men by an eclipse and on the sixth day of the week there was so great an earthquake that the ground appeared to sink down; a horrid noise being first heard beneath the surface'. In fact there can be no connection between an eclipse and a ground tremor, but William was again busy at the eclipse of 1140: 'It was feared that Chaos had come again . . . it was thought and said by many, not untruly, that the King [Stephen] would not continue a year in the government.' (In fact, Stephen reigned until 1154.) Several Scottish eclipses were given nicknames; Black Hour (1433), Black Saturday (1598), Mirk Monday (1652).

The eclipse of 1715 was well observed over much of England. Edmond Halley saw it, and gave a vivid description of the corona: 'A luminous ring of a pale whiteness, or rather pearl colour, a little tinged with the colours of the Iris, and concentric with the Moon.' He was also the first to see Baily's Beads – brilliant spots caused by the Sun's rays shining through valleys on the lunar limb immediately before and immediately after totality. They can sometimes be seen during an annular eclipse (as by Maclaurin, from Edinburgh, on 1 March 1737) but the first really detailed description of them was given in 1836, at the annular eclipse of 15 May, by Francis Baily, after whom they are named. (They were first photographed at the eclipse of 7 August 1869 by C. F. Hines and members of the Philadelphia Photographic Corps, observing from Ottuma in Iowa.)

The last British mainland totality before 1927 was that of 1724. Unfortunately the weather was poor and the only good report came from a Dr. Stukeley, from Haraden Hill near Salisbury. The spectacle, he wrote, 'was beyond all that he had ever seen or could picture to his imagination that most solemn'. The eclipse was much better seen from France.

In 1927 the track crossed parts of Wales and North England, but there was a great deal of cloud and the best results came from Giggleswick, where the Royal Astronomical Society party was stationed. Totality was brief – only 24 s – but the clouds cleared away at the vital moment, and useful photographs were obtained. On 30 June 1954 the track brushed the tip of Unst, northernmost of the Shetland Islands, but most observers went to Norway or Sweden. On 11 August 1999 the track crossed Devon and Cornwall, but most of the area was cloudy, though the partial phase was well seen from most of the rest of Britain. Turkey and Iran had good views; the prominences were particularly striking – not at all surprising as the Sun was rising to the peak of its 11-year cycle.

The maximum theoretical length of a British total eclipse is 5.5 min. That of 15 June 885 lasted for almost 5 min, and so will the Scottish total eclipse of 20 July 2381.

Another phenomenon seen at a total eclipse is that of shadow bands, wavy lines crossing the landscape just before and just after totality; they are, of course, produced in the Earth's atmosphere. They were first described by H. Goldschmidt at the eclipse of 1820.

The first attempt to photograph a total solar eclipse was made by the Austrian astronomer Majocci on 8 July 1842. He failed to record totality, although he did manage to photograph the partial phase. The first real success, showing the corona and prominences, was due to Berkowski on 28 July 1851, using the 6.25 Königsberg heliometer with an exposure time of 24 s. The flash spectrum was first photographed by the American astronomer C. Young on 22 December 1870. (The flash spectrum is the sudden change in the Fraunhofer lines from dark to bright, when the Moon blots out the photosphere in the background and the chromosphere is left shining 'on its own'.) The flash spectrum was first observed during an annular eclipse by Pogson, in 1872.

Nowadays, of course, total eclipses are shown regularly on television. The first attempt to show totality on television from several stations spread out along the track was made by the BBC at the eclipse of 15 February 1961. All went well, and totality was shown successively from France, Italy and Yugoslavia. There was, however, one bizarre incident. I was stationed atop Mount Jastrebač, in Yugoslavia, and with our party were several oxen used to haul the equipment up to the summit. It is quite true that animals tend to go to sleep as darkness falls, and, unknown to me, the Yugoslav director decided to show this as soon as totality began – so he trained the cameras on to the oxen and, just to make sure that the viewers were treated to a good view, he switched on floodlights!

The last total eclipse will probably occur in about 700 million years from now. By then the Moon will have receded to about 29 000 km further away from the Earth, and the disk will no longer appear large enough to cover the Sun.

EVOLUTION OF THE SUN

The Sun is a normal Main Sequence star. It is in orbit round the centre of the Galaxy; the period is of the order of 225 000 000 years – sometimes known as the 'cosmic year'. One cosmic year ago, the most advanced creatures on Earth were amphibians; even the dinosaurs had yet to make their entry. (It is interesting to speculate as to conditions here one cosmic year hence!) The apex of the Sun's way – i.e. the point in the sky toward which it is moving – is RA 18h, declination +34°, in Hercules; the antapex is at RA 6h, declination −34°, in Columba.

The age of the Earth is about 4.6 thousand million years, and the Sun is certainly older than this, so that perhaps 4800 million years to around 5000 million years is a reasonable estimate. The Sun was born inside a giant gas cloud, perhaps 50 light-years in diameter, which broke up into globules, one of which produced the Sun. The first stage was that of a protostar, surrounded by a cocoon of gas and dust which may be termed a solar nebula (an idea first proposed by Immanuel Kant as long ago as the year 1755). Contraction led to increased heat; there was a time when the fledgling star varied irregularly, and sent out an energetic 'wind' (the so-called T Tauri stage), but eventually the cocoon was dispersed, and the Sun became a true star. When the core temperature reached around 10 000 000 °C, nuclear reactions began. Initially the Sun was only 70% as luminous as it is now, but eventually it settled on to the Main Sequence, and began a long period of comparatively steady existence.

The supply of available hydrogen 'fuel' is limited, and as it ages the Sun is bound to change. Over the next thousand million years there will be a slow but inexorable increase in luminosity, and the Earth will become intolerably hot from our point of view. Worse is to come. Four thousand million years from now the Sun's luminosity will have increased threefold, so that the surface temperature of the Earth will soar to 100 °C and the oceans will be evaporated. Another thousand million years, and the Sun will leave the Main Sequence to become a giant star, with different nuclear reactions in the core. There will be a period of instability, with swelling and shrinking (the 'asymptotic giant' stage) but eventually the Sun's diameter will grow to 50 times its present size; the surface temperature will drop, but the overall luminosity will increase by a factor of at least 300, with disastrous results for the inner planets. The temperature at the solar core will reach 100 000 000 °C and helium will react to produce carbon and oxygen. A violent solar wind will lead to the loss of the outer layers, so that for a relatively brief period on the cosmical scale the Sun will become a planetary nebula. Finally, all that is left will be a very small, dense core made up of degenerate matter; the Sun will have become a white dwarf, with all nuclear reactions at an end. After an immensely long period – perhaps several tens of thousands of millions of years – all light and heat will depart, and the end product will be a cold, dead black dwarf, perhaps still circled by the ghosts of the remaining planets.

It does not sound an inviting prospect, but at least it need not alarm us. The Sun is no more than half-way though its career on the Main Sequence; it is no more than middle-aged.

3 THE MOON

The Moon is officially ranked as the Earth's satellite. Relative to its primary, it is however extremely large and massive, and it might well be more appropriate to regard the Earth–Moon system as a double planet. Data are given in Table 3.1.

Table 3.1. Data.

Distance from the Earth, centre to centre (km):
 Mean   384 400
 Max   406 697 (apogee)
 Min   356 410 (perigee)
Distance from the Earth, surface to surface (km):
 Mean   376 284
 Max   398 581 (apogee)
 Min   348 294 (perigee)
Revolution period: 27.321 661 days
Synodic period: 29.53 days (29d 12h 44m 2s.9)
Mean orbital velocity: 1.023 km s^{-1} (3682 km h^{-1})
Mean sidereal daily motion: $47434''.8899 = 13°.17636$
Mean transit interval: 24h 50m.47
Orbital eccentricity: 0.0549
Mean orbital inclination: 5°9'
Axial rotation period: 27.321661 days (synchronous)
Inclination of lunar equator: to ecliptic $1°32'30''$, to orbit 6°41'
Rate of recession from Earth: 3.8 cm/year
Diameter: equatorial   3476 km
          polar   3470 km
Oblateness: 0.002
Apparent diameter from Earth:
 Max   33'31''
 Mean   31'5''
 Min   29'22''
Reciprocal mass, Earth = 1: 81.301 ($=7.350 \times 10^{25}$ g)
Density, water = 1: 3.342
Escape velocity: 2.38 km s^{-1}
Volume, Earth = 1: 0.0203
Surface gravity, Earth = 1: 0.1653
Mean albedo: 0.067
Atmospheric density: 10^{-14} that of the Earth's atmosphere at sea level
Surface temperature range (°C): −184 to +101
Optical libration, selenocentric displacement: longitude ±7°.6
                                               latitude ±6°.7
Nutation period, retrograde: period 18.61 tropical years
Mean albedo: 0.067

Table 3.2. Legendary names of full moons.

January	Winter Moon, Wolf Moon
February	Snow Moon, Hunger Moon
March	Lantern Moon, Crow Moon
April	Egg Moon, Planter's Moon
May	Flower Moon, Milk Moon
June	Rose Moon, Strawberry Moon
July	Thunder Moon, Hay Moon
August	Grain Moon, Green Corn Moon
September	Harvest Moon, Fruit Moon
October	Hunter's Moon, Falling Leaves Moon
November	Frosty Moon, Freezing Moon
December	Christmas Moon, Long Night Moon

The *synodic period* (i.e. the interval between successive new moons or successive full moons) is 29d 12h 44m, so that generally there is one full moon every month. However, it sometimes happens that there are two full moons in a calendar month and one month (February) may have none. Thus in 1999 there were two full moons in January (on the 2nd and the 31st), none in February and two again in March (on the 2nd and the 31st, as with January). By tradition a second full moon in a month is known as a *blue moon*, but this has nothing whatsoever to do with a change in colour. (This is not an old tradition. It comes from the misinterpretation of comments made in an American periodical, the *Maine Farmers' Almanac*, in 1937.) Yet the Moon can occasionally look blue, due to conditions in the Earth's atmosphere. For example, this happened on 26 September 1950, because of dust in the upper air raised by vast forest fires in Canada. A blue moon was seen on 27 August 1883 caused by material sent up by the volcanic outburst at Krakatoa, and green moons were seen in Sweden in 1884 – at Kalmar, on 14 February, for 3 min, and at Stockholm on 12 January, also for 3 min.

Other full moons have nicknames (Table 3.2), but of these only two are in common use. In the northern hemisphere, the full moon closest to the autumnal equinox, which falls around 22 September, is called Harvest Moon because the ecliptic then makes its shallowest angle with the horizon, and the retardation – that is to say, the time lapse between moonrise on successive nights – is at its minimum; it may be no more than 15 min, although for most of the year it amounts to at least 30 min. It was held that this was useful to farmers gathering in their crops. Harvest Moon looks the same as any other full moon – and it is worth noting that the full moon looks no larger when low down than when high in the sky. Certainly it does give this impression, but the 'Moon Illusion' *is* an illusion and nothing more.

In Islam, the calendar follows a purely lunar cycle, so that over a period of about 33 years the months slowly regress through the seasons. Each month begins with the first sighting of the crescent Moon, and this is important in Islamic religion. An early sighting was made on 15 March 1972 by R. Moran of California, who used 10×50 binoculars and glimpsed the Moon 14h 53m past conjunction; on 21 January 1996 P. Schwann, from Arizona, used 25×60 binoculars to glimpse the Moon only 12h 30m after conjunction.

(As an aside: in 1992 a British political party, the Newcastle Green Party, announced that it would meet at new moon to discuss ideas and at full moon to act upon them. They have not, so far, won any seats in Parliament!)

There is no conclusive evidence of any link between the lunar phases and weather on Earth, or of any effect upon living things – apart from aquatic creatures, since the Moon is the main agent in controlling the ocean tides.

During the crescent stage the 'night' part of the Moon can usually be seen shining faintly. This is known as the *Earthshine* and is due solely to light reflected on to the Moon by the Earth – as was first realized by Leonardo da Vinci (1452–1519).

Moon legends and Moon worship

Every country has its own Moon legends – and who has not heard of the Man in the Moon? According to a German tale, the Old Man was a villager caught stealing cabbages, and was placed in the Moon as a warning to others; he was also a thief in Polynesian lore. Frogs and toads have also found their way there, and stories about the hare in the Moon are widespread. From China comes a delightful story. A herd of elephants made a habit of drinking at the Moon Lake, and trampled down many of the local hare population. The chief hare then had an excellent idea; he told the elephants that by disturbing the waters they were angering the Moon-Goddess, by destroying her reflection. The elephants agreed that this was most unwise, and made a hasty departure.

To the people of Van, in Turkey, the Moon was a young bachelor who was engaged to the Sun. Originally the Moon had shone in the daytime and the Sun at night, but the Sun, being feminine, was afraid of the dark – and so they changed places. In many mythologies the Sun is female and the Moon male, although this is not always the case. For example, in Greenland it is said that the Sun and Moon were brother and sister, Anninga and Malina. When Malina smeared her brother's face with soot, she fled to avoid his anger; reaching the sky, she became the Sun. Anninga followed and became the Moon, but he cannot fly equally high, and so he flies round the Sun hoping to surprise her. When he becomes tired at the time of lunar First Quarter, he leaves his house on a sled towed by four dogs, and hunts seals until he is ready to resume the chase.

There were many Moon gods, such as Artemis (Greece), Diana (Rome), Isis (Egypt) and Tsuki-yomi-no-kami (Japan). Moon worship continued until a surprisingly late stage, at least in Britain; from the Confessional of Ecgbert, Archbishop of York, in the 8th century it seems that homage was still being paid to the Moon as well as to the Sun.

ROTATION OF THE MOON

The Moon's rotation is synchronous (captured); i.e. the axial rotation period is the same as the orbital period. This means that the same area of the Moon is turned Earthward all the time, although the eccentricity of the lunar orbit leads to libration zones which are brought alternately in and out of view. From Earth, 59% of the Moon's surface can be studied at one time or another; only 41% is permanently out of view. There is no mystery about this behaviour; tidal forces over the ages have been responsible. Most other planetary satellites also have synchronous rotation with respect to their primaries.

The barycentre, or centre of gravity of the Earth–Moon system, lies 1707 km beneath the Earth's surface, so that the statement that 'the Moon moves round the Earth' is not really misleading.

The fact that the Moon has synchronous rotation was noted by Cassini in 1693; Galileo may also have realized it. The libration zones are so foreshortened that from Earth they are difficult to map, and good maps were not possible until the advent of space-craft. The first images of the averted 41% were obtained in 1959 by the Russian vehicle Luna (or Lunik) 3.

Because of tidal effects, the Moon is receding from the Earth at a rate of 3.83 cm/year; also, the Earth's rotation period is lengthening, on average, by 0.000 000 2 s/day, although motions of material inside the Earth mean that there are slight irregularities superimposed on the tidal increase in period.

ORIGIN OF THE MOON

Many theories have been advanced to explain the origin of the Moon. The attractive theory due to G. H. Darwin in 1881 – that the proto-Earth rotated so rapidly that it threw off a large piece of material, which became the Moon – is mathematically untenable. H. C. Urey proposed that the Moon accreted from the solar nebula in the same way as the Earth and became gravitationally linked later, but this would require a set of very special circumstances, and does not account for the Moon's lower density compared with the Earth. Then, in 1974, a completely different idea was put forward in America by W. Hartmann and D. R. Davis. This involves a collision between the Earth and a large impacting body, comparable in size with Mars, about 4000 million years ago. According to this theory, the cores of the Earth and the impactor merged, and mantle débris ejected during the collision accreted to form the Moon. This picture may not be accurate, but at least it seems more plausible than any of the other theories. (Urey once made the caustic comment that because all theories of the lunar origin seemed unlikely, science had proved that the Moon does not exist!)

MINOR SATELLITES

No minor Earth satellites of natural origin seem to exist. Careful searches have been made for them, notably in 1957 by Clyde Tombaugh (discoverer of the planet Pluto), but without result. A small satellite reported in 1846 by F. Pettit, Director of the Toulouse Observatory in France, was undoubtedly an error in observation – although

Jules Verne found it very useful in his great novel *From the Earth to the Moon* and its sequel *Round the Moon* (1865 and 1871). Clouds of loose material on the Moon's orbit, at the Lagrangian points, were reported by the Polish astronomer K. Kordylewski in 1961, but remain unconfirmed, and efforts made on various occasions to photograph them have been unsuccessful.

MAPPING THE MOON

It is possible that the first map of the Moon dates back 5000 years! A rudimentary etching found on a tomb at Knowle in County Meath (Ireland) does give an impression of a map of the lunar surface. Dr. Philip Brooke, who has made a careful study of it, estimates that it was made around 3000 BC.

The first suggestion that the Moon is mountainous was made by the Greek philosopher Democritus (460–370 BC). Earlier, Xenophanes (c 450 BC) had supposed that there were many suns and moons according to the regions, divisions and zones of the Earth! Certainly the main maria and some other features can be seen with the naked eye, and the first map which has come down to us was that of W. Gilbert, drawn in 1600, although it was not published until 1651 (Gilbert died in 1603).

Telescopes became available in the first decade of the 17th century. The first known telescopic map was produced in July 1609 by Thomas Harriot, one-time tutor to Sir Walter Raleigh. It shows a number of identifiable features, and was more accurate than Galileo's map of 1610. Another very early telescopic observer of the Moon was Sir William Lower, an eccentric Welsh baronet. His drawings, made in or about 1611, have not survived, but he compared the appearance of the Moon with a tart that his cook had made – 'Here some bright stuffe, there some darke, and so confusedlie all over'.

Galileo did at least try to measure the heights of some of the lunar mountains, from 1611, by the lengths of their shadows. He concentrated on the lunar Apennines, and although he over-estimated their altitudes his results were of the right order. Much better results were obtained by J. H. Schröter, from 1778.

The first systems of nomenclature were introduced in 1645 by van Langren (Langrenus) and in 1647 by Hevelius, but few of these names have survived; for example, the crater we now call Plato was named by Hevelius 'the Greater Black Lake'. At that time, of course, it was widely although not universally believed that the bright areas were lands, and the dark areas were watery. The modern-type system was introduced in 1651 by the Jesuit astronomer G. Riccioli, who named the features in honour of scientists – plus a few others. He was not impartial; for instance, he allotted a major formation to himself and another to his pupil Grimaldi, and he did not believe in the Copernican theory, that the Earth moves round the Sun – so he 'flung Copernicus into the Ocean of Storms'. Riccioli's principle has been followed since, although clearly all the major craters were used up quickly and later distinguished scientists had to be given formations of lesser importance, at least until it became possible to map the Moon's far side by using space research methods.

Other maps followed, some of which are listed in Table 3.3. Tobias Mayer in 1775 was the first to introduce a system of lunar coordinates, although the first accurate measurements with a heliometer were not made until 1839, by the German astronomer F. W. Bessel.

Undoubtedly the first really great lunar observer was J. H. Schröter, whose astronomical career extended from 1778, when he set up his private observatory at his home in Lilienthal, near Bremen in Germany, until 1813, when his observatory was destroyed by invading French troops (the soldiers even plundered his telescopes, which were brass-tubed and were taken to be made of gold). Schröter made many drawings of lunar features and was also the first to give a detailed description of the rills[1], although some of these had been seen earlier by the Dutch observer Christiaan Huygens.

In 1837–8 came the first really good map of the Moon, drawn by W. Beer and J. H. Mädler from Berlin. Although they used a small telescope (Beer's 3.75 inch or 9.50 cm refractor) their map was a masterpiece of careful, accurate work, and it remained the standard for several decades. They also published a book, *Der Mond*, which was a detailed description of the whole of the visible surface. A larger map completed in 1878 by Julius Schmidt was based

[1] Often spelled rilles; I have kept to the original spelling. They can also be known as clefts.

Table 3.3. Selected list of pre-Apollo lunar maps.

Date	Diameter (cm)	Author
1610	7	Galileo
1634	21	Mellan
1645	40	van Langren
1647	29	Hevelius
1651	11.3	Riccioli
1662	38	Montanari
1680	53	Cassini
1775	21	Mayer
1797	30 (globe)	Russell
1824	95	Lohrmann (unfinished)
1837	95	Beer and Mädler
1859	30	Webb
1873	30	Proctor
1876	61	Neison
1878	187	Schmidt
1895	46	Elger
1898	43	König
1910	196	Goodacre
1927	46	Lamèch
1930	508	Wilkins
1934	156	Lamèch
1935	100	I.A.U.
1936	86	Fauth
1946	762	Wilkins[a]

[a] Revised and re-issued to one-third scale in 1959.

on that of Beer and Mädler; so too was the 1876 map and book written by E. Neison (real name, Nevill). Other useful atlases were those of Elger (1895) and Goodacre (1910, revised 1930); in 1930 the Welsh observer H. Percy Wilkins published a vast map, 300 inches (over 500 cm) in diameter; it was re-issued, to one-third the scale, in 1946.

The first good photographic atlas was published in 1899 by the Paris astronomers Loewy and Puiseux, but the first actual photographs date back much further; a Daguerreo-type was taken on 23 March 1840 by J. W. Draper, using a 12.0 cm reflector, but the image was less than 3 cm across and required an exposure time of 20 min. Nowadays, of course, there are photographic atlases of the entire surface, obtained by space-craft, and it is fair to say that the Moon is better charted than some regions of the Earth. However, special mention should be made of an Earth-based photographic atlas produced by H. R. Hatfield, using his 32 cm reflector. It was re-issued in 1999, and is ideal for use by the amateur observer, as it shows all areas of the Moon under different conditions of illumination.

There were, of course, some oddities. No less a person than the great Sir William Herschel, who died in 1822, never wavered in his belief that the Moon must be inhabited, and in 1822 the German astronomer F. von Paula Gruithuisen described a structure with 'dark gigantic ramparts', which he was convinced was a true city built by the local populace – although in fact the area shows nothing but low, haphazard ridges. There was also the famous Lunar Hoax of 1835, when a daily paper, the New York *Sun*, published some quite fictitious reports of discoveries made by Sir John Herschel from the Cape of Good Hope. The reports were written by a reporter R. A. Locke, and included descriptions of bat-men and quartz mountains. The first article appeared on 25 August, and was widely regarded as authentic; only on 16 September did the *Sun* confess to a hoax. One religious group in New York City even started to make plans to send missionaries to the Moon in an attempt to convert the bat-men to Christianity.

This sounds very strange, but as late as the 1930s one eminent astronomer, W. H. Pickering, was maintaining that certain dark patches on the Moon might be due to the swarms of insects or even small animals. Only since the start of the Space Age have we been sure that the Moon is, and always has been, totally sterile.

SURFACE FEATURES

The most prominent surface features are of course the maria (seas). Although they have never contained water (as one eminent authority, H. C. Urey, believed as recently as 1966), they are undoubtedly old lava plains, and there are many 'ghost' craters whose walls have been virtually levelled by the lava. Of similar type are the 'lakes', 'marshes' and 'bays' (lacus, palus, sinus). Some of the maria, such as Imbrium and Crisium, are more or less regular in outline; others, such as Frigoris, are very irregular. Details of the Mare-type regions are given in Table 3.12 (page 47).

The largest of the 'regular' seas is the Mare Imbrium, with a diameter of over 1000 km; it is bounded in part by the mountain ranges of the Apennines, Alps and Carpathians. Its area is about the same as that of Pakistan, but the irregular

Oceanus Procellarum is considerably larger, and in area it is in fact greater than our Mediterranean. Most of the main seas make up a connected system; the exception is the distinct Mare Crisium. It is worth noting that although foreshortening makes it seem elongated in a north–south direction, the east–west diameter is actually greater (590 km, as against 490 km). Its area is about the same as that of the American state of Kansas. In general, the regular maria are the more depressed; thus the Mare Crisium lies about 4 km below the mean sphere, whereas the depth of the Oceanus Procellarum is on average only about 1 km.

There are no comparable seas on the far side of the Moon; the Mare Moscoviense and Mare Ingenii are smaller than some of the formations which are classed as craters. However, it is true that one major sea, the Mare Orientale, does extend on to the far side. Only its extreme eastern boundary is accessible from Earth. The main central area has a diameter of over 300 km; the outer rings extend for much further – out to more than 900 km[2].

The smaller mare-type features extend from the main seas. The Sinus Iridum (Bay of Rainbows) is particularly beautiful. It leads off the Mare Imbrium, and when the Sun is rising over it the western mountain border is first to catch the solar rays, so that for a brief period we see the appearance known popularly as the 'jewelled handle'. The 'seaward' wall has been levelled; only vague, discontinuous traces of it remain.

The largest and deepest basin on the surface is the South Pole-Aitken Basin, which is 2500 km across and lies around 12 km below the mean sphere. It was surveyed by the Clementine space-craft in 1994; it covers almost a quarter of the Moon's circumference. Smaller multi-ring basins (Apollo, Orientale and Korolev) lie in the same general area.

Craters are listed in Table 3.13 (page 48) and Table 3.14 (page 66). They are of many types; very often 'walled plains' would be a better term. In profile, a crater

---

[2] I discovered this formation in 1939, with the modest telescope in my observatory in Sussex; libration was at maximum, and I assumed that I was seeing the boundary of a minor limb-sea of the Humboldtianum type. I suggested its name – Mare Orientale, the Eastern Sea – but later the International Astronomical Union decided to reverse lunar east and west, so that the Eastern Sea is now on what is termed the Moon's *western* limb!

is more like a shallow saucer than a mine-shaft; the walls rise to only a modest height above the outer surface, while the floors are depressed. Central mountains or mountain groups are very common, but never attain the height of the outer ramparts. The depths of some craters are given in Table 3.4, but it must be remembered that these are at best no more than approximate. Some craters near the poles are so deep that their floors are always in shadow and therefore remain bitterly cold; one of these craters, Newton, has an depth of over 8 km below the crest of the wall.

Some craters, such as Grimaldi and Plato, have dark floors which make them identifiable under any conditions of illumination. There are also some very bright craters; the most brilliant of these is Aristarchus, which often appears prominent even when lit only by Earthshine. It has terraced walls and a prominent central peak. Some craters are the centres of systems of bright rays which stretch for long distances over the surface; the most prominent of these ray systems are associated with Copernicus, in the Mare Nubium, and Tycho in the southern uplands. Other important ray centres include Kepler, Olbers, Anaxagoras and Thales. The rays are not visible under low illumination, but near the time of full moon they dominate the entire scene.

One remarkable crater, Wargentin, is lava-filled, so that it is a large plateau. There are other plateaux here and there, but none even remotely comparable with Wargentin.

The most conspicuous rills (Table 3.5) on the Moon are those on the Mare Vaporum area (Hyginus, Ariadæus) and the Hadley Rill in the Apennine area, visited by the Apollo 15 astronauts. The Hadley Rill is 135 km long, 1–2 km wide and 370 m deep. Other famous rills are associated with Sirsalis, Bürg, Hesiodus, Triesnecker, Ramsden and Hippalus. Extending from Herodotus, near Aristarchus, is the imposing valley known as Schröter's Valley in honour of its discoverer; in a way this is misleading, since the crater named after Schröter is in a completely different area. It is worth noting that some rills are in part crater chains; the Hyginus Rill is an example of this, as it consists of a chain of small confluent craters. A much larger crater valley is to be found near Rheita, in the south-east quadrant of the Moon.

Domes, up to 80 km in diameter, are found in various parts of the Moon – for instance near Arago in the Mare Tranquillitatis, in the Aristarchus area, and on the floor of Capuanus. Many domes have symmetrical summit pits; their slopes are gentle.

Mountain ranges are merely the ramparts of the large maria; the Apennines, bordering the Mare Imbrium, are

Table 3.4. Crater depths. Depth values for lunar craters may carry large standard errors, and the figures given here are no more than approximate. The following are some typical mean values of the ramparts above the floor. Formations whose walls are particularly irregular in height are marked *. Depths are in metres.

4850	Tycho	2530	Pytheas
4730	Maurolycus	2510	Halley
4400	Theophilus	2450	Almanon
5220	Werner	2400	Ptolemæus*
4130	Walter	2400	Proclus
3900	Alpegragius	2320	Plinius
3850	Theon Junior	2300	Posidonius
3830	Alfraganus	2200	Hell
3770	Herschel	2150	Archimedes
3770	Copernicus	2100	Le Verrier
3750	Stiborius	2080	Campanus
3730	Abenezra	2000	Brayley
3650	Aristillus	1960	Fauth
3620	Arzachel	1860	Aratus
3570	Eratosthenes	1850	C Herschel
3510	Bullialdus	1850	Ammonius
3470	Theon Senior	1850	Gassendi
3430	Autolycus	1810	Feuillée
3320	Hipparchus*	1770	Boscovich*
3270	Thebit	1760	Mercator
3200	Godin	1740	Bessel
3140	Catharina	1730	Regiomontanus*
3130	Cayley	1730	Vitello
3120	Kant	1700	Manners
3110	Timocharis	1650	Beer
3110	Abulfeda	1550	Vitruvius
3110	Lansberg	1490	D'Arrest
3050	Manilius	1240	Cassini*
3010	Menelaus	1230	Tempel
3000	Aristarchus	1180	Agatharchides*
2980	Purbach	1040	Birt
2970	Diophantus	850	Kunowsky
2890	Lambert	750	Encke
2830	Theætetus	750	Hyginus
2800	Ukert	650	Stadius
2760	Stöfler*	600	Linné
2760	Mösting	380	Kies
2760	Triesnecker	310	Spörer
2720	Thebit A	0	Wargentin
2570	Kepler		

Table 3.5. Rills and valleys. (a) Valleys (valles), (b) rills (rimæ) and (c) rill systems (rimæ) (* = within crater).

(a)

Name	Lat.	Long.	Length (km)	
Alpine Valley	49N	3E	166	Very prominent; cuts through Alps; there is a delicate central rill.
Capella	7S	35E	49	Really a crater chain; cuts through Capella.
Reichenbach	31S	48E	300	SE of Reichenbach, narrowing to the S. Really a crater chain, much less prominent than that of Rheita.
Rheita	43S	51E	445	Major crater chain, starting in Mare Nectaris and abutting on Rheita.
Schröter's Valley	20N	51W	168	Great winding valley, extending from Herodotus. It starts at a 6 km crater N and widens to 10 km to form what is nicknamed the Conra-Head. The maximum depth is about 1000 m. The crater named after Schröter is nowhere near; it lies in quite another part of the Moon. (Mare Nubium area).
Snellius	31S	56E	592	Very long valley, directed toward the centre of the Nectaris basin.

(b)

Name	Lat.	Long.	Length (km)
Agatharcides	20S	28W	50
Archytas	53N	3E	90
Ariadæus	6N	14E	250
Birt	21S	9W	50
Brayley	21N	37W	311
Cauchy	10N	38E	140
Conon	19N	2E	30
Gay-Lussac	13N	22W	40
Hadley	25N	3E	80
Hesodius	30S	20W	256
Hyginus	7N	8E	219
Marius	17N	49W	121
Sheepshanks	58N	24E	200

(c)

Name	Lat.	Long.	Length (km) (total)
Alphonsus*	14S	2W	80
Archimedes	27N	4W	169
Arzachel*	18S	2W	50
Atlas*	47N	46E	60
Boscovich*	10N	11E	40
Bürg	44N	24E	147
Doppelmayer	26S	45W	162
Gassendi*	18S	40W	70
Gutenberg	5S	38E	330
Hevelius	1N	68W	182
Hippalus	25S	29W	191
Hypatia	0S	22E	206
Janssen*	46S	40E	114
Littrow	22N	30E	115
Menelaus	17N	18E	131
Mersenius	21S	49W	84
Petavius*	26S	59E	80
Pitatus	18N	24E	94
Posidonius*	32N	29E	70
Prinz	27N	43W	80
Ramsden	34S	31W	108
Repsold	51N	82W	166
Riccioli	2N	74W	400
Ritter	3N	18E	100
Sirsalis	16S	62W	426
Taruntius*	6N	46E	25
Triesnecker	4N	5E	215
Zupus	15S	53W	120

Table 3.6. Mountain ranges (Montes).

Name	Mid-Lat.	Mid-Long.	Length (km)	
Alps	46N	1W	281	Borders Mare Imbrium. Contains Mont Blanc, Alpine Valley.
Apennines	19N	4W	401	Borders Mare Imbrium. Contains Hadley, Huygens, Ampere, Serao, Wolf.
Carpathians	14N	24W	361	Borders Mare Imbrium.
Caucasus	38N	10E	445	Borders Mare Serenitatis. Fairly high peaks.
Cordilleras	17S	82W	574	Forms outer wall of Orientale.
Hæmus	20N	9E	560	Part of the border of Mare Serenitatis, separating it from Mare Vaporum.
Harbingers	17N	41W	90	Clumps of peaks E of Aristarchus.
Juras	47N	34W	422	Borders Sinus Iridum ('Jewelled Handle' effect).
Pyrenees	16S	41E	164	Not a true range, but a collection of moderate hills, roughly between Gutenberg and Bohnenberger.
Riphæans	8S	28W	189	Low range on the Mare Nubium, close to the bright crater Euclides. The northern section is sometimes called the Ural Mountains.
Rook	21S	82W	791	One of the inner circular mountain chains surrounding the Orientale basin.
Straight Range	48N	20W	90	Remarkable line of peaks in the Mare Imbrium, W of Plato (Montes Recti).
Spitzbergen	35N	5W	60	Bright little hills N of Archimedes, lying on the edge of a ghost ring. So named because in shape they resemble the terrestrial island group.
Taurus	28N	41E	172	Not a true range; mountainous region near Rømer.
Teneriffes	47N	12W	182	Mountainous region between Plato and the Straight Range.

Scarps (Rupes).

Name	Lat.	Long.	Length (km)	
Altai	24S	23E	427	Often called the Altai Mountains; really a scarp, on the edge of the Nectaris basin.
Rupes Cauchy	9N	37E	120	A fault, changing into a rill; in some ways not too unlike the Straight Wall.
Straight Wall	22S	8W	134	In Mare Nubium, W of Thebit; very prominent, appearing dark before full moon because of the shadow and bright after full. The angle of slope is no more than 40°.

Mons (Mountain)

Name	Lat.	Long.	Length (km)	
Ampère	19N	4W	30	Mountain massif in the Apennines.
Bradley	22N	1E	30	Mountain massif in the Apennines, close to Conon.
Hadley	26.5N	5E	25	Mountain massif in the northern Apennines.
Huygens	20N	3W	40	5400 m mountain massif in the central part of the Apennines.
La Hire	28N	25W	25	Isolated mountain in Mare Imbrium, NW of Lambert.
Mont Blanc	45N	1E	25	3600 m mountain in the Alps, SW of Cassini.
Pico	46N	9W	25	Triple-peaked mountain on the Mare Imbrium, S of Plato, over 2400 m high. It is bright and prominent. The area between it and Plato is occupied by a ghost ring, once named Newton although this name has since been transferred to a deep crater in the far south of the Moon.
Piton	41N	1W	25	Prominent mountain in Mare Imbrium, between Cassini and Piazzi Smyth.

Capes (Promontoria)

Name	Lat.	Long.	Length (km)	
Agarum	14N	66E	70	On the E border of Mare Crisium.
Agassiz	42N	2E	20	Edge of the Alps, NW of Cassini.
Archerusia	17N	22E	10	Edge of Mare Serenitatis, between Plinius and Tacquet.
Deville	43N	1E	20	Edge of the Alps, between Cape Agassiz and Mont Blanc.
Fresnel	29N	5E	20	Northern cape of the Apennines.
Heraclides	40N	33W	50	Western cape of Sinus Iridium.
Kelvin	27S	33W	50	In Mare Humorum, SW of Hippalus.
Laplace	46N	26W	50	Eastern cape of Sinus Iridium.
Tænarium	19S	8W	70	Edge of Mare Nubium, N of the Straight Wall. Sometimes spelled Ænarium.

Table 3.7. Lunar systems.

System	Age (thousand million years)	Events
pre-Nectarian	>3.92	Basins and craters formed before the Nectaris basin (multi-ring basins), e.g. Grimaldi.
Nectarian	3.92–3.85	Post-Nectaris, pre-Imbrian; includes some multi-ring basins, e.g. Clavius.
Imbrian	3.85–3.1	Extends from the formation of the Imbrian basin to the youngest mare lavas (Orientale, Schrödinger most basaltic maria, craters such as Archimedes and Plato).
Eratosthenian	3.1–1.0	Youngest craters and mare lavas (e.g. Eratosthenes).
Copernican	1.0–present	Begins with formation of Copernicus. Youngest craters (e.g. Tycho); ray systems.

particulary impressive. Isolated peaks and clumps of peaks are to be found all over the surface (Table 3.6).

In 1945 the American geologist and selenographer J. E. Spurr drew attention to the 'lunar grid' system, made up of families of linear features aligned in definite directions. It is also obvious that the distribution of the craters is not random; they form groups, chains and pairs, and when one crater intrudes into another it is almost always the smaller feature which breaks into the larger.

ORIGIN OF THE LUNAR FORMATIONS

This is a problem which has caused considerable controversy – and to a certain extent still does. Eccentric theories have not been lacking; for example P. Fauth, who died in 1943, supported the idea that the Moon is covered with ice. Weisberger, who died in 1952, denied the existence of any mountains or craters, and attributed the effects to storms and cyclones in a dense lunar atmosphere. The Spanish engineer Sixto Ocampo, in 1951, claimed that the craters were the result of an atomic war between two races of Moon men (the fact that some craters have central peaks while others have not proves, of course, that the two sides used different types of bombs; the last detonations on the Moon fired the lunar seas, which fell back to Earth and caused the Biblical Flood). However, in modern times the only serious question has been as to whether the craters were produced by internal action – that is to say, vulcanism – or whether they were due to impact. By now almost all authorities support the impact theory, which was proposed by Franz von Paula Gruithuisen in 1824, revived by G. K. Gilbert in 1892 and put into its present-day form by Ralph Baldwin in 1949.

According to this scenario, the sequence of events may have been more of less as follows (Table 3.7). The Moon was formed at about the same time as the Earth (4600 million years ago). The heat generated during the formation made the outer layers molten down to a depth of several hundred kilometres; less dense materials then separated out to the surface and in the course of time produced a crust. Then, between 4400 million and about 4000 million years ago, came the Great Bombardment, when meteorites rained down to produce the oldest basins such as the Mare Tranquillitatis and the Mare Fœcunditatis. The Imbrium basin dates back perhaps 3850 million years, and as the Great Bombardment ceased there was widespread vulcanism, with magma pouring out from below the crust and flooding the basins to produce structures such as the Mare Orientale and ringed formations of the Schrödinger type. Craters with dark floors, such as Plato, were also flooded at this time. The lava flows ended rather suddenly, by cosmical standards, and for the last few thousand million years the Moon has seen little activity, apart from the formation of occasional impact craters such as Copernicus and Tycho. It has been claimed that Copernicus is no more than at thousand million years old, and Tycho even younger. The ray systems are certainly latecomers, since the rays cross all other formations.

The Moon has experienced synchronous rotation since early times, and there are marked differences between the Earth-turned and the averted hemispheres. The crust is thicker on the far side and some of the basins are unflooded, which is why they are not classed as maria (palimpsests). The prominent feature Tsiolkovskii seems to be between a mare-type structure and a crater; it has a flooded, mare-type floor, but high walls and a central peak. It adjoins a formation of similar size, Fermi, which is unflooded.

One thing is certain; the Moon is today essentially inert. On 4 May 1783, and again on 19 and 20 April 1787, Sir William Herschel reported seeing active volcanoes, but there is no doubt that he observed nothing more significant than bright areas (such as Aristarchus) shining by earthlight. In modern times, transient lunar phenomena (TLP) have been reported on many occasions; they take the form of localized obscurations and glows. On 3 November 1958, N. A. Kozyrev, at the Crimean Astrophysical Observatory, obtained a spectrum of an event inside the crater Alphonsus, and on 30 October 1963 a red event in the Aristarchus area was recorded from the Lowell Observatory by J. Greenacre and J. Barr. In 1967 NASA published a comprehensive catalogue of the many TLP reports, compiled by Barbara Middlehurst and Patrick Moore; Moore issued a subsequent supplement. Over 700 TLP events were listed and although no doubt many of these are due to observational error it seems that others are genuine, presumably due to gaseous emissions from below the crust. They occur mainly round the peripheries of the regular maria and in areas rich in rills, and are commonest near lunar perigee, when the Moon's crust is under maximum strain.

Until recently the reality of TLP was questioned, perhaps because so many of the reports (though by no means all) came from amateur observers. However, full professional confirmation has now been obtained. Using the 83-cm telescope at the Observatory of Meudon, Audouin Dollfus has detected activity in the large crater Langrenus. He wrote: 'Illuminations have been photographed on the surface of the Moon. They appeared unexpectedly on the floor of Langrenus. Their shape and brightness was considerably modified in the following days, and they were simultaneously recorded in polarized light. They are apparently due to dust grain levitation above the lunar surface, under the effect of degassing from the interior ... The Langrenus observations indicate that the Moon is not a completely dead body. Degassing occasionally occurs in areas particularly fractured or fissured. Clouds of dust are lifted off the ground by the gas pressure.'

The brilliant crater Aristarchus has long been known to be particularly subject to TLP phenomena, and images from the Clementine space-craft in the 1990s have shown that there have been recent colour changes in the area; patches of ground have darkened and reddened. Winifred Cameron of the Lowell Observatory in Arizona, who has made a long study of TLP, considers that these changes are due to gaseous outbreaks stirring up the ground material, and it is indeed difficult to think of any other explanation. By terrestrial standards the lunar outbreaks are of course very mild indeed, but there is no longer any serious doubt that they do occur.

The most recent claim concerning the formation of a large impact crater related to a report dating from 18 July 1178, by Gervase of Canterbury. The crescent Moon was seen 'to split in two ... a flaming torch sprung up, spewing out over a considerable distance fire, hot coals and sparks. Meanwhile, the body of the Moon which was below, writhed ... and throbbed like a wounded snake'. This indicates a terrestrial cloud phenomenon, if anything; nevertheless, it has been seriously suggested that the phenomenon was the result of an impact on the far side which led to the formation of the ray-crater Giordano Bruno. In fact this is absurd, and in any case the altitude of the Moon at the time of the observation, as seen from Canterbury, was less than 5°. Therefore there is no doubt that the claim must be dismissed as merely a 'Canterbury tale'.

Major structural changes do not now occur. There have been two cases which have caused widespread discussion, but neither stands up to close examination. In the Mare Fœcunditatis there are two small craters, Messier and Messier A, which were said by Beer and Mädler, in 1837, to be exactly alike, with a curious comet-like ray extending from them to the west; in fact A is the larger of the two and is differently-shaped, but changes in solar illumination mean that they can often appear identical. In 1866 J. Schmidt, from Athens Observatory, reported that a small, deep crater on the Mare Serenitatis, Linné, had been transformed into a white patch; many contemporary astronomers, including Sir John Herschel, believed that a moonquake had caused the crater walls to collapse. Today Linné is a small impact crater standing on a white nimbus, and there seems no possibility of any real change having occurred – particularly as Mädler observed it in the 1830s and again after 1866, and reported that it looked exactly the same as it had always done.

Presumably there is a certain amount of exfoliation, because the temperature range is very great. Surface temperatures were first measured with reasonable accuracy by the fourth Earl of Rosse, from Birr Castle in Ireland. His papers, from 1869, indicated that near noon the temperature rose to about 100 °C, although Langley later erroneously concluded that the temperature never rose above 0°. It is now known that the noon equatorial temperature is about 101 °C, falling to −184 °C at night; the poles remain fairly constant at −96 °C.

It is widely believed that some meteorites found on Earth have come from the Moon. Over a dozen 'lunar meteorites' have been listed; two were found in Libya and the rest in Antarctica. For example, there are the two stones, MAC 88104 and 88105, found in the MacAlpine Hills region of Antarctica in 1990. The combined mass is 724 g; the stones are breccias (fused collections of rock fragments). It has been estimated that they were blasted away from the Moon by a huge impact and took about 100 000 years to reach the Earth, and have been awaiting discovery for at least 30 000 years. Most of the 'lunar meteorites' are breccias, though a few are different, such as Yamato 793169 (6.1 g) which appears to be of the same type as mare basalts. The evidence in favour of a lunar origin for these meteorites is not conclusive, but it is certainly strong.

MISSIONS TO THE MOON

The idea of reaching the Moon is very old; as long ago as the second century AD a Greek satirist, Lucian of Samosata, wrote a story about a lunar voyage (his travellers were propelled on to the Moon by the force of a powerful waterspout!) The first serious idea was due to Jules Verne, in his novel *From the Earth to the Moon* (1865); he planned to use a space-gun, but neglected the effects of friction against the atmosphere, quite apart from the shock of starting at a speed great enough to break free from Earth (escape velocity: 11.2 km s^{-1}). Before the end of the ninth century the Russian theoretical rocket pioneer, K. E. Tsiolkovskii, realized that the only way to achieve space travel is by using the power of the rocket.

The Space Age began on 4 October 1957, with the launch of Russia's first artificial satellite, Sputnik 1. Less than a year later the Americans made their first attempt to send a rocket vehicle to the Moon. It failed, as did others in the succeeding months, and the Russians took the lead; on 4 January 1959 their probe Luna 1 flew past the Moon at less than 6000 km and sent back useful information (such as the fact that the Moon has no detectable overall magnetic field). The first lander, again Russian (Luna 2), came down on the Moon on 13 September 1959, and in the following month the Soviet scientists achieved a notable triumph by sending Luna 3 round the Moon, obtaining the first pictures of the areas which are always turned away from Earth.

In the period from 1961 to 1965 the Americans launched their Ranger probes, which impacted the Moon and sent back valuable data before crash-landing. These were followed by the Surveyors (1966–1968), which made controlled landings and sent back a great deal of information as well as images. However, the first controlled landing was made by Russia's Luna 9, on 31 January 1966, which came down in the Oceanus Procellarum, and finally disposed of a curious theory according to which the maria, at least, would be covered by a deep layer of soft, treacherous dust.

Both the USA and the USSR were making efforts to achieve manned lunar landings. The Russian plans had to be abandoned when it became painfully obvious that their rockets were not sufficiently reliable, but the American Apollo programme went ahead, and culminated in July 1969 when Neil Armstrong and Edwin Aldrin stepped out on to the bleak rocks of the Mare Tranquillitatis. By the end of the Apollo programme, in December 1972, our knowledge of the Moon had been increased beyond all recognition. Meanwhile the Russians had used unmanned sample-and-return probes and had also dispatched two movable vehicles, the Lunokhods, which could crawl around the lunar surface under guidance from their controllers on Earth.

There followed a long hiatus in the programme of lunar exploration, but new probes were sent to the Moon during the 1990s, including one Japanese vehicle (Hagomoro, carrying the small satellite Hiten). The American Clementine (1994) and Prospector (1998) provided maps of the entire surface which were superior to any previously obtained as well as making some surprising claims; such as the possibility of locating ice inside the deep polar craters whose floors are always in shadow.

Details of the lunar missions are given in Table 3.8.

Table 3.8. Missions to the Moon.

(a) American

Name	Launch date	Landing date	Lat.	Long.	Results
Pre-ranger					
Pioneer 0 (Able 1)	17 Aug 1958	—	—	—	Failed after 77 s (explosion of lower stage of launcher).
Pioneer 1	11 Oct 1958	—	—	—	Reached 113 000 km. Failed to achieve escape velocity.
Pioneer 2	9 Nov 1958	—	—	—	Failed when third stage did not ignite.
Pioneer 3	6 Dec 1958	—	—	—	Reached 106 000 km. Failed to achieve escape velocity.
Pioneer 4	3 Mar 1959	—	—	—	Passed within 60 000 km of the Moon on 5 March. Now in solar orbit.
Able 4	26 Nov 1959	—	—	—	Failure soon after take-off.
Able 5A	25 Oct 1960	—	—	—	Total failure.
Able 5B	15 Dec 1960	—	—	—	Exploded 70 s after take-off.
Ranger (intended hard landers)					
Ranger 1	23 Aug 1961	—	—	—	Launch vehicle failure.
Ranger 2	18 Nov 1961	—	—	—	Launch vehicle failure.
Ranger 3	26 Jan 1962	—	—	—	Missed Moon by 37 000 km on 28 Jan. No images returned. Now in solar orbit.
Ranger 4	23 Apr 1962	26 Apr 1962	?	?	Landed on night side; instruments and guidance failure.
Ranger 5	18 Oct 1962	—	—	—	Missed Moon by over 630 km. No data received. Now in solar orbit.
Ranger 6	30 Jan 1964	2 Feb 1964	0.2N	21.5E	Landed in Mare Tranquillitatis. Camera failed; no images received.
Ranger 7	28 July 1964	31 July 1964	10.7S	20.7W	Landed in Mare Nubium. 4306 images returned.
Ranger 8	17 Feb 1965	20 Feb 1965	2.7N	24.8E	Landed in Mare Tranquillitatis. 7137 images returned.
Ranger 9	21 Mar 1965	24 Mar 1965	12.9S	2.4W	Landed in Alphonsus. 5814 images returned.
Surveyors (controlled landers)					
Surveyor 1	30 May 1966	2 June 1966	2.5S	43.2W	Landed in Mare Nubium, near Flamsteed. 11 150 images returned. Transmitted until 13 July; contact regained until January 1967.
Surveyor 2	20 Sept 1966	22 Sept 1966	SE of Copernicus		Guidance failure; crash-landed, site uncertain; no images returned.
Surveyor 3	17 Apr 1967	19 Apr 1967	2.9S	23.3W	Landed in Oceanus Procellarum, 612 km E of Surveyor 1, close to site of later Apollo 12 landing. 6315 images returned. Soil physics studied.
Surveyor 4	14 July 1967	16 July 1967	0.4N	1.3W	Crashed in Sinus Medii. No data returned.
Surveyor 5	8 Sept 1967	10 Sept 1967	1.4N	23.2E	Landed in Mare Tranquillitatis, 25 km from later Apollo 11 site. 18 000 images returned; soil physics studied. Contact lost on 16 December.
Surveyor 6	7 Nov 1967	9 Nov 1967	0.5N	1.4W	Landed in Sinus Medii. 29 000 images returned, soil physics studied. Re-started and moved 3 m. Contact lost on 14 December.
Surveyor 7	7 Jan 1968	9 Jan 1968	40.9S	11.5W	Landed on N rim of Tycho. 21 274 images returned and much miscellaneous information. Contact lost on 20 Feb 1968.

Table 3.8. (Continued)

Name	Launch date	Landing date	Lat.	Long.	Results
Orbiters (Mapping vehicles; no data attempted from the lunar surface)					
Orbiter 1	10 Aug 1966	29 Oct 1966	6.7N	162E	207 images returned. Controlled impact on far side at end of mission.
Orbiter 2	7 Nov 1966	11 Oct 1967	4.5N	98E	422 images returned. Controlled impact on far side at end of mission.
Orbiter 3	4 Feb 1967	9 Oct 1967	14.6N	91.7W	307 images returned. Controlled impact on far side at end of mission.
Orbiter 4	4 May 1967	6 Oct 1967	Far side		326 images returned; first images of polar regions. Impact on far side at end of mission; location uncertain.
Orbiter 5	1 Aug 1967	31 Jan 1968	0	70W	212 images returned. Controlled impact at end of mission.

Apollo (manned missions)

No	CM name	LM name	Launch	Land	Splash-down	Lat.	Long.	Area	Crew	EVA	Schedule
7	—	—	11 Oct 1968	—	22 Oct 1968	—	—	—	W Schirra, D Eisele, R Cunningham	—	Test orbiter (10 days 20 h)
8			21 Dec 1968	—	27 Dec 1968	—	—	—	F Borman, J Lovell, W Anders	—	Flight round Moon (6 days 3 h)
9	Gumdrop	Spider	3 Mar 1969	—	13 Mar 1969	—	—	—	J McDivett, D Scott, W Schweickart	—	LM test in Earth orbit (10 days 2 h)
10	Charlie Brown	Snoopy	18 May 1969	—	26 May 1969	—	—	—	T Stafford, J Young, E Cernan	—	LM test in lunar orbit (8 days 0 h)
11	Columbia	Eagle	16 July 1969	19 July 1969	24 July 1969	0°40′N	23°49′E	Mare Tranquillitatis	N Armstrong, E Aldrin, J Collins	2.2 h	Landed; ALSEP
12	Yankee Clipper	Intrepid	14 Nov 1969	19 Nov 1969	24 Nov 1969	3°12′S	23°24′W	Oceanus Procellarum, near Surveyor 3	C Conrad, A Bean, R Gordon	7.6 h (1.4 km)	Landing; ALSEP
13	Odyssey	Aquarius	11 Apr 1970	—	17 Apr 1970	—			J Lovell, F Haise, J Swigert	—	Aborted landing
14	Kitty Hawk	Antares	31 Jan 1971	5 Feb 1971	9 Feb 1971	3°40′S	17°28′W	Fra Mauro formation, Mare Nubium	A Shepard, E Mitchell, S Roosa	9.2 h (3.4 km)	Exploration; lunar cart
15	Endeavour	Falcon	26 July 1971	30 July 1971	7 Aug 1971	26°06′N	3°39′E	Hadley–Apennine region, near Hadley Rill	D Scott, J Irwin, A Worden	18.3 h (28 km)	Exploration; LRV
16	Casper	Orion	16 Apr 1972	21 Apr 1972	27 Apr 1972	8°36′S	15°31′E	Descartes formation, 50 km W of Kant	J Young, C Duke, T Mattingly	20.1 h (26 km)	Various experiments; LRV
17	America	Challenger	7 Dec 1972	11 Dec 1972	19 Dec 1972	20°12′N	30°45′E	Taurus–Littrow in Mare Serenitatis, 750 km E of Apollo 15	E Cernan, H Schmitt, R Evans	22 h (29 km)	Geology; LRV

Apollo 1 (21 Feb 1967) exploded on the ground, killing the crew (G Grissom, E White, R Chaffee). Apollos 2 and 3 were not used. Apollos 4 (9 Nov 1967), 5 (22 Jan 1968) and 6 (4 Apr 1968) were unmanned test Earth orbiters.

Post-Apollo missions

Clementine	Launch 25 Jan 1994	Entered lunar orbit 19 Feb 1994; mapping programme, surveying 38 000 000 square km of the Moon at 11 different wavelengths. Left lunar orbit on 3 May to rendezvous with asteroid Geographas, but failed to achieve this (on-board malfunction).
Prospector	Launch 6 Jan 1998	Extensive and prolonged lunar mapping and analysis programme; crashed into polar crater 31 July 1999 – unsuccessful search for water ice.

Table 3.8. (Continued)

(b) Russian

Name	Launch	Landing	Lat.	Long.	
Luna 1	2 Jan 1959	—	—	—	Passed Moon at 5955 km on 4 Jan, proving that the Moon lacks a magnetic field. Contact lost after 62 h. Studied solar wind. Now in solar orbit.
Luna 2	12 Sept 1959	13 Sept 1969	30N	1W	Crash-landed in Mare Imbrium, probably near Archimedes (uncertain).
Luna 3	4 Oct 1959	—	—	—	Went round Moon, imaging the far side. Approached Moon to 6200 km.
Luna 5	9 May 1965	12 May 1965	1.5S	25W	Unsuccessful soft lander. Crashed in Mare Nubium.
Luna 6	8 June 1965	—	—	—	Passed Moon at 161 000 km on 11 July; failure. Now in solar orbit.
Zond 3	18 July 1965	—	—	—	Approached Moon to 9219 km. Photographic probe; 25 images returned, including some of the far side. Images returned on 27 July from 2 200 000 km. Now in solar orbit.
Luna 7	4 Oct 1965	7 Oct 1965	9.8N	47.8W	Unsuccessful soft-lander. Crashed in Oceanus Procellarum.
Luna 8	3 Dec 1963	6 Dec 1963	9.6N	62W	Unsuccessful soft-lander. Crashed in Oceanus Procellarum.
Luna 9	31 Jan 1966	3 Feb 1966	7.1N	64W	Successful soft-lander; 100 kg capsule landed in Oceanus Procellarum. Images returned. Contact lost on 7 Feb.
Luna 10	31 Mar 1966	—	—	—	Lunar satellite; approached Moon to 350 km; gamma-ray studies of lunar surface layer. (Entered lunar orbit on 3 Apr.) Contact lost on 30 May, after 460 orbits. Now in lunar orbit.
Luna 11	24 Aug 1966	—	—	—	Lunar satellite; entered lunar orbit on Aug 28, and approached Moon to 159 km. Radiation, meteoritic and gravitational studies. Contact lost on 1 Oct.
Luna 12	22 Oct 1966	—	—	—	Lunar satellite. Entered lunar orbit on 25 Oct, and imaged craters to a resolution of 15 m. Contact lost on 19 Jan 1967.
Luna 13	21 Dec 1966	23 Dec 1966	18.9N	63W	Soft landing in Oceanus Procellarum. Images returned; soil and chemical studies. Contact lost on 27 December.
Luna 14	7 April 1968	—	—	—	Lunar satellite; approached Moon to 160 km. Valuable data obtained.
Zond 5	14 Sept 1968	—	—	—	Went round the Moon, approaching to 1950 km, and returned to Earth on 21 September. Plants, seeds, insects and tortoises carried.
Zond 6	10 Nov 1968	—	—	—	Went round the Moon, approaching to 2420 km, and filmed the far side. Returned to Earth on 17 November.
Luna 15	13 July 1969	21 July 1969	17N	60E	Unsuccessful sample and return probe. Crashed in Mare Crisium.
Zond 7	7 August 1969	—	—	—	Went round the Moon, approaching to 2000 km, and took colour images of both Moon and Earth. Returned to Earth.
Luna 16	12 Sept 1970	15 Sept 1970	0.7S	56.3E	Landed in M. Fœcuniditatis, secured 100 gr of material, and after $26\frac{1}{2}$ hours lifted off and returned to Earth (24 September).

Table 3.8. (Continued)

(b) Russian

Name	Launch	Landing	Lat.	Long.	
Zond 8	20 Oct 1970	—	—	—	Circum-lunar flight; colour pictures of Earth and Moon. Returned to Earth on 27 October, splashing down in Indian Ocean.
Luna 17	10 Nov 1970	17 Nov 1970	30.2N	35W	Carried Lunokhod I to Mare Imbrium.
Luna 18	2 Sept 1971	15 Sept 1971	3.6N	56.5E	Unsuccessful soft-lander. Contact lost during descent manœuvre to Mare Fœcunditatis.
Luna 19	28 Sept 1971	—	—	—	Lunar satellite. Contact kept for 4000 orbits. Studies of mascons, lunar gravitational field, solar flares, etc.
Luna 20	14 Feb 1972	17 Feb 1972	3.5N	56.6E	Landed near Apollonius, S. of Mare Crisium (120 km N. of Luna 16 site), drilled into the lunar surface, and returned with samples on 25 February.
Luna 21	8 Jan 1973	16 Jan 1973	25.9N	30.5E	Carried Lunokhod 2 to a site near Le Monnier, 180 km from Apollo 17 site.
Luna 22	29 May 1974	—	—	—	Lunar satellite. TV images, gravitation and radiation studies. Contact maintained until 6 November 1975.
Luna 23	28 Oct 1974	—	Mare Crisium	—	Unsuccessful sample and return mission; crashed.
Luna 24	9 Aug 1976	18 Aug 1976	12.8N	62.2E	Landed in Mare Crisium, drilled down to 2 metres, collected samples, lifting off on 19 August and landing back on Earth on 22 August.

Lunokhods

Name	Carrier	Weight (kg)	Site	
Lunokhod 1	Luna 7	756	Mare Imbrium	Operated for 11 months after arrival on 17 Nov 1970. Area photographed exceeded 80 000 m^2. Over 200 panoramic pictures and 20 000 images returned. Distance travelled, 10.5 km.
Lunokhod 2	Luna 21	850	Le Monnier	Operated until mid-May 1973. 86 paranormic pictures and 80 000 TV images obtained. Distance travelled, 37 km. On 3 June the Soviet authorities announced that the programme had ended.

(c) Japanese

Name	Launch	
Hagomoro	24 Jan 1990	Launched by Muses-A vehicle. The satellite Hiten ('Flyer') was ejected and put into lunar orbit on 15 Feb 1992; it had a mass of 180 kg and carried a micrometeoroid detector. Hiten crash-landed on the Moon on 10 Apr 1993, at lat. 34S, long. 55E, near Furnerius.
Nozomi	3 July 1998	Mars mission. Imaged Moon from 514 000 km on 18 July 1998.

(d) US/European

Galileo	Launch 18 Oct 1989. Galileo, en route to Jupiter, flew past the Earth–Moon system on 8 Dec 1990; surveyed the Moon's far side, and on 9 Dec imaged the far hemisphere from 550 000 km. The closest approach to Earth was 960 km. A second flyby occurred on 8 Dec 1992, when Galileo passed Earth at 302 km, and on the 9th imaged the Earth and Moon together.
NEAR–Shoemaker	(Near Earth Asteroid Rendevous Spacecraft) named in honour of E. Shoemaker. Swung past Earth on 23 Jan 1998, en route for the asteroid Eros, and obtained a view of Earth and Moon from above their south poles.

On 18 August 1999 the US Saturn probe Cassini flew past the Earth–Moon system, and imaged the Moon from a range of 377 000 km.

STRUCTURE OF THE MOON

The results from the Apollo missions and the various unmanned probes have led to a change in many of our ideas about the Moon. Moreover, one professional geologist has been there – Dr Harrison ('Jack') Schmitt, with Apollo 17 – and his expertise was naturally invaluable.

The upper surface is termed the regolith. This is a loose layer or débris blanket, continually churned by the impacts of micrometeorites. (It is often referred to as 'soil', but this is misleading, because there is nothing organic about it.) It is made up chiefly of very small particles ('dust'), but with larger rocks, a few metres across, here and there; it contains many different ingredients. In the maria it is around 2–8 m deep, but it is thicker over the highlands, and may in places go down to 10 m or even more.

The highland crust averages 61 km in depth, but ranges from an average of 55 km on the near or Earth-turned side of the Moon to up to 67 km on the far side. The maria are of course volcanic; they cover 17% of the surface, mainly on the near side (they are much less common on the far side, because of the greater thickness of the crust), and they are in general no more than 1–2 km deep, except near the centres of the large basins. At a fairly shallow level there are areas of denser material, which have been located because an orbiting space-craft will speed up when affected by them; these are known as mascons (*mass con*centrations). They lie under the large basins, such as Imbrium and Orientale.

On the highlands, the rock fragments are chiefly anorthosites, with minerals such as plagioclase, pyroxene and ilmenite. Details of these materials are given in Table 3.9.

Much of our knowledge about the lunar interior comes from seismic investigations – in fact, moonquakes – just as we depend upon earthquakes for information about the interior of our own world. Of course, moonquakes are very mild by terrestrial standards, and never exceed a value of 3 on the Richter scale, and they are of two main types. Most originate from a zone 800–1000 km below the surface, and are common enough; there is a definite correlation between moonquake frequency and lunar perigee. Shallow moonquakes, at depths of 50–200 km, also occur, although they are much less frequent. It is worth noting that the epicentres of moonquakes seem

Table 3.9. Materials in the lunar crust.

Pyroxene, a Ca–Mg–Fe silicate, is the most common mineral in lunar lavas, making up about half of most specimens; it forms yellowish–brown crystals up to a few centimetres in size.

Plagioclase or feldspar, a Na– or Ca–Al silicate, forms elongated white crystals.

Anorthosite is a rock type containing the minerals plagioclase, pyroxene and/or olivine in various proportions.

Basalt is a rock type containing the minerals plagioclase, pyroxene and ilmenite in varying proportions.

Olivine, a Mg–Fe silicate, is made up of pale green crystals a few millimetres in size; it is not uncommon in the anorthosites.

Ilmenite, present in the basalts, is an Fe–Ti oxide.

to be linked with areas particularly subject to TLP – mainly although not entirely around the peripheries of the regular maria.

There have also been man-made moonquakes, caused by the impacts of discarded lunar modules. These show that the outer few kilometres of the Moon are made up of cracked and shattered rock, so that signals can echo to and fro; it was even said that after the impact of an Apollo module the Moon 'rang like a bell'!

Below the crust comes the mantle, the structure of which seems to be relatively uniform. A 1 ton meteorite which hit the Moon in July 1972 indicated, from its seismic effects, that there is a region 1000–1200 km below the surface where the rocks are hot enough to be molten (Apollo measurements disposed of an earlier theory that the Moon's globe could be cold and solid all the way through). Finally, there may be a metallic core, although its existence has not been definitely proved, and it cannot be much more than 1000 km in diameter. Results from the Lunar Prospector mission of 1998/9 have led to an estimate of an iron-rich core between 440 and 900 km in diameter. Certainly the Moon's core is much smaller than that of the Earth, both relatively and absolutely. It is significant that there is no overall magnetic field now, although the remnant magnetism of some rocks indicates that between 3.6 and 3.9 thousand million years a definite field existed – which was not evident either before or after that period. There are however locally magnetized areas, notably the curious Reiner Gamma, a 'swirl' in the near side, and the crater Van de Graaff, on the far side.

All the rocks brought home for analysis are igneous, or breccias produced by impact processes; the Apollo missions recovered 2196 samples, with a total weight of 381.69 kg, now divided into 35 600 samples. The youngest basalt (No 12022) was given an age of 3.08 thousand million years, while the oldest (No 10003) dated back 3.85 thousand million years. There were no sedimentary or metamorphic rocks. The famous 'orange soil', found by the Apollo 17 astronauts and at first thought to indicate recent vulcanism, proved to be small glassy orange beads, sprayed out some 3.7 thousand million years ago in erupting fountains of basaltic magmas.

In the lavas, basalts are dominant. They contain more titanium than terrestrial lavas; over 10% in the Apollo 11 samples, for example, as against 1–3% in terrestrial basalts. Small amounts of metallic iron were found. Many lunar rocks have much less sodium and potassium than do terrestrial rocks. A new mineral – an opaque oxide of iron, titanium and magnesium, not unlike ilmenite – has been named armalcolite, in honour of *Arm*strong, *Al*drin and *Col*lins. There is also a different type of basalt, KREEP; this name comes from the fact that it is rich in potassium (chemical symbol K), rare earth elements and phosphorus. The average age of the highland rocks is from 4 to 4.2 thousand million years; in fact over 99% of the surface dates back for over 3 thousand million years and 90% goes back for more than 4 thousand million years. One interesting anorthosite rock, 4 thousand million years old, was collected by the Apollo 15 astronauts; it is white and was at once nicknamed the Genesis Rock.

The Apollo 12 sample 12013 (collected by Conrad from the Oceanus Procellarum) is unique. It is about the size of a lemon, and contains 61% of SiO_2, whereas the associated lavas have only 35–40% of SiO_2. It also contains 40 times as much potassium, uranium and thorium, making it one of the most radioactive rocks found anywhere on the Moon. It is composed of a dark grey breccia, a light grey breccia and a vein of solidified lava.

Unexpected results came from one of the most recent lunar probes, Clementine, named after the character in the old mining song who was 'lost and gone forever'. Although Clementine was among the cheapest of all probes (it cost $55 000 000) it was remarkably successful insofar as the Moon is concerned. It was a joint NASA–USAF venture, and was launched not from Canaveral, but from the Vandenberg Air Force Base, on 23 January 1995. On 21 February it entered lunar orbit and continued mapping until 23 April; by the time it left lunar orbit, on 4 May, the whole of the surface had been mapped. Unfortunately a fault developed, making it impossible to go on to an encounter with an asteroid (Geographos) as had been hoped. The last lunar flyby was on 20 July, the 25th anniversary of the Apollo 11 landing, after which Clementine entered a solar orbit. The minimum distance from the Moon had been 425 km.

Surprisingly, some investigators claimed that the neutron spectrometer on Clementine had been used to detect ice in some of the deep polar craters, whose floors are always in shadow and where the temperature is always very low. This seemed to be inherently unlikely, since none of the materials brought home by the astronauts had shown any sign of hydrated substances, and in any case it was not easy to see how the ice could have got there. It could hardly have been deposited by an impacting comet, because the temperature at the time of the collision would have been too high; and there is no evidence of past water activity, as there is for instance upon Mars. Yet it was even suggested that there might be enough ice to provide a useful water supply for future colonists; a thousand million gallons of water was one estimate.

Then came Prospector, launched on 6 January 1998, and put into a stable orbit which took it round the Moon once in every 118 min at a distance of 96 km from the surface. Prospector was designed not only to continue with the mapping programme, but also to make a deliberate search for ice deposits – and before long the results seemed to confirm those of Clementine. Yet it was not claimed that ice, as such, had been detected. All that had been found were apparent indications of hydrogen, which could be interpreted in several different ways.

Prospector, like Clementine, carried a neutron spectrometer. Neutrons are ejected when cosmic rays from space strike atoms in the Moon's crust, and these neutrons can be detected from the space-craft. Collisions between cosmic-ray particles and heavy atoms produce 'fast' neutrons; if hydrogen atoms are hit, the 'slow' neutrons are much less energetic and the spectrometers can distinguish between the two types. It was found that the neutron energy coming from the polar regions was reduced and from this the presence of hydrogen was inferred, which in turn could suggest the presence of water ice. On the other hand, the hydrogen could be due to the solar wind, which bombards the lunar surface all the time.

Efforts were made to confirm the presence of ice by using the large radio telescope at Arecibo in Puerto Rico. Radar studies did indeed give the same indications – but these were also found in regions which are not in permanent shadow and where frozen material could not possibly exist, so that very rough ground might well be responsible. From the outset there were many sceptics about the 'ice' idea. I had no faith in it; how could the ice have arrived there? All the samples brought home so far are completely lacking in hydrated materials. A major sceptic was Harrison Schmitt, the only professional geologist who has been to the Moon.

Finally, on 31 July 1999, a test was made. At the end of its active career Prospector was deliberately crashed on to the Moon, landing inside a polar crater where ice, if it existed at all, would be present. It was hoped that the cloud of material thrown up would show traces of water. In fact the results were completely negative. No traces of water were found, and all in all it seems that the whole idea of ice inside polar craters must be abandoned.

ATMOSPHERE

The Moon's low escape velocity means that it cannot be expected to retain much in the way of atmosphere. Initially it was believed that the atmosphere must be dense; this was the view of Schröter (1796) and also Sir William Herschel, who believed that habitability of the Moon to be 'an absolute certainty'. W. H. Pickering (1924) believed the atmosphere to be dense enough to support insects or even small animals, but in 1949 B. Lyot searched for lunar twilight effects and concluded that the atmosphere must have a density less than 1/10 000 of that of the Earth at sea-level. In what was then the USSR, V. Fesenkov and Y. N. Lipski made similar investigations and came to the final conclusion that the density was indeed in the region 1/10 000 that of our air.

The first reliable results came from the Apollo missions. The orbiting sections of Apollos 15 and 16 traced small quantities of radon and polonium seeping out from below the surface, and this was no surprise, because these gases are produced by the radioactive decay of uranium, which is not lacking in the lunar rocks. LACE, the Lunar Atmospheric Composition Experiment, taken to the Moon by Apollo 17, did detect an excessively tenuous atmosphere, mainly helium (due to the solar wind) and argon (seeping out from below the crust). Later D. Potter and T. Morgan, at the McDonald Observatory in Texas, identified two more gases, sodium and potassium. Traces of silicon, aluminium and oxygen have also been detected in the excessively tenuous upper atmosphere. The sodium seems to surround the Moon rather in the manner of a cometary corona. The lunar atmosphere seems to be in the nature of a collisionless gas; the total weight of the lunar atmosphere can be no more than about 30 tons. The density is of the order of 10^{-14} that of the Earth's atmosphere.

ECLIPSES OF THE MOON

Eclipses of the Moon are caused by the Moon's entry into the cone of shadow cast by the Earth. At the mean distance of the Moon, the diameter of the shadow cone is approximately 9170 km; on average the shadow is 1367 650 km long. Totality may last for up to 1 h 44 min.

Lunar eclipses may be either total or partial. If the Moon misses the main cone and merely enters the zone of 'partial shadow' or penumbra to either side, there is slight dimming, but a penumbral eclipse is not easy to detect with the naked eye. Of course, the Moon must pass through the penumbra before entering the main cone or umbra (Table 3.10).

During an eclipse the Moon becomes dim, often coppery. The colour and brightness during an eclipse

Table 3.10. Lunar eclipses. Lunar eclipses occurred/will occur on the following dates: * = total, for other partial eclipses the maximum phase is given. For penumbral eclipses (P) the maximum percentage is given in brackets.

(a) 1960–2000

Date		Type	Date		Type
1960	Mar 13	*	1981	July 17	58
	Sept 5	*	1982	Jan 9	*
1961	Mar 2	81		July 6	*
	Aug 26	99		Dec 30	*
1963	July 6/7	71	1983	June 25	34
	Dec 30	*	1985	May 4	*
1964	Dec 19	*		Oct 28	*
1965	June 13/4	18	1986	Apr 24	*
1967	Apr 24	*		Oct 17	*
	Oct 18	*	1987	Oct 1	1
1968	Apr 13	*	1988	Aug 27	29
	Oct 6	*	1989	Feb 20	*
1970	Feb 21	5		Aug 17	*
	Aug 17	41	1990	Feb 9	*
1971	Feb 10	*		Aug 6	68
	Aug 6	*	1991	Dec 21	9
1972	Jan 30	*	1992	June 15	68
	July 26	55		Dec 9	*
1973	Dec 10	11	1993	June 4	*
1974	June 4	83		Nov 29	*
	Nov 29	*	1994	May 25	24
1975	May 25	*	1995	Apr 15	11
	Nov 18/9	*	1996	Apr 4	*
1976	May 13	13		Sept 27	*
1977	Apr 4	21	1997	Mar 24	92
1978	Mar 24	*		Sept 16	*
	Sept 16	*	1999	July 28	40
1979	Mar 13	89			
	Sept 6	*			

Table 3.10. (Continued)

(b) 2000–2008

Date		Type	Time of mid eclipse (GMT)		Duration of eclipse Totality		Duration of eclipse Partial	
			h	m	h	m	h	m
2000	Jan 21	*	04	45	1	16	3	22
2000	July 16	*	13	57	1	0	3	16
2001	Jan 9	*	20	22	0	30	1	38
2001	July 5	49	14	57	—	—	1	19
2001	Dec 30	P(89)	10	30	—	—	—	—
2002	May 6	P(69)	12	05	—	—	—	—
2002	June 24	P(21)	21	29	—	—	—	—
2002	Nov 20	P(86)	01	47	—	—	—	—
2003	May 16	*	03	41	0	26	1	37
2003	Nov 9	*	01	20	0	11	1	45
2004	May 4	*	20	32	0	38	1	41
2004	Oct 28	*	03	05	0	40	1	49
2005	Apr 24	P(87)	09	57	—	—	—	—
2005	Oct 17	6	12	04	—	—	0	28
2006	Mar 14	P(100)	23	49	—	—	—	—
2006	Sept 7	18	18	52	—	—	0	45
2007	Mar 3	*	23	22	0	37	1	50
2008	Feb 21	*	03	27	0	24	1	42
2008	Aug 16	81	21	11	—	—	1	34

(c) 2008–2020

Date		Type	Date		Type
2009	Dec 31	8	2015	Apr 4	*
2010	June 26	54		Sept 28	*
	Dec 21	*	2017	Aug 7	25
2011	June 15	*	2018	Jan 31	*
	Dec 10	*		July 27	*
2012	June 4	37	2019	Jan 21	*
2013	Apr 25	1		July 16	65
2014	Apr 15	*			
	Oct 8	*			

Table 3.11. The Danjon scale for lunar eclipses.

0	Very dark; Moon almost invisible.
1	Dark; grey or brownish colour; details barely identifiable.
2	Dark or rusty-red, with a dark patch in the middle of the shadow; brighter edges.
3	Brick-red; sometimes a bright or yellowish border to the shadow.
4	Coppery or orange-red; very bright, with a bluish cast and varied hues.

depend upon the conditions in the Earth's atmosphere; thus the eclipse of 19 March 1848 was so 'bright' that lay observers refused to believe that an eclipse was happening at all. On the other hand, it is reliably reported that during the eclipses of 18 May 1761 and 10 June 1816 the Moon became completely invisible with the naked eye. The French astronomer A. Danjon has given an 'eclipse scale' from 0 (dark) to 4 (bright), and has attempted to correlate this with solar activity, although the evidence is far from conclusive. The Danjon scale is given in Table 3.11.

The Greek astronomer Anaxagoras (c 500–428 BC) gave a correct explanation of lunar eclipses, but in early times eclipses caused considerable alarm. For example, the Californian Indians believed that a monster was attacking the Moon, and had to be driven away by making as much noise as possible (as with the Chinese at the time of a solar eclipse), while in an old Scandinavian poem, the *Edda*, it is said that the monster Managarmer is trying to swallow the Moon, and staining the air and ground with blood. During an eclipse the Orinoco Indians would take their hoes and labour energetically in their cornfields, as they felt that the Moon was showing anger at their laziness.

Ancient eclipse records are naturally uncertain. It has been claimed that an eclipse seen in the Middle East can be dated back to 3450 BC; the eclipse of 1361 BC is more definite. Ptolemy gives the dates of observed eclipses as 721 BC and 720 BC. In *The Clouds*, the Greek playwright Aristophanes alludes to an eclipse seen from Athens on 9 October 425 BC. There was certainly a lunar eclipse in August 413 BC, which had unfortunate results for Athens, since is persuaded Nicias, the commander of the Athenian expedition to Sicily, to delay the evacuation of his army; the astrologers advised him to stay where he was 'for thrice nine days'. When he eventually tried to embark his forces, he found that he had been blockaded by the Spartans. His fleet was destroyed and the expedition annihilated – a reverse which led directly to the final defeat of Athens in the Peloponnesian War. According to Polybius, an eclipse in September 218 BC so alarmed the Gaulish mercenaries in the service of Attalus I of Pergamos that they refused to continue a military advance. On the other hand, Christopher Columbus turned the eclipse of 1504 AD to his advantage. He was anchored off Jamaica and the local inhabitants refused to supply his men with food; he threatened to extinguish the Moon, and when the eclipse took place the natives were so alarmed that there was no further trouble.

Obviously, a lunar eclipse can happen only at full moon. In the original edition of the famous novel *King Solomon's Mines*, H. Rider Haggard described a full moon, a solar eclipse and another full moon on successive days. When the mistake was pointed out he altered the second edition, turning the solar eclipse into a lunar one!

OCCULTATIONS

Occultations of stars by the Moon is common, and were formerly of great value for positional purposes. When a star is occulted, it shines steadily up to the instant of immersion; the lunar atmosphere is far too rarefied to have an effect. A planet will take some time to disappear. W. H. Pickering maintained that the lunar atmosphere could produce effects during an occultation of Jupiter, but this has long since been discounted.

On 23 April 1998 four Brazilian observers, led by E. Karkoschka, witnessed an exceptional event. Using a 10 cm refractor, they saw Venus and Jupiter occulted simultaneously; the site was 100 km N of Recife. The last occultation when Venus and Jupiter were occulted at the same time was 567 AD.

Table 3.12. Named mare, lacus, palus and sinus areas. Since these cover wide areas, the coordinates are given for their centres, but are of course approximate only. Diameter values are also approximate, since many of the maria have very irregular boundaries.

Name		Lat.	Long.	Diam. (km)	
Mare Anguis	Serpent Sea	23N	67E	150	Area 10 000 km^2. Narrow darkish area NE of Mare Crisium.
Mare Australe	Southern Sea	40S	93E	603	Irregular, patchy area in the SE. Area about 149 000 km^2.
Mare Cognitum	Known Sea	10S	23W	376	Part of Mare Nubium, E of the Riphæans. Landing site of Ranger 7 in 1964.
Mare Crisium	Sea of Crises	17N	59E	505	Well-defined; separate from main system. Area about 200 000 km^2 (similar to Great Britain).
Mare Fœcunditatis	Sea of Fertility	8S	31E	909	Irregular; confluent with Mare Tranquillitatis. Area 344 000 km^2.
Mare Frigoris	Sea of Cold	56N	1E	1596	Elongated, irregular; in places narrow. Area about 441 000 km^2 (including Lacus Mortis). Bounded in part by the Alps.
Mare Humboldtianum	Humboldt's Sea[a]	57N	81E	273	Limb sea, beyond Endymion; fairly regular.
Mare Humorum	Sea of Humours	24S	39W	389	Regular; leads off Mare Nubium. Area 118 000 km^2.
Mare Imbrium	Sea of Showers	33N	16W	1123	Largest regular sea; area 863 000 km^2 (equal to Britain and France combined). Bounded by the Alps, Apennines and Carpathians. Contains Palus Nebularum and Palus Putredinis, as well as major craters such as those of the Archimedes group.
Mare Insularum	Sea of Islands	7N	31W	513	Part of Mare Nubium; ill-defined; south of Copernicus.
Mare Marginis	Marginal Sea	13N	86E	420	Limb sea beyond Mare Crisium; fairly well-defined. Area 62 000 km^2.
Mare Nectaris	Sea of Nectar	15S	36E	333	Leads off Mare Tranquillitatis. Area 100 000 km^2 regular. Central part of a very ancient basin, whose border is marked by the Altai Scarp.
Mare Nubium	Sea of Clouds	21S	17W	715	Ill-defined N border. Area (with Mare Cognitum) 265 000 km^2.
Mare Orientale	Eastern Sea	19S	93W	327	Limb sea, beyond Corderillas. Vast ringed structure. Only the E part visible from Earth under favourable libration.
Oceanus Procellarum	Ocean of Storms	18S	57W	2568	Area 2 290 000 km^2. Irregular; contains Aristarchus.
Mare Serenitatis	Sea of Serenity	28N	17E	707	Regular; area 314 000 km^2. Contains few conspi- cuous craters; Bessel is the most prominent and Linné is also on the Mare. Crossed by a long ray coming from the south, and wrinkle-ridges also cross it.
Mare Smythii	Smyth's Sea[b]	1N	87E	373	Well-defined limb sea; area 104 000 km^2.
Mare Spumans	The Foaming Sea	1N	65E	139	Darkish area S of Mare Crisium. Area 16 000 km^2.
Mare Tranquillitatis	Sea of Tranquillity	9N	31E	873	Confluent with Mare Serenitatis, but is lighter, patchier and less regular. Area 440 000 km^2; rather irregular. Nectaris and Fœcunditatis lead off it.
Mare Undarum	Sea of Waves	7N	64E	243	Darkish irregular area near Firmicus. Area 21 000 km^2.
Mare Vaporum	Sea of Vapours	13N	4E	245	Area 55 000 km^2; SE of the Apennines. Contains some very dark patches; also the Hyginus Rill and part of the Ariadæus Rill.
Mare Moscoviense	Moscow Sea	27N	148E	277	Far side.
Mare Ingenii	Sea of Ingenuity	34S	163E	318	Far side.
Lacus Æstatis	Summer Lake	15S	69W	90	Two dark areas N of Crüger. Combined area about 1000 km^2.
Lacus Autumni	Autumn Lake	10S	84W	183	Dark patches in the Corderillas.
Lacus Bonitatis	Lake of Goodness	23N	44E	92	Small darkish area near Macrobius.
Lacus Doloris	Lake of Grief	17N	9E	110	Darkish area N of Manilius.
Lacus Excellentiæ	Lake of Excellence	35S	44W	184	Vague darkish area near Clausius.
Lacus Felicitatis	Lake of Happiness	19N	5E	90	Small darkish area N of Mare Vaporum.
Lacus Gaudii	Lake of Joy	16N	13E	50	Darkish area between Manilius and the Hæmus Mountains.
Lacus Hiemis	Winter Lake	15N	14E	50	Small darkish area SW of Menelaus.
Lacus Lenitatis	Lake of Tenderness	14N	12E	80	Darkish area E of Manilius.
Lacus Mortis	Lake of Death	45N	27E	151	Dark, adjoining Lacus Somniorum. Area 21 000 km^2. Contains the Bürg rills.
Lacus Odii	Lake of Hate	19N	7E	70	Darkish patch W of the Hæmus Mountains.
Lacus Perseverantiæ	Lake of Perseverance	8N	62E	70	Darkish patch adjoining Firmicus to the W.
Lacus Somniorum	Lake of the Dreamers	38N	29E	384	Irregular darkish area leading off the Mare Serenitatis. Area 70 000 km^2.
Lacus Spei	Lake of Hope	43N	65E	80	Dark strip between Messala and Zeno.
Lacus Temporis	Lake of Time	45N	58E	117	Irregular area between Mercurius and Atlas.
Lacus Timoris	Lake of Fear	39S	27W	117	Narrow darkish patch E of Hainzel.
Lacus Veris	Spring Lake	16S	86W	396	Narrow irregular area in Rook Mountains. Total area 12 000 km^2.
Lacus Luxuriæ	Lake of Luxury	19N	176E	50	Far side.
Lacus Oblivionis	Lake of Forgetfulness	21S	168W	50	Far side.
Lacus Solitudinis	Lake of Solitude	28S	104E	384	Far side.
Palus Epidemiarum	Marsh of Epidemics	32S	28W	286	Darkish area adjoining Mercator and Campanus. Area 27 000 km^2.
Palus Nebularum	Marsh of Clouds	140N	6W	150	E part of Mare Imbrium. Name deleted from some maps.
Palus Putredinis	Marsh of Decay	16N	0.4E	161	Part of Mare Imbrium, near Archimedes.
Palus Somnii	Marsh of Sleep	14N	45E	143	Curiously-coloured area bounded by bright rays from Proclus.
Sinus Æstuum	Bay of Heats	11N	9W	290	Fairly regular dark area leading off the Mare Nubium, E of Copernicus; area 40 000 km^2.
Sinus Amoris	Bay of Love	18N	39E	130	Part of Mare Tranquillitatis, E of Maraldi.
Sinus Asperitatis	Bay of Asperity	4S	27E	206	Part of Mare Nectaris; rough area between Theophilus and Hypatia.
Sinus Concordiæ	Bay of Harmony	11N	42E	142	Bay S of Palus Somnii.
Sinus Fidei	Bay of Faith	18N	2E	70	Outlet of Mare Vaporum, to the N.
Sinus Honoris	Bay of Honour	12N	18E	109	Edge of Mare Tranquillitatis, NW of Maclear.
Sinus Iridum	Bay of Rainbows	44N	31W	236	Beautiful bay, extending from Mare Imbrium.
Sinus Lunicus	Luna Bay	32N	1W	50	In Mare Imbrium, between Aristillus and Archimedes; probable landing site of Luna 2 in 1959.
Sinus Medii	Central Bay	2N	2E	260	Small bay near the apparent centre of the disk; area 22 000 km^2.
Sinus Roris	Bay of Dew	54N	57W	400	Area joining Mare Frigoris to Oceanus Procellarum.
Sinus Successus	Bay of Success	1N	59E	132	Ill-defined darkish area W of Mare Spumans.

[a] Alexander von Humboldt, German natural historian (1769–1859).

[b] Admiral William Henry Smyth, British astronomer (1768–1865).

Table 3.13. Selected craters on the near side of the Moon. c.p. = central peak.

Name	Lat.	Long.	Diameter (km)	Notes	Name	
Abbot	5.6N	54.8E	10	Uplands, E of Taruntius (Apollonius K).	Charles; American	1872–1973
Abel	34.5S	87.3E	122	Flooded walled plain, E of Furnerius.	Niels; Norwegian mathematician	1802–1829
Abenezra	21.0S	11.9E	42	W of Nectaris; well-formed pair with Azophi.	Abraham ben Ezra; Spanish–Jewish astronomer	1092–1167
Abetti	19.9N	27.7E	65	Obscure; dark floor; in Serenitatis, NW of Argæus.	Antonio; Italian astronomer	1846–1928
Abulfeda	13.8S	13.9E	65	Abenezra area; pair with Almanon	Abu'L fida, Ismail; Syrian geographer	1273–1331
Acosta	5.6S	60.1E	13	N of Langrenus (Langrenus C).	Cristobal; Portuguese natural historian	1515–1580
Adams	31.9S	68.2E	66	E of Vendelinus; irregular walls.	John Couch; English astronomer	1819–1892
Agatharchides	19.8S	30.9W	48	Humorum area; remains of c.p.; irregular walls.	Greek geographer	?–150 BC
Agrippa	4.1N	10.5E	44	Vaporum area; regular walls; pair with Godin.	Greek astronomer	c 92 AD
Airy	18.1S	5.7E	36	Pair with Argelander; irregular walls.	George Biddell; English Astronomer Royal	1810–1892
Al-Bakri	14.3N	20.2E	12	NW of Plinius (Tacquet A).	Al-Bakri; Spanish–Arab astronomer	1010–1094
Al-Biruni	17.9N	92.5E	77	Libration zone; Marginis area.	Persian mathematician/geographer	973–1048
Al-Marrakushi	10.4S	55.8E	8	W of Langrenus (Langrenus D).	Moroccan geographer/astronomer	c 1261
Albategnius	11.7S	4.3E	114	Companion to Hipparchus; irregular, terraced walls.	Al-Battani; Iraqi astronomer	850–929
Aldrin	1.4N	22.1E	3	On Tranquillitatis, E of Sabine (Sabine B).	Buzz; American astronaut; Apollo 11	1930–
Alexander	40.3N	13.5E	81	N end of Caucasus; darkish floor; low walls.	Alexander the Great of Macedon	356–323
Alfraganus	5.4S	19.0E	20	NW of Theophilus; v bright; minor ray-centre.	Al Fargani; Persian astronomer	?–840
Alhazen	15.9N	71.8E	32	Near border of Crisium (not Schröter's Alhazen).	Abu Ali Ibn Al Haitham; Iraqi mathematician	987–1038
Aliacensis	30.6S	5.2E	79	Pair with Werner; Walter area; high walls.	D'Ailly Pierre; French geographer	1350–1420
Almanon	16.8S	15.2E	49	Regular; pair with Abulfeda.	Al Mamun; Persian astronomer	786–833
Alpetragius	16.0S	4.5W	39	Outside Alphonsus; high, terraced walls; huge c.p. with summit pit.	Nur Ed-Din Al Betrugi; Moroccan astronomer	?–c 1100
Alphonsus	13.7S	3.2W	108	Ptolemæus chain; low c.p.; rills on floor.	Alfonso X; Spanish astronomer	1223–1284
Ameghino	3.3N	57.0E	9	Uplands SW of Apollonius.	Fiorino; Italian natural historian	c 1854–1911
Ammonius	8.5S	0.8W	8	Prominent; in Ptolemæus (Ptolemæus A).	Greek philosopher	?–c 517
Amontons	5.3S	46.8W	2	Crateret of Fœcunditatis, S of Messier.	Guillaume; French physicist	1663–1705
Amundsen	84.3S	85.6S	101	Libration area; well-formed; pair with Scott.	Roald; Norwegian explorer	1872–1928
Anaxagoras	73.4N	10.1W	50	N. polar area; distorts Goldschmidt; ray-centre.	Greek astronomer	500–428 BC
Anaximander	66.9N	51.3W	67	Pythagoras area; pair with Carpenter; no c.p.	Greek astronomer	c 611–547 BC
Anaximenes	72.5N	44.5W	80	Near Philolaus; rather low walls.	Greek astronomer	585–528 BC
Anděl	10.4S	12.4E	35	Highlands W of Theophilus; low, irregular walls.	Karel; Czech astronomer	1884–1947
Andersson	49.7S	95.3W	13	Libration zone; S of Guthnick and Rydberg.	Leif; American astronomer	1943–1979
Ångström	29.9N	41.6W	9	In Imbrium, N of Harbingers.	Anders; Swedish physicist	1814–1874
Ansgarius	12.7S	79.7E	94	Distinct; E of Fœcunditatis; pair with La Peyrouse.	St. Ansgar; German theologian	801–864
Anuchin	49.0S	101.3E	57	Libration zone; beyond Australe.	Dimitri; Russian geographer	1843–1923
Anville	1.9N	49.5E	10	E of Secchi, In Fœcunditatis (Taruntius G).	Jean-Baptiste; French cartographer	1697–1782
Apianus	26.9S	7.9E	63	Aliacensis area; high walls.	Bienewitz; German astronomer	1495–1552
Apollonius	4.5N	61.1E	53	Uplands; S of Crisium; well-formed.	Greek mathematician	3rd century BC
Arago	6.2N	21.4E	26	On Tranquillitatis; domes nearby.	François; French astronomer	1786–1853
Aratus	23.6N	4.5E	10	In Apennines; v bright; not regular in outline.	Greek astronomer	c 315–245 BC
Archimedes	29.7N	4.0W	82	On Imbrium; v regular; darkish floor; no c.p.	Greek mathematician/physicist	c 287–212 BC
Archytas	58.7N	5.0E	31	On Frigoris; bright, distinct; c.p.	Greek mathematician	c 428–347 BC
Argelander	16.5S	5.8E	34	Albategnius area; pair with Airy; c.p.	Friedrich; German astronomer	1799–1875
Ariadæus	4.6N	17.3E	11	Vaporum area; associated with great rill.	King of Babylon; chronologist	?–317 BC
Aristarchus	23.7N	47.4W	40	Brilliant; terraced walls; c.p.; inner bands.	Greek astronomer	?310–230 BC
Aristillus	33.9N	1.2E	55	Archimedes group; fine c.p.	Greek astronomer	c 280 BC
Aristoteles	50.2N	17.4E	87	High walls; pair with Eudoxus.	Greek astronomer	383–322 BC
Armstrong	1.4N	25.0E	4	Tranquillitatis; E of Sabine (Sabine E).	Neil; American astronaut (Apollo)	1930–
Arnold	66.8N	35.9E	94	NW of Democritus, N of Frigoris; low walls.	Christoph; German astronomer	1650–1695
Arrhenius	55.6S	91.3W	40	Libration zone; beyond Inghirami.	Svante; Swedish chemist	1859–1927
Artemis	25.0N	25.4W	2	Pair with Verne; between Euler and Lambert.	Greek moon goddess	
Artsimovich	27.6N	36.6W	8	On Imbrium (Diophantus A).	Lev; Russian physicist	1909–1973
Aryabhāta	6.2N	35.1E	22	Between Maskelyne and Cauchy; flooded; irregular (Maskelyne E).	Indian astronomer	476–c 550
Arzachel	18.2S	1.9W	96	Ptolemæus group; high walls; c.p.	Al Zarkala; Spanish–Arab astronomer	c 1028–1087
Asada	7.3N	49.9E	12	NW of Taruntius; edge of Fœcunditatis; distinct (Taruntius A).	Goryu; Japanese astronomer	1734–1799
Asclepi	55.1S	25.4E	42	S uplands; W of Hommel; distinct.	Giuseppe; Italian astronomer	1706–1776
Aston	32.9N	87.7W	43	Limb; beyond Ulugh Beigh.	Francis; British chemist	1877–1945
Atlas	46.7N	44.4E	87	Pair with Hercules; high walls; much interior detail.	Mythological	Greek; Titan
Atwood	5.8S	57.7E	29	NW of Langrenus; trio with Bihara and Naonobu (Langrenus K).	G; British mathematician	1745–1807
Autolycus	30.7N	1.5E	39	Archimedes group; regular, distinct.	Greek astronomer	?–c 330 BC
Auwers	15.1N	17.2E	20	Foothills of Hæmus; not very bright.	Georg Friedrich; German astronomer	1838–1915
Auzout	10.3N	64.1E	32	Outside Crisium; low c.p.	Adrien; French astronomer	1622–1691
Avery	1.4S	81.4E	9	Limb; N Smythii (Gilbert U).	Oswald; Canadian doctor	1877–1955
Avicenna	39.7N	97.2W	74	Libration zone; N of Lorentz.	Abu Ali Ibn Sina; Persian doctor	980–1037
Azophi	22.1S	12.7E	47	W of Nectaris; well-formed pair with Abenezra.	Al-Sûfi; Persian astronomer	903–986
Baade	44.8S	81.8W	55	Limb; beyond Schickard and Inghirami.	Walter; German astronomer	1893–1960
Babbage	59.7N	57.1W	143	Irregular enclosure near Pythagoras.	Charles; British mathematician	1792–1871
Babcock	4.2N	93.9E	99	Libration zone; Smythii area, beyond Neper.	Harold; American astronomer	1882–1968
Back	1.1N	80.7E	35	Limb; S of Schubert (Schubert B).	Ernst; German astronomer	1881–1959
Bacon	51.0S	19.1E	69	Licetus area; high walls; low c.p.	Roger; British natural philosopher	1214–1294
Baillaud	74.6N	37.5E	89	Uplands NE of Meton; rather low walls.	Benjamin; French astronomer	1848–1934
Bailly	66.5S	69.1W	287	'Field of ruins'; uplands in the far S.	Jean Sylvain; French astronomer	1736–1793
Baily	49.7N	30.4E	26	Frigoris area, N of Bürg.	Francis; British astronomer	1774–1844
Balboa	19.1N	83.2W	69	SE of Otto Struve.	Vasco de; Spanish explorer	1475–1517
Ball	35.9S	8.4W	41	On edge of Deslandres; high walls.	William; British astronomer	?–1690

Table 3.13. (Continued)

Name	Lat.	Long.	Diameter (km)	Notes	Name	
Balmer	20.3S	69.8E	138	Ruined wall plain W of Hekatæus.	Johann; Swiss mathematician	1825–1898
Banachiewicz	5.2N	80.1E	92	Libration zone; N of Schubert.	Tadeusz; Polish astronomer	1882–1954
Bancroft	28.0N	6.4W	13	Deep; NW of Archimedes (Archimedes A).	W D; American chemist	1867–1953
Banting	26.6N	16.4E	5	In Serenitatis, E of Linné (Linné E).	Frederick Grant; Canadian doctor	1891–1941
Barkla	10.7S	67.2E	42	Complex between Langrenus and Kapteyn (Langrenus A).	Charles; British physicist	1877–1944
Barnard	29.5S	85.6E	105	Limb; beyond Ansgarius and Legendre.	Edward; American astronomer	1857–1923
Barocius	44.9S	16.8E	82	Outside Maurolycus; high but broken walls.	Francesco; Italian mathematician	c 1570
Barrow	71.3N	7.7E	92	NW of W C Bond; low, broken walls.	Isaac; British mathematician	1630–1677
Bartels	24.5N	89.8W	55	Limb; beyond Otto Struve.	Julius; German geophysicist	1899–1964
Bayer	51.6S	35.0W	47	Outside Schiller; high, terraced walls.	Johann; German astronomer	1572–1625
Beals	37.3S	86.5E	48	Limb; beyond Gauss.	Carlyle F; Canadian astronomer	1899–1979
Beaumont	18.0S	28.8E	53	Bay on Nectaris.	Leonce; French geologist	1798–1874
Beer	27.1N	9.1W	9	On Imbrium; twin with Feuillée.	Wilhelm; German selenographer	1797–1850
Behaim	16.5S	79.4E	55	E of Vendelinus; high walls; c. craters.	Martin; German geographer	1436–1506
Belkovich	61.1N	90.2E	214	Humboldtianum area; high walls with 2 craters; c. peaks.	Igor; Russian astronomer	1904–1949
Bell	21.8N	96.4W	86	Libration zone; beyond Einstein.	Alexander; Scottish inventor	1847–1922
Bellot	12.4S	48.2E	17	Edge of Fœcunditatis; bright floor.	Joseph; French explorer	1826–1853
Bernouilli	35.0N	60.7E	47	E of Geminus; fairly regular.	Jacques; Swiss mathematician	1667–1748
Berosus	33.6N	69.9E	74	E of Cleomedes; terraced walls. Pair with Hahn.	Babylonian astronomer	c 250 BC
Berzelius	36.6N	50.9E	50	Taurus area; darkish area; c.p.	Jons; Swedish chemist	1779–1848
Bessarion	14.9N	37.3W	10	Deep craterlet on Imbrium; S of Brayley.	Johannes; Greek scholar	c 1369–1472
Bessel	21.8N	17.9E	15	On Serenitatis; associated with a long ray.	Friedrich Wilhelm; German astronomer	1784–1846
Bettinus	63.4S	44.8W	71	Bailly area; one of a line; high walls.	Mario; Italian mathematician	1582–1657
Bianchini	48.7N	34.3W	38	In Jura Mts; c.p.; rather irregular.	Francesco; Italian astronomer	1662–1729
Biela	54.9S	51.3E	76	Vlacq area; high walls; c.p.	Wilhelm von; Austrian astronomer	1782–1856
Biharz	5.8S	56.3E	43	Trio with Naonobu and Atwood (Langrenus F).	T; German doctor	1825–1862
Billy	13.8S	50.1W	45	Pair with Hansteen; very dark floor; S edge of Procellarum.	Jacques de; French mathematician	1602–1679
Biot	22.6S	51.1E	12	In Fœcunditatis; very bright.	Jean-Baptiste; French astronomer	1774–1862
Birmingham	65.1N	10.5W	92	N of Frigoris; low-walled, irregular.	John; Irish astronomer	1829–1884
Birt	22.4S	8.5W	16	On Nubium, W of Straight Wall; rill to the W; profile irregular.	William; British selenographer	1804–1881
Black	9.2S	80.4E	18	NE of La Peyrouse (Kästner F).	Joseph; French chemist	1728–1799
Blagg	1.3N	1.5E	5	Quite distinct; in Sinus Medii.	Mary; British astronomer	1858–1944
Blancanus	63.8S	21.4W	117	Near Clavius; pair with Scheiner; high walls.	Giuseppe Biancani; Italian mathematician	1566–1624
Blanchard	58.5S	94.4W	40	Libration zone; N of Hausen, beyond Pingré.	J P; French aeronaut	1753–1809
Blanchinus	25.4S	2.5E	61	N of Werner; uneven walls; rough floor.	Giovanni Blanchini; Italian astronomer	c 1458
Bobillier	19.6N	15.5E	6	In Serenitatis; E of Sulpicius Gallus.	E; French geometer	1798–1840
Bode	6.7N	2.4W	18	Outside Æstuum; bright; minor ray-centre.	Johann Elert; German astronomer	1747–1826
Boethius	5.6N	72.3E	10	Well-formed (Dubiago U).	Greek physicist	c 480–524
Boguslawsky	72.9S	43.2E	97	Southern uplands; high walls.	Palon von; German astronomer	1789–1851
Bohnenberger	16.2S	40.0E	33	Edge of Nectaris; low walls.	Johann von; German astronomer	1765–1831
Bohr	12.4N	86.6W	71	Limb; beyond Vasco da Gama.	Niels; Danish physicist	1885–1962
Boltzmann	74.9S	90.7W	76	Libration zone; closely N of Drygalski.	Ludwig; Austrian physicist	1844–1906
Bombelli	5.3N	56.2E	10	E of Apollonius (Apollonius T).	R; Italian mathematician	1526–1572
Bond, G P	33.3S	35.7W	23	E of Posidonius; fairly regular.	George P; American astronomer	1825–1865
Bond, W C	65.4N	3.7E	156	N of Frigoris; old and broken.	William Cranch; American astronomer	1789–1859
Bonpland	8.3S	17.4W	60	In Nubium; Fra Mauro group; fairly regular.	Aimé; French botanist	1773–1858
Boole	63.7N	87.4W	63	Limb; beyond Pythagoras.	George; British mathematician	1815–1864
Borda	25.1S	46.6E	44	W of Petavius; low walls.	Jean; French astronomer	1733–1799
Borel	22.3N	26.4E	4	In Serenitatis; SW of Le Monnier (Le Monnier C).	Felix; French mathematician	1871–1956
Born	6.0S	66.8E	14	Distinct; SE of Langrenus (Maclaurin Y).	Max; German physicist	1882–1970
Boscovich	9.8N	11.1E	46	Edge of Vaporum; low walls; irregular; very dark floor.	Ruggiero; Italian physicist	1711–1787
Boss	45.8N	89.2E	47	Limb; beyond Mercurius.	Lewis; American astronomer	1846–1912
Bouguer	52.3N	35.8W	22	In Jura uplands; very distinct.	Pierre; French hydrographer	1698–1758
Boussingault	70.2S	54.6E	142	S uplands; made up of 3 large rings.	Jean; French chemist	1802–1887
Bowen	17.6N	9.1E	8	N of Manilius; flattish floor (Manilius A).	Ira; American astronomer	1898–1973
Brackett	17.9N	23.6E	8	In Serenitatis; N of Plinius; inconspicuous.	Frederick; American physicist	1896–
Bragg	42.5N	102.9W	84	Libration zone; beyond Gerard.	William; Australian physicist	1862–1942
Brayley	20.9N	36.9W	14	On Procellarum; low c.p.	Edward; British geographer	1801–1870
Breislak	48.2S	18.3E	49	SE of Maurolycus; fairly regular; c.p.	Scipione; Italian chemist	1748–1826
Brenner	39.0S	39.3E	97	Broken; adjoins Janssen (S uplands).	'Leo' (assumed name); Austrian astronomer	1855–1928?
Brewster	23.3N	34.7E	10	Between Rømer and Littrow (Rømer L).	David; Scottish optician	1781–1868
Brianchon	75.0N	86.2W	134	Large c.p. enclosure; beyond Carpenter.	Charles; French mathematician	1783–1864
Briggs	26.5N	69.1W	37	On Procellarum; Otto Struve area; similar to Seleucus.	Henry; British mathematician	1556–1630
Brisbane	49.1S	68.5E	44	Australe area; fairly regular.	Sir Thomas; Scottish geographer.	1770–1860
Brown	46.4S	17.9W	34	NE of Longomontanus; irregular.	Ernest; British mathematician	1866–1938
Bruce	1.1N	0.4E	6	Distinct; in Sinus Medii.	Catherine Wolfe; American philanthropist	1816–1900
Brunner	9.9S	90.9E	53	Libration zone; beyond Kästner; Hirayama area.	William; Swiss astronomer	1878–1958
Buch	38.8S	17.7E	53	Maurolycus area; adjoins Büsching.	Christian von; German geologist	1774–1853
Bullialdus	20.7S	22.2W	60	On Nubium; massive walls; terraced; c.p.	Ismael Bouillaud; French astronomer	1605–1694
Bunsen	41.4N	85.3W	52	Between Gerard and La Voisier.	Robert; German physicist	1811–1899
Burckhardt	31.1N	56.5E	56	N of Cleomedes; member of a complex group.	Johann; German astronomer	1773–1825
Burnham	13.9S	7.3E	24	Albategnius area; low walls.	Sherburne; American astronomer	1838–1921
Bürg	45.0N	28.2E	39	N of Lacus Mortis; large c.p. with pit; major rill system nearby.	Johann; Austrian astronomer	1766–1834
Büsching	38.0S	20.0E	52	Adjoins Buch, but is less regular.	Anton; German geographer	1724–1793

Table 3.13. (Continued)

Name	Lat.	Long.	Diameter (km)	Notes	Name	
Byrd	85.3N	9.8E	93	N polar area; walled plain adjoining Gioja.	Richard; American explorer	1888–1957
Byrgius	24.7S	65.3W	87	W of Humorum; Byrgius A, on its E crest, is a ray-centre.	Joost Burgi; Swiss horologist	1552–1632
Cabæus	84.9S	35.5W	98	S polar uplands; fairly high walls.	Niccolo Cabeo; Italian astronomer	1586–1650
Cajal	12.6N	31.1E	9	On Tranquillitatis, E of Jansen (Jansen F).	Santiago; Spanish doctor	1852–1934
Calippus	38.9N	10.7E	32	N end of Caucasus; irregular.	Greek astronomer	c 330 BC
Cameron	6.2N	45.9E	10	Intrudes into NW wall of Taruntius (Taruntius C).	Robert; American astronomer	1925–1972
Campanus	28.0S	27.8W	48	W edge of Nubium; pair with Mercator, but with a lighter floor.	Giovanni Campano; Italian astronomer	c 1200–?
Cannizarro	55.6N	99.6W	56	Libration zone; in Poczobut.	Stanislao; Italian chemist	1826–1910
Cannon	19.9N	81.4E 3	56	Limb N of Marginis; light floor.	Annie Jump; American astronomer	1863–1941
Capella	7.5S	35.0E	90	Uplands N of Nectaris; large c.p.; cut by crater valley.	Martianus; Roman astronomer	c 400–?
Capuanus	34.1S	26.7W	59	Edge of Epidemiarum; domes on floor.	Francesco; Italian astronomer	c 1400–?
Cardanus	13.2S	72.4W	49	On Procellarum; c.p.; pair with Krafft.	Girolamo Cardano; Italian mathematician	1501–1576
Carlini	33.7N	24.1W	10	Bright crateret on Imbrium.	Francesco; Italian astronomer	1783–1862
Carmichael	19.6N	40.4E	20	Well-formed; W of Macrobius, on Tranquillitatis.	Leonard; American psychologist	1898–1973
Carpenter	69.4N	50.9W	59	N of Frigoris; adjoins Anaximander.	James; British astronomer	1840–1899
Carrel	10.7N	26.7E	15	On Tranquillitatis; c. crater; SW of Jansen (Jansen B).	Alexis; French doctor	1873–1944
Carrillo	2.2S	80.9E	16	N of Kästner.	Flores; Mexican soil engineer	1911–1967
Carrington	44.0N	62.1E	30	Messala group.	Richard; British astronomer	1826–1875
Cartan	4.2N	59.3E	15	Regular; closely W of Apollonius (Apollonius D).	Elié; French mathematician	1869–1951
Casatus	72.8S	29.5W	108	S of Clavius; high walls; intrudes into Klaproth.	Paolo Casani; Italian mathematician	1617–1707
Cassini	40.2N	4.6E	56	Edge of Nebularum; low walls; contains a deep crater, A.	Giovanni; Italian astronomer	1625–1712
Cassini, J J	68.0N	16.0W		Near Philolaus; irregular ridge-bounded area.	Jacques J; Italian–French astronomer	1677–1756
Catalán	45.7S	87.3W	25	Limb; beyond Schickard and Baade.	Miguel; Spanish spectroscopist	1894–1957
Catharina	18.1S	23.4E	104	Theophilus group; rough floor; no c.p.	Greek theologian (St Catherine)	?–c 307
Cauchy	9.6N	38.6E	12	Bright crater in Tranquillitatis.	Augustin; French mathematician	1789–1857
Cavalerius	5.1N	66.8W	57	In Hevel group; central ridge.	Buonaventura Cavalieri; Italian mathematician	1598–1647
Cavendish	24.5S	53.7W	56	W of Humorum; fairly high walls.	Henry; British chemist	1731–1810
Caventou	29.8N	29.4W	3	On Imbrium; fairly prominent (Lahire D).	Joseph; French chemist	1795–1877
Cayley	4.0N	15.1E	14	V bright; uplands W on Tranquillitatis.	Arthur; British astronomer	1821–1895
Celsius	34.1S	20.1E	36	Rabbi Levi group; rather elliptical.	Anders; Swedish astronomer	1701–1744
Censorinus	0.4S	32.7E	3	Brilliant; uplands SE of Tranquillitatis.	Roman astronomer	238–?
Cepheus	40.8N	45.8E	39	Somniorum area; forms a pair with Franklin.	Mythological character	—
Chacornac	29.8N	31.7E	51	Edge of Serenitatis; adjoins Posidonius.	Jean; French astronomer	1823–1873
Chadwick	52.7N	101.3W	30	Libration zone; beyond Pingré, near de Roy.	James; British physicist	1891–1974
Challis	79.5N	9.2E	55	N polar area; contact pair with Main.	James; British astronomer	1803–1862
Chamberlin	58.9S	95.7E	58	Libration zone; beyond Hanno.	Thomas; American geologist	1843–1928
Chapman	50.4N	100.7W	71	Libration zone; beyond Galvani.	Sydney; British geophysicist	1888–1970
Chappe	61.2S	91.5W	59	Libration zone.	Jean-Baptiste d'Auteroche; French astronomer	1728–1769
Chevallier	44.9N	51.2E	52	Atlas area; low walls.	Temple; British astronomer	1794–1873
Ching-te	20.0N	30.0E	4	Craterlet on Tranquillitatis, SW of Littrow.	Chinese male name	—
Chladni	4.0N	1.1E	13	Sinus Medii area; bright; abuts on Murchison.	Ernst; German physicist	1756–1827
Cichus	33.3S	21.1W	40	Just S of Nubium; well-formed.	Francesco; Italian astronomer	1257–1327
Clairaut	47.7S	13.9E	75	Maurolycus area; broken walls.	Alexis; French mathematician	1713–1765
Clausius	36.9S	43.8W	24	Schickard area; bright and distinct.	Rudolf; German physicist	1822–1888
Clavius	58.8S	14.1W	245	S uplands; massive walls; no c.p.; curved line of craters on floor.	Christopher; German mathematician	1537–1612
Cleomedes	27.7N	56.0E	125	Crisium area; distorted by Tralles.	Greek astronomer	?–c 50 BC
Cleostratus	60.4N	77.0W	62	Pythagoras area; distinct.	Greek astronomer	?–c 500 BC
Clerke	21.7N	29.8E	6	On Tranquillitatis; W of Littrow (Littrow B).	Agnes; British astronomer	1842–1907
Collins	1.3N	23.7E	2	On Tranquillitatis; E of Sabine (Sabine D).	Michael; American astronaut (Apollo 11)	1930–
Colombo	15.1S	45.8E	76	In Fœcunditatis; irregular; broken by large crater, A.	Christopher Columbus; Spanish explorer	1446–1506
Compton	55.3N	103.8E	182	Libration zone; beyond Humboldtianum; crossed by rill.	Arthur Holly; American physicist	1892–1962
Condamine	53.4N	28.2W	48	Edge of Frigoris; fairly regular.	Charles de la; French physicist	1701–1774
Condon	1.9N	60.4E	34	Dark floor; low walls; N of Webb (Webb R).	Edward; American physicist	1902–1974
Condorcet	12.1N	69.6E	74	Outside Crisium; regular; no c.p.	Jean; French mathematician	1743–1794
Conon	21.6N	2.0E	21	In Apennine uplands; fairly distinct.	Greek astronomer	c 260 BC
Cook	17.5S	48.9E	46	Edge of Fœcunditatis; darkish floor.	James; British explorer	1728–1779
Copernicus	9.7N	20.1W	107	Great ray centre; massive terraced walls; c.p.	Mikołaj Kopernik; Polish astronomer	1473–1543
Couder	4.8S	92.4W	21	Libration zone; N of Orientale.	Andre; French astronomer	1897–1978
Cremona	67.5N	90.6W	85	Libration zone; beyond Pythagoras.	Luigi; Italian mathematician	1830–1903
Crile	14.2N	46.0E	9	Distinct; S of Proclus (Proclus F).	George; American doctor	1864–1943
Crozier	13.5S	50.8E	22	Fœcunditatis area; SE of Colombo; c.p.	Francis; British explorer	1796–1848
Crüger	16.7S	66.8W	45	SW of Prodellarum; regular; v dark floor; no c.p.	Peter; German mathematician	1580–1639
Curie	22.9S	91.0E	151	Libration zone; beyond W Humboldt; pair with Sklodowska.	Pierre; French physicist	1859–1906
Curtis	14.6N	56.6E	2	On Crisium; E of Picard (Picard Z).	Heber; American astronomer	1872–1942
Curtius	67.2S	4.4E	95	Moretus area; massive, terraced walls; regular.	Albert Curtz; German astronomer	1600–1671
Cusanus	72.0N	70.8E	63	Limb beyond Democritus; well-formed; no c.p.	Nikolas Krebs; German mathematician	1401–1464
Cuvier	50.3S	9.9E	75	Licetus area; high walls; c.p.	Georges; French palæontologist	1769–1832
Cyrillus	13.2S	24.0E	98	Theophilus group; low c.p.; rather irregular.	St Cyril; Egyptian theologian	?–444
Cysatus	66.2S	6.1W	48	Moretus area; high walls.	Jean-Baptiste Cysat; Swiss astronomer	1588–1657
D'Arrest	2.3N	14.7E	30	E of Godin; low, broken walls.	Heinrich; German astronomer	1822–1875
da Vinci	9.1N	45.0E	37	N of Taruntius; irregular; low, broken walls.	Leonardo; Italian artist and inventor	1452–1519
Daguerre	11.9S	33.6E	46	In Nectaris; very low walls.	Louis; French photographer	1789–1851

Table 3.13. (Continued)

Name	Lat.	Long.	Diameter (km)	Notes	Name	
Dale	9.6S	82.9E	22	NE of La Peyrouse; trio with Black and Kreiken.	Sir Henry; British physiologist	1875–1968
Dalton	17.1N	84.3W	60	Limb crater; E of Einstein and W of Krafft.	John; British chemist	1766–1844
Daly	5.7N	59.6E	17	Well-formed; twin with Apollonuis F (Apollonius P).	Reginald; Canadian geologist	1871–1957
Damoiseau	4.8S	61.1W	36	E of Grimaldi; very irregular.	Marie; French astronomer	1768–1846
Daniell	35.3N	31.1E	29	Adjoins Posidonius; no c.p.	John; British physicist/meteorologist	1790–1845
Darney	14.5S	23.5W	15	Bright crater in Nubium; Fra Mauro area.	Maurice; French astronomer	1882–1958
Darwin	20.2S	69.5W	120	Grimaldi area; low walls; contains large dome.	Charles; British naturalist	1809–1882
Daubrée	15.7N	14.7E	14	SW of Menelaus; darkish floor (Menelaus S).	Gabriel; French geologist	1814–1896
Davy	11.8S	8.1W	34	Edge of Nubium; irregular walls.	Humphry; British physicist	1778–1829
Dawes	17.2N	26.4E	18	Distinct; between Serenitatis and Tranquillitatis.	William Rutter; British astronomer	1799–1868
De Gasparis	25.9S	50.7W	30	W of Humorum, S of Mersenius; fairly regular.	Annibale; Italian astronomer	1819–1892
De la Rue	59.1N	52.3E	134	Very low, broken walls; adjoins Endymion to the NW.	Warren; British astronomer	1815–1889
De Morgan	3.3N	14.9E	10	Bright craterlet; uplands W of Tranquillitatis.	Augustus; British mathematician	1806–1871
De Roy	55.3S	99.1W	43	Libration zone; beyond Pingré; pair with Chadwick.	Felix; Belgian astronomer	1883–1942
De Sitter	80.1N	39.6E	64	Libration zone; N of Euctemon.	Willem; Dutch astronomer	1872–1934
De Vico	19.7S	60.2W	20	Deep crater; W of Gassendi.	Francesco; Italian astronomer	1805–1848
Debes	29.5N	51.7E	30	Outside Cleomedes; fusion of 2 rings.	Ernst; German cartographer	1840–1923
Dechen	46.1N	68.2W	12	In NW of Procellarum; not bright.	Ernst von; German geologist	1800–1889
Delambre	1.9S	17.5E	51	Tranquillitatis area; high walls.	Jean-Baptiste; French astronomer	1749–1822
Delaunay	22.2S	2.5E	46	Albategnius area, near Faye; irregular.	Charles; French astronomer	1816–1872
De l'Isle	29.9N	34.6W	25	On Procellarum; pair with Diophantus; c.p.	Joseph; French astronomer	1688–1768
Delmotte	27.1N	60.2E	32	N of Crisium; not prominent.	Gabriel; French astronomer	1876–1950
Deluc	55.0S	2.8W	46	Clavius area; walls of moderate height.	Jean; Swiss geologist	1727–1817
Dembowski	2.9N	7.2E	26	Low walls; E of Sinus Medii.	Baron Ercole; Italian astronomer	1815–1881
Democritus	62.3N	35.0E	39	Highlands N of Frigoris; very deep.	Greek astronomer	c 460–360 BC
Demonax	77.9S	60.8E	128	Boguslawski area; fairly regular.	Greek philosopher	?–c 100 BC
Desargues	70.2N	73.3W	85	Limb formation; beyond Anaximander.	Gerard; French mathematician	1593–1662
Descartes	11.7S	15.7E	48	NE of Abulfeda; low, broken walls.	Rene; French mathematician	1596–1650
Deseilligny	21.1N	20.6E	6	Distinct craterlet on Serenitatis.	Jules; French selenographer	1868–1918
Deslandres	33.1S	4.8W	256	Ruined enclosure W of Walter.	Henri; French astrophysicist	1853–1948
Dionysius	2.8N	17.3E	18	Brilliant crater on edge of Tranquillitatis.	Greek astronomer	9–120
Diophantus	27.6N	34.3W	17	On Procellarum; c.p.; pair with de l'Isle.	Greek mathematician	?–c 300
Doerfel	69.1S	107.9W	68	Libration zone; beyond Hansen.	Georg; German astronomer	1643–1688
Dollond	10.4S	14.4E	11	W of Theophilus; borders a large 'ghost'.	John; British optician	1706–1761
Donati	20.7S	5.2E	36	S of Albategnius; irregular; c.p.; pair with Faye.	Giovanni; Italian astronomer	1826–1873
Donner	31.4S	98.0E	58	Libration zone; beyond W Humboldt.	Anders; Finnish astronomer	1873–1949
Doppelmeyer	28.5S	41.4W	63	Bay in Humorum; remnant of c.p.	Johann; German astronomer	1671–1750
Dove	46.7S	31.5E	30	Janssen area; low walls.	Heinrich; German physicist	1803–1879
Draper	17.6N	21.7W	8	In Imbrium; S of Pytheas; one of a pair.	Henry; American astronomer	1837–1882
Drebbel	40.9S	49.0W	30	Schickard area; well-formed.	Cornelius; Dutch inventor	1572–1634
Dreyer	10.0N	96.9E	61	Libration zone; beyond Marginis.	Johann Ludwig Emil; Danish astronomer	1852–1926
Drude	38.5S	91.8W	24	Libration zone; beyond Orientale.	Paul; German physicist	1863–1906
Drygalski	79.3S	84.9W	149	Cabæus area; irregular.	Erich von; German geophysicist	1865–1949
Dubiago	4.4N	70.0E	51	Smythii area; regular.	Dimitri; Russian astronomer	1850–1918
Dugan	64.2N	103.3E	50	Libration zone; beyond Humboldtianum; beyond Belkovich.	Raymond; American astronomer	1878–1940
Dunthorne	30.1S	31.6W	15	Edge of Epidemiarum; broad walls.	Richard; British astronomer	1711–1775
Dziewulski	21.2N	98.9E	63	Libration zone; beyond Marginis.	Wladyslaw; Polish astronomer	1878–1962
Eckert	17.3N	58.3E	2	Craterlet on Crisium; NE of Picard.	Wallace; American astronomer	1902–1971
Eddington	21.3N	72.2W	118	Flooded plain between Seleucus and Otto Struve.	Sir Arthur; British astronomer	1882–1944
Edison	25.0N	99.1E	62	Libration zone; Marginis area; adjoins Lomonosov.	Thomas; American inventor	1847–1931
Egede	48.7N	10.6E	37	Near Alpine Valley; low walls; lozenge-shaped.	Hans; Danish natural historian	1686–1758
Eichstädt	22.6S	78.3W	49	Orientale area; regular.	Lorentz; German mathematician	1596–1660
Eimmart	24.0N	64.8E	46	Near edge of Crisium; regular.	Georg; German astronomer	1638–1705
Einstein	16.3N	88.7W	198	Otto Struve area; contains central crater.	Albert; German physicist/mathematician	1879–1955
Elger	35.3S	29.8W	21	Edge of Epidemiarum; low walls; imperfect.	Thomas Gwyn; English selenographer	1838–1897
Ellison	55.1N	107.5W	36	Libration zone; between Xenophanes and Poczobut.	Mervyn; Irish astronomer	1909–1963
Elmer	10.1S	84.1E	16	Limb crater; beyond La Peyrouse.	Charles; American astronomer	1872–1954
Encke	4.7N	36.6W	28	On Procellarum; Kepler area.	Johann; German mathematician/astronomer	1791–1865
Endymion	53.9N	57.0E	123	Humboldtianum area; darkish floor; no c.p.	Greek mythological character	—
Epigenes	67.5N	4.6W	55	N of Frigoris; broad walls.	Greek astronomer	?–c 200 BC
Epimenides	40.9S	30.2W	27	One of a pair E of Hainzel.	Greek philosopher	c 596 BC
Eppinger	9.4S	25.7W	6	Deep craterlet NE of Riphæans (Euclides D).	H; Czech doctor	1879–1946
Eratosthenes	14.5N	11.3W	58	End of Apennines; terraced; very deep; c.p.	Greek astronomer/geographer	c 276–196 BC
Erro	5.7N	98.5E	61	Libration zone; Smythii area; beyond Babcock.	Luis; Mexican astronomer	1897–1955
Esclangon	21.5N	42.1E	15	On Tranquillitatis; W of Macrobius; low walls (Mercurius L).	Ernest; French astronomer	1876–1954
Euclides	7.4S	29.5W	11	Near Riphæans; lies on bright nimbus.	Euclid; Greek mathematician	?–c 300 BC
Euctemon	76.4N	31.3E	62	Walled plain beyond Meton.	Greek astronomer	?–c 432 BC
Eudoxus	44.3N	16.3E	67	S of Frigoris; pair with Aristoteles.	Greek astronomer	c 408–355 BC
Euler	23.3N	29.2W	27	On Imbrium; minor ray-centre.	Leonhard; Swiss mathematician	1707–1783
Fabbroni	18.7N	29.2E	10	Edge of Mare; NW of Vitruvius (Vitruvius E).	Giovanni; Italian chemist	1752–1822
Fabricius	42.9S	42.0E	78	Intrudes into Janssen; rough floor; c.p.	David; Dutch astronomer	1564–1617

Table 3.13. (Continued)

Name	Lat.	Long.	Diameter (km)	Notes	Name	
Fabry	42.9N	100.7E	184	Libration zone; in Harkhebi.	Charles; French physicist	1867–1945
Fahrenheit	13.1N	61.7E	6	On Crisium; W of Agarum (Picard X).	Gabriel; Dutch physicist	1686–1736
Faraday	42.4S	8.7E	69	Intrudes into Stöfler; irregular.	Michael; British chemist	1791–1867
Faustini	87.3S	77.0E	39	Libration zone.	Arnaldo; Italian polar geographer	1874–1944
Fauth	6.3N	20.1W	12	On Procellarum; S of Copernicus; double crater.	Philipp; German selenographer	1867–1941
Faye	21.4S	3.9E	36	S of Albategnius; irregular; c.p.; pair with Donati.	Hervé; French astronomer	1814–1902
Fedorov	28.2N	37.0W	6	On Imbrium; W of Diophantus.	A; Russian rocket engineer	1872–1920
Fényi	44.9S	105.1W	38	Libration zone; beyond Rydberg and Guthnick.	Gyula; Hungarian astronomer	1845–1927
Fermat	22.6S	19.8E	38	Altai area; distinct.	Pierre de; French mathematician	1601–1665
Fernelius	38.1S	4.9E	65	N of Stöfler; rather irregular.	Jean; French astronomer/doctor	1497–1558
Feuillée	27.4N	9.4W	9	On Imbrium; twin with Beer.	Louis; French natural scientist	1660–1732
Finsch	23.6N	21.3E	4	In Serenitatis; NE of Bessel; darkish floor.	Otto; German zoologist	1839–1917
Firmicus	7.3N	63.4E	56	S of Crisium; dark floor; no c.p.	Julius; Latin astronomer	?–c 330
Flammarion	3.4S	3.7W	74	NW of Ptolemæus; irregular enclosure.	Camille; French astronomer	1842–1925
Flamsteed	4.5S	44.3W	20	On Procellarum; associated with 100 km 'ghost'.	John; British astronomer	1646–1720
Focas	33.7S	93.8W	22	Libration zone; well-formed; beyond Rook Mts.	Ionnas; Greek astronomer	1908–1969
Fontana	16.1S	56.6W	31	Between Billy and Crüger; well-formed c.p.	Francesco; Italian astronomer	c 1585–1656
Fontenelle	63.4N	18.9W	38	N edge of Frigoris; deep and distinct.	Bernard de; French astronomer	1657–1757
Foucault	50.4N	39.7W	23	Jura area; bright and deep.	Leon; French physicist	1819–1868
Fourier	30.3S	53.0W	51	W of Humorum; terraced, with central crater.	Jean-Baptiste; French mathematician	1768–1830?
Fox	0.5N	98.2E	24	Libration zone; Smythii area; regular.	Philip; American astronomer	1878–1944
Fra Mauro	6.1S	17.0W	101	On Nubium; low walls; trio with Bonpland and Parry.	Italian geographer	?–1459
Fracastorius	21.5S	33.2E	112	Great bay at S of Nectaris.	Girolamo Fracastoro; Italian astronomer	1483–1553
Franck	22.6N	35.5E	12	On Mars; E of Rømer (Rømer K).	James; German physicist	1882–1964
Franklin	38.8N	47.7E	56	Regular; SE of Atlas.	Benjamin; American inventor	1706–1790
Franz	16.6N	40.2E	25	Edge of Somnii; low walls.	Julius; German astronomer	1847–1913
Fraunhofer	39.5S	59.1E	56	S of Furnerius; NW wall broken by craters.	Joseph von; German optician	1787–1826
Fredholm	18.4N	46.5E	14	Highlands S of Macrobius (Macrobius D).	Erik; Swedish mathematician	1866–1927
Freud	25.8N	52.3W	2	Craterlet adjoining Schröter's Valley.	Sigmund; Austrian psychoanalyst	1856–1939
Froelich	80.3N	109.7W	58	Libration zone; beyond Mouchez; pair with Lovelace.	Jack; American rocket scientist	1921–1967
Fryxell	21.3S	101.4W	18	Libration zone; beyond Orientale.	Roald; American geologist	1934–1974
Furnerius	36.0S	60.6E	135	In Petavius chain; rather broken walls.	Georges Furner; French mathematician	c 1643
Galen	21.9N	5.0E	10	S of Aratus, E of Conon (Aratus A).	Claudius; Greek doctor	c 129–200
Galilaei	10.5N	62.7W	15	On Procellarum; obscure.	Galileo; Italian scientist	1564–1642
Galle	55.9N	22.3E	21	On Frigoris; distinct.	Johann; German astronomer	1812–1910
Galvani	49.6N	84.6W	80	Limb; beyond Repsold.	Luigi; Italian physicist	1737–1798
Gambart	1.0N	15.2W	25	On Procellarum, SSE of Copernicus; regular; low walls.	Jean; French astronomer	1800–1836
Ganswindt	79.6S	110.3E	74	Libration zone; beyond Demonax; adjoins Schrödinger.	Hermann; German rocket inventor	1856–1934
Gardner	17.7N	34.6E	18	Distinct; E of Vitruvius (Vitruvius A).	Irvine; American physicist	1889–1972
Gärtner	59.1N	34.6E	115	Bay on Frigoris; 'seaward' wall barely traceable.	Christian; German geologist	c 1750–1813
Gassendi	17.6S	40.1W	101	Edge of Humorum; 'seaward' wall low; c.p.; many rills on floor.	Pierre; French astronomer	1592–1655
Gaudibert	10.9S	37.8E	34	Edge of Nectaris, low walls.	Casimir; French astronomer	1823–1901
Gauricus	33.8S	12.6W	79	Pitatus group; irregular outline.	Luca Gaurico; Italian astronomer	1476–1558
Gauss	35.7N	79.0E	177	High walls; c.p.; along limb from Humboldtianum.	Karl; German mathematician	1777–1855
Gay-Lussac	13.9N	20.8W	26	N of Copernicus; irregular enclosure.	Joseph; French physicist	1778–1850
Geber	19.4S	13.9E	44	Regular; between Almanon and Abulfeda.	Jabir ben Aflah; Arab astronomer	c 1145
Geissler	2.6S	76.5E	16	Small crater W of Smythii (Gilbert D).	Heinrich; German physicist	1814–1879
Geminus	34.5N	56.7E	85	Crisium area; broad, terraced walls; c. hill.	Greek astronomer	?–c 70 BC
Gemma Frisius	34.2S	13.3E	87	N of Maurolycus; high but broken walls.	Reinier; Dutch doctor	1508–1555
Gerard	44.5N	80.0W	90	W of Sinus Roris; fairly distinct.	Alexander; Scottish explorer	1792–1839
Gernsback	36.5S	99.7E	48	Libration zone; beyond Australe.	Hugo; American writer	1884–1967
Gibbs	18.4S	84.3E	76	Limb; NW of Hekatæus.	Josiah; American physicist	1839–1903
Gilbert	3.2S	76.0E	112	Walled plain E of Smythii; NW of Kästner.	Grove; American geologist	1843–1918
Gill	63.9S	75.9E	66	Beyond Rosenberger.	Sir David; Scottish astronomer	1843–1914
Ginzel	14.3N	97.4E	55	Libration zone; beyond Marginis.	Friedrich; Austrian astronomer	1850–1926
Gioja	83.3N	2.0E	41	Polar crater; fairly regular and distinct.	Flavio; Italian inventor	c 1302
Glaisher	13.2N	49.5E	15	W border of Crisium; obscure.	James; British meteorologist	1809–1903
Goclenius	10.0S	45.0E	72	Edge of Fœcunditatis; lava-flooded; low c.p.	Rudolf Göckel; German mathematician	1572–1621
Goddard	14.8N	89.0E	89	Limb beyond Marginis; dark floor.	Robert; American rocket scientist	1882–1945
Godin	1.8N	10.2E	34	S of Vaporum; c.p.; pair with Agrippa.	Louis; French astronomer	1704–1760
Goldschmidt	73.2N	3.8W	113	E of Anaxagoras; N of Frigoris; low, broken walls.	Hermann; German astronomer	1802–1866
Golgi	27.8N	60.0W	5	Small distinct craterlet N of Schiaparelli.	Camillo; Italian doctor	1843–1926
Goodacre	32.7S	14.1E	46	E of Aliacensis; low c.p.	Walter; British selenographer	1856–1938
Gould	19.2S	17.2W	34	'Ghost' in Nubium, E of Bullialdus.	Benjamin; American astronomer	1824–1896
Graff	42.4S	88.6W	36	Limb formation, beyond Schickard.	Kasimir; Polish astronomer	1878–1950
Greaves	13.2N	52.7E	13	Deep craterlet on Crisium; N of Lick (Lick D).	William; British astronomer	1897–1955
Grimaldi	5.5S	68.3W	172	W of Procellarum; no c.p.; very dark floor; irregular, low walls.	Francesco; Italian physicist/astronomer	1618–1663
Grove	40.3N	32.9E	28	In Somniorum; bright and deep.	Sir William; British physicist	1811–1896
Gruemberger	66.9S	10.0W	93	Moretus area; high walls.	Christoph; Austrian astronomer	1561–1636
Gruithuisen	32.9N	39.7W	15	Bright craterlet on Procellarum contains a very deep crater.	Franz von; German astronomer	1774–1852
Guericke	11.5S	14.1W	63	Fra Mauro group; broken, irregular walls.	Otto von; German physicist	1602–1686
Gum	40.4S	88.6E	54	Australe area beyond Marinus; shallow, flooded.	Colin; Australian astronomer	1924–1960
Gutenberg	8.6S	41.2E	74	Edge of Fœcunditatis; near Goclenius; irregular.	Johann; German inventor	c 1398–1468

Table 3.13. (Continued)

Name	Lat.	Long.	Diameter (km)	Notes	Name	
Guthnick	47.7S	93.9W	36	Libration zone; pair with Rydberg.	Paul; German astronomer	1879–1947
Gyldén	5.3S	0.3E	47	N of Ptolemæus; partly lava-filled; crater valley to W.	Hugo; Swedish astronomer	1841–1896
Hagecius	59.8S	46.6E	76	Vlacq group; wall broken by craters.	Thaddeus Hayek; Czech astronomer	1525–1600
Hahn	31.3N	73.6E	84	Crisium area; regular; c.p.; pair with Berosus.	Friedrich von; German astronomer	1879–1968
Haidinger	39.2S	25.0W	22	S of Epidemiarum; inconspicuous.	Wilhelm von; Austrian geologist	1795–1871
Hainzel	41.3S	33.5W	70	N of Schiller; compound, 2 coalesced rings.	Paul; German astronomer	c 1570
Haldane	1.7S	84.1E	37	Limb formation; Smythii area.	John; British biochemist	1892–1964
Hale	74.2S	90.8E	83	Libration zone; beyond Boussingault; terraced.	George Ellery; American astronomer	1868–1938
Hall	33.7N	37.0E	35	Somniorum area; E of Posidonius.	Asaph; American astronomer	1829–1907
Halley	8.0S	5.7E	36	Hipparchus group; regular; pair with Hind.	Edmond; British astronomer	1656–1742
Hamilton	42.8S	84.7E	57	Limb, beyond Oken; deep and regular.	Sir William; Irish mathematician	1805–1865
Hanno	56.3S	71.2E	56	Australe area; darkish floor.	Roman explorer	c 500 BC
Hansen	14.0N	72.5E	39	Regular; similar to Alhazen; Crisium area near Agarum.	Peter; Danish astronomer	1795–1874
Hansky	9.7S	97.0E	43	Libration zone; Smythii area, SE of Hirayama.	Alexei; Russian astronomer	1870–1908
Hansteen	11.5S	52.0W	44	Regular; S edge of Procellarum; pair with Billy.	Christopher; Norwegian astronomer	1784–1873
Harding	43.5N	71.7W	22	Sinus Roris; low walls.	Karl; German astronomer	1765–1834
Hargreaves	2.2S	64.0E	16	Irregular; E of Maclaurin (Maclaurin S).	Frank; British optician	1891–1970
Harkhebi	39.6N	98.3E	237	Libration; eroded, incomplete; N of Marginis; contains Fabry.	Egyptian astronomer	c 300 BC
Harpalus	52.6N	43.4W	39	Edge of Frigoris; deep, prominent.	Greek astronomer	c 460 BC
Hartwig	6.1S	80.5W	79	W of Grimaldi; adjoins Schlüter to the E.	Carl; German astronomer	1851–1923
Hase	29.4S	62.5E	83	S of Petavius; rather irregular	Johann; German mathematician	1684–1742
Hausen	65.0S	88.1W	167	Bailly area; c.p.	Christian; German astronomer	1693–1743
Hayn	64.7N	85.2E	87	Limb; beyond Strabo	Friedrich; German astronomer	1863–1928
Hédervári	81.8S	84.0E	69	Libration zone.	Peter; Hungarian astronomer	1931–1984
Hekatæus	21.8S	79.4E	167	SE of Vendelinus; irregular walls; c.p.	Greek geographer	c 476 BC
Heinrich	24.8N	15.3W	6	In Imbrium; NW of Timocharis (Timocharis A).	Wladimir; Czech astronomer	1884–1965
Heinsius	39.5S	17.7W	64	Tycho; irregular; 3 craters on its S wall.	Gottfried; German astronomer	1709–1769
Heis	32.4N	31.9W	14	Bright craterlet on Imbrium.	Eduard; German astronomer	1806–1877
Helicon	40.4N	23.1W	24	In Imbrium; pair with Le Verrier.	Greek astronomer	?–c 400 BC
Hell	32.4S	7.8W	33	In W of Deslandres; low c.p.	Maximilian; Hungarian astronomer	1720–1792
Helmart	7.6S	87.6E	26	Libration zone; Smythii area; adjoins Kao.	Friedrich; German astronomer	1843–1917
Helmholtz	68.1S	64.1E	94	Fairly regular; S uplands; Boussingault area.	Hermann von; German scientist	1821–1894
Henry, Paul	23.5S	58.9W	42	W of Humorum; distinct pair with Prosper Henry.	Paul Henry; French astronomer	1848–1905
Henry, Prosper	23.5S	58.9W	42	Pair with Paul Henry.	Prosper; French astronomer	1849–1903
Heraclitus	49.2S	6.2E	90	Very irregular; BC Cuvier–Licetus group, S of Stöfler.	Greek philosopher	c 540–480
Hercules	46.7N	39.1E	69	Bright walls; deep interior crater; pair with Atlas.	Greek mythological hero	—
Herigonius	13.3S	33.9W	15	NE of Gassendi; bright; c.p.	Pierre Hérigone; French astronomer	c 1644
Hermann	0.9S	57.0W	15	On Procellarum; E of Lohrmann; bright.	Jacob; Swiss mathematician	1678–1833
Hermite	86.0N	89.9W	104	Well-formed; limb; beyond Anaxagoras.	Charles; French mathematician	1822–1901
Herodotus	23.2N	49.7W	34	Fairly regular; pair with BC Aristarchus; great valley nearby.	Greek historian	c 484–408
Herschel	5.7S	2.1W	40	N of Ptolemæus; terraced walls; large c.p.	William; Hanoverian/British astronomer	1738–1822
Herschel, Caroline	34.5N	31.2W	13	On Imbrium; bright; group with Carlini and de l'Isle.	Caroline; Hanoverian/British astronomer	1750–1848
Herschel, John	62.0N	42.0W	165	N of Frigoris; ridge-bordered enclosure.	John; British astronomer	1792–1871
Hesiodus	29.4S	16.3W	42	Companion to Pitatus; rill runs SW from it.	Hesiod; Greek author	c 735 BC
Hevel	2.2N	67.6W	115	Grimaldi chain; convex floor; with low c.p. and several rills.	Johann Hewelcke; Polish astronomer	1611–1687
Heyrovsky	39.6S	95.3W	16	Libration zone; beyond Cordilleras.	Jaroslav; Czech chemist	1890–1967
Hill	20.9N	40.8E	16	Edge of Mare, W of Macrobius (Macrobius B).	George; American astronomer	1838–1914
Hind	7.9S	7.4E	29	Hipparchus group; pair with Halley; regular.	John Russell; British astronomer	1823–1895
Hippalus	24.8S	30.2W	57	Bay on Humorum; remnant of c.p.; associated with rills.	Greek explorer	?–c 120
Hipparchus	5.1S	5.2E	138	Low-walled, irregular; pair with Albategnius.	Greek astronomer	c 140 BC
Hirayama	6.1S	93.5E	132	Libration zone; Smythii area; regular.	Kiyotsugu; Japanese astronomer	1874–1943
Hohmann	17.9S	94.1W	16	Libration zone; small crater in Orientale.	Walter; German space engineer	1880–1945
Holden	19.1S	62.5E	47	S of Vendelinus; deep.	Edward; American astronomer	1846–1914
Hommel	54.7S	33.8E	126	S uplands; 2 large craters in floor.	Johann; Greek astronomer	1518–1562
Hooke	41.2N	54.9E	36	W of Messala; fairly regular.	Robert; British scientist	1635–1703
Hornsby	23.8N	12.5E	3	Craterlet in Serenitatis; between Linné and Sulpicius Gallus.	Thomas; British astronomer	1733–1810
Horrebow	58.7N	40.8W	24	Pair with Robinson; outside J Herschel; deep.	Peder; Danish astronomer	1679–1764
Horrocks	4.0S	5.9E	30	Within Hipparchus; regular.	Jeremiah; British astronomer	1619–1641
Hortensius	6.5N	28.0W	14	On Procellarum; bright; domes to N.	Hove, Martin van den; Dutch astronomer	1605–1639
Houtermans	9.4S	87.2E	29	Limb; SE of Kästner.	Friedrich; German physicist	1903–1966
Hubble	22.1N	86.9E	80	Limb; SE of Plutarch; partly flooded.	Edwin; American astronomer	1889–1953
Huggins	41.1S	1.4W	65	Between Nasireddin and Orontius; irregular.	Sir William; British astronomer	1824–1910
Humason	30.7N	56.6W	4	On Procellarum; E of Lichtenberg (Lichtenberg G).	Milton; American astronomer	1891–1972
Humboldt, Wilhelm	27.0S	80.9E	189	E of Petavius; rills on floor.	Wilhelm von; German philologist	1767–1835
Hume	4.7S	90.4E	23	Libration zone; Smythii area; bordering Hirayama.	David; Scottish philosopher	1711–1776
Huxley	20.2N	4.5W	4	Apennine region; E of Imbrium (Wallace B).	Thomas; British biologist	1825–1895
Hyginus	7.8N	6.3E	9	Depression in Vaporum; great crater-rill.	Caius; Spanish astronomer	?–100?
Hypatia	4.3S	22.6E	40	S of Tranquillitatis; low walls; irregular.	Egyptian mathematician	?–415
Ibn Battuta	6.9S	50.4E	11	Prominent; NE of Goclenius (Goclenius A).	Moroccan geographer	1304–1377
Ibn Rushd	11.7S	21.7E	32	Between Cyrillus and Kant (Cyrillus B).	Spanish astronomer/philosopher	1126–1198
Ibn Yunis	14.1N	91.1E	58	Libration zone; Marginis area; adjoins Goddard.	Averrdes; Egyptian astronomer	950–1009

Table 3.13. (Continued)

Name	Lat.	Long.	Diameter (km)	Notes	Name	
Ideler	49.2S	22.3E	38	SE of Autolycus; distinct.	Christian; German astronomer	1766–1846
Idelson	81.5S	110.9E	60	Libration zone; Demonax area.	Naum; Russian astronomer	1855–1951
Ilyin	17.8S	97.5W	13	Libration zone; in Orientale.	N; Russian rocket scientist	1901–1937
Inghirami	47.5S	68.8W	91	Schickard area; regular; c.p.	Giovanni; Italian astronomer	1779–1851
Isidorus	8.0S	33.5E	42	Outside Nectaris; deep; pair with Capella.	St Isidore; Roman astronomer	570–636
Ivan	26.9N	43.3W	4	Small craterlet NE of Prinz (Prinz B).	Russian male name	—
Jacobi	56.7S	11.4E	68	Heraclitus group; fairly distinct.	Karl; German mathematician	1804–1851
Jansen	13.5N	28.7E	23	In Tranquillitatis; low walls, darkish floor.	Janszoon; Dutch optician	1580–c 1638
Janssen	45.4S	40.3E	199	S uplands; great ruin, broken in the N by Fabricius.	Pierre Jules; French astronomer	1842–1907
Jeans	55.8S	91.4E	79	Beyond Hanno; libration zone; between Chamberlin and Lyot.	Sir James; British astronomer	1877–1946
Jehan	20.7N	31.9W	5	On Imbrium; SW of Euler (Euler K).	Turkish female name	—
Jenkins	0.3N	78.1E	38	W of Smythii.	Louise; American astronomer	1888–1970
Jenner	42.1S	95.9E	71	Libration zone; Australe area.	Edward; British doctor	1749–1823
Joliot	25.8N	93.1E	164	Libration zone; NE of Marginis; interior detail.	Fréderic Joliot-Curie; French physicist	1900–1958
Joy	25.0N	6.6E	5	Craterlet in Hæmus foothills.	Alfred; American astronomer	1882–1973
Julius Cæsar	9.0N	15.4E	90	Vaporum area; low, irregular walls; v dark floor.	Roman emperor	c 102–44 BC
Kaiser	36.5S	6.5E	52	N of Stöfler; well-marked; no c.p.	Frederik; Dutch astronomer	1808–1872
Kane	63.1N	26.1E	54	N of Frigoris; fairly regular.	Elisha; American explorer	1820–1857
Kant	10.6S	20.1E	33	W of Theophilus; large c.p. with summit pit.	Immanuel; German philosopher	1724–1804
Kao	6.7S	87.6E	34	Adjoins Helmert; between Kästner and Kiess.	Ping-Tse; Taiwan astronomer	1888–1970
Kapteyn	10.8S	70.6E	49	E of Langrenus; inconspicuous.	Jacobus; Dutch astronomer	1851–1922
Kästner	6.8S	78.5E	108	Smythii area; distinct.	Abraham; German mathematician	1719–1800
Keldysh	51.2N	43.6E	33	In Hercules; bright, regular (Hercules A).	Mstislav; Russian mathematician	1911–1978
Kepler	8.1N	38.0W	31	In Procellarum; pair with Encke; major ray-centre.	Johannes; German mathematician/astronomer	1571–1630
Kies	26.3S	22.5W	45	On Nubium; Bullialdus area; very low walls.	Johann; German mathematician/astronomer	1713–1781
Kiess	6.4S	84.0E	63	Limb; beyond Kästner.	Carl; American astrophysicist	1887–1967
Kinau	60.8S	15.1E	41	Jacobi group; high walls; c.p.	C A; German selenographer/botanist	?–1850
Kirch	39.2N	5.6W	11	Bright craterlet in Imbrium.	Gottfried; German astronomer	1639–1710
Kircher	67.1S	45.3W	72	Bettinus chain; Bailly area; very high walls.	Athanasius; German humanitarian	1601–1680
Kirchhoff	30.3N	38.8E	24	Cleomedes area; larger of 2 craters W of Newcomb.	Gustav; German physicist	1824–1887
Klaproth	69.8S	26.0W	119	Darkish floor; S uplands; contact pair with Casatus.	Martin; German mineralogist	1743–1817
Klein	12.0S	2.6E	44	In Albategnius; regular; c.p.	Hermann; German selenographer	1844–1914
Knox-Shaw	5.3N	80.2E	12	Abuts on Banachiewicz (Banachiewicz F).	Harold; British astronomer	1855–1970
Kopff	17.4S	89.6W	41	Limb; Orientale area; W of Crüger.	August; German astronomer	1882–1960
Krafft	16.6N	72.6W	51	On Procellarum; pair with Cardanus; darkish floor.	Wolfgang; German astronomer	1743–1814
Kramarov	2.3S	98.8W	20	Libration zone; Rook area.	G M; Russian space scientist	1887–1970
Krasnov	29.9S	79.6W	40	Irregular; limb; W of Doppelmayer.	Aleksander; Russian astronomer	1866–1907
Kreiken	9.0S	84.6E	23	NE of La Peyrouse; trio with Black and Dale.	E A; Dutch astronomer	1896–1964
Krieger	29.0N	45.6W	22	Aristarchus area; distinct, but walls broken by craterlets.	Johann; German selenographer	1865–1902
Krishna	24.5N	11.3E	3	Craterlet in Serenitatis; S of Linné.	Indian male name	—
Krogh	9.4N	65.7E	19	Well-formed; SE of Auzout (Auzout B).	Schack; Danish zoologist	1874–1949
Krusenstern	26.2S	5.9E	47	Werner area; not prominent.	Adam; Russian explorer	1770–1846
Kugler	53.8S	103.7E	65	Libration zone; beyond Brisbane and Jeans.	Franz Xaver; German chronologist	1862–1929
Kuiper	9.8S	22.7W	6	Prominent; on Mare Cognitum; W of Bonpland (Bonpland E).	Gerard; Dutch astronomer	1905–1973
Kundt	11.5S	11.5W	10	Prominent; Guericke and Davy (Guericke C).	August; German physicist	1839–1894
Kunowsky	3.2N	32.5W	18	On Procellarum; E of Encke; low central ridge.	Georg; German astronomer	1786–1846
La Peyrouse	10.7S	76.3E	77	E of Fœcunditatis; well-formed; pair with Ansgarius.	Jean, Comte de; French explorer	1741–1788
Lacaille	23.8S	1.1E	67	Werner area; irregular.	Nicholas de; French astronomer	1713–1762
Lallemand	14.3S	84.1W	18	W of Rocca (Kopff A).	Andre; French astronomer	1904–1978
Lacroix	37.9S	59.0W	37	N of Schickard; regular; c.p.	Sylvestre; French mathematician	1765–1843
Lade	1.3S	10.1E	55	Highlands S of Godin; low walls.	Heinrich von; German astronomer	1817–1904
Lagalla	44.6S	22.5W	85	Abuts on Wilhelm I; low-walled; irregular.	Giulio; Italian philosopher	1571–1624
Lagrange	32.3S	72.8W	225	W of Humorum; low walls; rough floor.	Joseph; Italian mathematician	1736–1813
Lalande	4.4S	8.6W	24	Ptolemæus area; low c.p.	Joseph de; French astronomer	1732–1807
Lamarck	22.9S	69.8W	100	Ruined plain; S of Darwin and W of Byrgius.	Jean; French natural historian	1744–1829
Lamb	42.9S	100.1E	106	Libration zone; Australe area.	Sir Horace; British mathematician	1849–1934
Lambert	25.8N	21.0W	30	In Imbrium; bright, central crater.	Johann; German astronomer	1728–1777
Lamé	14.7S	64.5E	84	On NE wall of Vendelinus; well-formed.	Gabriel; French mathematician	1795–1870
Lamèch	42.7N	13.1E	13	Eudoxus area; very low, irregular walls.	Felix; French selenographer	1894–1962
Lamont	4.4N	23.7E	106	On Tranquillitatis, Arago area; low walls.	John; Scottish astronomer	1805–1879
Landsteiner	31.3N	14.8W	6	Craterlet on Imbrium; E of Carlini.	Karl; Austrian pathologist	1868–1943
Langley	51.1N	86.3W	59	Limb beyond Repsold; N of Galvani.	Samuel; American astronomer/physicist	1834–1906
Langrenus	8.9S	61.1E	127	Petavius area; massive walls; c.p.	Michel van Langren; Belgian selenographer	c 1600–1675
Lansberg	0.3S	26.6W	38	On Nubium; massive walls; c.p.	Philippe van; Belgian astronomer	1561–1632
Lassell	15.5S	7.9W	23	In Nubium area; Alphonsus area; low walls.	William; British astronomer	1799–1880
Laue	28.0N	96.7W	87	Libration zone; beyond Ulugh Beigh; intrudes into Lorentz.	Max von; German physicist	1879–1960
Lauritsen	27.6S	96.1E	52	Beyond Humboldt; abuts on Curie.	Charles; Danish physicist	1892–1968
La Voisier	38.2N	81.2W	70	W of Procellarum; well-formed.	Antoine; French chemist	1743–1794
Lawrence	7.4N	43.2E	24	Flooded; NW of Taruntius (Taruntius M).	Ernst; American physicist	1901–1958

Table 3.13. (Continued)

Name	Lat.	Long.	Diameter (km)	Notes	Name	
Le Monnier	26.6N	30.6E	60	Bay on Serenitatis; smooth floor.	Pierre; French astronomer	1715–1799
Le Verrier	40.3N	20.6W	20	Distinct crater on Imbrium; pair with Helicon.	Urbain; French astronomer	1811–1877
Leakey	3.2S	37.4E	12	Obscure; near Censorinus (Censorinus F).	Louis; British archæologist	1903–1972
Lebesque	5.1S	89.0E	11	Smythii area; near Warner.	Henri; French mathematician	1875–1941
Lee	30.7S	40.7W	41	S edge of Humorum; damaged by lava.	John; British astronomer	1783–1866
Legendre	28.9S	70.2E	78	W Humboldt area; central ridge.	Adrien; French mathematician	1752–1833
Legentil	74.4S	76.5W	113	S of Bailly; fairly distinct.	Guillaume; French astronomer	1725–1792
Lehmann	40.0S	56.0W	53	Schickard area; irregular.	Jacob; German astronomer	1800–1863
Lepaute	33.3S	33.6W	16	Distinct small crater at edge of Epidemiarum.	Nicole Reine; French astronomer	1723–1788
Letronne	10.8S	42.5W	116	Bay at S edge of Procellarum; low c.p.; N wall destroyed by lava.	Jean; French archæologist	1787–1848
Lexell	35.8S	4.2W	62	Edge of Deslandres; N wall reduced; remnant of c.p.	Anders; Finnish astronomer	1740–1784
Licetus	47.1S	6.7E	74	Cuvier-Heraclitus group, near Stöfler; fairly regular.	Fortunio Liceti; Italian physicist	1577–1657
Lichtenberg	31.8N	67.7W	20	Edge of Procellarum; minor ray-centre.	Georg; German physicist	1742–1799
Lick	12.4N	52.7E	31	Edge of Crisium; incomplete.	James; American benefactor	1796–1876
Liebig	24.3S	48.2W	37	W of Humorum; moderate walls.	Justus, Baron von; German chemist	1803–1873
Lilius	54.5S	6.2E	61	Jacobi group; high walls; c.p.	Luigi; Italian philosopher	?–1576
Lindbergh	5.4S	52.9E	12	Distinct; on Fœcunditatis; SE of Messier (Messier G).	Charles; American aviator	1902–1974
Lindblad	70.4N	98.8W	66	Libration zone; beyond Pythagoras; SW of Brianchon.	Bertil; Swedish astronomer	1895–1965
Lindenau	32.3S	24.9E	53	Rabbi Levi group; terraced walls.	Bernhard von; German astronomer	1780–1854
Lindsay	7.0S	13.0E	32	In highlands; N of Åndel (Dollond C).	Eric; Irish astronomer	1907–1974
Linné	27.7N	11.8E	2	In Serenitatis; craterlet on a light nimbus.	Carl von; Swedish botanist	1707–1778
Liouville	2.6N	73.5E	16	Distinct crater W of Schubert (Dubiago S).	Joseph; French mathematician	1809–1882
Lippershey	25.9S	10.3W	6	In Nubium; Pitatus area; distinct craterlet.	Hans (Jan); Dutch optician	?–1619
Littrow	21.5N	31.4E	30	Edge of Serenitatis; irregular.	Johann; Czech astronomer	1781–1840
Lockyer	46.2S	36.7E	34	Intrudes into Janssen; bright walls.	Sir (Joseph) Norman; British astronomer	1836–1920
Loewy	22.7S	32.8W	24	Edge of Humorum; distinct.	Moritz; French astronomer	1833–1907
Lohrmann	0.5S	67.2W	30	Between Grimaldi and Hevel; fairly regular; c.p.	Wilhelm; German selenographer	1796–1840
Lohse	13.7S	60.2E	41	On W wall of Vendelinus; deep; c.p.	Oswald; German astronomer	1845–1915
Lomonosov	27.3N	98.0E	92	Libration zone; N of Marginis; Joliot group.	Mikhail; Russian astronomer	1711–1765
Longomontanus	49.6S	21.8W	157	Clavius area; complex walls; much floor detail.	Christian; Danish astronomer	1562–1647
Lorentz	32.6N	95.3W	312	Libration zone; huge enclosure; contains Nernst, Röntgen.	Hendrik; Dutch mathematician	1853–1928
Louise	28.5N	34.2W	2	Craterlet between Diophantus and de l'Isle.	French female name	—
Louville	44.0N	46.0W	36	In Jura Mtns; low walls, darkish floor.	Jacques; French astronomer	1671–1732
Lovelace	82.3N	106.4W	54	Libration zone; pair with Froelich; S of Hermite.	William; American space scientist	1907–1965
Lubbock	3.9S	41.8E	13	W edge of Fœcunditatis; fairly bright.	Sir John; British astronomer	1803–1865
Lubiniezky	17.8S	23.8W	43	In Nubium; Bullualdus area; low walls.	Stanislaus; Polish astronomer	1623–1675
Lucian	14.3N	36.7E	7	On Tranquillitatis; W of Lyell (Maraldi B).	Greek writer	125–190
Ludwig	7.7S	97.4E	23	Libration zone; beyond Smythii and Hirayama.	Carl; German physiologist	1816–1895
Luther	33.2N	24.1E	9	Distinct craterlet in N of Serenitatis.	Robert; German astronomer	1822–1900
Lyell	13.6N	40.6E	32	W edge of Somnii; darkish floor.	Sir Charles; Scottish geologist	1797–1875
Lyot	49.8S	84.5E	132	Flooded; irregular wall; dark floor; N of Australe.	Bernard; French astronomer	1897–1952
Maclaurin	1.9S	68.0E	50	W of Smythii; concave floor; uneven walls.	Colin; Scottish mathematician	1698–1746
Maclear	10.5N	20.1E	20	On Tranquillitatis; darkish floor.	Thomas; Irish astronomer	1794–1879
MacMillan	24.2N	7.8W	7	In Imbrium; SW of Archimedes.	William; American astronomer	1871–1948
Macrobius	21.3N	46.0E	64	Crisium area; high walls; compound c.p.	Ambrosius; Roman writer	?–c 140
Mädler	11.0S	29.8E	27	On Nectaris; irregular.	Johann von; German selenographer	1794–1874
Maestlin	4.9N	40.6W	7	On Procellarum; near Encke; obscure.	Michael; German mathematician	1550–1631
Magelhæns	11.9S	44.1E	40	Edge of Fœcunditatis; pair with A; darkish floor.	Fernao de (Magellan); Portuguese explorer	1480–1521
Maginus	50.5S	6.3W	194	Clavius area; irregular walls; obscure near full moon.	Giovanni Magini; Italian astronomer	1555–1617
Main	80.8N	10.1E	46	N polar area; contact twin with Challis.	Robert; British astronomer	1808–1878
Mairan	41.6N	43.4W	40	Jura area; bright, regular.	Jean de; French geophysicist	1678–1771
Malapert	84.9S	12.9E	69	Irregular form; near S. pole, E of Cabæus.	Charles; Belgian astronomer	1581–1639
Mallet	45.4S	54.2E	58	Rheita Valley area; inconspicuous.	Robert; Irish seismologist	1810–1881
Manilius	14.5N	9.1E	38	On edge of Vaporum; brilliant walls; c.p.	Marcus; Roman writer	c 50 BC
Manners	4.6N	20.0E	15	On Tranquillitatis; Arago area; bright.	Russell; British astronomer	1800–1870
Manzinus	67.7S	26.8E	98	Boguslawsky area; high, terraced walls.	Carlo Manzini; Italian astronomer	1599–1677
Maraldi	19.4N	34.9E	39	In N of Tranquillitatis; distinct.	Giovanni; Italian astronomer	1709–1788
Marco Polo	15.4N	2.0W	28	Apennines area; irregular; darkish floor.	Italian explorer	1254–1324
Marinus	39.4S	76.5E	58	Australe area; distinct; c.p.	Greek geographer	c 100
Markov	53.4N	62.7W	40	On Sinus Roris; sharp rim.	Aleksandr; Russian astrophysicist	1897–1968
Marth	31.1S	29.3W	6	In Epidemiarum; concentric crater.	Albert; German astronomer	1828–1897
Maskelyne	2.2N	30.1E	23	In Tranquillitatis; low c.p.	Nevil; British astronomer	1732–1811
Mason	42.6N	30.5E	33	Bürg area; pair with Plana.	Charles; British astronomer	1730–1787
Maunder	14.6S	93.8W	55	Libration zone on N edge Orientale; regular; c.p.	Annie; British astronomer	1868–1947
					Edward; British astronomer	1851–1928
Maupertuis	49.6N	27.3W	45	In Juras; irregular mountain enclosure.	Pierre de; French mathematician	1698–1759
Maurolycus	42.0S	14.0E	114	E of Stöfler; rough floor; central mountain group.	Francesco Maurolico; Italian mathematician	1494–1575
Maury	37.1N	39.6E	17	Atlas area; bright and deep.	Matthew; American oceanographer	1806–1873
Maxwell	30.2N	98.9E	107	Libration zone; beyond Gauss; intrudes into Richardson.	James Clerk; Scottish physicist	1831–1879
Mayer, C	63.2N	17.3E	38	N of Frigoris; rhomboidal.	Charles; German astronomer	1719–1783
Mayer, T	15.6N	29.1W	50	In Carpathians; c.p.	Johann Tobias; German astronomer	1723–1762
McAdie	2.1N	92.1E	45	Libration zone; E of Smythii; low walls.	Alexander; American meteorologist	1863–1943
McClure	15.3S	50.3E	23	Edge of Fœcunditatis; E of Colombo; regular.	Robert; British explorer	1807–1873

Table 3.13. (Continued)

Name	Lat.	Long.	Diameter (km)	Notes	Name	
McDonald	30.4S	20.9W	7	On Imbrium; SE of Carlini (Carlini B).	Thomas; Scottish selenographer	?–1973
McLaughlin	47.1N	92.9W	79	Libration zone; beyond Galvani; rather irregular.	Dean; American astronomer	1901–1965
Mee	43.7S	35.3W	126	Abuts on Hainzel; low, broken walls.	Arthur; Scottish astronomer	1860–1926
Mees	13.6N	96.1W	50	Libration zone; beyond Einstein.	Kenneth; English photographer	1882–1960
Mendel	48.8S	109.4W	138	Libration zone; beyond Orientale.	Gregor; Austrian biologist	1822–1884
Menelaus	16.3N	16.0E	26	In Hæmus Mtns; brilliant; c.p.	Greek astronomer	c 98
Menzel	3.4N	36.9E	3	On Tranquillitatis; E of Maskelyne.	Donald; American astronomer	1901–1976
Mercator	29.3S	26.1W	46	Pair with Campanus; dark floor.	Gerard de; Belgian cartographer	1512–1594
Mercurius	46.6N	66.2E	67	Humboldtianum area; low c.p.	Roman messenger of the gods	
Merrill	75.2N	116.3W	57	Libration zone; beyond Brianchon.	Paul; American astronomer	1887–1961
Mersenius	21.5S	49.2W	84	W of Humorum; convex floor; rills nearby.	Marin Mersenne; French mathematician	1588–1648
Messala	39.2N	60.5E	125	Humboldtianum area; oblong; broken walls.	Ma-Sa-Allah; Jewish astronomer	762–815
Messier	1.9S	47.6E	11	On Fœcunditatis; twin with A; 'comet' to the W.	Charles; French astronomer	1730–1817
Metius	40.3S	43.3E	87	Janssen group; distinct; pair with Fabricius.	Adriaan; Dutch astronomer	1571–1635
Meton	73.6N	18.8E	130	N polar area; smooth floor; compound formation.	Greek astronomer	c 432 BC
Milichius	10.0N	30.2W	12	On Procellarum; bright; dome to the W.	Jacob Milich; German mathematician	1501–1559
Miller	39.3S	0.8E	61	Orontius group; fairly distinct.	William; British chemist	1817–1870
Mitchell	49.7N	20.2E	30	Distinct; abuts on Aristoteles.	Maria; American astronomer	1818–1889
Moigno	66.4N	28.9E	36	W of Arnold; contains central crater; no c.p.	François; French mathematician	1804–1884
Möltke	0.6S	24.2E	6	Distinct; N edge of Tranquillitatis.	Helmuth; German benefactor	1800–1891
Monge	19.2S	47.6E	36	Edge of Fœcunditatis; rather irregular.	Gaspard; French mathematician	1746–1818
Montanari	45.8S	20.6W	76	Longomontanus area; distorted.	Geminiano; Italian astronomer	1633–1687
Moretus	70.6S	5.8W	111	S uplands; very high walls; massive c.p.	Theodore Moret; Belgian astronomer	1602–1667
Morley	2.8S	64.6E	14	NW of Maclaurin (Maclaurin R).	Edward; American chemist	1838–1923
Moseley	20.9N	90.1N	90	Libration zone; beyond Einstein.	Henry; British physicist	1887–1915
Mösting	0.7S	5.9W	24	Medii In; A, to the N, is used as a reference point.	Johan; Danish benefactor	1759–1843
Mouchez	78.3N	26.6W	81	Ruined plain near Philolaus.	Ernest; French astronomer	1821–1892
Moulton	61.1S	97.2E	49	Libration zone; beyond Hanno.	Forest Ray; American astronomer	1871–1952
Müller	7.6S	2.1E	22	Ptolemæus area; fairly regular.	Karl; Czech astronomer	1866–1942
Murchison	5.1N	0.1W	57	Edge of Medii; low-walled; irregular.	Sir Roderick; Scottish geologist	1792–1871
Mutus	63.6S	30.1E	77	S uplands; 2 large craters on floor.	Vincente; Spanish astronomer	?–1673
Nansen	80.9N	05.3E	104	Libration zone; beyond Einstein.	Fridtjof; Norwegian explorer	1861–1930
Naonobu	4.6S	57.8E	34	Trio with Biharz and Atwood (Langrenus B).	Ajima; Japanese mathematician	c 1732–1798
Nasireddin	41.0S	0.2E	52	Orontius group; fairly distinct.	Nasir al-Din; Persian astronomer	1201–1274
Nasmyth	50.5S	56.2W	76	Phocylides group; fairly regular; no c.p.	James; Scottish engineer	1808–1890
Natasha	20.0N	31.3W	12	On Imbrium; SW of Euler (Euler P).	Russian female name	—
Naumann	35.4N	62.0W	9	In N of Procellarum; bright walls.	Karl; German geologist	1797–1873
Neander	31.3S	39.9E	50	Rheita Valley area; well-formed.	Michael; German mathematician	1529–1581
Nearch	58.5S	39.1E	75	Vlacq area; craterlets on floor.	Greek explorer	c 325 BC
Neison	68.3N	25.1E	53	Meton area; regular; no c.p.	Edmond Neville; British selenographer	1849–1940
Neper	8.5N	84.6E	137	Marginis/Smythii area; deep.	John; Scottish mathematician	1550–1617
Neumayer	71.1S	70.7E	76	Boussingault area; distinct.	Georg; German meteorologist	1826–1909
Newcomb	29.9N	43.8E	41	Cleomedes area; S wall broken by crater.	Simon; Canadian astronomer	1835–1909
Newton	76.7S	16.9W	78	Moretus area; very deep; irregular.	Sir Isaac; British mathematician	1643–1727
Nicholson	26.2S	85.1W	38	In Rook Mtns.	Seth; American astronomer	1891–1963
Nicolai	42.4S	25.9E	42	Janssen area; regular.	Friedrich; German astronomer	1793–1846
Nicollet	21.9S	12.5W	15	In Nubium; W of Birt; distinct.	Jean; French astronomer	1788–1843
Nielsen	31.8N	51.8W		On Procellarum; between Lichtenberg and Gruithuisen (Wollaston C).	Axel; Danish astronomer	1902–1980
Niépce	71.7N	119.1W	57	Libration zone; beyond Brianchon.	Joseph; French photographer	1765–1833
Nobile	85.2S	53.5E	73	Libration zone.	Umberto; Italian explorer	1885–1978
Nobili	0.2N	75.9E	42	Trio with Jenkins and X (Schubert Y).	Leopoldo; Italian physicist	1784–1835
Nöggerath	48.8S	45.7W	30	Schiller area; low walls.	Johann; German geologist	1788–1877
Nonius	34.8S	3.8E	69	Stöfler area; fairly regular.	Pedro Nunez; Portuguese mathematician	1492?–1578
Nunn	4.6N	91.1E	19	Libration zone; on N edge of Smythii; low walls.	Joseph; American engineer	1905–1968
Œnopides	57.0N	64.1W	67	Limb area near Sinus Roris; high walls.	Greek astronomer	?500–430 BC
Œrsted	43.1N	47.2E	42	Edge of Somniorum; rather irregular.	Hans; Danish chemist	1777–1851
Oken	43.7S	75.9E	71	Australe area; prominent, darkish floor.	Lorenz Okenfuss; German biologist	1779–1851
Olbers	7.4N	75.9W	74	Grimaldi area; major ray-centre.	Heinrich; German astronomer/doctor	1758–1840
Omar Khayyám	58.0N	102.1W	70	Libration zone; in Poczobut.	Al-Khayyami; Persian astronomer/poet	c 1050–1123
Opelt	16.3S	17.5W	48	'Ghost' in Nubium; E of Bullialdus.	Friedrich; German astronomer	1794–1863
Oppolzer	1.5S	0.5W	40	Edge of Medii; low walls.	Theodor von; Czech astronomer	1841–1886
Orontius	40.6S	4.6W	105	Irregular; one of a group NE of Tycho.	Finnaeus Oronce; French mathematician	1494–1555
Palisa	9.4S	7.2W	33	Alphonsus area; on edge of larger ring.	Johann; Czech astronomer	1848–1925
Palitzsch	28.0S	64.5E	64 × 91 × 32	Outside Petavius to the E; really a crater-chain.	Johann; German astronomer	1723–1788
Pallas	5.5N	1.6W	46	Medii area; adjoins Murchison; c.p.	Peter; German geologist	1741–1811
Palmieri	28.6S	47.7W	40	Humorum area; darkish floor.	Luigi; Italian physicist	1807–1896
Paneth	63.0N	94.8W	65	Libration zone; beyond Xenophanes; N of Smoluchowski.	Friedrich; German chemist	1887–1958
Parkhurst	33.4S	103.6E	96	Libration zone; beyond Australe; irregular.	John; American astronomer	1861–1925
Parrot	14.5S	3.3E	70	Albategnius area; v irregular, compound structure.	Johann; Russian physicist	1792–1840

Table 3.13. (Continued)

Name	Lat.	Long.	Diameter (km)	Notes	Name	
Parry	7.9S	15.8W	47	On Nubium; Fra Mauro group; fairly regular.	William; British explorer	1790–1855
Pascal	74.6N	70.3W	115	Limb formation beyond Carpenter.	Louis; French mathematician	1623–1662
Peary	88.6N	33.0E	73	N polar area; beyond Gioja.	Robert; American explorer	1856–1920
Peirce	18.3N	53.5E	18	In Crisium; conspicuous.	Benjamin; American astronomer	1809–1880
Peirescius	46.5S	67.6E	61	Australe area; rather irregular.	Nicolas Peiresc; French astronomer	1580–1637
Pentland	64.6S	11.5E	56	S uplands; near Curtius; high walls.	Joseph; Irish geographer	1797–1873
Petermann	74.2N	66.3E	73	Limb beyond Arnold.	August; German geographer	1822–1878
Petavius	25.1S	60.4E	188	Great crater; c.p.; major rill on floor.	Denis Petau; French chronologist	1583–1652
Peters	68.1N	29.5E	15	Plain; no c.p.; between Neison and Arnold.	Christian; German astronomer	1806–1880
Petit	2.3N	63.5E	5	Small crater E of Spumans (Apollonius W).	Alexis; French physicist	1771–1820
Petrov	61.4S	88.0E	49	Flooded; beyond Pontécoulant.	Evgenii; Russian rocket scientist	1900–1942
Pettit	27.5S	86.6W	35	Limb formation; pair with Nicholson.	Edison; American astronomer	1889–1962
Petzval	62.7S	110.4W	90	Libration zone; beyond Hausen.	Joseph von; Austrian optician	1807–1891
Phillips	26.6S	75.3E	122	W of W Humboldt; central ridge.	John; British astronomer/geologist	1800–1874
Philolaus	72.1S	32.4W	70	Frigoris area; deep, regular; pair with Anaximenes.	Greek astronomer	c 400
Phocylides	52.7S	57.0W	121	Schickard group; much interior detail.	Johannes Holwarda; Dutch astronomer	1618–1651
Piazzi	36.6S	67.9W	134	Schickard area; very broken walls.	Giuseppe; Italian astronomer	1746–1826
Piazzi Smyth	41.9N	3.2W	13	On Imbrium; bright.	Charles; Scottish astronomer	1819–1900
Picard	14.6N	54.7E	22	Largest crater on Crisium; central hill.	Jean; French astronomer	1620–1682
Piccolomini	29.7S	32.2E	87	End of Altaiscarp; high walls; c.p.	Alessandro; Italian astronomer	1508–1578
Pickering	2.9S	7.0E	15	Hipparchus area; distinct.	Edward; American astronomer	1846–1919
					William; American astronomer	1858–1938
Pictet	43.6S	7.4W	62	Closely E of Tycho; fairly regular.	Marc Pictet-Turretin; Swiss physicist	1752–1825
Pilâtre	60.2S	86.9W	50	Beyond Pingré; low, irregular walls.	de Rozier; French aeronaut	1753–1785
Pingré	58.7S	73.7W	88	Limb formation beyond Phocylides; no c.p.	Alexandre; French astronomer	1711–1796
Pitatus	29.9S	13.5W	106	S edge of Nubium; passes connect it with Hesiodus.	Pietri Pitati; Italian astronomer	?–c 1500
Plana	42.2N	28.2E	44	Bürg area; pair with Mason; darkish floor.	Baron Giovanni; Italian astronomer	1781–1864
Plaskett	82.1N	174.3E	109	Libration zone; large walled plain; N polar area.	John; Canadian astronomer	1865–1941
Plato	51.6N	9.4W	109	Edge of Imbrium; very regular; v dark floor.	Greek philosopher	c 428–c 347 BC
Playfair	23.5S	8.4E	47	Abenezra area; fairly regular.	John; Scottish mathematician/geologist	1748–1819
Plinius	15.4N	23.7E	43	Between Serenitatis and Tranquillitatis; central craters.	Gaius; Roman natural scientist	23–79
Plutarch	24.1N	79.0E	68	NE of Crisium; distinct; c.p.	Greek biographer	c 46–c 120
Poczobut	57.1N	98.8W	195	Libration zone; large plain broken by several craters.	Martin; Polish astronomer	1728–1810
Poisson	30.4S	10.6E	42	Aliacensis area; compound; very irregular.	Simeon; French mathematician	1781–1840
Polybius	22.4S	25.6E	41	Theophilus/Catharina area.	Greek historian	?204–?122 BC
Pomortsev	0.7N	66.9E	23	Distinct (Dubiago P).	Mikhail; Russian rocket scientist	1851–1916
Poncelet	75.8N	54.1W	69	Limb, beyond Anaximenes and Philolaus.	Jean; French mathematician	1788–1867
Pons	25.3S	21.5E	41	Altai area; very thick walls.	Jean; French astronomer	1761–1831
Pontanus	28.4S	14.4E	57	Altai area; regular; no c.p.	Giovanni Pontano; Italian astronomer	1427–1503
Pontécoulant	58.7S	66.0E	91	S uplands; high walls.	Philippe; Comte de; French mathematician	1795–1874
Popov	17.2N	99.7E	65	Libration zone; beyond Marginis.	Aleksandr; Russian physicist	1859–1905
					Cyril; Bulgarian astronomer	1880–1966
Porter	56.1S	10.1W	51	On wall of Clavius; well-formed; c.p.	Russell; American telescope designer	1871–1949
Posidonius	31.8N	29.9E	95	Edge of Serenitatis; narrow walls; much interior detail.	Greek geographer	?135–?51 BC
Priestley	57.3S	108.4E	52	Libration zone; beyond Hanno and Chamberlin.	Joseph; British chemist	1733–1804
Prinz	25.5N	44.1W	46	NE of Aristarchus; incomplete; domes nearby.	Wilhelm; Belgian astronomer	1857–1910
Proclus	16.1N	46.8E	28	W of Crisium; brilliant; low c.p.; minor ray-centre.	Greek mathematician/astronomer	410–485
Proctor	46.4S	5.1W	52	Maginus area; fairly regular.	Mary; British astronomer	1862–1957
Protagoras	56.0N	7.3E	21	In Frigoris; bright and regular.	Greek philosopher	?481–?411 BC
Ptolemæus	9.3S	1.9W	164	Trio with Alphonsus and Arzachel; has Ammonius; darkish floor.	Greek astronomer, geographer/mathematician	c 120–180
Puiseux	27.8S	39.0W	24	On Humorum; near Doppelmeyer; v low walls.	Pierre; French astronomer	1855–1928
Pupin	23.8N	11.0W	2	Small but distinct; SE of Timocharis (Timocharis K).	Michael; Jugoslav physicist	1858–1935
Purbach	25.5S	2.3W	115	Walter group; rather irregular.	Georg von; Austrian mathematician	1423–1461
Purkyně	1.6S	94.9E	48	Libration zone; beyond Smythii.	Jan; Czech doctor	1787–1869
Pythagoras	63.5S	63.0W	142	NW of Iridum; high, massive walls; high c.p.	Greek philosopher and mathematician	c 532 BC
Pytheas	20.5N	20.6W	20	On Imbrium; bright; c.p; minor ray-centre.	Greek navigator and geographer	c 308 BC
Rabbi Levi	34.7S	23.6E	81	One of a group SW of Piccolomimi.	Ben Gershon; Jewish philosopher/astronomer	1288–1344
Raman	27.0N	55.1W	10	Irregular; NW of Herodotus (Herodotus D).	Chandrasekhara; Indian physicist	1888–1970
Ramsden	0.0N	0.0E	24	Edge of Epidemiarum; rills nearby.	Jesse; British instrument maker	1735–1800
Rankine	3.9S	71.5E	8	Craterlet E of Gilbert.	William; Scottish physicist	1820–1872
Rayleigh	29.3N	89.6E	114	Limb walled plain NE of Seneca.	John (Lord); British physicist	1842–1919
Réaumur	2.4S	0.7E	30	Medii area; low walls.	René; French physicist	1683–1757
Regiomontanus	28.3S	1.0W	129 × 105	Between Walter and Purbach; distorted; c.p.	Johann Muller; German astronomer	1436–1476
Regnault	54.1N	88.0W	46	Limb; near Xenophanes.	Henri; French chemist	1810–1878
Reichenbach	30.3S	48.0E	71	Rheita area; irregular; crater valley to SE.	Georg von; German optician	1722–1826
Reimarus	47.7S	60.3E	48	Rheita Valley area; irregular.	Nicolai Reymers; German mathematician	1550–c 1600
Reiner	7.0N	54.9W	29	On Procellarum; pair with Marius; c.p.	Vincento Reinieri; Italian astronomer	?–1648
Reinhold	3.3N	22.8W	42	SW of Copernicus; pair with B to the NE.	Erasmus; German astronomer	1511–1553
Repsold	51.3N	78.6W	109	W of Roris; one of a group.	Johann; German inventor	1770–1830
Respighi	2.8N	71.9E	18	SE of Dubiago; distinct (Dubiago C).	Lorenzo; Italian astronomer	1824–1890
Rhæticus	0.0N	4.9E	45	Medii area; low walls.	Georg von; Hungarian astronomer	1514–1576
Rheita	37.1S	47.2E	70	Sharp crests; associated with great crater valley.	Anton; Czech astronomer	1597–1660
Riccioli	3.3S	74.6W	139	Companion to Grimaldi; low walls; very dark patches on floor.	Giovanni; Italian astronomer	1598–1671

Table 3.13. (Continued)

Name	Lat.	Long.	Diameter (km)	Notes	Name	
Riccius	36.9S	26.5E	71	Rabbi Levi group; broken walls; rough floor.	Matteo Ricci; Italian mathematician	1552–1610
Richardson	31.1N	180.5E	141	Limb beyond Gauss, Vestine; broken by Maxwell.	Sir Owen; British physicist	1879–1959
Riemann	38.9N	86.8E	163	Ruined walled plain; beyond Gauss.	Georg; German mathematician	1826–1866
Ritchey	11.1S	8.5E	24	E of Albategnius; broken walls.	George; American astronomer/optician	1864–1945
Rittenhouse	74.5S	106.5E	26	Libration zone beyond Neumayer W of Schrödinger.	David; American astronomer	1732–1796
Ritter	2.0N	19.2E	29	On Tranquillitatis; c.p.; pair with Sabine.	Karl; German geographer	1779–1859
Ritz	15.1S	92.2E	51	Libration zone; beyond Ansgarius.	Walter; Swiss physicist	1878–1909
Robinson	59.0N	45.9W	24	Frigoris area; distinct; similar to Horrebow.	(John) Romney; Irish astronomer	1792–1882
Rocca	12.7S	72.8W	89	S of Grimaldi; irregular walls.	Giovanni; Italian mathematician	1607–1656
Rocco	28.0N	45.0W	4	Craterlet E of Krieger (Krieger D).	Italian male name	—
Rosenberger	55.4S	43.1E	95	Vlacq group; darkish floor; c.p.	Otto; German mathematician	1800–1890
Ross	11.7N	21.7E	24	On Tranquillitatis; c.p.	⎰ James Clark; British explorer ⎱ Frank; American astronomer	1800–1862 / 1874–1966
Rosse	17.9S	35.0E	11	On Nectaris; bright.	3rd Earl of Rosse; Irish astronomer	1800–1867
Röst	56.4S	33.7W	48	Schiller area; regular; pair with Weigel.	Leonhard; German astronomer	1688–1727
Rothmann	30.8S	27.7E	42	Altai area; fairly deep and regular.	Christopher; German astronomer	?–1600
Rozhdestvensky	85.2N	155.4W	177	Libration zone; polar area; contains 2 craters.	Dimitri; Russian astronomer	1876–1940
Rømer	25.4N	36.4E	39	Taurus area; massive c.p. with summit pit.	Ole; Danish astronomer	1644–1710
Röntgen	33.0N	91.4W	126	Libration zone; in Lorentz.	Wilhelm; German physicist	1845–1923
Rümker	40.8N	58.1W	70	Very irregular structure; part-plateau; near Harding.	Karl; German astronomer	1788–1862
Runge	2.5S	86.7E	38	Smythii area.	Carl; German mathematician	1856–1927
Russell	26.5N	75.4W	103	Extension of Otto Struve.	Henry Norris; American astronomer	1877–1957
Ruth	28.7N	45.1W	3	Craterlet adjoining Krieger to the NE.	Hebrew female name	—
Rutherfurd	60.9S	12.1W	48	On wall of Clavius; distinct c.p.	Lewis; American astronomer	1816–1892
Rydberg	46.5S	96.3W	49	Libration zone; pair with Guthnick.	Johannes; Swedish physicist	1854–1919
Rynin	47.0N	103.5W	75	Libration zone, beyond Galvani and McLaughlin.	Nikolai; Russian rocket scientist	1877–1942
Sabatier	13.2N	79.0E	10	E of Condorcet; low walls.	Paul; French chemist	1854–1941
Sabine	1.4N	20.1E	30	On Tranquillitatis; c.p.; pair with Ritter.	Sir Edward; Irish physicist/astronomer	1788–1883
Sacrobosco	23.7S	16.7E	98	Altai area; irregular.	Johannes Sacrobuschus; British astronomer	c 1200–1256
Sampson	29.7N	16.5W	1	On Imbrium; distinct craterlet NW of Timocharis.	Ralph; British astronomer	1866–1939
Santbech	20.9S	44.0E	64	E of Fracastorius; darkish floor.	Daniel; Dutch mathematician	c 1561
Santos–Dumont	27.7N	4.8E	8	S end of Apennines (Hadley B).	Alberto; Brazilian aeronaut	1873–1932
Sarabhai	24.7N	21.0E	7	On Serenitatis; NE of Bessel (Bessel A).	Vikram; Indian astrophysicist	1919–1971
Sasserides	39.1S	9.3W	90	Irregular enclosure N of Tycho.	Gellius Sascerides; Danish astronomer	1562–1612
Saunder	4.2S	8.8E	44	E of Hipparchus; low walls.	Samuel; British selenographer	1852–1912
Saussure	43.4S	3.8W	54	N of Maginus; interrupts larger ring.	Horace de; Swiss geologist	1740–1799
Scheele	9.4S	37.8W	4	Distinct; on Procellarum; S of Wichmann (Letronne D).	Carl; Swedish chemist	1742–1786
Scheiner	60.5S	27.5W	110	Clavius area; high walls; floor craterlet; pair with Blancanus.	Christopher; German astronomer	1575–1650
Schiaparelli	23.4N	58.8W	24	On Procellarum; distinct.	Giovanni; Italian astronomer	1835–1910
Schickard	44.3S	55.3W	206	Great walled plain; rather low walls.	Wilhelm; German mathematician/astronomer	1592–1635
Schiller	51.9S	39.0W	180 × 97	Schickard areal fusion of 2 rings.	Julius; German astronomer	c 1627
Schlüter	5.9S	83.3W	89	Beyond Grimaldi; prominent; terraced walls.	Heinrich; German astronomer	1815–1844
Schmidt	1.0N	18.8E	11	In Tranquillitatis; Sabine/Ritter area; bright.	⎰ Julius; German astronomer ⎸ Bernhard; Estonian optician ⎱ Otto; Russian astronomer	1825–1884 / 1879–1935 / 1891–1956
Schömberger	76.7S	24.9E	85	Regular; Boguslawsky area.	Georg; Austrian astronomer	1597–1845
Schönfeld	44.8N	98.1W	25	Libration zone; regular; beyond Gerard.	Eduard; German astronomer	1828–1891
Schorr	19.5S	89.7E	53	Limb formation; beyond Gibbs.	Richard; German astronomer	1867–1951
Schrödinger	67.0S	132.4E	312	Libration zone; beyond Hanno; associated with great valley.	Erwin; Austrian physicist	1887–1961
Schröter	2.6N	7.0W	35	Medii area; low walls.	Johann; German astronomer	1745–1816
Schubert	2.8N	81.0E	54	In Smythii area; distinct.	Theodor; Russian cartographer	1789–1865
Schumacher	42.4N	60.7E	60	Messala area; fairly distinct.	Heinrich; German astronomer	1780–1850
Schwabe	65.1N	45.6E	25	NE of Democritus; dark floor.	Heinrich; German astronomer	1789–1875
Schwarzschild	70.1N	121.2E	212	Libration zone; beyond Petermann; interior detail.	Karl; German astronomer	1873–1916
Scoresby	77.7N	14.1E	55	Polar uplands; deep and prominent; c.p.	William; British explorer	1789–1857
Scott	82.1S	48.5E	103	Beyond Schömberger; pair with Nansen.	Robert Falcon; British explorer	1868–1912
Secchi	2.4N	43.5E	22	In Fœcunditatis; bright walls; c.p.	(Pietro) Angelo; Italian astronomer	1818–1878
Seeliger	2.2S	3.0E	8	N of Hipparchus; irregular.	Hugo von; German astronomer	1849–1924
Segner	58.9S	48.3W	67	Schiller area; well-formed; prominent; with Zucchius.	Johann; German mathematician	1704–1777
Shackleton	89.9S	0.0E	19	South polar.	Ernest; British explorer	1874–1922
Shaler	32.9S	85.2W	48	Limb beyond Lagrange; pair with Wright.	Nathaniel; American geologist	1841–1906
Shapley	9.4N	56.9E	23	Off N edge of Crisium; dark floor (Picard H).	Harlow; American astronomer	1885–1972
Sharp	45.7N	40.2W	39	In Jura Mtns; deep; small c.p.	Abraham; British astronomer	1651–1742
Sheepshanks	59.2N	16.9E	25	N of Frigoris; fairly regular.	Anne; British benefactor	1789–1876
Shi Shen	76.0N	104.1E	43	Libration zone; beyond Nansen.	Chinese astronomer	c 300 BC
Short	74.6S	7.3W	70	Moretus group; deep; high walls.	James; Scottish mathematician/optician	1710–1768
Shuckburgh	41.6N	52.8E	38	Cepheus area; fairly regular.	Sir George; British geographer	1751–1804
Shuleykin	27.1S	92.5W	15	Libration zone; just beyond Orientale; Rook area.	Mikhail; Russian radio engineer	1884–1939
Sikorsky	66.1S	103.2E	98	Libration zone; crossed by Schrödinger Valley.	Igor; Russian aeronautical engineer	1889–1972
Silberschlag	6.2N	12.5E	13	Near Ariadæus; bright.	Johann; German astronomer	1721–1791
Simpelius	73.0S	15.2E	70	Moretus area; deep.	Hugh Sempill; Scottish mathematician	1596–1654
Sinas	8.8N	31.6E	11	On Tranquillitatis; not prominent.	Simon; Greek benefactor	1810–1876
Sirsalis	12.5S	60.4W	42	Contact pair with A; associated with rill.	Gerolamo Sersale; Italian astronomer	1584–1654

Table 3.13. (Continued)

Name	Lat.	Long.	Diameter (km)	Notes	Name	
Sklodowska	18.2S	95.5E	127	Libration zone; well-formed, beyond Hekatæus.	Marie (Curie); Polish physicist/chemist	1867–1934
Slocum	3.0S	89.0E	13	Smythii area.	Frederick; American astronomer	1873–1944
Smithson	2.4N	53.6E	5	In E of Tranquillitatis (Taruntius N).	James; British chemist	1765–1829
Smoluchowski	60.3N	96.8W	83	Libration zone; intrudes into Poczobut.	Marian; Polish physicist	1872–1917
Snellius	29.3S	55.7E	82	Furnerius area; high walls; c.p.; pair with Stevinus.	Willibrord Snell; Dutch mathematician	1591–1626
Somerville	8.3S	64.9E	15	Distinct; E of Langrenus (Langrenus J).	Mary; Scottish physicist/mathematician	1780–1872
Sömmering	0.1N	7.5W	28	Medii area; low walls.	Samuel; German doctor	1755–1830
Sosigenes	8.7N	17.6E	17	Edge of Tranquillitatis; bright, low c.p.	Greek astronomer/chronologist	c 46 BC
South	58.0N	50.8W	104	Frigoris area; ridge-bounded enclosure.	Sir James; British astronomer	1785–1867
Spallanzani	46.3S	24.7E	32	W of Janssen; low walls.	Lazzaro; Italian biologist	1729–1799
Spörer	4.3S	1.8W	27	N of Ptolemæus; partly lava-filled.	Friedrich; German astronomer	1822–1895
Spurr	27.9N	1.2W	11	In Putredinis; SE of Archimedes.	Josiah; American geologist	1870–1950
Stadius	10.5N	13.7W	60	Pitted ghost ring E of Copernicus.	Jan Stade; Belgian astronomer	1527–1579
Steinheil	48.6S	46.5E	67	SW of Janssen; contact twin with Watt.	Karl von; German astronomer	1801–1870
Stevinus	32.5S	54.2E	74	Furnerius area; high walls; c.p.; pair with Snellius.	Simon Stevin; Belgian mathematician	1548–1620
Stewart	2.2N	67.0E	13	Distinct; SW of Dubiago (Dubiago Q).	John Quincy; American astrophysicist	1894–1972
Stiborius	34.4S	32.0E	43	Altai area; S of Piccolomini; c.p.	Andreas Stoberl; German astronomer	1465–1515
Stöfler	41.1S	6.0E	126	S of Walter; dark floor; broken by Faraday.	Johann; German astronomer	1452–1531
Stokes	52.5N	88.1W	51	Limb; Regnault–Repsold area.	Sir George; British mathematician	1819–1903
Strabo	61.9N	54.3E	55	Near De la Rue; minor ray-centre.	Greek geographer	54 BC–24 AD
Street	46.5S	10.5W	57	S of Tycho; fairly regular; no c.p.	Thomas; British astronomer	1621–1689
Struve	43.0N	65.0E	18	Adjoins Messala; lies on dark patch.	Friedrich G. W.; Russian astronomer	1793–1864
Struve, Otto	22.4N	77.1W	164	W edge of Procellarum; pair of 2 ancient rings.	Otto; Russian astronomer	1819–1905
					Otto; American astronomer	1897–1963
Suess	4.4N	47.6W	8	On Procellarum; W of Encke; obscure.	Eduard; Austrian geologist	1831–1914
Sulpicius Gallus	19.6N	11.6E	12	On Serenitatis; very bright.	Gaius; Roman astronomer	c 166 BC
Sundman	10.8N	91.6W	40	Libration zone; beyond Vasco da Gama and Bohr.	Karl; Finnish astronomer	1873–1949
Sven Hedin	2.0N	76.8W	150	W of Hevel; irregular; broken walls.	Swedish explorer	1865–1952
Swasey	5.5S	89.7E	23	Smythii area.	Ambrose; American inventor	1846–1937
Swift	19.3N	53.4E	10	On Crisium; N of Peirce; prominent (Peirce B).	Lewis; American astronomer	1820–1913
Sylvester	82.7N	79.6W	58	Libration zone; beyond Philolaus.	James; British mathematician	1814–1897
Tacchini	4.9N	85.8E	40	Limb beyond Banachiewicz (Neper K).	Pietro; Italian astronomer	1838–1905
Tacitus	16.2S	19.0E	39	Catharina area; polygonal; 2 floor craters.	Cornelius; Roman historian	c 55–120
Tacquet	16.6N	19.2E	7	Just on Serenitatis, near Menelaus; bright.	Andre; Belgian mathematician	1612–1660
Talbot	2.5S	85.3E	11	Smythii area.	William Fox; British photographer	1800–1877
Tannerus	56.4S	22.0E	28	S. uplands, near Mutus; c.p.	Adam Tanner; Austrian mathematician	1572–1632
Taruntius	5.6N	46.5E	56	On Fœcunditatis; concentric crater; c.p.; low walls.	Lucius Firmanus; Roman philosopher	c 86 BC
Taylor	5.3S	16.7E	42	Delambre area; rather elliptical.	Brook; British mathematician	1685–1731
Tebbutt	9.6N	53.6E	31	Flooded; foothills of Crisium (Picard G).	John; Australian astronomer	1834–1916
Tempel	3.9N	11.9E	45	Uplands W of Tranquillitatis; bright.	(Ernst) Wilhelm; German astronomer	1821–1889
Thales	61.8N	50.3E	31	Near Strabo; major ray-centre.	Greek philosopher/astronomer	c 636–546 BC
Theætetus	37.0N	6.0E	24	On Nebularum; low c.p.	Greek geometrician	c 380 BC
Thebit	22.0S	4.0W	56	Edge of Nubium; wall broken by A, which is itself broken by F.	Thabit ibn Qurra; Arab astronomer	836–901
Theiler	13.4N	83.3E	7	Craterlet W of Marginis.	Max; S African bacteriologist	1899–1972
Theon Junior	2.3S	15.8E	17	Near Delambre; bright; pair with Theon Senior.	Greek astronomer	?–c 380 BC
Theon Senior	0.8S	15.4E	18	Near Delambre; bright; pair with Theon Junior.	Greek mathematician	?–c 100
Theophilus	11.4S	26.4E	110	Very deep; massive walls, c.p. complex; trio with Cyrillus and Catharina.	Greek astronomer	?–412
Theophrastus	17.5N	39.0E	9	On Tranquillitatis; NW of Franz (Maraldi M).	Greek botanist	c 372–287 BC
Timæus	62.8N	0.5W	32	Edge of Frigoris; bright.	Greek astronomer	?–c 400 BC
Timocharis	26.7N	13.1W	33	On Imbrium; central crater; minor ray-centre.	Greek astronomer	c 280 BC
Timoleon	35.0N	75.0E	130	Adjoins Gauss; fairly distinct.	Greek general and statesman	c 337 BC
Tisserand	21.4N	48.2E	36	Crisium area; regular; no c.p.	François; French astronomer	1845–1896
Titius	26.8S	100.7E	73	Libration zone; beyond W Humboldt and Lauritsen.	Johann; German astronomer	1729–1796
Tolansky	9.5S	16.0W	13	Between Parry and Guericke; flat floor (Parry A).	Samuel; British physicist	1907–1973
Torricelli	4.6S	28.5E	22	On Nectaris; irregular, compound structure.	Evangelista; Italian physicist	1608–1647
Toscanelli	27.4N	47.5W	7	Distinct; in highlands N of Aristarchus (Aristarchus C).	Paolo; Italian cartographer/doctor	1397–1482
Townley	3.4N	63.3E	18	S of Apollonius; distinct (Apollonius G).	Sidney; American astronomer	1867–1946
Tralles	28.4N	52.8E	43	On wall of Cleomedes; very deep.	Johann; German physicist	1763–1822
Triesnecker	4.2N	3.6E	26	Vaporum area; associated with great rill system.	Franz; Austrian astronomer	1745–1817
Trouvelot	49.3N	5.8E	9	Alpine Valley area; rather bright.	Etienne; French astronomer	1827–1895
Tucker	5.6S	88.2E	7	Craterlet in Smythii area.	Richard; American astronomer	1859–1952
Turner	1.4S	13.2W	11	On Nubium, near Gambart; deep.	Herbert Hall; British astronomer	1861–1930
Tycho	43.4S	11.1W	102	Terraced walls; c.p.; brightest ray-centre.	Tycho Brahe; Danish astronomer	1546–1601
Ukert	7.8N	1.4E	23	Edge of Vaporum; rills nearby.	Friedrich; German historian	1780–1851
Ulugh Beigh	32.7N	81.9W	54	W of Procellarum; high walls; c.p.	Mongolian astronomer	1394–1449
Urey	27.9N	87.4E	38	Limb formation; beyond Seneca.	Harold; American chemist	1893–1981
Väisälä	25.9N	47.8W	8	Distinct craterlet; N of Aristarchus.	Yrjo; Finnish astronomer	1891–1971
Van Albada	9.4N	64.3E	21	Distinct; closely S of Auzout (Auzout A).	Gale; Dutch astronomer	1912–1972
Van Biesbroeck	28.7N	45.6W	9	Closely S of Krieger (Krieger B).	Georges; Belgian astronomer	1880–1974

Table 3.13. (Continued)

Name	Lat.	Long.	Diameter (km)	Notes	Name	
Van Vleck	1.9S	78.3E	31	Dark floor; N of Kästner (Gilbert M).	John; American astronomer	1833–1912
Vasco da Gama	13.6N	83.9W	83	W of Procellarum; central ridge.	Portuguese navigator	1469–1524
Vashakidze	43.6N	93.3E	44	Libration zone; outside Harkhebi.	Mikhail; Russian astronomer	1909–1956
Vega	45.4S	63.4E	75	Australe area; deep.	Georg von; German mathematician	1756–1802
Vendelinus	16.4S	61.6E	131	Petavius chain; broken and irregular.	Godefroid Wendelin; Belgian astronomer	1580–1667
Vera	26.3N	43.7W	2	Aristarchus area; origin of long rill (Prinz A).	Latin female name	—
Verne	24.9N	25.3W	2	Between Euler and Lambert; pair with Artemis.	Latin male name	—
Very	25.6N	25.3E	5	Craterlet in Serenitatis; W of Le Monnier (Le Monnier B).	Frank; American astronomer	1852–1927
Vestine	33.9N	93.9E	61	Libration zone; beyond Gauss.	Ernest; American physicist	1906–1968
Vieta	29.2S	56.3W	87	W of Humorum; low; c.p.	François; French mathematician	1540–1603
Virchow	9.8N	83.7E	16	Adjoins Neper to the N (Neper G).	Rudolph; German pathologist	1821–1902
Vitello	30.4S	37.5W	42	S edge of Humorum; concentric crater.	Erazmus Witelo; Polish physicist	1210–1285
Vitruvius	17.6N	31.3E	29	Between Serenitatis and Tranquillitatis; low but rather bright walls.	Marcus; Roman engineer	c 25 BC
Vlacq	53.3S	38.8E	89	Janssen area; deep; c.p.; one of a group of 6.	Adriaan; Dutch mathematician	c1600–1667
Vögel	15.1S	5.9E	26	SE of Albategnius; chain of 4 craters.	Hermann; German astronomer	1841–1907
Volta	53.9N	84.4W	123	Limb formation near Repsold.	Count Allessandro; Italian physicist	1745–1827
von Behring	7.8S	71.8E	38	Distinct; W of Kästner (Maclaurin F).	Emil; German bacteriologist	1854–1917
Voskresensky	28.0N	88.1W	49	Flooded; beyond Otto Struve.	Leonid; Russian rocket scientist	1913–1965
Wallace	20.3N	8.7W	26	In Imbrium; imperfect ring; very low walls.	Alfred Russel; British natural historian	1823–1913
Walter	33.1S	1.0E	128	Massive walls; interior peak and craters; trio with Regiomontanus and Purbach.	Bernard Walther; German astronomer	1430–1504
Wargentin	49.6S	60.2W	84	Schickard group; the famous plateau.	Per; Swedish astronomer	1717–1783
Warner	4.0S	87.3E	35	Regular; Smythii area.	Worcester; American inventor	1846–1929
Watt	49.5S	48.6E	66	SW of Janssen; contact twin with Steinheil.	James; Scottish inventor	1736–1819
Watts	8.9N	46.3E	15	N of Taruntius; low walls; darkish floor (Taruntius D).	Chester; American astronomer	1889–1971
Webb	0.9S	60.0E	21	Edge of Fœcunditatis; darkish floor; c.p.; minor ray-centre.	Thomas; British astronomer	1806–1885
Weierstrass	1.3S	77.2E	33	Fairly regular; E of Maclaurin (Gilbert N).	Karl; German mathematician	1815–1897
Weigel	58.2S	38.8W	35	Schillar area; fairly regular; pair with Röst.	Erhard; German mathematician	1625–1699
Weinek	27.5S	37.0E	32	NE of Piccolomimo; darkish floor.	Ladislaus; Czech astronomer	1848–1913
Weiss	31.8S	19.5W	66	Pitatus group; very irregular.	Edmund; German astronomer	1837–1917
Werner	28.0S	3.3E	70	Very regular; high walls; c.p.; pair with Aliacensis.	Johann; German mathematician	1468–1528
Wexler	69.1S	90.2E	51	Libration zone; beyond Neumayer; regular.	Harry; American meteorologist	1911–1962
Whewell	4.2N	13.7E	13	Bright craterlet in Tranquillitatis uplands.	William; British astronomer	1794–1866
Wichmann	7.5S	38.1W	10	On Procellarum; associated with large 'ghost'.	Moritz; German astronomer	1821–1859
Widmanstätten	6.1S	85.5E	46	Limb; E of Maclauri.	Aloys; German physicist	1753–1849
Wildt	9.0N	75.8E	11	Distinct; W of Neper (Condorcet K).	Rupert; German astronomer	1905–1976
Wilhelm I	43.4S	20.4W	106	Lomgomintanus area; uneven walls.	Landgrave of Hesse; German astronomer	1532–1592
Wilkins	29.4S	19.6E	57	Rabbi Levi group; irregular.	(Hugh) Percy; Welsh selenographer	1896–1960
Williams	42.0N	37.2E	36	Somniorum area; not prominent.	Arthur; British astronomer	1861–1938
Wilson	69.2S	42.4W	69	Bettinus chain; near Bailly; deep, regular; no c.p.	Alexander; Scottish astronomer / Charles; Scottish physicist	1714–1786 / 1869–1959
Winthrop	10.7S	44.4W	17	Ruined crater on W wall of Letronne (Letronne P).	John; American astronomer	1714–1779
Wöhler	38.2S	31.4E	27	E of Riccius; fairly regular.	Friedrich; German chemist	1800–1882
Wolf	22.7S	16.6E	25	In S of Nubium; irregular; low walls.	Maximilian; German astronomer	1863–1932
Wollaston	30.6N	46.9W	10	Bright crater in Harbinger Mountains.	William Hyde; British physicist/chemist	1766–1828
Wright	31.6S	86.6W	39	Beyond Lagrange; pair with Shaler.	Thomas; British philosopher / William; American astronomer / Frederick; American astronomer	1711–1786 / 1871–1959 / 1878–1953
Wrottesley	23.9S	56.8E	57	Petavius group; twin-peaked central mountain.	John (Baron); British astronomer	1798–1867
Wurzelbauer	33.9S	15.9W	88	Pitatus group; irregular walls, much floor detail.	Johann von; German astronomer	1651–1725
Wyld	1.4S	98.1E	93	Libration zone; beyond Smythii.	James; American rocket scientist	1913–1953
Xenophanes	57.5N	82.0W	125	Limb near Sinus Roris; high walls; c.p.	Greek philosopher	?560–?478 BC
Yakovkin	54.5S	78.8W	37	Limb; beyond Phocylides (Pingré H).	A A; Russian astronomer	1887–1974
Yangel	17.0N	4.7E	8	In highlands NW of Manilius (Manilius F).	Mikhail; Russian rocket scientist	1911–1971
Yerkes	14.6N	51.7E	36	On W edge of Crisium; low walls; irregular.	Charles; American benefactor	1837–1905
Young	41.5S	50.9E	71	Rheita Valley area; irregular.	Thomas; British doctor/physicist	1773–1829
Zach	60.9S	5.3E	70	E of Clavius; fairly deep and regular.	Freiherr von; Hungarian astronomer	1754–1832
Zagut	32.0S	22.1E	84	Rabbi Levi group; irregular.	Abraham; Jewish astronomer	?–c 1450
Zähringer	5.6N	40.2E	11	Deep; W of Taruntius (Taruntius F).	Josef; German physicist	1929–1970
Zasyadko	3.9N	94.2E	11	Libration area; Smythii area; in Babcock.	Alexander; Russian rocket scientist	1779–1837
Zeeman	75.2S	133.6W	190	Libration zone; beyond Drygalski.	Pieter; Dutch physicist	1865–1943
Zeno	45.2N	72.9E	65	E of Mercurius; deformed.	Greek philosopher	c 335–263 BC
Zinner	26.6N	58.8W	4	Distinct craterlet; N of Schiaparelli (Schiaparelli B).	Ernst; German astronomer	1886–1970
Zöllner	8.0S	18.9E	47	NW of Theophilus; elliptical.	Johann Karl; German astronomer	1834–1882
Zsigmondy	59.7N	104.7W	65	Libration zone; beyond Poczobut.	Richard; Austrian chemist	1865–1929
Zucchius	61.4S	50.3W	64	Schiller area; distinct; pair with Segner.	Niccolo Zucchi; Italian astronomer	1586–1670
Zupus	17.2S	52.3W	38	S of Billy; low walls; irregular; very dark floor.	Giovanni Zupi; Italian astronomer	1590–1650

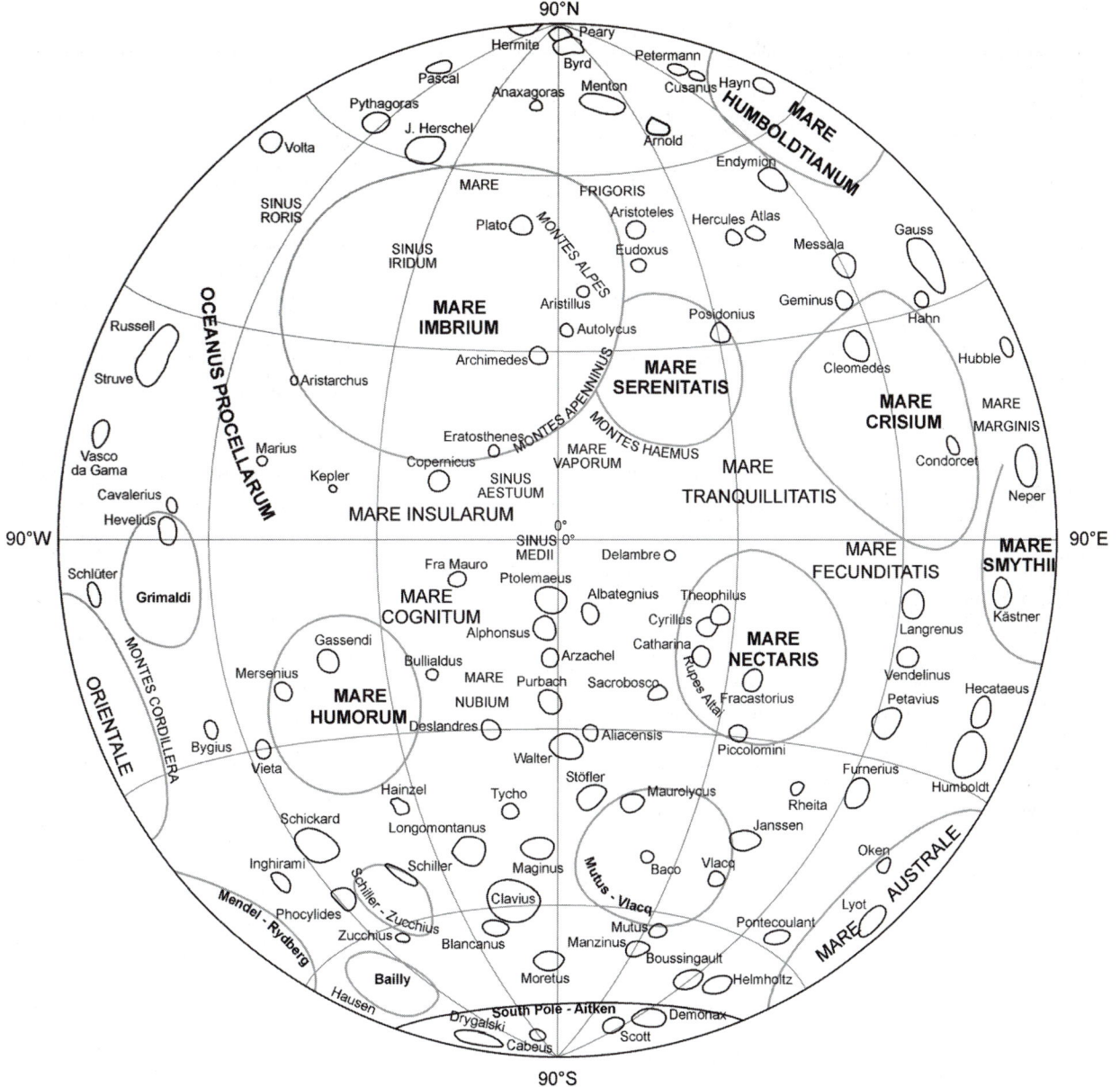

Figure 3.1. Outline map of the Moon.

Figure 3.2. The Moon – first quadrant. (Courtesy: Philip's.)

Figure 3.3. The Moon – second quadrant. (Courtesy: Philip's.)

Figure 3.4. The Moon – third quadrant. (Courtesy: Philip's.)

Figure 3.5. The Moon – fourth quadrant. (Courtesy: Philip's.)

Table 3.14. Selected craters on the far side of the Moon.

Name	Lat.	Long.	Diameter (km)	Name origin	
Abbe	57.3S	175.2E	66	Ernst; German astronomer	1840–1905
Abul Wafa	1.0N	116.6E	55	Persian astronomer	940–998
Aitken	16.8S	173.4E	135	Robert; American astronomer	1864–1951
Alden	23.6S	110.8E	104	Harold; American astronomer	1890–1964
Alekhin	68.2S	131.3W	70	Nikolai; Russian rocket scientist	1913–1964
Alter	18.7N	107.5W	64	Dinsmore; American astronomer	1888–1968
Amici	9.9S	172.1W	54	Giovanni; Italian astronomer	1787–1863
Anders	41.3S	142.9W	40	William; American astronaut	1933–
Anderson	15.8N	171.1E	109	John; American astronomer	1876–1959
Antoniadi	69.7S	172.0W	143	Eugenios; Greek astronomer	1870–1944
Apollo	36.1S	151.8W	537	Named in honour of Apollo lunar missions	
Appleton	37.2N	158.3E	63	Sir Edward; British physicist	1892–1965
Artamonov	25.5N	101.5E	60	Nikolai; Russian rocket scientist	1906–1965
Artemev	10.8N	144.4W	67	Vladimir; Russian rocket scientist	1885–1962
Ashbrook	81.4S	112.5W	156	Joseph; American astronomer	1918–1980
Avicenna	39.7N	97.2W	74	Ibn Sina; Persian doctor	980–1037
Avogadro	63.1N	164.9E	139	Amedeo (Comte de); Italian physicist	1776–1856
Backlund	16.0S	103.0E	75	Oscar; Russian astronomer	1846–1916
Baldet	53.3S	151.1W	55	François; French astronomer	1885–1964
Barbier	23.8S	157.9E	66	Daniel; French astronomer	1907–1965
Barringer	28.9S	149.7W	68	Daniel; American engineer	1860–1929
Becquerel	40.7N	129.7E	65	Antoine; French physicist	1852–1908
Bečvar	1.9S	125.2E	67	Antonin; Czech astronomer	1901–1965
Beijerinck	13.5S	151.8E	70	Martinus; Dutch botanist	1851–1931
Bellingshausen	60.6S	164.6W	63	Faddey; Russian explorer	1778–1852
Belopolsky	17.2S	128.1W	59	Aristarch; Russian astronomer	1854–1934
Belyayev	23.3N	143.5E	54	Pavel; Russian cosmonaut	1925–1970
Bergstrand	18.8S	176.3E	43	Carl; Swedish astronomer	1873–1948
Berkner	25.2N	105.2W	86	Lloyd; American geophysicist	1905–1967
Berlage	63.2N	162.8W	92	Hendrik; Dutch geophysicist	1896–1968
Bhabba	55.1S	164.5W	64	Homi; Indian physicist	1909–1966
Birkeland	30.2S	173.9E	82	Olaf; Norwegian physicist	1867–1917
Birkhoff	58.7N	146.1W	345	George; American mathematician	1884–1944
Bjerknes	38.4S	113.0E	48	Vilhelm; Norwegian physicist	1862–1951
Blazhko	31.6N	148.0W	54	Sergei; Russian astronomer	1870–1956
Bobone	26.9N	131.8W	3	Jorge; Argentine astronomer	1901–1958
Boltzmann	74.9S	90.7W	76	Ludwig; Austrian physicist	1844–1906
Bolyai	33.6S	125.9E	135	Janos; Hungarian mathematician	1802–1860
Borman	38.8S	147.7W	50	Frank; American astronaut	1928–
Bose	53.5S	168.6W	91	Jagadis; Indian botanist	1858–1937
Boyle	53.1S	178.1E	57	Robert; British chemist	1627–1691
Brashear	73.8S	170.7W	55	John; American astronomer	1840–1920
Bredikin	17.3N	158.2W	59	Fedor; Russian astronomer	1831–1904
Brianchon	75.0N	86.2W	134	Charles; French mathematician	1783–1864
Bridgman	43.5N	137.1E	80	Percy; American physicist	1882–1961
Brouwer	36.2S	126.0W	158	Dirk; American astronomer	1902–1966
Brunner	9.9S	90.9E	53	William; Swiss astronomer	1878–1958
Buffon	40.4S	133.4W	106	Georges; French natural historian	1707–1788
Buisson	1.4S	112.5E	56	Henri; French astronomer	1873–1944
Butlerov	12.5N	108.7W	40	Alexander; Russian chemist	1828–1886
Buys-Ballot	20.8N	174.5E	55	Christoph; Dutch meteorologist	1817–1890
Cabannes	60.9S	169.6W	80	Jean; French physicist	1885–1959
Cajori	47.4S	168.8E	70	Florian; American mathematician	1859–1930
Campbell	45.3N	151.4E	219	Leon; American astronomer	1862–1938
Cantor	38.2N	118.6E	80	Georg; German mathematician	1845–1918
Carnot	52.3N	143.5W	126	Nicholas; French physicist	1796–1832
Carver	43.0S	126.9E	59	George; American botanist	1864?–1943
Cassegrain	52.4S	113.5E	55	Giovanni; French astronomer	1625–1712
Ceraski	49.0S	141.6E	56	Witold; Polish astronomer	1849–1925
Chaffee	38.8S	153.9W	49	Roger; American astronaut	1935–1967
Champollion	37.4N	175.2E	58	Jean; French Egyptologist	1790–1832
Chandler	43.8N	171.5E	85	Seth; American astronomer	1846–1913
Chang Heng	19.0N	112.2E	43	Chinese astronomer	78–139
Chant	40.0S	109.2W	33	Clarence; Canadian astronomer	1865–1956
Chaplygin	6.2S	150.3E	137	Sergei; Russian mathematician	1869–1942
Chapman	50.4N	100.7W	71	Sydney; British geophysicist	1888–1970
Chappell	54.7N	177.0W	80	James; American astronomer	1891–1964
Charlier	36.6N	131.5W	99	Carl; Swedish astronomer	1862–1934
Chaucer	3.7N	140.0W	45	Geoffrey; British writer/astronomer	c 1340–1400
Chauvenet	11.5S	137.0E	81	William; American astronomer	1820–1870

Table 3.14. (Continued)

Name	Lat.	Long.	Diameter (km)	Name origin	
Chebyshev	33.7S	131.1W	178	Pafnutif; Russian astronomer	1821–1894
Chrétien	45.9S	162.9E	88	Henri; French astronomer/mathematician	1870–1956
Clark	38.4S	118.9E	49	Alvan; American astronomer/optician	1804–1887
				Alvan G; American astronomer/optician	1832–1897
Coblentz	37.9S	126.1E	33	William; American astronomer	1873–1962
Cockcroft	31.3N	162.6W	93	Sir John; British nuclear physicist	1897–1967
Comrie	23.3N	112.7W	59	Leslie; New Zealand astronomer	1893–1950
Comstock	21.8N	121.5W	72	George; American astronomer	1855–1934
Congreve	0.2S	167.3W	57	Sir William; British rocket pioneer	1772–1828
Cooper	52.9N	175.6E	36	John; American humanitarian	1887–1967
Coriolis	0.1N	171.8E	78	Gaspard de; French physicist	1792–1843
Coulomb	54.7N	114.6W	89	Charles de; French physicist	1736–1806
Crocco	47.5S	150.2E	75	Gaetano; Italian aeronautical engineer	1877–1968
Crommelin	68.1S	146.9W	94	Andrew; Irish astronomer	1865–1939
Crookes	10.3S	164.5W	49	Sir William; British physicist	1832–1919
Cyrano	20.5S	156.6E	80	Cyrano de Bergerac; French writer	1615–1655
Dædalus	5.9S	179.4E	93	Greek mythological character	—
D'Alembert	50.8N	163.9E	248	Jean; French mathematician	1717–1783
Danjon	11.4S	124.0E	71	Andre; French astronomer	1890–1967
Dante	25.5N	180.0E	54	Alighieri; Italian poet	1265–1321
Das	26.6S	136.8W	38	Smil; Indian astronomer	1902–1961
Davisson	37.5S	174.6W	87	Clinton; American physicist	1881–1958
Dawson	67.4S	134.7W	45	Bernhard; Argentinian astronomer	1890–1960
Debye	49.6N	176.2W	142	Peter; Dutch physicist	1884–1966
De Forest	77.3S	162.1W	57	Lee; American inventor/physicist	1873–1961
Dellinger	6.8S	140.6E	81	John; American physicist	1886–1962
Delporte	16.0S	121.6E	45	Eugene; Belgian astronomer	1882–1955
Denning	16.4S	142.6E	44	William; British astronomer	1848–1931
Deutsch	24.1N	110.5E	66	Armin; American astronomer	1918–1969
de Vries	19.9S	176.7W	59	Hugo; Dutch botanist	1848–1935
Dewar	2.7S	165.5E	50	Sir James; British chemist	1842–1923
Dirichlet	11.1N	151.4W	47	Peter; German mathematician	1805–1859
Doppler	12.6S	159.6W	110	Christian; Austrian physicist	1803–1853
Douglass	35.9N	122.4W	49	Andrew; American astronomer	1867–1962
Dryden	33.0S	155.2W	51	Hugh; American physicist	1898–1965
Dufay	5.5N	169.5E	39	Jean; French astronomer	1896–1967
Dugan	64.2N	103.3E	50	Raymond; American astronomer	1878–1940
Dunér	44.8N	179.5E	62	Nils; Swedish astronomer	1839–1914
Dyson	61.3N	121.2W	63	Sir Frank; British astronomer	1868–1939
Ehrlich	40.9N	172.4W	30	Paul; German doctor	1854–1915
Eijkman	63.1S	143.0W	54	Christian; Dutch doctor	1858–1930
Einthoven	4.9S	109.6E	69	Willem; Dutch physiologist	1879–1955
Ellerman	25.3S	120.1W	47	Ferdinand; American astronomer	1869–1940
Elvey	8.8N	100.5W	74	Christian; American astronomer	1899–1970
Emden	63.3N	177.3W	111	J Robert; Swiss astrophysicist	1862–1940
Engelhardt	5.7N	159.0W	43	Vasili; Russian astronomer	1828–1915
Eötvös	35.5S	133.8E	99	Roland von; Hungarian physicist	1848–1919
Esnault-Pelterie	47.7N	141.4W	79	Robert; French rocket engineer	1881–1957
Espin	28.1N	109.1E	75	Thomas; British astronomer	1858–1934
Evans	9.5S	133.5W	67	Sir Arthur; British archæologist	1851–1941
Evdokimov	34.8N	153.0W	50	Nikolai; Russian astronomer	1868–1940
Evershed	35.7N	150.5W	66	John; British astronomer	1864–1956
Fechner	59.0S	124.9E	63	Gustav; German physicist	1801–1887
Feoktistov	30.9N	140.7E	23	Konstantin; Russian cosmonaut	1926–
Fermi	19.3S	122.6E	183	Enrico; Italian physicist	1901–1954
Fersman	18.7N	126.0E	151	Alexander; Russian geochemist	1883–1945
Firsov	4.5N	112.2E	51	Georgi; Russian rocket engineer	1917–1960
Fitzgerald	27.5N	171.7W	110	George; Irish physicist	1851–1901
Fizeau	58.6S	133.9W	111	Armand; French physicist	1819–1896
Fleming	15.0N	109.6E	106	Alexander; British doctor	1881–1955
				Williamina; American astronomer	1857–1911
Foster	23.7N	141.5W	33	John; Canadian physicist	1860–1964
Fowler	42.3N	145.0W	146	Alfred; British astronomer	1868–1940
Freundlich	25.0N	171.0E	85	Erwin Finlay; German astronomer	1885–1964
Fridman	12.6S	126.0W	102	Alexander; Russian physicist	1885–1925
Frost	37.7N	118.4W	75	Edwin; American astronomer	1866–1935

Table 3.14. (Continued)

Name	Lat.	Long.	Diameter (km)	Name origin	
Gadomski	36.4N	147.3W	65	Jan; Polish astronomer	1889–1966
Gagarin	20.2S	149.2E	265	Yuri; Russian cosmonaut	1934–1968
Galois	14.2S	151.9W	222	Evariste; French mathematician	1811–1832
Gamow	65.3N	145.3E	129	George; Russian astronomer/physicist	1904–1968
Ganswindt	79.6S	110.3E	74	Hermann; German inventor	1856–1934
Garavito	47.5S	156.7E	74	José; Colombian astronomer	1865–1920
Geiger	14.6S	158.5E	34	Johannes; German physicist	1882–1945
Gerasimovič	22.9S	122.6W	86	Boris; Russian astronomer	1889–1937
Giordano Bruno	35.9N	102.8E	22	Italian astronomer/philosopher	1548–1600
Glasenapp	1.6S	137.6E	43	Sergei; Russian astronomer	1848–1937
Golitzyn	25.1S	105.0W	36	Boris; Russian physicist	1862–1916
Golovin	39.9N	161.1E	37	Nicholas; American rocket scientist	1912–1969
Grachev	3.7S	108.2W	35	Andrei; Russian rocket scientist	1900–1964
Green	4.1N	132.9E	65	George; British mathematician	1793–1841
Gregory	2.2N	127.3E	67	James; Scottish astronomer	1638–1675
Grigg	12.9N	129.4W	36	John; New Zealand astronomer	1838–1920
Grissom	47.0S	147.4W	58	Virgil; American astronaut	1926–1967
Grotrian	66.5S	128.3E	37	Walter; German astronomer	1890–1954
Gullstrand	45.2N	129.3W	43	Állvar; Swedish opthalmologist	1862–1930
Guyot	11.4N	117.5E	92	Arnold; Swiss geographer	1807–1884
Hagen	48.3S	135.1E	55	Johann; Austrian astronomer	1847–1930
Harriot	33.1N	114.3E	56	Thomas; British astronomer/mathematician	1560–1621
Hartmann	3.2N	135.3E	61	Johannes; German astronomer	1865–1936
Harvey	19.5N	146.5W	60	William; British doctor	1578–1657
Heaviside	10.4S	167.1E	165	Oliver; British physicist/mathematician	1850–1925
Helberg	22.5N	102.2W	62	Robert; American aeronautical engineer	1906–1967
Henderson	4.8N	152.1E	47	Thomas; Scottish astronomer	1798–1844
Hendrix	46.6S	159.2W	18	Don; American optician	1905–1961
Henyey	13.5N	151.6W	63	Louis; American astronomer	1910–1970
Hertz	13.4N	104.5E	90	Heinrich; German physicist	1857–1894
Hertzsprung	2.6N	129.2W	591	Ejnar; Danish astronomer	1873–1967
Hess	54.3S	174.6E	88	Victor; Austrian physicist	1883–1964
Heymans	75.3N	144.1W	50	Corneille; Belgian physiologist	1892–1968
Hilbert	17.9S	168.2E	55	Johann; Austrian astronomer	1847–1930
Hippocrates	70.7N	145.9W	60	Greek doctor	c 140 BC
Hoffmeister	15.2N	136.9E	45	Cuno; German astronomer	1892–1968
Hogg	33.6N	121.9E	38	Arthur; Australian astronomer	1903–1966
				Frank; Canadian astronomer	1904–1951
Holetschek	27.6S	150.9E	38	Johann; Austrian astronomer	1846–1923
Houzeau	17.1S	123.5W	71	Jean; Belgian astronomer	1820–1888
Hutton	37.3N	168.7E	50	James; Scottish geologist	1726–1797
Icarus	5.3S	173.2W	96	Greek mythical aviator	—
Idelson	81.5S	110.9E	60	Naum; Russian astronomer	1885–1951
Ingalls	26.4N	153.1W	37	Alnert; American optician	1888–1958
Innes	27.8N	119.2E	42	Robert; Scottish astronomer	1861–1933
Izsak	23.3S	117.1E	30	Imre; Hungarian astronomer	1929–1965
Jackson	22.4N	163.1W	71	John; Scottish astronomer	1887–1958
Joffe	14.4S	129.2W	86	Abram; Russian physicist	1880–1960
Joule	27.3N	144.2W	96	James; British physicist	1818–1889
Jules Verne	35.0S	147.0E	143	French writer	1828–1905
Kamerlingh Onnes	15.0N	115.8W	66	Heike; Dutch physicist	1853–1926
Karpinsky	73.3N	166.3E	92	Alexei; Russian geologist	1846–1936
Kearons	11.4S	112.6W	23	William; American astronomer	1878–1948
Keeler	10.2S	161.9E	160	James; American astronomer	1857–1900
Kekulé	16.4N	138.1W	94	Friedrich; German chemist	1829–1896
Khwolson	13.8S	111.4E	54	Orest; Russian physicist	1852–1934
Kibaltchich	3.0N	146.5W	92	Nikolai; Russian rocket scientist	1853–1881
Kidinnu	35.9N	122.9E	56	Or Cidenas; Babylonian astronomer	?–c 343 BC
Kimura	57.1S	118.4E	28	Hisashi; Japanese astronomer	1870–1943
King	5.0N	120.5E	76	Arthur; American physicist	1876–1957
				Edward; American astronomer	1861–1931
Kirkwood	68.8N	156.1W	67	Daniel; American astronomer	1814–1895
Kleimenov	32.4S	140.2W	55	Ivan; Russian rocket scientist	1898–1938
Klute	37.2N	141.3W	75	Daniel; American rocket scientist	1921–1964
Koch	42.8S	150.1E	95	Robert; German doctor	1843–1910
Kohlschütter	14.4N	154.0E	53	Arnold; German astronomer	1883–1969

Table 3.14. (Continued)

Name	Lat.	Long.	Diameter (km)	Name origin	
Kolhörster	11.2N	114.6W	97	Werner; German physicist	1887–1946
Komarov	24.7N	152.2E	78	Vladimir; Russian cosmonaut	1927–1967
Kondratyuk	14.9S	115.5E	108	Yuri; Russian rocket pioneer	1897–1942
Konstantinov	19.8N	158.4E	66	Konstantin; Russian rocket scientist	1817–1871
Korolev	4.0S	157.4W	437	Sergei; Russian rocket scientist	1906–1966
Kostinsky	14.7N	118.8E	75	Sergei; Russian astronomer	1867–1937
Kovalevskaya	30.8N	129.6W	115	Sofia; Russian mathematician	1850–1891
Kovalsky	21.9S	101.0E	49	Marian; Russian astronomer	1821–1884
Kramers	53.6N	127.6W	61	Hendrik; Dutch physicist	1894–1952
Krasovsky	3.9N	175.5W	59	Feodosii; Russian geodetist	1878–1948
Krylov	35.6N	165.8W	49	Alexei; Russian mathematician	1863–1945
Kulik	42.4N	154.5W	58	Leonid; Russian mineralogist	1883–1942
Kuo Shou Ching	8.4N	133.7W	34	Chinese astronomer	1231–1316
Kurchatov	38.3N	142.1E	106	Igor; Russian nuclear physicist	1903–1960
Lacchini	41.7N	107.5W	58	Giovanni; Italian astronomer	1884–1967
Lamarck	22.9S	69.8W	100	Jean; French natural historian	1744–1829
Lamb	42.9S	100.1E	106	Sir Horace; British mathematician	1849–1934
Lampland	31.0S	131.0E	65	Carl; American astronomer	1873–1951
Landau	41.6N	118.1W	214	Lev; Russian physicist	1908–1968
Lane	9.5S	132.0E	55	Jonathan; American astrophysicist	1819–1880
Langemak	10.3S	118.7E	97	Georgi; Russian rocket scientist	1898–1938
Langevin	44.3N	162.7E	58	Paul; French physicist	1872–1946
Langmuir	35.7S	128.4W	91	Irving; American physicist	1881–1957
Larmor	32.1N	179.7W	97	Sir Joseph; British mathematician	1857–1942
Leavitt	44.8S	139.3W	66	Henrietta; American astronomer	1868–1921
Lebedev	47.3S	107.8E	102	Petr; Russian physicist	1866–1912
Lebedinsky	8.3N	164.3W	62	Alexander; Russian astrophysicist	1913–1967
Leeuwenhoek	29.3S	178.7W	125	Antony van; Dutch microscopist	1632–1723
Leibnitz	38.3S	179.2E	245	Gottfried; German mathematician	1646–1716
Lemaître	61.2S	149.6W	91	Georges; Belgian mathematician	1894–1966
Lenz	2.8N	102.1W	21	Heinrich Emil; Estonian physicist	1804–1865
Leonov	19.0N	148.2E	33	Alexei; Russian cosmonaut	1934–
Leucippus	29.1N	116.0W	56	Greek philosopher	c 440 BC
Levi-Civita	23.7S	143.4E	121	Tullio; Italian mathematician	1873–1941
Lewis	18.5S	113.8W	42	Gilbert; American chemist	1875–1946
Ley	42.2N	154.9E	79	Willy; German rocket scientist	1906–1969
Lobachevsky	9.9N	112.6E	84	Nikolai; Russian mathematician	1793–1856
Lodygin	17.7S	146.8W	62	Alexander; Russian inventor	1847–1923
Lomonsov	27.3N	98.0E	92	Mikhail; Russian cartographer	1711–1765
Lorentz	32.6N	95.3W	312	Hendrik; Dutch physicist	1853–1928
Love	6.3S	129.0E	84	Augustus; British mathematician	1863–1940
Lovelace	82.3N	106.4W	54	William; American space scientist	1907–1965
Lovell	36.8S	141.9W	34	James; American astronaut	1928–
Lowell	12.9S	103.1W	66	Percival; American astronomer	1855–1916
Lucretius	8.2S	120.8W	63	Titus; Roman philosopher	c 95–55 BC
Lundmark	39.7S	152.5E	106	Knut; Swedish astronomer	1889–1958
Lyman	64.8S	163.6E	84	Theodore; American physicist	1874–1954
Mach	18.5N	149.3W	180	Ernst; Austrian physicist	1838–1916
Maksutov	40.5S	168.7W	83	Dimitri; Russian optician	1896–1964
Malyi	21.9N	105.3E	41	Alexander; Russian rocket scientist	1907–1961
Mandelstam	5.4N	162.4E	197	Leonid; Russian physicist	1879–1944
Marci	22.6N	167.0W	25	Jan; Czech physicist	1595–1667
Marconi	9.6S	145.1E	73	Guglielmo; Italian radio pioneer	1874–1937
Mariotte	28.5S	139.1W	65	Edme; French physicist	1620–1684
McKellar	15.7S	170.8W	51	Andrew; Canadian astronomer	1910–1960
McMath	17.3N	165.6W	86	Francis; American engineer/astronomer	1867–1938
				Robert; American astronomer	1891–1962
McNally	22.6N	127.2W	47	Paul; American astronomer	1890–1955
Mees	13.6N	96.1W	50	Kenneth; British photographer	1882–1960
Meggers	24.3N	123.0E	52	William; American physicist	1888–1968
Meitner	10.5S	112.7E	87	Lise; Austrian physicist	1878–1968
Mendeleev	5.7N	140.9E	313	Dimitri; Russian chemist	1834–1907
Merrill	75.2N	116.3W	57	Paul; American astronomer	1887–1961
Mesentsev	72.1N	128.7W	89	Yuri; Russian rocket scientist	1929–1965
Meshcerski	12.2N	125.5E	65	Ivan; Russian mathematician	1859–1935
Michelson	7.2N	120.7W	123	Albert; German physicist	1852–1931
Milanković	77.2N	168.8E	101	Milutin; Jugoslav astronomer	1879–1958
Millikan	46.8N	121.5E	98	Robert; American physicist	1868–1953
Mills	8.6N	156.0E	32	Mark; American physicist	1917–1958
Milne	31.4S	112.2E	272	Arthur; British mathematician/astronomer	1896–1950

Table 3.14. (Continued)

Name	Lat.	Long.	Diameter (km)	Name origin	
Mineur	25.0N	161.3W	73	Henri; French mathematician/astronomer	1899–1954
Minkowski	56.5S	146.0W	113	Hermann; German mathematician	1864–1909
				Rudolph; American astronomer	1895–1976
Mitra	18.0N	154.7W	92	Sisir Kumar; Indian physicist	1890–1963
Möbius	15.8N	101.2E	50	August; German mathematician	1790–1868
Mohorovičič	19.0S	165.0W	51	Andrija; Jugoslav geophysicist	1857–1936
Moiseev	9.5N	103.3E	59	Nikolai; Russian astronomer	1902–1955
Montgolfier	47.3N	159.8W	88	Jacques; French balloonist	1745–1799
				Joseph; French balloonist	1740–1810
Moore	37.4N	177.5W	54	Joseph; American astronomer	1878–1949
Morozov	5.0N	127.4E	42	Nikolai; Russian natural scientist	1854–1945
Morse	22.1N	175.1W	77	Samuel; American inventor	1791–1872
Nagaoka	19.4N	154.0E	46	Hantaro; Japanese physicist	1865–1940
Nassau	24.9S	177.4E	76	Jason; American astronomer	1892–1965
Nernst	35.3N	94.8W	116	Walther; German physical chemist	1864–1941
Neujmin	27.0S	125.0E	101	Grigori; Russian astronomer	1885–1946
Nièpce	72.7N	119.1W	57	Joseph N; French photographer	1765–1833
Nijland	33.0N	134.1E	35	Albertus; Dutch astronomer	1868–1936
Nikolayev	35.2N	151.3E	41	Andrian; Russian cosmonaut	1929–
Nishina	44.6S	170.4W	65	Yoshio; Japanese physicist	1890–1951
Nobel	15.0N	101.3W	48	Alfred; Swedish inventor	1833–1896
Nöther	66.6N	113.5W	67	Emmy; German mathematician	1882–1935
Numerov	70.7S	160.7W	113	Boris; Russian astronomer	1891–1941
Nušl	32.3N	167.6E	61	Frantisek; Czech astronomer	1867–1925
Obruchev	38.9S	162.1E	71	Vladimir; Russian geologist	1863–1956
O'Day	30.6S	157.5E	71	Marcus; American physicist	1897–1961
Ohm	18.4N	113.5W	64	Georg; German physicist	1787–1854
Olcott	20.6N	117.8E	81	William; American astronomer	1873–1936
Omar Kháyyám	58.0N	102.1W	70	Al Khayyami; Persian mathematician, poet	c 1050–1123
Oppenheimer	35.2S	166.3W	208	J Robert; American physicist	1904–1967
Oresme	42.5S	169.2E	76	Nicole; French mathematician	1323?–1382
Orlov	25.7S	175.0W	81	Alexander; Russian astronomer	1880–1954
				Sergei; Russian astronomer	1880–1958
Ostwald	10.4N	121.9E	104	Wilhelm; German chemist	1853–1932
Pannekoek	4.2S	140.5E	71	Antonie; Dutch astronomer	1873–1960
Papaleski	10.2N	164.0E	97	Nikolai; Russian physicist	1880–1947
Paracelsus	23.0S	163.1E	83	Theopnrastus von Hohenheim; Swiss chemist	1493–1541
Paraskevopoulos	50.4N	149.9W	94	John; Greek astronomer	1889–1951
Parenago	25.9N	108.5W	93	Pavel; Russian astronomer	1906–1960
Parkhurst	33.4S	103.6E	96	John; American astronomer	1861–1925
Parsons	37.3N	171.2W	40	John; American astronomer	1913–1952
Paschen	13.5S	139.8W	124	Friedrich; German physicist	1865–1940
Pasteur	11.9S	104.6E	224	Louis; French chemist	1822–1895
Pauli	44.5S	136.4E	84	Wolfgang; Austrian physicist	1900–1958
Pavlov	28.8S	142.5E	148	Ivan; Russian physiologist	1849–1936
Pawsey	44.5N	143.0E	60	Joseph; Australian radio astronomer	1908–1962
Pease	12.5N	106.1W	38	Francis; American astronomer	1881–1938
Perelman	24.0S	106.0E	46	Yakov; Russian rocket scientist	1882–1942
Perepelkin	10.0S	129.0E	97	Evgeny; Russian astrophysicist	1906–1940
Perkin	47.2N	175.9W	62	Richard; American telescope maker	1906–1969
Perrine	45.2N	127.8W	86	Charles; American astronomer	1867–1951
Petrie	45.3N	108.4E	33	Robert; Canadian astronomer	1906–1966
Petropavlovsky	37.2N	114.8W	63	Boris; Russian rocket engineer	1898–1933
Petzval	62.7S	110.4W	90	Joseph von; Austrian optician	1807–1891
Pirquet	20.3S	139.6E	65	Baron Guido von; Austrian space scientist	1867–1936
Pizzetti	34.9S	118.8E	44	Paolo; Italian geodetist	1860–1918
Planck	57.9S	136.8E	324	Max; German physicist	1858–1947
Plaskett	82.1N	174.3E	109	John; Canadian astronomer	1865–1941
Plummer	25.0S	155.0W	73	Henry; British astronomer	1875–1946
Pogson	42.2S	110.5E	50	Norman; British astronomer	1829–1891
Poincaré	56.7S	163.6E	319	Jules; French mathematician	1854–1912
Poinsot	79.5N	145.7W	68	Louis; French mathematician	1777–1859
Polzunov	25.3N	114.6E	67	Ivan; Russian heat engineer	1728–1766
Poynting	18.1N	133.4W	128	John; British physicist	1852–1914
Prager	3.9S	130.5E	60	Richard; German astronomer	1884–1945
Prandtl	60.1S	141.8E	91	Ludwig; German physicist	1875–1953
Priestley	57.3S	108.4E	52	Joseph; British chemist	1733–1804

Table 3.14. (Continued)

Name	Lat.	Long.	Diameter (km)	Name origin	
Quetelet	43.1N	134.9W	55	Lambert; Belgian astronomer	1796–1874
Racah	13.8S	179.8W	63	Giulio; Israeli physicist	1909–1965
Raimond	14.6N	159.3W	70	J J; Dutch astronomer	1903–1961
Ramsay	40.2S	144.5E	81	Sir William; British chemist	1852–1916
Rasumov	39.1N	114.3W	70	Vladimir; Russian rocket engineer	1890–1967
Rayet	44.7N	114.5E	27	George; French astronomer	1839–1906
Rayleigh	29.3N	89.6E	114	John, Lord Rayleigh; British physicist	1842–1919
Riccò	75.6N	176.3E	65	Annibale; Italian astronomer	1844–1911
Riedel	48.9S	139.6W	47	Klaus; German rocket scientist	1907–1944
				Walter; German rocket scientist	1902–1968
Riemann	38.9N	86.8E	163	Georg; German mathematician	1826–1866
Rittenhouse	74.5S	106.5E	26	David; American astronomer/inventor	1732–1796
Roberts	71.1N	174.5W	89	Alexander; South African astronomer	1857–1938
				Isaac; British astronomer	1829–1904
Robertson	21.8N	105.2W	88	Howard; American physicist	1903–1961
Roche	42.3S	136.5E	160	Edouard; French astronomer	1820–1883
Rowland	57.4N	162.5W	171	Henry; American physicist	1848–1901
Rozhdestvensky	85.2N	155.4W	177	Dimitri; Russian physicist	1876–1940
Rumford	28.8S	169.8W	61	Benjamin, Count Rumford; British physicist	1753–1814
Šafařik	16.6N	176.9E	27	Vojtech; Czech astronomer	1829–1902
Saha	1.6S	102.7E	99	Meghnad; Indian astrophysicist	1893–1956
Sänger	4.3N	102.4E	75	Eugen; Austrian rocket engineer	1905–1964
St John	10.2N	150.2E	68	Charles; American astronomer	1857–1935
Sanford	32.6N	138.9W	55	Roscoe; American astronomer	1883–1958
Sarton	49.3N	121.1W	69	George; Belgian historian of science	1844–1956
Scaliger	27.1S	108.9E	84	Joseph; French chronologist	1540–1609
Schaeberle	26.7S	117.2E	62	John; American astronomer	1853–1924
Schjellerup	69.7N	157.1E	62	Hans Carl; Danish astronomer	1827–1887
Schlesinger	47.4N	138.6W	97	Frank; American astronomer	1871–1943
Schliemann	2.1S	155.2E	80	Heinrich; German archæologist	1822–1890
Schneller	41.8N	163.6W	54	Herbert; German astronomer	1901–1967
Schrödinger	67.0S	132.4E	312	Erwin; Austrian physicist	1887–1961
Schuster	4.2N	146.5E	108	Sir Arthur; British mathematician	1851–1934
Schwarzschild	70.1N	121.2E	212	Karl; German astronomer	1873–1916
Seares	73.5N	145.8E	110	Frederick; American astronomer	1873–1964
Sechenov	7.1S	142.6W	62	Ivan; Russian physiologist	1829–1905
Segers	47.1N	127.7E	17	Carlos; Argentine astronomer	1900–1967
Seidel	32.8S	152.2E	62	Ludwig von; German astronomer	1821–1896
Seyfert	29.1N	114.6E	110	Carl; American astronomer	1911–1960
Shajn	32.6N	172.5E	93	Grigori; Russian astrophysicist	1892–1956
Sharanov	12.4N	173.3E	74	Vsevolod; Russian astronomer	1901–1964
Shatalov	24.3N	141.5E	21	Vladimir; Russian cosmonaut	1927–
Shi Shen	76.0N	104.1E	43	Chinese astronomer	?–c 300 BC
Siedentopf	22.0N	135.5E	61	Heinrich; German astronomer	1906–1963
Siepinski	27.2S	154.5E	69	Waclaw; Polish mathematician	1882–1969
Sisakian	41.2N	109.0E	34	Norai; Russian doctor	1907–1966
Sklodowska	18.2S	95.5E	127	Marie Curie; Polish physicist	1867–1934
Slipher	49.5N	160.1E	69	Earl; American astronomer	1883–1964
				Vesto; American astronomer	1875–1969
Sniadecki	22.5S	168.9W	43	Jan; Polish astronomer	1756–1830
Sommerfeld	65.2N	162.4W	169	Arnold; German physicist	1868–1951
Spencer Jones	13.3N	165.6E	85	Sir Harold; British astronomer	1890–1960
Stark	25.5S	134.6E	49	Johannes; German physicist	1874–1957
Stebbins	64.8N	141.8W	131	Joel; American astronomer	1878–1966
Stefan	46.0N	108.3W	125	Josef; Austrian physicist	1835–1893
Stein	7.2N	179.0E	33	Johan; Dutch astronomer	1871–1951
Steklov	36.7S	104.9W	36	Vladimir; Russian mathematician	1864–1926
Steno	32.8N	161.8E	31	Nicolaus; Danish doctor	1638–1686
Sternfeld	19.6S	141.2W	100	Ari; Russian space scientist	1905–1980
Stetson	39.6S	118.3W	64	Harlan; American astronomer	1885–1964
Stoletov	45.1N	155.2W	42	Alexander; Russian physicist	1839–1896
Stoney	55.3S	156.1W	45	(George) Johnstone; Irish physicist	1826–1911
Størmer	57.3N	146.3E	69	Carl; Norwegian astronomer	1874–1957
Stratton	5.8S	164.6E	70	Frederick; British astronomer	1881–1960
Strömgren	21.7S	132.4W	61	Elis; Danish astronomer	1870–1947
Subbotin	29.2S	135.3E	67	Mikhail; Russian astronomer	1893–1966
Sumner	37.5N	108.7E	50	Thomas; American geographer	1807–1876
Swann	52.0N	112.7E	42	William; British physicist	1884–1962
Szilard	34.0N	105.7E	122	Leo; Hungarian physicist	1898–1964

Table 3.14. (Continued)

Name	Lat.	Long.	Diameter (km)	Name origin	
Teisserenc de Bort	32.2N	135.9W	62	Leon; French meteorologist	1855–1913
ten Bruggencate	9.5S	134.4E	59	Paul; German astronomer	1901–1961
Tereshkova	28.4N	144.3E	31	Valentina; Russian cosmonaut	1937–
Tesla	38.5N	124.7E	43	Nikola; Jugoslav inventor	1856–1943
Thiel	40.7N	134.5W	32	Walter; German space scientist	1910–1943
Thiessen	75.4N	169.0W	66	Georg; German astronomer	1914–1961
Thomson	32.7S	166.2E	117	Sir Joseph; British physicist	1856–1940
Tikhomirov	25.2N	162.0E	65	Nikolai; Russian chemical engineer	1860–1930
Tikhov	63.3N	171.7E	83	Gavriil; Russian astronomer	1875–1960
Tiling	53.1S	132.6W	38	Reinhold; German rocket scientist	1890–1933
Timiryazev	5.5S	147.0W	53	Kliment; Russian botanist	1843–1920
Titov	28.6N	150.5E	31	German; Russian cosmonaut	1935–
Trümpler	29.3N	167.1E	77	Robert; Swiss astronomer	1866–1956
Tsander	6.2N	149.3W	181	Friedrich; Russian rocket scientist	1887–1933
Tsiolkovskii	21.2S	128.9E	185	Konstantin; Russian rocket engineer	1857–1935
Tsu Chung-chi	17.3N	145.1E	28	Chinese mathematician	430–501
Tyndall	34.9S	117.0E	18	John; British physicist	1820–1893
Valier	6.8N	174.5E	67	Max; German rocket engineer	1895–1930
Van de Graaff	27.4S	172.2E	233	Robert; American physicist	1901–1967
Van den Bergh	31.3N	159.1W	42	G; Dutch astronomer	1890–1966
Van der Waals	43.9S	119.9E	104	Johannes; Dutch physicist	1837–1923
Van Gent	15.4N	160.4E	43	H; Dutch astronomer	1900–1947
Van Maanen	35.7N	128.0E	60	Adriaan; Dutch astronomer	1884–1946
Van Rhijn	52.6N	146.4E	46	Pieter; Dutch astronomer	1886–1960
Van't Hoff	62.1N	131.8W	92	Jacobus; Dutch astronomer	1852–1911
Van Wijk	62.8S	118.8E	32	Uco; Dutch astronomer	1924–1966
Vavilov	0.8S	137.9W	98	Nicolai; Russian botanist	1887–1943
Vening Meinesz	0.3S	162.6E	87	Felix; Dutch geophysicist	1887–1966
Ventris	4.9S	158.0E	95	Michael; British archæologist	1922–1956
Vernadsky	23.2N	130.5E	91	Vladimir; Russian mineralogist	1863–1945
Vesalius	3.1S	114.5E	61	Andreas; Belgian doctor	1514–1564
Vetchinkin	10.2N	131.3E	98	Vladimir; Russian physicist engineer	1888–1950
Vilev	6.1S	144.4E	45	Mikhail; Russian chemist	1893–1919
Volterra	50.8N	132.2E	52	Vito; Italian mathematician	1890–1940
von der Pahlen	24.8N	132.7W	56	Emanuel; German astronomer	1882–1952
von Kármán	44.8S	175.9E	180	Theodor; Hungarian aeronautical scientist	1881–1963
von Neumann	40.4N	153.2E	78	John; American mathematician	1903–1957
von Zeipel	42.6N	141.6W	83	E H; Swedish astronomer	1873–1959
Walker	26.0S	162.2W	32	American pilot	1921–1966
Waterman	25.9S	128.0E	76	Alan; American physicist	1892–1967
Watson	62.6S	124.5W	62	James; American astronomer	1838–1880
Weber	50.4N	123.4W	42	Wilhelm; German astronomer	1804–1891
Wegener	45.2N	113.3W	88	Alfred, Austrian meteorologist	1880–1930
Wells	40.7M	122.8E	114	H G (Herbert) Wells; English writer	1866–1946
Weyl	17.5N	126.2W	108	Hermann; German mathematician	1885–1955
White	44.6S	158.3W	39	Edward; American astronaut	1930–1967
Wiechert	84.5S	163.0E	41	Emil; German geophysicist	1861–1928
Wiener	40.8N	146.6E	120	Norbert; American mathematician	1894–1964
Wilsing	21.5S	155.2W	73	Johannes; German astronomer	1856–1943
Winkler	42.2N	179.0W	22	Johannes; German rocket scientist	1897–1947
Winlock	35.6N	105.6W	64	Joseph; American astronomer	1826–1875
Woltjer	45.2N	159.6W	46	Jan; Dutch astronomer	1891–1946
Wood	43.0N	120.8W	78	Robert; American physicist	1868–1955
Xenophon	22.8S	122.1E	25	Greek natural philosopher	c 430–354 BC
Yablochkov	60.9N	128.3E	99	Pavel; Russian electrical engineer	1847–1894
Yamamoto	58.1N	160.9E	76	Issei; Japanese astronomer	1889–1959
Zeeman	75.2S	133.6W	190	Pieter; Dutch physicist	1865–1943
Zelinsky	28.9S	166.8E	53	Nikolai; Russian chemist	1860–1953
Zernike	18.4N	168.2E	48	Frits; Dutch physicist	1888–1966
Zhiritsky	24.8S	120.3E	35	Georgi; Russian rocket scientist	1893–1966
Zhukovsky	7.8N	167.0W	81	Nikolai; Russian physicist	1847–1921
Zsigmondy	59.7N	104.7W	65	Richard; Austrian chemist	1865–1929
Zwicky	15.4S	168.1E	150	Swiss astrophysicist	1898–1974

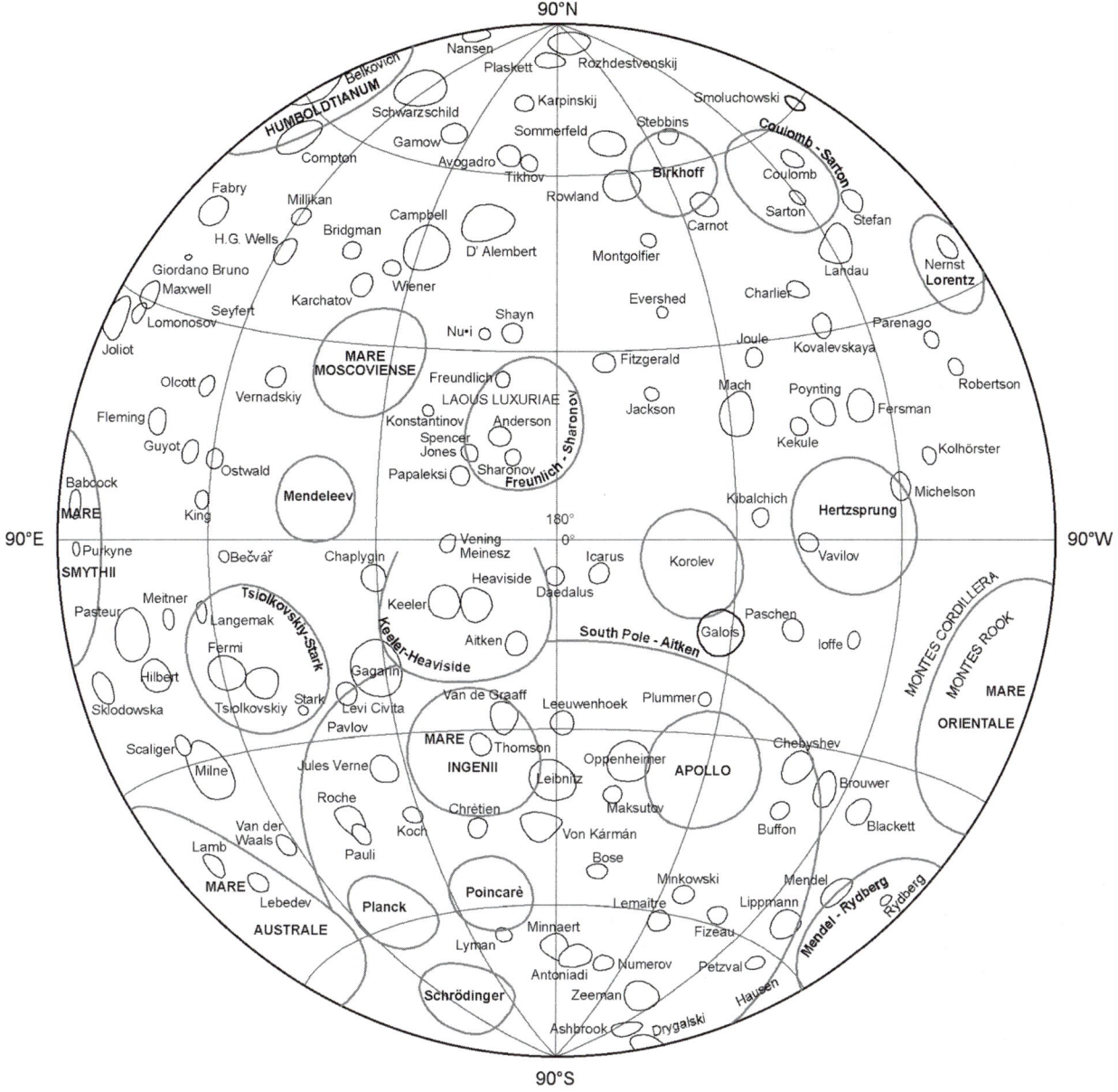

Figure 3.6. Outline map of the far side of the Moon.

4 MERCURY

The innermost planet, Mercury, is also the smallest of the principal planets – apart from Pluto, which may be in a completely different category. It was once thought that a still closer-in planet existed and it was even given a name: Vulcan. It would be wellnigh impossible to observe except during a total solar eclipse or a transit across the Sun's disk, but various claims were made, notably by a French amateur, Lescarbault, who stated that he had observed a transit on 26 March 1859. One believer was U. J. J. Le Verrier, whose mathematical work had led to the identification of Neptune in 1846. Vulcan's distance from the Sun was said to be about 21 000 000 km, with a period of 19d 17h. However, it is now certain that Vulcan does not exist. The only bodies which may move within the orbit of Mercury are comets, plus some asteroids such as Phæthon.

Data for Mercury are given in Table 4.1.

MOVEMENTS

The planet must have been known in prehistoric times, even if its nature were not realized. The oldest observation which has come down to us is dated 15 November 265 BC, when the planet was one lunar diameter away from a line joining the stars Delta and Beta Scorpii. This information was given by the last great astronomer of Classical times, Ptolemy (c 120–180 AD). Plato (*Republic*, X 14) commented upon the yellowish colour of Mercury, although most naked-eye observers will describe it as white. Mercury can actually become brighter than any star, but can never be seen against a really dark sky.

There is an oft-quoted story that the great astronomer Copernicus never saw Mercury at all, because of mists arising from the River Vistula, near his home, but the story is certainly false. Mercury is by no means hard to identify when well placed, and the skies were much less polluted in Copernicus' time than they are now (he died in 1543). The maximum angular elongation from the Sun is 28°. Elongation dates for the period 2000–2015 are given in Table 4.2.

The phases of Mercury are easy to see with a small telescope. They were probably suspected in the

Table 4.1. Data.

Distance from the Sun:
  mean 57.9 million km (0.387 a.u.)
  max 69.7 million km (0.467 a.u.)
  min 45.9 million km (0.306 a.u.)
Sidereal period: 87.969 days
Synodic period: 115.88 days
Rotation period: 58.6461 days
Mean orbital velocity: 47.87 km s^{-1}
Axial inclination: negligible
Orbital eccentricity: 0.206
Orbital inclination: 7°00′15″.5
Diameter: 4878 km
Surface area: 7.475×10^7 km^2
Apparent diameter from Earth:
  max 12″.9
  min 4″.5
Reciprocal mass, Sun = 1:6000 000
Density, water = 1:5.44
Mass: 3.3×10^{26} g
Mass, Earth = 1:0.055
Volume, Earth = 1:0.056
Escape velocity: 4.25 km s^{-1}
Surface gravity, Earth = 1:0.38
Mean surface temperature:
  day +350 °C
  night −170 °C
Extremes of surface temperature:
  day +427 °C
  night −183 °C
Oblateness: negligible
Albedo: 0.06
Maximum magnitude: −1.9
Mean diameter of Sun, as seen from Mercury:
  1°22′40″

first half of the 17th century by Galileo, Marius and Martin van den Hove (=Hortensius), but we cannot be sure. They were definitely seen by Giovanni Zupus in 1639, and confirmed by Hevelius in 1644.

Mercury, like Venus, can pass in transit across the face of the Sun and does so more frequently than with Venus, although during a transit it is not visible with the naked eye.

Table 4.2. Elongations of Mercury, 2000–2015.

Eastern

2000	15 Feb, 9 June, 6 Oct
2001	28 Jan, 22 May, 18 Sept, 26 Dec
2002	11 Jan, 4 May, 1 Sept
2003	16 Apr, 14 Aug, 9 Dec
2004	29 Mar, 27 July, 21 Nov
2005	12 Mar, 9 July, 3 Nov
2006	24 Feb, 20 June, 17 Oct
2007	7 Feb, 2 June, 29 Sept
2008	22 Jan, 14 May, 11 Sept
2009	4 Jan, 26 Apr, 24 Aug, 18 Dec
2010	8 Apr, 7 Aug, 1 Dec
2011	23 Mar, 20 July, 14 Nov
2012	5 Mar, 4 July, 26 Oct
2013	16 Feb, 12 June, 9 Oct
2014	31 Jan, 25 May, 21 Sept
2015	14 Jan, 7 May, 4 Sept, 29 Dec

Western

2000	28 Mar, 27 July, 25 Nov
2001	11 Mar, 9 July, 29 Oct
2002	21 Feb, 21 June, 13 Oct
2003	4 Feb, 3 June, 26 Sept
2004	17 Jan, 14 May, 9 Sept, 29 Dec
2005	26 Apr, 23 Aug, 12 Dec
2006	8 Apr, 7 Aug, 25 Nov
2007	22 Mar, 20 July, 8 Nov
2008	3 Mar, 1 July, 22 Oct
2009	13 Feb, 13 July, 6 Oct
2010	27 Jan, 26 May, 19 Sept
2011	9 Jan, 7 May, 3 Sept, 23 Dec
2012	18 Apr, 16 Aug, 4 Dec
2013	31 Mar, 30 July, 18 Nov
2014	14 Mar, 12 July, 1 Nov
2015	24 Feb, 24 Jun, 16 Oct

Table 4.3. Transits of Mercury.
(a) 1631–2000

1631 Nov 7	1832 May 5
1644 Nov 9	1835 Nov 7
1651 Nov 3	1845 May 8
1661 May 3	1848 Nov 9
1664 Nov 4	1861 Nov 12
1677 Nov 7	1868 Nov 5
1690 Nov 10	1878 May 6
1697 Nov 3	1881 Nov 8
1707 May 5	1891 May 10
1710 Nov 6	1894 Nov 10
1723 Nov 9	1907 Nov 14
1736 Nov 11	1914 Nov 7
1740 May 2	1924 May 8
1743 Nov 5	1927 Nov 10
1753 May 6	1937 May 11
1756 Nov 7	1940 Nov 11
1769 Nov 9	1953 Nov 14
1776 Nov 2	1957 May 6
1782 Nov 12	1960 Nov 7
1786 May 4	1970 May 9
1789 Nov 5	1973 Nov 10
1799 May 7	1986 Nov 13
1802 Nov 9	1993 Nov 6
1815 Nov 12	1999 Nov 15
1822 Nov 5	

(b) 2000–2100

Date	Mid-transit (GMT)
2003 May 7	07.53
2006 Nov 8	21.42
2016 May 9	14.59
2019 Nov 11	15.21
2032 Nov 13	08.55
2039 Nov 7	08.48
2049 May 7	14.26
2052 Nov 9	02.32
2062 May 10	21.39
2065 Nov 11	20.08
2078 Nov 14	13.44
2085 Nov 7	13.37
2095 May 8	21.09

The first prediction of a transit was made by Kepler, for the transit of 7 November 1631, and this enabled Gassendi to observe the transit. Transit dates between 1631 and 2100 are given in Table 4.3.

Transits can occur only in May and November. May transits occur with Mercury near aphelion; at November transits Mercury is near perihelion, and November transits are the more frequent in the ratio of seven to three. The longest transits (those of May) may last for almost 9 h. During a transit, Mercury appears much blacker than any sunspot. The Mercurian atmosphere is too tenuous to produce any observable effects, as happens with Venus.

The transit of 6 November 1993 was observed at X-ray wavelengths. The observations were made from the Japanese satellite Yohkoh. Mercury blocked X-ray emissions from the corona, so appearing as a tiny dark hole

Table 4.4. Planetary occultations and close conjunctions involving Mercury.

	Date	GMT	Separation (")	Elongation
Neptune	1914 Aug 10	08.11	−30	18W
Mars	1942 Aug 19	12.36	−20	16E
Mars	1985 Sept 4	21.00	−46	16W
Mars	1989 Aug 5	21.54	+47	18E
Mars	2032 Aug 23	04.26	+16	13W
Saturn	2037 Sept 15	21.32	+18	15W
Neptune	2039 May 5	09.42	−39	13W
Neptune	2050 June 4	06.46	−46	17W
Neptune	2067 July 15	12.04	+13	18W
Mars	2079 Aug 11	01.31	Occ.	11W
Venus	2084 Dec 24	05.11	+48	17W

Mercury will occult Jupiter on 27 October 2088 and 7 April 2094, but the elongations will be less than 6 degrees.

in the X-ray corona. The transit of 15 November 1999 was exceptional. It was a 'grazing transit'; Mercury followed a short chord across the Sun's NE limb. In fact, over some parts of the Earth there was only a partial transit, as Mercury did not pass wholly on to the solar disk. The transit was total from Papua New Guinea, NE Australia, Hawaii, western South America and most of North America; partial from Antarctica and most of Australia. In New Zealand, the transit was total from North Island but partial from South Island. This situation will not recur for several centuries.

Occasionally Mercury may occult another planet; this last happened on 9 December 1808, when Mercury passed in front of Saturn. The next occasion will be on 11 August 2079, when Mercury will occult Mars. Close planetary conjunctions involving Mercury are given in Table 4.4, for the period 1900–2100; all these occur more than 10° from the Sun, and the separations are below 60 arcsec.

MAPS OF MERCURY

Mercury is a difficult object to study from Earth (and, incidentally, it cannot be studied at all with the Hubble Space Telescope, because it is too close to the Sun in the sky). The first serious telescopic observations were made in the late 18th century by Sir William Herschel, who, however, could make out no surface detail. At about the same time observations were made by J. H. Schröter, who recorded some surface patches and who believed that he had detected high mountains. It seems certain that these results were illusory.

The first attempt to produce a proper map was made by G. V. Schiaparelli, from Milan, using 21.8 cm and 49 cm refractors between 1881 and 1889. His method was to study the planet in daylight, when both it and the Sun were high above the horizon. Schiaparelli recorded various dark markings, and concluded that the rotation period must be synchronous – that is to say, equal to Mercury's orbital period. This would mean that part of the planet would be in permanent sunlight and another part in permanent darkness, with an intervening 'twilight zone' over which the Sun would rise and set, always keeping fairly close to the horizon. Percival Lowell, at the Flagstaff Observatory in Arizona, drew a map in 1896 showing 'canal-like' linear features, but these were completely non-existent.

The best pre-Space Age map was drawn by E. M. Antoniadi, using the 83 cm refractor at Meudon, near Paris; like Schiaparelli, he observed in daylight. The map was published in 1934, together with a book dealing with all aspects of the planet[1]. He agreed with Schiaparelli that the rotation period must be synchronous, and he also believed the atmosphere to be dense enough to support obscurations. Both these conclusions are now known to be wrong.

Antoniadi's map showed various dark and bright features, and to these he gave names; there was a degree of agreement with Schiaparelli's map, although there were very marked differences. Antoniadi's names are given in Table 4.5. Yet they referred only to albedo features, and the map is not accurate enough for the names to be retained – which was not Antoniadi's fault; he was certainly the best planetary observer of his time.

The modern map is based entirely upon the results from Mariner 10, the only space-craft to have flown past Mercury. Mariner 10 was launched from Cape Canaveral on 3 November 1973. It by-passed Venus on 5 February 1974, at a range of 4200 km, after which it went on to encounter Mercury, making three active passes: on 29 March 1974 (at 705 km from the surface), 21 September

[1] Surprisingly, the book was not translated into English until 1974, when I did so – by which time its interest was mainly historical. Although Antoniadi was Greek, he spent much of his life in France and wrote his book in French.

Table 4.5. Albedo features on Antoniadi's map.

Name	Lat.	Long. W
Apollonia	45.0N	315.0
Aurora	45.0N	90.0
(Victoria Rupes region)		
Australia	72.5S	0.0
(Bach region)		
Borea	75.0N	0.0
(Borealis region)		
Caduceata	45.0N	135.0
(Shakespeare region)		
Cyllene	41.0S	270.0
Heliocaminus	40.0N	170.0
Hesperis	45.0S	355.0
Liguria	45.0N	225.0
Pentas	5.0N	310.0
Phæthontias	0.0N	167.0
(Tolstoy region)		
Pieria	0.0N	270.0
Pleias Gallia	25.0N	130.0
Sinus Argiphontæ	10.0S	335.0
Solitudo Admetei	55.0N	90.0
Solitudo Alarum	15.0S	290.0
Solitudo Atlantis	35.0S	210.0
Solitudo Criophori	0.0N	230.0
Solitudo Helii	10.0S	180.0
Solitudo Hermæ Trismegisti	45.0S	45.0
(Discovery Rupes region)		
Solitudo Horarum	25.0N	115.0
Solitudo Iovis	0.0N	0.0
Solitudo Lycaonis	0.0N	107.0
(Beethoven region)		
Solitudo Maiæ	15.0S	155.0
Solitudo Martis	35.0S	100.0
Solitudo Neptuni	30.0N	150.0
Solitudo Persephones	41.0S	225.0
Solitudo Phœnicis	25.0N	225.0
Solitudo Promethei	45.0S	142.5
(Michelangelo region)		
Tricrena	0.0N	36.0
(Kuiper region)		

1974 (at 47 000 km) and on 16 March 1975 (at 327 km). By that time the equipment was deteriorating, and contact was finally lost on 24 March 1975. No doubt the probe is still in solar orbit, and still making periodical approaches to Mercury, but we have no hope of locating it again. Unfortunately the same areas of the planet were in sunlight at all three active passes, and so we have good maps of less than half the total surface, although there is no reason to expect that the remaining areas will be basically different from those available to Mariner.

Mercury has proved to be a world of craters, mountains, low plains (planitiæ), scarps (rupes), dorsa (ridges) and valleys. The craters are named after famous artists, musicians, painters and authors; planitiæ after the names for Mercury in different languages; rupes after ships of discovery or scientific expeditions, and valleys after radio telescopes. Only three astronomers are commemorated on Mercury; Antoniadi and Schiaparelli, and also G. P. Kuiper, who was closely concerned with planetary space research. A selected list of named features on Mercury is given in Table 4.7.

ATMOSPHERE

As might be expected from its low escape velocity, Mercury has only an excessively tenuous atmosphere; its total weight is probably no more than about 8 tons. It was first positively detected by the instruments on Mariner 10 (earlier spectroscopic reports of an atmosphere, initially by H. Vogel in 1871, were erroneous). Its density is so low that it may be regarded as an exosphere. The ground density is of the order of 10^{-10} mbar.

The main constituents are sodium, oxygen, helium, potassium and some argon. Hydrogen and helium may well originate from the solar wind, while sodium, potassium and oxygen are probably gases released by the vaporization of the surface material (the regolith) by the impact of micrometeorites. But there is no chance that the atmosphere is dense enough to support 'clouds' of any type, as Antoniadi believed – and it is very clear that life of any kind on Mercury is out of the question.

INTERIOR AND MAGNETIC FIELD

Rather surprisingly, Mercury does have a definite magnetic field, first identified by Mariner 10. The encounter yielded a value of 350 gammas (0.3 G) at the surface, or about 1% of the Earth's magnetic field. This means that the magnetic field of Mercury is stronger than those of Venus, Mars or the Moon. The field is dipolar, with two equal magnetic poles of opposite polarity, inclined by 11° to the rotational axis of the planet. The polarity is the same as that of the Earth's field (i.e. a compass needle would point north). The magnetic field is just sufficient to deflect the solar wind away from the planet's surface, and a magnetosphere is detectable, with a bow shock at 1.5 Mercurian radii from the centre of the globe.

This indicates the presence of an iron-rich core, which may well be molten. The overall density of Mercury is greater than that of any other planet apart from the Earth, and the core is likely to be about 3600 km in diameter – larger than the whole globe of the Moon. Overlying the core are a solid mantle and crust of silicates with a combined thickness of around 600 km; the surface is covered with a layer of porous silicate 'dust' forming the top of a regolith, which goes down to a few metres or a few tens of metres. Compared with that of the Moon, the crust itself may be rather deficient in iron and titanium-bearing minerals, which block or absorb microwaves. There have been suggestions that in the early history of the Solar System Mercury was struck by a large body and had its outer layers ripped off, leaving behind the heavy iron-rich core. This sounds plausible, but of course there is no proof. By weight, Mercury is 70% iron and only 30% rocky material, so that is contains twice as much iron per unit volume as any other planet or satellite.

AXIAL ROTATION

For many years the synchronous rotation period, favoured by Schiaparelli and Antoniadi, was generally accepted, but in 1962 W. E. Howard and his colleagues at Michigan measured the long-wavelength radiations from Mercury, and found that the dark side was much warmer than it would be if it never received any sunlight. In 1965 the non-synchronous period was confirmed by radar methods, largely by R. Dyce and G. Pettengill, using the large radio telescope at Arecibo in Puerto Rico. The true period is 58.6 days, which is two-thirds of the orbital period, and when Mercury is best placed for observation from Earth it is always the same area which is presented.

There is no area of permanent daylight, no region of permanent night and no twilight zone, but the Mercurian calendar is decidedly peculiar. The axial inclination is negligible, so that Mercury spins in an almost 'upright'

sense relating to its orbital plane. When Mercury is near perihelion, the orbital angular velocity exceeds the constant spin angular velocity, so that an observer on Mercury would see the Sun slowly move in a retrograde direction for eight Earth days around each perihelion passage. The Sun would then almost hover over what may be called a 'hot pole', one of which is the site of the largest basin found on the planet – aptly named the Caloris Basin.

It is interesting to follow the course of a Mercurian 'day'. To an observer at a hot pole, the Sun will be at the zenith at the time of perihelion, so that its apparent diameter will be at its greatest. As it nears the zenith it will stop and move retrograde for eight Earth days before resuming its original direction of motion. As it drops toward the horizon it will shrink, finally setting 88 Earth days after having risen – not to be seen again for another 88 Earth days. But to an observer 90° away, the Sun will be at its largest when rising, as Mercury will then be at perihelion; sunrise will be protracted, because the Sun will appear, almost vanish again and finally climb toward the zenith, when it will be at its minimum apparent size. It will swell as it drops in the sky; it will set, rise again briefly and then depart. The interval between one sunrise and the next is 176 Earth days.

No doubt Mercury once rotated much faster than it does now; it has been 'braked' by the gravitational pull of the Sun, and it has been suggested that it had reached its present slow rotation rate as early as 500 million years after its formation.

The temperature at the hot pole reaches 427 °C at maximum, but the night temperature drops to −183 °C. The temperature range is greater than for any other planet in the Solar System. Near the poles there are some craters whose floors are always in shadow, and which remain bitterly cold; in 1991 radar measurements made with the VLA (Very Large Array) in New Mexico led to the suggestion that ice might exist in these craters. Results from Arecibo, and from Mariner 10, were later quoted in support. However, the evidence is at best very suspect – as in the case of the Moon – and the existence of ice on a world such as Mercury would be very surprising indeed. Moreover, the same radar results have been found in areas on Mercury which do receive sunlight, and where ice could not possibly survive.

Table 4.6. Mercurian systems.

Name	Age (thousands of millions of years)	Features	Lunar counterparts
Pre-Tolstoyan	4	Intercrater plains, multi-ring basins	Pre-Nectarian
Tolstoyan		Plains materials, smaller basins, craters	Nectarian
Calorian		Basins, plains materials, craters	Imbrian
Mansurian		Craters	Eratosthenian
Kuiperian	1	Craters, ray-systems	Copernican

SURFACE FEATURES

Craters are widespread in Mercury, and some of them are ray-centres; indeed, the first crater to be identified during the approach of Mariner 10 was a ray-centre, and was named Kuiper, in honour of the Dutch astronomer Gerard Kuiper (1905–1973). The south pole of Mercury is marked by the crater Chao Meng-Fu. It has been agreed that the 20th-degree meridian passes through the centre of the 1.5 km crater Hun Kal, 0°58′ south of the Mercurian equator (the name stands for 20 in the language of the Maya, who used a base-20 number system). Craters less than 20 km in diameter are bowl-shaped, with depths of about one-fifth of their diameters; craters between 20 and 90 km across have flatter floors, often with central peaks and terraced walls. The general distribution of the craters is of the lunar type, so that if one crater breaks into another it is almost always the smaller formation which is the intruder.

The most imposing feature on Mercury is the Caloris Basin (Caloris Planitia), at one of the two 'hot poles'. It is 1300 km in diameter, and is bounded by a ring of smooth mountain blocks rising 1–2 km above the surrounding surface; a second, weaker scarp lies 100–160 km beyond the main one. Unfortunately only half of Caloris was within range of Mariner 10. Antipodal to it is an area which is officially termed 'hilly and lineated terrain', although often called 'weird terrain'. It covers 360 000 km², and consists of hills, depressions and valleys which have destroyed older features. Some of the hills rise to 2 km. It may well be that the formation of this terrain has been due to the vast impact which produced Caloris.

There is an obvious resemblance between the surface of Mercury and that of the Moon, but there are important differences in detail. In particular, about 45% of the Mercurian surface mapped from Mariner 10 is occupied by 'intercrater plains', which are very old and are unlike anything on the Moon. There are also lobate scarps, cliffs from 20 to over 500 km long and up to 3 km high; they seen to be essentially thrust faults, cutting through other features and displacing older landforms. Again there is no lunar counterpart. It seems probable that they formed in response to a 1–2 km shrinkage in the planet's diameter early in its history – predicted by thermal models. There was undoubtedly much more general melting on Mercury than there ever was on the smaller Moon, and this accounts for the significant differences between the two bodies. Moreover, even the most densely-cratered regions of Mercury do not contain as many formations as the most densely-cratered areas of the Moon.

Apart from Caloris, the largest circular structure is Beethoven, with a diameter of 643 km. Ray-craters include Kuiper, Copley, Mens, Tansen and Snorri.

Obviously we know much less about the past history of Mercury than we do about the Moon. A tentative timescale has been drawn up, and is given in Table 4.6, but it may not be very accurate.

Mercury has no satellite. This seems certain. There was however an alarm on 27 March 1974, two days before the Mariner 10 flyby of the planet; one instrument began recording bright emissions in the extreme ultra-violet. They vanished, but then reappeared, and a satellite was suspected. However, it turned out that the object was an ordinary star – 31 Crateris. If Mercury had a satellite of appreciable size, it would almost certainly have been found by now.

No doubt more space-craft will be sent to Mercury during the 21st century, but manned landings there seem to be most unlikely. Mercury is a fascinating world, but is not a welcoming one.

Table 4.7. (a) Craters on Mercury.

Name	Lat.	Long. W	Diameter
Abu Nuwas	17.4N	20.4	116
Africanus Horton	51.5S	41.2	135
Ahmed Baba	58.5N	126.8	127
Al-Akhtal	50.2N	07.0	102
Alencar	63.5S	103.5	120
Al-Hamadhani	38.8N	89.7	186
Al-Jāhiz	1.2N	21.5	91
Amru Al-Qays	12.3N	175.6	50
Andal	47.7S	37.7	108
Aristoxenus	82.0N	11.4	69
Asvaghosa	10.4N	21.0	90
Bach	68.5S	103.4	214
Balagtas	22.6S	13.7	98
Balzac	10.3N	144.1	80
Barma	41.3S	162.8	128
Bartok	29.6S	134.6	112
Bashō	32.7S	169.7	80
Beethoven	20.8S	123.6	643
Belinskii	76.0S	103.4	70
Bello	18.9S	120.0	129
Bernini	79.2S	136.5	146
Bjornson	73.1N	109.2	88
Boccaccio	80.7S	29.8	142
Boethuis	0.9S	73.3	129
Botticelli	63.7N	109.6	143
Brahms	58.5N	176.2	96
Bramante	47.5S	61.8	159
Brontë	38.7N	125.9	60
Bruegel	49.8N	107.5	75
Brunelleschi	9.1S	22.2	134
Burns	54.4N	115.7	45
Byron	8.5S	32.7	105
Callicrates	66.3S	32.6	70
Camões	70.6S	69.6	70
Carducci	36.6S	89.9	117
Cervantes	74.6S	122.0	181
Cezanne	8.5S	123.4	75
Chao Meng-Fu	87.3S	134.2	167
Chekhov	36.2S	61.5	199
Chiang K'ui	13.8N	102.7	35
Chong Ch'ol	46.4N	116.2	162
Chopin	65.1S	123.1	129
Chu Ta	2.2N	105.1	110
Coleridge	55.9S	66.7	110
Copley	38.4S	85.2	30
Couperin	29.8N	151.4	80
Dario	26.5S	10.0	151
Degas	37.4N	126.4	60
Delacroix	44.7S	129.0	146
Derzhavin	44.9N	35.3	159
Desprez	80.8N	90.7	50
Dickens	72.9S	153.3	78
Donne	2.8N	13.8	88
Dostoevskii	45.1S	176.4	411
Dowland	53.5S	179.5	100
Dürer	21.9N	119.0	180
Dvorák	9.6S	11.9	82
Echegaray	42.7N	19.2	75
Eitoku	22.1S	156.9	100
Equiano	40.2S	30.7	99
Fet	4.9S	179.9	24
Flaubert	13.7S	72.2	95
Futabatei	16.2S	83.0	66
Gainsborough	36.1S	183.3	100
Gauguin	66.3N	96.3	72
Ghiberti	48.4S	80.2	123
Giotto	12.0N	55.8	150
Gluck	37.3N	18.1	105
Goethe	78.5N	44.5	383
Gogol	28.1S	146.4	87

Table 4.7. (Continued)

Name	Lat.	Long. W	Diameter
Goya	7.2S	152.0	135
Grieg	51.1N	14.0	65
Guido d'Arezzo	38.7S	18.3	66
Hals	54.8S	115.0	100
Han Kan	71.6S	143.8	50
Handel	3.4N	33.8	166
Harunobu	15.0N	140.7	110
Hauptmann	23.7S	179.9	120
Hawthorne	51.3S	115.1	107
Haydn	27.3S	71.6	270
Heine	32.6N	124.1	75
Hesiod	58.5S	35.0	107
Hiroshige	13.4S	26.7	138
Hitomaro	16.2S	15.8	107
Holbein	35.6N	28.9	113
Holberg	67.0S	61.1	61
Homer	1.2S	36.2	314
Horace	68.9S	52.0	58
Hugo	38.9N	47.0	198
Hun Kal	1.6S	21.4	13
Ibsen	24.1S	35.6	159
Ictinus	79.1S	165.2	119
Imhotep	18.1S	37.3	159
Ives	32.9S	111.4	20
Janáček	56.0N	153.8	47
Jōkai	72.4N	135.3	106
Judah ha-Levi	10.9N	17.7	80
Kālidāsā	18.1S	179.2	107
Keats	69.9S	154.5	115
Kenkō	21.5S	16.1	99
Khansa	59.7S	51.9	111
Kōshō	60.1N	138.2	65
Kuan Han-Ch'ing	29.4N	52.4	151
Kuiper	11.3S	31.1	62
Kusosawa	53.4S	21.8	159
Leopardi	73.0S	180.1	72
Lermontov	15.2N	48.1	152
Lessing	28.7S	89.7	100
Li Ch'ing-Chao	77.1S	73.1	61
Li Po	16.9N	35.0	120
Liang K'ai	40.3S	182.8	140
Liszt	16.1S	168.1	85
Lu Hsun	0.0N	23.4	98
Lysippus	0.8S	132.5	140
Ma Chih-Yuan	60.4S	78.0	179
Machaut	1.9S	82.1	106
Mahler	20.0S	18.7	103
Mansart	73.2N	118.7	95
Mansur	47.8N	162.6	100
March	31.1N	175.5	70
Mark Twain	11.2S	137.9	149
Marti	75.6S	164.6	68
Martial	69.1N	177.1	51
Matisse	24.0S	89.8	186
Melville	21.5N	10.1	154
Mena	0.2S	124.4	25
Mendes Pinto	61.3S	17.8	214
Michelangelo	45.0S	109.1	216
Mickiewicz	23.6N	103.1	100
Milton	25.2S	174.8	186
Mistral	4.5N	54.0	110
Mofolo	37.7S	28.2	114
Molière	15.6N	16.9	132
Monet	44.4N	10.3	303
Monteverdi	63.8N	77.3	138
Mozart	8.0N	190.5	270
Murasaki	12.6S	30.2	130
Mussorgsky	32.8N	96.5	125
Myron	70.9N	79.3	31

Table 4.7. (Continued)

Name	Lat.	Long. W	Diameter
Nampeyo	40.6S	50.1	52
Nervo	43.0N	179.0	63
Neumann	37.3S	34.5	120
Nizāmi	71.5N	165.0	76
Ōkyo	69.1S	75.8	65
Ovid	69.5S	22.5	44
Petrarch	30.6S	26.2	171
Phidias	8.7N	149.3	160
Philoxenus	8.7S	111.5	90
Pigalle	38.5S	9.5	154
Po Chü-i	7.2S	165.1	68
Po Ya	46.2S	20.2	103
Polygnotus	0.3S	68.4	133
Praxiteles	27.3N	59.2	182
Proust	19.7N	46.7	157
Puccini	65.3S	46.8	70
Purcell	81.3N	146.8	91
Pushkin	66.3S	22.4	231
Rabelais	61.0S	62.4	141
Rajnis	4.5N	95.8	82
Rameau	54.9S	37.5	51
Raphael	19.9S	75.9	343
Ravel	12.0S	38.0	75
Renoir	18.6S	51.5	246
Repin	19.2S	63.0	107
Riemenschneider	52.8S	99.6	145
Rilke	45.2S	12.3	86
Rimbaud	62.0S	148.0	85
Rodin	21.1N	18.2	229
Rubens	59.8N	74.1	175
Rublev	15.1S	156.8	132
Rūdaki	4.0S	51.1	120
Rude	32.8S	79.6	75
Rumi	24.1S	104.7	75
Sadi	78.6S	56.0	68
Saikaku	72.9N	176.3	88
Sarmiento	29.8S	187.7	145
Sayat-Nov	28.4S	122.1	158
Scarlatti	40.5N	100.0	129
Schoenberg	16.0S	135.7	29
Schubert	43.4S	54.3	185
Scopas	81.1S	172.9	105
Sei	64.3S	89.1	113
Shakespeare	49.7N	150.9	370
Shelley	47.8S	127.8	164
Shevchenko	53.8S	46.5	137
Sholem-Aleichem	50.4N	87.7	200
Sibelius	49.6S	144.7	90
Simonides	29.1S	45.0	95
Sinan	15.5N	29.8	147
Smetana	48.5S	70.2	190
Snorri	9.0S	82.9	19
Sophocles	7.0S	145.7	150
Sor Juana	49.0N	23.9	93
Sōseki	38.9N	37.7	90
Sōtatsu	49.1S	18.1	165
Spitteler	68.6S	61.8	68
Stravinsky	50.5N	73.5	190
Strindberg	53.7N	135.3	190
Sullivan	16.9S	86.3	145
Sūr Dās	47.1S	93.3	132
Surikov	37.1S	124.6	120
Takanobu	30.8N	108.2	80
Takayoshi	37.5S	163.1	139
Tansen	3.9N	70.9	34
Tchaikovsky	7.4N	50.4	165
Thākur	3.0S	63.5	118
Theophanes	4.9S	142.4	45
Thoreau	5.9N	132.3	80
Tintoretto	48.1S	22.9	92

Table 4.7. (Continued)

Name	Lat.	Long. W	Diameter
Titian	3.6S	42.1	121
Tolstoy	16.3S	163.5	390
Ts'ai Wen-Chi	22.8S	22.2	119
Ts'ao Chan	13.4S	142.0	110
Tsurayuki	63.0S	21.3	87
Tung Yüan	73.6N	55.0	64
Turgenev	65.7N	135.0	116
Tyagaraja	3.7N	148.4	105
Unkei	31.9S	62.7	123
Ustad Isa	32.1S	165.3	136
Vālmiki	23.5S	141.0	221
van Dijck	76.7N	163.8	105
van Eyck	43.2N	158.8	282
van Gogh	76.5S	134.9	104
Velazquez	37.5N	53.7	129
Verdi	64.7N	168.6	163
Vincente	56.8S	142.4	98
Vivaldi	13.7N	85.0	213
Vlaminck	28.0N	12.7	97
Vyāsa	48.3N	81.1	290
Wagner	67.4S	114.0	140
Wang Meng	8.8N	103.8	165
Wergeland	38.0S	56.5	42
Whitman	41.1N	110.4	70
Wren	24.3N	35.2	221
Yeats	9.2N	34.6	100
Yun Sŏn-Do	72.5S	109.4	68
Zeami	3.1S	147.2	120
Zola	50.1N	177.3	80

Table 4.7. (b) Other named features on Mercury.

Dorsum		
Antoniadi	25.1N	30.5
Schiaparelli	23.0N	264.2

Mons		
Caloris Montes	39.4N	187.2

Planitia		
Borealis	73.4N	79.5
Budh	22.0N	150.9
Caloris	30.5N	189.8
Odin	23.3N	171.6
Sobkou	39.9N	129.9
Suisei	59.2N	150.8
Tir	0.8N	176.1

Rupes		
Adventure	65.1S	65.5
Astrolabe	41.6S	70.7
Discovery	56.3S	38.3
Endeavour	37.5N	31.3
Fram	56.9S	93.3
Gjoa	66.7S	159.3
Heemskerck	2.9N	125.3
Hero	58.4S	171.4
Mirni	37.3S	39.9
Pourquoi-Pas	58.1S	156.0
Resolution	63.8S	51.7
Santa Maria	5.5N	19.7
Victoria	50.9N	31.1
Vostok	37.7S	19.5
Zarya	42.8S	20.5
Zeehaen	51.0N	157.0

Vallis		
Arecibo	27.5S	28.4
Goldstone	15.8S	31.7
Haystack	4.7N	46.2
Simeiz	13.2S	64.3

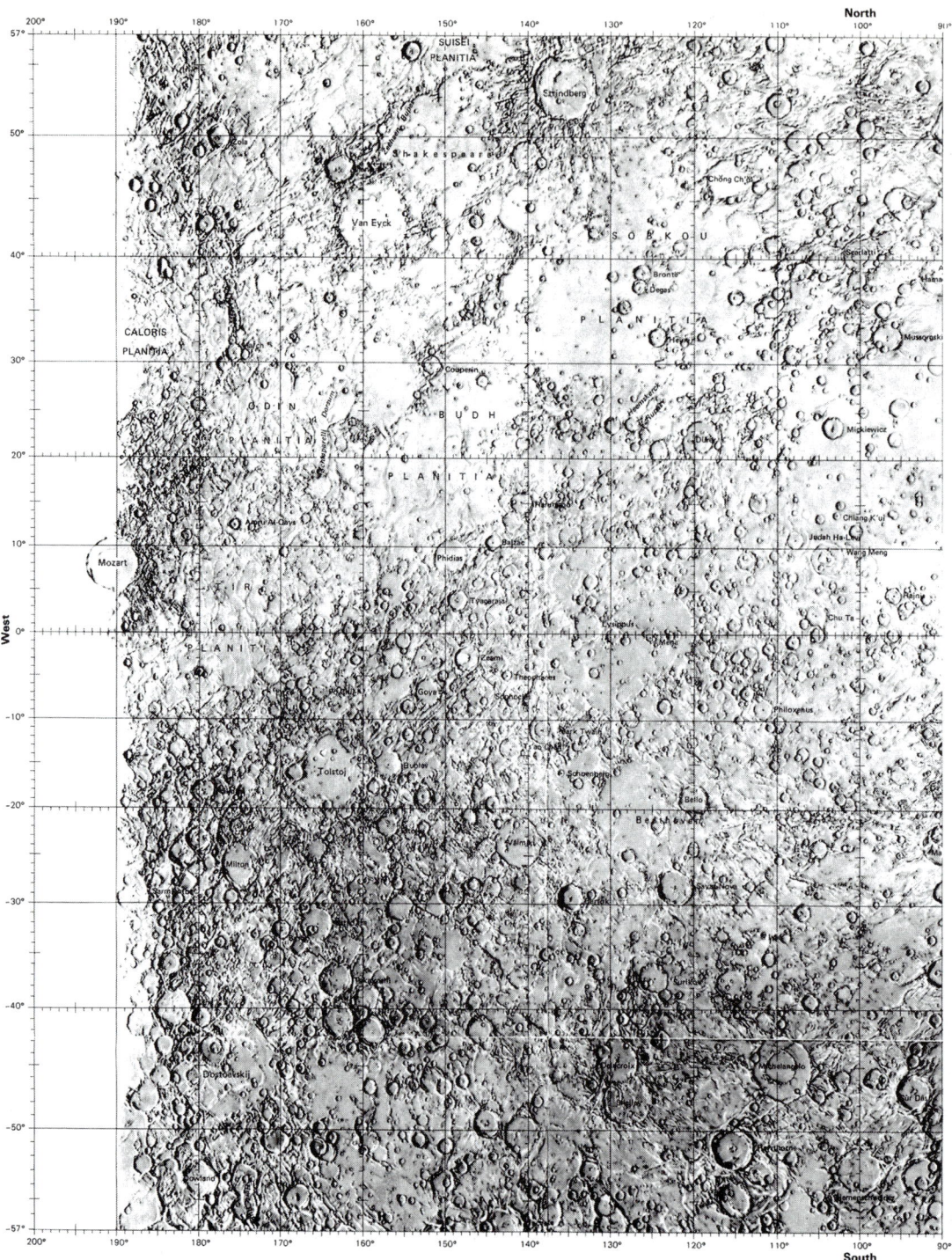

Figure 4.1. Mercury.

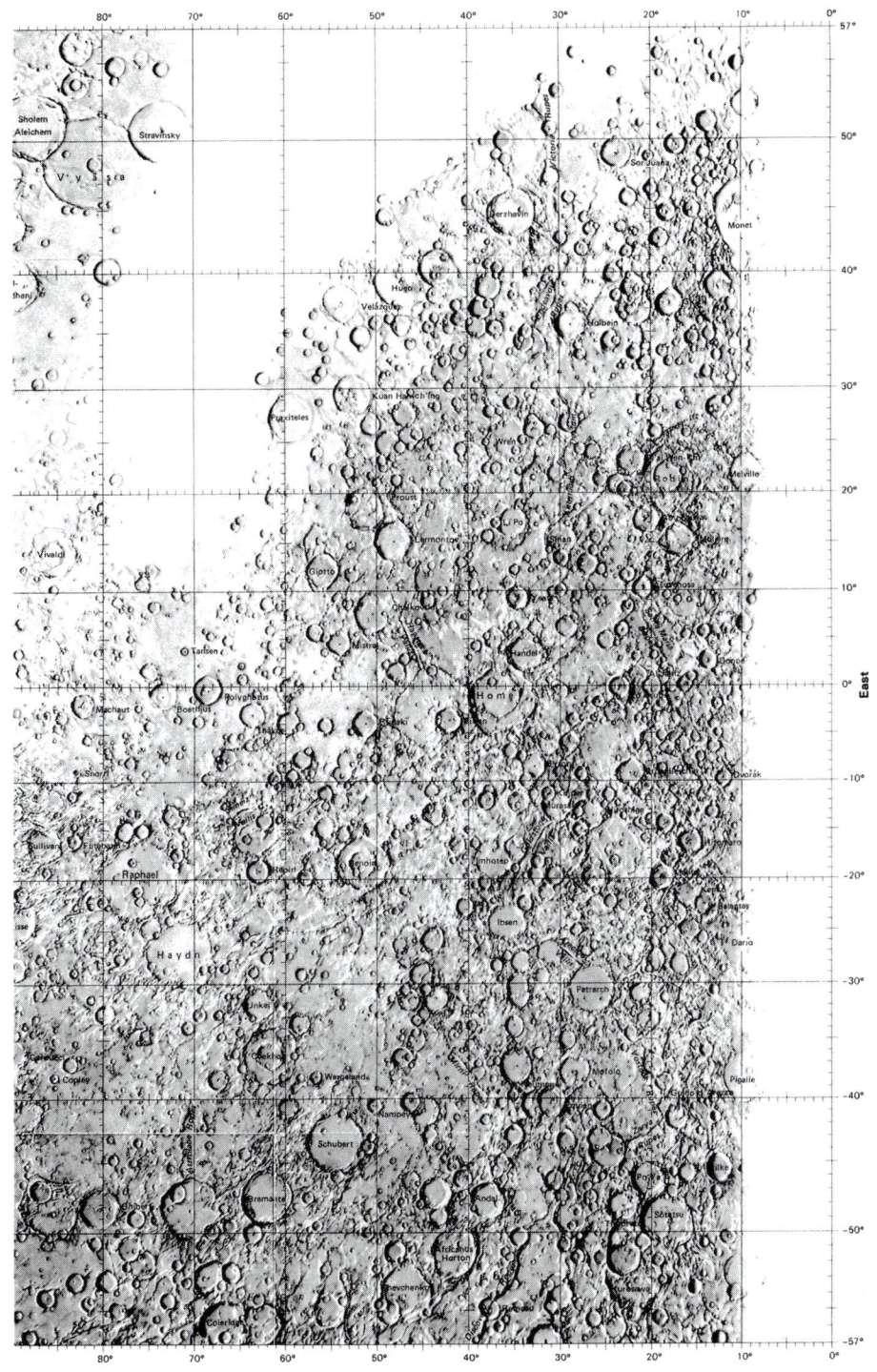

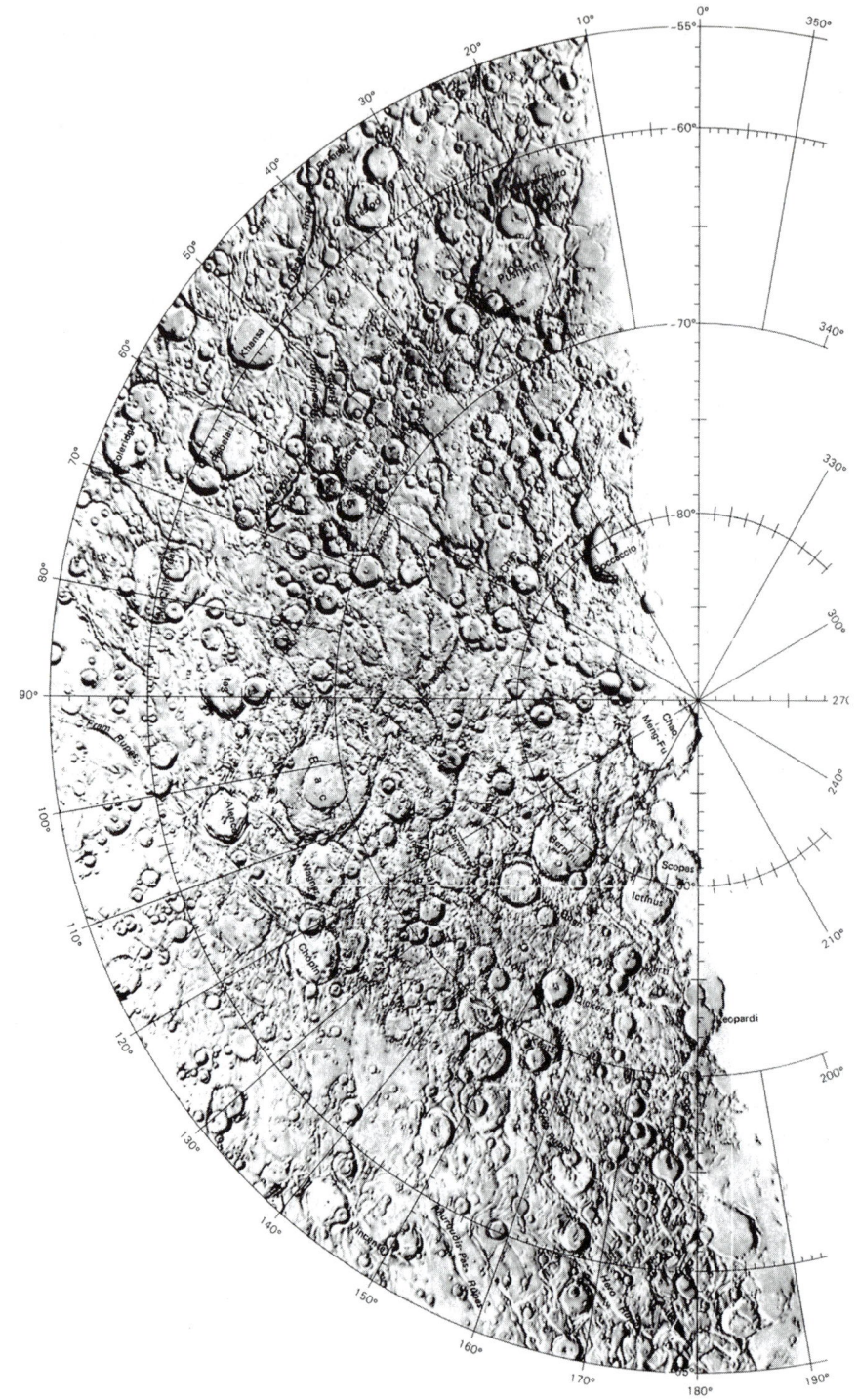

Figure 4.2. Mercury – south polar region.

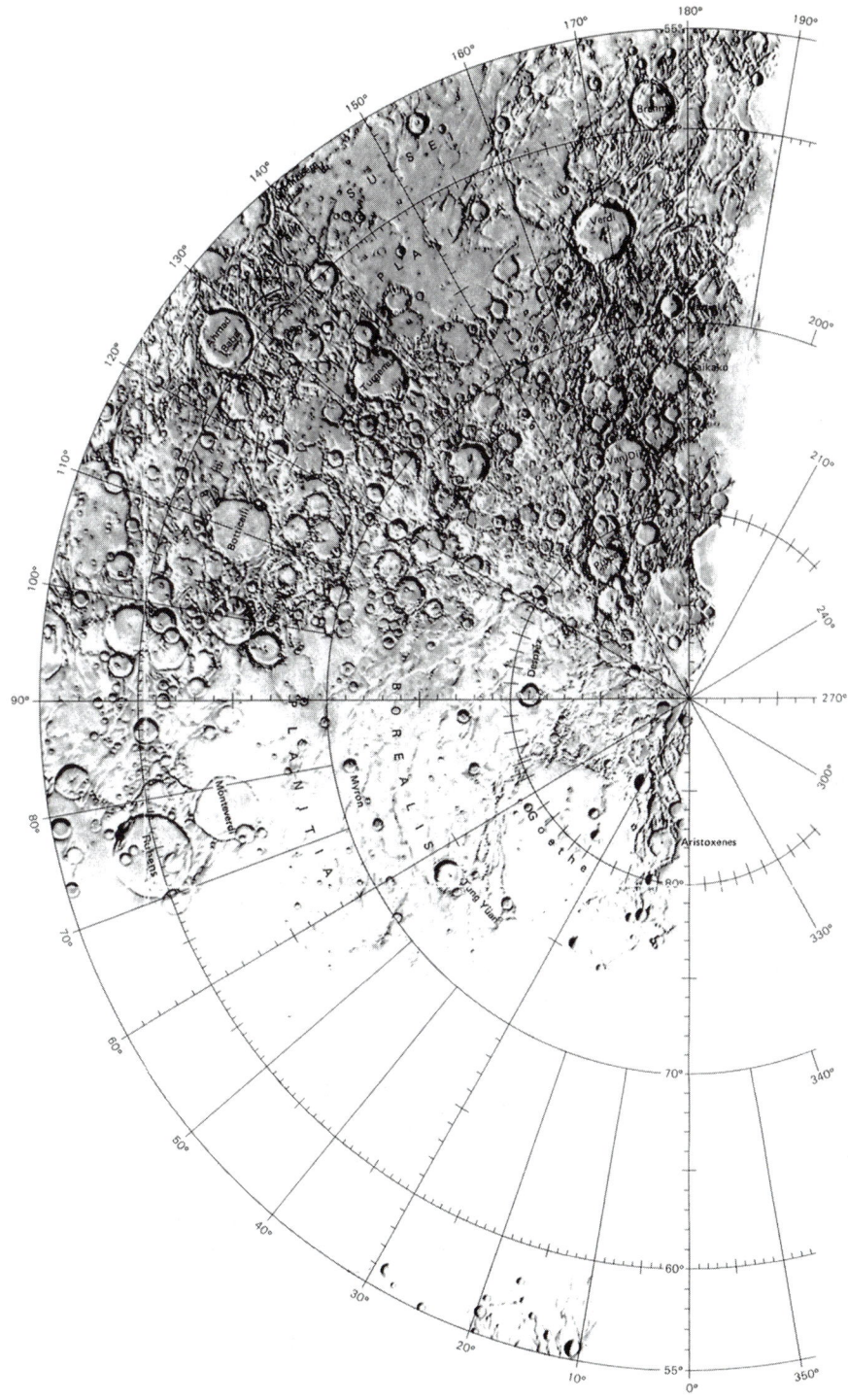

Figure 4.3. Mercury – north polar region.

5 Venus

Venus, the second planet in order of distance from the Sun, is almost a twin of the Earth in size and mass; it is only very slightly smaller and less dense. However, in all other respects it is quite unlike the Earth. Only during the past 40 years have we been able to find out what Venus is really like; its surface is permanently hidden by its thick, cloudy atmosphere, and before the Space Age Venus was often referred to as 'the planet of mystery'. Data are given in Table 5.1.

Venus is the brightest object in the sky apart from the Sun and the Moon. At its best it can even cast shadows – as was noted by the Greek astronomer Simplicius, in his *Commentary on the Heavens of Aristotle*, and by the Roman writer Pliny around 60 AD. Venus must have been known since prehistoric times. The most ancient observations which have come down to us are Babylonian, and are recorded on the Venus Tablet found by Sir Henry Layard at Konyunjik, now to be seen in the British Museum. Homer (*Iliad*, XXII, 318) refers to Venus as 'the most beautiful star set in the sky' and the name is, of course, that of the Goddess of Love and Beauty.

(As an interesting aside, Venus was once referred to by Napoleon Bonaparte. According to the French astronomer F. Arago, Napoleon was visiting Luxembourg when he saw that the crowd was paying more attention to the sky than to him; it was noon, but Venus was easily visible and Napoleon saw it. Not surprisingly, his followers referred to it as being the star 'of the Conqueror of Italy'. In more recent times, Venus has been responsible for innumerable UFO reports – one of them, indeed, from President Carter of the United States!)

Movements

Venus moves round the Sun in a practically circular orbit. Its elongation can be as much as $47°$, so that it can be above the horizon from as much as $5\frac{1}{2}$ hours after sunset or before sunrise; phenomena for the period 2000–2015 are given in Table 5.2.

Table 5.1. Data.

Distance from the Sun:
  mean 108.2 million km = 0.723 a.u.
  max 109.0 million km = 0.728 a.u.
  min 107.4 million km = 0.718 a.u.

Sidereal period: 224.701 days

Synodic period: 583.9 days

Rotation period: 243.018 days

Mean orbital velocity: 35.0 km s^{-1}

Axial inclination: $177°.33$

Orbital inclination: $3°23'39''.8$

Orbital eccentricity: 0.0167

Diameter: 12.104 km

Oblateness: negligible

Apparent diameter from Earth:
  mean $37''.3$
  max $65''.2$
  min $9''.5$

Mass: 4.868×10^{24} kg

Reciprocal mass, Sun = 1: 408.520

Mass, Earth = 1: 0.815

Density, water = 1: 5.25

Volume, Earth = 1: 0.86

Escape velocity: 10.36 km s^{-1}

Surface gravity, Earth = 1: 0.909

Mean surface temperature:
  cloud tops $-33\,°C$
  surface $467\,°C$

Albedo: 0.76

Maximum magnitude: -4.4

Mean diameter of Sun, as seen from Venus: $44'15''$.

In 1721, Edmond Halley was the first to note that Venus, unlike Mercury, is at its brightest during the crescent stage, when about 30% of the daylight hemisphere is turned in our direction. When full, Venus is of course on the far side of the Sun; at inferior conjunction, when it is closest

Table 5.2. Phenomena of Venus, 2000–2015.

E elongation	Inferior conjunction	W elongation	Superior conjunction
2001 Jan 17	2001 Mar 29	2001 June 8	2000 June 11
2002 Aug 22	2002 Nov 1	2003 Jan 11	2002 Jan 14
2004 Mar 29	2004 June 8	2004 Aug 17	2003 Aug 18
2005 Nov 3	2006 Jan 13	2005 Mar 25	2005 Mar 31
2007 June 9	2007 Aug 18	2007 Oct 28	2006 Oct 27
2009 Jan 14	2009 Mar 27	2009 June 5	2008 June 9
2010 Aug 20	2010 Oct 29	2011 Jan 8	2010 Jan 11
2012 Mar 27	2012 June 5	2012 Aug 15	2011 Aug 16
2013 Nov 1	2014 Jan 10	2014 Mar 22	2013 Mar 28
2015 June 6	2015 Aug 16	2015 Oct 26	2014 Oct 25

The maximum elongation of Venus during this period is $47°07'$, but all elongations range from $45°23'$ to $47°07'$. At the superior conjunctions of 11 June 2000 and 9 June 2008 Venus will actually be occulted by the Sun.

to the Earth, its dark side faces us, and the planet cannot be seen at all except during the rare occasions of a transit.

The observed and theoretical phases do not always agree. This is particularly evident during the time of dichotomy or half-phase. When Venus is waning, in the evening sky, dichotomy is earlier than predicted; when Venus is waxing, in the morning sky, dichotomy is late. The discrepancy may amount to several days, although it is true that timing the exact time of observed dichotomy is not easy. This effect was first noted by J. H. Schröter in 1793 – and is now generally referred to as the Schröter Effect, a term which I introduced about 40 years ago. It is due to the effects of Venus' atmosphere.

The phases were first observed telescopically by Galileo, in 1610. This was important, because according to the old Ptolemaic theory, with the Earth in the centre of the planetary system, Venus could never show a full cycle of phases. The observation strengthened Galileo's conviction in the Copernican or Sun-centred system. (The phases had not previously been mentioned specifically, although very keen-sighted people can see the crescent form with the naked eye.)

TRANSITS

Transits occur in pairs, separated by eight years, after which no more occur for over a century. Dates of past and future transits are given in Table 5.3. Unlike Mercury, Venus is

easy so see with the naked eye during transit – but since the last opportunity was in 1882, there can at present (2000) be no living person who can remember one.

The first prediction of a transit was made by Kepler, who in 1627 found that a transit was due on 6 December 1631. It was not actually observed, because it happened during night over Europe. The first transit to be observed was that of 1639, 24 November O.S. (4 December N.S.) It was seen by two amateurs, Jeremiah Horrocks and William Crabtree; independent calculations had been made by Horrocks (Kepler had not predicted a transit for 1639). It has been claimed that the Eastern scholar Al-Farabi saw a transit in 910 AD, from Kazakhstan, and this may well be true, but there is no proof, and neither is it clear that Al-Farabi had any idea about the cause – even if he did really seen Venus against the Sun. (It might even have been a large sunspot.)

Early in the 18th century, Edmond Halley suggested using transits of the inferior planets to measure the length of the astronomical unit or Earth–Sun distance. Transits of Mercury could not be observed with sufficient accuracy, but those of Venus seemed more promising, and Halley, following up an earlier comment by James Gregory, recommended careful studies of the transits of 1761 and 1769. Unfortunately the method proved to be disappointing and is now obsolete, so that future transits will be regarded as of academic interest only. The trouble was due to an effect termed the Black Drop. As Venus passes on to the solar disk it seems to draw a strip of blackness after it, and when this strip disappears the transit has already begun; again the atmosphere of Venus is responsible. However, the 1769 transit had one other important consequence. Captain Cook was detailed to take the astronomer Charles Green to Tahiti, to observe the transit; the observations were duly made – and Cook then continued upon his voyage of discovery to Australia.

OCCULTATIONS

Venus can of course be occulted by the Moon, and can itself occult stars and, occasionally, planets. An occultation of Mars by Venus was observed on 3 October 1590 by M. Möstlin, from Heidelberg, and Mercury was occulted by Venus on 17 May 1737; this was seen by J. Bevis from Greenwich. The last occultation of a planet by Venus was on 3 January 1818, when Venus passed in front of Jupiter. The next occasion will be on 22 November 2065, when Venus will again occult Jupiter, but the elongation from the Sun will be only 8°W. Close planetary conjunctions involving Venus for the period 2000–2100 are given in Table 5.4.

When Venus occults a star, the light from the star dims appreciably as it passes through Venus' atmosphere before the actual occultation takes place. This was very evident when Regulus was occulted on 7 July 1959, and proved to be very useful in estimating the density of Venus' atmosphere, which was not then well known. (I was able to make good estimates with a 30 cm reflector at Selsey, in Sussex.)

TELESCOPIC OBSERVATIONS

Early telescopic observers were unable to see any genuine details on Venus, but on 9 January 1643 G. Riccioli recorded the Ashen Light, or faint visibility of the night side of Venus. It was formerly dismissed as a mere contrast effect, but it is now believed to be due to electrical phenomena in the planet's upper atmosphere.

Dark markings on the disk were reported by F. Fontana in 1645, but he was using a small-aperture, long-focus refractor, and there is no doubt that his 'markings' on Venus were illusory. In 1727 F. Bianchini, from Rome, went so far as to produce a map of the surface, and even gave names to

Table 5.3. Transits of Venus, 1631–2200.

Date	Mid-transit (GMT)
1631 Dec 7	05.21 (not observed)
1639 Dec 4	18.27
1761 June 6	05.19
1769 June 3	22.26
1874 Dec 9	04.07
1882 Dec 6	17.06
2004 June 8	08.21
2012 June 6	01.31
2117 Dec 11	02.52
2125 Dec 8	16.06

These are followed by transits on 2247 June 11, 2255 June 8, 2360 Dec 13 and 2368 Dec 10.
Earlier transits occurred in 1032, 1040, 1153, 1275, 1283, 1396, 1518 and 1526.

the features he believed that he had recorded – such as 'the Royal Sea of King John', 'the Sea of Prince Constantine' and 'the Strait of Vasco da Gama'. Again these markings were illusory; Bianchini's telescope was of small aperture, and as the focal length was about 20 m it must have been very awkward to use.

J. H. Schröter, using better telescopes (including one made by William Herschel) observed Venus from 1779, at his observatory at Lilienthal, near Bremen. He recorded markings, which he correctly interpreted as being atmospheric, but also claimed to have seen high mountains protruding above the atmosphere.

In fact no Earth-based optical telescope will show surface details; all that can be made out are vague, impermanent, cloudy features. Neither is conventional photography more helpful, but in 1923 F. E. Ross, at Mount Wilson, took good photographs at infra-red and ultra-violet wavelengths. The infra-red pictures showed no detail, but vague features were shown in ultra-violet, indicating high-altitude cloud phenomena.

Table 5.4. Planetary conjunctions involving Venus.
(a) 2000–2015.

Planet	Date	Closest approach (GMT)	Distance (″)
Uranus	2000 Mar 4	00.37	234
Jupiter	2000 May 17	10.30	42
Uranus	2003 Mar 28	12.46	157
Mercury	2005 June 27	16.01	233
Uranus	2006 Feb 14	15.31	83
Saturn	2006 Aug 26	23.36	257
Uranus	2015	18.41	317

(b) Close conjunctions, 2000–2100 (separations below 60″, elongation at least 10° from the Sun).

Planets	Date	GMT	Separation (″)	Elongation (°)
Venus–Neptune	2022 Apr 27	19.21	−26	43W
Venus–Neptune	2023 Feb 15	15.35	−42	28E
Venus–Uranus	2077 Jan 20	20.01	−43	11W
Mercury–Venus	2084 Dec 24	05.11	+48	17W

Venus occulted Regulus on 7 July 1959, at 14.28 GMT, and will do so again on 1 October 2044, at 22.02 GMT. On 17 November 1981 Venus occulted the second-magnitude star Nunki (Sigma Sagittarii).

ROTATION PERIOD

Until fairly recent times the axial rotation period of Venus was unknown. Efforts were made to determine it by observing the drifts of surface markings across the disk, as is easy enough with Mars of Jupiter, but fails for Venus because the markings are too ill-defined. The first attempt, by G. D. Cassini in 1666, gave 23h 21m. Many other estimates followed, by visual, photographic and spectroscopic methods, but were no better. In a monograph published in 1962, I listed all the estimates of rotation periods published between 1666 and 1960. There were over 100 of them – and every one turned out to be wrong.

In 1890 G. V. Schiaparelli proposed a synchronous rotation period. This would mean that the rotation period and the orbital period would be equal at 224.7 days, and Venus would keep the same hemisphere turned toward the Sun all the time. However, this did not seem to fit the facts, and in 1954 G. P. Kuiper proposed a period of 'a few weeks'.

An estimate of the rotation period found by spectroscopic methods (the Doppler shift) was made by R. S. Richardson, at Mount Wilson, in 1956. He concluded that the rotation was very slow and retrograde – that is to say, opposite in sense to that of the Earth. During the 1950s French observers, using photography, claimed that the upper clouds had a retrograde rotation period of four days. Surprisingly, both these results proved to be correct.

Venus was first contacted by radar in 1961, by a team at the Lincoln Laboratory in the United States (an earlier result, in 1958, proved to be erroneous). Several other groups made radar contact with the planet at about the same time, and it became possible to obtain a reliable value for the rotation period.

The true period is 243.02 days, retrograde, so that Venus is the only planet to have a rotation period longer than its orbital period. The solar day on Venus is equal to 118 Earth-days, and if it were possible to see the Sun from the surface it would rise in the west and set in the east. (In fact, the Sun could never be seen through the clouds.) However, the upper clouds do indeed rotate in only four days, retrograde, so that the atmospheric structure is very unusual. The reason for this curious state of affairs is not known. It has been suggested that Venus was struck by a large impactor; this does not seem very plausible, but it is difficult to think of anything better.

ATMOSPHERE

The atmosphere of Venus was first reported by the Russian astronomer, M. V. Lomonosov, during the transit of 1761; he saw that the outline of the planet was hazy rather than clear-cut, and correctly interpreted this as being due to a dense atmosphere. Subsequently the existence of a substantial atmosphere was not seriously questioned except by Percival Lowell, at Flagstaff, who in 1897 published a map showing linear features radiating from a dark central patch which he named 'Eros'. These features were, however, illusory.

In 1923–28 E. Pettit and S. B. Nicholson, using a thermocouple attached to the 2.5 m (100 inch) Hooker reflector at Mount Wilson, made the first reliable measurements of the temperatures of the upper clouds of Venus. They gave a value of $-38\,°C$ for the day side and $-33\,°C$ for the dark side, which is in excellent agreement with modern values. Then, in 1932, W. S. Adams and T. Dunham, also at Mount Wilson, used spectroscopic methods to analyze the upper atmosphere and identified carbon dioxide – now known to make up almost the whole of the atmosphere. Carbon dioxide has a strong greenhouse effect, and it followed that the surface of Venus must be very hot indeed. A high surface temperature was also indicated by the first radio measurements of Venus at centimetre wavelengths, made by Mayer and his team in the United States in 1958.

The composition of the clouds remained uncertain. In 1937 R. Wildt suggested that they were made up of formaldehyde, and this remained the favoured theory until space-craft results showed it to be incorrect. Neither was the nature of the surface known. In 1954 F. L. Whipple and D. H. Menzel proposed that Venus was mainly water-covered, and that the clouds were composed chiefly of H_2O,

but this attractive 'marine' theory was disproved by the Mariner 2 results of 1962. Venus is far too hot for liquid water to exist on its surface.

Virtually all our detailed knowledge of Venus' atmosphere, beneath the cloud tops, comes from the various space-craft launched since 1961; a list of these is given in Table 5.5. Some of the earlier Venera probes designed to land on the surface were actually crushed during their descent, because the atmosphere was even thicker than had been expected.

Below an altitude of 80 km the atmosphere is made up of over 96% of carbon dioxide and about 3% of nitrogen (N_2), which does not leave much room for anything else; there are minute traces of carbon monoxide, helium, argon, sulphur dioxide, oxygen and water vapour and other gases such as krypton and xenon. The clouds are rich in sulphuric acid, and at some levels there must be sulphuric acid 'rain', which however evaporates well before reaching the surface. The troposphere extends from the surface to an altitude of 65 km; above, the stratosphere and mesosphere extend to 95 km, and then comes the upper atmosphere, which reaches out to at least 400 km. Venus has no detectable overall magnetic field (it must be at least 25 000 times weaker than that of the Earth), but the dense atmosphere and magnetic eddy currents induced in its ionosphere produce a well-marked bow shock, and prevent the solar wind particles from reaching the surface.

The wind structure is remarkable; the whole atmosphere may be said to be super-rotating. The winds decrease from 100 m s^{-1} at the cloud-top level to only 50 m s^{-1} at 50 km, and only a few metres per second at the surface, although even a slow wind will have tremendous force in that dense atmosphere. It is notable that there is little wind erosion on the surface, although there are obvious signs of æolian depositional activity. The atmospheric pressure at the surface is about 90 times greater than that of the Earth's air at sea-level – roughly equivalent to being under water on the sea floor of Earth at a depth of 9 km. The greenhouse effect of the carbon dioxide leads to a surface temperature of around $467\,°C$, and this is practically the same for the day and night hemispheres of the planet.

There are various cloud layers. The upper clouds lie at 70 km, and at a height of 63 km the temperature is $13\,°C$, with an outside pressure of 0.5 atmospheres. At an altitude of 50 km above the surface the temperature is 20 °C; below lies a clear layer, and then a layer of denser cloud. Beneath this layer, at 47 km, there is a second clear region. The cloud deck ends at 30 km above the ground, and at the surface there is almost complete calm. The light level is low, and following the successful landings of the first Soviet probes it was said that the illumination was roughly the same as that at Moscow on a cloudy winter day. There was no need to use the searchlights with which the probes had been equipped.

Very valuable information was obtained from two balloons dropped into Venus' atmosphere by the Russian Vega space-craft in June 1985, en route to rendezvous with Halley's Comet. (Small landers were also dropped, and sent back data from the actual surface.) The balloon from Vega 1 entered the atmosphere at 11 km s^{-1} on 10 June, over the night side, and was tracked for 46 hours as it drifted over into the day side; the Vega 2 balloon (15 June) was equally successful. One revelation was that at the level of the balloons there were stronger vertical atmospheric up-currents than had been predicted.

On 10 February 1990 the Galileo space-craft flew past Venus, en route for Jupiter, and scanned the night side; it appeared that the cloud deck, 50–58 km high, was very turbulent. It may be transporting heat upward from below via very large convention cells.

SPACE-CRAFT TO VENUS

The first of all interplanetary probes was Russia's Venera 1, launched on 12 February 1961. It lost contact when it had receded to 7 500 000 km from Earth, and we cannot be sure what happened to it; it may have by-passed Venus in May 1961 at around 100 000 km, and is presumably still in solar orbit. In July 1962 the Americans made their first attempt, with Mariner 1, but the result was disastrous – Mariner 1 plunged into the sea, apparently because someone had forgotten to feed a minus sign into a computer (a slight mistake which cost approximately $4 280 000). But then came the triumphant Mariner 2, and the era of direct planetary exploration had well and truly begun.

During its encounter with Venus, Mariner 2 revolutionized many of our ideas about the planet.

The Whipple–Menzel marine theory was killed at once; the high surface temperature and long rotation were confirmed, as was the absence of any detectable magnetic field. Venus and Earth were indeed non-identical twins.

Great interest in the exploration of Venus was maintained during the 1960s and 1970s. The Russians concentrated upon controlled landings, and after several initial failures they succeeded; in October 1975 Veneras 9 and 10 were able to transmit for 55 and 65 minutes respectively after arrival before being put permanently out of action by the hostile environment. The landing procedures were obviously difficult; everything had to be automatic, and the space-craft had to be chilled before beginning their descent through the dense, fiercely hot atmosphere.

American efforts were concentrated upon fly-by missions and orbiters. Mariner 5 by-passed Venus in 1967 and sent back data, but it was aimed essentially at Mercury, and passed Venus in what is termed a 'gravity assist' manœuvre. The Pioneer Venus mission in 1978 was complex; it consisted of a 'bus' carrying four smaller probes which were released well before the rendezvous and made 'hard' landings on the surface, leaving the 'bus' to burn away in Venus' upper atmosphere. The landers were not designed to transmit after arrival, although in fact one of them (the 'Day' probe, which came down in the area now called Themis Regio on the sunlit hemisphere) did so for over an hour.

Most of our detailed knowledge of the surface comes from the Magellan orbiter, which operated well for over four years (1990–1994). Since then there have been no deliberate Venus missions, although data were obtained from Galileo, bound for Jupiter, in 1990 and a certain amount from the Cassini space-craft, which passed Venus in 1998 and 1999 on its way to a rendezvous with Saturn.

It is probably true to say that since we have established the unfriendly nature of Venus, interest in the planet has to a certain extent waned, and has been transferred to Mars. It is very clear that no manned missions to Venus are likely to be undertaken in the foreseeable future.

Table 5.5. Missions to Venus, 1961–2000.

Name	Launch date	Encounter date	Closest approach (km)	Capsule landing area Lat.	Long.		Results
Venera 1	12 Feb 1961	19 May 1961	100 000	—	—		Contact lost at 7 500 000 km from Earth.
Mariner 1	22 July 1962	—	—	—			Total failure; fell in sea.
Mariner 2	27 Aug 1962	14 Dec 1962	34 833	—	—		Fly-by. Contact lost on 4 Jan 1963.
Zond 1	2 Apr 1964	?	100 000?	—	—		Contact lost in a few weeks.
Venera 2	12 Nov 1965	27 Feb 1966	24 000	—	—		In solar orbit.
Venera 3	16 Nov 1965	1 Mar 1966	Landed	?	?		Lander crushed during descent.
Venera 4	12 June 1967	18 Oct 1967	Landed	+19	038	Eistla Regio	Data transmitted during descent.
Mariner 5	14 June 1967	19 Oct 1967	4100	—	—		Fly-by. Data transmitted.
Venera 5	5 Jan 1969	16 May 1969	Landed	−03	018	E of Navka Planitia	Lander crushed during descent.
Venera 6	10 Jan 1969	17 May 1969	Landed	−05	023	E of Navka Planitia	Lander crushed during descent.
Venera 7	17 Aug 1970	15 Dec 1970	Landed	−05	351	E of Navka Planitia	Transmitted for 23 min after landing.
Venera 8	26 Mar 1972	22 July 1972	Landed	−10	335	E of Navka Planitia	Transmitted for 50 min after landing.
Mariner 10	3 Nov 1973	5 Feb 1974	5800	—	—		Data transmitted. En route to Mercury.
Venera 9	8 June 1975	21 Oct 1975	Landed	+31.7	290.8	Beta Regio	Transmitted for 55 min after landing. 1 picture.
Venera 10	14 June 1975	25 Oct 1975	Landed	+16	291	Beta Regio	Transmitted for 65 min after landing. 1 picture.
Pioneer Venus 1	20 May 1978	4 Dec 1978	145	—	—		Orbiter, 145 to 66 000 km, period 24 h. Contact lost, 9 Oct 1992.
Pioneer Venus 2	8 Aug 1978	4 Dec 1978	Landed (9 Dec)				Multiprobe. 'Bus' and 4 landers.
			Large Probe	+04.4	304.0	Beta Regio	No transmission after landing.
			North Probe	+59.3	004.8	Ishtar Regio	No transmission after landing.
			Day Probe	+31.7	317.0	Themis Regio	Transmitted for 67 min after landing.
			Night Probe	−28.7	056.7	N of Aino Planitia	No transmission after landing.
			Bus	−37.9	290.9	Themis Regio	Crash-landing.
Venera 11	9 Sept 1978	25 Dec 1978	Landed	−14	299	Navka Planitia	Transmitted for 95 min after landing.
Venera 12	14 Sept 1978	22 Dec 1978	Landed	−07	303.5	Navka Planitia	Transmitted for 60 min after landing.
Venera 13	30 Oct 1981	1 Mar 1982	Landed	−07.6	308	Navka Planitia	Transmitted for 60 min. Soil analysis.
Venera 14	4 Nov 1981	5 Mar 1982	Landed	−13.2	310.1	Navka Planitia	Transmitted for 60 min. Soil analysis.
Venera 15	2 June 1983	10 Oct 1983	1000	—	—		Polar orbiter, 1000–65 000 km. Radar mapper.
Venera 16	7 June 1983	16 Oct 1961	1000	—	—		Polar orbiter, 1000–65 000 km. Radar mapper.
Vega 1	15 Dec 1984	11 June 1985	8890	—	—		Fly-by; en route to Halley's Comet.
			Lander	+08.5	176.9	Rusalka Planitia	Lander transmitted for 20 min after arrival. Balloon dropped into Venus' atmosphere.
Vega 2	20 Dec 1984	15 June 1985	8030	—	—		Fly-by; en route to Halley's Comet.
			Lander	−07.5	179.8	Rusalka Planitia	Lander transmitted for 21 min after arrival. Balloon dropped into Venus' atmosphere.
Magellan	5 May 1989	10 Aug 1990	294	—	—		Orbiter, 294–8450 km. Radar mapper. Burned away in Venus' atmosphere 11 Oct 1994.
Galileo	18 Oct 1989	10 Feb 1990	16 000	—	—		Fly-by. En route to Jupiter.
Cassini	18 Oct 1997	26 Apr 1998	284	—	—		Fly-by. En route for Saturn.
		20 June 1999	598	—	—		

SURFACE FEATURES

In every way Venus is an intensely hostile planet. The first pictures from the surface, sent back by Veneras 9 and 10, were obtained under a pressure of about 90 000 mbars and an intolerably high temperature – which had been expected, in view of the greenhouse effects of the atmospheric carbon dioxide. The Venera 9 landscape was described as 'a heap of stones', several dozen centimetres in diameter and with sharp edges; the Venera 10 landing site was smoother, as through it were an older plateau.

The first attempts at analysis of the surface materials were made in March 1982 by Veneras 13 and 14, which landed in the general region of the area now known as Phœbe Regio. Venera 13 dropped a lander which continued to transmit for a record 127 minutes after arrival; the temperature was +457 °C and the pressure 89 atmospheres. Venera 14 came down in a plain near Navka Planitia; there were fewer of the sharp, angular rocks of the Venera 13 site. The temperature was given as 465 °C, and the pressure 94 atmospheres. In both cases it was reported that highly alkaline potassium basalts were much in evidence.

The first attempts at mapping Venus were made by using Earth-based radar, but only with the Pioneer mission in 1978 did it become possible to obtain really reliable information. Then came the Magellan mission, which was launched in May 1989 and which proved to be completely successful; it remained fully operative until it burned away in Venus' atmosphere on 11 October 1994. Magellan could resolve features down to 120 m; the orbital period was 3.2 h, and the high inclination of the orbit meant that the polar zones could be studied as well as the rest of the planet. The main dish (3.7 m across) transmitted downwards a pulse at an oblique angle to the space-craft, striking the surface below much as a beam of sunlight will do on Earth. The surface rocks modified the pulse before it was reflected back to the antenna; rough areas are radar-bright, while smooth areas are radar-dark. A smaller antenna sent down a vertical pulse, and the time lapse between transmission and return gave the altitude of the surface below to an accuracy of 30 m. Altogether, Magellan studied about 98% of the total surface of Venus.

Venus is a world of volcanic plains, highlands and lowlands. The plains cover 65–70% of the surface, with lowlands accounting for less than 30% and highlands for only 8%. About 60% of the surface lies within 500 m of Venus' mean radius, with only 5% at more than 2 km above it; the total range of elevations is 13 km. The highest mountains are the Maxwell Mountains, which rise to 11 km above the mean radius or 8.2 km above the adjacent plateau in Ishtar Terra. The lowest point is Diana Chasma, in the Aphrodite area, 2 km below the mean radius.

The features now mapped on Venus are given in Table 5.6, but this is of course a selected list, since many more features have not been given names. It was laid down that all names on the planet should be female – the only exception being the Maxwell Mountains, named after the 19th-century Scottish physicist James Clerk Maxwell; this name was given before the official policy was formulated. Many of the names are familiar, such as Florence Nightingale and Marie Curie, but not everyone will know that, for example, Auralia was Julius Cæsar's mother, Heng-O was a Chinese Moon goddess, Marie Vigier Lebrun a French painter and Vellamo a Finnish mermaid!

Vulcanism dominates the surface, and there are lava flows everywhere. There are two very large highland areas, Ishtar Terra in the northern hemisphere and Aphrodite area, which lies mainly in the south but is crossed by the equator. Ishtar is about the size of Australia (2900 km in diameter) and consists of western and eastern components separated by the Maxwell Mountains, the highest elevations on Venus and which have steep slopes of up to 35° in places. Maxwell forms the eastern edge of a high plateau, Lakshmi Planum, which is bounded to the south, west and north respectively by the Freyja Akna and Danu Mountains. Lakshmi is relatively smooth, covered with lava which has flowed from the caldera-like structure Colette; Colette itself has collapsed to 3 km below the adjacent surface.

Aphrodite is larger – 9700 × 3200 km – and consists of eastern and western elevated areas separated by a lower area. Western Aphrodite is made up of two regions, Thetis and Ovda, which lie from 3 to 4 km above the mean radius of Venus. These regions are dominated by what used to be called 'parquet' terrain; this term was abandoned as being too unscientific, and was replaced by 'tesseræ'. Tessera terrain is characterized by extreme roughness, and covers

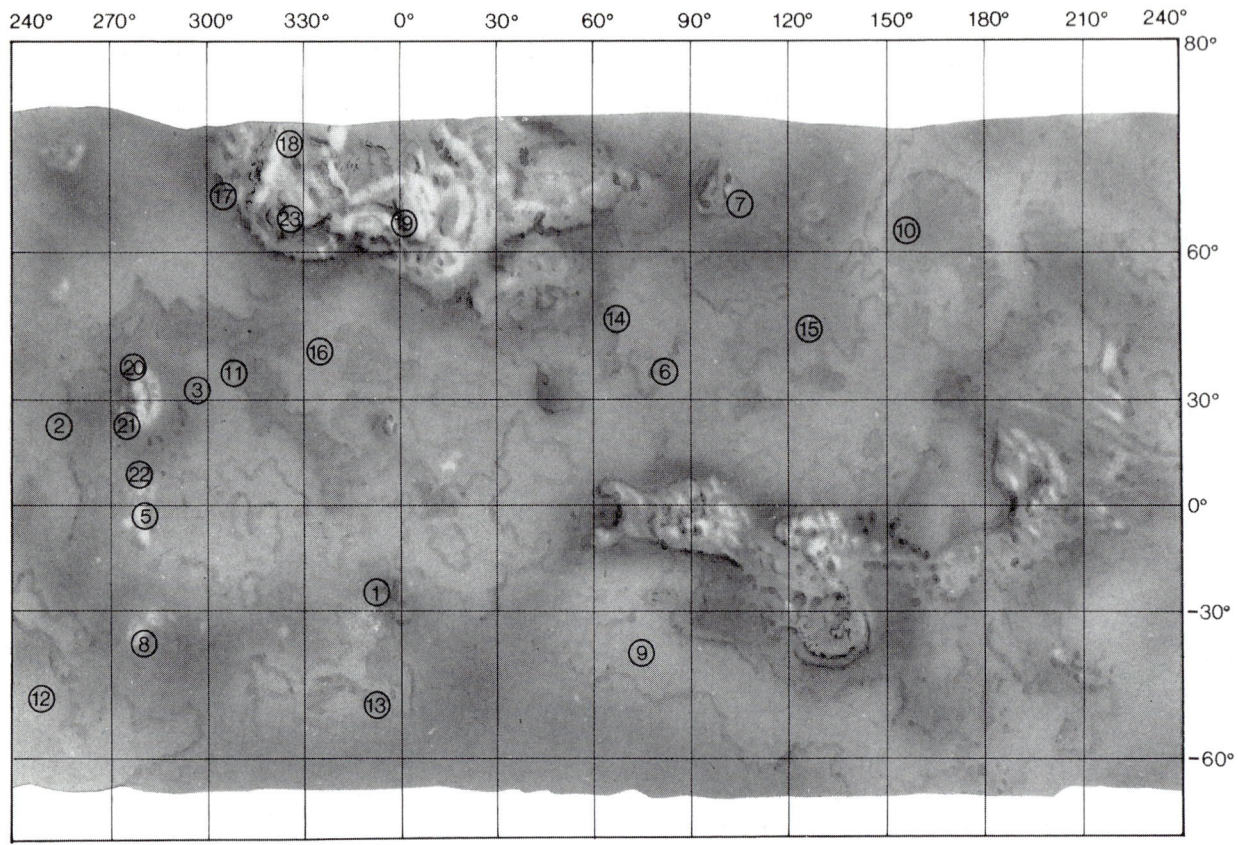

Figure 5.1. Venus.

about 8% of the total surface of Venus. It seems to be unique to Venus – at least so far as we know; this also applies to the coronæ and arachnoids. Eastern Aphrodite is dominated by chasmata, which are deep, narrow canyons. One end of Aphrodite has been nicknamed the Scorpion's Tail, although known officially as Atla Regio; it is thought to be one of the main volcanic areas.

Also of note is the highland area of Beta Regio and Phœbe Regio. Here we have what is partly a large shield volcano and partly a tessera-type highland, cut by a rift valley similar to a large scale version of the terrestrial East African Rift. There is also the southern highland of Lada Terra, first noted by the US probe Magellan.

Volcanic activity has produced features ranging from huge shield volcanoes, such as Sapas Mons (base 40 km across, height 1.5 km, with a large summit caldera) and Maat Mons, down to small structures, domes and what are still called pancake structures, probably in the nature of low, gentle domes. There are coronæ, complex features usually more or less circular, up to 2 km high and 400 km, across, surrounded by ridges and troughs, and there are the curious 'arachnoids', so named because of their outward resemblance to spiders' webs. The main volcanoes are of the shield type, as in our Hawaii. The high atmospheric pressure on Venus inhibits the flow of gas coming out in solution as the molten rock rises, making Hawaiian-type fire fountains rather unlikely; in general the magma is less dense than the rock through which it rises. Magellan recorded 140 large volcanoes with bases more than 100 km across, as well as many others of smaller size.

The crust of Venus seems to be less mobile than that of the Earth, so that terrestrial-type plate tectonics do not apply; a volcano which forms over a 'hot spot' will not drift away, as Mauna Kea in Hawaii has done, but will remain active over a very long period. There has been extensive flooding from lavas sent out both from volcanic calderæ and from fissures at lower levels, probably from 300 to 500 million years ago. Current activity is probable, perhaps in the smaller highland area of Beta Regio, which has two massive peaks – Theia Mons, which is certainly a shield volcano, and its neighbour Rhea Mons.

Impact craters abound, but are rather different from those on Mercury or the Moon; the dense atmosphere means that no meteorite more than 30 km across can hit the surface with sufficient force to produce a crater, and impact craters below 3 km across are absent, although there are some vast structures. Mead, the largest, is 280 km in diameter. The smaller craters are not lunar-type bowls, but are less regular. For the last few hundred million years the surface activity has been dominated by rift-associated vulcanism, and the fact that impact craters are less crowded than those of the Moon or Mercury indicates that the overall age of the surface features cannot be more than about 750 million years – perhaps considerably less. The largest circular lowland is Atalanta Regio, east of Ishtar; it is on average 1.4 km below the level of the mean radius and is about the area of the Gulf of Mexico. There are long lava channels, such as Hildr Chasm which is longer than the Nile.

THE INTERIOR OF VENUS

The lithosphere of Venus seems to be predominantly basaltic, and may go down to around 20–40 km, although in some areas (mainly associated with tesseræ), it may be more – perhaps down to 60 km. Below this comes the mantle and the core, about which our knowledge is very limited. The lack of a detectable magnetic field may be significant.

SATELLITE

Venus has no satellite. This now seems quite definite. A satellite was reported by G. D. Cassini in 1666, when on 18 August he saw what he regarded as a genuine attendant; other reports followed, the last coming from Montbaron, at Auxerre, on 29 March 1974. It is certain that the observers were deceived by 'telescopic ghosts'.

Table 5.6. Features on Venus. (Bold numbers indicate map references.)

Name	Lat.	Long. E	Diameter (km)
Craters: selected list			
Addams	56.1S	98.0	85
Aglaonice	26.51S	339.9	65
Alcott	59.5S	354.5	63
Andreianova	3.0S	68.8	70
Aurelia	20.3N	331.8	31
Baker	62.6N	40.5	105
Barsova	61.3N	223.0	79
Barton	27.4N	337.5	54
Boleyn	24.5N	220.0	70
Bonnevie	36.1S	127.0	91
Boulanger	26.5S	99.3	62
Cleopatra	65.9N	7.0	105
Cochran	51.8N	143.2	100
de Beauvoir	2.0N	96.1	58
Dickinson	74.3N	177.3	69
Dix	36.9S	329.1	68
Ermolova	60.2N	154.2	64
Erxleben	59.9S	39.4	30
Fedorets	59.6N	65.1	54
Gautier	26.5N	42.8	60
Graham	6.0S	6.0	75
Greenaway	22.9N	145.0	92
Henie	51.9S	145.8	70
Hepworth	5.1N	94.6	61
Isabella	29.8S	204.2	165
Jhirad	16.8S	105.6	50
Joliot-Curie	1.6S	62.5	100
Kenny	44.3S	271.1	50
Klenova	78.1N	104.2	140
Langtry	17.0S	155.0	50
Marie Celeste	23.5N	140.2	95
Markham	4.1S	155.6	69
Mead	12.5N	57.2	280
Meitner	55.6S	321.6	150
Millay	24.4N	111.1	50
Mona Lisa	25.5N	25.1	86
Nevelson	35.3S	307.8	75
O'Keeffe	24.5N	228.7	72
Ponselle	63.0S	289.0	52
Potanina	31.6N	53.1	90
Sanger	33.8N	288.6	85
Sayers	67.5S	230.0	90
Seymour	18.2N	326.5	65
Stanton	23.3S	199.1	104
Stowe	43.3S	233.2	78
Tubman	23.6N	204.5	50
Vigier Lebrun	17.3N	141.1	59
Warren	11.8S	176.5	50
Wheatley	16.6N	268.1	72
Yablochkina	48.2N	195.1	63
Zhilova	66.3N	125.5	56
Chasma			Length (km)
Aranyani	69.3N	74.4	718
Artemis	41.2S	138.5	3087
Baba-Jaga	53.2N	49.5	580
Dali	17.6S	167.0	2077
Daura	72.4N	53.8	729
Devana **22**	9.6N	284.4	1616
Diana	14.8S	154.8	938
Ganis	16.3N	196.4	615
Hecate	18.2N	254.3	3145
Heng-O	6.6N	355.5	734
Ix Chel	10.0S	73.4	503
Juno	30.5S	111.1	915
Kaygus	49.6N	52.1	503
Kottravey	30.5N	76.8	744
Kozhla-Ava	56.2N	50.6	581
Kuanja	12.0S	99.5	890
Lasdona	69.3N	34.4	697
Medeina	46.2N	89.3	606
Mežas Mate	51.0N	50.7	506
Misne	77.1N	316.5	610

Table 5.6. (Continued)

Name	Lat.	Long. E	Length (km)
Chasma (Continued)			
Morana	68.9N	24.0	317
Mots	51.9N	56.1	464
Parga	24.5S	271.5	1870
Quilla	23.7S	127.3	973
Varz	71.3N	27.0	346
Vir-Ava	14.7S	124.1	416
Vires-Akka	75.6N	341.6	742

Name	Lat.	Long. E	Diameter (km)
Colles			
Akkruva	46.1N	115.5	1059
Jurate	56.8N	153.5	418
Mena	52.5S	160.0	850

Name	Lat.	Long. E	Length (km)
Corona			
Artemis	35.0S	135.0	2600
Atete	16.0S	243.5	600
Beiwe	52.6N	306.5	600
Ceres	16.0S	151.5	675
Copia	42.5S	75.5	500
Heng-O	2.0N	355.0	1060
Lilwani	29.5S	271.5	500
Maram	7.5S	221.5	600
Quetzalpetlatl	64.0S	354.5	400
Tacoma	37.0S	288.0	500

Name	Lat.	Long. E	Length (km)
Dorsum			
Ahsonnutli	47.9N	194.8	1708
Aušrä	49.4N	25.3	859
Bezlea	30.4N	36.5	807
Breksta	35.9N	304.0	700
Dennitsa	85.6N	205.9	872
Frigg	51.2N	148.9	896
Hera	36.4N	29.5	813
Iris	52.7N	221.3	2050
Juno	31.0S	95.6	1652
Laŭma	64.8N	190.4	1517
Mardezh-Ava	32.4N	68.6	906
Nambi	72.5S	213.0	1125
Nephele	39.7N	139.0	1937
Okipeta	66.0N	238.5	1200
Saule	58.0S	206.0	1375
Sel-Anya	79.4N	81.3	975
Semuni	75.9N	8.0	514
Tezan	81.4N	47.1	1079
Uni	33.7N	114.3	800
Varma-Ava	62.3N	267.7	767
Vedma	49.8N	170.5	3345
Zorile	39.9N	338.4	1041

Name	Lat.	Long. E	Length (km)
Fluctus			
Eriu	35.0S	358.0	1200
Kaiwan	48.0S	1.5	1200
Mylitta	56.0S	353.5	1250

Name	Lat.	Long. E	Diameter (km)
Fossæ			
Arionrod	37.0S	239.9	715
Bellona	38.0N	222.1	855
Enyo	61.0S	344.0	900
Hildr	45.4N	159.4	677
Nike	62.0S	347.0	850

Name	Lat.	Long. E	Length (km)
Linea			
Antiope	40.0S	350.0	1240
Guor	17.0N	2.6	1050
Kalaipahoa	60.5S	338.0	2400

Name	Lat.	Long. E	Length (km)
Linea			
Kara	44.0S	306.0	700
Molpadia	48.0S	359.0	1350
Morrigan	54.5S	311.0	3200
Penardun	54.0S	344.0	975

Table 5.6. (Continued)

Name	Lat.	Long. E	Diameter (km)
Mons			
Gula	21.9N	359.1	276
Hathor	38.7S	324.7	333
Innini	34.6S	328.5	339
Maat	0.5N	194.6	395
Mbokomu	15.1S	215.2	460
Melia	62.8N	119.3	311
Nephthys	33.0S	317.5	350
Ozza	4.5N	201.0	507
Rhea **20**	32.4N	282.2	217
Sapas	8.5N	188.3	217
Sekmet	44.2N	240.8	338
Sif	22.0N	352.4	200
Tefnut	38.6S	304.0	182
Theia **21**	22.7N	281.0	226
Tuulikki	10.3N	274.7	520
Ushas	24.3S	324.6	413
Venilia	32.7N	238.8	320
Xochiquetzal	3.5N	270.0	80

Name	Lat.	Long. E	Diameter (km)
Montes			
Akna **17**	68.9N	318.2	830
Danu	58.5N	334.0	808
Freyja **18**	74.1N	333.8	579
Maxwell	65.2N	3.3	797
Nokomis	18.9N	189.9	486

Name	Lat.	Long. E	Diameter (km)
Patera			
Anning	66.5N	57.8	135
Aspasia	56.4N	189.1	150
Boadicea	56.0N	96.0	220
Colette	66.5N	322.8	149
Eliot	39.0N	79.0	116
Hatshepsut	28.1N	64.5	118
Hiei Chu	48.3N	97.4	139
Hroswitha	35.8N	34.8	163
Kottauer	36.7N	39.6	136
Nzingha	69.0N	206.0	143
Raskova	51.0S	222.8	80
Razia	46.2N	197.8	157
Sacajawea	64.3N	335.4	233
Sand	42.0M	15.5	181
Sappho	14.1N	16.5	92
Schumann-Heink	74.0N	215.0	120
Stopes	42.5N	47.0	169
Tarbell	58.2S	351.5	80
Tipporah	38.9N	43.0	99
Tituba	42.5N	214.0	163
Trotula	41.3N	18.9	146
Yaroslavna	38.8N	21.2	112

Name	Lat.	Long. E	Diameter (km)
Planitia			
Aino **9**	40.5S	94.5	4983
Atalanta **10**	45.6N	165.8	2048
Audra	61.5N	71.5	1861
Bereghinya	28.6N	23.6	3902
Ganiki	25.9N	189.7	5158
Guinevere **11**	21.9N	325.0	7519
Helen **12**	51.7S	263.9	4362
Kawelu	32.8N	246.5	3910
Lavinia **13**	47.3S	347.5	2820
Leda **14**	44.0N	65.1	2890
Louhi	80.5N	120.5	2441
Navka	8.1S	317.6	2100
Niobe **15**	21.0N	112.3	5008
Nsomeka	55.0S	170.0	7000
Rusalka	9.8N	170.1	3655
Sedna **16**	42.7N	340.7	3572
Snegurochka	86.6N	328.0	2773
Vellamo	45.4N	149.1	2154
Vinmara	53.8N	207.6	1634

Name	Lat.	Long. E	Diameter (km)
Planum			
Lakshmi **23**	68.6N	339.3	2343

Table 5.6. (Continued)

Name	Lat.	Long. E	Diameter (km)
Regio			
Alpha **1**	25.5S	1.3	1897
Asteria **2**	21.6N	267.5	1131
Atla	9.2N	200.1	3200
Bell	32.8N	51.4	1778
Beta **3**	25.3N	282.8	2869
Dione	31.5S	328.0	2300
Eistla	10.5N	21.5	8015
Hyndla	22.5N	294.5	2300
Imdr	43.0S	212.0	1611
Metis **4**	72.0N	256.0	729
Ovda	2.8S	85.6	5280
Phœbe **5**	6.0S	282.8	2852
Tethus **7**	66.0N	120.0	2410
Themis **8**	37.4S	284.2	1811
Thetis	11.4S	129.9	2801
Ulfrun	20.5N	223.0	3954
Rupes			
Fornax	30.3N	201.1	729
Gabie	67.5N	109.9	350
Hestia	6.0N	71.1	588
Uorsar	76.8N	341.2	820
Ut	55.3N	321.9	676
Vesta	58.3N	323.9	788
Terra			
Aphrodite	5.8S	104.8	9999
Ishtar	70.4N	27.5	5609
Lada	54.4S	342.5	8614
Tessera			
Ananke	53.3N	133.3	1060
Atropos	71.5N	304.0	469
Clotho	56.4N	334.9	289
Dekla	57.4N	71.8	1363
Fortuna	69.9N	45.1	2801
Itzpapalotl	75.7N	317.6	380
Kutue	39.5N	108.8	653
Lachesis	44.4N	300.1	664
Laima	55.0N	48.5	971

Table 5.6. (Continued)

Name	Lat.	Long. E	Diameter (km)
Tessera (Continued)			
Manzan-Gurme	39.5N	359.5	1354 (Tesseræ)
Meni	48.1N	77.9	454
Meshkenet	65.8N	103.1	1056
Moira	58.7N	310.5	361
Nemesis	45.9N	192.6	355
Shimti	31.9N	97.7	1275
Tellus **6**	42.6N	76.8	2329
Virilis	56.1N	239.7	782
Tholus			
Ale	68.2N	247.0	87
Ashtart	48.7N	247.0	138
Bast	57.8N	130.3	83
Brigit	49.0N	246.0	115
Mahuea	37.5S	164.7	110
Nertus	61.2N	247.9	66
Semele	64.3N	202.0	194
Upunusa	66.2N	242.4	223
Wurunsemu	40.6N	209.9	83
Zorya	9.4S	335.3	22
Undæ			
Al-Uzza	67.7N	90.5	150
Menat	24.8S	339.4	25
Ningal	9.0N	60.7	225

Name	Lat.	Long. E	Length (km)
Vallis			
Anuket	66.7N	8.0	350
Avfruvva	2.0N	70.0	70
Baltis	37.3N	161.4	6000
Bayara	45.6N	16.5	500
Belisama	50.0N	22.5	220
Bennu	1.3N	341.2	710
Citlalpul	57.4S	185.0	2350
Kallistos	51.1S	21.5	900
Lo Shen	12.8S	80.6	224
Samundra	24.1S	347.1	85
Sati	3.2N	334.4	225
Saga	76.1N	340.6	450
Sinann	49.0S	270.0	425
Ta'urua	80.2S	247.5	525
Vakarine	5.0N	336.4	625
Ymoja	71.6S	204.8	390

6 EARTH

The Earth is the largest and most massive of the inner group of planets. Data are given in Table 6.1. In the Solar System, only the Earth is suited for advanced life of our kind; it lies in the middle of the 'ecosphere', the region round the Sun where temperatures are neither too high nor too low. Venus lies at the extreme inner edge of the ecosphere, and Mars at the extreme outer edge.

Table 6.1. Data.

Distance from Sun:
  mean 149.5979 million km (1 a.u.)
  max 152.0962 million km (1.0167)
  min 147.0996 million km (0.9833)

Perihelion (2000): 3 January

Aphelion (2000): 4 July

Equinoxes (2000): 20 March, 07h 35m;
  22 September, 17h 27m

Solstices (2000): 21 June, 01h 48m;
  21 December, 13h 37m

Obliquity of the ecliptic:
  23°.43942 (2000), 23°.43929 (2000)

Sidereal period: 365.256 days

Rotation period: 23h 56m 04s

Mean orbital velocity: 29.79 km s^{-1}

Orbital inclination: 0° (by definition)

Orbital eccentricity: 0.01671

Diameter: equatorial 12 756 km; polar 12 714 km

Oblateness: 1/298.25

Circumference: 40 075 km (equatorial)

Surface area: 510 565 500 km^2

Mass: 5.974 × 10^{24} g

Reciprocal mass, Sun = 1: 328 900.5

Density, water = 1: 5.517

Escape velocity: 11.18 km s^{-1}

Albedo: 0.37

Mean surface temperature: 22 °C

The Earth–Moon system is often regarded as a double planet rather than as a planet and a satellite. The effect of tidal friction increases the Earth's axial rotation period by an average of 1.7 ms per century.

STRUCTURE

The rigid outer crust and the upper mantle of the Earth's globe make up what is termed the lithosphere; below this comes the asthenosphere, where rock is partially melted. Details of the Earth's structure are given in Table 6.2.

Table 6.2. Structure of the Earth.

Depth (km)		% of Earth's mass
0–50	Continental crust	0.374
0–10	Oceanic crust	0.099
10–400	Upper mantle	10.3
400–650	Transition zone	7.5
650–2890	Lower mantle	49.2
2890–5150	Outer core	30.8
5150–6370	Inner core	1.7

The crust has an average depth of 10 km below the oceans, but down to around 50 km below the continents; the base of the crust is marked by the Mohorovičič Discontinuity (the 'Moho', named after the Jugoslav scientist Andrija Mohorovičič, who discovered that the velocity of seismic waves changes abruptly at this depth, indicating a sudden change in density). Between 50 and 100 km below the surface the lithospheric rocks become hot and structurally weak. The outer shell is divided into eight major 'plates' and over 20 minor ones; the boundaries between the plates are associated with transform faults, subduction zones, earthquakes, volcanoes and mountain ranges. The continents drift around relative to each other; this was first proposed in 1915 by the Austrian meteorologist Alfred Wegener, and has led on to the science of plate tectonics.

Below the lithosphere comes the mantle, which extends down to 2890 km and contains 67% of the Earth's

Table 6.3. Geological periods. Ages in millions of years.

	From	To		
Pre-Cambrian era				
Archæan		>3490	Formation of the crust. Oldest rocks.	No life.
Proterozoic	3490	590	Shallow seas widespread.	First life. Stromatolites. Marine algæ. Jellyfish.
Palæozoic era				
Cambrian	590	505	Climate probably fairly mild in the northern hemisphere.	Graptolites, trilobites.
Ordovician	505	438	Probably moderate to warm. Volcanic activity.	Brachiopods, trilobites. Shelled invertebrates.
Silurian	438	408	Probably warm. Continental movements.	Armoured fishes. Scorpions.
Devonian	408	360	Warm, sometimes arid. Greenland, NW Scotland, and N America probably joined.	Land plants, Amphibians. Insects. Spiders. Graptolites die out.
Carboniferous	360 (Mississippian) 320 (Pennsylvanian)	320 286	Africa moves against a joined Europe and N America. Climate warm. Swamps and shallow seas.	Amphibians, winged insects. Coal measures laid down. First reptiles at end of the period.
Permian	286	248	Widespread deserts. Gondwanaland (S America, Africa, India, Australia, Antarctica) near South Polar regions. Formation of great continent of Pangæa.	Last trilobites. Spread of reptiles. Conifers. Cold period at end; major extinction of species.
Mesozoic era				
Triassic	248	213	Pangæa comprised Eurasia to the N and Gondwanaland to the S, separated by the Tethys Ocean. Hot climate.	Ammonites. Large marine reptiles. First dinosaurs. First small and primitive mammals.
Jurassic	213	144	Pangæa breaking up; rupture between Africa and N America began in the Gulf of Mexico.	Dinosaurs. Archæopteryx; Ammonites. Small mammals.
Cretaceous	144	65	Continental shifts; separation of S Africa and S America. Cooler than in the Jurassic; icecap over Antarctica.	Dinosaurs, dying out at the end of the period – K–T extinction.
Cenozoic era (Tertiary)				
Palæocene	65	55	Continental drifting. Climate warm.	Rise of mammals. First modern-type plants.
Eocene	55	38	Continental drifting. Australia separates. Climate warm. Volcanic activity.	Widespread forests. Mammals. Snakes.
Oligocene	38	25	North Europe moves northward. Climate warm to temperate.	Rise of modern-type mammals.
Miocene	25	5	Spread of grasslands, at the expense of forests. Climate temperate.	Grazing mammals. Whales. First primates.
Pliocene	5	2.5	Continents approaching present form. Climate cooler.	Primates. Apes.
(Quaternary)				
Pleistocene	2.5	10 000 yr	Periodical Ice Ages, with inter-glacials	First men.
Holocene	10 000 yr	Present	End of Ice Ages. Modern world.	Civilization.

mass. Partial melting of mantle material produces basalt, which issues from volcanic vents on the ocean floors. The base of the mantle is marked by the Gutenberg Discontinuity, where the rock composition changes from silicate to metallic and its state from solid to liquid. The outer liquid core extends down to 5150 km, and contains 31% of the Earth's mass. The inner core, down to the centre of the globe, is solid and has 1.7% of the total mass; it has been said to 'float' in the surrounding liquid core. The core is iron-rich. The solid inner core, approximately 2400 km in diameter, is thought to have a central temperature of about 4530 °C, with a density of 13.1 g cm^{-3}. Currents in the liquid core, involving iron, are responsible for the Earth's magnetic field. The outer boundary of the solid core is known as the Lehmann Discontinuity, first identified by the Danish scientist Inge Lehmann in 1936.

Most of our knowledge of the Earth's interior comes from studies of earthquakes. Seismic waves are of three types; surface, primary (P-waves) and secondary (S-waves). P-waves ('push-waves') can travel through liquid; S-waves ('shake-waves') cannot, and it was this which gave the first definite proof that the Earth does have a liquid core.

GEOLOGICAL EVOLUTION

The age of the Earth is approximately 4.6 thousand million years. Palæontology – the study of past life forms through fossil remains – has enabled us to draw up a fairly reliable picture of the Earth's evolution; details are given in Table 6.3. There have been periodical ice ages, the last of which ended only 10 000 years ago, and no doubt the Earth has been struck by massive bodies from space; it is often maintained that a violent impact about 65 000 000 years ago caused a dramatic change in climate, leading to the extinction of the dinosaurs. This is known as the K-T extinction, separating the Cretaceous Period (K) from the the Triassic Period (T). Evidence is said to come from the amount of iridium in rocks laid down at that period – although it must be added that an earlier and even more widespread extinction of species occurred at the end of the Permian Period, and rocks of that age are not enriched in iridium.

ATMOSPHERE

The Earth's atmosphere is divided into various layers. The structure is given in Table 6.4, the composition in Table 6.5.

The lowest layer (troposphere) includes all normal clouds and all 'weather'. On average the temperature falls by 1.6 °C per 300 m altitude (the 'lapse rate'). The upper boundary, the tropopause, is considerably higher over the equator than over the polar regions. Above comes the stratosphere, first studied from 1904 by T. de Bort, using unmanned balloons; the temperature is stable up to 25 km, but then increases – bearing in mind that scientific 'temperature' is defined by the speeds at which the atoms and molecules move around, and is the not the same as what we ordinarily refer to as 'heat'; it rises to 470–490 °C at the stratopause, which is the upper boundary of the stratosphere. The ozone layer lies in the stratosphere; it is this which absorbs solar ultra-violet radiation and increases the temperature. Next comes the mesosphere (a term introduced by S. Chapman in 1950), where the temperature decreases with height; noctilucent clouds are found here. Above comes the thermosphere, also known as the ionosphere because it contains the layers which reflect some radio waves back to Earth and make long-range radio communication possible. It consists of electrically charged particles produced by the ionization of atmospheric atoms and molecules by solar and galactic radiation; it is markedly affected by changes in the solar wind, and it is here that we find auroræ as well as meteor trails. Finally there is the exosphere, which is a collisionless gas and is very tenuous; it has no definite boundary, but simply thins out until the density is no greater than that of the interplanetary medium.

MAGNETOSPHERE

The Earth has a fairly strong magnetic field; at the equator it is 0.305 G. The magnetic axis is at present offset to the rotational axis by 10.8° and the north magnetic pole is located at Ellef Ringnes Island, off north Canada. Magnetic field lines run between the magnetic poles, and charged particles become trapped, forming the magnetosphere.

The impact of the solar wind on the magnetopause (the outer boundary of the magnetosphere) compresses the magnetosphere on the day side of the Earth, while the field lines facing away from the Sun stream back to form the magnetotail. On the day side, the magnetosphere extends from 80 to 60 000 km, while on the night side it trails out to over 300 000 km.

Table 6.4. Structure of the atmosphere.

	Height (km)	
Troposphere	0 to 8–15 (high and middle latitudes)	Normal
	0 to 16–18 (low latitudes)	clouds
Tropopause	Upper boundary of the troposphere	
Stratosphere	18 to 50	Ozone layer
Stratopause	Upper boundary of the stratosphere	
Mesosphere	50 to 80	Meteors
Mesopause	Upper boundary of the mesosphere	
Ionosphere	80 to 1000	Aurora
(Thermosphere)		
Exosphere	Over 1000	Collisionless gas

Table 6.5. Composition of the lower atmosphere.

		Volume (%)
Nitrogen	N_2	78.08
Oxygen	O_2	20.95
Argon	Ar	0.93
Carbon dioxide	CO_2	0.03
Neon	Ne	18.18×10^{-4}
Helium	He	5.24×10^{-4}
Krypton	Kr	1.14×10^{-4}
Xenon	Xe	0.09×10^{-4}
Hydrogen	H_2	0.5×10^{-4}
Methane	CH_4	2.0×10^{-4}
Nitrous oxide	N_2O	0.5×10^{-4}

Very slight, variable traces of sulphur dioxide (SO_2) and carbon monoxide (CO).

The amount of water vapour is variable; in the range of 1%.

In the magnetosphere are the Van Allen Belts, discovered by the American astronomer J. Van Allen whose equipment was carried in the first successful US satellite, Explorer 1 of 1958. There are two belts. One is centred at about 3000 km above the Earth and has a thickness of 5000 km; it consists of energetic protons and electrons, probably originating from interactions between cosmic-ray particles and the upper atmosphere. The outer belt, centred 15 000 to 20 000 km above the Earth, is from 6000 to 10 000 km thick; it is made up of less energetic protons and electrons, believed to come mainly from the solar wind.

Because the Earth's magnetic axis is offset to the axis of rotation, there is an area where the inner Van Allen belt dips down toward the Earth's surface; this happens above the South Atlantic, off the Brazilian coast, and is known as the South Atlantic Anomaly. It allows charged particles to penetrate deeper into the atmosphere, and this can affect artificial satellites.

7 MARS

Mars, the fourth planet in order of distance from the Sun, must have been known since very ancient times, since when at its best it can outshine any other planet or star apart from Venus. Its strong red colour led to its being named in honour of the God of War, Ares (Mars); the study of the Martian surface is still officially known as 'areography'.

Mars was recorded by the old Egyptian, Chinese and Assyrian star-gazers, and the Greek philosopher Aristotle (384–322 BC) observed an occultation of Mars by the Moon, although the exact date of the phenomenon is not known. According to Ptolemy, the first precise observation of the position of Mars dates back to 27 January 272 BC, when the planet was close to the star β Scorpii.

Data for Mars are given in Table 7.1. Oppositions occur at a mean interval of 779.9 days, so that in general they fall in alternate years (Table 7.2). The closest oppositions occur when Mars is at or near perihelion, as in 2003, when the minimum distance will be only 56 000 000 km. The greatest distance between Earth and Mars, with Mars at superior conjunction, may amount to 400 000 000 km. The last favourable oppositions occur with Mars at aphelion, as in 1995 (minimum distance 101 000 000 km).

Mars shows appreciable phases, and at times only 85% of the day side is turned toward us. At opposition, the phase is of course virtually 100%. At times Mars may be occulted by the Moon, and there are also close conjunctions with other planets (Table 7.3). Planetary occultations of or by Mars are very rare; the next occasion will be on 11 August 2079, when Mars will be occulted by Mercury. Occultations of Mars by the Moon are reasonably frequent (Table 7.4).

THE MARTIAN SEASONS

The seasons on Mars are of the same general type as those of Earth, since the axial tilt is very similar and the Martian day (sol) is not a great deal longer (1 sol = 1.029 days). The lengths of the seasons are given in Table 7.5.

Southern summer occurs near perihelion. Therefore, climates in the southern hemisphere of Mars show a wider range of temperature than those in the north. The effects are

Table 7.1. Data.

Distance from the Sun:
  max 249 100 000 km (1.666 a.u.)
  mean 227 940 000 km (1.524 a.u.)
  min 206 700 000 km (1.381 a.u.)

Sidereal period: 686.980 days (= 668.60 sols)

Synodic period: 779.9 days

Rotation period: 24h 37m 22.6s (=1 sol)

Mean orbital velocity: 24.1 km s^{-1}

Axial inclination: 23°59′

Orbital inclination: 1°50′59″

Orbital eccentricity: 0.093

Diameter: equatorial 6794 km
              polar 6759 km

Apparent diameter from Earth:
  max 25″.7
  min 3″.3

Reciprocal mass, Sun = 1: 3098 700

Mass, Earth = 1: 0.107

Mass: 6.421 × 10^{26} g

Density, water = 1: 3.94

Volume, Earth = 1: 0.150

Escape velocity: 5.03 km s^{-1}

Surface gravity, Earth = 1: 0.380

Oblateness: 0.009

Albedo: 0.16

Surface temperature:
  max +26 °C
  mean −23 °C
  min −137 °C

Maximum magnitude: −2.8

Mean diameter of Sun, seen from Mars: 21′

Maximum diameter of Earth, seen from Mars: 46″.8

much greater than for Earth, partly because there are no seas on Mars and partly because of the greater eccentricity of the Martian orbit. At perihelion, Mars receives 44% more solar radiation than at aphelion.

Table 7.2. Oppositions of Mars 1999–2005.

Date	Closest approach to Earth	Minimum distance (millions of km)	Apparent diameter ($''$)	Magnitude	Constellation
1999 Apr 24	1999 May 1	87	16.2	−1.5	Virgo
2001 June 13	2001 June 21	67	20.8	−2.1	Sagittarius
2003 Aug 28	2003 Aug 27	56	25.1	−2.7	Capricornus
2005 Nov 7	2005 Oct 30	69	20.2	−2.1	Aries

There will then be oppositions on 2007 Dec 24, 2010 Jan 29, 2012 Mar 3, 2014 Apr 8 and 2016 May 22.

Between 1999 and 2005 Mars is at perihelion on 1999 Nov 25, 2001 Oct 12, 2003 Aug 30 and 2005 July 17. Aphelion is reached on 2000 Nov 2, 2002 Sept 21 and 2004 Aug 7.

The interval between successive oppositions of Mars is not constant; it may be as much as 810 days or as little as 764 days. Oppositions between 1900 and 2000 occurred on the following dates:

1901 Feb 22	1918 Mar 15	1935 Apr 6	1952 Apr 30	1969 May 31	1986 July 10
1903 Mar 29	1920 Apr 21	1937 May 19	1954 June 24	1971 Aug 10	1988 Sept 28
1905 May 8	1922 June 10	1939 July 23	1956 Sept 11	1973 Oct 25	1990 Nov 27
1907 July 6	1924 Aug 23	1941 Oct 10	1958 Nov 17	1975 Dec 15	1993 Jan 7
1909 Sept 24	1926 Nov 4	1943 Dec 5	1960 Dec 30	1978 Jan 22	1995 Feb 12
1911 Nov 25	1928 Dec 21	1946 Jan 13	1963 Feb 4	1980 Feb 25	1997 Mar 17
1914 Jan 5	1931 Jan 27	1948 Feb 17	1965 Mar 9	1982 Mar 31	1999 Apr 24
1916 Feb 9	1933 Mar 1	1950 Mar 23	1967 Apr 15	1984 May 11	

Table 7.3. Close planetary conjunctions involving Mars, 1900–2100.

	Date	UT	Separation ($''$)	Elongation ($°$)
Mercury–Mars	1942 Aug 19	12.36	−20	16E
Mars–Uranus	1947 Aug 6	01.59	+43	48W
Mercury–Mars	1985 Sept 4	21.00	−46	16W
Mars–Uranus	1988 Feb 22	20.48	+40	63W
Mercury–Mars	1989 Aug 5	21.54	+47	18E
Mercury–Mars	2032 Aug 23	04.26	+16	13W
Mercury–Mars	2079 Aug 11	01.31	Occultation	11W

TELESCOPIC OBSERVATIONS

The best pre-telescopic observations of the movements of Mars were made by the Danish astronomer Tycho Brahe, from his island observatory on Hven between 1576 and 1596. It was these observations which enabled Kepler, in 1609, to publish his first Laws of Planetary Motion, showing that the planets move round the Sun in elliptical rather than circular orbits.

The first telescopic observations of Mars were made by Galileo, in 1610. No surface details were seen. However, Galileo did detect the phase, as he recorded in a letter written to Father Castelli on 30 December of that year. The first telescopic drawing of the planet was made by F. Fontana, in Naples, in 1636 (the exact date has not been recorded); Mars was shown as spherical and 'in its centre was a dark cone in the form of a pill'. This feature was, of course, an

Table 7.4. Occultations of Mars by the Moon.

Date	UT (h)
2000 July 30	12
2000 Aug 28	03
2001 Oct 23	20
2002 May 14	19
2002 June 12	12
2002 Dec 30	02
2003 Jan 27	15
2003 July 17	08
2003 Sept 9	09
2003 Oct 6	16
2004 Feb 26	02
2004 Mar 25	23
2004 Oct 13	09
2004 Nov 11	04
2005 May 11	10
2005 Dec 12	04

Table 7.5. The Martian seasons.

	Days	Sols
S spring (N autumn)	146	142
S summer (N winter)	160	156
S autumn (N spring)	199	194
S winter (N summer)	182	177
Total	687	669

optical effect. Fontana's second drawing (24 August 1638) was similar.

On 28 November 1659, at 7 pm, Christiaan Huygens made the first telescopic drawing to show genuine detail. His sketch shows the Syrtis Major in easily recognizable form, although exaggerated in size. This sketch has been very useful in confirming the constancy of Mars' rotation period. It was Huygens himself who gave the first reasonably good estimate of the length of the rotation period; on 1 December 1659 he recorded that the period was 'about 24 hours'. In 1666, G. D. Cassini gave a value of 24h 40m, which is very close to the truth; in the same year he made the first record of the polar caps. It has been claimed that a cap was seen by Huygens in 1656, but his surviving drawing is very inconclusive.

(However, Huygens undoubtedly saw the south polar cap in 1672.)

The discovery that the polar caps do not coincide with the rotational (areographical) poles was made in 1719 by G. Maraldi – a year in which Mars was at perihelic opposition and was so bright that it caused a mild panic; some people mistook it for a red comet which was about to collide with the Earth! In 1704 Maraldi had made a series of observations of the caps, and had given a value for the rotation period of 24h 39m.

William Herschel observed Mars between 1777 and 1783, and suggested that the polar caps were made of ice and snow. Herschel also measured the rotation period, and his observations were later re-worked by W. Beer and J. H. Mädler, yielding a period of 24h 37m 23s.7, which is only one second in error.

Herschel also made the first good determination of the axial inclination of Mars; he gave a value of 28°, which is only 4° too great. At present the north pole star of Mars is Deneb (α Cygni), but the inclination ranges between 14.9° and 35.2° over a cycle of 51 000 Earth years, which has important long-term effects. In 25 000 years from now it will be the northern hemisphere which is turned sunward when Mars is at perihelion. These greater precessional effects are due to the fact that the globe of Mars is considerably more oblate than that of the Earth.

Useful drawings of Mars were made by J. H. Schröter, from Lilienthal in Germany, between 1785 and 1814, but Schröter never produced a complete map, and the first reasonably good map was due to Beer and Mädler, from Berlin, in 1830–32; the telescope used was Beer's 9.5 cm refractor. Beer and Mädler were also the first to report a dark band round the periphery of a shrinking polar cap. This band was seen by almost all subsequent observers, and in the late 19th century Percival Lowell attributed it – wrongly – to moistening of the ground by the melting polar ice.

Telescopic observations also revealed the presence of a Martian atmosphere. In 1783 Herschel observed the close approach of Mars to a background star, and from this concluded that the atmosphere could not be very extensive. Clouds on Mars were first reported by the French astronomer H. Flaugergues in 1811, who also suggested that the southern polar cap must have a greater range of

Table 7.6. Martian nomenclature.

Proctor	Schiaparelli
Beer Continent	Aeria and Arabia
Herschel II Strait	Sinus Sabæus
Arago Strait	Margaritifer Sinus
Burton Bay	Mouth of the Indus canal
Mädler Continent	Chryse
Christie Bay	Auroræ Sinus
Terby Sea	Solis Lacus
Kepler Land and Copernicus Land	Thaumasia
Jacob Land	Noachis and Argyre I
Phillips Island	Deucalionis Regio
Hall Island	Protei Regio
Schiaparelli Sea	More Sirenum, Lacus Phœnicis
Maraldi Sea	Mare Cimmerium
Hooke Sea and Flammarion Sea	Mare Tyrrhenum and Syrtis Minor
Cassini Land and Dreyer Island	Ausonia and Iapygia
Lockyer Land	Hellas
Kaiser Sea (or the Hourglass Sea)	Syrtis Major

size than that in the north, because of the more extreme temperature range – a comment verified observationally in 1811 by F. Arago. 'White' clouds in the Martian atmosphere were first seen by Angelo Secchi (Italy) in 1858. Secchi's sketches also show surface features, notably the Syrtis Major, which he called the 'Atlantic Canal' – an inappropriate name, particularly because there was at that time no suggestion that it might be artificial.

During the 1850s good drawings were made by the British amateur Warren de la Rue, using his 33 cm reflector, and useful maps were subsequently compiled by Sir Norman Lockyer, F. Kaiser, R. A. Proctor and others, although it was not until the work of G. V. Schiaparelli, from 1877, that really detailed maps were produced.

For some time it was tacitly assumed that the bright areas on Mars must be lands, while the dark areas were regarded as seas (although, strangely, Schröter believed that all the observed features were atmospheric in nature). Then, in 1860, E. Liais, a French astronomer living in Brazil, suggested that the dark areas were more likely to be vegetation tracts than oceans, and in 1863 Schiaparelli pointed out that the dark areas did not show the Sun's reflection, as they would be expected to do if they were made of water. Agreement was by no means universal; Secchi wrote that 'the existence of continents and seas has been conclusively proved', and in 1865 Camille Flammarion wrote that 'in places the water must be very deep', although he later modified this view and suggested that the dark areas might be composed of material in an intermediate state, neither pure liquid nor pure vapour. It was only it the late 19th century that the concept of major oceans on Mars was definitely abandoned.

NOMENCLATURE

With the compilation of better maps, thought was given to naming the various markings. In 1867 the British astronomer R. A. Proctor produced a map in which he named the features after famous observers – Cassini Land, Mädler Land and so on. His system was followed by other British observers, but was widely criticized. Various modifications were introduced, but in 1877 the whole system was overthrown in favour of a new one by G. V. Schiaparelli. The Proctor and Schiaparelli names are compared in Table 7.6.

Basically, Schiaparelli's system has been retained, although the space-probe results have meant that in recent years it has had to be drastically amended. The old and new systems are compared in Table 7.7. The system has also

Table 7.7. Old and new nomenclature.

Old	New
Mare Acidalium	Acidalia Planitia
Amazonis	Amazonis Planitia
Aonium Sinus	Aonium Terra
Arabia	Arabia Terra
Arcadia	Arcadia Planitia
Argyre I	Argyre Planitia
Ascræus Lacus	Ascræus Mons
Auroræ Sinus	Auroræ Planum
Mare Australe	Australe Planum
Mare Boreum	Boreum Planum
Mare Chronium	Chronium Planum
Chryse	Chryse Planitia
Mare Cimmerium	Cimmeria Terra
Elysium	Elysium Planitia
Mare Hadriacum	Hadriaca Patera
Hellas	Hellas Planitia
Hesperia	Hesperia Planum
Icaria	Icaria Planum
Isidis Regio	Isidis Planitia
Lunæ Lacus	Lunæ Planum
Margaritifer Sinus	Margaritifer Terra
Meridiani Sinus	Meridiani Terra
Nix Olympica	Olympus Mons
Noachis	Noachis Terra
Nodus Gordii	Arsia Mons
Ophir	Ophir Planum
Pavonis Lacus	Pavonis Mons
Promethei Sinus	Promethei Terra
Mare Sirenum	Sirenum Terra
Solis Lacus	Solis Planum
Syria	Syria Planum
Syrtis Major	Syrtis Major Planum
Tempe	Tempe Terra
Tharsis	Tharsis Planum
Mare Tyrrhenum	Tyrrhenum Terra
Utopia	Utopia Planitia
Xanthe	Xanthe Terra

been extended to take into account features such as craters, which are not identifiable with Earth-based telescopes.

THE CANALS

Who has not heard of the canals of Mars? Not so very many decades ago they were regarded as well-established features, quite possibly of artificial origin.

The first detection of an alleged canal network was due to Schiaparelli in 1877, when he recorded 40 features; he called them *canali* (channels), but this was inevitably translated as canals. Streaks had been recorded earlier by various observers, including Beer and Mädler (1830–32) and W. R. Dawes (1864), but Schiaparelli's work marked the beginning of the 'canal controversy'. Schiaparelli himself maintained an open mind, and wrote 'The suggestion has been made that the channels are of artificial origin. I am very careful not to combat this suggestion, which contains nothing impossible'. In 1879 he reported the twinning, or gemination, of some canals. The first observers to support the network were Perrotin and Thollon, at the Nice Observatory, in 1886, and subsequently canals became fashionable; they were widely reported even by observers using small telescopes. Percival Lowell, who built the observatory at Flagstaff in Arizona mainly to observe Mars, was convinced of their artificiality, and wrote 'That Mars is inhabited by beings of some sort or other is as certain as it is uncertain what these beings may be'. In 1892 W. H. Pickering observed canals and other features in the dark regions as well as the bright areas, and thus more or less killed the theory that the dark areas might be seas, but it was often claimed that a canal was a narrow water-course, possibly piped, surrounded to either side by strips of irrigated land. As recently as 1956 G. de Vaucouleurs, a leading observer of Mars, still maintained that though the canals were certainly not artificial, they did have 'a basis of reality'.

In fact, this is not so. The canals were due to tricks of the eye, and do not correspond to any true features on Mars. Equally illusory is the 'wave of darkening'; it had been claimed that when a polar cap shrank, releasing moisture, the plants near the cap became more distinct, and that this effect spread steadily from the polar regions to the Martian equator. Only since the Space Age has it been shown that the dark areas are not old sea-beds, and are not coated with organic material[1].

Lowell's brilliant Martian civilization has long since been banished to the realm of myth, but as an aside it

[1] From 1953, when I was engaged in mapping the Moon, I was able to make extensive use of the Lowell refractor, and I also turned it toward Mars. I saw nothing remotely resembling a canal, and neither could I follow the alleged 'wave of darkening'. Under the circumstances, I am delighted that I failed.

is interesting to look back at some of the suggestions made about signalling to the Martians. The first idea seems to have been due to the great German mathematician K. F. Gauss, about 1802; his plan was to draw vast geometrical patterns in the Siberian tundra. In 1819 J. von Littrow, of Vienna, proposed to use signal fires lit in the Sahara. Later, in 1874, Charles Cros, in France, put forward a scheme to focus the Sun's heat on to the Martian deserts by means of a huge burning-glass; the glass could be swung around to write messages in the deserts!

The first (and only) prize offered for communicating with extra-terrestrial beings was the Guzman Prize, announced in Paris on 17 December 1900. The sum involved was 100 000 francs – but Mars was excluded, because it was felt that calling up the Martians would be too easy.

Radio then came into the story. In 1906, when Marconi set up a telegraph station at Cape Clear, British operators reported a strange regular signal of three dots (the Morse letter S) which they could not explain, while in 1921 Marconi himself reported receiving the Morse letter V at 150 000 m. Not to be outdone, the astronomer David Todd and balloonist L. Stevens planned to take a sensitive radio receiver up in a balloon, reducing terrestrial interference and making it easier to detect messages from Mars. At about the same time (1909) the well-known physicist R. W. Wood suggested building a large array of cylinders, blacked on one side, so that they could be set up in a desert and swung round to send signals to the Martians.

A concerted effort was made in August 1924, when Mars was at its closest. Radio transmitters in various parts of the United States were temporarily shut down so that signals from Mars could be picked up, and ready to translate them was W. F. Friedman, head of the code section of the US Army Signals Corps. At Dulwich, in Outer London, radio listeners reported possible Martian signals at 30 000 m. In 1926 a Dr Mansfield Robinson went to the Central Telegraph Office in London and dispatched a telegram to Mars – for which he was charged the standard 18 pence per word. Tactfully, the postal authorities noted it as 'Reply not guaranteed'.

Perhaps the last word was said in 1992 by Mr. C. Cockell, who fought the General Election on behalf of the Forward to Mars Party. The constituency he selected was Huntingdon, where the sitting member was the then Prime Minister, Mr. John Major. It is sad to relate that Mr. Cockell lost his deposit!

EARTH-BASED OBSERVATIONS, PRE-1964

Energetic observations of Mars were continued during the years before the opening of the Space Age. The behaviour of the polar caps was carefully followed; it was generally believed that the caps were very thin – probably no more than a few millimetres thick, so that they would be in the nature of hoar-frost; there was considerable support for a theory due to A. C. Ranyard and Johnstone Stoney (1898) that the material was solid carbon dioxide rather than water ice. The first useful measurements of the surface temperature of the planet were made in 1909 by Nicholson and Petit at Mount Wilson and by Coblentz and Lampland at Flagstaff; they found that Mars has a mean surface temperature of $-28\,°C$, as against $+15\,°C$ for Earth. Lampland also announced the detection of what became known as the 'violet layer', supposed to block out short-wave solar radiation, and prevent them from reaching the Martian surface, except on occasions when it temporarily cleared away; we now know that it does not exist – it is as unreal as the canal network and the wave of darkening.

In 1933 W. S. Adams and T. Dunham analyzed the Martian atmosphere by using the Doppler method. When Mars is approaching the Earth, the spectral lines should be shifted to the short-wave end of the band; when Mars is receding, the shift should be to the red. It was hoped that in this way any lines due to gases in the Martian atmosphere could be disentangled from the lines produced by these same gases in the Earth's atmosphere. From their results, Adams and Dunham concluded that the amount of oxygen over Mars was less than 0.1% of the amount existing in the atmosphere of the Earth. In 1947 G. P. Kuiper reported spectroscopic results indicating carbon dioxide in the Martian atmosphere, but it was generally believed that most of the atmosphere was made up of nitrogen – whereas we now know that it is mainly composed of carbon dioxide. In 1934 the Russian astronomer N. Barabaschev estimated the atmospheric pressure at the surface, giving a value of 50 mbars; he later increased this to between 80 and 90 mbars – a gross overestimate, since the real pressure is nowhere as high as 10 mbars. In 1954 W. M. Sinton claimed to have detected organic matter in the spectra of the dark areas, but these results were later found to be spurious. One intriguing theory was due to E. J. Öpik, an Estonian astronomer resident at Armagh in Northern Ireland. He maintained that the dark areas had to be made up of material which could grow and push aside the red, dusty stuff blown from the 'deserts' – otherwise they would soon be covered up. Certainly there were major dust storms, usually when Mars was near perihelion, as in 1924, 1929, 1941 and 1956.

Before 1964 it was believed that Mars was a world without major mountains or valleys; that the caps were thin, and very probably made up of solid carbon dioxide; that the atmosphere was composed chiefly of nitrogen, with a ground pressure of the order of 87 mbars; that the dark areas were old sea-beds, covered with primitive vegetation; and that the red regions were 'deserts', not of sand but of reddish minerals such as felsite or limonite. Then came the flight of Mariner 4, and in a very short time all these conclusions, apart from the last, were found to be completely wrong.

SPACE MISSIONS TO MARS

It is believed that the Soviet authorities made several unsuccessful attempts to send probes to Mars between 1960 and 1962, but no details have ever been released. The first space-craft to Mars of which we have definite information was the Soviet Mars 1, launched on 1 November 1962; contact with it was lost on 21 March 1963, at a range of about 105 000 km, and though it may have passed fairly close to Mars contact was never regained. The first successful mission was America's Mariner 4, which by-passed Mars in July 1965. Since then many probes have been launched – all American or Russian apart from the Japanese Nozomi ('Hope') sent up in 1998. Surprisingly, almost all the information has come from American vehicles; the Russians have had very little success. A list of the Mars missions between 1962 and the present time is given in Table 7.8.

Dealing first with the Russian probes, it has to be admitted that the story is not a happy one. Between 1971 and 1974 six space-craft were launched, but all were either total or partial failures; little was learned from them. Two missions to the inner satellite, Phobos, were sent up in

1988, but the first of these went out of contact during the outward journey – human error was responsible – and the second 'went silent' before it had started the main part of its mission, for unknown reasons. Even more disappointing was the loss of the elaborate Mars 96 probe, which carried a number of experiments. Unfortunately the fourth stage of the rocket launcher failed, and Mars 96 fell Earthward, burning away in the atmosphere. Japan's first and only attempt, Nozomi, went on its way in July 1998 and is intended to orbit the planet, sending back miscellaneous data. It is expected to reach the orbit of Mars in December 2003. Otherwise, the field has been left to the United States.

Mariner 3 (5 November 1964) was a prompt failure; control was lost, and although the probe must have entered solar orbit there is no hope of contacting it. However, its twin, Mariner 4, was a triumphant success. It flew past Mars, making its closest approach at 01h 0m 57s on 15 July 1965 and sent back the first close-range images, showing unmistakable craters. Many years earlier, it had been claimed that craters had been seen from Earth, by E. E. Barnard in 1892, using the Lick Observatory refractor, and by J. Mellish in 1917, with the Yerkes refractor; but these observations were never published, and are decidedly dubious. Craters had been predicted in 1944 by D. L. Cyr (admittedly for the wrong reasons), but it was not until the flight of Mariner 4 that they were definitely found. Mariner 4 also confirmed the thinness of the atmosphere, and demonstrated that the dark areas were not depressed sea-beds; indeed some, such as Syrtis Major, are plateaux. The idea of vegetation tracts was abandoned, and the final nail driven into the coffin of the canals. Mariner 4 remained in contact until 20 December 1967; it is now in solar orbit, with a period of 587 days. Its perihelion distance from the Sun is 165 000 000 km, while at aphelion it swings out to 235 000 000 km. Of course, all track of it has since been lost.

Mariners 6 and 7, of 1969, were also successful and sent back good images – notably of Hellas, a bright feature which was once thought to be a snow-covered plateau but now known to be a deep basin; when filled with cloud it can become so brilliant that it can easily be mistaken for an extra polar cap. Yet it was only later that the true character of Mars was revealed. By ill chance, Mariners 4, 6 and 7 passed over the least spectacular areas, and it was only

Table 7.8. Missions to Mars.

Name	Nationality	Launch date	Encounter date	Closest approach or orbiter (km)	Landing site of capsule	Results
Mars 1	USSR	1 Nov 1962	?	190 000?	—	Contact lost at 106 000 km.
Mariner 3	USA	5 Nov 1964	—	—	—	Shroud failure. In solar orbit, but contact lost soon after launch.
Mariner 4	USA	28 Nov 1964	14 July 1965	9789	—	Returned 21 images, plus miscellaneous data. Contact lost on 21 Dec 1967.
Zond 2	USSR	30 Nov 1964	Aug 1965?	?	—	Contact lost on 2 May 1965.
Mariner 6	USA	24 Feb 1969	31 July 1969	3392	—	Returned 76 images; flew over Martian equator. In solar orbit.
Mariner 7	USA	27 Mar 1969	4 Aug 1969	3504	—	Returned 126 images, mainly over the S hemisphere. In solar orbit.
Mariner 8	USA	8 May 1971	—	—	—	Total failure: fell in the sea.
Mars 2	USSR	19 May 1971	27 Nov 1971	In orbit, 2448 × 24 400	44S, 213W (Eridania)	Capsule landed, with Soviet pennant, but no images received.
Mars 3	USSR	28 May 1971	2 Dec 1971	In orbit, 1552 × 212 800	45S, 158W (Phæthontis)	Orbiter returned data. Contact with lander lost 20 s after arrival.
Mariner 9	USA	30 May 1971	13 Nov 1971	In orbit, 1640 × 16 800	—	Returned 7329 images. Contact lost on 27 Oct 1972.
Mars 4	USSR	21 July 1973	10 Feb 1974	Over 2080	—	Missed Mars; some fly-by data returned. Failed to orbit.
Mars 5	USSR	25 July 1973	12 Feb 1974	In orbit, 1760 × 32 500	—	Failure; contact lost.
Mars 6	USSR	5 Aug 1973	12 Mar 1974	?	?24S, 25W (Erythræum)	Contact lost during landing sequence.
Mars 7	USSR	9 Aug 1973	9 Mar 1974	1280	—	Failed to orbit; missed Mars.
Viking 1	USA	20 Aug 1975	19 June 1976	In orbit	22.4N, 47.5W (Chryse)	Landed 20 July 1976.
Viking 2	USA	9 Sept 1975	7 Aug 1976	In orbit	48N, 226W (Utopia)	Landed 3 Sept 1976.
Phobos 1	USSR	7 July 1988	?	?	—	Contact lost, 29 Aug 1988.
Phobos 2	USSR	12 July 1988	—	In orbit, 850 × 79 750	—	Contact lost, 27 Mar 1989. Some images and data from Mars and Phobos returned.
Mars Observer	USA	25 Sept 1992	24 Aug 1993	—	—	Contact lost, 25 Aug 1993.
Mars 96	Russia	16 Nov 1996	—	—	—	Total failure. Fell in the sea.
Pathfinder	USA	4 Dec 1996	4 July 1997	—	19.33N 33.55W	Landed in Ares Vallis. Carried Sojourner rover. Contact lost 6 Oct 1997.
Global Surveyor	USA	7 Nov 1996	11 Sept 1997	In orbit	—	Data returned.
Nozomi	Japan	3 July 1998	—	—	—	Intended orbiter.
Mars Climate Orbiter	USA	11 Dec 1998	—	—	—	Intended orbiter. Contact lost, 23 Sept 1999.
Mars Polar Lander	USA	3 Jan 1999	3 Dec 1999	—	76.25, 195.3W (probable)	Intended orbiter/lander. Contact lost, 3 Dec 1999.

in 1971 that the picture changed. Mariner 8 failed – the second stage of the rocket launcher failed to ignite – but Mariner 9 entered a closed path round Mars and for the first time provided views of the towering volcanoes and the deep valleys. When it first reached the neighbourhood of Mars, a major dust storm was in progress, but this soon cleared, and the full-scale photographic coverage of the surface could begin.

The most widespread dust-storms occur when Mars is near perihelion, as was the case when Mariner 9 reached the planet (perihelion had fallen in the previous September). If the windspeed exceeds a certain critical value – 50 to 100 m s^{-1} – grains of surface material, about 100 μm across, are whipped up and given a 'skipping' motion, known technically as saltation. On striking the surface they propel smaller grains, a few micrometres across, into the atmosphere, where they may remain suspended for weeks. Over 100 localized storms and 'dust devils' occur every Martian year, and at times a storm becomes global, so that for a while all surface details are hidden. The maximum windspeeds may reach 400 km h^{-1}, but in that tenuous atmosphere they will have little force.

Next came the Vikings, which were launched in 1975 and reached Mars in 1976. Each Viking consisted of an orbiter and a lander; the orbiter continued the surveys started by Mariner 9 and also acted as a relay for the lander. The lander, separated when the probe had entered Martian orbit, came down gently, braked partly by parachute and partly by retro-rockets; the touchdown speed was no more than 9.6 km h^{-1}. Fortunately both landers came down clear of the rocks which are strewn all over the planet's surface. By the end of 1976 our knowledge of Mars had been improved beyond all recognition.

ATMOSPHERE

The main constituent is indeed carbon dioxide, which accounts for more than 95% of the total; nitrogen accounts for 2.7% and argon for 1.6%, which does not leave much room for other gases (Table 7.9). The highest atmospheric pressure so far measured is 8.9 mbars, on the floor of the deep impact basin Hellas, while the pressure at the top of the lofty Olympus Mons is below 3 mbars. When Viking 1 landed in the golden plain of Chryse (latitude 22°.4N) the pressure was approximately 7 mbars. A decrease of 0.012 mbar per sol was subsequently measured, due to carbon dioxide condensing out of the atmosphere to be deposited on the south polar cap, but this is a seasonal phenomenon, and ceased while Vikings 1 and 2 were still operating. The maximum temperature of the aeroshell of Viking 1 during the descent to Mars was 1500 °C; that of Viking 2 was similar.

The atmospheric pressure is at present too low for liquid water to exist on the surface, but there is no doubt

Table 7.9. Composition of the Martian atmosphere, at the surface. (ppm = parts per million).

Carbon dioxide, CO_2	95.32%
Nitrogen, N_2	2.7%
Argon, ^{40}Ar	1.6%
Oxygen, O_2	0.03%
Carbon monoxide, CO	0.07%
Water vapour, H_2O	0.03% (variable)
Neon, Ne	2.5 ppm
Krypton, Kr	0.3 ppm
Xenon, Xe	0.08 ppm
Ozone, O_3	0.03 ppm

that water did once exist; Mariner 9 and the Viking orbiters showed clear evidence of old riverbeds and even islands, while confirmation was obtained from the Pathfinder mission of 1997 that the area where Pathfinder landed (Ares Vallis) was once water-covered. Mars may well go through very marked climatic changes. This may be due to the effects of the changing axial inclination (between 14.9° and 35.5° over a cycle of 51 000 years) and the changing orbital eccentricity (from 0.004 to 0.141 in a cycle of 90 000 years). When one of the poles is markedly tilted sunward when Mars is at perihelion, some of the volatiles may sublime, temporarily thickening the atmosphere and even causing rainfall. It has also been suggested that every few tens of millions of years Mars goes through spells of intense volcanic activity, when tremendous quantities of gases and vapours (including water vapour) are sent out from beneath the crust.

Quite apart from the dust storms, ice-crystal clouds are found on Mars; they are composed of water ice, and lie at around 10–15 km above the surface. Localized white clouds may be seen anywhere, together with sunrise and sunset fogs and hazes.

Because the Martian atmosphere is heated from below, temperatures in the troposphere decrease with altitude, as on Earth; the top of the Martian troposphere lies at about 40 km, with an average temperature lapse rate of 2.5° per km. Above the troposphere comes the mesosphere, where the temperatures become nearly isothermal. The mesosphere extends to about 80 km, and above this comes the excessively tenuous ionosphere, composed of ions and electrons. It extends from about 120 km up to several hundred kilometres; unlike the Earth's ionosphere it is not shielded from the solar wind by a strong magnetic field.

Thin though it is, the Martian ionosphere was used to brake the Global Surveyor space-craft and place it into a virtually circular orbit.

POLAR CAPS

The seasonal variations of the polar caps are striking. It is now clear that at each pole there is a residual cap which is overlaid by a seasonal coating of solid carbon dioxide. The carbon dioxide condenses out of the atmosphere in the autumn of each hemisphere, producing clouds which accumulate over the poles and preventing us from seeing just how the caps develop. When these 'hoods' disperse the caps below are revealed. With the onset of spring and summer, temperatures rise and the seasonal caps disappear, leaving only the permanent residual caps.

The caps are not identical, because of the differences in climate between the two hemispheres. The northern seasonal cap is smaller and darker than its counterpart in the south, because it is laid down at a time in the Martian year when the atmosphere contains a large amount of dust; this dust is precipitated on to the surface together with the carbon dioxide, whereas the southern cap is laid down when the atmosphere is much less dust-laden. The northern residual cap is the larger of the two (diameter 1000 km), as against 400 km for the southern cap. The temperatures differ; the residual southern cap is, predictably, the colder of the two, with temperatures going down to below $-130\,°C$, while temperatures above the residual northern cap have been known to rise to $-68\,°C$, which is well above the frost point of carbon dioxide and not far from the frost point of water in a thin atmosphere which contains only a small amount of precipitable H_2O. Finally, there are important differences in composition. The northern residual cap is almost certainly water ice, while the southern is a mixture of water ice and carbon dioxide ice. The thickness of the caps is considerable; recent measures indicate that the depth of the northern cap is of the order of 5 km. It has been estimated that if all the water in the atmosphere were condensed it would cover the Martian surface with a layer only 1/100 mm deep, but the release of all the water in the ice-caps would produce a layer 10 m deep.

GENERAL TOPOGRAPHY

There are high peaks and deep valleys on Mars. On Earth, altitudes are reckoned from sea-level, but there are no seas on Mars and it has been agreed to use a datum line where the average atmospheric pressure is 6.2 mbars. This means that all values of altitudes and depressions must be regarded as somewhat arbitrary.

The two hemispheres are not alike. Generally speaking, the southern part of the planet is heavily cratered and much of it lies up to 3 km above the datum line; the northern part is lower – mainly below the datum line – and is more lightly cratered, so that it is presumably younger. The very ancient craters which once existed there have been eroded away, and in general the slopes are lower than those in the south. However, the demarcation line does not follow the Martian equator; instead, it is a great circle inclined to the equator at an angle of 35°.

Rather surprisingly, the two deepest basins on Mars, Hellas and Argyre, are in the south; the floor of Hellas is almost 5 km below the datum line, Argyre about 3 km below. Both are relatively smooth, and both can become brilliant when cloud-filled.

The main volcanic area is the Tharsis bulge. This is a crustal upwarp, centred at latitude 14°S longitude 101°W; it straddles the demarcation line between the two 'hemispheres' and rises to a general altitude of around 9 km. Along it lie the three great volcanoes of Arsia Mons, Pavonis Mons and Ascræus Mons, which are spaced out at intervals of between 650 and 720 km; only Arsia Mons is south of the equator. Olympus Mons lies 1500 km to the west of the main chain, and is truly impressive, with a 600 km base and a complex 80 km caldera. The slopes are fairly gentle (6° or less); lava flows are much in evidence, and there is an extensive 'aureole', made up of blocks and ridges, around the base. It is just over 24 km high – three times the height of our Everest – and is a shield volcano, essentially similar to those of Hawaii, but on a much grander scale. Summit calderæ are also found on Arsia Mons (diameter 110 km), Ascræus Mons (65 km) and Pavonis Mons (45 km). On the northern flank of the Tharsis bulge is the unique Alba Patera, which rises to no more than 3.2 km, but is over 1500 km across, with a central caldera.

The second major volcanic area is Elysium, centred at latitude 25°N, longitude 210°W, It is smaller than Tharsis,

Table 7.10. Altitudes of some Martian volcanoes. (These altitudes are bound to be rather arbitrary, as there is no sea-level on Mars. They are reckoned from the 'datum line' where the average atmospheric pressure is 6.2 mb.).

Name	Altitude (m)
Olympus Mons	24 000
Ascræus Mons	18 000
Pavonis Mons	18 000
Arsia Mons	9100
Elysium Mons	9000
Tharsis Tholus	6000
Hecates Tholus	6000
Albor Tholus	5000
Uranius Tholus	3000
Ceraunius Tholus	2000

Table 7.11. Estimated volcano ages.

	Age (thousands of millions of years)
Tempe Patera	3.4
Ceraunius Tholus	2.4
Uranius Tholus	2.3
Elysium Mons	2.2
Alba Patera	1.7
Hecates Tholus	1.7
Tharsis Tholus	1.4
Arsia Mons	0.7
Pavonis Mons	0.3
Ascræus Mons	0.1
Olympus Mons	0.03

but is still over 1900 km across; on average it reaches about 4 km above the datum line. Volcanoes there include Elysium Mons, Albor Tholus and Hecates Tholus. Isolated volcanoes are also found elsewhere – for example the Hellas area, as well as near Syrtis Major and Tempe. The 'tholi' (domes) are smaller and steeper than the 'montes', so that the material from which they formed may have been relatively viscous; like the montes, most of them have summit calderæ. There are also the 'pateræ', scalloped, collapsed shields with shallow slopes and complex summit calderæ; some are symmetrical, others less regular, with radial channels running down their flanks. They may be composed of relatively loose material, with ash flows very much in evidence. The altitudes of some of the Martian volcanoes are given in Table 7.10.

Plate tectonics do not apply to Mars, so that when a volcano forms over a 'hot spot' it remains there for a very long time – which accounts of the great size of the major structures. They are generally assumed to be extinct, although a certain amount of doubt must remain. Some, such as Uranius Tholus and Elysium Mons, seem to be over 2000 million years old, but some of the Tharsis volcanoes may have been active much more recently and Olympus Mons may have last erupted a mere 30 million years ago. Estimated volcano ages are given in Table 7.11.

The greatest canyon system is the Valles Marineris, a huge gash in the surface over 4500 km long, with a maximum width of 600 km and a greatest depth of about 7 km below the rim. It begins at the complex Noctis Labyrinthus, often nicknamed the Chandelier, where we find canyons which dwarf our own Grand Canyon of the Colorado. The Valles Marineris extends eastward toward Auroræ Planum, ending in the blocky terrain not far from the well-known Margaritifer Terra (once known as Margaritifer Sinus, the Gulf of Pearls). For most of its course it runs roughly parallel to the line of demarcation.

Craters abound, all over Mars; many of them have central peaks similar to those of the lunar craters, and the distribution laws are much the same. They have in general been named after astronomers (Table 7.17). There are also features which look so like dry riverbeds that they can hardly be anything else. There is evidence of past flash-floods, so that some of the old craters have been literally sliced in half. Raging torrents must have carried rocks down into the low-lying areas, and some of the 'rivers', such as the Kasei Vallis, are hundreds of kilometres long.

SURFACE EVOLUTION

Our knowledge of the evolution of the Martian surface is far from complete, but there are several fairly well-defined epochs, listed in Table 7.12. The main bombardment came during the Noachian epoch; cratering declined during the Hesperian, and virtually ceased toward the end of the Amazonian.

Table 7.12. Martian epochs. Ages are given in thousands of millions of years, but are bound to be somewhat uncertain.

Epoch	From	To	
Early Noachian	4.5	4.4	Intense bombardment. Impact basins (Hellas, Argyre).
Middle Noachian	4.4	4.3	Cratering; highland vulcanism.
Late Noachian	4.3	3.8	Intercrater plains; lava flows; sinuous channels.
Early Hesperian	3.8	3.7	Initial faulting of Valles Marineris. Complex ridged plains; lava flows; declining cratering rate.
Late Hesperian	3.7	3.6	Vulcanism in Tharsis. Long sinuous channels.
Early Amazonian	3.6	2.3	Smooth plains such as Acidalia. Extensive vulcanism. Lava flows.
Middle Amazonian	2.3	0.7	Continued vulcanism; major lava flows (Tharsis).
Late Amazonian	0.7	Present	Residual ice-caps. Major vulcanism in Tharsis and elsewhere, dying out at a relatively late stage; disappearance of surface water.

INTERNAL STRUCTURE

As yet we do not have a complete picture of the structure of Mars, but it is thought that there is an iron-rich core, perhaps 2900 km in diameter, overlaid by a 3500 km mantle and a crust which can hardly be more than around 100 km deep. There is a weak magnetic field, with a strength of no more than 1/800 of the Earth's field, but which nevertheless supports the idea of a core rich in iron; whether this core still acts in the fashion of a dynamo is not clear. The overall field has an orientation similar to that of the Earth, so that in theory a compass needle would point north, but there are localized magnetic anomalies in the crust, so that using a magnetic compass on Mars would be highly unreliable.

VIKING LANDERS; THE SEARCH FOR LIFE

Both Viking landers came down in the red parts of Mars; Viking 1 in Chryse and Viking 2 in the more northerly Utopia. The first picture from Viking 1, taken immediately after touchdown, showed a rockstrewn landscape and the overall impression was that of a barren, rocky desert, with extensive dunes as well as pebbles and boulders. The colour was formed by a thin veneer of red material, probably limonite (hydrated ferric oxide) covering the dark bedrock. The sky was initially said to be salmon-pink, although later pictures modified this to yellowish pink. Temperatures were low, ranging between $-96\,^{\circ}$C after dawn to a maximum of $-31\,^{\circ}$C near noon. Winds were light. They were strongest at about 10 am local time, but even then were no more than 22 km h^{-1} breezes; later in the sol they dropped to around 7 km h^{-1}; coming from the southwest rather than the east. The pattern was fairly regular from one sol to the next.

The Viking 2 site, in Utopia, was not unlike Chryse, but there were no large boulders, and the rocks looked 'cleaner'. There were no major craters in sight; the nearest large formation, Mie, was over 200 km to the west. Small and medium rocks were abundant, most of them vesicular. (Vesicles are porous holes, formed as a molten rock cools at or near the surface of a lava flow, so that internal gas bubbles escape.) There were breccias and one good example of a xenolith – a 'rock within a rock' probably formed when a relatively small rock was caught in the path of a lava flow and was coated with a molten envelope which subsequently solidified. The temperatures were very similar to those at Chryse.

One main aim of the Viking missions was to search for life. Soon after arrival Viking 1 collected samples by using a scoop, drew them inside the space-craft, analyzed them and transmitted the results to Earth. There were three main experiments:

- *Pyrolytic Release.* Pyrolysis is the breaking up of organic compounds by heat. The experiment was based on the assumption that any Martian life would contain carbon, one species of which, carbon-14, is radioactive, so that when present it is easy to detect.

The sample was heated sufficiently to break up any organic compounds which were present.

- *Labelled Release.* This also involved carbon-14, and assumed that the addition of water to a Martian sample would trigger off biological processes if any organisms were present.

- *Gas Exchange.* It was assumed that any biological activity on Mars would involve the presence of water, and the idea was to see whether providing a sample with suitable nutrients would persuade any organisms to release gases, thereby altering the composition of the artificial atmosphere inside the test chamber.

However, the results of all three experiments were inconclusive, and this was also the case when they were repeated from Viking 2. The investigators had to admit that they still could not say definitely whether there was or was not any trace of life on Mars.

Pathfinder And Sojourner

The next successful lander was Pathfinder, which came down on Mars on 4 July 1997 – America's Independence Day. This time there was no attempt to make a 'soft' landing or to put the space-craft into an initial orbit round Mars. Pathfinder was encased in airbags, and on impact it literally bounced in the manner of a beach-ball. The landing speed was over 90 km h^{-1}, and the first bounce sent Pathfinder up to more than 160 m; altogether there were 15 to 17 bounces before the space-craft came to rest in an upright position, having rolled along for about a minute after the final touchdown. On Sol 2 – the second day on Mars – the tiny Sojourner rover could emerge, crawling down a ramp on to the Martian surface. Like Viking 1, Pathfinder had landed in Chryse; the distance from the Viking 1 lander was 800 km. On arrival, the main base was renamed the Sagan Memorial Station, in honour of the American astronomer Carl Sagan.

The site had been carefully chosen. It lay at the mouth of a large outflow channel, Ares Vallis, which had been carved by a violent flood in the remote past. Ares Vallis had once been a raging torrent of water, and it was thought that rocks of many different types would have been swept down into the area, which did indeed prove to be the case; there was no reasonable doubt that the whole region had

once been a flood-plain of standing water. Mars had not always been as arid as it is now.

Sojourner itself was a miniature vehicle, 65 cm long by 18 cm high; there were six 13 cm wheels. Sojourner could be guided from Earth, and was able to analyze the surface rocks and make general observations to supplement the panoramic views from the Sagan station. Sojourner could climb over small rocks and skirt round larger ones. As it moved around it left a track in the 'soil', exposing the darker material below; the soil itself was finer than talcum powder, and was likened to the fine-grained silt found in regions such as Nebraska in the United States. The soil density was 1.2–2.0 g cm^{-3}, much the same as dry soils on Earth.

Rock analysis was carried out by an instrument on Sojourner known as the Alpha Proton X-ray Spectrometer, or APXS for short. It carried a small quantity of radioactive curium-244, which emits alpha-particles (helium nuclei). When Sojourner came up to a rock, the alpha-particles from APXS bombarded the rock; in some cases the particles interacted with the rock and bounced back, while in other cases protons or X-rays were generated. The backscattered alpha-particles, protons and X-rays were counted, and their energies determined. The numbers of particles counted at each energy level gave a clue as to the abundance of the various elements in the rocks and also to the rock types.

Most of the rocks were, as expected, volcanic. The first to be examined, because it was nearest to the Sagan Station, was nicknamed Barnacle Bill, because of its outward appearance[2], and proved to be similar to terrestrial volcanic rocks known as andesites; another rock, Yogi, was basaltic and less rich in silicon. However, several Martian rocks proved to have a higher silicon content than the common basalt, and it was reasoned that the original molten lava must have modestly differentiated, so that the heavy elements such as iron sank to the bottom and the lighter silicon compounds rose to the top of the crust, which was the source of the lavas. There was evidence

---

[2] Nicknames included Hassock, Sausage, Mermaid, Squash, Wedge, Stripe, Soufflé and Desert Princess. Nobody can accuse the NASA teams of being lacking in imagination. Yogi was so named because from some angles it bore a slight resemblance to the celebrated Yogi Bear!

of layering or bedding, suggesting a sedimentary origin, and the pebbles and cobbles found in hollows of some of the rocks also suggested conglomerates formed in running water. Windspeeds were light during the operational life of Pathfinder, seldom exceeding 10 m s^{-1}. Rather surprisingly, dust devils – miniature tornadoes – were common, although in that tenuous atmosphere they had little force. Ice-crystal clouds were recorded, at altitudes of around 13 km.

Altogether, the lander sent back 16 000 images, and 550 were received from Sojourner before the mission ended. The least really good transmission was received on 27 September 1997. There were brief fragmentary signals on 2 and 8 October, but then Pathfinder lost contact – and of course Sojourner no longer had a relay to send data back to Earth.

The next US probe, Mars Global Surveyor (MGS), entered Mars orbit on 11 September 1997. It was purely an orbiter. The initial orbit round Mars was highly elliptical, but at each closest approach the probe dipped into the upper Martian atmosphere and was slowed by friction, so that successive closest approaches would be lower and lower; this would continue, until eventually MGS would be moving in a circular path at an altitude of around 400 km above the surface of the planet. Achievement of the final orbit was delayed because of technical problems on the space-craft itself, but the method proved in the end to be very satisfactory.

A weak magnetic field (with a strength about 1/800 of that of the Earth's field) was confirmed on 15 September 1997, and before long excellent images were being received. (One picture, sent back on 5 April 1998, showed a rock which had earlier been imaged by the Viking orbiters, and gave a strange resemblance to a human face, although the MGS images showed, once and for all, that the rock was quite ordinary and the curious appearance had been due to nothing more significant than light and shadow patterns. Another crater, Galle, imaged by MGS, has been nicknamed the 'Happy Smiling Face', because of the arrangement of details on its floor.) Views of the canyons in the Noctis Labyrinthus area showed clear evidence of layering; flooded craters were identified, and it was also evident that many of the surface features had been shaped by winds. The polar caps were found to be thick; the northern ice-cap

goes down to a depth of at least 5 km, and large areas of it were surprisingly smooth.

Another orbiter, Nozomi or 'Hope' – Japan's first attempt at sending a probe to Mars – began its journey on 3 July 1998; then came the US, Mars Climate Orbiter, which would, it was hoped, operate for a full Martian year after entering its final orbit 240 km above the surface of the planet. Unfortunately, two teams were involved in the approach manœuvre; one team was working in Imperial units and the other in Metric. The engine was programmed to fire using thrust data in pounds rather than in Metric newtons. Five minutes after the firing of the orbital insertion engine, the space-craft entered the denser atmosphere, at an altitude 80 to 90 km lower than had been intended – with the result that it either burned up or else crashed on to the surface. It was a blunder which cost 125 million dollars.

Mars Polar Lander was launched on 3 January 1999, and was designed to come down on the following 3 December at latitude 76°S, longitude 195°W, 800 km from the Martian south pole. It carried microphones, so that it was hoped to pick up actual sounds from Mars. All went well until ten minutes before the scheduled landing on 3 December. Two microprobes (Amundsen and Scott) were released; these were intended to send down penetrators and search for subsurface water. They would impact at over 600 kilometres per hour, and penetrate to a depth of several metres. Unfortunately no signals were received after landing, either from the main probe or from the microprobes. The cause of this failure is unclear.

All these missions are leading up to the dispatch of a sample-and-return probe, which will bring Martian samples back to Earth for analysis. Then, and only then, will we know whether Mars is sterile. Lowly organisms may well exist perhaps below ground level, but one thing is certain on Mars; there can be no life as advanced as a blade of grass.

At least we already have good maps of the entire Martian surface. A selected list of formations is given in Table 7.16.

Table 7.13. SNC ('Martian') meteorites.

Name	Location found	Date found	Mass (g)
Chassigny	Chassigny, France	3 Oct 1815	±4000
Shergotty	Shergotty, India	25 Aug 1865	±5000
Nakhla	Nakhla, Egypt	28 June 1911	±40 000
Lafayette	Lafayette, Indiana	1931	±800
Governador Valadares	Governador Valadares, Brazil	1958	158
Zagami	Zagami, Nigeria	3 Oct 1962	±18 000
ALHA 77005	Allan Hills, Antarctica	19 Dec 1977	482
Yamato 793605	Yamato Mtns, Antarctica	1979	16
EETA 79001	Elephant Moraine, Antarctica	13 Jan 1980	7900
ALH 84001	Allan Hills, Antarctica	27 Dec 1984	1940
LEW 88516	Lewis Cliff, Antarctica	22 Dec 1988	13
QUE 94201	Queen Elizabeth Range, Antarctica	16 Dec 1994	12
—	Northern Africa	May 1998	2200
Los Angeles	Mojave Desert, California	Oct 1999	452.5 and 254.4

METEORITES FROM MARS?

It has been claimed that certain meteorites, known as SNC meteorites (because the first three to be identified came from Shergotty in India, Nakhla in Egypt and Chassigny in France) are of Martian origin. Of special note is the meteorite ALH 84001, found in the Allan Hills of Antarctica. It is shaped rather like a potato, and measures 15 cm × 10 cm × 7.6 cm; when found, it was covered with a fusion crust made of black glass. Its age seems to be around 4.5 thousand million years. Claims for a Martian origin have also been made for other SNC meteorites; a list is given in Table 7.13.

Careful examination of ALH 84001 showed the presence of tiny features which some investigators regarded as being due to primitive forms of microscopic life. They were moreover accompanied by certain minerals which sometimes accompany microbacteria on Earth. These features were however very small indeed; the largest of them was no more than 500 nm long (1 nm is one thousand-millionth of a metre), and the general opinion at present is that they are either inorganic or are due to terrestrial contamination.

On the 'Martian' theory, ALH 84001 was blasted away from Mars about 16 000 000 years ago by a violent impact. It could not have been put initially into an Earth-crossing path, but entered an orbit around the Sun. At first this orbit was similar to that of Mars itself, and there were various encounters; finally the two bodies separated, and ALH 84001 continued in its own orbit, until by chance it encountered the Earth. About 13 000 years ago it plumped conveniently down in Antarctica, where the meteorite collectors found it.

This may or may not be the true picture. Whether or not the SNC meteorites really come from Mars remains an intriguing possibility, but it certainly cannot be regarded as definitely proved.

THE SATELLITES OF MARS

Mars has two dwarf satellites, Phobos and Deimos, named after the attendants of the mythological war god.

The first mention of possible Martian satellites was fictional – by Jonathan Swift, in Gulliver's *Voyage to Laputa* (1727), Swift described two satellites, one of which had a revolution period shorter than the rotation period of its primary, but at that time there was no telescope which could have shown either Phobos or Deimos. Two satellites were also described in another novel, Voltaire's *Micromégas* (1750). The reasoning was, apparently, that since Earth had one satellite and Jupiter was known to have four, Mars could not possibly manage with less than two!

A satellite reported telescopically in 1645 by A. Schyrle was certainly nothing more than a faint star. The first systematic search for satellites was made in 1783 by William Herschel. The result was negative, and H. D'Arrest, from Copenhagen, in 1862 and 1864, was similarly unsuccessful.

The first satellite to be discovered was Deimos, by Asaph Hall at Washington on 10 August 1877, using the 66 cm Clark refractor at Washington. On 16 August he discovered Phobos. The names of the satellites were suggested to Hall by Mr. Madan of Eton. Both are decidedly elusive, because of their small size and their closeness to Mars. Data are given in Table 7.14. E. M. Antoniadi, using the 83 cm Meudon refractor in 1930, reported that Phobos was white and Deimos bluish, but these results were certainly spurious. Much more recently – in 1952, 1954 and 1956 – G. P. Kuiper, using the McDonald 208 cm reflector, made a careful search and concluded – rightly – that no new satellite as much as a kilometre across could exist.

The origin of the satellites is not certainly known. It is tempting to believe that they are ex-asteroids, which were captured by Mars long ago, and certainly they do seem to be similar in nature to the asteroids which have been surveyed from close range by space probes – although it is true that the capture would involve some very special circumstances. Note, however, that there is one asteroid, 647 Eureka, which is a Martian 'Trojan' and moves in the same orbit as Mars, although it keeps well clear of the planet and is in no danger of collision.

Tidal effects indicate that the orbit of Phobos is slowly shrinking, so that the satellite may impact Mars at some time between 30 and 100 million years in the future. Deimos, however, is in a stable path and will not suffer a similar fate.

Neither satellite would be of much use in lighting up the Martian nights. Moreover, Phobos would be invisible to any observer on the planet above latitude 69° north or south; for Deimos the limiting latitude would be 82°. To a Martian observer, Phobos would appear less than half the diameter of the Moon as seen from Earth, and would give only about as much light as Venus does to us;

from Phobos, Mars would subtend a mean angle of 42°. The maximum apparent diameter of Deimos as seen from Mars would be only about twice that of Venus seen from Earth, and with the naked eye the phases would be almost imperceptible. Deimos would remain above the Martian horizon for $2\frac{1}{2}$ sols consecutively; Phobos would cross the sky in only $4\frac{1}{2}$ h, moving from west to east, and the interval between successive risings would be a mere 11 h. Total solar eclipses could never occur. Phobos would transit the Sun 1300 times a year, taking 19 s to cross the disk, while Deimos would show an average of 130 transits, each taking 1 min 48 s. To an observer on Mars, eclipses of the satellites by the shadow of the planet would be very frequent.

Predictably, both Phobos and Deimos have synchronous rotation periods. This leads to great ranges in temperature, particularly in the case of Phobos. During its $7\frac{1}{2}$ hour 'day' the temperature may rise to $-4\,°C$, while at night it falls to $-112\,°C$. There will, of course, be pronounced libration effects.

The satellites are irregular in shape; the first proof of this was obtained in 1969, when Mariner 7 photographed the shadow cast by Phobos on Mars and showed it to be elliptical. The first accurate size measurements were made by Mariner 9 in 1971–2; the craters on their surfaces were also first recorded from Mariner 9. The best views so far have come from Mars Global Surveyor. Earlier, in 1988, the Russians launched two probes with the intention of making controlled landings on Phobos, but both failed.

Both satellites have low albedoes, of the order of 5%. Phobos shows one major crater, Stickney, which is 9.6 km in diameter; the impact which produced it must have come close to shattering Phobos completely. Grooves extending across the surface from Stickney appear to be surface fractures caused by the impact; near the crater, the grooves are around 700 m across and 90 m deep, although further away from the crater the widths and depths are more of the order of 100–200 m and 10–20 m, respectively. Surface details of the satellites are given in Table 7.15.

The wide temperature range on Phobos gives a clue as to the composition of its surface. The upper regolith must be made up of finely-ground powder at least 1 m deep – it

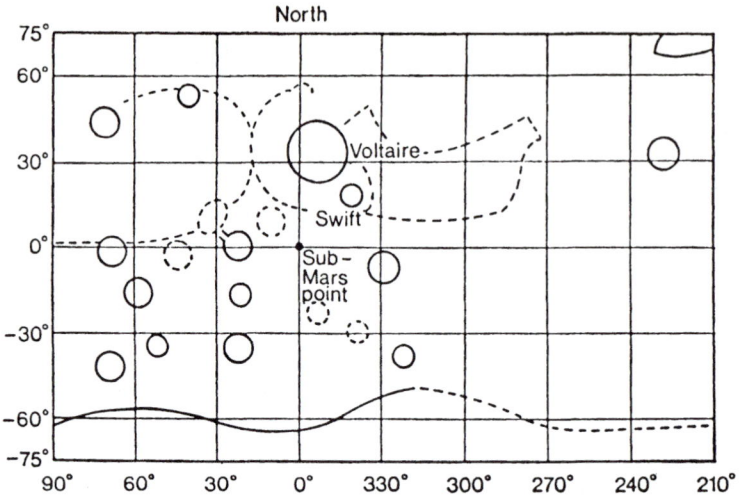

Figure 7.1. Deimos.

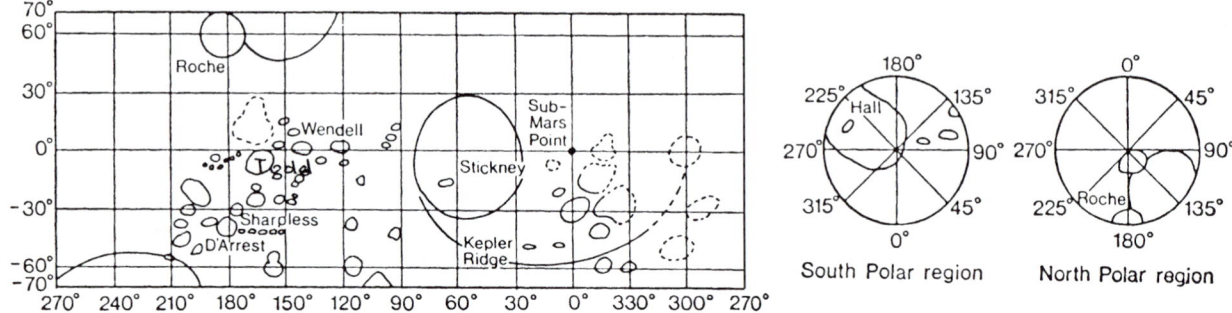

Figure 7.2. Phobos.

has been said that an astronaut would find himself 'hip-deep in dust'. The surface has been pounded by constant impacts of meteroids, some of which have started landslides leaving dark trails marking the steep slopes of the main craters. This is particularly evident in Stickney, where boulders on the rim measure up to 50 m across; rocks rolling down the slopes have left obvious tracks, showing that the gravity on Phobos, weak though it may be, is not inappreciable. The gravity field has about 1/1000 the strength of that of the Earth at sea-level.

Deimos, with its gravity field only about 1/3000 of that of the Earth, has a more subdued surface; the largest crater, Voltaire, is less than 3 km in diameter, though it has been suggested that an 11 km wide depression near the south pole may be an impact scar. There are nearly as many craters per unit area as there are on Phobos, but those of Deimos have been more eroded and filled in.

As an aside: in 1959 the eminent Russian astronomer Iosif Shklovsky published a paper in which he claimed that because Phobos was being 'braked' by the upper limits of

Table 7.14. The satellites of Mars.

	Phobos	Deimos
Mean distance from Mars (km)	9378	23 459
Mean angular distance from Mars, at mean opposition	24″.6	1′01″.8
Mean sidereal period (days)	0.3189	1.2624
Mean synodic period	7h 39m 26s.6	1d 5h 21m 15s.7
Orbital inclination ($°$)	1.068	0.8965
Orbital eccentricity	0.01515	0.0003
Diameter (km)	$27 \times 22 \times 18$	$15 \times 12 \times 10$
Density, water = 1	2.0	1.7
Mass (g)	1.08×10^{19}	1.8×10^{18}
Escape velocity (m s^{-1})	3–10	6
Magnitude at mean opposition	11.6	12.8
Maximum apparent magnitude, seen from Mars	−3.9	−0.1
Apparent diameter seen from Mars		
max	12′.3	2′
min	8′	1′.7

Table 7.15. Satellite features.

	Lat.	Long. W	Diameter (km)		
Phobos					
Craters					
D'Arrest	35.0S	185.0	2.3	Heinrich; German astronomer	1822–1875
Hall	75.0S	225.0	3	Asaph; American astronomer	1829–1907
Roche	60.0N	185.0	1.2	Edouard; French astronomer	1820–1883
Sharpless	25.0S	165.0	0.3	Bevan; American astronomer	1904–1950
Stickney	5.0S	55.0	4.6	Angeline; wife of Asaph Hall	?–1938
Todd	5.0S	160.0	1	David; American astronomer	1855–1939
Wendell	0.0	140.0	0.5	Oliver; American astronomer	1845–1912
Dorsum					
Kepler Dorsum	30N	250	4	Johannes; German astronomer	1571–1630
Deimos					
Craters					
Swift	10N	358	0.2	Jonathan; British writer	1667–1745
Voltaire	31N	2	0.3	François; French writer	1694–1778

the thin Martian atmosphere it must be of negligible mass and even hollow – in which case it would have to be regarded as a space-station launched by the Martians. Naturally, this idea was welcomed by flying saucer enthusiasts, who went so far as to suggest that Phobos and Deimos had not been discovered before 1877 because they did not then exist. Years later, when the satellites had been surveyed from close range, I asked Shklovsky whether he had changed his mind. He replied that his original paper had been nothing more than a practical joke. No comment!

Table 7.16. Selected list of Martian formations.

(a) Craters.

	Lat.	Long. W	Diameter (km)
Adams	31.3N	197.1	100
Agassiz	70.1S	88.4	104
Airy	5.2S	0.0	56
Alexei Tolstoy	47.6S	246.4	94
Alitus	34.5S	49.0	29
Aniak	32.1S	69.6	58
Antoniadi	21.7N	299.0	381
Arago	10.5N	330.2	154
Arandas	42.6N	15.1	22
Arkhangelsky	41.3S	24.6	119
Arrhenius	40.2S	237.0	132
Azul	42.5S	42.3	20
Azusa	5.5S	40.5	41
Babakin	36.4S	71.4	78
Bakhuysen	23.1S	344.3	162
Balboa	3.5S	34.0	20
Baldet	23.0N	294.5	195
Baltisk	42.6S	54.5	48
Bamba	3.5S	41.7	21
Bamberg	40.0N	3.0	57
Barabashov	47.6N	68.5	126
Barnard	61.3S	298.4	128
Becquerel	22.4N	7.9	167
Beer	14.6S	8.2	80
Bernard	23.8S	154.2	129
Berseba	4.4S	37.7	36
Bianchini	64.2S	95.1	77
Bjerknes	43.4S	188.7	89
Boeddicker	14.8S	197.6	107
Bond	33.3S	35.7	104
Bouguer	18.6S	332.8	106
Bozkir	44.4S	32.0	89
Brashear	54.1S	119.2	126
Briault	10.1S	270.2	100
Bunge	34.2S	48.4	78
Burroughs	72.5S	243.1	104
Burton	14.5S	156.3	137
Byrd	65.6S	231.9	122
Camiling	0.8S	38.1	21
Camiri	45.1S	41.9	20
Campbell	54.0S	195.0	123
Cartago	23.6S	17.8	33
Cassini	23.8N	327.9	415
Cerulli	32.6N	337.9	120
Chamberlin	66.1S	124.3	125
Charlier	68.6S	168.4	100
Chekalin	24.7S	26.6	87
Chia	1.6N	59.5	94
Chimbote	1.5S	39.8	65
Chincoteague	41.5N	236.0	35
Choctaw	41.5S	37.0	20
Clark	55.7S	133.2	93
Coblentz	55.3S	90.2	111
Cobres	12.1S	153.7	89
Columbus	29.7S	165.8	114
Comas Solà	20.1S	158.4	132
Concord	16.6N	34.1	20
Copernicus	50.0S	168.6	292
Crommelin	5.3N	10.2	111
Cruls	43.2S	196.9	83
Curie	29.2N	4.9	98
Da Vinci	1.5N	39.1	98
Daly	66.4S	23.0	99
Dana	72.6S	33.1	95
Darwin	57.2S	19.2	166
Dawes	9.3S	322.3	191
Dein	38.5N	2.4	24
Dejnev	25.7S	164.5	156
Denning	17.5S	326.6	165
Dia-Cau	0.3S	42.8	28
Dison	25.4S	16.3	20
Dokuchaev	60.8S	127.1	73
Douglass	51.7S	70.4	97
Du Martheray	5.8N	266.4	94

Table 7.16. (Continued)

	Lat.	Long. W	Diameter (km)
Du Toit	71.8S	49.7	75
Edam	26.6S	19.9	20
Eddie	12.5N	217.8	90
Eiriksson	19.6S	173.7	56
Escanalte	0.3N	244.8	83
Eudoxus	45.0S	147.2	92
Fesenkov	21.9N	86.4	86
Flammarion	25.7N	311.7	160
Flaugergues	17.0S	340.9	235
Focas	33.9N	347.2	82
Fontana	63.2S	71.9	78
Foros	34.0S	28.0	23
Fournier	4.25S	287.5	112
Gale	5.3S	222.3	172
Gali	44.1S	36.9	24
Galilaei	5.8N	26.9	124
Galle	50.8S	30.7	230
Gilbert	68.2S	273.8	115
Gill	15.8N	354.5	81
Glazov	20.8S	26.4	20
Gledhill	53.5S	272.9	72
Globe	24.0S	27.1	45
Graff	21.4S	206.0	157
Green	52.4S	8.3	184
Grójec	21.6S	30.6	37
Guaymas	26.2N	44.8	20
Gusev	14.6S	184.6	166
Hadley	19.3S	203.0	113
Haldane	53.0S	230.5	72
Hale	36.1S	36.3	136
Halley	48.6S	59.2	81
Hartwig	38.7S	15.6	104
Heaviside	70.8S	94.8	103
Heinlein	64.6S	243.8	83
Helmholtz	45.6S	21.1	107
Henry	11.0N	336.8	165
Herschel	14.9S	230.1	304
Hilo	44.8S	35.5	20
Hipparchus	45.0S	151.1	104
Holden	26.5S	33.9	141
Holmes	75.0S	293.9	109
Hooke	45.0S	44.4	145
Huancayo	3.7S	39.8	25
Huggins	49.3S	204.3	82
Hussey	53.8S	126.5	100
Hutton	71.9S	255.5	99
Huxley	62.9S	259.2	108
Huygens	14.0S	304.4	456
Ibragimov	25.9S	59.5	89
Innsbrück	6.5S	40.0	64
Janssen	2.8N	322.4	166
Jarry-Desloges	9.6S	276.1	97
Jeans	69.9S	205.5	71
Joly	74.6S	42.5	81
Kaiser	46.6S	340.9	201
Kakori	41.9S	29.6	25
Kansk	20.8S	17.1	34
Kantang	24.8S	17.5	64
Karpinsk	46.0S	31.8	28
Karshi	23.6S	19.2	22
Kashira	27.5S	18.3	68
Kasimov	25.0S	22.8	92
Keeler	60.7S	151.2	92
Kepler	47.2S	218.7	219
Kipini	26.0N	31.5	75
Knobel	6.6S	226.9	127
Korolev	72.9N	195.8	84
Kovalsky	30.0S	141.4	299
Krishtofovich	48.6S	262.6	111
Kuba	25.6S	19.5	25
Kufra	40.6N	239.7	32
Kuiper	57.3S	157.1	86
Kunowsky	57.0N	9.0	60
Kushva	44.3S	35.4	39
Labria	35.3S	48.0	60
Lamas	27.4S	20.5	21

Table 7.16. (Continued)

	Lat.	Long. W	Diameter (km)
Lambert	20.1S	334.6	87
Lamont	58.3S	113.3	72
Lampland	36.0S	79.5	71
Lassell	21.0S	62.4	86
Lasswitz	9.4S	221.6	122
Lau	74.4S	107.3	109
Le Verrier	38.2S	342.9	139
Lebu	20.6S	19.4	20
Li Fan	47.4S	153.0	103
Liais	75.4S	252.9	128
Libertad	23.3N	29.4	31
Liu Hsin	53.7S	171.4	129
Lockyer	28.2N	199.4	74
Lohse	43.7S	16.4	156
Lomonosov	64.8N	8.8	151
Lorica	20.1S	28.3	67
Loto	22.2S	22.3	22
Lowell	52.3S	81.3	201
Luga	44.6S	47.2	42
Luki	30.0S	37.0	20
Luzin	27.3N	328.8	86
Lyell	70.0S	15.6	134
Lyot	50.7N	330.7	220
Mädler	10.8S	357.3	100
Magadi	34.8S	46.0	57
Magelhæns	32.9S	174.5	102
Maggini	28.0N	350.4	146
Main	76.8S	310.9	102
Maraldi	62.2S	32.1	119
Marbach	17.9N	249.2	20
Marca	10.4S	158.2	83
Mariner	35.2S	164.3	151
Marth	13.1N	3.6	104
Martz	35.2S	215.8	91
Maunder	50.0S	358.1	93
McLaughlin	22.1N	22.5	90
Mellish	72.9S	24.0	99
Mena	32.5S	18.5	31
Mendel	59.0S	198.5	82
Mie	48.6N	220.4	93
Milankovič	54.8N	146.6	113
Millman	54.4S	149.6	82
Millochau	21.5S	274.7	102
Mitchel	67.8S	284.0	141
Molesworth	27.8S	210.6	175
Moreux	42.2N	315.5	138
Müller	25.9S	232.0	120
Murgoo	24.0S	22.3	24
Mutch	0.6N	55.1	200
Nansen	50.5S	140.3	82
Nardo	27.8S	32.7	23
Peridier	25.8N	276.0	99
Navan	26.2S	23.2	25
Newcomb	24.1S	359.0	259
Newton	40.8S	157.9	287
Nicholson	0.1N	164.5	114
Niesten	28.2S	302.1	114
Nitro	21.5S	23.8	28
Noma	25.7S	24.0	38
Nordenskiold	53.0S	158.7	87
Ochakov	42.5S	31.6	30
Oraibi	17.4N	32.4	31
Ostrov	26.9S	28.0	67
Ottumwa	24.9N	55.7	55
Oudemans	10.0S	91.7	121
Pãros	22.2N	98.1	40
Pasteur	19.6N	335.5	114
Perepelkin	52.8N	64.6	112
Perrotin	3.0S	77.8	95
Pettit	12.2N	173.9	104
Phillips	66.4S	45.1	183
Pickering	34.4S	132.8	112
Playfair	78.0S	125.5	68
Podor	44.6S	43.0	25
Polotsk	20.1S	26.1	23
Poona	24.0N	52.3	20

Table 7.16. (Continued)

	Lat.	Long. W	Diameter (km)
Porter	50.8S	113.8	113
Poynting	8.4N	112.8	80
Priestly	54.3S	229.3	40
Proctor	47.9S	330.4	168
Ptolemæus	46.4S	157.5	184
Pulawy	36.6S	76.7	51
Pylos	16.9N	30.1	29
Quenisset	34.7N	319.4	127
Rabe	44.0S	325.2	99
Radau	17.3N	4.7	115
Rayleigh	75.7S	240.1	153
Redi	60.6S	267.1	62
Renaudot	42.5N	297.4	69
Reuyl	9.5S	193.1	74
Revda	24.6S	28.3	26
Reynolds	75.1S	157.6	91
Richardson	72.6S	180.3	82
Ritchey	28.9S	50.9	82
Roddenberry	49.9S	4.5	140
Ross	57.6S	107.6	88
Rossby	47.8S	192.2	82
Ruby	25.6S	16.9	25
Rudaux	38.6N	309.0	52
Russell	55.0S	347.4	138
Rutherford	19.2N	10.6	116
Ruza	34.3S	52.8	20
Salaga	47.6S	51.0	28
Sangar	27.9S	24.1	28
Santa Fe	19.5N	48.0	20
Sarno	44.7S	54.0	20
Schaeberle	24.7S	309.8	160
Schiaparelli	2.5S	343.4	461
Schmidt	72.2S	77.5	194
Schöner	20.4N	309.5	185
Schröter	1.8S	303.6	337
Secchi	57.9S	257.7	218
Semeykin	41.8N	351.2	71
Seminole	24.5S	18.9	21
Shambe	20.7S	30.5	29
Sharanov	27.3N	58.3	95
Shatskii	32.4S	14.7	69
Sibu	23.3S	19.6	31
Sigli	20.5S	30.6	31
Sklodowska	33.8N	2.8	116
Slipher	47.7S	84.5	129
Smith	66.1S	102.8	71
Sögel	21.7N	55.1	29
Sokol	42.8S	40.5	20
Soochow	16.8N	28.9	30
South	77.0S	338.0	111
Spallanzani	58.4S	273.5	72
Spencer Jones	19.1S	19.8	85
Stege	2.6N	58.4	72
Steno	68.0S	115.3	105
Stokes	56.0N	189.0	70
Stoney	69.8S	138.4	177
Suess	67.1S	178.4	72
Sumgin	37.0S	48.6	83
Sytinskaya	42.8N	52.8	90
Tabor	36.0S	58.5	20
Tara	44.4S	52.7	27
Tarakan	41.6S	30.1	37
Taza	44.0S	45.1	22
Teisserenc de Bort	0.6N	315.0	118
Terby	28.2S	286.0	135
Tikhonravov	13.7N	324.1	390
Tikhov	51.2S	254.1	107
Timbuktu	5.7S	37.7	63
Timoshenko	42.1N	63.9	84
Trouvelot	16.3N	13.0	168
Trümpler	61.7S	150.6	77
Turbi	40.9S	51.2	25
Tuskegee	2.9S	36.2	69
Tycho Brahe	49.5S	213.8	108
Tyndall	40.1N	190.4	79
Valverde	20.3N	55.8	35

Table 7.16. (Continued)

	Lat.	Long. W	Diameter (km)
Verlaine	9.4S	295.9	42
Very	49.8S	176.8	127
Viana	19.5N	255.3	29
Vik	36.0S	64.0	25
Vinogradov	56.3S	216.0	191
Vishniac	76.7S	276.1	76
Vivero	49.4N	241.3	64
Voeykov	32.5S	76.1	67
Vogel	37.0S	13.2	124
von Kármán	64.3S	58.4	100
Wabash	21.5N	33.7	42
Wallace	52.8S	249.2	159
Waspam	20.7N	56.6	40
Wegener	64.3S	4.0	70
Weinbaum	65.9S	245.5	86
Wells	60.1S	237.4	94
Wicklow	2.0S	40.7	21
Wien	10.5S	220.1	105
Williams	18.8S	164.1	125
Windfall	2.1S	43.5	20
Wirtz	48.7S	25.8	128
Wislencius	18.3S	348.7	138
Wright	58.6S	150.8	106
Zilair	32.0S	33.0	43
Zongo	32.0S	42.0	23
Zulanka	2.3S	42.3	47
Zuni	19.3N	29.6	25

(b) The features listed here are not visible with ordinary Earth-based telescopes, apart from the main dark areas and some of the smaller features such as Olympus Mons (formerly Nix Olympica, the Olympic Snow). The length or diameter in km is given. Positions refer to the centres of the features.

	Lat.	Long. W	Length/Diameter (km)
Catena (chain of craters)			
Acheron Catena	38.2N	100.7	554
Alba Catena	35.2N	114.6	148
Ceraunius Catena	37.4N	108.1	51
Elysium Catena	18.0N	210.4	66
Ganges Catena	2.6S	69.3	221
Labeatis Catenæ	18.8N	95.1	318
Phlegethon Catena	40.5N	101.8	875
Tithoniæ Catena	5.5S	71.5	380
Tractus Catena	27.9N	103.2	1234
Chasma (large recilinear chain)			
Chasma Australe	82.9S	273.8	491
Chasma Boreale	83.2N	21.3	318
Capri Chasma	8.7S	42.6	1498
Candor Chasma	6.5S	71.0	816
Eos Chasma	12.6S	45.1	963
Gangis Chasma	8.4S	48.1	541
Hebes Chasma	1.1S	76.1	285
Ius Chasma	7.2S	84.6	1003
Juventæ Chasma	1.9S	61.8	495
Melas Chasma	10.5S	72.9	526
Ophir Chasma	4.0S	72.5	251
Tithonium Chasma	4.6S	86.5	904
Fossa (ditch)			
Acheron Fossæ	38.7N	136.6	1120
Alba Fossæ	43.4N	103.6	2077
Amenthes Fossæ	10.2N	259.3	1592
Ceraunius Fossæ	24.8N	110.5	711
Claritas Fossæ	34.8S	99.1	2033
Coloe Fossæ	37.1N	303.9	930

Table 7.16. (Continued)

	Lat.	Long. W	Length/Diameter (km)
Coracis Fossæ	34.8S	78.6	747
Elysium Fossæ	27.5N	219.5	1114
Fossa (ditch) (Continued)			
Icaria Fossæ	53.7S	135.0	2153
Ismeniæ Fossæ	38.9N	326.1	858
Labeatis Fossæ	20.9M	95.0	388
Mareotis Fossæ	45.0N	79.3	795
Memnonia Fossæ	21.9S	154.4	1370
Nili Fossæ	24.0N	283.0	709
Noctis Fossæ	3.3S	99.0	692
Olympica Fossæ	24.4N	115.3	573
Sirenum Fossæ	34.5S	158.2	2712
Tantalus Fossæ	44.5N	102.4	1990
Tempe Fossæ	40.2N	74.5	1553
Thaumasia Fossæ	45.9S	97.3	1118
Ulysses Fossæ	11.4N	123.3	720
Uranius Fossæ	25.8N	90.1	438
Zephyrus Fossæ	23.9N	214.4	452
Labyrinthus (valleu complex)			
Adamas Labyrinthus	36.5N	255.0	664
Noctis Labyrinthus	7.2S	101.3	976
Mensa (mesa)			
Aeolis Mensæ	3.7S	218.5	841
Baetis Mensa	5.4S	72.4	181
Cydonia Mensæ	37.0N	12.8	854
Deuteronilus	45.7N	337.9	627
Galaxias Mensæ	36.5N	212.7	327
Nepenthes Mensæ	9.3N	241.7	1704
Nilokeras Mensæ	32.9N	51.1	327
Nilosyrtis Mensæ	35.4N	293.6	693
Protonilus Mensæ	44.2N	309.4	592
Sacra Mensæ	25.0N	69.2	602
Zephyria Mensæ	10.1S	188.1	230
Mons (mountain or volcano)			
Arsia Mons	9.4S	120.5	485
Ascræus Mons	11.3N	104.5	462
Charitum Montes	58.3S	44.2	1412
Elysium Mons	25.0N	213.0	432
Erebus Montes	39.9N	170.5	594
Hellespontus Montes	45.5S	317.5	681
Libya Montes	2.7N	271.2	1229
Nereidum Montes	41.0S	43.5	1677
Olympus Mons	18.4N	133.1	624
Pavonis Mons	0.3N	112.8	375
Phlegra Montes	40.9N	197.4	1310
Tartarus Montes	25.1N	188.7	1011
Tharsis Montes	2.8N	113.3	2105
Patera (shallow crater, volcanic structure)			
Alba Patera	40.5N	109.9	464
Amphitrites Patera	59.1S	299.0	138
Apollinaris Patera	8.3S	186.0	198
Biblis Patera	2.3N	123.8	117
Diacria Patera	34.8N	132.7	75
Hadriaca Patera	30.6S	267.2	451
Meroe Patera	7.2N	291.5	60
Nili Patera	9.2N	293.0	70
Orcus Patera	14.4N	181.5	381
Peneus Patera	58.0S	307.4	123
Tyrrhena Patera	21.9S	253.2	597
Ulysses Patera	2.9N	121.5	112
Uranius Patera	26.7N	92.0	276
Planitia (low, smooth plain)			
Acidalia Planitia	54.6N	19.9	2791
Amazonis Planitia	16.0N	158.4	2816

Table 7.16. (Continued)

	Lat.	Long. W	Length/ Diameter (km)
Arcadia Planitia	46.4N	152.1	3052
Argyre Planitia	49.4S	42.8	868
Chryse Planitia	27.0N	36.0	1500
Elysium Planitia	14.3N	241.1	3899
Hellas Planitia	44.3S	293.8	2517
Isidis Planitia	14.1N	271.0	1238
Utopia Planitia	47.6N	277.3	3276

Planum (high plain or plateau)

Auroræ Planum	11.1S	50.2	590
Planum Australe	80.3S	155.1	1313
Planum Boreum	85.0N	180.0	1066
Bosporus Planum	33.4S	64.0	330
Planum Chronium	62.0S	212.6	1402
Dædalia Planum[a]	13.9S	138.0	2477
Hesparia Planum	18.3S	251.6	1869
Icaria Planum	42.7S	107.2	840
Lunæ Planum	9.6N	66.6	1845
Malea Planum	65.9S	297.4	1068
Ophir Planum	9.6S	62.2	1068
Planum Angustum	80.0S	83.9	208
Sinai Planum	12.5S	87.1	1064
Syria Planum	12.0S	103.9	757
Syrtis Major Planum	9.5N	289.6	1356
Thaumasia Planum	22.0S	65.0	930

Rupes (scarp)

Amenthes Rupes	1.8N	249.4	441
Argyre Rupes	63.3S	66.7	592
Arimanes Rupes	10.0S	147.4	203
Avernus Rupes	8.9S	186.4	200
Bosporus Rupes	42.8N	57.2	500
Cerberus Rupes	8.4N	195.4	1254
Chalcoporus Rupes	54.9S	338.6	380
Claritas Rupes	26.0S	105.5	—
Cydnus Rupes	59.6N	257.5	430
Elysium Rupes	25.4N	211.4	180
Morpheos Rupes	36.2S	234.1	321
Ogygis Rupes	34.1S	54.9	225
Olympus Rupes	17.2N	133.9	1819
Phison Rupes	26.6N	309.4	149
Pityusa Rupes	63.0S	328.0	290
Promethei Rupes	76.8S	286.8	1491
Rupes Tenuis	81.9N	65.2	—
Tartarus Rupes	6.6S	184.4	81
Thyles Rupes	73.2S	205.6	269
Ulyxis Rupes	63.2S	198.8	303

Scopulus (lobate or irregular scarp)

Charybdis Scopulus	24.9S	339.9	513
Coronæ Scopulus	33.9S	294.1	306
Eridania Scopulus	53.4S	217.9	510
Nilokeras Scopulus	31.5N	57.2	1064
Œnotria Scopulus	11.2S	283.2	1438
Scylla Scopulus	25.5S	342.0	474
Tartarus Scopulus	6.4S	181.7	127

Sulcus (subparallel furrows or ridges)

Amazonis Sulci	3.3S	145.1	243
Apollinaris Sulci	11.3S	183.4	300
Cyane Sulci	25.5N	128.4	286
Gigas Sulci	9.9N	127.6	467
Gordii Sulci	17.9N	125.6	316
Lycus Sulci	29.2N	139.8	1639
Medusæ Sulci	5.3N	159.8	91
Memnonia Sulci	6.5S	175.6	361

Terra (land mass)

Aonia Terra	58.2S	94.8	3372
Arabia Terra	25.0N	330.0	6000

[a] Formerly Erythræum Planum.

Table 7.16. (Continued)

	Lat.	Long. W	Length/ Diameter (km)
Terra Cimmeria	34.0S	215.0	2285
Margaritifer Terra	16.2S	21.3	2049
Terra Meridiani	7.2S	356.0	1622
Noachis Terra	45.0S	350.0	3500
Promethei Terra	52.9S	262.2	2761
Terra Sabæa	10.4S	330.6	1367
Terra Sirenum	37.0S	160.0	2165
Tempe Terra	41.3N	70.5	2055
Tyrrhena Terra	14.3S	278.5	2817
Xanthe Terra	4.2N	46.3	3074

Tholus (domed hills)

Albor Tholus	19.3N	209.8	165
Australis Tholus	59.0S	323.0	40
Ceraunius Tholus	24.2N	97.2	108
Hecates Tholus	32.7N	209.8	183
Iaxartes Tholus	72.0N	15.0	53
Jovis Tholus	18.4N	117.5	61
Kison Tholus	73.0N	358.0	50
Ortygia Tholus	70.0N	8.0	100
Tharsis Tholus	13.4N	90.8	153
Uranius Tholus	26.5N	97.8	71

Unda (dune)

Abalos Undæ	81.0N	83.1	80 090
Hyperboreæ Undæ	77.5N	46.0	80 050

Vallis (valley)

Al-Qahira Vallis	17.5S	196.7	546
Ares Vallis	9.7N	23.4	1690
Auqakuh Vallis	28.6N	299.3	195
Bahram Vallis	21.4N	58.7	403
Brazos Valles	6.3S	341.7	494
Dao Vallis	36.8S	269.4	667
Evros Vallis	12.6S	345.9	335
Granicus Valles	28.7N	227.6	445
Harmakhis Vallis	39.1S	267.2	585
Hebrus Valles	20.0N	233.9	299
Hrad Vallis	37.8N	221.6	719
Huo Hsing Vallis	31.5N	293.9	340
Indus Vallis	19.2N	322.1	253
Kasei Valles	22.8N	68.2	2222
Loire Valles	18.5S	16.4	670
Louros Valles	8.7S	81.9	423
Ma'adim Vallis	21.0S	182.7	861
Maja Valles	15.4N	56.7	1311
Mamers Vallis	41.7N	344.5	945
Mangala Valles	7.4S	150.3	880
Marti Vallis	11.0N	182.0	1700
Mawrth Vallis	22.4N	16.1	575
Naktong Vallis	5.2N	326.7	807
Nanedi Valles	5.5N	48.7	470
Nirgal Vallis	28.3S	41.9	511
Ravi Vallis	0.4S	40.7	169
Samara Valles	24.2S	18.9	575
Scamander Vallis	16.1N	331.1	272
Shalbatana Vallis	5.6N	43.1	687
Simud Vallis	11.5N	38.5	1074
Tinjar Valles	38.2N	235.7	390
Tisia Valles	11.6S	313.9	390
Tiu Vallis	8.6N	34.8	970
Uzboi Vallis	30.0S	35.6	340
Valles Marineris	11.6S	70.7	4128

Vastitas (widespread lowland)

Vastitas Borealis	67.5N	180.0	9999

Table 7.17. Martian crater names of astronomers.

Adams	Walter S.	American	1876–1956
Airy	George Biddell	British	1801–1892
Arago	Dominique François	French	1786–1853
Bakhuysen	Henricus van de Sande	Dutch	1838–1923
Baldet	Ferdinand	French	1885–1964
Barabashov	Nikolai	Russian	1894–1971
Barnard	Edward Emerson	American	1857–1923
Beer	Wilhelm	German	1797–1850
Bianchini	Francesco	Italian	1662–1729
Bond	George P.	American	1825–1865
Boeddicker	Otto	German	1853–1937
Briault	P.	French	?–1922
Burton	Charles E.	British	1846–1882
Campbell	William Wallace	American	1862–1938
Cassini	Giovanni	Italian	1625–1712
Cerulli	Vicenzio	Italian	1859–1927
Charlier	Carl V. L.	Swedish	1862–1934
Clark	Alvan	American	1804–1887
Coblentz	William	American	1873–1962
Comas Solá	Jose	Spanish	1868–1937
Copernicus	Nikolas (Mikołaj)	Polish	1473–1543
Crommelin	Andrew Claude de la C.	Irish	1865–1939
Darwin	George	British	1845–1912
Dawes	William Rutter	British	1799–1868
Denning	William Frederick	British	1848–1931
Douglass	Andrew E.	American	1867–1962
Du Martheray	Maurice	Swiss	1892–1955
Du Toit	Alexander	South African	1878–1948
Eddie	Lindsay	South African	1845–1913
Escalante	F.	Mexican	c 1930
Eudoxus	—	Greek	408–355 BC
Fesenkov	Vasili	Russian	1889–1972
Flammarion	Camille	French	1842–1925
Flaugergues	Honoré	French	1755–1835
Focas	Ioannas	Greek	1909–1969
Fontana	Francesco	Italian	1585–1685
Fournier	Georges	French	1881–1954
Gale	Walter F.	Australian	1865–1945
Galilei	Galileo	Italian	1564–1642
Galle	Johann	German	1812–1910
Gill	David	Scottish	1843–1914
Gledhill	Joseph	British	1836–1906
Graff	Kasimir	German	1878–1950
Green	Nathan	British	1823–1899
Gusev	Matwei	Russian	1826–1866
Hale	George Ellery	American	1868–1938
Halley	Edmond	British	1656–1742
Hartwig	Ernst	German	1851–1923
Henry	Paul and Prosper	French	1848–1905; 1849–1903
Herschel	William	Hanoverian/British	1738–1822
Hipparchus	—	Greek	c 160–125 BC
Holden	Edward	American	1846–1914
Hooke	Robert	British	1635–1703
Huggins	William	British	1824–1910
Ibragimov	Nadir	Russian	1932–1977
Janssen	Pierre	French	1824–1907
Jarry-Desloges	René	French	1868–1951
Jeans	James Hopwood	British	1877–1946
Kaiser	Frederik	Dutch	1808–1872
Keeler	James	American	1857–1900
Kepler	Johannes	German	1571–1630
Knobel	Edward	British	1841–1930
Kovalsky	Marian	Russian	1821–1884
Kunowsky	George	German	1786–1846
Lamont	Johann von	German	1805–1879
Lampland	Carl O.	American	1873–1951
Lau	Hans E.	Danish	1879–1918
Le Verrier	Urbain	French	1811–1877
Li Fan	—	Chinese	c 85 AD

Table 7.17. (Continued)

Liais	Emmanuel	French	1826–1900
Liu Hsin	—	Chinese	?–22
Lockyer	J. Norman	British	1836–1920
Lohse	Oswald	German	1845–1915
Lowell	Percival	American	1855–1916
Lomonosov	Mikhail	Russian	1711–1765
Lyot	Bernard	French	1897–1952
Mädler	Johann Heinrich von	German	1794–1874
Maggini	Mentore	Italian	1890–1941
Main	Robert	British	1808–1878
Maraldi	Giacomo	Italian	1665–1729
Marth	Albert	British	1828–1897
Maunder	Edward W.	British	1851–1928
McLaughlin	Dean B.	American	1901–1965
Mellish	John	American	1886–1970
Milankovič	Milutin	Jugoslav	1879–1958
Millman	Peter	Canadian	1906–1990
Millochau	Gaston	French	1866–?
Mitchel	Ormsby	American	1809–1862
Moreux	Theophile	French	1867–1954
Müller	Carl	German	1851–1925
Mutch	Thomas	American	1931–1980
Newcomb	Simon	American	1835–1909
Newton	Isaac	British	1643–1727
Nicholson	Seth	American	1891–1963
Niesten	Louis	Belgian	1844–1920
Oudemans	Jean	Dutch	1827–1906
Perepelkin	Evgenii	Russian	1906–1940
Perrotin	Henri	French	1845–1904
Pettit	Edison	American	1890–1962
Phillips	Theodore	British	1868–1942
Phillips	John	British	1800–1874
Pickering	Edward C.	American	1846–1919
Pickering	William H.	American	1858–1938
Porter	Russell W.	American	1871–1949
Poynting	John Henry	British	1852–1914
Proctor	Richard A.	British	1837–1888
Ptolemæus	Claudius	Greek	c 120–180
Quenisset	Ferdinand	French	1872–1951
Radau	Rudolphe	French	1835–1911
Rabe	Wilhelm	German	1893–1958
Renaudot	Gabrielle	French	1877–1962
Ritchey	George W.	American	1864–1945
Ross	Frank E.	American	1874–1966
Rudaux	Lucien	French	1874–1947
Russell	Henry Norris	American	1877–1957
Schaeberle	John	American	1853–1924
Schiaparelli	Giovanni	Italian	1835–1910
Schmidt	Johann	German	1825–1884
Schmidt	Otto	Russian	1891–1956
Schröter	Johann Hieronymus	German	1745–1816
Secchi	Angelo	Italian	1818–1878
Semeykin	Boris	Russian	1900–1937
Sharonov	Vesvolod	Russian	1901–1964
Slipher	Vesto	American	1875–1969
South	James	British	1785–1867
Spencer Jones	Harold	British	1890–1960
Sytinskaya	Nadezhda	Russian	1906–1974
Terby	François	Belgian	1846–1911
Tikhonravov	Gavriil	Russian	1851–1916
Trouvelot	Etienne	French	1827–1895
Tycho Brahe	—	Danish	1546–1601
Very	Frank W.	American	1852–1927
Vogel	Hermann Carl	German	1841–1907
Vishniac	Wolf	American	1922–1974
von Kármán	Theodor	Hungarian	1881–1963
Wirtz	Carl Wilhelm	German	1876–1939
Wislensius	Walter	German	1859–1905
Wright	William H.	American	1871–1959
Williams	Arthur Stanley	British	1861–1938

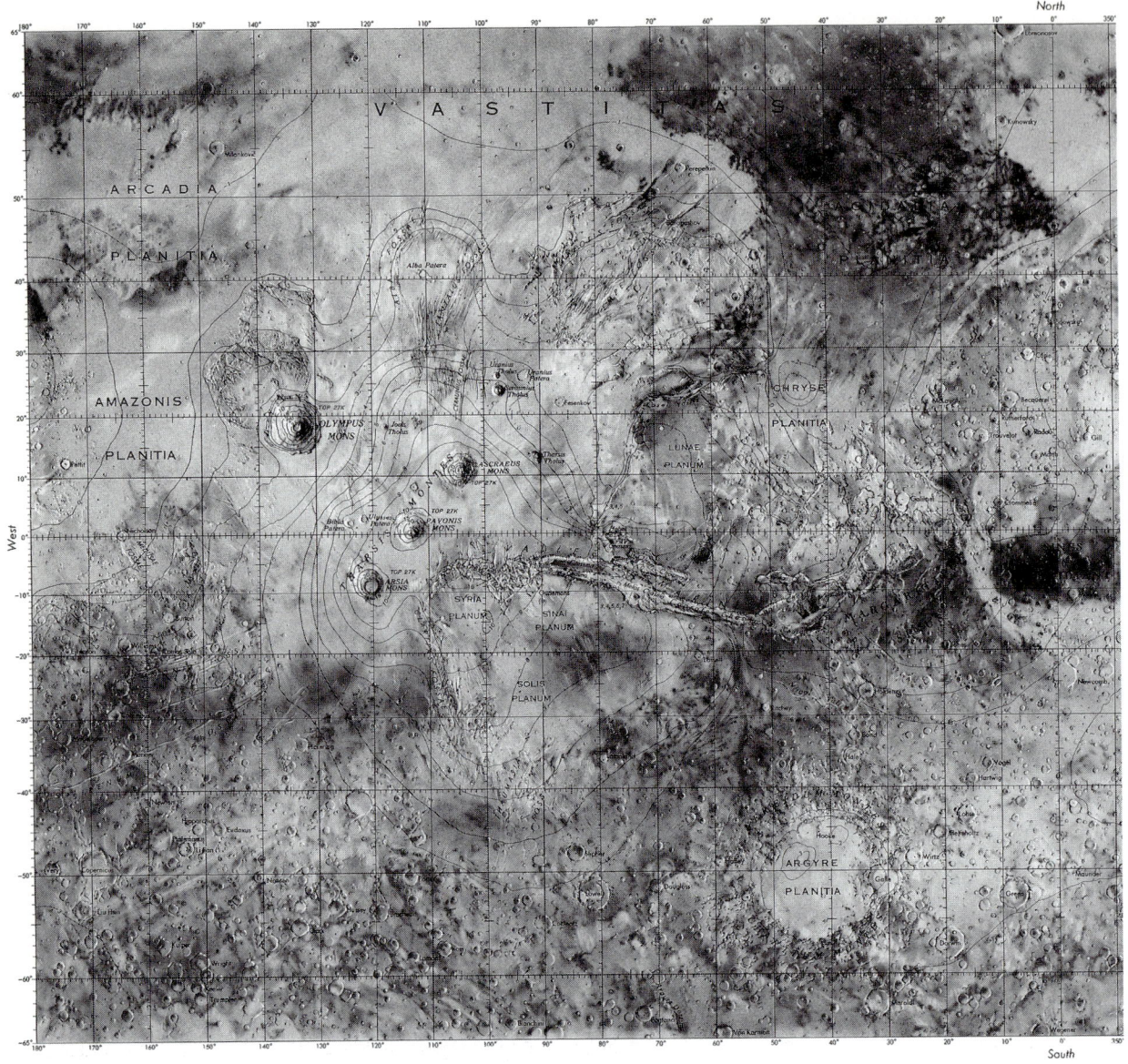

Figure 7.3. Mars. (Courtesy: NASA.)

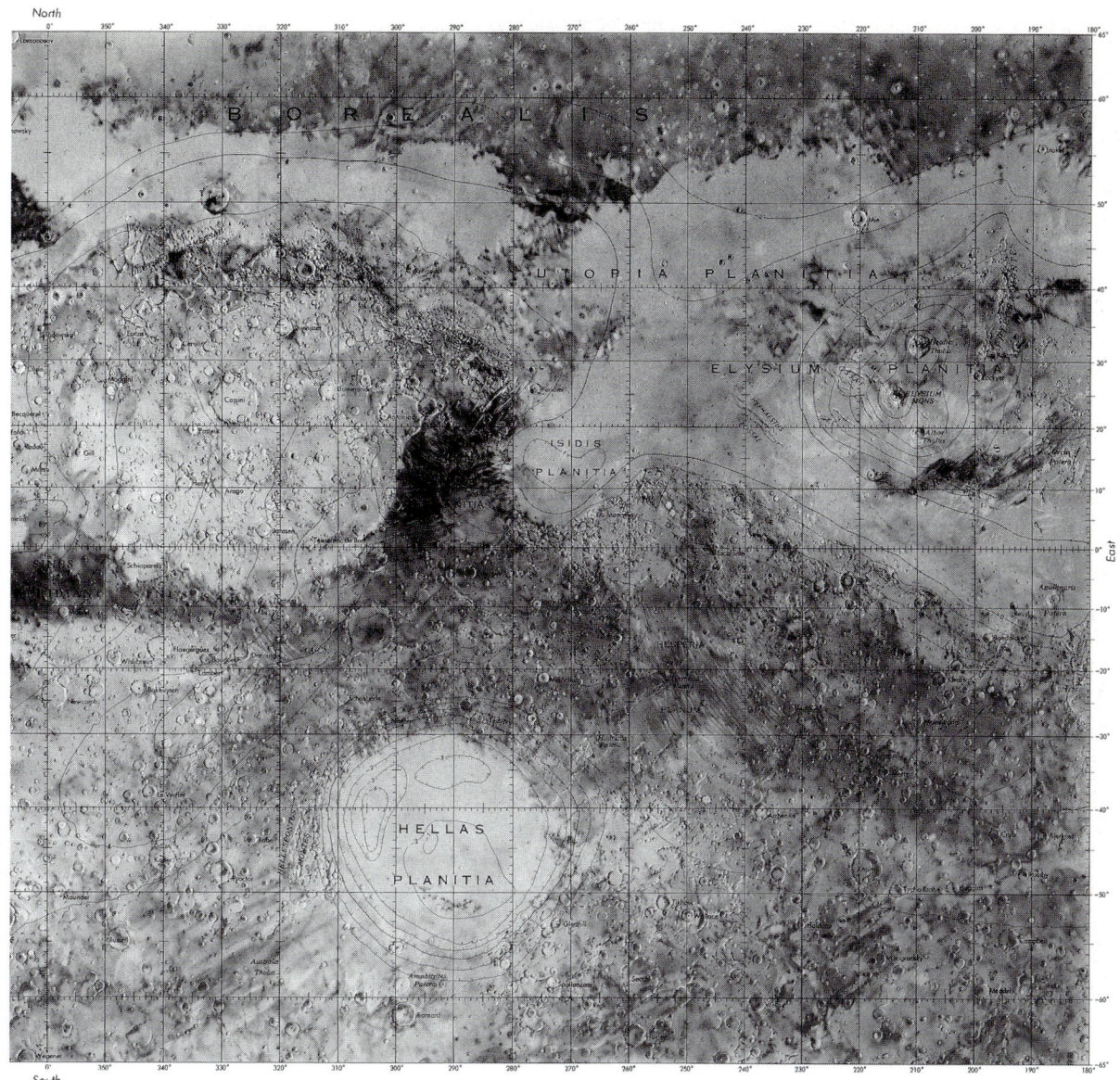

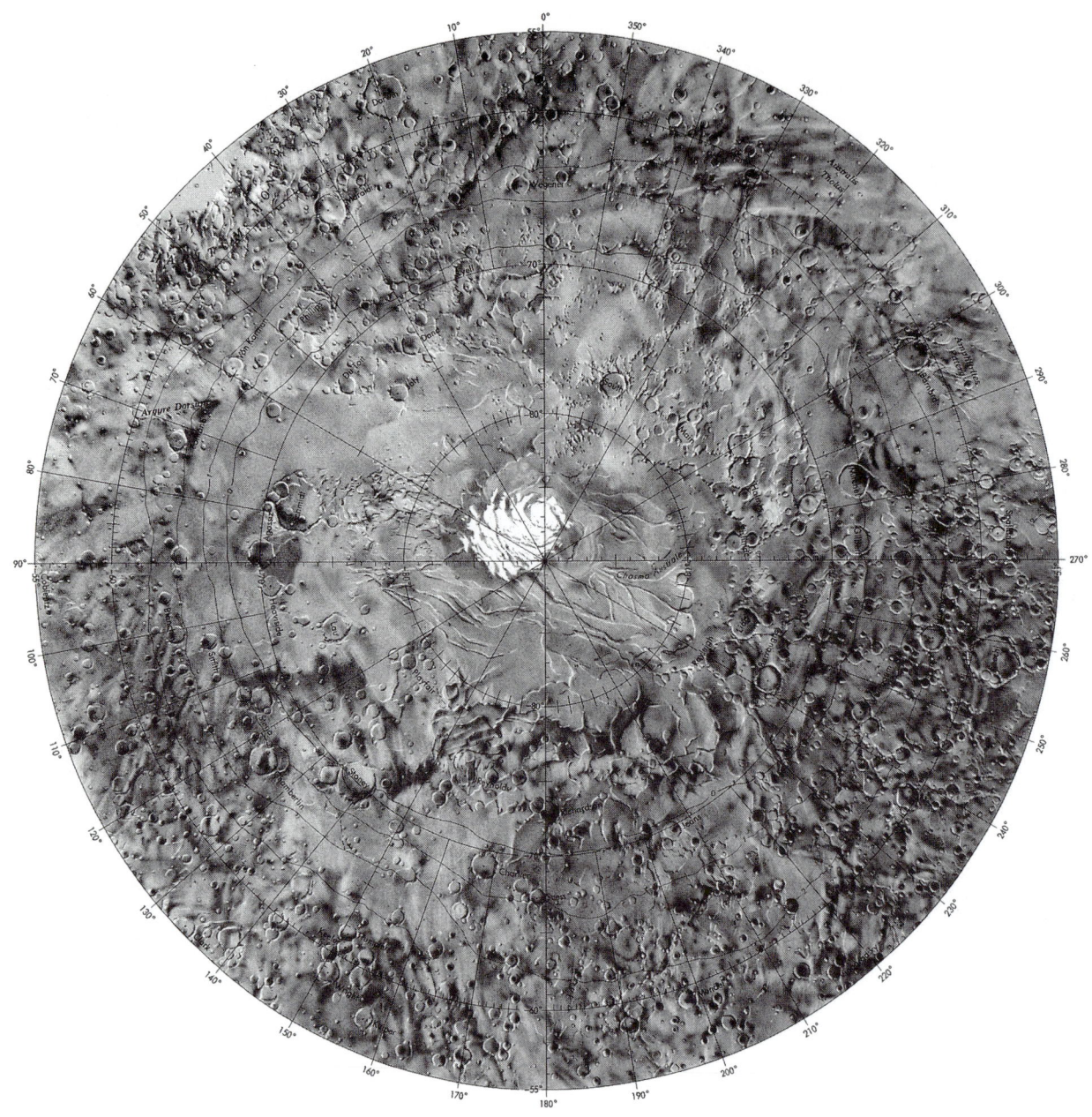

Figure 7.4. Mars – south polar region. (Courtesy: NASA.)

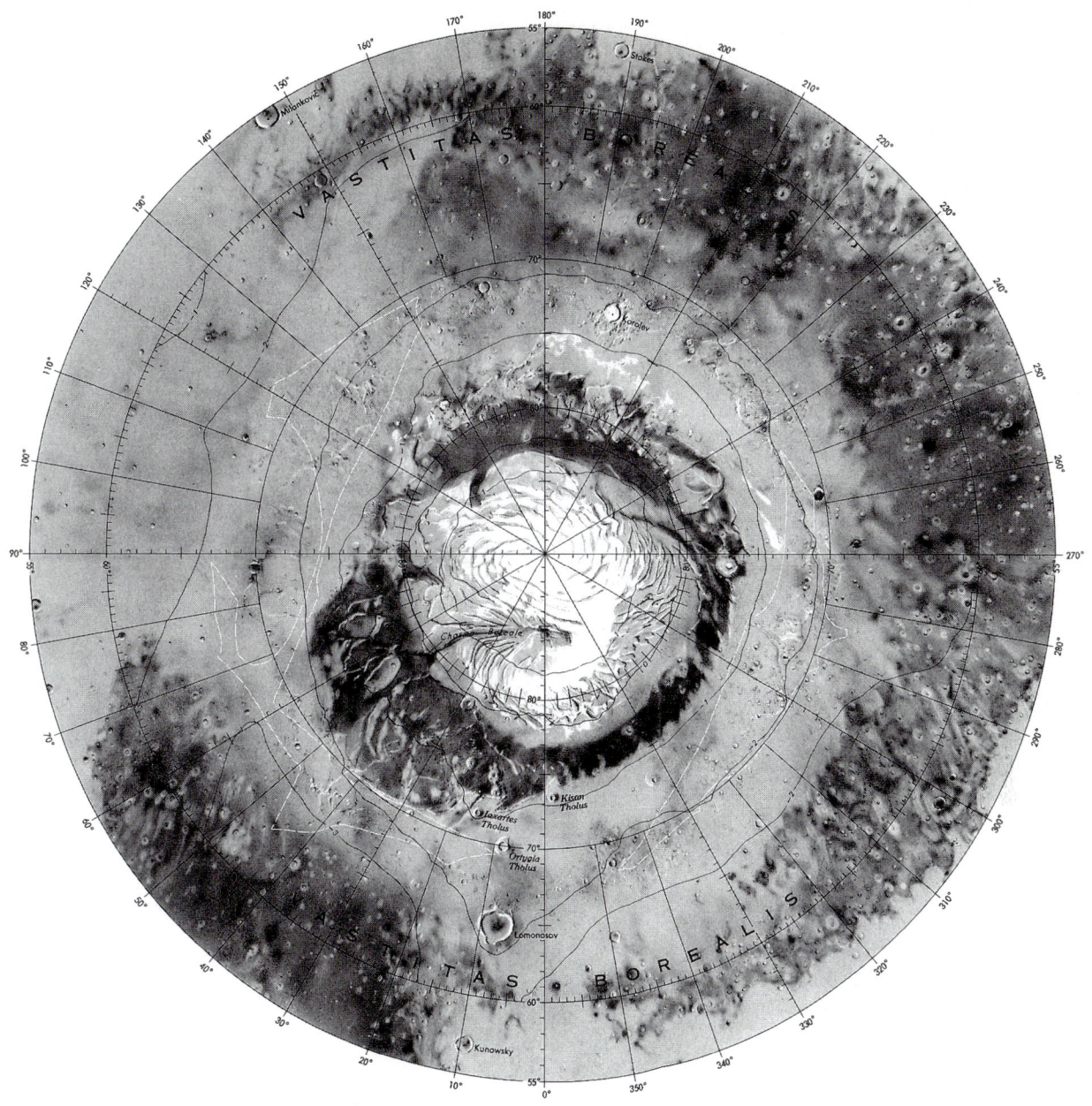

Figure 7.5. Mars – north polar region. (Courtesy: NASA.)

8 THE MINOR PLANETS

The Minor Planets or Asteroids are small bodies, all below 1000 km in diameter. Most of them occupy the 'Main Belt', between the orbits of Mars and Jupiter. The first asteroid, Ceres, was discovered in 1801, and three others (Pallas, Juno and Vesta) were found between 1801 and 1807. These are known popularly as 'the Big Four', although in fact Juno is not even among the dozen largest asteroids. Of the entire swarm, only 3 Vesta is visible with the naked eye.

DISCOVERY

A systematic search for a planet orbiting between the paths of Mars and Jupiter was initiated in 1800 by six astronomers meeting at Lilienthal, where Johann Schröter had his observatory. They based their search on Bode's law (actually first described by J. Titius of Wittemberg, but popularized by J. E. Bode in 1772). The law may be summed up as follows:

Take the numbers 0, 3, 6, 12, 24, 48, 96, 192 and 384, each of which (apart from the first) is double its predecessor. Add 4 to each. Taking the Earth's distance from the Sun as 10 units, the remaining figures give the mean distances of the planets with reasonable accuracy out to as far as Saturn, the outermost planet known in 1772. Uranus, discovered in 1781, fits in well. Neptune and Pluto do not, but these planets were not discovered until many years later. Bode's law is given in Table 8.1.

It is now believed that the 'law' is nothing more than a coincidence; it fails completely for Neptune, the third most massive planet in the Solar System, and also for Pluto which admittedly does not appear to be worthy of true planetary status. However, it was enough for the 'Celestial Police' to begin a hunt for a planet corresponding to the Bode number 28. Schröter became President of the association; the secretary was the Hungarian astronomer Baron Franz Xavier von Zach. They very logically concentrated their searches in the region of the ecliptic.

Ceres was discovered on 1 January 1801, the first day of the new century, by G. Piazzi from Palermo. (He was working on a new star catalogue and was not a member of

Table 8.1. Bode's law.

Planet	Distance by Bode's law	Actual distance
Mercury	4	3.9
Venus	7	7.2
Earth	10	10.0
Mars	16	15.2
—	28	—
Jupiter	52	52.0
Saturn	100	95.4
Uranus	196	191.8
Neptune	—	300.7
Pluto	388	394.6

the 'Police', although he joined later.) Ceres was found to have a Bode distance of 27.7. Pallas, Juno and Vesta were identified by the time that the 'Police' disbanded in 1813; the fifth, Astræa, was not found until 1845, as the result of painstaking searches by a German amateur, K. Hencke. Data for the first four asteroids are given in Table 8.2.

Since 1847 no year has passed without the discovery of at least one new asteroid. By 1868, 100 were known. by 1921 the number had risen to 1000; by 1946, to 2000; by 1985, to 3000; and by 1999 to over 10 000. These are the 'numbered' asteroids; many others – around 50 000 – have been seen, but have not been observed sufficiently for orbits to be calculated. The total number of asteroids must be at least 70 000, but the combined mass is only about 0.04% of the mass of the Earth or 3% of the mass of the Moon. Ceres is the giant of the asteroid swarm; all the rest of the main belt asteroids make up only about 1.5 times the mass of Ceres, and Ceres, Pallas and Vesta combined make up 55% of the total mass of the asteroid swarm making up the main belt.

The first asteroid to be discovered photographically was 323 Brucia, on 20 December 1891, by Max Wolf. Systematic searches are now being made with dedicated telescopes, the 0.91 m Spacewatch Telescope and the 1 m

Table 8.2. The first four asteroids.

Asteroid number and name	Diameter (km)	~Mass (10^{15} kg)	Rotation period (h)	Orbital period (years)	Type
1 Ceres	960×932	870 000	9.075	4.60	C
2 Pallas	$571 \times 525 \times 482$	318 000	7.811	4.61	U
3 Juno	288×230	20 000	7.210	4.36	S
4 Vesta	525	300 000	5.342	3.63	V

Linear Telescope. The Kleť Observatory, in the Czech Republic, is concerned entirely with asteroids.

NOMENCLATURE

When an asteroid is found, it is given a provisional designation, such as 1999 CE. The first four characters give the year of discovery; the fifth, the time of year reckoned in two-week periods in an alphabetical sequence – A the first two weeks in January, B the second two weeks in January, C the first two weeks in February and so on (I is omitted); the fifth character indicates the order of discovery. Thus 1999 CE indicates that the asteroid was the fifth to be discovered during the second half of February. If a month has over 24 discoveries, a subscript is added; thus the 25th asteroid to be found during the first half of February 1999 is 1999 CA$_1$. When a reliable orbit has been worked out, the asteroid is assigned a number. Data for the first hundred known asteroids are given in Table 8.3 and a selected list of data for other interesting asteroids is given in Table 8.4.

The discoverer of an asteroid is entitled to suggest a name, and this recommendation is almost always accepted by the International Astronomical Union. The first asteroid to be named was 1 Ceres, discovered by Piazzi from Palermo and named by him after the patron goddess of Sicily. All early names were mythological, but when the supply of deities became exhausted more 'modern' names were introduced, some of them decidedly bizarre; the first was that of 125 Liberatrix, which seems to have been given by the discoverer, P. M. Henry of Paris, in honour of Joan of Arc. Asteroids may be named after countries (136 Austria, 1125 China, 1197 Rhodesia) and people (1123 Shapleya, after Dr. Harlow Shapley; 1462 Zamenhof, after the inventor of Esperanto; 1486 Marilyn, after the daughter of Paul Herget, Director of the Cincinnati Observatory.

(I feel very honoured that 2602 Moore, discovered by E. Bowell from Flagstaff in Arizona, was named by him after me!)

The first asteroid to be named after an electronic calculator is 1625 The NORC – the Naval Ordnance Research Calculator at Dahlgre, Virginia. Plants are represented (978 Petunia, 973 Aralia), as are universities (694 Ekard – 'Drake' spelled backwards), musical plays (1047 Geisha), foods (518 Halawe – the discoverer, R. S. Dugan, was particularly fond of the Arabian sweet *halawe*), social clubs (274 Philagoria – a club in Vienna) and shipping lines (724 Hapag–Hamburg–Amerika line). One name has been expunged; 1267 was named Vladimirov in honour of a Soviet philanthropist, Anatoli Vladimirov, but was withdrawn when Vladimirov was exposed as a financial swindler. One name was auctioned: 250 Bettina, whose discoverer, Palisa, offered to sell his right of naming for £50. The offer was accepted by Baron Albert von Rothschild, who named the asteroid after his wife. Names of politicians and military leaders are in general not allowed, although we do have 1581 Abanderada – someone who carries a banner – associated with Eva Perón, wife of the former ruler of Argentina. Asteroid 1000 is, appropriately enough, named Piazzia.

(My own favourite names are 2309 Mr. Spock, named after a ginger cat which was itself named after the Vulcanian space-traveller in the fictional starship *Enterprise*, and 6042 Cheshirecat!)

719 Albert was discovered in 1911, and then lost. It was finally recovered in 2000, so that no numbered asteroids are now unaccounted for. Asteroid 330, Adalberta, never existed at all; it was recorded by Max Wolf in 1892, but was photographed only twice, and it was then found that the images were of two separate stars. One named asteroid,

Table 8.3. The first hundred asteroids. q = perihelion distance, Q = aphelion distance, P = orbital period, e = eccentricity, i = inclination, T = type, A = albedo, D = diameter (km), R = rotation period, m = magnitude.

No	Name	Discoverer	q (a.u.)	Q (a.u.)	P (years)	e	i (°)	T	A	D (km)	R (h)	m
1	Ceres	Piazzi, 1 Jan 1801	2.55	2.77	4.60	0.078	10.60	C	0.054	960 × 932	9.08	7.4
2	Pallas	Olbers, 28 Mar 1802	2.12	2.77	4.62	0.234	34.80	CU	0.074	571 × 525 × 482	7.81	8.0
3	Juno	Harding, 1 Sept 1804	1.98	2.67	4.36	0.258	13.00	S	0.151	288 × 230	7.21	8.7
4	Vesta	Olbers, 29 Mar 1807	2.15	2.37	3.63	0.090	7.14	V	0.229	525	5.34	6.5
5	Astræa	Hencke, 8 Dec 1845	2.08	2.57	4.13	0.190	5.36	S	0.140	120	16.81	9.8
6	Hebe	Hencke, 1 July 1847	1.94	2.43	3.77	0.202	14.79	S	0.164	204	7.28	8.3
7	Iris	Hind, 13 Aug 1847	1.84	2.39	5.51	0.229	5.51	S	0.154	208	7.14	7.8
8	Flora	Hind, 18 Oct 1847	1.86	2.20	3.27	0.156	5.89	S	0.144	162	12.35	8.7
9	Metis	Graham, 26 Apr 1848	2.10	2.39	3.69	0.121	5.59	S	0.139	158	5.08	9.1
10	Hygela	De Gasparis, 12 Apr 1849	2.76	3.13	5.54	0.120	3.84	C	0.041	430	17.50	10.2
11	Parthenope	De Gasparis, 11 May 1850	2.06	2.45	3.84	0.100	4.63	S	0.126	156	7.83	9.9
12	Victoria	Hind, 13 Sept 1850	1.82	2.33	3.56	0.220	8.38	S	0.114	136	8.65	9.9
13	Egeria	De Gasparis, 2 Nov 1850	2.36	2.58	4.14	0.087	16.50	C	0.041	244	7.05	10.8
14	Irene	Hind, 19 May 1851	2.16	2.59	4.16	0.016	9.11	S	0.162	150	9.35	9.6
15	Eunomia	De Gasparis, 29 July 1851	2.15	2.64	4.30	0.185	11.76	S	0.155	260	6.08	8.5
16	Psyche	De Gasparis, 17 Mar 1851	2.53	2.92	5.00	0.133	3.09	M	0.093	248	4.20	9.9
17	Thetis	Luther, 17 Apr 1852	2.13	2.47	3.88	0.138	5.60	S	0.103	98	12.28	10.7
18	Melpomene	Hind, 14 June 1852	1.80	2.30	3.48	0.217	10.14	S	0.144	162	11.57	8.9
19	Fortuna	Hind, 22 Aug 1852	2.05	2.44	3.82	0.159	1.57	C	0.032	198	7.45	10.1
20	Massalia	De Gasparis, 19 Sept 1852	2.06	2.41	3.74	0.145	0.70	S	0.164	134	8.09	9.2
21	Lutetia	Goldschmidt, 15 Nov 1852	2.04	2.44	3.80	0.161	3.07	M	0.093	108	8.17	10.5
22	Calliope	Hind, 16 Nov 1852	2.62	2.91	4.96	0.098	13.70	M	0.130	174	4.15	10.6
23	Thalia	Hind, 15 Dec 1852	2.02	2.63	4.26	0.231	10.15	S	0.164	116	12.31	10.1
24	Themis	De Gasparis, 5 Apr 1853	2.71	3.13	5.54	0.134	0.76	C	0.030	228	8.38	11.8
25	Phocæa	Chacornac, 6 Apr 1853	1.79	2.40	3.72	0.254	21.59	S	0.184	72	9.95	10.3
26	Proserpina	Luther, 5 May 1853	2.41	2.67	2.41	0.089	3.56	S	±0.2	88	12.00	11.3
27	Euterpe	Hind, 8 Nov 1853	1.95	2.35	3.60	0.171	1.59	S	0.147	116	8.50	9.9
28	Bellona	Luther, 1 Mar 1854	2.37	2.78	4.64	0.148	9.40	S	0.132	124	15.70	10.8
29	Amphitrite	Marth, 1 Mar 1854	2.37	2.56	4.08	0.073	6.09	S	0.140	200	5.39	10.0
30	Urania	Hind, 22 July 1854	2.07	2.37	3.63	0.126	2.09	S	0.144	94	13.69	10.6
31	Euphrosyne	Ferguson, 1 Sept 1854	2.43	3.15	5.58	0.228	26.35	C	0.030	270	5.53	11.0
32	Pomona	Goldschmidt, 26 Oct 1854	2.37	2.59	4.16	0.084	5.52	S	0.140	92	9.44	11.3
33	Polyhymnia	Chacornac, 28 Oct 1854	1.89	2.86	4.84	0.341	1.89	S	±0.2	62	18.60	11.0
34	Circe	Chacornac, 6 Apr 1855	2.40	2.69	4.40	0.107	5.49	C	0.039	112	15.00	12.3
35	Leucothea	Luther, 19 Apr 1855	2.34	3.00	5.20	0.220	8.04	C	?	±67	?	12.4
36	Atalantë	Goldschmidt, 5 Oct 1855	1.91	2.74	4.55	0.305	18.49	C	0.024	120	9.93	11.6
37	Fides	Luther, 5 Oct 1855	2.17	2.64	4.29	0.177	3.07	S	0.186	96	7.33	10.7
38	Leda	Chacornac, 12 Jan 1856	2.33	2.74	4.54	0.151	6.95	CU	±0.2	±118	13.00	12.2
39	Lætitia	Chacornac, 8 Feb 1856	2.45	2.77	4.60	0.114	10.38	S	0.169	156	5.13	10.3
40	Harmonia	Goldschmidt, 31 Mar 1856	2.16	2.67	3.41	0.047	4.26	S	0.123	116	9.14	10.5
41	Daphne	Goldschmidt, 22 May 1856	2.03	2.77	4.61	0.269	15.79	C	0.056	204	5.99	10.3
42	Isis	Pogson, 23 May 1856	1.89	2.43	3.81	0.226	8.54	S	0.125	94	13.59	10.2
43	Arladne	Pogson, 15 Apr 1857	1.83	2.20	3.27	0.169	3.47	S	0.113	84	5.75	10.2
44	Nysa	Goldschmidt, 27 May 1857	2.06	2.42	3.77	0.151	3.71	E	0.377	68	6.42	9.7
45	Eugenia	Goldschmidt, 27 June 1857	2.49	2.72	4.49	0.083	6.60	FC	0.030	216	5.70	11.5
46	Hestia	Pogson, 16 Aug 1857	2.09	2.52	4.01	0.172	2.33	F	0.028	164	21.04	11.1
47	Aglaia	Luther, 15 Sept 1857	2.50	2.88	4.89	0.133	4.98	C	0.027	158	13	12.1
48	Doris	Goldschmidt, 19 Sept 1857	2.90	3.11	5.49	0.068	6.54	C	±0.03	246	11.90	11.9
49	Pales	Goldschmidt, 19 Sept 1857	2.35	3.08	5.41	0.236	3.18	C	?	176	10.4	11.4
50	Virginia	Ferguson, 4 Oct 1857	1.88	2.65	4.31	0.289	2.84	C	?	±88	24	11.6
51	Nemausa	Laurent, 22 Jan 1858	2.21	2.37	3.63	0.066	9.97	CU	0.050	152	7.8	10.8
52	Europa	Goldschmidt, 4 Feb 1858	2.79	3.11	5.48	0.103	7.46	C	0.035	292	5.6	11.1
53	Calypso	Luther, 4 Apr 1858	2.09	2.62	4.24	0.203	5.16	C	?	110	26.6	11.7
54	Alexandra	Goldschmidt, 10 Sept 1858	2.12	2.71	4.47	0.197	11.78	C	0.030	176	7.0	10.9
55	Pandora	Searle, 10 Sept 1858	2.36	2.76	4.59	0.145	7.20	CMEU	?	112	4.8	11.6
56	Melete	Goldschmidt, 9 Sept 1859	2.00	2.60	4.19	0.232	8.09	P	0.026	144	16.0	11.5
57	Mnemosyne	Luther, 22 Sept 1859	2.78	3.15	5.59	0.118	15.12	S	0.140	116	12.3	12.2
58	Concordia	Luther, 24 Mar 1860	2.59	2.70	4.44	0.043	5.07	C	0.030	104	?	13.1
59	Elpis	Chacornac, 12 Sept 1860	2.40	2.71	4.47	0.117	8.64	C	0.05	104	13.7	11.4
60	Echo	Ferguson, 14 Sept 1860	1.96	2.40	3.70	0.183	3.60	S	0.154	52	26.2	11.5
61	Danaë	Goldschmidt, 9 Sept 1860	2.50	3.00	5.16	0.162	18.21	S	?	88	11.5	11.7
62	Erato	Förster, 14 Sept 1860	2.53	3.11	5.49	0.186	2.28	C	?	64	?	13.0
63	Ausonia	De Gasparis, 10 Feb 1861	2.09	2.40	3.71	0.126	5.78	S	0.128	92	8.8	10.2
64	Angelina	Tempel, 4 Mar 1861	2.35	2.69	4.40	0.125	1.31	E	0.342	60	8.8	11.3
65	Cybele	Tempel, 8 Mar 1861	3.07	3.43	6.36	0.105	3.55	CPF	0.022	308	6.1	11.8
66	Mala	Tuttle, 9 Apr 1861	2.19	2.65	4.30	0.174	3.05	C	0.029	92	?	12.5
67	Asia	Pogson, 17 Apr 1861	1.97	2.42	3.77	0.187	6.01	S	0.157	60	15.89	11.6
68	Leto	Luther, 29 Apr 1861	2.27	2.78	4.64	0.185	7.96	S	0.126	128	14.84	10.8
69	Hesperia	Schiaparelli, 29 Apr 1861	2.47	2.98	5.13	0.170	8.56	M	?	108	5.66	11.3
70	Panopæa	Goldschmidt, 5 May 1861	2.13	2.61	4.23	0.184	11.59	C	0.039	152	14.0	11.5
71	Niobe	Luther, 13 Aug 1861	2.27	2.75	4.57	0.175	23.29	S	0.140	106	28.8	11.3
72	Feronia	Safford, 29 May 1861	1.99	2.67	3.41	0.120	5.42	U	0.032	96	8.1	12.0
73	Clytia	Tuttle, 7 Apr 1862	2.55	2.66	4.35	0.042	2.38	?	?	±56	?	13.4
74	Galatea	Tempel, 29 Aug 1862	2.11	2.78	4.63	0.239	4.07	C	?	±108	9	12.0

Table 8.3. (Continued)

No	Name	Discoverer	q (a.u.)	Q (a.u.)	P (years)	e	i (°)	T	A	D (km)	R (h)	m
75	Eurydice	Peters, 22 Sept 1862	1.86	2.67	4.37	0.304	4.99	M	?	±48	8.9	11.1
76	Freia	D'Arrest, 21 Oct 1862	2.82	3.40	6.28	0.171	2.11	P	0.03	196	9.8	12.3
77	Frigga	Peters, 12 Nov 1862	2.31	2.67	4.36	0.133	2.43	M	0.113	66	9.0	12.1
78	Diana	Luther, 15 Mar 1863	2.09	2.62	4.25	0.205	8.66	C	?	144	7.2	10.9
79	Eurynome	Watson, 14 Sept 1863	1.97	2.44	3.28	0.193	4.62	S	0.137	80	5.8	10.8
80	Sappho	Pogson, 3 May 1864	1.84	2.30	3.48	0.200	8.67	SU	0.113	84	14.1	12.8
81	Terpsichore	Tempel, 30 Sept 1864	2.26	2.86	4.82	0.209	7.81	C	?	122	?	12.1
82	Alcmene	Luther, 27 Nov 1864	2.15	2.76	4.58	0.223	2.84	S	0.138	66	13.0	11.5
83	Beatrix	De Gasparis, 26 Apr 1865	2.23	2.43	3.79	0.084	4.98	C	0.030	118	10.2	12.0
84	Clio	Luther, 25 Aug 1865	1.80	2.36	3.63	0.236	9.32	C	0.037	88	?	11.2
85	Io	Peters, 19 Sept 1865	2.14	2.65	4.32	0.193	11.96	C	0.042	148	6.8	11.2
86	Semele	Tietjen, 4 Jan 1886	2.45	3.11	5.48	0.213	4.79	C	?	112	16.6	12.6
87	Sylvia	Pogson, 16 May 1866	3.19	3.48	6.50	0.083	10.87	P	?	282	5.2	12.6
88	Thisbe	Peters, 15 June 1866	2.31	2.77	4.61	0.163	5.21	C	0.045	210	6.0	10.7
89	Julia	Stéphan, 6 Aug 1866	2.09	2.55	4.08	0.180	16.12	S	0.086	168	11.4	10.3
90	Antiope	Luther, 1 Oct 1866	2.64	3.15	5.59	0.162	2.24	C	?	128	?	12.4
91	Ægina	Stéphan, 4 Nov 1866	2.32	2.59	4.17	0.106	2.12	C	0.031	104	6.0	12.2
92	Undina	Peters, 7 July 1867	2.91	3.20	5.71	0.087	9.88	U	±0.03	184	15.9	12.0
93	Minerva	Watson, 24 Aug 1867	2.37	2.76	4.56	0.141	8.56	C	0.039	168	6.0	11.5
94	Aurora	Watson, 6 Sept 1867	2.90	3.16	5.62	0.081	8.01	C	0.029	190	7.2	12.5
95	Arethusa	Luther, 23 Nov 1867	2.63	3.07	5.38	0.143	12.93	C	0.019	228	8.7	12.1
96	Ægle	Coggia, 17 Feb 1868	2.62	3.05	5.32	0.140	16.01	U	0.04	112	?	12.4
97	Clotho	Tempel, 17 Feb 1868	1.98	2.67	4.36	0.256	11.76	MP	0.121	108	35.0	10.5
98	Ianthe	Peters, 18 Apr 1868	2.18	2.68	4.40	0.188	15.56	C	?	106	?	12.6
99	Dike	Borrelly, 28 May 1868	2.14	2.66	4.35	0.195	13.88	C	?	80	30.0	12.1
100	Hekate	Watson, 11 July 1868	2.62	3.10	5.46	0.155	6.39	SU	?	84	10.0	12.1

Hermes, has been lost (1937 UB). It was discovered by K. Reinmuth, from Heidelberg, by-passing the Earth at only 800 000 km on 28 October 1937; it reached the eighth magnitude, and moved at 5° per hour, so that it crossed the sky in nine days. It has not been seen again and its recovery will be a matter of luck. It was, of course, well away from the main swarm.

In recent years asteroid-hunting has become a popular pastime, among amateurs as well as professionals. For example, the British amateur Brian Manning now has five numbered asteroids to his credit. Several 19th-century astronomers were particularly prolific; thus K. Reinmuth discovered 246 asteroids, Max Wolf 232 and J. Palisa 121.

ASTEROID ORBITS

Asteroid orbits are of several definite types.

- *Main Belt*. These orbits lie between those of Mars and Jupiter.
- *Near-Earth Asteroids (NEA)*
  (a) Aten class; average distance from the Sun less than 1 a.u., although some may cross the Earth's orbit. All Atens are very small, and no doubt there are a great many of them.
  (b) Apollo class; average distance from the Sun over 1 a.u., but orbits do cross that of the Earth.
  (c) Amor class; orbits cross that of Mars but not that of the Earth. Asteroid 251, Lick, has an orbit lying entirely between those of the Earth and Mars; its period is 1.63 years and the orbital inclination is over 39°.
- *Trojans*. These move in orbits which are the same as that of Jupiter, although they are on average either 60° ahead of (L4) or 60° behind (L5) of Jupiter, and are in no danger of being engulfed. There is one known Mars Trojan, 5261 Eureka, discovered on 20 June 1990 by H. E. Holt and D. H. Levy. It is very small, with a diameter of no more than 3 km. No doubt other Martian Trojans exist, but their faintness must make them very difficult objects. No Trojan type asteroids have been found in the orbits of Mercury, Venus or the Earth.
- *Centaurs*. These lie well beyond the main belt and beyond the orbit of Jupiter; their average distances from the Sun lie between the orbits of Jupiter and Neptune.
- *Kuiper Belt Objects (KBOs)*. Orbits close to or beyond that of Neptune. Most are small, with diameters of a few hundred kilometres, but it is now widely believed that Pluto is a KBO rather than a true planet.

Table 8.4. Some further asteroid data (symbols as Table 8.3. The following asteroids are of note in some way or other – either for orbital peculiarities, physical characteristics or because they are the senior members of 'families' – or even because they have very fast or very slow rotation periods, for example.

No	Name	Discoverer	q (a.u.)	Q (a.u.)	P (years)	e	i (°)	T	D (km)	R (h)	m
128	Nemesis	Watson, 25 Nov 1872	2.40	2.75	4.56	0.126	6.25	CEU	116	39.0	11.6
132	Æthra	Watson, 13 June 1873	1.61	2.61	4.22	0.384	25.07	SU	38	?	11.9
135	Hertha	Peters, 18 Feb 1874	1.93	2.42	3.78	0.204	2.29	M	80	8.4	10.5
153	Hilda	Palisa, 2 Nov 1875	3.40	3.97	7.91	0.142	7.83	P	222	?	13.3
158	Koronis	Knorre, 4 Jan 1876	2.71	2.87	4.86	0.053	1.00	S	36	14.2	14.0
221	Eos	Palisa, 18 June 1882	2.72	3.01	5.23	0.098	10.87	CEU	112	10.4	12.4
243	Ida	Palisa, 29 Sept 1884	2.74	2.86	4.84	0.042	1.14	S	58 × 23	4.6	14.6
253	Mathilde	Palisa, 12 Nov 1885	1.94	3.35	5.61	0.262	6.70	C	66 × 48 × 46	417.7	10.0
279	Thule	Palisa, 25 Oct 1888	4.22	4.27	8.23	0.011	2.33	D	130	7.4	15.4
288	Glauke	Luther, 20 Feb 1890	2.18	2.76	4.58	0.210	4.33	S	30	1500	13.2
311	Claudia	Charlotte, 11 Jan 1891	2.89	2.90	4.94	0.003	3.23	S	28	11.5	14.9
324	Bamberga	Palisa, 23 Feb 1892	1.77	2.68	4.39	0.341	11.14	C	252	29.4	9.2
349	Dembowska	Charlotte, 9 Dec 1892	2.66	2.92	5.00	0.092	8.26	R	164	4.7	10.6
423	Diotima	Charlotte, 7 Dec 1896	2.97	3.07	5.38	0.031	11.22	C	208	4.6	12.4
434	Hungaria	Wolf, 11 Sept 1898	1.80	1.94	2.71	0.074	22.51	E	11	26.5	13.5
444	Gyptis	Coggia, 31 Mar 1899	2.29	2.77	4.61	0.173	10.26	CS	166	6.2	11.8
451	Patientia	Charlotte, 4 Dec 1899	2.85	3.06	5.36	0.070	15.23	C	280	9.7	11.9
511	Davida	Dugan, 30 May 1903	2.61	3.18	5.66	0.177	15.93	C	324	5.2	10.5
532	Herculina	Wolf, 20 Apr 1904	2.28	2.77	4.62	0.176	16.35	S	220	9.4	10.7
704	Interamnia	Cerulli, 2 Oct 1910	2.61	3.06	5.36	0.148	17.30	F	338	8.7	11.0
747	Winchester	Metcalf, 7 Mar 1913	1.97	3.00	5.19	0.342	18.17	C	204	9.4	10.7
878	Mildred	Nicholson, 6 Sept 1916	1.82	2.36	3.63	0.231	2.02	?	26	?	15.0
903	Nealley	Palisa, 13 Sept 1918	3.16	3.24	5.84	0.025	11.69	?	42	?	15.1
944	Hidalgo	Baade, 31 Oct 1920	2.01	5.84	14.15	0.656	42.40	MEU	28	10.0	14.9
951	Gaspra	Neujmin, 30 July 1916	1.82	2.20	3.28	0.173	4.10	S	19 × 12 × 11	20.0	14.1
1269	Rollandia	Neujmin, 20 Sept 1930	2.76	3.90	7.69	0.097	2.76	D	110	55	15.3
1300	Marcelle	Reisa, 10 Feb 1934	2.76	2.78	4.64	0.009	9.54	?	21	?	15.9
1578	Kirkwood	Cameron, 10 Jan 1951	3.02	3.94	7.82	0.231	0.81	D	56	?	15.8
2602	Moore	Bowell, 24 Jan 1982	2.14	2.38	3.69	0.10	6.6	UX	14	?	17.0
9969	Braille	Helin and Laurence, 27 May 1992	1.32	3.35	3.58	0.43	—	V	2.2 × 0.6	6	—

There are also some bodies which have very exceptional orbits. Such is 1996 TL$_{66}$, discovered by Jane Luu and her colleagues in October 1966. Assuming that it has a darkish surface, it must be around 500 km in diameter. When found it seems to have been close to its perihelion, but the orbit is highly eccentric and the aphelion distance may be as much as 170 a.u. It is possible that 1996 PW, discovered by E. Helin on 9 August 1996, may be even more extreme, with a period of 5800 years and an aphelion distance of 645 a.u. – almost 100 000 million km; the diameter is probably about 10 km.

These curious bodies may well bridge the gap between the Kuiper Belt and the Oort Cloud of comets, and indeed the distinction between the various small members of the Solar System seems to be blurred. 1996 TL$_{66}$ and 1996 PW look like asteroids, but their orbits are cometary. On the other hand, Comet Wilson–Harrington, first seen in 1951 and which had a perceptible tail, was lost for many years and recovered in 1979, this time in the guise of an asteroid; it has been given an asteroid number, 4015, and now shows no sign of cometary activity. Comet 133P/Elst–Pizarro, discovered in August 1996, also has a tail, but moves entirely within the main asteroid belt and has been given an asteroid number, 7968. The NEA asteroid 3200 Phæthon, discovered on 11 December 1983 from the IRAS satellite, has an orbit so like that of the Geminid meteor stream that it could be the 'parent' of that stream – in which case Phæthon is an ex-comet which has lost all its volatiles. Even more significant perhaps is 2060 Chiron, which was the first Centaur to be discovered and which moves mainly between the orbits of Saturn and Uranus. It is around 180 km in diameter, and though it looked asteroidal when discovered it has since shown comet-like activity. It appears to be much too large to be classed as a comet, but its true nature remains unclear, and it may not be unique. All in all, the relationships between various types of objects seem to be much more complicated than was thought until very recently.

TYPES OF ASTEROIDS

Asteroids are divided into various types according to their physical and surface characteristics. The main classes are as follows.

- *C (carbonaceous).* These are the most numerous, increasing in number from 10% at a distance of 2.2 a.u. up to 80% at 3 a.u. They are of low albedo – often below 5%, darker than coal – and have flat, featureless spectra, similar to those of carbonaceous chondrites. Ceres, the largest asteroid, is of type C.

- *S (silicaceous).* These are most numerous in the inner part of the main zone, making up 60% of the total at 2.2 a.u., but only 15% at 3 a.u. Their albedoes are in the range 15–25%, and they seem to resemble the metal-bearing meteorites known as chrondites; generally they are *reddish.* Prominent S-type asteroids include 3 Juno and 5 Astræa.

- *M (metallic).* Moderate albedoes, and may be the metal-rich cores of larger 'parent asteroids' which have been broken up by collision. 16 Psyche is almost pure nickel–iron alloy in composition.

- *E (enstatite).* Relatively rare; high albedoes, sometimes over 40%. They may resemble some types of chondrites, in which enstatite ($MgSiO_3$) is a major constituent. 214 Aschera, 434 Hungaria and 1025 Riema are of this type.

- *D.* Low albedo; reddish. Their surfaces seem to be 90% clays, with magnetite and carbon-rich substances. Most of the D-type asteroids are remote from the Sun, including many Trojans, but there are also a few in the main belt, including 336 Lacadiera and 361 Bononia. The largest D-type asteroid is 704 Interamnia.

- *A.* Almost pure olivine. The best examples are 246 Asporina, 329 Nenetta and 446 Æternitas.

- *P.* Dark and reddish; not unlike type D.

- *Q.* Sometimes used for NEA asteroids; in composition they seem to resemble chrondites (see the chapter on Meteorites).

- *V.* Igneous rock surfaces. Very rare; 4 Vesta is the only large example.

- *U.* Asteroids which are regarded as unclassifiable. For example, 2201 Oljato, discovered in 1947 by H. Giclas and recovered in 1979 by E. Helin, has a spectrum unlike any other in the Solar System. Its diameter is no more than 3 km; it shows no trace of cometary activity.

Other classes sometimes used are B (similar to C, but slightly brighter and more neutral in colour), F (neutral in colour), G (C-like, but brighter), T (reddish, intermediate between D and S) and R (like S, but with greater indications of olivine in their spectra).

There was also the extraordinary object discovered on 5 December 1991 by J. V. Scotti, who was carrying out a routine patrol with the 0m.9 Spacewatch telescope. It passed within 460 000 km of the Earth, and varied in brightness; the estimated diameter was a mere 6 m. The orbital inclination was $0°.4$, and the period only slightly longer than that of the Earth. It was given an asteroidal designation, 1991 VG, but is widely believed to have been a piece of man-made satellite débris rather than an asteroid. It has not been seen again, and is unlikely to be recovered.

MAIN-BELT ASTEROIDS

Ceres is the giant of the swarm; it is the only asteroid more than 600 km in diameter, and of the rest only 2 Pallas, 4 Vesta and 10 Hygeia are over 400 km across. The largest and brightest asteroids are listed in Table 8.5. Several main-belt asteroids have been contacted by radar – initially 16 Psyche and 33 Klotho, with the Arecibo radio telescope in 1981 – and several, including 950 Gaspra, 243 Ida and 55 Mathilde, have been surveyed from close range by spacecraft from 1991.

It is generally assumed that no large planet could form in this region of the Solar System because of the powerful disruptive influence of Jupiter. Gaps in the main zone were predicted by D. Kirkwood in 1857, and confirmed in 1866; they too are due to the gravitational influence of Jupiter. For example, the Hilda asteroids have periods of approximately two-thirds that of Jupiter, so that the perturbations are cumulative. There are distinct 'families' of asteroids, whose members probably have a common origin in the disruption of a larger body. They are sometimes

Table 8.5. (a) The largest asteroids. (This does not include Centaurs and trans-Neptunians.) (b) The brightest asteroids.

(a)

No	Name	Mean diameter (km)	Type
1	Ceres	960 × 932	C
2	Pallas	571 × 525 × 482	CU
4	Vesta	525	V
10	Hygeia	430	C
704	Interamnia	338	D
511	Davida	324	C
65	Cybele	308	P
52	Europa	292	C
87	Sylvia	282	P
451	Patientia	280	C
31	Euphrosyne	260	S
15	Eunomia	260	S
324	Bamberga	252	C
3	Juno	288 × 230	S
16	Psyche	248	M
48	Doris	246	C
13	Egeria	244	C
624	Hektor	232	D
24	Themis	228	C
95	Arethusa	228	C

(b)

No	Name	Mean opposition magnitude
4	Vesta	6.4
1	Ceres	7.3
2	Pallas	7.5
7	Iris	7.8
3	Juno	8.1
6	Hebe	8.3
15	Eunomia	8.5
8	Flora	8.7
18	Melpomene	8.9

(433 Eros can reach magnitude 8.3 at closest approach)

called Hirayama families, because they were first recognized by K. Hirayama in 1918. The main families are listed in Table 8.6. Outside the most remote family (Hilda) is 279 Thule, of type D; its diameter is 130 km and its period

8.8 years. It may be said to mark the outpost of the main swarm, and could be one of several at about the same distance, but Thule itself is large by asteroidal standards, and smaller members would certainly be faint and difficult to identify.

The orbits of main-belt asteroids show a wide range of inclinations; for example 52°.0 for 1580 Betulia, only 0°.014 for 1383 Limburgia. Pallas, the second largest asteroid, has an inclination of almost 35°. Some orbits are almost circular; thus for 311 Claudia the eccentricity is only 0.0031, and for 903 Neally it is 0.0039. Mutual perturbations are measurable; thus the mass of Vesta was determined largely by its effects upon the movements of 197 Arete, an S-type asteroid 42 km in diameter and therefore very much smaller and less massive than Vesta.

Ordinary telescopes show asteroids as nothing more than dots of light, but surface details have now been mapped on some of them. Ceres, imaged in January 1998 with the NASA Infra-red Facility 3 m telescope on Mauna Kea in Hawaii, proved to be slightly football-shaped rather than perfectly spherical, with a more varied surface than had been expected. Vesta has been mapped with the Hubble Space Telescope, and has proved to be the most geologically diverse of the main asteroids. There are ancient lava flows; ground-based spectroscopy had already shown that there were basaltic regions, and it follows that the asteroid once had a molten interior. There are two distinct hemispheres, containing different types of solidified lava, and there is one huge impact crater, over 300 km across, with a central peak almost 13 km high. One side of the asteroid has what may be called quenched lava flows, while the other has characteristics of molten rock that cooled and solidified underground, being subsequently exposed by impacts. It has been suggested that Vesta may be the source of the eucrite meteorites which have fallen on Earth, although there is certainly no proof. It has also been suggested, again without proof, that the 3 km asteroid 9969 Braille may be a broken-off part of Vesta; on 29 July 1999 Braille was imaged from close range by the probe Deep Space 1, which had been launched on 24 October 1998.

Vesta has a rotation period of 5.3 h. Other main-belt asteroids spin at different speeds; the period is only 2 h

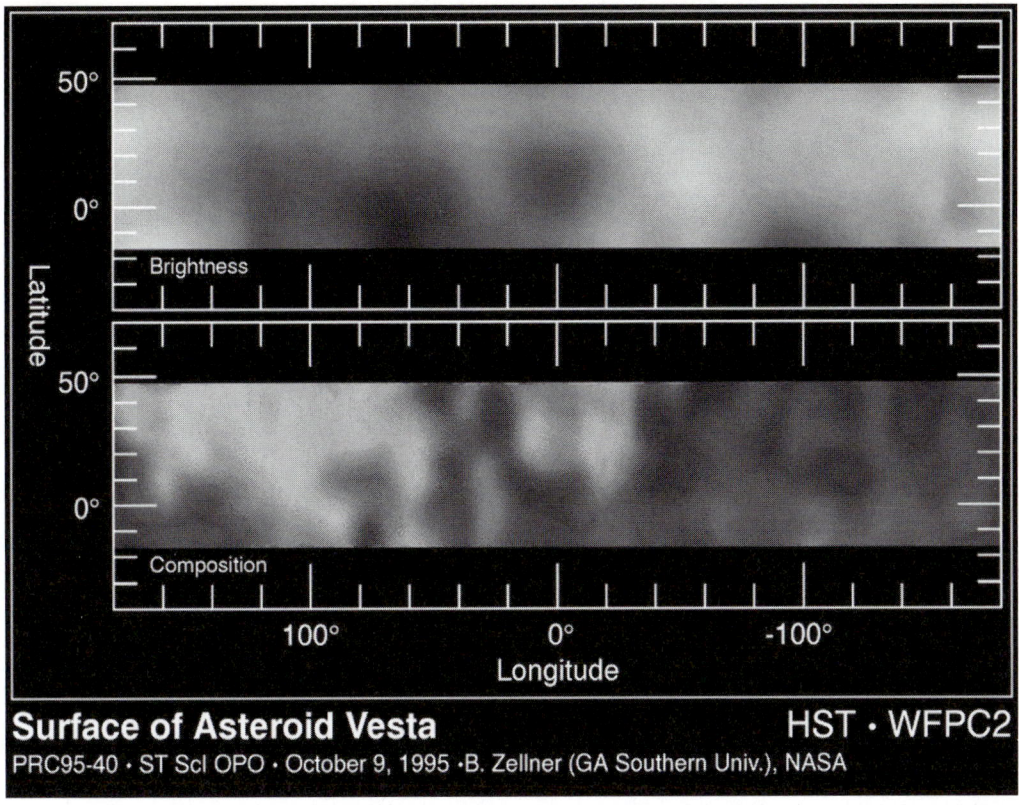

Figure 8.1. Vesta. (Courtesy: NASA.)

52 min for 321 Florentina, as much as 1500 h for 288 Glauke (although it has been suggested that Glauke may be a binary asteroid rather than a single slow-rotation body). Rotation rates for other asteroids may be obtained from their variations in brightness; thus 1864 Daedalus changes by 0.9 magnitude over a period of 8 h 34 min, while 1226 Crocus also varies by 0.9 magnitude, although in the longer period of 3.07 days.

The first three main-belt asteroids to be surveyed by space-craft were 951 Gaspra (13 November 1991), 243 Ida (28 August 1993) and 253 Mathilde (27 June 1997) – the first two by the Galileo probe en route for Jupiter and the third by the NEAR (Shoemaker) probe making for Eros. They proved to be very different from each other. Gaspra, imaged from a range of 16 000 km, proved to be wedge-shaped (not unlike a distorted potato) with a darkish, rocky surface pitted with craters. There were

also indications of three areas from which pieces had been broken off, so that Gaspra may well be the survivor from a series of major collisions; it is small, measuring 19 km × 12 km × 11 km. Craters on it have been named after famous spas such as Bath, Aix and Rotorua. Surprisingly, there were magnetic effects strong enough to create a 'bubble' in the solar wind. Some named features are listed in Table 8.7.

Ida is larger than Gaspra (58 km × 23 km) and is heavily cratered; many of these are larger than craters on Gaspra. Ida is a member of the Koronis family, with a rotation period of 4.6 h; like Gaspra it is of type S and is reddish, so that it is presumably composed of a mixture of pyroxene, olivine and iron minerals. It was found to have a tiny satellite, now named Dactyl, measuring 1.2 km × 1.4 km × 1.6 km, orbiting at a distance of about 100 km from Ida's centre. Strangely, it and Ida seem to be composed of decidedly

Table 8.6. The main asteroid families.

Family	Distance from the Sun (a.u.)	
Hungaria	1.8–2.0	Just inside the main belt. Most are of class E, with some of class S. Inclinations exceed 16°. Members include 434 Hungaria, 1103 Sequoia and 1025 Riema. Hungaria is 11.4 km in diameter; most of the rest are much smaller.
Flora	2.2	Most populous family; well over 150 members. Most are reddish. 8 Flora itself is of type S, 162 km in diameter and is virtually spherical.
Maria	2.25	Separated from the main belt by inclinations of around 15°. Most members are of type S, including 170 Maria itself, which is 40 km in diameter.
Phocæa	2.4	Rather ill-defined. 25 Phocæa is of type S, with a diameter of 72 km.
Koronis	2.8–2.9	The only rich family with low inclination and low eccentricity; about 60 members. Almost all are of type S. 158 Koronis has a diameter of 36 km.
Budrosa	2.9	Only 6 members; inclinations about 6°. 338 Budrosa, diameter 80 km, is of type M; most of the others have to be classified as U.
Eos	3.0	The most compact of all the families; high inclinations, 11° for Eos itself. Almost all are of type S. The largest members are 221 Eos (diameter 98 km), 579 Sidonia (80 km) and 639 Latona (68 km).
Themis	3.1	Over 100 members; mainly of type C. Inclinations and eccentricities are low. 24 Themis is 228 km in diameter – large by asteroidal standards. Other members include 90 Antiope (128 km), 222 Lucia (55 km) and 171 Ophelia (113 km).
Hilda	4.0	In the outer region; over 30 members mainly of classes P, C and D. 153 Hilda is 222 km in diameter, and of type P.

different materials. The largest crater is Afon, named after a Russian cave.

Mathilde proved to be surprising. It measured 66 km × 4.8 km × 46 km, but is very irregular in outline and is cratered; the largest crater is 30 km across. The albedo is only 3%, so that Mathilde is as black as charcoal; the surface may consist of carbon-rich material which has not been altered by planet-building processes. The mean density is only 1.3 g cm^{-2}. Fittingly, the craters are to be named after famous coal mines! The rotation period is very long indeed: 417.7 h, over 17 days. Clearly Mathilde has had a tortured history; it was even claimed that 'there are more huge craters than there is asteroid'.

In 1998 astronomers using the Canada–France–Hawaii 3.58 m telescope on Mauna Kea discovered a satellite of asteroid 45 Eugenia; the team was led by W. Merline. The satellite orbits Eugenia at a distance of 1200 km; the orbital inclination is 45 degrees. The satellite is 13 km in diameter. A satellite of asteroid 3671 Dionysus has been suspected, but not yet confirmed. In 1999 astronomers using the 3.6 m reflector at La Silla (Chile) found that asteroid 216 Kleopatra is a bifurcated shape with two lobes of similar size, 217 × 93 km; the separation was given as 0″.125. The shape is strikingly like that of a dog's bone!

Some 'target' asteroids for present and future space missions are listed in Table 8.8.

Table 8.7. Selected list of features on Gaspra.

Crater	Lat.	Long. W	Diameter (km)
Aix	47.9N	160.3	6
Bath	13.4N	122.0	10
Charax	8.6N	0.0	11
Lisdoonvara	16.5N	358.1	10
Ramlösa	15.0N	4.9	10
Rotorua	18.8N	30.7	6

There are three named regions: Dunne (15.0N, 15.0W), Neujmin (2.0N, 80.0W) and Yeates (65.0N, 75.0W). These are named after persons associated with Gaspra. Grigori Neujmin discovered the asteroid, on 30 July 1916; James Dunne was the planner of the Galileo mission and Clatne Yeates was project manager of the mission

ASTEROIDS CLOSER IN THAN THE MAIN BELT

Most of the asteroids which invade the inner reaches of the Solar System are very small indeed. Ganymed, much the largest of them, is only 40 km in diameter[1]. The absolute magnitude of an asteroid is the apparent magnitude that it would have if seen from unit distance (1 a.u.) at full phase. Table 8.9 (page 140) links absolute magnitude (H) with diameter, but the values are found to be very uncertain, because we do not have precise values for the asteroid albedoes.

[1] Not to be confused with Ganymede, the largest satellite of Jupiter.

Table 8.10. Selected list of Amor asteroids (symbols as Table 8.3, H = absolute magnitude).

No	Name	q (a.u.)	Q (a.u.)	P (years)	e	i (°)	H	D (km)	m	T	R (h)	Discoverer
433	Eros	1.133	1.783	1.76	0.223	10.83	11.1	33 × 13 × 13	11.9	S	5.270	Witt, 1898. Close in 1931, 1975
719	Albert	1.191	2.637	4.28	0.550	11.31	15.8	2.6	16.8	?	?	Palisa, 1911. Lost
887	Alinda	1.087	3.884	3.97	0.558	9.25	13.8	4	15.0	S	74	Wolf, 1918
1036	Ganymed	1.227	4.090	4.35	0.537	26.45	9.4	40	10.4	S	10.3	Baade, 1924
1221	Amor	1.083	2.755	2.66	0.435	11.90	17.7	1	19.1	?	?	Delporte, 1932. Well placed every 8 years
1580	Betulia	1.119	3.270	3.26	0.049	52.01	14.5	1	15.0	U	6.1	Johnson, 1950
1627	Ivar	1.124	2.603	2.54	0.397	8.44	13.2	6	14.2	S	?	Hertzsprung, 1929
1915	Quetzalcoatl	1.081	3.994	4.03	0.577	20.50	19.0	0.4	20.1	SU	4.9	Wilson, 1953. Recovered in 1974
1916	Boreas	1.250	3.295	3.43	0.450	12.84	14.9	3	16.1	S	?	Arend, 1953. Recovered in 1974
1917	Cuyo	1.067	3.235	3.15	0.505	23.99	13.9	3	16.5	?	?	Cesco and Samuel, 1968
1943	Anteros	1.064	1.796	1.71	0.256	8.70	15.7	4	16.5	S	?	Gibson, 1973
1951	Lick	1.304	1.390	1.63	0.062	39.09	15.3	2.2	17.2	?	?	Wirtanen, 1949
1980	Tezcatlipoca	1.085	2.334	2.23	0.365	26.85	13.9	6.2	15.1	U	?	Wirtanen, 1980
2059	Babuquivari	1.256	4.044	4.31	0.526	10.99	15.8	3.8	16.0	?	?	Goethe, 1963
2061	Anza	1.048	3.481	3.40	0.537	3.74	16.6	2.4	18.0	C	?	Giclas, 1960
2202	Pele	1.120	1.120	3.463	0.512	1.12	17.6	1.2	18.5	?	?	Lemola, 1972
2368	Beltrovata	1.234	2.976	3.06	0.413	5.26	15.2	4.8	16.8	DU	?	Wild, 1977
2608	Seneca	1.044	3.940	3.90	0.586	15.63	17.5	1.4	18.0	?	18.5	Schuster, 1978
3199	Nefertiti	1.128	2.021	1.98	0.283	32.97	14.8	4.4	16.3	?	?	Shoemaker, 1982
3271	Ul	1.271	2.933	3.05	0.394	25.00	16.7	2	18.0	?	?	Shoemaker, 1982
3288	Seleucus	1.103	2.962	2.90	0.457	5.93	15.3	4	16.5	?	75	Schuster, 1982
3352	McAuliffe	1.185	2.573	2.58	0.369	4.78	15.8	2.4	17.5	?	?	Thomas, 1981
3671	Dionysus	1.003	3.387	3.26	0.540	13.61	16.3	2	17	?	?	Shoemaker, 1984

Other named Amors include 3122 Florence, 3551 Verenia, 3552 Don Quixote, 3553 Mera, 4055 Magellan, 4401 Aditi, 4487 Pocohontas, 4503 Cleobulus, 4947 Ninkasi, 4954 Eric, 4957 Brucemurray, 5324 Lyapunov, 5370 Taranis, 5751 Zao, 5797 Bivoj, 5863 Tara, 5869 Tanith, 6489 Golevka, 7088 Ishtar and 7480 Norwan. Comet Wilson–Harrington has been assigned an asteroid number, 4015

Table 8.11. Selected list of Apollo asteroids (symbols as Table 8.3, H = absolute magnitude).

No	Name	q (a.u.)	Q (a.u.)	P (years)	e	i (°)	H	D (km)	m	T	R (h)	Discoverer
—	Hermes	0.617	2.662	2.10	0.624	6.2	18.0	1	—	?	?	Reinmuth, 1937. Lost
1566	Icarus	0.187	1.969	1.12	0.827	22.02	16.9	1.4	17.6	U	2.27	Baade, 1949
1620	Geographos	0.828	1.663	1.39	0.335	13.3	16.8	2 × 5	16.8	S	5.23	Wilson and Minkowski, 1951
1685	Toro	0.771	1.963	1.60	0.436	9.4	14.2	7.6	15.1	S	10.20	Wirtanen, 1948
1862	Apollo	0.647	2.295	1.78	0.560	6.3	16.2	1.4	17.0	S	3.07	Reinmuth, 1932
1863	Antinoüs	0.890	3.630	3.40	0.607	18.4	15.5	3	17.0	?	?	Wirtanen, 1948
1864	Dædalus	0.563	2.359	1.77	0.614	22.2	14.8	3.2	16.0	SU	8.57	Gehrels, 1971
1865	Cerberus	0.576	1.584	1.12	0.467	16.1	16.8	1.6	17.5	S	6.80	Kohoutek, 1971
1866	Sisyphus	0.873	2.914	2.61	0.539	41.1	13.0	7.6	14.5	U	?	Wild, 1972
1981	Midas	0.622	2.930	2.37	0.650	39.8	15.5	1.6	18.0	?	?	Kowal, 1973
2063	Bacchus	0.701	1.455	1.11	0.349	9.4	17.1	1.2	18.7	?	?	Kowal, 1977
2101	Adonis	0.441	3.307	2.57	0.764	1.4	18.7	2	19.5	?	?	Delporte, 1936
2102	Tantalus	0.905	1.676	1.47	0.298	64.0	16.2	2	17.5	?	?	Kowal, 1978
2135	Aristæus	0.794	2.405	2.02	0.503	23.0	17.9	0.8	19.2	?	?	Helin, 1977
2201	Oljato	0.628	3.723	3.20	0.711	2.5	15.2	2.8	16.7	?	24.0	Giclas, 1947
2212	Hephaistos	0.362	3.975	3.18	0.835	11.9	13.9	5.4	15.2	U	?	Chernykh, 1978
2329	Orthos	0.820	3.986	3.73	0.659	24.4	14.9	3.2	16.3	?	?	Schuster, 1976
3103	Eger	0.907	1.904	1.67	0.355	20.9	15.4	5	16.0	?	?	Lovas, 1982
3200	Phæthon	0.140	2.403	1.43	0.890	22.0	14.6	5	16.0	?	4?	IRAS, 1983
3360		0.632	4.296	3.85	0.744	22.0	16.3	1.4	18.0	?	?	Helin and Dunbar, 1981 (1981 VA)
3361	Orpheus	0.810	1.599	1.33	0.322	2.7	19.0	0.8	19.8	?	?	Torres, 1982
3752	Camillo	0.986	1.841	1.68	0.303	55.6	15.5	1.6	17	?	?	Helin and Narucci, 1985
4660	Nereus	0.953	2.027		0.360	1.4	15.5	1.4					Helin, 1982

Other named Apollos include 3838 Epona, 4179 Toutatis, 4183 Cuno, 4257 Ubasti, 4341 Poseidon, 4450 Pan, 4486 Mithra, 4544 Xanthus, 4581 Asclepius, 4769 Castalia, 5011 Ptah, 5143 Heracles, 5731 Zeus, 6063 Jason, 6239 Minos, 7092 Cadmus and 5786 Talos

Apollos which approach the Sun to within 0.2 a.u. are 5756 Talos (q = 0.187, Q = 1.976); 1566 Icarus (0.187, 1.969) and 3200 Phæthon (0.140, 2.403)

close approaches to Earth: for instance, 0.024 a.u. in 1992, 0.035 a.u. in 1996 and 0.073 a.u. in 2000. On 29 September 2004 Toutatis will pass Earth at 0.010 a.u. – the closest predicted approach of any asteroid or comet during the next 30 years. The minimum distance will then be a mere 1549 719 km.

Aten type. All these are very small indeed; 2340 Hathor is no more than about 200 m in diameter. 2100 Ra-Shalom has the shortest orbital period – 277 days. 3753 Cruithne ranks as an unusual companion of the Earth, since it has almost the same orbital period and describes a curious sort of 'horseshoe' path with respect to the Earth. There is

Table 8.12. Selected list of Aten asteroids (symbols as Table 8.3, H = absolute magnitude).

No	Name	q (a.u.)	Q (a.u.)	P (years)	e	i (°)	H	D (km)	m	T	R (h)	Discoverer
2062	Aten	0.790	1.143	0.95	0.183	18.9	16.8	0.9	18.4	S	?	Helin, 1976
2100	Ra-Shalom	0.469	1.195	0.76	0.450	15.8	16.0	1.6	17.2	U	19.80	Helin, 1978
2340	Hathor	0.464	1.224	0.77	0.450	5.9	19.2	0.2	21.5	U	?	Kowal, 1976
3362	Khufu	0.526	1.453	0.98	0.469	9.9	18.3	1.4	18.8	?	?	Dunbar and Barocci, 1984
3554	Amun	0.701	1.247	0.96	0.280	23.4	15.8	2	18	?	?	Shoemaker, 1986
3753	Cruithne	0.484	1.511	0.99	0.515	19.8	15.1	2.7	?	?	?	Waldron, 1986
5381	Sekhmet	0.667	1.228				16.5	1	18	?	?	Shoemaker, 1991

One Aten approaches the Sun to within 2 a.u.: 1995 CR ($q = 0.120$, $Q = 1.692$)

Table 8.13. Close approaches to Earth by asteroids.

Asteroid	Absolute magnitude, H	Date	Minimum distance (a.u.)
1994 XM$_1$	28.0	1994 Dec 9.8	0.0007 (= 112 000 km)
1993 KA$_2$	29.0	1993 May 20.9	0.0010
1994 ES$_1$	28.5	1994 Mar 15.7	0.0011
1991 BA	28.5	1991 Jan 18.7	0.0011
1995 FF	26.5	1995 Mar 27.2	0.0029[a]
1996 JA$_1$	20.5	1996 May 19.7	0.0030
1991 VG	28.8	1991 Dec 5.4	0.0031[b]
4851 Asclepius	20.5	1989 Mar 22.9	0.0046
1994 WR$_{12}$	22.0	1994 Nov 24.8	0.0048
Hermes	18	1937 Oct 30.7	0.0049
1995 UB	27.5	1995 Oct 17.2	0.0050
1998 KY$_{26}$	25.5	1998 June 8.2	0.0054
1993 UA	25.0	1993 Oct 18.8	0.0067
1994 GV	27.5	1994 Apr 12.1	0.0069
1993 KA	26.0	1993 May 17.9	0.0071
1997 UA$_{11}$	25.0	1997 Oct 26.2	0.0071
1997 CD$_{17}$	27.5	1997 Feb 9.8	0.0074
2340 Hathor	20.3	1976 Oct 20.7	0.0078
1988 TA	21.0	1988 Sept 29.0	0.0099

[a] On 1995 Mar 27.0, 1995 FF approached the Moon to 0.0013 a.u.

[b] 1991 VG may be a piece of man-made space débris rather than an asteroid.

no fear of collision, because Cruithne's orbit is inclined at an angle of almost 20°. A list of Aten asteroids is given in Table 8.12.

In 1998 D. Tholen and R. Whiteley, using the 2.24 m telescope on Mauna Kea, discovered asteroid 1998 DK$_{36}$, diameter 40 m, which was suspected to have an orbit lying wholly within that of the Earth. If this is so, then we might have to accept a new class of NEA, but so far confirmation is lacking.

POSSIBLE ASTEROID COLLISIONS

NEAs are much more plentiful than was believed before systematic searches were started, and the chances of a damaging impact are not nil. Table 8.13 is a list of observed close approaches by asteroids – but the dangers come not from asteroids which are known, but from those which are not! Table 8.14 lists what are called PHAs (Potentially Hazardous Asteroids). If one of these is seen to be on a collision course, there might be a chance of diverting it by a nuclear explosion on or near it.

Table 8.14. Selected list of potentially hazardous asteroids.

No	Name	Minimum distance from Earth (a.u.)	Perihelion distance, q (a.u.)	Aphelion distance, Q (a.u.)	Absolute magnitude, H
	Hermes	0.003	0.616	2.662	18
1566	Icarus	0.040	0.187	1.969	16.9
1620	Geographos	0.046	0.828	1.663	15.6
1862	Apollo	0.028	0.647	2.295	16.2
1981	Midas	0.000	0.622	2.930	15.5
2101	Adonis	0.012	0.441	3.308	18.7
2102	Tantalus	0.029	0.905	1.675	16.2
2135	Aristæus	0.015	0.795	2.405	17.9
2201	Oljato	0.001	0.623	3.721	15.2
2340	Hathor	0.006	0.464	1.223	19.2
3200	Phæthon	0.026	0.140	2.403	14.6
3361	Orpheus	0.013	0.819	1.599	19.0
3362	Khufu	0.018	0.526	1.453	18.3
3671	Dionysus	0.034	1.003	3.388	16.3
3757		0.026	1.017	2.653	18.9
4015	Wilson–Harrington	0.049	1.000	4.289	16.0
4034		0.023	0.023	1.530	18.1
4179	Toutatis	0.006	0.919	4.104	15.3
3183	Cuno	0.038	0.718	3.243	14.4
4450	Pan	0.027	0.596	2.287	17.2
4486	Mithra	0.045	0.743	3.658	15.6
4581	Asclepius	0.004	0.657	1.387	20.4
4660	Nereus	0.005	0.953	2.026	18.2
4769	Castalia	0.023	0.550	1.577	16.9
4953		0.040	0.555	2.687	14.1
5011	Ptah	0.026	0.818	2.453	17.1
5189		0.044	0.810	2.292	17.3
5604		0.037	0.551	1.303	16.4
5189		0.044	0.810	2.292	17.3
5604		0.037	0.551	1.303	16.4
5693		0.008	0.527	2.016	17.0
6037		0.024	0.636	1.904	18.7
6239	Minos	0.028	0.676	1.627	17.9
6489	Golovka	0.038	1.012	4.023	19.2
7335		0.042	0.913	2.628	17.0
7482		0.017	0.904	1.788	16.8
7753		0.005	0.761	2.175	18.6
7822		0.033	0.938	1.308	17.4
8014		0.018	0.951	2.543	18.7
8566		0.017	0.857	2.156	16.5
9856		0.033	0.844	3.647	17.4

Table 8.15. Selected list of Trojan asteroids (symbols as Table 8.3, H = absolute magnitude, m = apparent magnitude, at mean opposition). Asteroids marked * are east of Jupiter (L3); the others, west (L4).

No	Name	Year of discovery	q (a.u.)	Q (a.u.)	P (years)	e	i (°)	H	D (km)	R (h)	T	m
Jupiter Trojans												
588	Achilles*	1906	4.413	5.593	11.77	0.149	10.3	8.7	116	?	D	15.3
617	Patroclus	1906	4.501	5.957	11.97	0.139	22.1	8.2	164	?	P	15.2
624	Hektor*	1907	5.088	5.321	11.76	0.022	18.2	7.5	300 × 150	6.92	D	16.2
659	Nestor*	1908	4.624	5.812	12.01	0.114	4.5	9.0	110	?	C	15.8
884	Priamus	1917	4.522	5.786	11.71	0.123	8.9	8.8	94	?	D	16.0
911	Agamemnon*	1919	4.880	5.588	11.87	0.068	21.8	7.9	144	?	D	15.1
1143	Odysseus*	1930	4.771	5.743	12.01	0.092	3.1	7.9	179	?	D	15.6
1172	Æneas	1930	4.635	5.714	11.72	0.104	16.7	8.3	162	?	D	15.7
1173	Anchises	1930	4.596	6.055	12.21	0.137	6.9	8.9	162	?	C	16.0
1208	Troilus*	1931	4.744	5.698	11.85	0.091	33.6	9.0	124	?	C	16.0
1404	Ajax*	1936	4.685	5.897	12.01	0.115	18.0	9.0	92	?	?	16.0
1437	Diomedes*	1937	4.903	5.364	11.52	0.045	20.6	9.3	172	18.0	C	15.7
1583	Antilochus*	1950	4.825	5.376	11.55	0.054	28.6	8.6	158	?	D	16.3
1647	Menelaus*	1957	5.124	5.369	12.03	0.023	5.6	10.3	50	?	?	18.1
1749	Telamon*	1949	4.615	5.780	11.99	0.112	6.1	9.2	56	?	?	17.5
1867	Deiphobus*	1971	4.915	5.375	11.75	0.045	26.9	8.6	140	?	D	
1868	Thersites*	1960	4.708	5.876	12.07	0.110	16.8	10.7	104	?	CFPO	
1869	Philoctetes*	1960	4.957	5.647	12.24	0.065	4.0	11.0	28	?	?	
1870	Glaukos	1971	5.083	5.426	12.00	0.033	6.6	10.5	41	?	?	
1871	Astyanax	1971	5.126	5.497	12.33	0.035	8.6	11.0	35	?	?	
1872	Helenos	1971	5.003	5.500	11.76	0.047	14.7	11.2	50	?	?	
1873	Agenor	1971	4.780	5.743	12.09	0.092	21.8	10.5	45	?	?	
2146	Stentor*	1976	4.663	5.738	11.88	0.103	39.3	10.8	50	?	?	
2148	Epeios*	1976	4.892	5.503	11.02	0.059	9.2	11.1	38	?	?	
2207	Antenor	1977	5.048	5.211	11.67	0.016	6.8	8.9	122	?	D	
2223	Sarpedon	1977	5.095	5.261	11.69	0.016	16.0	9.4	96	?	D	
2241	Alcathous	1979	4.897	5.570	12.02	0.066	16.6	8.6	132	?	D	
2260	Neptolemus*	1975	4.953	5.426	11.82	0.046	17.8	9.3	98	?	D	
2357	Phereclos	1981	4.951	5.420	11.78	0.045	2.7	8.9	96	?	D	
2363	Cebriones	1977	4.954	5.336	11.92	0.037	32.2	9.1	100	?	?	
2456	Palamedes*	1966	4.758	5.550	11.95	0.077	13.9	9.6	80	?	?	
2594	Acamas	1978	4.672	5.551	11.70	0.086	5.5	11.5	26	?	?	
2674	Pandarus	1982	4.821	5.529	11.79	0.068	1.9	9.0	80	?	?	
2759	Idomeneus*	1980	4.828	6.604	11.73	0.965	22.0	9.8	51	?	?	
2797	Teucer*	1981	4.660	5.601	11.83	0.092	22.4	8.4	101	?	?	
2893	Peiroös	1975	4.793	5.597	11.95	0.077	14.6	9.2	80	?	?	
2895	Memnon	1981	4.965	5.484	11.88	0.050	27.2	9.3	64	?	?	
2920	Automedon*	1981	4.984	5.295	11.85	0.030	21.1	8.8	80	?	?	

Other named Jupiter Trojans include:

3063 Makhaon*	4007 Euryalos*	4722 Agelaos	7543 Prylis*
3420 Laocoon	4057 Demophon*	4754 Panthoös	7815 Dolon
3317 Paris	4060 Deipylos*	4791 Iphidamas	8060 Anius*
3391 Sinon*	4063 Euforbo*	4792 Lykaon	8125 Tyndareus*
3451 Mentor	4068 Menestheus*	4805 Asteropaios	
3540 Protesilaos*	4086 Podalirius*	4827 Dares	
3548 Eurybates*	4138 Kalchas*	4828 Misebus	
3564 Talthybius*	4348 Poulydamas	4829 Sergestus	
3596 Meriones*	4501 Eurypyloa*	4832 Palinurus	
3709 Polypoites*	4543 Phoinix*	4833 Meges*	
3793 Leonteus*	4707 Khryses	4834 Thoas*	
3794 Sthenelos*	4708 Polydoros		
3801 Thrasymedes*	4709 Ennomos		

No	Name	Year of discovery	q (a.u.)	Q (a.u.)	P (years)	e	i (°)	H	D (km)	R (h)	T	m
Martian Trojan												
5261	Eureka	1990	1.425	1.622	1.88	0.065	20.2	16.2	1.5	?	?	

THE JUPITER TROJANS

In 1906 Max Wolf, from Heidelberg, discovered asteroid 588, Achilles, which was found to move in the same orbit as Jupiter. It oscillated around the Lagrangian point, 60° ahead of Jupiter. Other members of the group were then found, some 60°E and others at the second Lagrangian point, 60°W; they were named after the participants in the Trojan War. They oscillate around their Lagrangian points, and some move from about 45° from Jupiter out to 80° and then back again. By asteroidal standards they are large, but their great distance means that they are faint. The senior member of the swarm is 624 Hektor, which seems to be cylindrical and to measure 300 km by 150 km; its magnitude varies over a range of 1.1 magnitudes in a period of 6.923 h, which is presumably the rotation period. It is even possible that Hektor, like the much smaller Toutatis, is a contact binary asteroid. Several hundreds of Jupiter Trojans are now known. A selected list is given in Table 8.15. Details of the one known Martian Trojan (5261 Eureka) are also given in Table 8.15.

HIDALGO AND DAMOCLES

On 31 October 1920 W. Baade discovered asteroid 944 Hidalgo, which proved to have an unusual orbit. It travels from the inner edge of the asteroid belt, at 2.0 a.u., out to 9.7 a.u., just beyond the orbit of Saturn; the eccentricity is 0.66 and the inclination 42°. It seems to be of type D, and magnitude variations indicate a rotation period of 10 hours. The diameter is probably about 50 km. The orbit appears cometary, but despite careful scrutiny Hidalgo has never been known to show comet-like activity. The period is 14.15 years.

Another asteroid which ranges far out into the Solar System is 5335 Damocles; it moves from 1.6 a.u. out to 22.2 a.u. in a period of 40.9 years, so that its orbit crosses those of Mars, Jupiter, Saturn and Uranus. However, the high inclination (60.9°) means that it is safe from collision at the present epoch. It is very small – no more than about 15 km in diameter.

CENTAURS

On 1 November 1977 C. Kowal, using the Schmidt telescope at Palomar, discovered the remarkable object 2060 Chiron. Its path lies mainly between those of Saturn

Table 8.16. Chiron.

Designation:	Asteroid 2060, cometary designation 95/P.
Perihelion date:	1996 Feb 14, 18.06 UT.
Perihelion distance:	8.463 0422 a.u.
Aphelion distance:	18.943 14 a.u.
Orbital period:	50.7 years.
Eccentricity:	0.3831.
Inclination:	6.935 degrees.
Mass:	2×10^{19} to 2×10^{19} g.
Rotation period:	5.9 hours.
Diameter:	148 to 208 km.
Discoverer:	C. Kowal, 1977 Nov 1 (on a plate taken 18 October).

and Uranus. At perihelion (as on 14 February 1996) it comes to within 1278 million km of the Sun, and about one-sixth of its orbit lies within that of Saturn, but at aphelion it recedes to 2827 million km, greater than the minimum distance between the Sun and Uranus. The period is 50.9 years. The orbit is unstable over a time scale of some millions of years. In 1664 BC Chiron approached Saturn to a distance of 16 000 000 km, which is not much greater than the distance between Saturn and its outermost satellite, Phœbe, which is almost certainly a captured asteroid. At discovery the magnitude was 18, but at perihelion it rose to 15. Light-curve studies give a rotation period of just under 6 h. Estimates of the diameter range from 148 to over 200 km. Chiron's image can be traced back on plates taken as long ago as 1895, so that its orbit is very well known.

Preliminary spectroscopic results indicated a fairly low albedo with a dusty or rocky surface, but there was a major surprise in 1988, when Chiron was found to be brightening – not spectacularly, but appreciably. Inevitably there were suggestions that it might be a huge comet rather than an asteroid, and this idea was strengthened in April 1990, when K. Meech and J. S. Belton, using electronic equipment on the 4 m reflector at Kitt Peak in Arizona, photographed Chiron and found that it appeared to be 'fuzzy'; in other words, it had developed a coma.

Table 8.17. Selected list of Centaurs (symbols as Table 8.3).

No	Name	q (a.u.)	Q (a.u.)	P (years)	e	i (°)	D (km)	R (h)
2060	Chiron	8.46	18.79	50.9	0.38	6.9	208	5.9
5145	Pholus	8.66	31.78	92.1	0.9	24.7	240?	?
7066	Nessus	11.81	37.18	124	0.52	15.6	80	?
8405		6.83	29.29				60	?

Table 8.18. The first six Kuiper Belt objects.

Object	q (a.u.)	Q (a.u.)	i (°)	D (km)	P (years)	Discoverers
1 1992 QB$_1$	40.8	43.9	2.2	283	290.2	Jewitt and Luu, 30 Aug 1992
2 1993 FW	42.1	43.9	7.7	286	291.2	Jewitt and Luu, 28 Mar 1993
3 1993 RO		39.3	3.7	139		Jewitt and Luu, 14 Sept 1993
4 1993 RP		39.3	2.6	96		Jewitt and Luu, 15 Sept 1993
5 1993 SB	26.8	39.4	1.9	188		Williams *et al*, 17 Sept 1993
6 1993 SC	32.4	39.5	5.2	319	319	Williams *et al*, 17 Sept 1993

Using the 2.24 m telescope on Mauna Kea on 29 January 1990, D. Jewitt and J. X. Luu found that the coma extended for 80 000 km and was elongated away from the Sun in comet-like fashion. The ejected material was thought to be vaporized carbon monoxide carrying away dust grains.

This is certainly comet-like behaviour, and on occasion the diameter of the coma has been known to reach almost 2 000 000 km; the brightness can vary by a factor of four over a few hours, and a gravitationally-bound 'dust atmosphere' appears to be suspended in the inner 1200 km of the coma. Moreover this dust shows evidence of structure, indicating that there may be particle plumes issuing from the nucleus. Yet Chiron is 40 times larger and 50 000 times more massive than any known comet. It may well be an escapee from the Kuiper Belt.

Other bodies with orbits crossing those of the giant planets were found later, and have been named after mythological centaurs. (Hidalgo and Damocles are not officially classed as Centaurs, because they are very small and have perihelia much closer to the Sun.) 5145 Pholus has a much more eccentric orbit than Chiron, crossing those of Saturn, Uranus and Neptune; while Chiron is greyish and active (see Table 8.16), Pholus is inert and very red. Half a dozen Centaurs are now known; those which have been allotted numbers are listed in Table 8.17. No doubt many more exist.

KUIPER BELT OBJECTS

In January 1943 K. E. Edgeworth suggested the possibility of a swarm of minor bodies orbiting the Sun in the outermost part of the planetary system. The suggestion was made independently by G. P. Kuiper in 1951. A belt of asteroidal-sized objects does in fact exist and, perhaps rather unfairly, it is now always known as the Kuiper Belt.

The first Kuiper Belt object (KBO) was discovered on 31 August 1992 by D. Jewitt and J. X. Luu, using the 2.2 m telescope on Mauna Kea. The magnitude was 23; it was given the provisional designation of 1992 QB$_1$. Its orbit keeps it well beyond that of Neptune, and by asteroidal standards it is large, with an estimated diameter of 283 km. Other KBOs were soon found, with aphelion distances in the range 35–45 a.u.; it is clear that the swarm is very populous indeed. It has even been suggested that there may be at least 35 000 KBOs more than 100 km in diameter – making it more massive than the main-belt asteroid zone.

Table 8.18 lists data for some of the early KBO discoveries.

Most KBOs seem to be dark and reddish; two, 1994 JQ$_{11}$ and 1994 VK$_8$, are almost 400 km in diameter – larger than any main-belt asteroid apart from Ceres, Pallas, Vesta and Hygeia. It has been suggested that Pluto may be simply the largest member of the KBO swarm, and that short-period comets also come from there, but so far the evidence is not conclusive.

9 JUPITER

Jupiter is much the largest and most massive planet in the Solar System; its mass is greater than those of all the other planets combined. It has been suggested that it may have been responsible for preventing approaching comets invading the inner Solar System, and thereby protecting the Earth from bombardment. Data are given in Table 9.1.

Table 9.1. Data.

Distance from the Sun:
    max 815 700 000 km (5.455 a.u.)
    mean 778 340 000 km (5.203 a.u.)
    min 740 900 000 km (4.951 a.u.)

Sidereal period: 11.86 years = 4332.59 days

Synodic period: 398.88 days

Rotation period:
    System I (equatorial) 9h 50m 30s
    System II (rest of planet) 9h 55m 41s
    System III (radio methods) 9h 55m 29s

Mean orbital velocity: 13.07 km s^{-1}

Axial inclination: 3°4′

Orbital inclination: 1°18′15″.8

Diameter: equatorial 142 884 km
            polar 133 708 km

Oblateness: 0.065

Apparent diameter from Earth: max 50″.1
                       min 30″.4

Reciprocal mass, Sun = 1: 1047.4

Density, water = 1: 1.33

Mass, Earth = 1: 317.89 (1.899 × 10^{24} kg)

Volume, Earth = 1: 1318.7 (143.128 × 10^{10} km^3)

Escape velocity: 60.22 km s^{-1}

Surface gravity, Earth = 1: 2.64

Mean surface temperature: −150 °C
Albedo: 0.43

Maximum magnitude: −2.6

Mean diameter of Sun, as seen from Jupiter: 6°9″

Distance from Earth: max 968 100 000 km
                  min 588 500 000 km

MOVEMENTS

Jupiter is well placed for observation for several months in every year. The opposition brightness has a range of only about 0.5 magnitude. Generally speaking it 'moves' about one constellation per year; thus the 1998 opposition was in Pisces, that of 1999 in Aries and that of 2000 in Taurus. Opposition data for the period 2000–2005 are given in Table 9.2. Some years pass without an opposition, as in 2001 – because the opposition of 28 November 2000 is followed by the next on 1 January 2002; the next 'missed year' will be 2013.

The perihelion years are 1987, 1999, 2011; the aphelion years are 1993, 2005, 2017.

Generally Jupiter is the brightest of the planets apart from Venus; its only other rival is Mars at perihelic opposition.

OCCULTATIONS AND CONJUNCTIONS

Occultations of and by Jupiter involving other planets are rare. The last occasion when Jupiter was occulted by a planet was on 3 January 1818, when Venus occulted Jupiter; this will happen again on 22 November 2065, at 12.46 UT, but the elongation from the Sun will be only 8°W. Jupiter last occulted a planet on 15 August 1623, when it occulted Uranus, but on 10 May 1955, at 20.39 UT, Jupiter and Uranus were only 46″ apart; elongation from the Sun was 65°E, so that the event was easily observed (some unwary observers thought that Jupiter had acquired an extra satellite, although in fact Uranus appeared appreciably larger and dimmer than the Galilean satellites). Data for occultations and close conjunctions are given in Tables 9.3 and 9.4.

EARLY OBSERVATIONS

Since Jupiter is generally the brightest object in the sky apart from the Sun, the Moon and Venus, it must have been known since the dawn of human history. The first telescopic observations were made in 1610 by pioneers such as Galileo and Marius. The four main satellites were discovered, but at first no details were seen on Jupiter itself. Using the current

Table 9.2. Oppositions of Jupiter 2000–2005.

Date	Diameter (arcsec)	Constellation	Declination at opposition (°)	Magnitude
2000 Nov 28	48.5	Taurus	+20	−2.4
2002 Jan 1	47.1	Gemini	+23	−2.3
2003 Feb 2	45.5	Cancer	+18	−2.0
2004 Mar 4	44.5	Leo	+8	−2.0
2005 Apr 3	44.2	Virgo	−4	−2.0

There follow oppositions on 2006 May 4 (Dec −15°), 2007 June 5 (−22°), 2008 July 9 (−23°), 2009 August 14 (−15°) and 2010 September 21 (−2°).

Table 9.3. Occultations of Jupiter 1900–2100.

Occulting planet	Date	UT	Elongation (°)
Venus	2065 Nov 22	12.46	8W
Mercury	2088 Oct 27	13.44	5W
Mercury	2074 Apr 7	10.49	2W

Obviously, these last two events will be very difficult to observe.

Table 9.4. Planetary conjunctions, 2000–2005. The following conjunctions involve Jupiter. The elongation is over 10° from the Sun. On 2000 May 17 Venus and Jupiter are only 42″ apart at 11h UT, but the elongation is only 7°W.

Date	Planet	UT	Separation (°) (′)		Elongation (°)
2000 Apr 6	Mars	23	1	06	23E
2000 May 31	Saturn	10	1	11	17W
2001 May 16	Mercury	17	2	47	21E
2001 July 12	Mercury	22	1	56	21W
2001 Aug 5	Venus	24	1	12	38W
2002 June 3	Venus	18	1	39	34E
2002 July 3	Mars	06	0	49	12E

small-aperture, long-focus refractors, N. Zucchi may have seen the main equatorial belts in 1630, and F. Fontana definitely recorded three belts in 1633. In 1648 F. Grimaldi showed that the belts are parallel with the Jovian equator. In 1659 C. Huygens published a good drawing showing the two equatorial belts.

More detailed drawings were made from 1665 by G. D. Cassini, first at Milan and then from Paris. He found the globe of Jupiter to be appreciably oblate, and recorded over half a dozen 'bands'; by watching the drift of the surface features, including one well-marked spot, he gave a rotation period of just over 9h 55m, which was very near the truth. Other observers of the period included G. Campani and Robert Hooke.

Careful studies of Jupiter were made in the latter part of the 16th century by William Herschel and J. H. Schröter, but detailed results were delayed until the 17th century, with observers such as J. H. Mädler, W. de la Rue, W. Lassell, W. R. Dawes and the Earl of Rosse. Rotation periods were measured; Sir George Airy, the Astronomer Royal, gave 9h 55m 21s, but it became clear that the rotation is differential, with different latitudes having different periods. Studies

of Jupiter were pioneered in America by William Bond, at Harvard College. A famous drawing by Warren de la Rue, on 25 October 1865, showed that different features show differences in colour (de la Rue used a home-made 13 inch reflector). In 1890 the British Astronomical Association was founded, and ever since then its Jupiter Section has monitored the surface features. These observations are invaluable, particularly since they go back to the era before good planetary photographs became available.

Some early theories sound bizarre today. In 1698 Huygens maintained that Jupiter must be a moist, life-bearing world and that the belts were strips of vegetation, no doubt supporting animals. Even in the early 17th century W. Whewell believed Jupiter to be a globe of 'ice and water', with a cindery nucleus; he described 'huge gelatinous monsters languidly floating in icy seas'. However, it was later assumed that Jupiter must be a miniature sun, able to warm its satellite system – a theory which was still generally accepted until less than 80 years ago.

BELTS AND ZONES

The surface is dominated by the dark belts and bright zones, all of which are variable, although in general their latitudes do not change much. There are also various striking features, notably the Great Red Spot, described below. The main belts are listed in Table 9.5.

Jupiter has the shortest 'day' insofar as the principal planets are concerned (some of the asteroids rotate much more quickly). It is not possible to give an overall value for the rotation period, because Jupiter does not spin in the way that a solid body would do. The equatorial zone has a shorter period than the rest of the planet. Conventionally, System I refers to the region between the north edge of the South Equatorial Belt and the south edge of the North Equatorial Belt; the mean period here is 9h 50m 30s, although individual features may have periods which differ perceptibly from this. System II, comprising the rest of the surface of the planet, has a period of 9h 55m 41s, although again different features have their own periods; that of the Great Red Spot varies between 9h 55m 36s and 9h 55m 42s. In addition there is System III, which relates not to optical features but to the bursts of decametre radio radiation; the period is 9h 55m 29s.7. There are no true 'seasons' on Jupiter, because the axial inclination to the perpendicular of the orbital plane is only just over 3° – less than for any other planet.

The most prominent belt is generally the North Equatorial, while the South Equatorial Belt is very variable and may become obscure at times; on the other hand, at times during 1962–3 the two Equatorial Belts appeared to merge, and during 1988 the South Equatorial was fully equal to the North Equatorial in width and intensity. Revivals of the South Equatorial Belt are the most spectacular phenomena seen on Jupiter; they involve sudden outbreaks of bright and dark clouds, with intense turbulence and many spots moving on rapid currents. Other belts are also subject to marked variations in intensity, although the North Temperate and South Temperate Belts are usually

Table 9.5. Average latitudes of the belts [a]. These latitudes are subject to slight variation and are given here in round numbers.

		Lat. (°)
South Polar Region	SPRn	−53
South South Temperate Belt	SSTBs	−47
	SSTBn	−42
South Temperate Belt	STBs	−33
	STB	−30
	STBn	−27
South Tropical Band	STropB	−25
South Equatorial Belt	SEBs	−21
	SEBn	−7
North Equatorial Belt	NEBs	+7
	NEBn	+18
North Tropical Band	NTropB	+23
North Temperate Belt	NTBs	+24
	NTBn	+31
North North Temperate Belt	NNTBs	+36
	NNTB	+38
	NNTBn	+39
North North North Temperate Belt	NNNTBs	+43
	NNNTB	+45
	NNNTBn	+47
North North North North Temperate Belt	NNNNTB	+49

[a] Detailed values are given by J. J. Rogers 1995 *The Giant Planet Jupiter* (Cambridge: Cambridge University Press). I thank Dr. Rogers for allowing me to give these data here.

well-marked. Around latitude 16°N may be seen brown ovals which are known as 'barges'; they are low-lying, and it may be that their colour is due to the fact that they are slightly warmer than the adjacent regions, so that some of the ammonia ice particles begin to melt.

We now have a sound knowledge of the way in which the Jovian winds blow. In the equatorial region, they blow west–east at about 100 m s^{-1} relative to the core, reaching maximum speed 6 to 7° north and south of the equator. In the northern hemisphere the east wind decreases with increasing latitude, until at 18°N the clouds are moving westward at 25 m s^{-1}. North of this, the windspeed falls to zero, and then shifts in an eastward direction, reaching a maximum of about 170 m s^{-1} at 24°N. In the southern hemisphere conditions are not quite the same, due probably to the presence of the Great Red Spot. But in both hemispheres there is an alternating pattern of eastward and westward jet streams, which mark the boundaries of the visible belts. Spots such as the Great Red Spot circulate cyclonically or anticyclonically as if rolling between the jet streams.

THE GREAT RED SPOT

The Great Red Spot is undoubtedly the most famous feature on Jupiter. Together with its characteristic 'Hollow', it has certainly been in existence for many centuries. It may have been recorded by Cassini as long ago as 1665, although the identification is not certain. A sketch made by R. Hooke on 26 June 1666 may also show it, and it is also possible that he made an observation of it in 1664. It was seen several times during the 19th century, following the first observation of the Hollow in 1831 by H. Schwabe; it was seen to encroach into the southern part of the South Equatorial Belt. The Spot itself was seen in 1858–9 by W. Huggins as a dark ring with a light interior, and also by Lord Rosse, with his great Birr Castle reflector. Suddenly, in 1878, it became very prominent, and brick-red in colour. It remained very conspicuous until 1882, but subsequently faded. Since then it has shown variations in both intensity and colour, and at times it has been invisible, but it always returns, so that during telescopic times it has been to all intents and purposes a permanent feature – unlike any other of the spots. At its greatest extent it was said to measure 40 000 km in an east–west direction and 14 000 km north–south, giving it a greater surface area than that of the Earth, although more recently the dimensions have been 24 000 km by 12 000 km. Whether or not this shrinkage will continue remains to be seen. The latitude varies little from a value of 22.4°S, but the Spot drifts around in longitude, and over the past century the total longitude drift has amounted to about 1200°.

It was once thought that the Red Spot might be the top of a glowing volcano, but this was soon shown to be untenable. It was then suggested that it might be a solid or semi-solid body floating in Jupiter's outer gas, in which case it would be expected to disappear if its level sank for any reason (possibly a decrease in the density of the outer gas). In 1963 R. Hide suggested that it might be the top of a 'Taylor column', a sort of standing wave above a mountain or a depression below the gaseous layer. However, the space-probe results have given us quite a different picture. The Red Spot is a phenomenon of Jovian meteorology – a high-level anticyclonic vortex, with wind speeds of up to 360 km h^{-1}. To the south the Spot is bounded by an east wind, while to the north it is bounded by a strong west wind. This means that as the winds are reflected round the Spot, they set up anti-clockwise rotation, with a period of 12 days at the outer edge and nine days inside. The vortex is a high-pressure area elevated 8 km above the adjacent cloud deck by the upward convection of warmer gases from below; smaller clouds to the north-east and north-west, beautifully shown on a 1997 image from the Galileo probe, look very like Earth's towering thunderstorms. The cause of the red colour is not definitely known. It may be due to the condensation of phosphorus at the cloud tops. Certainly there is a great deal of interior structure.

WHITE OVALS

Also of great importance are (or were) the three white ovals on the edge of the South Temperate Belt, which are similar to the Red Spot in shape and which also drift around in longitude; all have dark borders. They have been under observation since 1939 and are longer-lived than any other spots on Jupiter with the exception of the Great Red Spot itself, and may also rotate in an anticyclonic sense. In February 1998 observations from professionals (at the Pic du Midi) and amateurs indicated that two of the ovals had merged, forming a single, larger oval; the temperature has been given as −157 °C, which is perceptibly colder than the surrounding regions, as with other atmospheric ovals.

THE SOUTH TROPICAL DISTURBANCE

Also in the latitude of the Great Red Spot was a feature known as the South Tropical Disturbance (STD), discovered by P. Molesworth on 28 February 1901 and last recorded by many observers during the apparition of 1939–40. It took the form of a shaded zone between white spots. The rotation period of the STD was shorter than that of the Red Spot, so that periodically the STD caught up the Red Spot and passed it; the two were at the same latitude, and the interactions were of great interest. Nine conjunctions were observed, and possibly the beginning of the tenth in 1939–40, although by then the STD had practically vanished. Its average rotation period was 9h 55m 27s.6. It has not reappeared, but there have been several smaller, shorter-lived disturbances of the same type; the most notable of these lasted from 1979 to 1981.

INTERNAL STRUCTURE OF JUPITER

In 1923 and 1924 a classic series of papers by H. Jeffreys finally disposed of the idea that Jupiter is a miniature sun, giving off vast amounts of heat. Jeffreys proposed a model in which Jupiter would have a rocky core, a mantle composed of solid water ice and carbon dioxide, and a very deep, tenuous atmosphere. Methane and ammonia – both hydrogen compounds – were identified in the atmosphere by R. Wildt in 1932, and it was proposed that Jupiter must consist largely of hydrogen. In 1934 Wildt proposed a model giving Jupiter a rocky core 60 000 km in diameter, overlaid by an ice shell 27 000 m thick, above which lay the hydrogen-rich atmosphere. (This was certainly more plausible than a strange theory proposed by E. Schoenberg in 1943. Schoenberg believed Jupiter to have a solid surface, with volcanic rifts along parallels of latitude; heated gases rising from these rifts would produce the belts!)

New models were proposed independently in 1951 by W. Ramsey in England and W. DeMarcus in America. According to Ramsey, the 120 000 km diameter core was composed of hydrogen, so compressed that it assumed the characteristics of a metal. The core was overlaid by an 8000 km deep layer of ordinary solid hydrogen, above which came the atmosphere. Today it is believed that Jupiter is mainly liquid (a suggestion made long ago, in 1871, by G. W. Hough, who also believed the Red Spot to be a floating island). The latest models are based on work carried out by J. D. Anderson and W. B. Hubbard in the United States. There is no reason to think that they are very far from the truth, although it would be idle to pretend that our knowledge is at all complete.

It seems that there is a relatively small, rocky core made up of iron and silicates, at a temperature of around 20 000 °C (perhaps rather more). Above this is a thick shell of liquid metallic hydrogen. At about 46 000 km from the centre of the planet there is a transition from liquid metallic hydrogen to liquid molecular hydrogen; in the transition region the temperature is assumed to be around 11 000 °C, with a pressure about three million times that of the Earth's air at sea level. Above the liquid molecular hydrogen comes the gaseous atmosphere, which is about 1000 km deep. The change in site is gradual; there is no hard, sharp boundary, so that we cannot say definitely where the 'atmosphere' ends and the actual body of the planet begins.

Jupiter radiates 1.7 times more energy than it would do if it depended only upon radiation received from the Sun. Probably this excess heat is nothing more than what remains of the heat generated when Jupiter was formed. It has been suggested that the globe is slowly contracting, with release of energy, but this explanation is not now generally favoured.

Note, incidentally, that Jupiter's core is not nearly hot enough to trigger off stellar-type nuclear reactions. Jupiter is not a 'failed star' or even a brown dwarf; it is definitely a planet.

ATMOSPHERE

What may be termed the atmosphere of Jupiter has a depth of approximately 1000 km, although of course no absolutely precise figure can be given. Most of it is hydrogen (H_2). According to one recent analysis, hydrogen accounts for 80.4% of the total and helium for 13.6%, which does not leave much room for anything else. Methane (CH_4) may account for up to 0.2%, and there are traces of ammonia (NH_3), hydrogen sulphide (H_2S) and ethane (C_2H_6). The amount of water (H_2O) is very small, and certainly no more than 0.1%.

Gases warmed by the internal heat of the planet rise into the upper atmosphere and cool, producing clouds of ammonia crystals floating in gaseous hydrogen. These clouds form the bright zones on Jupiter, which are both higher and colder than the dark belts. It had been assumed that below the ammonia ice clouds came a layer of ammonium hydrosulphide, and below that a layer of water ice or liquid water droplets, although the results from the Galileo space-craft indicate that some of our long-held ideas may be in need of revision.

COMET COLLISION, 1994

A remarkable event, unique in our experience, occurred in July 1994, when a comet was observed to hit Jupiter.

The comet was discovered on 26 March 1993 by Eugene and Carolyn Shoemaker, working in collaboration with David Levy; because it was this team's ninth discovery the comet was known as Shoemaker–Levy 9 (S/L 9). The image was found on a plate taken three nights earlier with the Schmidt telescope at Palomar. Carolyn Shoemaker described it as a 'squashed comet' quite unlike anything previously seen. It was in orbit not round the Sun, but round Jupiter, and had probably been in this sort of path ever since 1929 – perhaps even earlier. Calculations showed that on 7 July 1992 it had passed only 21 000 km from Jupiter, and had been disrupted, literally torn apart by the Giant Planet's powerful gravitational pull. A year later, in July 1993, it reached apojove – its furthest point from Jupiter – and solar perturbations put it into a collision course. It was calculated that the chain of fragments would impact Jupiter in July 1994, and this is precisely what happened. Over 20 fragments were identified, strung out in the manner of a pearl necklace, and were lettered from A to W (I and O being omitted). The first fragment (A) was due to impact on 16 July, at 20h 11m UT and the last (W) on 22 July, at 8h 5m UT.

The largest fragments were G and Q, while J and M soon faded out altogether. Subsequently P and Q split in two; P2 then split again, while P1 disappeared. By the time they reached Jupiter, the fragments were stretched out over about 29 000 000 km, with separate 'tails' and dusty 'wings' extending ahead of and behind the main swarm.

All the fragments landed in about the same latitude: around 50°S, well south of the Red Spot and in the area of the South South Temperate Belt. Unfortunately all the impacts occurred on the side of Jupiter turned away from the Earth, but the planet's quick rotation brought the affected areas into view after only a few minutes, and the results of the impacts were very marked[1]. Great dark spots were produced; the most impressive was G, which impacted at 7h 32m on 18 July. Fragment G, which may have been 3–4 km across, created a fireball at least 3000 km high and left a multi-ringed scar on the cloud deck. It was estimated that if fragment G had hit the Earth it would have made a crater 60 km in diameter; the large fragments produced vast clouds of 'smoke' which remained visible from Earth for many months. At one stage the scars seemed to link up, producing what gave the impression of an extra belt on the planet.

The impacts were observed from the Hubble Space Telescope, and also from the Galileo probe, then on its way to Jupiter. Hubble results showed the presence of ammonia and hydrogen sulphide in the G fireball as it cooled, but there was no sign of the expected water or ice layer beneath the cloud tops.

There were no permanent effects on Jupiter, but the whole area was violently disturbed, and dark material in the Jovian stratosphere produced by the impacts could still be traced well into 1996. There had been suggestions that the impactor might have been an asteroid rather than a comet, but it now seems definite that S/L 9 really was a comet which had spent most of its career in the Kuiper Belt, beyond the orbit of Neptune.

Could there have been any previous observations of cometary impacts? In 1690 G. D. Cassini recorded suspicious dark spots, as did Johann Schröter in 1785 and 1786, using an excellent 13 cm reflector made by William Herschel; but it is quite impossible to decide whether or not these were due to an impact of a comet.

THE GALILEO ENTRY PROBE

On 18 October 1989 the Galileo space-craft was launched. It was made up of an orbiter, designed to assume a closed path round Jupiter and transmit data, and an entry probe, to plunge into the clouds and send back results until being destroyed. The experiment was highly successful. At 22.04 UT on 7 December 1995 the entry probe met the Jovian atmosphere, at latitude 6.5°N, longitude 4.5°W, and transmitted for 57.6 min before losing contact. By then it had penetrated to a depth of about 600 km below the tenuous upper reaches of the Jovian atmosphere. The results were in some ways decidedly unexpected. Only one

---

[1] I was observing with the 66 cm refractor as Herstmonceux, then the site of the Royal Greenwich Observatory. The scar left by Impact A was far more prominent than I had expected. Later scars were easily visible in the 5 cm finder of my small portable refractor, and were much blacker than anything I had ever previously seen on Jupiter apart from satellite shadows.

well-defined, distinct cloud structure was found, apparently corresponding to the previously predicted cloud layer of ammonium hydrosulphide. There was much less lightning activity than had been expected. However, one major surprise concerned the winds. It had been assumed that the strong Jovian winds, about 380 km h^{-1} at the entry level, were more or less confined to the upper atmosphere, but Galileo showed that this is not the case; the velocity increased to over 500 km h^{-1} below the visible level. This seems to indicate that the Jovian winds are not produced by solar heating, as on Earth, or by the condensation of water vapour; it is now more likely that the cause is heat escaping from the deep interior of the globe.

The atmosphere was found to be much dryer than had been expected; it cannot contain more than one-fifth to one-tenth the percentage of water contained in the Sun. It may well be that Galileo plunged into the Jovian equivalent of a desert region on Earth; it entered at a point on the edge of the North Equatorial Belt, which is atypical of the planet as a whole. Certainly we have to admit that our knowledge of conditions below the Jovian cloud tops is very far from complete.

RADIO EMISSIONS

Radio radiation from Jupiter was detected by B. F. Burke and K. L. Franklin, in the United States, in 1955. (It has to be admitted that the discovery was accidental.) The main emissions are in wavelengths 10–500 m (decametric) and 0.1–3 m (decimetric); there are also kilometric emissions (0.3–5 km) and millimetric (below 10 cm) due to plasma in Jupiter's powerful magnetic field, outside the atmosphere and strongly influenced by the volcanic satellite Io.

MAGNETOSPHERE

Jupiter has a very powerful magnetic field – much the strongest in the entire Solar System. The strength is 4.2 G at the Jovian equator and 10–14 G at the magnetic poles; by contrast, the strength of the Earth's magnetic field at the equator is a mere 0.3 G. With Jupiter, the magnetic axis is inclined to the rotational axis at an angle of 9.6°. The polarity is opposite to that of the Earth, so that if it were possible to use a magnetic compass on Jupiter the needle would point south.

With regard to magnetic phenomena, distances from the centre of Jupiter are usually reckoned in terms of the planet's radius, R_J. The volcanic satellite Io, which has such a profound effect upon these phenomena, lies at a distance of 5.9R_J, corresponding to about 422 000 km.

The magnetic field is generated inside Jupiter, near the outer boundary of the shell of metallic hydrogen. The field is not truly symmetrical, but beyond a distance of a few R_J it more or less corresponds to a dipole.

There is a huge magnetosphere; if it could be seen with the naked eye from Earth, its apparent diameter would exceed that of the full moon. The outer boundary is formed at what is known as the magnetopause, where the incoming solar wind particles are deflected and produce a bow shock, at about 10R_J ahead of the actual magnetopause. The region between the bow shock and the true magnetopause is termed the magnetosheath. On the sunward side of Jupiter the field tends to be compressed; on the night side it is stretched out into a 'magnetotail' which may be up to 650 000 000 km long, so that at times it can even engulf Saturn.

There are zones of intense radiation (protons and electrons), at least 10 000 times more powerful than the Van Allen zones surrounding the Earth. Pioneer 10, the first space-probe to encounter these zones (in December 1972), received a total of over 250 000 rads. Since a dose of 500 rads is fatal to a man, future astronauts will be well advised to keep well clear of the danger zone.

In 1964 K. E. Bigg realized the orbital position of Io has a marked effect upon Jupiter's decametric radiation, and it is now known that the satellite is connected to Jupiter by a very strong flux tube setting up a potential difference of 400 000 V. Molecules are sputtered off Io's surface by particles in the magnetosphere, producing a 'torus' tilted to Io's orbit by 7°, so that Io passes through it twice for each rotation of Jupiter. There is also a tenuous sodium cloud round Io which extends all round the orbit of the satellite.

LIGHTNING AND AURORÆ

Lightning is very intense on Jupiter; for example enormous bursts were recorded on the planet's night side by the Galileo probe in November 1996, with individual flashes hundreds of kilometres across. The flashes are of the 'cloud-to-cloud' variety, and no doubt there is thunder as well. Auroræ are also intense; they were first detected in 1977, and were recorded by Voyager 1 during its passage across the night side of the planet in March 1979.

Jupiter is also a source of cosmic radiation; cosmic rays from it have been detected as far away as the orbit of Mercury. Truly the Giant Planet is an energetic world!

THE RINGS OF JUPITER

Jupiter's ring system was discovered on 4 March 1979, on a single image sent back by the Voyager 1 probe as it passed through the equatorial plane of the planet. It is now known that there are three detectable rings. Details are given in Table 9.6.

Table 9.6. The rings of Jupiter. The Halo and Gossamer rings are extremely tenuous. Metis and Adrastea lie in the Main Ring, and Amalthea at the outer edge of the Gossamer Ring. Traces of the Gossamer Ring extend out almost as far as the orbit of Thebe.

Name	Distance from centre of Jupiter (km)	Width (km)	Thickness (km)
Halo	92 000–122 500	30 500	~1000
Main	122 500–128 940	6400	~30
Gossamer	181 000–222 000	41 000	?

The rings are very dark, and are quite unlike the bright, icy rings of Saturn. They are caused by material coming from the small inner satellites Metis, Adrastea, Amalthea and perhaps Thebe. This material is released when the satellites are struck by interplanetary meteoroids at speeds greatly magnified by Jupiter's powerful gravitational field – the situation has been compared with the cloud of chalk dust produced when two erasers are banged together. Metis and Adrastea, with their low escape velocities and their closeness to Jupiter, are probably the most important contributors. Metis and Adrastea lie inside the Main Ring, with Amalthea and Thebe further out.

There seem to be no ice particles in the rings, and the ring material is more dark, reddish soot.

The outer (Gossamer) ring is actually composed of two faint and more or less uniform rings, one enclosing the other; they extend from the outer boundary of the Main Ring

(122 500 km) and extend out to over 222 000 km, although the ring is so tenuous that it is difficult to give a precise boundary. The fainter of the two extends radially inward from the orbit of Thebe, while the denser of the two – the enclosed ring – extends radially inward from the orbit of Amalthea. In each case the centres of the rings are fainter than the edges. The Main Ring extends from the orbit of Adrastea to the edge of the Halo Ring, while the Halo Ring is toroidal, extending radially from 122 000 km to 92 000 km.

Voyager and Galileo results show that the rings are much brighter in forward-scattered light than in back-scattered or reflected light. This indicates that the ring particles are in general only 1–2 μm across. Such particles have relatively short lifetimes in stable orbits, so that the rings must be continually replenished by material produced by the small satellites.

SPACE-CRAFT TO JUPITER

Six space-craft have now encountered Jupiter; two Pioneers, two Voyagers, Galileo and the solar polar probe Ulysses, which by-passed Jupiter and used the powerful Jovian gravity to put it into its planned orbit well out of the perpendicular. Details are given in Table 9.7.

The Pioneers were virtual twins; both were successful. Pioneer 10 carried out studies of the Jovian atmosphere and magnetosphere, and returned over 300 images. It showed that the radiation zones are far stronger than had been previously believed. The first energetic particles were detected when Pioneer was still over 20 000 000 km from Jupiter, and the radiation level increased steadily as Pioneer moved inward; at the minimum distance from the upper clouds (131 400 km) the instruments were almost saturated, and if the minimum distance had been much less the mission would have failed. The proposed orbit of Pioneer 2 was hastily altered to a different trajectory which would carry it quickly over Jupiter's equatorial zone, where the danger is at its worst. Pioneer 10 is now on its way out of the Solar System; it carries a plaque to give a clue to its planet of origin – although whether any other beings would be able to decipher the message seems rather debatable.

Pioneer 11 confirmed the earlier findings, and was then put into a path which took it out to a rendezvous with Saturn. It too is now leaving the Solar System permanently.

Table 9.7. Missions to Jupiter, 1972–2000.

Name	Launch date	Encounter date	Nearest approach (km)	Remarks
Pioneer 10	2 Mar 1972	3 Dec 1973	131 400	Complete success; new images and data. Now on its way out of the Solar System; still contactable.
Pioneer 11	5 Apr 1973	2 Dec 1974	46 400	Complete success. Went on to rendezvous with Saturn (1979 Sept 1). Now on its way out of the Solar System; still contactable.
Voyager 1	5 Sept 1977	5 Mar 1979	350 000	Detailed information about Jupiter and the Galilean satellites Io, Ganymede and Callisto; volcanoes on Io discovered. Went on to rendezvous with Saturn (12 Nov 1980). Now on its way out of the Solar System; still contactable.
Voyager 2	20 Aug 1977	9 July 1979	714 000	Complemented Voyager 1. Went on to fly by Saturn (1981), Uranus (1986) and Neptune (1989). Now on its way out of the Solar System; still contactable.
Galileo	18 Oct 1989	7 Dec 1995	Entry	Orbiter and entry probe; fly-bys of Venus (10 Feb 1990) and Earth (8 Dec 1990 and 8 Dec 1992); images of asteroids Gaspra (10 Oct 1989) and Ida (8 Aug 1993). Orbiter still moving round Jupiter and sending data.
Ulysses	6 Oct 1990	8 Feb 1992	378 000	Studies of Jupiter's magnetosphere, radiation zones and general environment. Went on to survey the poles of the Sun.

The third Jupiter probe, Voyager 1, was actually launched a few days later than its twin Voyager 2, but travelled in a more economical path. (To confuse matters still further, initial faults detected in the first space-craft caused a switch in numbers, so that the original Voyager 1 became Voyager 2 and *vice versa*.) The Voyagers were much more elaborate than the Pioneers, and the results obtained were of far higher quality. Voyager 1 also surveyed the satellites Io, Ganymede and Callisto. Voyager 2 followed much the same programme, and was also sent close to Europa, the only Galilean satellite not well studied by its predecessor. Voyager 1 went on to survey Saturn, while Voyager 2 was able to encounter Uranus and Neptune as well.

Galileo was made up of an orbiter and an entry probe. After a six-year journey through the Solar System, Galileo approached Jupiter in 1995; on 13 July of that year the orbiter was separated from the entry probe, and the two reached Jupiter on different trajectories. The entry probe dived into the Jovian clouds on 7 December 1995, and survived for 57.6 min, by which time it had penetrated to a depth of about 600 km. Six hours before entry it also detected a new, very intense radiation zone round Jupiter.

The orbiter acted as a relay, and after the demise of the entry probe began a long-continued survey of the satellite system.

Ulysses was essentially a solar probe, and Jupiter was incidental – it had to be encountered to put Ulysses into its planned path, well out of the ecliptic. However, during its pass of Jupiter, Ulysses did send back some useful data about the magnetosphere, radiation zones and general environment.

SATELLITES
Jupiter's satellite family is unlike any other in the Solar System. There are four main satellites, of planetary size, always known as the Galileans, although probably Marius saw them slightly before Galileo did so. Of these, three are larger than our Moon and the fourth (Europa) only slightly smaller, while Ganymede is actually larger than Mercury, although less massive. There are 12 smaller satellites, four close in and eight further out than the Galileans; the outer satellites are probably asteroidal, and are so perturbed by the Sun that their orbits are not even approximately circular. The outer four have retrograde motion. Data for the satellites are given in Table 9.8.

Table 9.8. Satellites of Jupiter.

Satellite	Discoverer	Mean distance from Jupiter (km)	Mean angular distance from Jupiter, at mean opposition distance	
			(')	(")
XVI Metis	S Synott, 1979	127 969		
XV Adrastea	D Jewitt and E Danielson, 1979	128 971		
V Amalthea	E Barnard, 1892	181 300	0	59.4
XIV Thebe	S Synott, 1979	221 895		
I Io	Galileo and Marius, 1610	421 600	2	18.4
II Europa	Galileo and Marius, 1610	670 900	3	40.1
III Ganymede	Galileo and Marius, 1610	1 070 000	5	51.2
IV Callisto	Galileo and Marius, 1610	1 880 000	10	17.6
XIII Leda	C Kowal, 1974	11 094 000	60	45
VI Himalia	C Perrine, 1904	11 480 000	62	45
X Lysithea	S Nicholson, 1938	11 720 000	64	05
VII Elara	C Perrine, 1905	11 737 000	64	10
XII Ananke	S Nicholson, 1951	21 200 000	116	
XI Carme	S Nicholson, 1938	22 600 000	128	
VIII Pasiphaë	P Melotte, 1908	23 500 000	129	
IX Sinope	S Nicholson, 1914	23 700 000	130	
S1999/J1	1999	24 000 000		

	Sidereal period (days)	Mean synodic period				Orbital inclination (°)	Orbital eccentricity
		(d)	(h)	(m)	(s)		
Metis	0.294 779					0.0000	0.0000
Adrastea	0.298 260					0.0000	0.0000
Amalthea	0.498 179	0	11	57	17.6	0.40	0.003
Thebe	0.674 536					1.0659	0.0183
Io	1.769 138	1	18	28	35.9	0.040	0.004
Europa	3.551 181	3	13	17	53.7	0.470	0.009
Ganymede	7.154 553	7	03	59	35.9	0.195	0.002
Callisto	16.689 02	16	18	05	06.9	0.281	0.007
Leda	238.72					26.07	0.148
Himalia	250.5662					27.63	0.158
Lysithea	259.22					29.02	0.107
Elara	259.653					24.77	0.207
Ananke	631					147	0.169
Carme	692					163	0.207
Pasiphaë	735					147	0.378
Sinope	758					153	0.275
S1999/J1	±730						

Table 9.8. (Continued)

	Rotation period (days)	Mean orbital velocity (km s^{-1})	Diameter (km)	Density, water = 1	Reciprocal mass, Jupiter = 1
Metis	?	31.57	60 × 28	2.8	
Adrastea	?	31.45	26 × 20 × 16	?	
Amalthea	0.498 179	26.47	262 × 146 × 143	1.8	
Thebe	0.674 536	23.93	110 × 90	1.5	
Io	1.769 138	17.34	3660 × 3637 × 3631	3.55	21 300
Europa	3.551 181	13.74	3130	3.01	39 000
Ganymede	7.154 553	10.88	5268	1.94	12 700
Callisto	16.689 02	8.21	4806	1.86	17 800
Leda	?	3.38	16	2.7	
Himalia	0.4	3.34	186	2.8	
Lysithea	?	3.29	36	3.1	
Elara	0.5	3.29	76	3.3	
Ananke	?	2.44	30	2.7	
Carme	?	2.37	40	2.8	
Pasiphaë	?	2.32	50	2.9	
Sinope	?	2.27	36	3.1	
S1999/J1	?		5		

	Escape velocity (km s^{-1})	Visual geometric albedo	Apparent diameter as seen from Jupiter (′)	Apparent diameter as seen from Jupiter (″)	Magnitude at mean opposition distance
Metis	0.0253	0.05			17.5
Adrastea	0.0143	0.05			19.1
Amalthea	0.0842	0.05	7	24	14.1
Thebe	0.0434	0.05			15.7
Io	2.56	0.61	35	40	5.0
Europa	2.02	0.64	17	30	5.3
Ganymede	2.74	0.42	18	06	4.6
Callisto	2.45	0.20	9	30	5.6
Leda	0.0097	low	0	0.15	20.2
Himalia	0.117	0.03	0	8.2	14.8
Lysithea	0.0240	low	0	0.03	18.4
Elara	0.0522	0.03	0	0.14	16.8
Ananke	0.0184	low	0	0.2	18.9
Carme	0.0253	low	0	0.2	18.0
Pasiphaë	0.0319	low	0	0.2	17.0
Sinope	0.0240	low	0	0.2	18.3
S1999/J1					

The Galileans can be seen with almost any telescope or even with good binoculars. Very keen-sighted people have even reported naked-eye sightings, and there is considerable evidence that one of them (probably Ganymede, or else two satellites close together) was seen from China by Gan De as long ago as 364 BC. The first attempted maps of the Galileans were due to A. Dollfus and his colleagues at the Pic du Midi Observatory in 1961. Some features were recorded, but, predictably, the maps were not very accurate. Today we have detailed maps obtained by spacecraft, and details can also be followed with the Hubble Space Telescope.

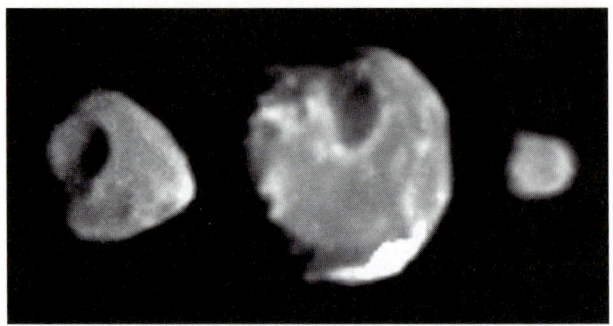

Figure 9.1. Thebe, Amalthea and Metis. (Courtesy: NASA.)

The Small Inner Satellites

Metis. Named after the daughter of Zeus (Jupiter) by his first consort, Oceanus. It was identified on the Voyager images, and lies within the Main Ring. Galileo images taken in 1996 and 1997 show that it is rather elliptical, with a longest diameter of about 60 km; the albedo is low.

Adrastea. Named after a daughter of Jupiter and Ananke (equated with Nemesis, the goddess of rewards and punishments). It too was discovered on Voyager images. It lies in the outer part of the Main Ring and is, with Metis, one of the main sources of the ring particles. It has low albedo, but nothing much else is known about it.

Amalthea. A mythological name; possibly the goat which suckled the infant Jupiter (Zeus) or possibly the daughter of Melisseus, King of Crete, who brought up the infant on a diet of goat's milk. It was discovered in 1892 by E. E. Barnard, using the 36 inch (91 cm) refractor at the Lick Observatory; this was the last satellite discovery to be made visually.

The satellite is irregular in form; the surface is very red, due probably to contamination from Io. It has synchronous rotation, with its longest axis pointing toward Jupiter, and is heavily cratered. The two largest craters, Gaea and Pan, are of immense size relative to the overall diameter of Amalthea. Pan is 90 km in diameter. It was well imaged in January 2000 by the Galileo probe, from a range of 351 km. Gaea seems to have a depth of between 10 and 20 km; if the latter figure is correct, the slope angle of the wall will be 30°, making it the steepest known scarp

Table 9.9. Features on Amalthea.

	Lat.	Long. W	Diameter (km)
Craters			
Gaea	80.0S	90.0	80
Pan	55.0N	35.0	100
Faculæ			
Ida Facula	20.0N	175.0	
Lyctos Facula	20.0S	120.0	

in the Solar System. (It would be interesting to watch a piece of material fall from the crest to the floor; the descent time would be about 10 min!) Both Pan and Gaea are deeper, relatively, than craters of similar size on the Moon. Between them, from longitude 0–60° W, is a complex region of troughs and ridges, tens of kilometres long and up to at least 20 km wide. The two bright patches, Ida and Lyctos, are each about 15 km across and are presumably mountains (Table 9.9).

Amalthea is exposed to the Jovian radiation field, and also to energetic ions, protons and electrons produced in Jupiter's magnetosphere; it is also bombarded by micrometeorites, and by sulphur, oxygen and sodium ions that have been blasted away from Io.

Thebe. Named after the daughter of the river-god Asopus; discovered on the Voyager images. It moves beyond the main part of the Gossamer Ring. It has synchronous rotation and low albedo. The surface is dominated by a 40 km crater, imaged by the Galileo probe in 2000.

The Galilean Satellites

Io. Named after the daughter of Inachus, King of Argos; Jupiter was enamoured of her and Juno, Jupiter's wife, ill-naturedly changed Io into a white heifer.

The satellite is slightly larger than our Moon, and is the densest of the four Galileans. Before the space missions it was tacitly assumed to be a rocky, cratered world, but in the event nothing could have been further from the truth; it is the most volcanically active of any body in the Solar System.

In March 1979, S. Peale and his colleagues in America suggested that since Io's orbit is not perfectly circular, the interior might be 'flexed' by the gravitational pulls of Jupiter and the other Galileans, heating it sufficiently to produce active surface volcanoes. A week later, on 9 March, this prediction was dramatically verified. Linda Morabito, a member of the Voyager imaging team, was looking for a faint star, AGK-10021, as a check on Io's position when she saw what was undoubtedly a volcanic plume rising from the limb of the satellite. Subsequently nine plumes were detected, together with volcanic craters and numerous calderæ; impact craters were absent. The lack of impact craters means that the surface cannot be more than a million years old, and there must be constant 'resurfacing', with deposition of a layer 1 mm thick each year.

The surface is made up of vent regions, plains regions and mountains; the average surface temperature is $-143\,°C$. Mountains are appreciable; the highest, Hæmus, rises to 13 km, and its steep slopes means that it cannot be solid sulphur, even though sulphur and sulphur dioxide cover most of Io's surface. The mountains are presumably siliceous, with an outer coating of sulphur sent out by the volcanoes. The plains are crossed by yellow and brownish-yellow flows; the original Voyager images made them look redder than they really are. There are extensive deposits of sulphur dioxide (SO_2); the gas is vented from the volcanic areas and is frozen out when it reaches the bitterly cold surface.

The volcanoes seem to be of two main types. The sulphur volcanoes, such as Pele, Surt and Aten, send out material at up to 1 km s^{-1}; eruptions last for days or months (for example Pele was erupting at the time of the Voyager 1 pass, but was quiescent when Voyager 2 flew past the planet). The sulphur dioxide volcanoes, such as Prometheus, Amirani and Volund, have lower vent velocities, but eruptions go on for months or years consecutively. (Loki, one of the most violent centres, seems to be of a hybrid type.) The temperatures of the volcanoes are very high, and the Galileo probe recorded that Pillan Patera reached over 2000 $°C$. This is too hot for the material to be sulphur, and it now seems that intensely heated lava, in the form of silica enriched by magnesium and sodium, may be responsible for much or all of Io's vulcanism. However, the plumes are more in the nature of geysers, emitting sulphur dioxide particles and gas rather than water as in terrestrial geysers. They rise to hundreds of kilometres, although the vent velocities are not sufficient to expel material from Io altogether. The black patches seen round the geysers are due to sulphur dioxide frost. Over 200 calderæ have been identified, although by no means all are active; there are probably many lava lakes. Observations from the Galileo probe in late 1999 showed over 100 active centres. Pele volcano showed a red ring of sulphur, over 1200 km in diameter, deposited by a plume of material emerging from the volcano. Loki is the most powerful volcano in the Solar System, emitting more heat than all the Earth's active volcanoes combined; the temperature of the lava attains 1027 $°C$. A selected list of surface features is given in Table 9.10.

Io seems to have a dense core, rich in iron and iron sulphide, which extends half-way from the centre of the globe to the surface, and is overlaid by a mantle of partly molten rock; above this comes the relatively thin, rocky, sulphur-coated crust. The atmosphere of sulphur dioxide is excessively tenuous, and corresponds to what we usually call a good laboratory vacuum; it may also be very variable in both density and distribution. Its highest pressure is one-thousand millionth of that of the Earth's air at sea level.

On Io, the active volcanic areas and pateræ have been named after gods of fire, thunder, volcanoes, mythical blacksmiths and solar deities (for example, Pele is the Hawaiian goddess of fire), catenæ after Sun-gods and the other features after people and places associated with myths involving Io.

Io is indeed a strange, colourful place, but since it moves well within Jupiter's radiation zones it must be just about the most lethal world in the entire Solar System. On 25 November 1999 the Galileo probe passed only 299 km from its surface.

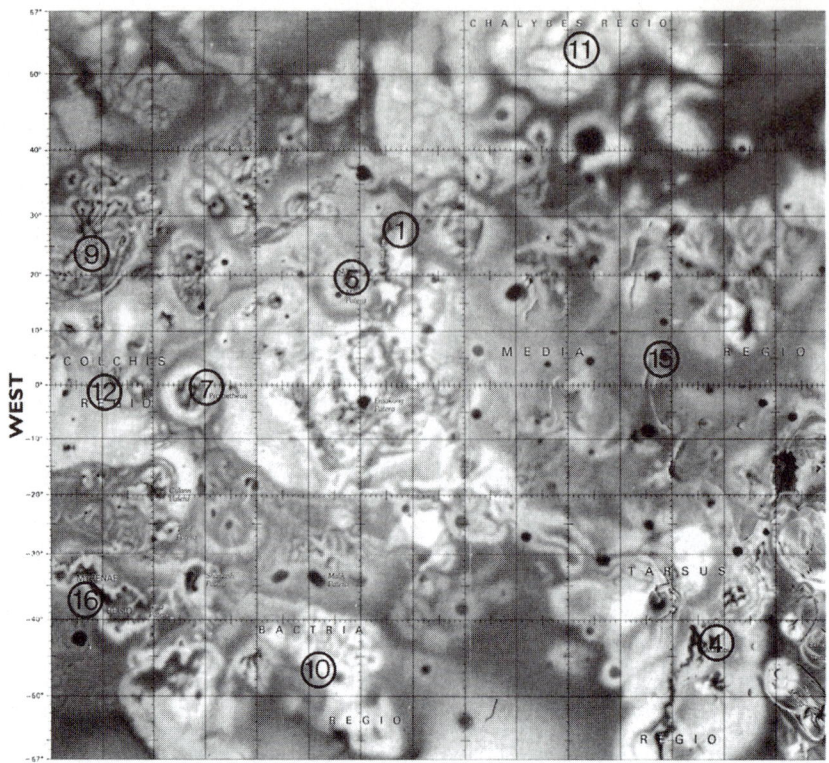

Figure 9.2. Io.

Table 9.10. Selected list of features on Io. Heights and widths are derived from observations from space-craft and the Hubble Space Telescope. They are no doubt very variable. (Bold numbers indicate map references.)

	Lat.	Long. W	Height (km)	Width (km)
Eruptive sites				
Amirani **1**	25.9N	114.5	95	220 (Plume 5)
Aten	47.9S	310.1	300	1200
Culann Patera	19.9S	158.7	—	—
Kanehekili	18.0S	037.0	—	—
Loki **2**	17.9N	302.6	200	400 (Plume 2)
Malik Patera	34.2S	128.5	—	—
Marduk **3**	27.1S	207.5	70	195 (Plume 7)
Masubi **4**	46.3S	54.7	—	— (Plume 8)
Maui **5**	16.5N	124.0	90	230 (Plume 6)
Pele **6**	18.6S	257.8	400	1200 (Plume 1)
Pillan Patera	12.0S	244.0	140	400
Prometheus **7**	1.6S	153.0	75	270 (Plume 3)
Ra Patera	8.6S	325.3	—	400
Surt **8**	45.5N	337.9	300	1200
Volund **9**	25.0N	184.3	100	125 (Plume 4)
Zamama	18.0N	173.0	—	—

NORTH

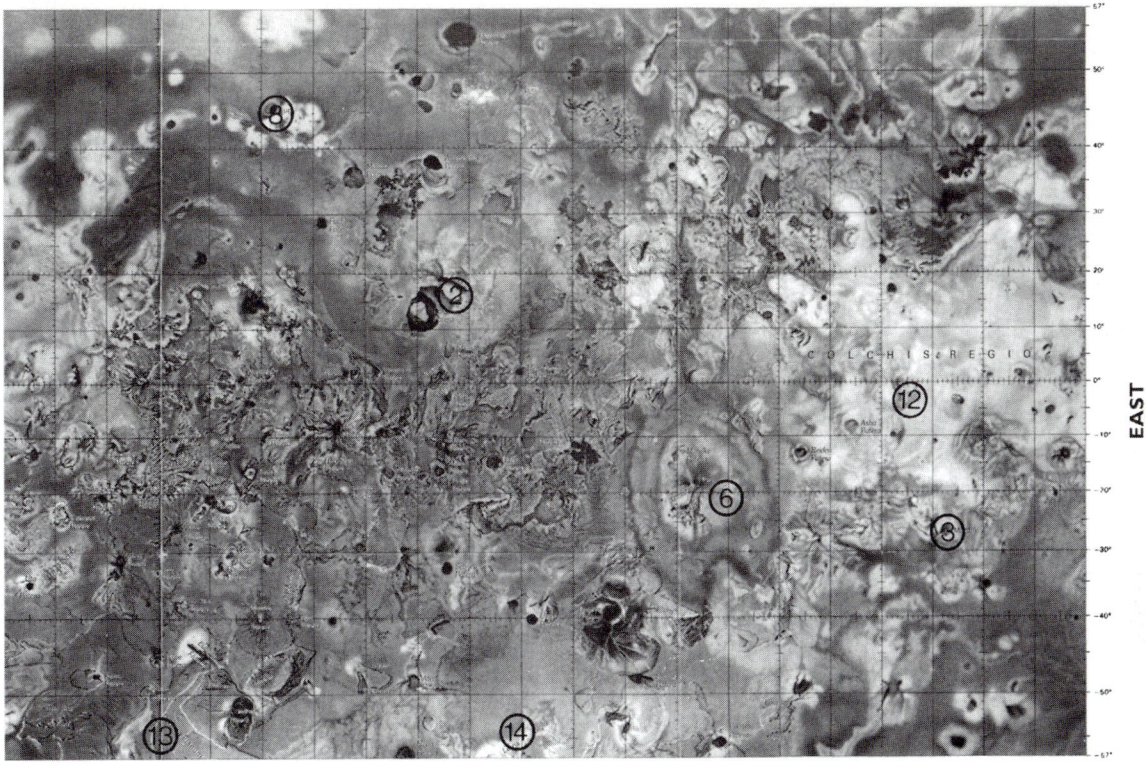

SOUTH

Table 9.10. (Continued)

	Lat.	Long.	Diameter (km)		Lat.	Long.	Diameter (km)		Lat.	Long.	Diameter (km)
Pateræ				Reiden	13.4S	235.7	70	Hæmus	68.9S	46.6	
Amaterasu	37.7N	306.6	100	Ruwa	0.4N	3.0	50	Silpium	52.6S	272.9	
Aten	47.9S	310.0	40	Shakuru	23.6N	266.4	70	*Mensæ*			
Atar	30.2N	278.9	125	Shamash	33.7S	152.1	110	Echo	79.6S	357.4	
Babbar	39.5S	272.1	95	Svarog	48.3S	267.5	70	Epaphus	53.5S	241.3	
Cataquil	24.2S	18.7	125	Taranis	70.8S	28.6	105	Iynx	61.1S	304.6	
Creidne	52.4S	343.5	125	Tol-Ava	1.7N	322.0	70	Pan	49.5S	35.4	
Daedalus	19.1N	274.3	40	Tupan	18.0S	141.0	50	*Planum*			
Discura	37.0N	119.0	70	Ülgen	40.4S	288.0	49	Argos	47.0S	318.2	140
Emakong	3.2S	119.1	80	Vahagn	23.8S	351.7	70	Danube	20.9S	258.7	150
Galai	10.7S	288.3	90	Viracocha	61.2S	281.7	55	Dodona **13**	56.8S	352.9	390
Gibil	14.9S	294.9	95	Zal	42.0N	76.0	130	Ethiopia	44.9S	27.0	105
Gish Bar	17.0N	90.0	150	*Catenæ*				Hybristes	54.0S	21.1	150
Heinseb	29.7N	244.8	60	Mazda	8.6S	313.5		Iopolis	34.5S	333.5	125
Horus	9.6S	338.6	125	Reshet	0.8N	305.6		Lyrcea	40.3S	269.3	310
Huo Shen	15.1S	329.3	90	*Fluctus*				Nemea	73.3S	275.5	500
Isum	29.0N	208.0	100	Eubœa	45.1S	351.3		*Regiones*			
Kane	47.8S	13.4	115	Fjorgynn	17.5N	358.0	300	Bactria **10**	45.8S	123.4	
Khalla	6.0N	303.4	80	Ionian	5.0N	250.0		Chalybes **11**	45.5N	83.2	
Loki	12.6N	308.8	250	Kanehekili	16.0S	38.0	250	Colchis **12**	5.3N	199.8	
Lu Huo	38.4S	354.1	90	Lei-Kung	38.0N	204.0	400	Illyrikon	72.0S	160.0	700
Mafuike	13.9S	260.0	110	Marduk	27.0S	209.0	150	Lerna **14**	64.0S	292.6	
Malik	34.2S	128.5	85	Masubi	48.0S	60.4	800	Media **15**	4.6N	58.8	
Masaya	22.5S	348.1	125	Tung Yo	16.4S	357.8		Mycenæ **16**	37.3S	165.9	
Mihr	16.4S	305.6	40	Uta	32.6S	19.2		Tarsus	43.7S	61.4	
Nina	38.3S	164.2	425	*Montes*				*Tholus*			
Nusku	64.7S	4.6	90	Boösaule	4.4S	270.1	590	Apis	11.2S	348.8	
Nyambe	0.6N	343.9	50	Eubœa	46.3S	339.9		Inachus	15.9S	348.9	

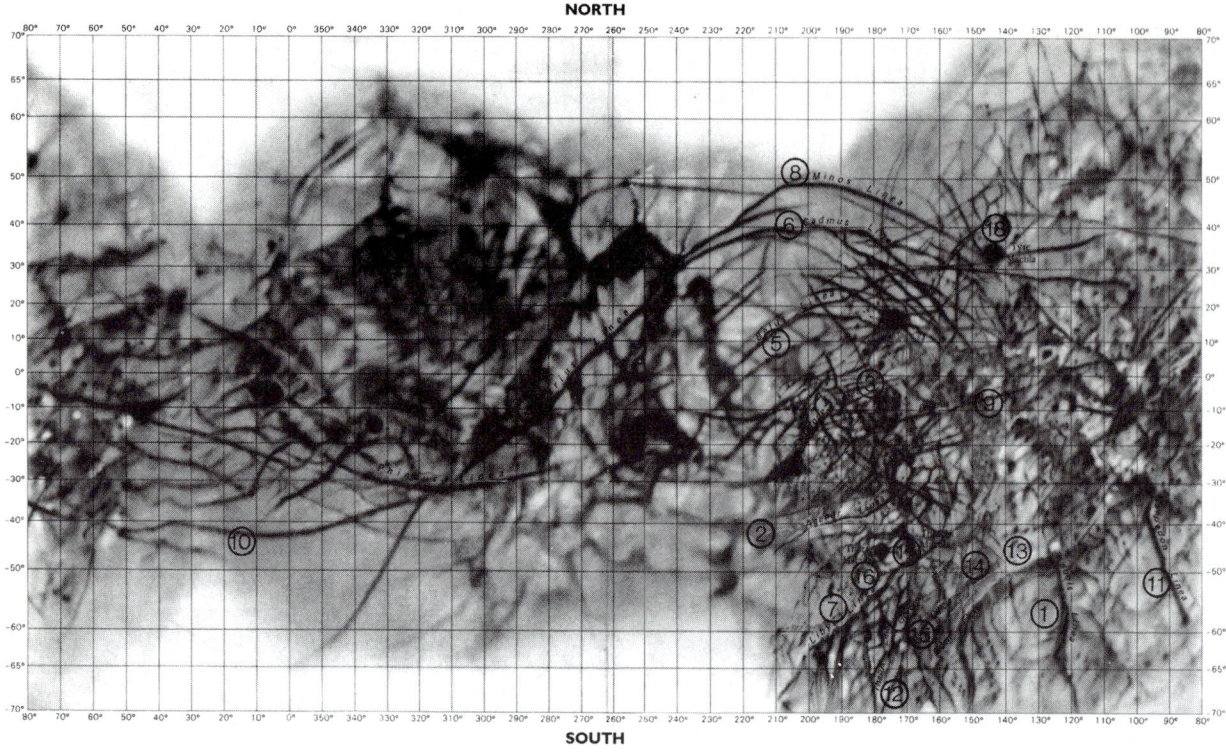

Figure 9.3. Europa.

Europa. Europa is the second and smallest of the four Galileans. In mythology she was the daughter of King Agenor of Tyre and sister of Cadmus; Jupiter (Zeus) assumed the form of a bull and carried her across the sea to Crete, where she bore him several children.

Europa is as different from Io as it could possibly be. Its surface is smooth and white, covered with water ice or snow; the average albedo is 0.7 for the white regions and 0.5–0.6 for the slightly darker areas, so that Europa is particularly reflective. It is also very cold, with a mean surface temperature of −145 °C. There are few impact craters, showing that the surface must be young – perhaps only a few millions of years old, so that there must be constant resurfacing.

The surface features are unlike any found elsewhere in the Solar System, and when they were first seen, on the Voyager images, new terms had to be introduced (flexus, linea, macula). There is little surface relief, and no hills as

much as 1 km high. The trailing hemisphere is somewhat darker than the leading part of the satellite, due no doubt to contamination from Io.

Europa has been described as a map-maker's nightmare; one region looks very much like another. The criss-crossing ridges and linear features extend for thousands of kilometres; there are ridges and narrow grooves, and shallow pits a few kilometres across, as well as darker patches with diameters of from 50 to 500 km.

There is absolutely no doubt that the surface material is water ice, and in 1982 R. Reynolds and S. Squyres suggested that below the crust there might be an ocean of liquid water. This is now regarded as a real possibility. Io has marked effects upon Jupiter's magnetic field, but the same would not be expected of Europa; yet this does apparently happen. Water, particularly salty water, is a good conductor of electricity, and if Jupiter's field sets up a current in an underground ocean on Europa this will make its presence felt.

Table 9.11. Selected list of features on Europa. (Bold numbers indicate map references.)

Name	Lat.	Long. W	Diameter/ length (km)
Craters			
Cilix	1.2N	181.9	23
Govannan	37.5S	302.6	10
Manann'an	20.0N	240.0	30
Morvran	5.7S	152.2	25
Pwyll	26.0	271.0	26
Rhiannon	81.8S	199.7	25
Taliesin	23.2S	137.4	48
Tegid	0.6S	164.0	29
Flexus			
Cilicia **13**	47.6S	142.6	639
Delphi	69.7S	172.3	1125
Gortyna **14**	42.4S	144.6	1261
Phocis	48.6S	197.2	298
Sidon **15**	64.5S	170.4	1216
Linea			
Adonis **1**	51.8S	113.2	758
Agenor **2**	43.6S	208.2	1326
Alphesibœa	28.0S	182.6	1642
Argiope **3**	8.2S	202.6	934
Asterius **4**	17.7N	265.6	2735
Astypalœa	76.5S	220.3	1030
Belus **5**	11.8N	228.3	2580
Cadmus **6**	27.8N	173.1	1212
Echion	13.1S	184.3	1217
Ino	5.0S	163.0	1400
Katreus	39.5S	215.5	245
Libya **7**	56.2S	183.3	452
Minos **8**	45.3N	195.7	2134
Pelagon	34.0N	170.0	800
Pelorus **9**	17.1S	175.9	1770
Phœnix	14.5N	184.7	732
Phineus **10**	33.0S	269.2	1984
Rhadamanthys	18.5N	200.8	1780
Sarpedon **11**	42.2S	89.4	940
Tectamus	17.9N	181.9	719
Telephassa	2.8S	178.8	800
Thasus **12**	68.7S	187.4	1027
Thynia	57.9S	148.6	398
Maculæ			
Boestia	54.0S	166.0	22
Cycleides	64.0S	192.0	105
Thera **16**	47.7S	180.9	78
Thrace **17**	46.6S	171.2	173
Large ring features			
Callanish	16.0S	333.4	100
Tyre **18**	31.7N	147.0	148

According to one model, Europa has an iron-rich core about 1250 km across, overlaid by a silicate mantle and then an ice-water crust up to 150 km thick; the depth of the ocean – if it exists at all – has been given as anything from a few metres up to as much as 150 km.

The plains are broken into plates a few kilometres across, and it has even been proposed that they drift about above the liquid or mushy material below; along fracture lines, warm material may well up to produce the linear features (a process termed cryovulcanism). One very young crater, Pwyll, shows bright rays extending in all directions and crossing all other features; possibly the impactor penetrated the crust through to the darker material below, and the crater floor is now at the same level as the outer terrain, so that it may be filled with slushy material. There are even features which look uncannily like icebergs. In 1999 the Galileo space-probe detected sulphuric acid on the surface of Europa. Hydrogen peroxide has also been detected.

Observations from the Hubble Space Telescope and the Galileo space-craft have shown that Europa has an excessively tenuous oxygen atmosphere, with a density no more than one hundred thousand millionth of that of the Earth's air at sea-level. If all of it were compressed to the density of our air, it would just about fill the Royal Festival Hall. The icy surface of the satellite is subject to impacts from dust and charged particles, and these processes cause the surface ice to produce water vapour as well as gaseous fragments of water molecules. These are then broken up into hydrogen and oxygen. The hydrogen escapes, while the oxygen is retained to form a thin atmosphere extending up to perhaps 200 km; obviously it must be continuously replenished from below.

Craters on Europa are named after Celtic heroes, and other features after people associated with myths involving Europa. A selected list of surface features is given in Table 9.11.

There are many cycloid-shaped cracks known as *flexi*. G. Hoppa and B. Randall (University of Arizona) suggested in 1999 that they were formed as Europa's icy crust responded to tidal forces induced by Jupiter. There is certainly a tidal bulge 30 m high, and this shifts location during each revolution, since the orbit of the satellite is slightly eccentric. According to Hoppa and Randall, this causes tension cracks to open and propagate along the

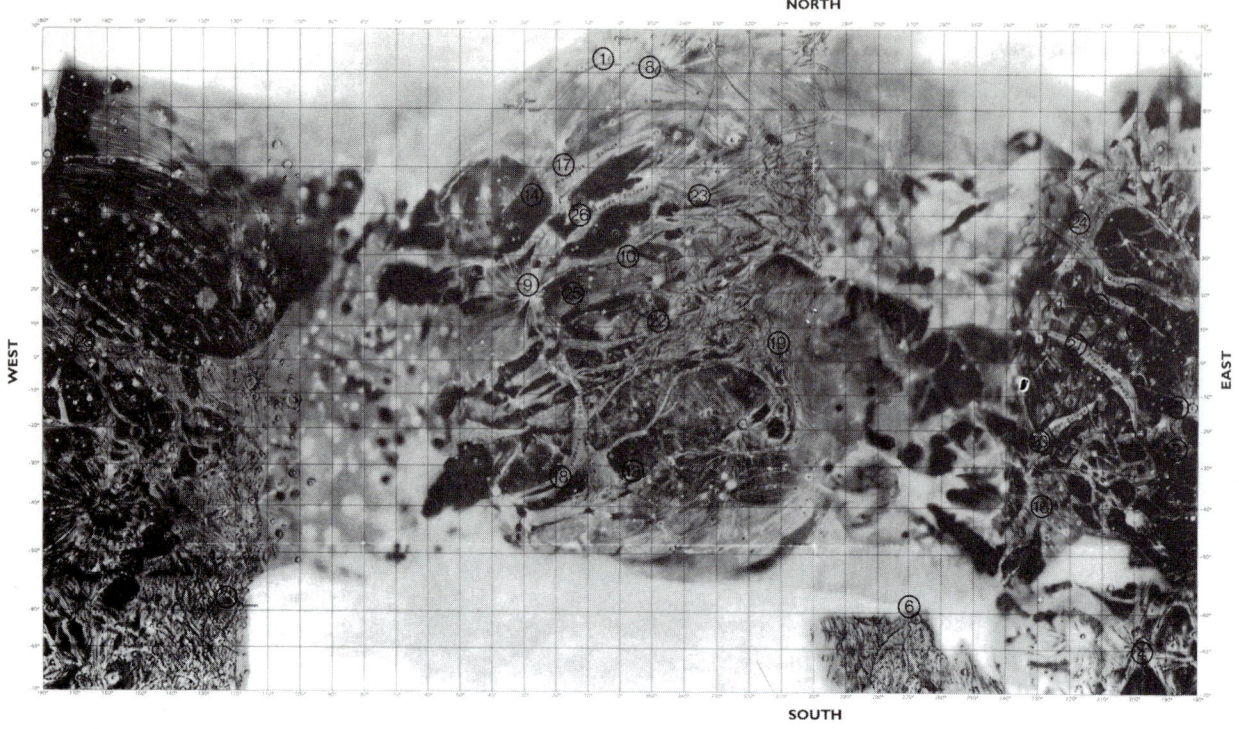

Figure 9.4. Ganymede.

surface at a rate of around $3 \, \text{km} \, \text{h}^{-1}$. This, of course, would indicate that the tidal bulge is sliding freely over the interior, and is further evidence in support of an underground ocean (in which it has even been suggested that life might exist, though this is, to put it mildly, highly speculative).

On 3 January 2000 the Galileo probe passed Europa at 351 km, and magnetic effects added credibility to the idea of an underground ocean. Water, particularly salty water, conducts electric currents well, but ice does not, so that an ice-shell seems unlikely.

Ganymede. Ganymede is the largest satellite in the Solar System. It is named after a handsome shepherd boy summoned by Jupiter to become cup-bearer to the gods. (One has to admit that Jupiter's motives were not entirely altruistic!)

Ganymede is much less dense than Io or Europa, and is of quite different type; the overall density is less than twice that of water. The interior structure is not well known, but there must be an iron-rich, partially molten core; above this comes the mantle, of which the lower part is siliceous and the upper part icy. The mantle is overlaid by the thin, icy crust. All in all, it seems that the globe is made up of a combination of rocky materials (60%) and ice (40%).

There are two types of surface: dark areas and bright regions. The dark regions are well-defined; the most prominent has been appropriately named Galileo Regio. They are heavily cratered, and are presumably the oldest parts of the surface. Crossing them are dark furrows (sulci), from 5 to 10 km wide; very often they indicate the outlines of distorted circles, and may well have originated from massive impacts early in Ganymede's history. There are also light-floored, rough features without walls; these are termed palimpsests, and are very ancient indeed. Their floors are relatively flat; some of them are as much as 200 km across.

The bright regions are characterized by grooves (sulci), which run in some cases for thousands of kilometres, although the vertical relief does not exceed a few hundred metres; they are often fist-topped, with gentle slopes of up

Table 9.12. Selected list of features on Ganymede. The latitudes and longitudes are central values. (Bold numbers indicate map references.)

Name	Lat.	Long. W	Diameter/ length (km)	Name	Lat.	Long. W	Diameter/ length (km)
Regiones				*Craters* (Continued)			
Barnard **10**	0.8N	1.0	2547	Ta-Urt	26.5N	306.5	85
Galileo **11**	35.7N	137.6	3142	Thoth	42.4S	146.0	107
Marius **12**	12.1N	199.3	3572	Tros **9**	11.0N	31.1	109
Nicholson **13**	34.0S	356.7	3719	Zaqar	57.5N	41.3	52
Perrine **14**	38.8N	30.0	2145				
				Faculæ			
				Abydos	34.1N	154.0	165
Craters				Busiris	14.9N	216.1	348
Achelous **1**	60.3N	13.5	51	Buto	12.6N	204.3	236
Agreus	15.2N	225.4	72	Coptos	9.4N	209.8	332
Amon	33.4N	223.3	102	Dendera	0.0	257.0	114
Anubis	82.7S	118.5	97	Edfu	26.8N	147.7	187
Ashima	37.7S	122.4	82	Memphis	15.4N	132.5	344
Bau	24.1N	53.3	81	Ombos	3.8N	238.6	90
Enkidu	27.9S	328.4	121	Punt	26.1S	242.2	228
Eshmun **2**	17.8S	191.5	99	Sais	37.9N	14.2	137
Gilgamesh **3**	61.7S	123.9	175	Siwah	7.5N	143.2	220
Halieus	35.2N	168.0	90	Tettu	38.6N	160.9	86
Hapi	31.3S	212.4	85	Thebes	4.8N	202.4	475
Irkilla	31.1S	114.7	116				
Ishkur	0.1M	11.5	83	*Fossæ*			
Isimu	8.1N	2.5	90	Lakhamu Fossa	12.5S	228.3	392
Isis **4**	67.9S	197.2	68	Lakhmu Fossæ	30.3N	142.3	2871
Kadi	48.8N	181.0	94	Zu Fossæ	53.0N	129.4	1386
Khonsu	38.0S	189.7	86				
Kingu	35.7S	227.4	91	*Sulci*			
Kulla	34.8N	115.0	82	Aquarius Sulcus **17**	50.0N	11.5	1341
Melkart **5**	10.0S	185.8	111	Arbela Sulcus	22.3S	353.6	1896
Misharu	5.3S	338.3	95	Bubastis Sulci	79.8S	263.1	2197
Mush	13.5S	115.0	97	Dardanus Sulcus **18**	39.3S	20.2	2559
Neith	28.9N	9.0	93	Elam Sulci	57.4N	205.5	1866
Nidaba	19.0N	123.8	188	Mashu Sulcus **21**	31.1N	209.2	3030
Ninki	6.6S	120.9	170	Mysia Sulci **22**	9.6S	28.6	4221
Ninlil	7.6N	118.7	91	Nippur Sulcus	40.9N	191.5	2158
Nunsum	13.3S	140.1	91	Phrygia Sulcus **25**	12.4N	19.3	3205
Nut	60.1S	268.0	93	Sippar Sulcus	15.8S	191.0	1539
Osiris **7**	37.8N	165.2	109	Tiamat **27**	3.2N	209.2	1310
Sati	30.5N	14.9	98	Ur Sulcus	48.0N	178.0	950
Sebek **8**	59.5N	178.9	70	Uruk Sulcus **28**	8.4N	169.0	2456
Seker	40.8S	351.0	117	Xibaltia	35.0N	80.0	2000
Selket	16.7N	107.4	140				

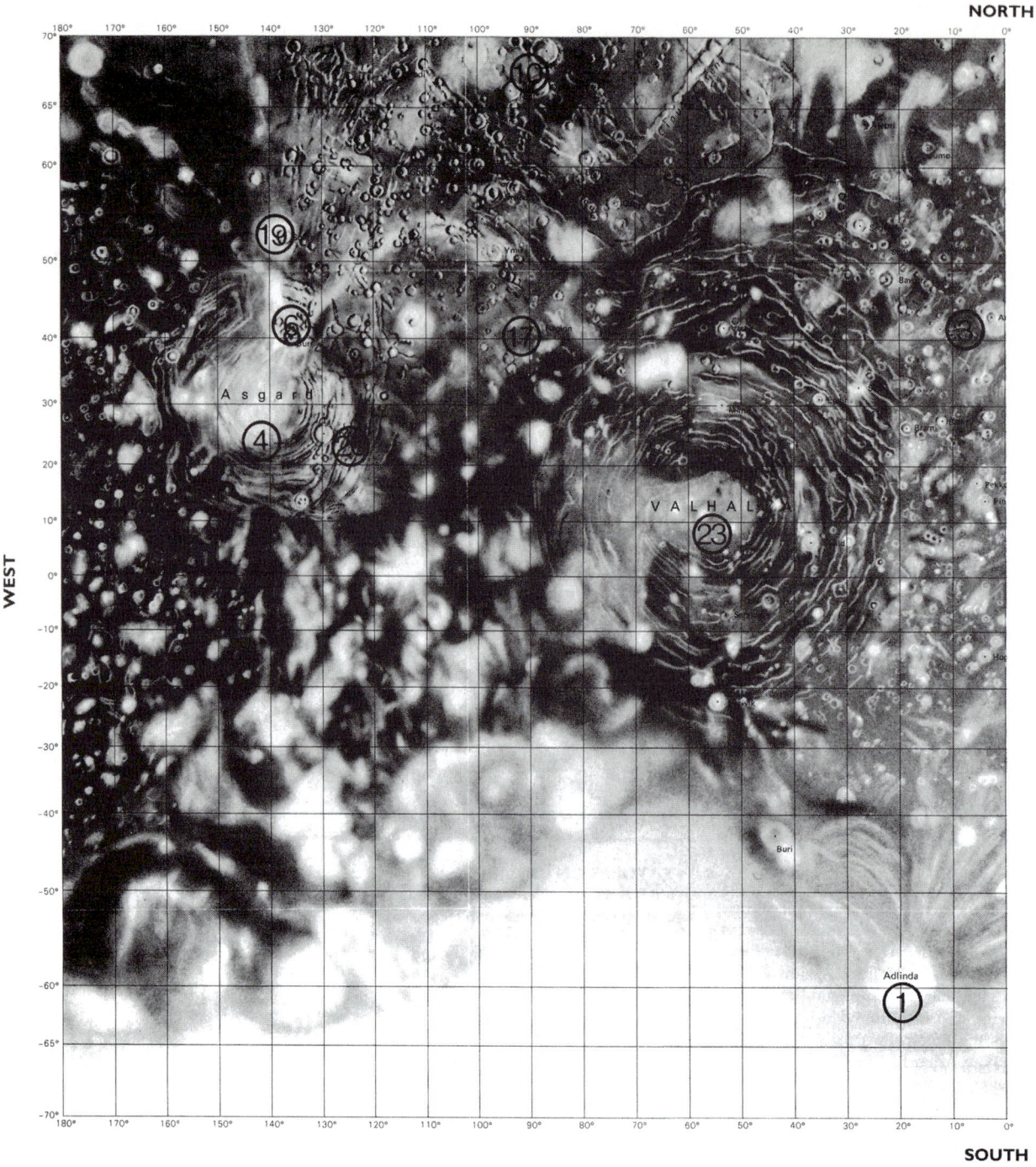

Figure 9.5. Callisto.

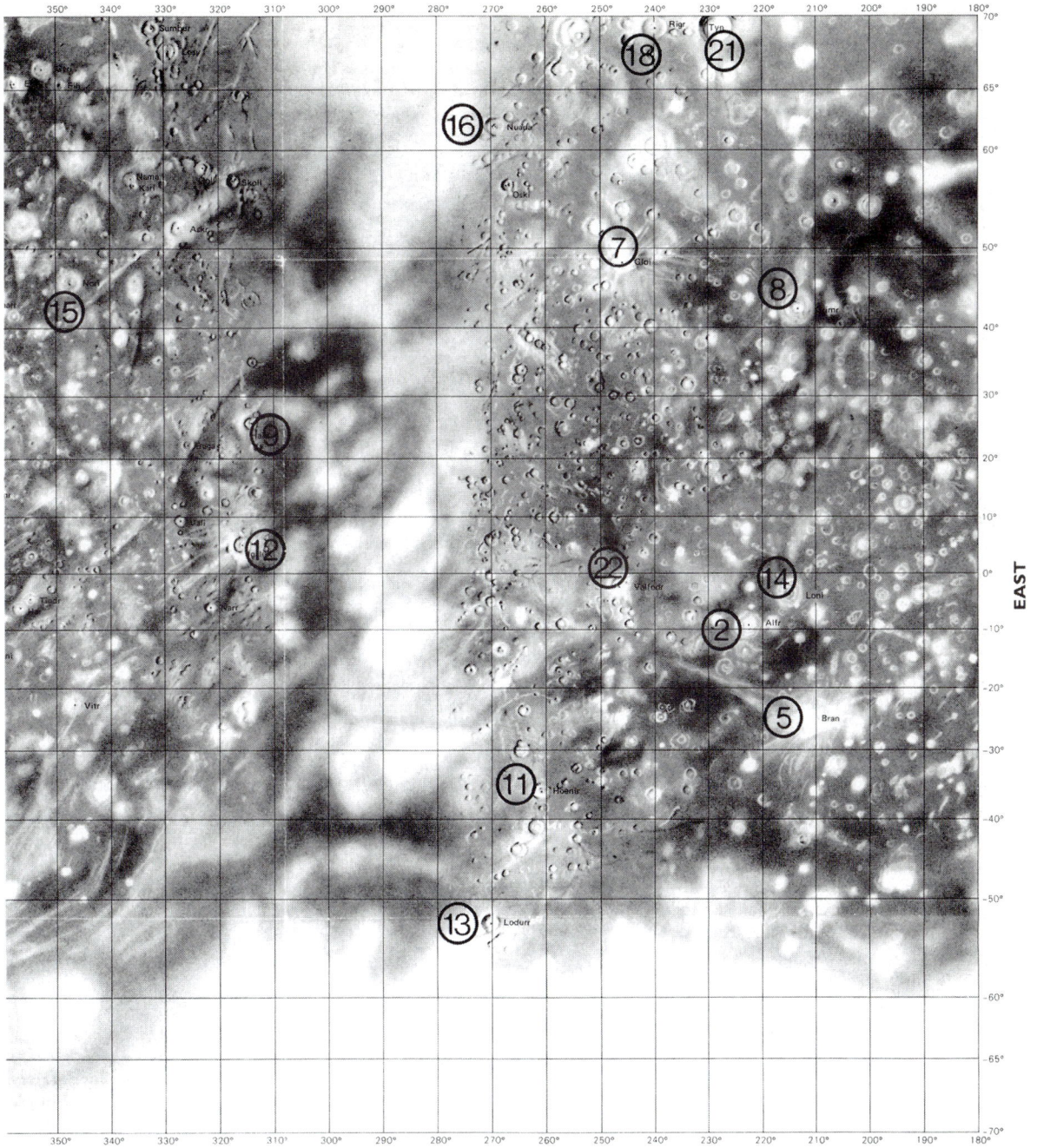

to 20°. These areas are essentially icy. Of the faculæ (bright spots) the most prominent is Memphis, which contains dark-floored craters which have been punched through to the darker material below – indicating that much of the bright palimpsest is a thin sheet.

Impact craters abound; the largest well-marked crater is the 175 km Gilgamesh, which is surrounded by outlying concentric escarpments with an overall diameter of 800 km. Small craters tend to have central peaks, while with larger craters (over 35 km across) central pits are more common. There are also ray-craters, such as Osiris, whose brilliant rays stretch out for over 1000 km. Unquestionably there has been marked tectonic activity in past ages, although Ganymede today is to all intents and purposes inert.

One major surprise, due to the Galileo probe, is that Ganymede has a magnetic field. By terrestrial standards it is weak, but it is sufficient to produce a well-defined magnetosphere – so that we have a magnetosphere within a magnetosphere. The magnetic axis is inclined to the rotational axis by about 10°. There is even vague evidence of polar auroræ.

The Hubble Space Telescope detected ozone on the surface, caused by the disruption of icy particles by bombardment from charged particles. There is also a very tenuous oxygen atmosphere, no denser than that of Europa and presumably of the same type.

On Ganymede, craters and fossæ are named after gods and heroes of the ancient Fertile Crescent peoples, faculæ after places associated with Egyptian myths, sulci after places associated with other ancient myths, and regions after astronomers who have discovered Jovian satellites (Galileo, Simon Marius, E. E. Barnard, S. B. Nicholson and C. D. Perrine). A selected list of formations on Ganymede is given in Table 9.12.

Callisto. Callisto, the fourth Galilean, is named after the daughter of King Lycaon of Arcadia, who was turned into a bear by Juno and subsequently placed in the sky as Ursa Major.

Callisto is almost as large as Mercury, but its relatively low albedo means that it is the faintest of the Galileans. It is also much further away from Jupiter, so that eclipse, transit and occultation phenomena are less frequent than with Io, Europa or Ganymede. It is also the least dense of

the four, although its escape velocity is still higher than that of Europa.

Although Callisto is almost equal to Ganymede in size, it is different in many respects. It seems to be comparatively undifferentiated, and until recently it was thought to lack any substantial iron-rich core; we are not yet certain whether such a core exists. The surface is icy, and is saturated with craters. The dominant features are two large ringed basins, Valhalla and Asgard. Valhalla is a complex structure; around the central 600 km palimpsest there are concentric rings, and the palimpsest itself is less heavily cratered than the surrounding areas, showing that Valhalla is young by Callistan standards – although the entire surface is very ancient, and there is no definite evidence of past tectonic activity, as there is on Ganymede. Asgard is similar in type, although smaller; the third basin is partly obscured by rays from Adlinda, the crater after which the basin itself is named. Elsewhere there are craters all over the surface, some of which have dark floors with bright rims and central peaks, but very small craters seem to be less numerous than was thought before the data sent back by the Galileo space probe. There are various catenæ (crater-chains), due probably to the impacts of comets or asteroids which were broken up before impacting. For example, Gipul Catena is a crater-chain 620 km long; the largest crater is 40 km across. Svol, crossing the 81 km crater Skul, is another good example of a catena. Callisto shows no Ganymede-type grooves, and there are not many craters over 100 km in diameter.

Ringed basins are named after the homes of the Norse gods and heroes; craters from heroes and heroines from Northern myths, and catenæ from mythological places in Scandinavia. A selected list of features on Callisto is given in Table 9.13.

Before the Galileo results it was generally assumed that Callisto was solid through to its centre, and that it had been totally inert since the very early days of the Solar System. The globe seems to consist of a mixture of rock and ice in equal proportions, with the percentage of rock increasing with increasing depth. The main surprise has been that as Callisto moves through Jupiter's magnetic field it seems to produce the same sorts of effects as Europa, and this has led to a change of opinion; beneath the 200 km thick crust there may be a salty ocean, up to 10 km deep. This may sound inherently improbable, but it is

Table 9.13. Selected list of features on Callisto. (Bold numbers indicate map references.)

Name	Lat.	Long. W	Diameter/length (km)	Name	Lat.	Long. W	Diameter/length (km)
Basins				*Craters* (Continued)			
Adlinda **1**	56.6S	23.1	900	Hödr **10**	69.0N	91.0	76
Asgard **4**	32.0N	139.8	1347	Högni	13.5S	4.5	65
Valhalla **23**	15.9N	56.6	2748	Igaluk **12**	5.6N	315.9	105
				Ivarr	6.1S	321.5	68
Catena				Lodurr **13**	51.2S	270.8	76
Gipul Catena	70.2N	48.2	588	Loni **14**	3.6S	214.9	86
				Nār	1.7S	46.4	63
Craters				Nori **15**	45.4N	343.5	86
Adal	75.4N	80.8	40	Nuada **16**	62.1N	273.2	66
Ägröi	43.3N	11.0	55	Reginn **17**	39.7N	90.8	51
Ahti	41.8N	103.1	52	Rigr **18**	70.9N	245.0	54
Akycha	72.5N	318.6	67	Sequinek	55.5N	25.5	80
Ali	59.3N	56.2	61	Sköll	55.6N	315.3	55
Anarr **3**	44.1N	0.6	47	Skuld	10.1N	37.7	81
Aningan	50.5N	8.2	287	Sudri **19**	55.4N	137.1	69
Askr	51.7N	324.1	64	Tindr	2.5S	355.5	64
Aziren	35.4N	178.3	64	Tornarsuk **20**	28.7N	128.6	104
Balkr	29.1N	11.9	64	Tyll	43.3N	165.4	65
Bavörr	49.2N	20.3	84	Tyn **21**	70.8N	233.6	60
Brami	28.9N	19.2	67	Valfödr **22**	1.2S	247.8	81
Bran **5**	24.3S	207.7	89	Vanapagan	38.1N	158.0	62
Buri	38.7S	46.2	98	Veralden	33.2N	96.1	75
Burr **6**	42.5N	135.5	74	Vestri	43.3N	52.8	75
Fadir	56.4N	12.7	81	Vidarr	11.9N	193.6	84
Finnr	15.5N	4.3	65	Vitr	22.4S	349.3	76
Gloi **7**	49.0N	245.7	112	Vu-Mart	22.9N	170.9	79
Grimr **8**	41.6N	215.2	90	Vutash	31.9N	102.9	55
Haki **9**	24.9N	315.1	69	Ymir	51.4N	101.3	77

not easy to account for the magnetic effects in any other way. According to M. Kivelson, principal investigator for Galileo's magnetometer, 'The new data certainly suggest that something is hidden below Callisto's surface, and that something may well be a salty ocean.'

Galileo has also detected an excessively tenuous atmosphere, which seems to be made up of carbon dioxide. The density is so low that it ranks as an exosphere, i.e. a collisionless gas. It is easily lost because of the effects of ultra-violet radiation from the Sun, and so it must be constantly replenished, perhaps by venting from Callisto's interior – although this again mitigates against the assumption that the satellite is completely inert. Altogether, Callisto has provided some unexpected results.

The Outer Small Satellites

The outer satellites are very small and probably asteroidal. They fall into two well-defined groups; the inner (Leda, Himalia, Lysithea and Elara) and the outer (Ananke, Carme, Pasiphaë and Sinope). The outer four satellites have retrograde motion.

Leda. Named after the wife of King Tyndareos of Sparta; she was the mother of the 'Heavenly Twins', Castor and Pollux.

Leda was discovered by C. Kowal in 1974. (At about the same time Kowal suspected the presence of another satellite, similar in brightness, but this has never been confirmed.) Because of its faintness and very small size, little is known about Leda.

Himalia. Named after one of Jupiter's many consorts – by whom she had three sons. It is much the largest of the asteroidal satellites, and the only one over 100 km in diameter. Variations in brightness have enabled its rotation period to be determined ($9\frac{1}{2}$ h) and it may be elliptical in shape, but nothing else is known about it.

Lysithea. Named after a daughter of the sea-god Oceanus. Again we know little about it.

Elara. One of Perrine's discoveries in 1905, shortly after the identification of Himalia; Elara is much the smaller of the two, and no details are available. In mythology Elara was one of Jupiter's countless lovers, and mother of the giant Titius.

Ananke. In mythology Ananke was the mother of Adrastea, by Jupiter. It is the smallest of the outer group of retrograde satellites, and we have no information about its surface. It was discovered photographically by Nicholson in 1951.

Carme. One of Jupiter's innumerable partners, and mother of Britomartis, a Cretan goddess of happiness. Like Lysithea, it was discovered by Nicholson in 1938. Also like Lysithea, we know nothing about its surface features.

Pasiphaë. Discovered by P. J. Melotte in 1908, Pasiphaë was 'lost' until 1914, and again between 1941 and 1955; like the other asteroidal outer satellites, it is so perturbed by the Sun that its orbit is very variable, and no two revolutions are alike. Of course, modern equipment ensures that there is no fear of its being mislaid again, but about its physical details we are totally ignorant.

Sinope. The outermost asteroidal satellite was discovered by accident. On 21 July 1914, S. B. Nicholson set out to photograph Pasiphaë; when he developed the plate, he found not only Pasiphaë but also the newcomer. Little is known about Sinope. It is quite possible that the four outer satellites are the débris from a larger body which was disrupted for some reason. No doubt further very small attendants await discovery.

10 SATURN

Saturn, the outermost planet known in ancient times, is sixth in order of distance from the Sun. It was named in honour of the first ruler of Olympus, who was succeeded by his son Jupiter (Zeus). It moves across the sky more slowly than any of the other planets known before the invention of the telescope, and has thus been associated with the passage of time. Data are given in Table 10.1.

Table 10.1. Data.

Distance from the Sun:
   max 1506.4 million km (10.069 a.u.)
   mean 1426.8 million km (9.359 a.u.)
   min 1347.6 million km (9.008 a.u.)
Distance from Earth:
   max 1658.5 million km
   min 1195.5 million km
Sidereal period: 10 746.94 days (29.4235 years)
Mean synodic period: 378.09 days
Rotation period: equatorial 10h 13m 59s
   internal 10h 39m 25s
Mean orbital velocity: 9.65 km s^{-1}
Axial inclination: 26°44' (26°.73)
Orbital eccentricity: 0.055 55
Orbital inclination: 2°29'21"
Diameter: equatorial 120 536 km
         polar 108 728 km
Apparent diameter seen from Earth: max 20".1
                       min 14".5
Oblateness: 0.098
Mass, Earth = 1: 95.17 (=1.899 × 10^{27} kg)
Reciprocal mass, Sun = 1: 3498.5
Density, water = 1: 0.70
Volume, Earth = 1: 752
Escape velocity: 35.26 km s^{-1}
Surface gravity, Earth = 1: 1.19
Geometric albedo: 0.47
Mean surface temperature: −180 °C
Opposition magnitude: max −0.3
                     min +0.8
Mean diameter of Sun, as seen from Saturn: 3'22".
Number of confirmed satellites: 18

MOVEMENTS

Saturn reaches opposition about 13 days later every year. Opposition dates for the period 2000–2005 are given in Table 10.2.

The opposition magnitude is affected both by Saturn's varying distance and by the angle of presentation of the rings. At its best, Saturn may outshine any star apart from Sirius and Canopus, but at the least favourable oppositions from this point of view – when the rings are edgewise-on, as in 1995 – the maximum magnitude may be little brighter than Aldebaran. A list of edgewise presentations is given in Table 10.3.

The intervals between successive edgewise presentations are 13 years 9 months and 15 years 9 months. During the shorter interval, the south pole is sunward; the southern ring-face is seen, and Saturn passes through perihelion. Perihelion fell in 1944 and 1974; the next will be in 2003. The aphelion dates are 1959, 1988 and 2018.

At edgewise presentations, the rings almost disappear, particularly when the Earth passes through the plane of the rings and also when the Sun does so. This is an obvious proof that the rings are very thin.

Occultations of Saturn by the moon are not uncommon; it is then interesting to note how small Saturn appears when compared with a lunar crater! There will be close conjunctions of Saturn and Venus on 26 August 2006 and 9 January 2016, and on 15 September 2037, at 21.30 UT, Mercury and Saturn will be separated by only 18 arcsec.

EARLY RECORDS

The first observations of Saturn must have been made in prehistoric times. The first recorded observations seem to have been those made in Mesopotamia in the mid-7th century BC. About 650 BC there is a record that Saturn 'entered the Moon', which is presumably a reference to an occultation of the planet. Very careful observations were made by the Greeks and, later, the Arabs; of course it was generally assumed that Saturn, like all other celestial bodies, moved round the Earth.

Table 10.2. Oppositions, 1999–2005.

Date	Magnitude	Diameter (″)	Declination at opposition	Constellation
1999 Nov 6	0.0	20.3	+13	Aries
2000 Nov 19	−0.1	20.5	+17	Taurus
2001 Dec 3	−0.3	20.6	+20	Taurus
2002 Dec 17	−0.3	20.7	+22	Taurus
2003 Dec 31	−0.3	20.7	+23	Taurus
2005 Jan 13	−0.2	20.6	+24	Taurus/Gemini

Copernicus recorded an observation of Saturn on 26 April 1514, when the planet lay in line with the stars 'in the forehead of Scorpio'; he also noted the position of Saturn on 5 May 1514, 13 July 1520 and 10 October 1527. Tycho Brahe noted that on 18 August 1563 Saturn was in conjunction with Jupiter. However, the first telescopic observation was delayed until July 1610, when Galileo examined it with his early telescope, using a magnification of ×32. He noted that 'the planet Saturn is not one alone, but is composed of three, which almost touch one another and never move nor change with respect to one another. They are arranged in a line parallel to the Zodiac, and the middle one is about three times the size of the lateral ones'. Galileo's telescope was not powerful enough to show the rings in their true guise, and subsequently he was puzzled to find that the lateral bodies had disappeared; in fact the rings were edgewise-on in December 1612. Later, Galileo again saw the lateral bodies, but was unable to interpret them.

A drawing made by the French astronomer Pierre Gassendi on 19 June 1633 shows what seems to be a ring, but Gassendi also failed to interpret it. The correct explanation was given by Christiaan Huygens in 1659, in his book *Systema Saturnium*. He had started his telescopic observations in 1655, using a magnification of ×50; in his book he gave the answer in an anagram which he had published to ensure priority. In translation, the anagram reads: 'The planet is surrounded by a thin flat ring, nowhere touching the body of the planet, and inclined to the ecliptic'.

Some earlier explanations were very wide of the mark. For instance, the French mathematician Gilles de Roberval believed Saturn to be surrounded by a torrid

Table 10.3. Edgewise presentations of Saturn's rings (Sun crossing).

1936 Dec 28
1950 Sept 21
1956 June 15
1980 Mar 3
1995 Nov 19
2009 Aug 10
2025 May 5
2039 Jan 22

zone giving off vapours, transparent and in small quantity but reflecting sunlight off the edges if of medium density, and producing an elongated appearance when thicker. Huygens' explanation was not universally accepted until 1665.

Subsequent telescopic improvements meant that the rings could be examined in more detail. The division between the two main rings (A and B) was discovered by G. D. Cassini in 1675, and is named after him (claims that W. Ball had seen the Division ten years earlier have been discounted). A narrower division, in Ring A, was discovered by J. F. Encke from Berlin on 28 May 1837, while the Crêpe or Dusky Ring (Ring C) was detected by W. Bond from Harvard in November 1850, using the 15 in (38 cm) Merz refractor there; independent confirmation came from W. Lassell and W. R. Dawes in England. The transparency of Ring C was discovered in 1852, independently by Lassell and by C. Jacob from Madras. In fact the Crêpe Ring is not a difficult object when Saturn is suitably placed; indications of it may have been shown by Campani as long ago as 1664.

Extra rings were later reported; in 1907 G. Fournier, using an 11 in (28 cm) refractor at Mont Revard in France, claimed to have seen a dusky ring outside the main system. It is now known that there are in fact three exterior rings, although they are very difficult to observe from Earth. An inner structure, D, extends down towards the cloud tops, but can hardly be regarded as a true ring.

Originally it was assumed that the rings must be solid; the first suggestion that this cannot be so was made by J. Cassini in 1705. Theoretical confirmation was provided in 1875 by James Clerk Maxwell, who showed that no solid ring could exist; it would be disrupted by Saturn's powerful gravitational pull. Spectroscopic proof was given in 1895 by J. E. Keeler, who showed that the rings do not rotate as a solid mass would do; the inner sections have the fastest rotation. Using the 13 in (33 cm) refractor at the Allegheny Observatory in Pittsburgh, Keeler measured the Doppler shifts of the spectral lines at the opposite ends of the rings (ansæ) and was able to measure the rotational speeds.

A very interesting observation was made on 9 February 1917 by two amateurs, M. A. Ainslie and J. Knight; Ainslie used a 9 in (23 cm) refractor at Blackheath, in Outer London, and Knight a 5 in (13 cm) refractor at Rye in Sussex. The star B.D.$+21°1714$ passed behind the rings, and the way in which the star was dimmed by the various rings and gaps provided valuable information. Yet in 1954 G. P. Kuiper, using the 200 in (5 m) Hale reflector at Palomar, claimed that the Cassini Division was the only genuine gap, Encke's Division being a mere 'ripple' and all the other reported gaps non-existent.

The fact that the rings were made up of small particles was well established during the 20th century. Radar reflections from the rings were obtained in 1972, and indicated that the ring particles were icy, with diameters of between 4 and 30 cm.

OBSERVATIONS OF THE GLOBE

Saturn's globe shows belts, not unlike those of Jupiter, but much less prominent, and sensibly curved. Predictably, the globe is flattened; this was first noted in 1789 by William Herschel. He gave the ratio of the equatorial to the polar diameter as 11:10, which is approximately correct. Apart from Jupiter, Saturn has the shortest rotation period of any planet in the Solar System.

Well-defined spots were seen in 1796 by J. H. Schröter and his assistant, K. Harding, from Schröter's observatory at Lilienthal, near Bremen. Prominent white spots are found occasionally. One was seen on 8 December 1876 by A. Hall; there was another in 1903, observed by A. S. Williams; and then, in 1933, a really spectacular outbreak, discovered on 3 August by W. T. Hay, using a 6 in (15 cm) refractor. (Hay may be better remembered today as Will Hay, the stage and screen comedian.) The spot remained identifiable until 13 September 1933. W. H. Wright considered it to have been of 'an eruptive nature'. Another spot was seen in 1960 and a major outbreak in 1990; on 25 September an American amateur, S. Wilber, detected a brilliant white spot in much the same latitude as the earlier ones. Within a few days the spot had been spread out by Saturn's strong equatorial winds, and by 10 October had been transformed into a bright zone all round the equator. Extra outbreaks were seen in it, clearly indicating an uprush of material from below.

There is an interesting periodicity here. The white spots were seen in 1876, 1903, 1933, 1960 and 1990; the intervals between outbreaks have been 27, 30, 27 and 30 years. This is very close to Saturn's orbital period of $29\frac{1}{2}$ years. It could be coincidental, but observers will be on watch for a new white spot around 2020.

COMPOSITION OF SATURN

Initially it was assumed that Saturn must be a sort of miniature Sun. As recently as 1882 R. A. Proctor, in his book *Saturn and its System*, wrote:

'Over a region hundreds of thousands of square miles in extent, the glowing surface of the planet must be torn by subplanetary forces. Vast masses of intensely hot vapour must be poured forth from beneath, and rising to enormous heights, must either sweep away the enwrapping mantle of cloud which had concealed the disturbed surface, or must itself form into a mass of cloud, recognizable because of its enormous extent. . . .'

The first modern-type model of Saturn was proposed in 1938 by R. Wildt. Wildt believed that there was a rocky core, overlaid by a thick layer of ice which was itself overlaid by the gaseous atmosphere.

Today it seems that there are strong resemblances between the compositions of Jupiter and Saturn, although

Saturn has a much smaller mass and lower density (less than that of water, so that in theory it would float if dropped into a vast ocean). There is presumably a silicate core at a high temperature, perhaps as high as 15 000 °C although possibly rather less; then comes a layer of liquid metallic hydrogen, succeeded by a layer of liquid molecular hydrogen and then by the atmosphere. Hydrogen and helium combined make up about 70% of the total mass of the planet.

Like Jupiter and Neptune (but unlike Uranus), Saturn sends out more energy than it would do if it depended entirely upon what it receives from the Sun; the excess amounts to 1.8% (as against 1.7% for Jupiter). However, the cause is different. Saturn must by now have lost the heat produced during its formation, whereas the more massive Jupiter has not had time to do so. It is likely that the excess energy of Saturn is gravitational, produced by helium droplets, generated in the upper regions, falling through the lighter hydrogen toward the centre of the globe. This also explains why Saturn has less high-altitude helium than Jupiter. The rotation period of the core is given as 10h 39m.4, which is longer than that of the upper clouds. Dynamo action in the layer of metallic hydrogen is responsible for Saturn's strong magnetic field.

ATMOSPHERE AND CLOUDS

Saturn's atmosphere contains relatively more hydrogen and less helium than in the case of Jupiter. Details are given in Table 10.4.

Table 10.4. Composition of the atmosphere of Saturn.

Major
Molecular hydrogen (H_2), 96.3% (uncertainty 2.4%)
Helium (He), 3.25% (uncertainty 3.25%)

Minor (parts per million)
Methane (CH_4), 4500
Ammonia (NH_3), 125
Deuterium (HD), 110
Ethane (C_2H_6), 7
(All these values are subject to uncertainty)

Aerosols
Ammonia ice, water ice, ammonia hydrosulphide.

There are thought to be three main cloud decks. The uppermost consists of ammonia ice, about 100 km below the tropopause, which lies at the top of Saturn's troposphere; the temperature here is of the order of −120 °C. Next come the clouds of ammonium hydrosulphide, about 170 km below the tropopause, where the temperature may be around −70 °C. Below come the H_2O clouds, around 230 km below the tropopause, where the temperature has risen to about 0 °C, although as yet we cannot claim that our knowledge of these regions is at all reliable. At about 120 km above the H_2O clouds the pressure is probably about the same as the pressure of the Earth's atmosphere at sea level.

Windspeeds in the atmosphere are high; the zonal winds along the belts, in a prograde direction (blowing toward the east) may attain over 600 km h^{-1}, although most are less violent. There is an equatorial jet stream; at higher latitudes there are alternating streams, but mostly prograde and rather weaker. As yet we know little about the wind-speeds well below the visible surface.

Saturn is colder than Jupiter, so that ammonia crystals form at higher levels, covering the planet with 'haze' and giving it a somewhat bland appearance. Features over 1000 km in diameter are uncommon, and even the largest ovals are no more than half the size of Jupiter's Red Spot. Seasonal effects occur; for example, the Sun crossed into Saturn's northern hemisphere in 1980, and during the Voyager missions the northern hemisphere was still the colder of the two. The south pole was 10° warmer than the north pole.

MAGNETIC FIELD AND AURORÆ

Saturn has a strong magnetic field, first detected in 1979 by the Pioneer 11 probe. It is 850 times stronger than that of the Earth, although 30 times weaker than that of Jupiter. Saturn is unique inasmuch as its rotational axis and magnetic axis are almost coincident; the magnetic field is therefore reasonably straightforward and axisymmetric, and the magnetosphere is much less active than that of Jupiter. The field seems to be slightly stronger at the north pole than at the south, and the centre of the field is displaced about 2400 km northward along the planet's axis.

The magnetosphere is about one-fifth the size of that of Jupiter; the bow shock lies at a range of about 60 300 km from the planet, and has a thickness of about 2000 km. Saturn does have radiation zones; the magnetosphere traps radiation belt particles, which extend as far as the outer edge of the main ring system. The numbers of electrons fall off quickly at the outer edge of Ring A, because the electrons are absorbed by the ring particles; the region between Ring A and the globe is probably the most radiation-free zone in the entire Solar System. There is no doubt a very extensive 'magnetotail'.

All the inner satellites are embedded in the magnetosphere. The outer boundary is somewhat variable, and is located close to the orbit of Titan, so that Titan is sometimes just inside the main magnetosphere and sometimes just outside it.

Auroræ occur on Saturn, and because the magnetic and the rotational poles are almost coincidental the 'auroral ovals' are centred on the poles. Initial observations were made from Pioneer 11 in 1979, and from 1980 auroræ were detected from a series of spectroscopic observations from the IUE (International Ultra-violet Explorer) satellite; the northern oval was first imaged from the Hubble Space Telescope in June 1992. A huge auroral curtain was found to rise as far as 2000 km above the cloud tops on 9 October 1994, when Saturn was 1 300 000 000 km from the Earth.

RINGS

Telescopically, Saturn's rings make it probably the most beautiful object in the entire sky. A small instrument will show them well when they are suitably placed, although most of our detailed knowledge about them has been obtained from space probes. Essentially, the rings are made up of innumerable particles and 'icebergs', ranging in size from fine dust to small houses. They are composed of loosely packed snowballs of water ice, although slight colour effects indicate that there may be a small amount of rocky material.

The total mass of the ring system is surprisingly small; if all the particles could be combined, they would make up a satellite no more than 300 km across at most. Although they are very extensive – the main system has a total diameter of over 270 000 km – they are very thin; the thickness cannot be more than 200 m, and is probably less. It has been said that if a model of the system were made from material the

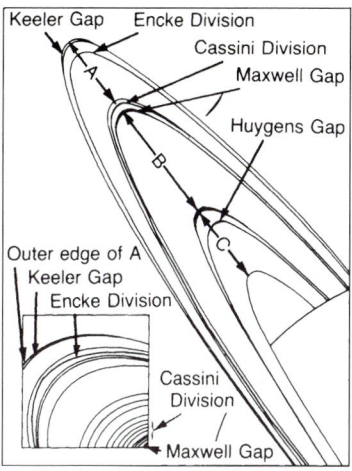

Figure 10.1. The rings of Saturn.

thickness of a 10p coin, the overall diameter would be of the order of 15–20 km. The origin of the ring system is uncertain. It may have been produced from material 'left over', so to speak, when Saturn itself was formed, or it may have been due to the break-up of a former icy satellite which wandered too close to the planet and was literally torn apart. Certainly the rings lie wholly within the Roche limit from Saturn.

The rings are much more complex than was believed before the Pioneer and Voyager missions. Instead of being more or less homogeneous, they have been found to be made up of thousands of small ringlets and narrow gaps. An ingenious experiment was carried out from Voyager 2, in 1981. As seen from the space-craft, the third-magnitude star Delta Scorpii was occulted by the rings. It had been expected that there would be comparatively few 'winks', as the starlight would shine through any gaps but would be blocked by ring material. In fact thousands of 'blips' were recorded, and it seems that there is very little empty space anywhere in the main ring system. Details of the rings are given in Table 10.5.

The so-called Ring D is not a true ring, and has no sharp inner edge; the particles may extend downward almost to the cloud tops. Ring C, the Crêpe or Dusky Ring, is over 17 000 km wide and has various gaps, including two with widths of 200 and 300 km respectively; the outer gap contains a narrow, denser and slightly eccentric bright ring,

Table 10.5. Ring data.

Feature	Distance from centre of Saturn (km)	Period (h)	Radial width (km)
(Cloud tops)	60 330	10.66	—
Inner edge of Ring D	66 900	4.91	7150
Outer edge of Ring D	73 150		
Inner edge of Ring C	74 510	5.61	
Maxwell Gap	87 500	6	270
Outer edge of Ring C	92 000	7.9	C: 17,500
Inner edge of Ring B	92 000	7.93	25,500
Outer edge of Ring B	117 500	11.41	
Inner edge of Cassini Division	117 500	11.4	
Centre of Cassini Division	119 000		4700
Outer edge of Cassini Division	122 200		
Huygens Gap	117 680	11.4	285–440
Inner edge of Ring A	122 200	11.4	
(Pan)	133 583	13.7	—
Centre of Encke Division	135 700	13.82	(Width of Division, 325 km)
Centre of Keeler Gap	136 530	14	(Width of Gap, 35 km)
Outer edge of Ring A	136 800	14.14	(Width Ring A, 14 600 km)
(Atlas)	137 640	14.4	—
(Prometheus)	139 350	14.7	
Centre of Ring F	140 210	14.94	30–500
(Pandora)	141 700	15.07	—
(Epimetheus)	151 422	16.65	—
(Janus)	151 472	16.7	—
Inner edge of Ring G	164 000		
Centre of Ring G	168 000	19.90	8000
Outer edge of Ring G	172 000		
Inner edge of Ring E	180 000	21	
(Mimas)	185 520	22.6	—
Brightest part of Ring E	230 000	31.3	
(Enceladus)	238 020	32.9	—
(Tethys/Telesto/Calypso)	294 660	44.9	
(Dione/Helene)	377 400	65.9	—
Outer edge of Ring E	480 000	80	(Width of Ring E, 300 000 km)
(Rhea)	527 040	108.4	—

90 km wide. On average, the C-ring particles seem to be about 2 m in diameter.

The main ring is, of course, Ring B, where the particles range in size from 10 cm to about 1 m. The particles are redder than those of the C ring and D region; the temperatures range from −180 °C in sunlight down to −200 °C in shadow. It has also been found that there is a cloud of neutral hydrogen extending to 60 000 km above and below the ring plane. Voyager images showed strange radial spokes in the B ring – incidentally, shown many years earlier in drawings made by E. E. Barnard and E. M. Antoniadi – and these are decidedly puzzling, because logically no such features should form; the difference in rotation period between the inner and the outer edges of the ring is over

3 h – and yet the spokes persisted for hours after emerging from the shadow of the globe; when they were distorted and broken up, new spokes emerged from the shadow to replace them. Presumably they are due to particles of a certain definite size elevated away from the ring plane by magnetic or electrostatic forces. They are confined wholly to Ring B.

The 4700 km wide Cassini Division is not empty, but contains many rings a few hundred kilometres wide, one of which is markedly eccentric. The particles are less red than those of Ring B, and more closely resemble those of Ring C.

Ring A is made up of particles ranging from fine grains up to about 10 m across. Encke's Division, in Ring A, is notable because it contains a tiny satellite, Pan; the satellite Atlas moves close to the outer edge of Ring A, and is responsible for the sharp border of the ring. (Incidentally, the Encke Division has been referred to as the Keeler Gap, but it has now, very properly, been decided that the old, familiar name would be retained; three other narrow gaps in the ring system have been named Huygens, Maxwell and Keeler.)

Outside the main system comes the irregular Ring F, which has been described as a convoluted tangle of narrow strands. It is stabilized by two small satellites, Prometheus and Pandora, to either side of its centre. The satellite slightly closer to Saturn (Prometheus, at the present time) will be moving faster than the ring particles, and will speed up a particle which strays away from the main system; the outer satellite, moving more slowly, will drag any errant particles back and return them to the central region. For obvious reasons, these two tiny moons are referred to as 'shepherd satellites'.

Beyond Ring F and its shepherds come the two co-orbital satellites, Epimetheus and Janus. Next there is the very tenuous Ring G. The final ring, E, is 300 000 km wide, but very tenuous indeed, and it has even been said that it is made up of slight condensations of débris in the orbital plane of the satellites. Its brightest part is slightly closer-in than the orbit of Enceladus, and it is possible that material ejected from Enceladus may be concerned in the formation of the ring. Ring E extends out beyond the orbit of Tethys, and almost as far as the orbit of Dione.

Why are the rings so complex? The Cassini Division had been explained as being due to the satellite Mimas;

a particle moving in the Division will have an orbital period half that of Mimas, and cumulative perturbations should drive it away from the 'forbidden zone', leaving the Division swept clear. Yet the Voyagers have shown that this explanation is clearly inadequate, and it now seems more likely that we are dealing with a density-wave effect. A satellite such as Mimas can alter the orbit of a ring particle and make it elliptical. This causes 'bunching' in various areas; a spiral density wave will be created, and particles in it will collide, moving inward toward the planet and leaving a gap just outside the resonance orbit. This theory, due to Peter Goldreich and Scott Tremaine, does sound plausible, but it would be wrong to claim that we yet really understand the mechanics of the ring system.

SPACE MISSIONS TO SATURN

Four space-craft have so far been launched to Saturn. Data are given in Table 10.6.

The first probe was Pioneer 11. Its prime target had been Jupiter, but the success of its predecessor, Pioneer 10, and the fact that there was adequate power reserve meant that it could be swung back across the Solar System to a rendezvous with Saturn in 1979. The results were preliminary only, but involved several important discoveries. Perhaps the most significant fact was that Pioneer was not destroyed by a ring particle collision; at that time nobody knew what conditions near Saturn would be like and estimates of the probe's chances of survival ranged from 1% to 99%.

Voyager 1, launched in 1977, surveyed Jupiter in 1979 and then went on to Saturn. It was a complete success, and obtained new data about the rings and satellites as well as the planet itself. One of its prime targets was Titan, already known to have an atmosphere; it was a surprise to find that the main constituent of this atmosphere was nitrogen. Voyager 1 then began a never-ending journey out of the Solar System; as I write these words (2000) it is still in touch.

Voyager 2 came next. This time Titan was not surveyed in detail; this would have meant that the probe would have been unable to go on to Uranus and Neptune. Again the mission was a complete success, and both the outer planets were also encountered. Like its predecessor,

Table 10.6. Space missions to Saturn.

Name	Launch date	Encounter date	Nearest approach (km)	Remarks
Pioneer 11	1973 Apr 5	1979 Sept 11	20 880	Preliminary results.
Voyager 1	1977 Sept 5	1980 Nov 12	124 200	Success. Good images of Titan, Rhea, Dione and Mimas.
Voyager 2	1977 Aug 20	1981 Aug 25	101 300	Success. Good images of Iapetus, Hyperion, Tethys and Enceladus. Went on to Uranus and Neptune.
Cassini/Huygens	1997 Oct 15	2004 July 1		Cassini orbiter; Huygens scheduled to land on Titan, Nov 2004.

Voyager 2 is now leaving the Solar System permanently. Its success was all the more striking because one of the main instruments failed early in the mission, and Voyager 2 operated throughout on its back-up system.

The Cassini–Huygens mission was launched on 15 October 1997, from Cape Canaveral, using a Titan-4B/Centaur rocket. The gravity-assist technique has to be used, involving swing-bys of Venus (21 April 1998), Venus again (20 June 1999), Earth (16 August 1999) and Jupiter (30 December 2000). The probe will be put into Saturn orbit on 1 July 2004, only 20 000 km above the cloud tops, to begin the first of 74 planned orbits. On 6 November 2004 the space-craft will be manœuvred into an impact trajectory with Titan, and on 8 November the Huygens lander will be released; Cassini itself will be deflected away from a collision course, leaving Huygens to land on Titan 18.4° north of the satellite's equator. The descent through Titan's atmosphere will take $2\frac{1}{2}$ h; on landing, data will, it is hoped, be received for up to half an hour. After that the Cassini orbiter, being used as a relay, will pass out of range – and before it is again suitably positioned, Huygens will be dead. Cassini itself will continue to orbit Saturn until July 2008, sending back data of all kinds. It is indeed an ambitious mission; let us hope that it will be successful.

SATELLITES

Saturn has a wealth of satellites. Eighteen are definitely known, and others have been suspected. In April 1861 H. Goldschmidt announced that he had found a satellite moving between the orbits of Hyperion and Iapetus, and named it Chiron; in 1904 W. H. Pickering claimed that he had found a satellite moving between the orbits of Titan and Hyperion, and named it Themis. Neither of these satellites was confirmed, and neither seems to exist, but no doubt other very small attendants await discovery. Details of the known satellites are given in Table 10.7.

Pan. This tiny satellite was found on photographs taken by the Voyager 2 probe; it was discovered by M. Showalter in 1990. It moves within the Encke Division in Saturn's A ring. It is presumably icy, but nothing definite is known about its composition. It was named after the half-man/half-goat son of Hermes and Dyope.

Atlas. This satellite was named after the Titan who carried the sky on his shoulders. It lies near the edge of Ring A, and may act as a 'shepherd'. It was identified, by R. Terrile, on Voyager images. Nothing is known about its physical condition, but again it is presumably icy.

Prometheus and Pandora. These F-ring shepherd satellites were identified, by S. Collins and his team, on Voyager images. Both seem to be of low density, so that ice is a major constituent. Both are cratered; Prometheus shows a number of ridges and valleys, as well as craters up to 20 km across, while Pandora lacks visible ridges, but does show two 30 km craters. Prometheus was the brother of Atlas, who gave humanity the gift of fire; Pandora, made of clay by Hephæstus at the request of Zeus (Jupiter), opened a box which set free all the ills which plague mankind today.

Epimetheus and Janus. Epimetheus (named after the Greek backward-looking god) and Janus (named for the two-faced Roman god who could look backward and forward at the same time) are co-orbital, and separated by only about

Table 10.7. Satellites of Saturn.

Name	Discoverer	Mean distance from Saturn (km)	Sidereal period (days)	Mean Synodic Period (d)	(h)	(m)	(s)	Mean angular distance from Saturn, opposition (')	('')	Orbital inclination (°)	Orbital eccentricity	Orbital velocity (km s⁻¹)	Rotation period (days)
Pan	M Showalter, 1990	133 583	0.575							0.0	0.0	16.90	?
Atlas	R Terrile, 1980	137 640	0.6019							0.0	0.0	16.63	?
Prometheus	S Collins and others, 1980	139 350	0.6130							0.0	0.003	16.54	?
Pandora	S Collins and others, 1980	141 700	0.6285							0.0	0.004	16.40	?
Epimetheus	R Walker and others, 1980	151 422	0.6942							0.34	0.009	15.87	0.6942
Janus	A Dollfus, 1966	151 472	0.6945							0.14	0.007	15.87	0.6945
Mimas	W Herschel, 1789	185 520	0.9424	0	22	37	12.4	0	30.0	1.53	0.020	14.32	0.9424
Enceladus	W Herschel, 1789	283 020	1.3702	1	8	53	22	0	38.4	0.02	0.005	12.64	1.3702
Tethys	G Cassini, 1684	294 660	1.8878	1	21	18	55	0	47.6	1.09	0.000	11.36	1.8878
Telesto	B Smith and others, 1980	294 660	1.8878	1	21	18	55	0	47.6	0.0	0.000	11.36	?
Calypso	D Pascu and others, 1980	294 660	1.8878	1	21	18	55	0	47.6	0.0	0.000	11.36	?
Dione	G Cassini, 1684	377 400	2.7369	2	17	42		1	01.1	0.02	0.002	10.03	2.7369
Helene	P Laques and J Lecacheus, 1980	377 400	2.7360	2	17	42		1	01.1	0.2	0.005	10.03	?
Rhea	G Cassini, 1672	527 040	4.5175	4	12	18		1	25.1	0.35	0.001	8.49	4.5175
Titan	C Huygens, 1655	1 221 850	15.9454	15	23	16		3	17.3	0.33	0.029	5.58	15.945
Hyperion	W Bond, 1848	1 481 100	21.2777	21	7	39		3	59.4	0.43	0.104	5.07	chaotic
Iapetus	G Cassini, 1671	3 561 300	79.3302	79	22	05		9	35	14.72	0.028	3.27	79.3302
Phœbe	W H Pickering, 1898	12 952 000	550.48	523	13			34	51	175.3	0.163	1.71	0.4

Name	Diameter (km)	Mass (g)	Reciprocal mass, Saturn = 1	Density, water = 1	Escape velocity (km s⁻¹)	Magnitude	Albedo	Apparent diameter seen from Saturn (')	('')
Pan	19.3	?	?	?	Low	19	0.5		
Atlas	37 × 34 × 27	?	?	?	Low	18.0	0.9		
Prometheus	145 × 82 × 62	1.4 × 10²⁰	—	0.27	0.022	15.8	0.6		
Pandora	114 × 84 × 62	1.3 × 10²⁰		0.7	0.227	16.5	0.9		
Epimetheus	144 × 108 × 98	5.5 × 10²⁰		0.63	0.322	15.7	0.8		
Janus	196 × 192 × 150	2.0 × 10²¹		0.67	0.052	14.5	0.8		
Mimas	421 × 395 × 385	3.7 × 10²²	15 000 000	1.17	0.161	12.9	0.5	10	54
Enceladus	512 × 495 × 488	6.5 × 10²²	7 000 000	1.24	0.212	11.7	0.99	10	36
Tethys	1058	6.1 × 10²³	910 000	0.98	0.436	10.2	0̂.9	17	36
Telesto	34 × 28 × 26	?	?	?	Low	18.5	0.5		
Calypso	30 × 16 × 16	?	?	?	Low	18.7	0̂.5		
Dione	1120	1.1 × 10²⁴	490 000	1.49	0.500	10.4	0.7	12	24
Helene	36 × 32 × 30	?		0.9	0.011	18.4	0̂.7		
Rhea	1528	2.3 × 10²⁴	250 000	1.33	0̂.659	9.7	0.7	10	42
Titan	5150	1.34 × 10²⁶	4150	1.88	2.65	8.3	0̂.21	17	10
Hyperion	410 × 260 × 220	?	~5 000 000	1.4?	0.11	14.2	0.3	0	43
Iapetus	1460̂	1.6 × 10²⁴	300 000	1.21	0.59	10.2–11.9	0.2(mean)	1	48
Phœbe	220	?	?	0.7?	0.07	16.4	0.06	0	3.2

50 km. Periodically they approach each other, and exchange orbits; this happens every four years. Both are irregular in shape, and may be the remnants of a larger body which was broken up long ago. Janus is heavily cratered, with some formations 30 km in diameter; Epimetheus is also cratered, with valleys and ridges. Both were identified on Voyager images, but Janus had been previously detected by A. Dollfus in 1966, during the edgewise presentation of Saturn's rings – thereby facilitating observations of the small inner satellites. (Subsequently I found it on a sketch made at the same time, but as I failed to identify it I can claim absolutely no credit!)

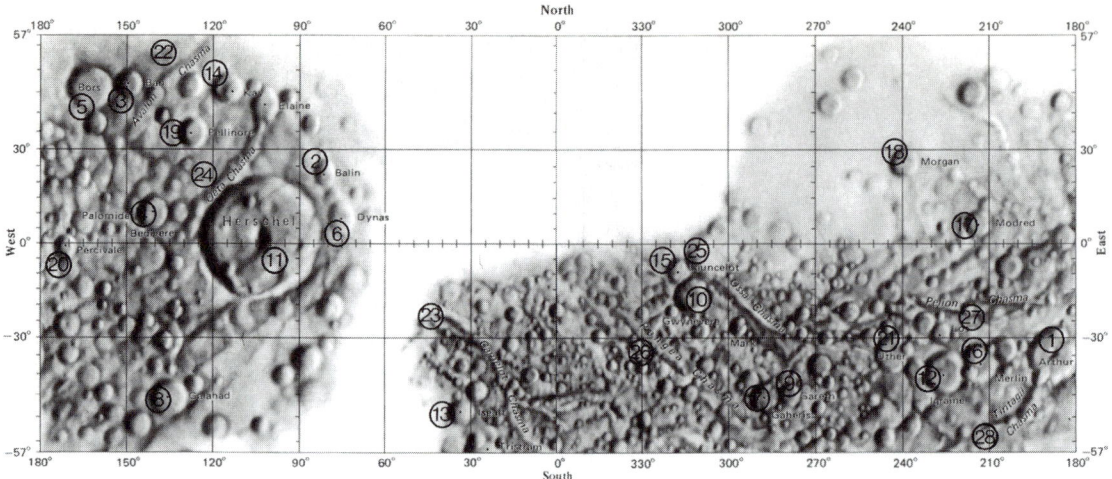

Figure 10.2. Mimas.

Mimas. Discovered by William Herschel, whose son, John, named it after one of the Titans, who was killed by Ares (Mars). The surface features have been given names from Arthurian legend – apart from the main crater, which is named after William Herschel himself.

Mimas is only slightly denser than water, and may be made up mainly of ice all the way through to its centre, although there could be some rock as well. Crater Herschel is 130 km across, one-third the diameter of Mimas itself; if it were produced by an impact, the whole satellite would have been in danger of disruption. Herschel is 10 km deep, with a 6 km central peak whose base measures 30 km × 20 km. There are many other craters, such as Modred and Bors, but few are as much as 50 km across, and larger craters are lacking in the south polar region. Grooves (chasmata) are much in evidence, and some are more or less parallel, indicating that the icy crust has been subjected to considerable strain. Like all the icy satellites, Mimas is very cold; the surface temperature is given as about −200 °C. A selected list of surface features is given in Table 10.8.

Table 10.8. Selected list of features on Mimas. (Bold numbers indicate map references.)

Name	Lat.	Long. W
Craters		
Arthur **1**	33.2S	195.6
Balin **2**	17.1N	86.8
Ban **3**	39.2N	156.4
Bedivere **4**	9.5N	152.3
Bors **5**	39.0N	166.0
Dynas **6**	3.8N	82.9
Gaheris **7**	40.9S	298.4
Galahad **8**	47.0S	135.0
Gareth **9**	40.9S	288.8
Gwynevere **10**	16.8S	324.0
Herschel **11**	2.9N	109.5
Igraine **12**	42.1S	231.1
Iseult **13**	46.4S	36.4
Kay **14**	47.0N	126.4
Launcelot **15**	10.0S	328.2
Merlin **16**	37.7S	219.5
Modred **17**	5.0N	213.0
Morgan **18**	23.6N	242.6
Pellinore **19**	29.5N	139.5
Percivale **20**	1.0S	180.2
Uther **21**	34.9S	251.1
Chasma		
Avalon **22**	39.6N	147.8
Camelot **23**	38.4S	27.6
Oeta **24**	32.2N	117.5
Ossa **25**	20.6S	307.8
Pangea **26**	22.6S	348.8
Pelion **27**	24.0S	248.3
Tintagel **28**	49.7S	205.0

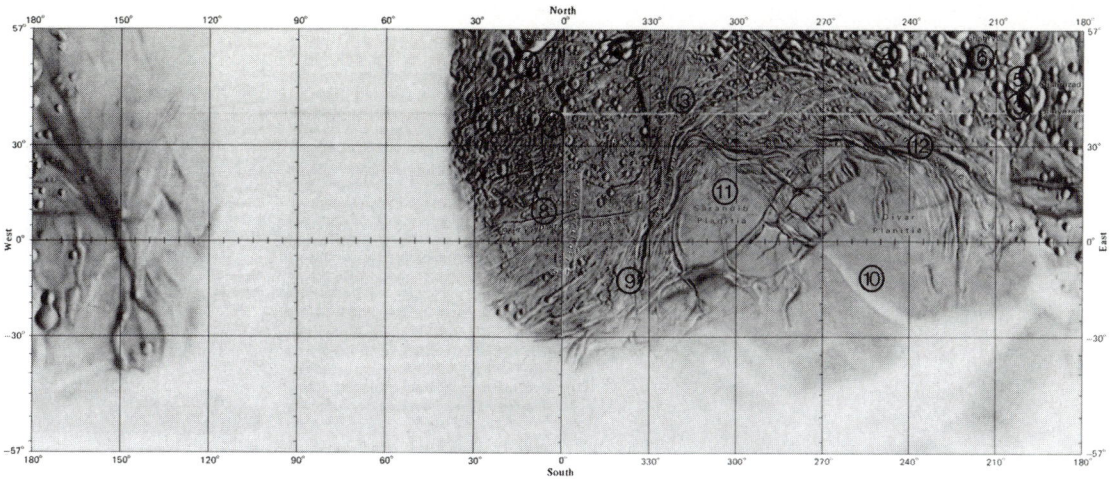

Figure 10.3. Enceladus.

Enceladus. Named by John Herschel after a Titan, who was crushed in a battle with the Olympian gods; earth was piled on top of him and became the island of Sicily.

Enceladus is very different from Mimas – or from any other Saturnian satellite. Instead of one huge crater, we find several completely different types of terrain. Craters exist in many areas, but give the impression of being fairly young and sharp, while there is an extensive plain which is almost crater-free and is instead dominated by long grooves. Its density is very low, and it is more reflective than any other satellite, so that the surface is exceptionally cold ($-201\,°C$).

Surprisingly, Enceladus may be active, and we may well have an example of 'cryovulcanism' – the icy equivalent of what we always call volcanic action. Certainly the surface seems to be very young, so that presumably it has been re-surfaced, and any large craters obliterated. We have to explain the paucity of craters over wide areas, and it may be that the interior is flexed by the tidal forces of Saturn and the outer, much more massive satellite Dione, whose orbital period is twice that of Enceladus. If so, there may be periods when soft ice or even water wells up over the surface. Selected features, named from the Arabian Nights, are listed in Table 10.9.

It is possible that Enceladus may be the main source of Saturn E ring, so that it is in every way exceptional.

Table 10.9. Selected list of features on Enceladus. (Bold numbers indicate map references.)

Name	Lat.	Long. W	Diameter/ length (km)
Craters			
Ali Baba **1**	57.2N	12.0	35
Dalilah **2**	52.9N	246.4	14
Dunyazad **3**	42.6N	196.5	30
Julnar **4**	54.2N	342.0	20
Shahrazad **5**	48.2N	195.1	20
Shahryar **6**	59.7N	225.0	21
Sindbad **7**	68.9N	211.4	23
Fossæ			
Bassorah **7**	45.4N	6.3	131
Daryabar **8**	9.7N	359.1	201
Isbanir **9**	12.6N	354.0	132
Planitia			
Diyar **10**	0.5N	239.7	311
Sarandib **11**	4.4N	298.0	200
Sulci			
Harran **12**	26.7N	237.6	276
Samarkand **13**	20.5N	326.8	383

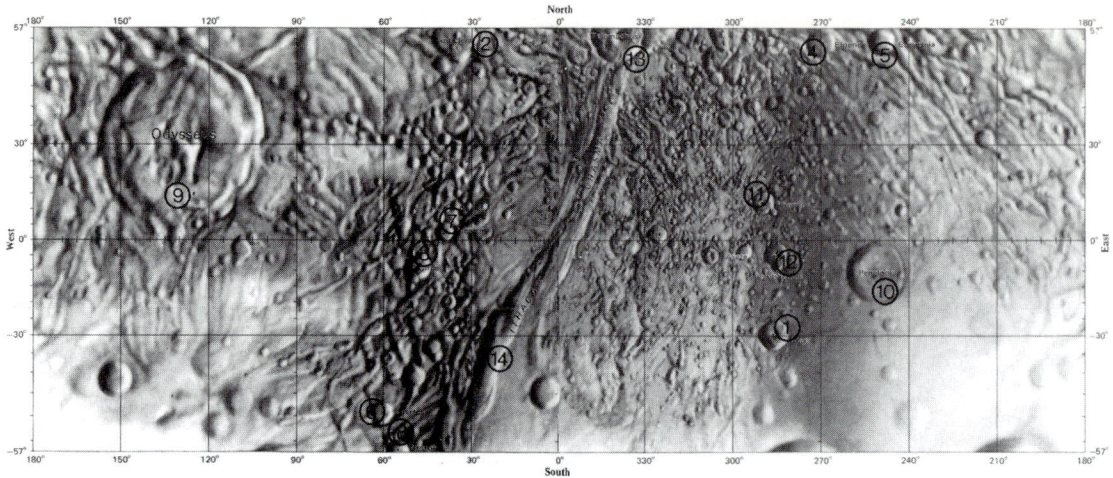

Figure 10.4. Tethys.

Tethys. Named after a Titaness, wife of Oceanus and mother of the Oceanids. Surface features, named after people and places in Homer's *Odyssey*, are listed in Table 10.10.

Most of our detailed knowledge of Tethys comes from Voyager 2 (Voyager 1 did not make a close approach), and it is unfortunate that some data were lost when Voyager 2 developed a temporary fault after it had started its journey from Saturn to Uranus. Tethys seems to be made up of almost pure ice; the surface temperature is −187 °C. There is one huge crater, Odysseus, with a diameter of 400 km – larger than Mimas. It is not very deep, and the curve of its floor follows the general shape of the globe. There are many other craters, notably Penelope, but the main feature is a huge trench, Ithaca Chasma, 2000 km long, running from near the north pole across the equator and along to the south polar region. Nothing similar is known in the Solar System. It was presumably formed when the water inside Tethys froze, expanding as it did so and fracturing the crust. It extends three-quarters of the way round Tethys; it is over 60 km wide and 4–6 km deep, with a rim rising to 0.5 km above the surrounding terrain.

Table 10.10. Selected list of features on Tethys. (Bold numbers indicate map references.)

Name	Lat.	Long. W
Craters		
Ajax **1**	29.1S	282.0
Anticleia **2**	52.3N	34.4
Circe **3**	12.1S	53.7
Elpenor **4**	54.8N	163.3
Eurycleia **5**	52.7N	245.9
Laertes **6**	47.6S	66.4
Mentor **7**	1.3S	45.0
Nausicaa	82.3N	357.3
Nestor **8**	54.6S	61.7
Odysseus **9**	30.0N	130.0
Penelope **10**	11.5S	248.0
Phemius **11**	12.0N	285.8
Polyphemus **12**	4.6S	282.8
Telemachus **13**	54.0N	338.7
Chasma		
Ithaca **14**	60S–35N	030–340

Telesto and Calypso. These are Tethys 'Trojans'; Telesto 60° ahead of Tethys, Calypso 60° behind. In mythology Telesto was one of the Oceanids, and Calypso a daughter of Atlas. Presumably they are icy, but as yet we have little further information about them, except that they are irregular in shape; Telesto is slightly the larger of the two.

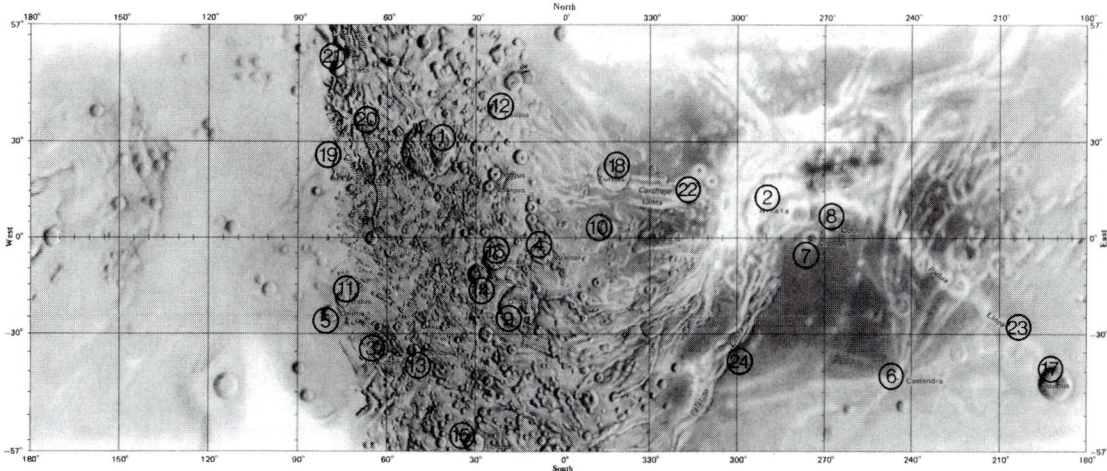

Figure 10.5. Dione.

Dione. Dione, named after the sister of Cronos and mother (by Zeus) of Aphrodite, is of special interest, because although little larger than Tethys it is much denser and more massive. It is indeed denser than any other Saturnian satellite apart from Titan, and there are suspicions that it may have an effect on Saturn's radio emission, since there seems to be a radio cycle of 2.7 days – which is also the orbital period of Dione. Moreover, as noted, Dione may also play a rôle in flexing the interior of Enceladus, whose revolution period is almost exactly half that of Dione. Surface features are named from Virgil's Æneid; a selected list is given in Table 10.11.

The surface is not uniform. The trailing hemisphere is relatively dark, with an albedo of 0.3, while the brightest features on the leading hemisphere have an albedo of 0.6. Most of the heavily cratered terrain lies on the trailing hemisphere. One very prominent feature is Amata, which may be either a crater or a basin; its precise nature is uncertain, but it is associated with a system of bright wispy features which extend over the trailing hemisphere and are accompanied by narrow linear troughs and ridges. These wispy features have been produced by bright, new ice which has seeped out from the interior, so that in the past Dione seems to have been much more active than Tethys or Rhea. There are large, well-marked craters, some, such as Æneas, with central peaks; another large crater, on the opposite hemisphere to Æneas, is Dido. Latium Chasma is rimmed, with a flattish

floor; although over 300 km long it is only 8–12 km wide, with a depth of less than 1 km. Other valleys are also seen. Dione's surface temperature has been given as −186 °C.

Table 10.11. Selected list of features on Dione. (Bold numbers indicate map references.)

Name	Lat.	Long. W	Diameter/ length (km)
Craters			
Æneas **1**	26.1N	46.3	166
Amata **2**	7.7N	285.3	231
Anchises **3**	33.7S	66.1	42
Antenor **4**	6.5S	10.4	82
Caieta **5**	23.3S	80.5	70
Cassandra **6**	39.5S	244.1	36
Catillus **7**	1.6S	273.0	35
Coras **8**	0.6N	266.4	37
Dido **9**	23.7S	18.5	118
Ilia **10**	0.1N	346.0	51
Italus **11**	18.1S	77.5	40
Lausus **12**	36.2N	23.2	28
Massicus **13**	34.8S	56.0	43
Remus **14**	13.2S	31.1	69
Ripheus **15**	56.1S	35.5	32
Romulus **16**	7.3S	26.5	81
Sabinus **17**	47.8S	175.6	79
Turnus **18**	16.2N	344.6	97
Chasma			
Larissa **19**	30.2N	71.1	315
Latium **20**	21.2N	69.5	381
Palatine	75.6S	25.1	394
Tibur **21**	57.2N	69.1	156
Linea			
Carthage **22**	12.7N	321.9	318
Padua **23**	20.0S	210.7	780
Palatine **24**	40.6S	305.4	645

Helene. A Dione Trojan, moving 60° ahead of Dione; it is named after a daughter of Jupiter (Zeus) and Leda. It is icy, and one relatively large crater has been recorded.

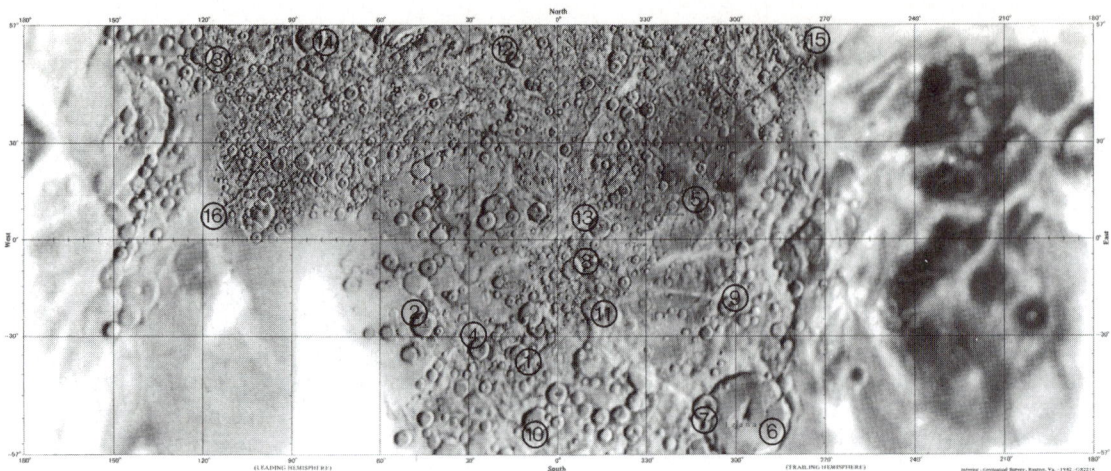

Figure 10.6. Rhea.

Rhea. This is the largest Saturnian satellite apart from Titan. It is named after the mother of Jupiter (Zeus); she was the sister of Cronos (Saturn) and also his wife – the morals of the ancient Olympians left a great deal to be desired! Surface features are named after creation myths. A selected list is given in Table 10.12.

Rhea is heavily cratered, but there are few really large formations, and craters tend to be rather irregular. As with Dione, the trailing hemisphere is the darker of the two, and there are wispy features, although not nearly so prominent as those of Dione. It has been found that there are two distinct types of terrain; the first contains craters over 40 km across, while the second area, in parts of the polar and equatorial regions, is characterized by craters of smaller size. Possibly an early cratering period produced the larger structures; there was then a period of resurfacing, presumably by material welling up from below, and then a second cratering era. On the leading hemisphere there is one ray-centre. Rhea seems to have a rocky core, around which most of the material is ice. There has certainly been no activity for a very long time.

A small telescope is capable of showing Rhea. An interesting observations was made on 8 April 1921 by six English observers independently: A. E. Levin, P. W. Hepburn, L. J. Comrie, E. A. L. Attkins, F. Burnerd and C. J. Spencer. Rhea was eclipsed by the shadow of Titan,

Table 10.12. Selected list of features on Rhea. (Bold numbers indicate map references.)

Name	Lat.	Long. W
Craters		
Bulagat **1**	38.2S	15.2
Djuli **2**	31.2S	46.7
Faro **3**	45.3N	114.0
Haik **4**	36.6S	29.3
Heller **5**	10.1N	315.1
Izanagi **6**	49.4S	310.3
Izanami **7**	46.3S	313.4
Kiho **8**	11.1S	358.7
Leza **9**	21.8S	309.2
Melo **10**	53.2S	7.1
Qat **11**	23.8S	351.6
Thunapa **12**	45.6N	21.3
Xamba **13**	2.1N	349.7
Chasma		
Kun Lun **15**	46.0N	307.5
Pu Chou **16**	26.1N	95.3

and according to the first three observers Rhea vanished completely for over half an hour.

Titan. Titan, discovered and named by Huygens in 1655, is the largest of Saturn's satellites, and the largest satellite in the Solar System apart from Ganymede. It is the only satellite to have a substantial atmosphere.

The first suggestion of an atmosphere came in 1908, when the Spanish astronomer J. Comas Solà reported limb darkening effects (essentially the same as those of the Sun). The existence of the atmosphere was proved spectroscopically in 1944 by G. P. Kuiper, but it was tacitly assumed that the main constituent would be methane, and that the density would be low; after all, the escape velocity is only 2.65 km s^{-1}, little more than that of our virtually airless Moon. As Voyager 1 approached Saturn's system, in November 1980, astronomers were still divided as to whether any surface details would be seen on Titan, or whether there would be too much cloud. In the event, no details were seen; Titan is permanently shrouded beneath its orange clouds. The northern hemisphere was observed to be the darker of the two, and there were vague indications of 'banding', but that was all. However, important discoveries were made. The atmosphere proved to be mainly composed of nitrogen, with a little methane and traces of other compounds. The ground pressure was given as 1.5 times that of the Earth's air at sea level (the latest value is 1.6), and the temperature on the surface was −178 °C.

Voyager 2 did not image Titan from close range, because of the need to go on to Uranus and Neptune, but from 1994 images have been obtained from the Hubble Space Telescope, in the near infra-red, showing brighter and darker areas; one prominent bright feature is 4000 km across – about the size of Australia. It must be some sort of solid surface feature. Other bright and dark regions could be continents or oceans. If there are seas or ponds on Titan, they will certainly not be of water; they will be chemical – made of ethane, methane or a mixture of both.

Excellent images of Titan have been obtained since 1996 by S. Gibbard and her colleagues, with the 10 m Keck I telescope on Mauna Kea in Hawaii. Using speckle interferometry techniques at 1.5 to 2.5 μm, Titan's hazy atmosphere was penetrated, revealing a mottled surface. It was suggested that the infra-red-dark areas might be basins filled with methane, ethane, or other hydrocarbons that precipitate out of the atmosphere; the bright regions could be a mixture of rock and water ice, washed clear of organic material.

Nitrogen makes up about 90% of the atmosphere, and most of the rest is methane, plus a little argon and traces of hydrogen, hydrocarbons and nitrogen and oxygen compounds. At the top of Titan's troposphere, at an altitude of 40 km above the surface, there are clouds of methane; above this, in the stratosphere, comes the 'haze', and this extends to about 200 km, with a detached and more tenuous second layer 100 km higher.

It is not likely that there is a global ocean, but liquid areas are very probable. When the Huygens probe lands there, in 2004, it may come down on solid ground or alteratively splash down in a chemical ocean which could even have waves. The surface temperature is close to the triple point of methane, so that there may even be methane cliffs, seas of liquid methane and a methane rain dripping down from the orange clouds above. At this low temperature, H_2O ice will be as hard as conventional rock.

The globe is probably made up of a rocky core, surrounded by a mantle of liquid water with some dissolved ammonia and methane, above which comes the crust. However, we have to admit that our information is still fragmentary, and if Huygens is successful we may be prepared for many surprises. The one thing we can say with confidence is that Titan is unlike any other world in the Solar System.

Hyperion. Hyperion, named for one of the Titans, is in many ways unusual. It is irregular in shape; this is strange, since an object of this size would be expected to be regular in form. To make matters even more puzzling, the longer axis does not point directly at Saturn, as dynamically it ought to do, and the rotation period is chaotic, changing from one orbit to another. On average, the rotation period is of the order of 13 days. It may well be that Hyperion is part of a larger body which broke up, but there is no sign of the rest of such a body.

Hyperion is less reflective than the other icy satellites, and seems to have an old surface with what may be called 'dirty ice'. There are several craters, with diameters up to 120 km, and one long ridge or scarp, extending to 300 km, which has been named Bond-Lassell in honour of the discoverers of Hyperion in 1858. (Bond actually found the satellite first, but Lassell's confirmation shortly afterwards was independent.) Features on Hyperion are named after solar and lunar deities; a short selected list is given in Table 10.13.

Hyperion is not an easy telescopic object. The best time to locate it is when it is in conjunction with Titan.

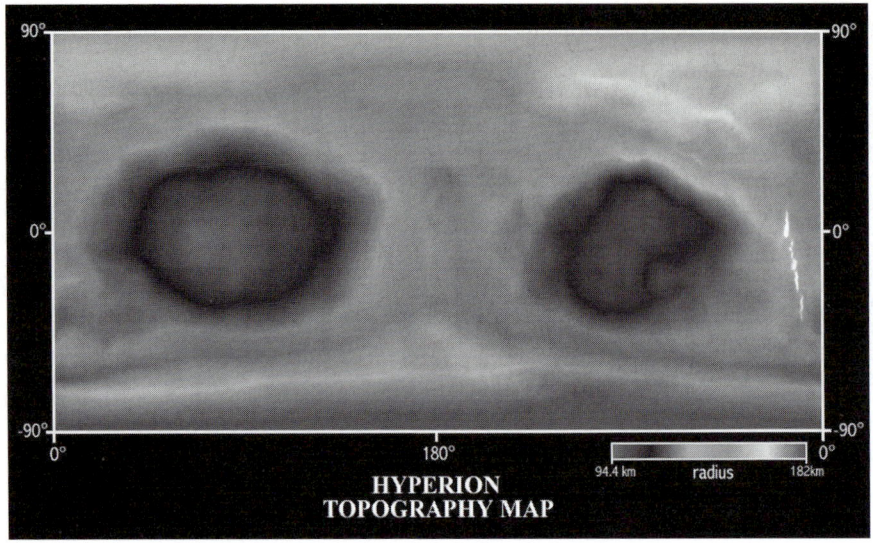

Figure 10.7. Hyperion. (Copyright by Calvin J. Hamilton.)

Table 10.13. Selected list of features on Hyperion.

Name	Lat.	Long. W
Craters		
Bahloo	36.0N	196.0
Helios	71.0N	132.0
Jarilo	61.0N	183.0
Meri	3.0N	171.0
Dorsum		
Bond-Lassell	48.0N	143.5

Iapetus. Iapetus, named for yet another of the Titans, was found by G. D. Cassini in 1671. He soon noticed that it is very variable. When west of Saturn it is an easy telescopic object, but when to the east it is much fainter, and at first Cassini assumed (wrongly) that it disappeared for a time during each of its 79-day orbits. In fact, the magnitude drops from almost 10 down to little brighter than 12.

The reason is that the two hemispheres have different albedoes. The leading hemisphere is as black as a blackboard, while the trailing hemisphere is bright and icy. The demarcation line is not abrupt; there is a 200–300 km transition zone. Some craters have dark floors, but unfortunately we do not know whether or not the material is the same as that which covers the leading hemisphere.

In view of the low density of Iapetus, it seems certain that the satellite itself is bright and icy, so that the dark material is a coating of some kind. There have been suggestions that the dark material has been wafted on to Iapetus from the outermost satellite, Phœbe, but this seems unlikely, partly because Phœbe is so small and distant and partly because the material is not quite the same colour. We also have to explain the dark floors of craters in the bright regions. Presumably, then, the dark material has welled out from below the surface. We know nothing about its depth. Carl Sagan once suggested that it might be thick and made up of organics, while others believe that it may be no more than a few millimetres deep. Obviously we know little about the dark areas, but the bright regions contain craters of the usual type, named from the Charlemagne period. A selected list is given in Table 10.14.

It is worth noting that future space-travellers will see Saturn well from Iapetus, because the rings will not always be edgewise-on. The orbital inclination is over 14°, and Saturn will indeed be a glorious object in the Iapetan sky. All the inner satellites move virtually in the plane of the rings.

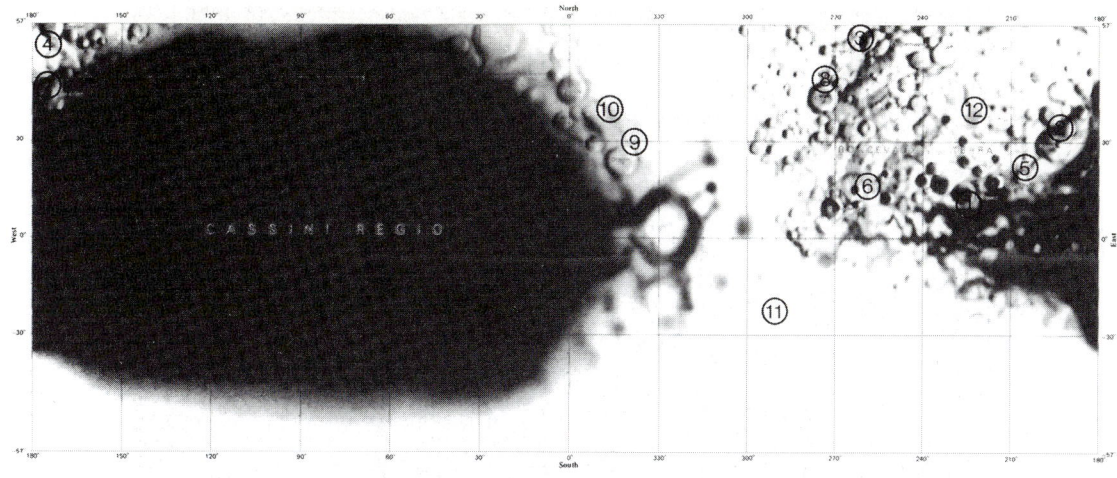

Figure 10.8. Iapetus.

Table 10.14. Selected list of features on Iapetus. (Bold numbers indicate map references.)

Name	Lat.	Long. W	Diameter/ length (km)
Craters			
Baligant **1**	16.4N	224.9	66
Basan **2**	33.3N	194.7	76
Charlemagne **3**	55.0N	258.8	95
Geboin **4**	58.6N	173.4	81
Grandoyne **5**	17.7N	214.5	65
Hamon **6**	10.6N	270.0	96
Marsilion **7**	39.2N	176.1	136
Ogier **8**	42.5N	275.1	100
Othon **9**	33.3N	347.8	86
Roland	73.3N	25.2	144
Turpin **10**	47.7N	1.4	87
Regio			
Cassini **11**	28.1S	92.6	?
Terra			
Roncevaux **12**	37.0N	239.5	1284

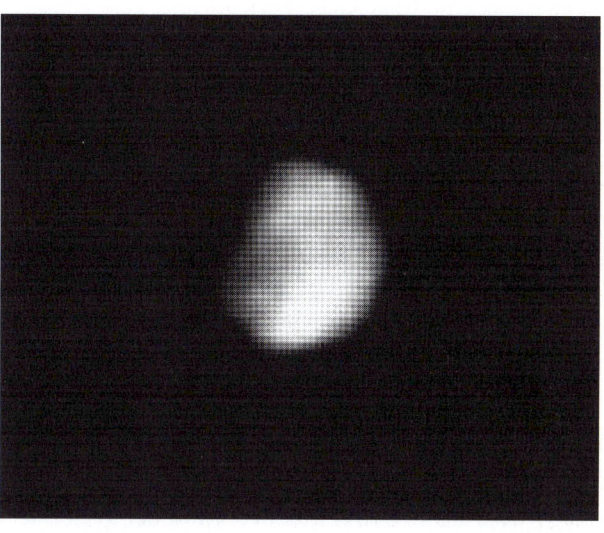

Figure 10.9. Phœbe from Voyager 2, 4 September 1981. (Courtesy: NASA.)

Phœbe. The outermost satellite, Phœbe, was discovered by W. H. Pickering in 1898 – the first satellite discovery made photographically. It takes over 550 days to complete one orbit, but its rotation period is only 9 h, so that it does not always keep the same face turned toward Saturn. Moreover, it has retrograde motion, and is almost certainly a captured asteroid rather than a bona-fide satellite. It is reddish and is roughly spherical in shape, but unfortunately neither Voyager went close to it, and so we know little about its surface features. It may have much in common with asteroids of the carbonaceous type, but we do not really know. Craters are indicated on an image acquired from Voyager 2 on 4 September 1981.

11 URANUS

Uranus, the seventh planet in order of distance from the Sun, was the first to be discovered in telescopic times, by William Herschel, in 1781. It is a giant world, but it and the outermost giant, Neptune, are very different from Jupiter and Saturn, both in size and in constitution. It is probably appropriate to refer to Jupiter and Saturn as gas giants and to Uranus and Neptune as ice giants.

Data for Uranus are given in Table 11.1.

Table 11.1. Data.

Distance from the Sun:
   max 3005 200 000 km (20.088 a.u.)
   mean 2869 600 000 km (19.181 a.u.)
   min 2734 000 000 km (18.275 a.u.)

Sidereal period: 84.01 years (30 685.4 days)

Synodic period: 369.66 days

Rotation period: 17.24 h (17h 14.4m)

Mean orbital velocity: 6.82 km s^{-1}

Axial inclination: 97°.86

Orbital inclination: 0°.773

Orbital eccentricity: 0.04718

Diameter: equatorial 51 118 km
         polar 49 946 km

Apparent diameter, seen from Earth: max 3″.7
                        min 3″.1

Reciprocal mass, Sun = 1: 22 869

Mass, Earth = 1: 14.6 (8.6978 × 10^{25} kg)

Volume, Earth = 1: 64

Escape velocity: 21.1 km s^{-1}

Surface gravity, Earth = 1: 1.17

Density, water = 1: 1.27

Oblateness: 0.023

Albedo: 0.51

Mean surface temperature: −214 °C

Maximum magnitude: −5.6

Mean diameter of Sun, seen from Uranus: 1′41″

Distance from Earth: max 3157 300 000 km
               min 2581 900 000 km

MOVEMENTS

Since Uranus' synodic period is less than five days longer than our year, Uranus comes to opposition every year; opposition dates for the period 2000–2005 are given in Table 11.2. The opposition magnitude does not vary a great deal; the planet can just be seen with the naked eye under good conditions. The most recent aphelion passage was that of 1 April 1925; Uranus was at its maximum distance from the Earth (21.09 a.u.) on 13 March of that year. The next aphelion will be that of 27 February 2009. The last perihelion passage was on 20 May 1966; Uranus was at its closest to the Earth (17.29 a.u.) on 9 March of that year. The next perihelion will be that of 13 August 2050.

Table 11.2. Oppositions of Uranus.

Date	Declination	Magnitude	Apparent diameter (″)
2000 Aug 11	−15°52′	6.0	3.62
2001 Aug 15	−14°37′	6.0	3.62
2002 Aug 20	−13°18′	6.0	3.61
2003 Aug 24	−11°55′	6.0	3.60
2004 Aug 27	−10°30′	6.1	3.60
2005 Sept 1	−9°02′	6.1	3.60

During this period Uranus passes from Sagittarius into Capricornus.

In June 1989 Uranus reached its greatest southerly declination (−23°.7). Greatest northern declination had been reached in March 1950.

Close planetary conjunctions involving Uranus are listed in Table 11.3. It is interesting to note that between January and March 1610 Uranus was within 3° of Jupiter. This was the time when Galileo was making his first telescopic observations of Jupiter, but Uranus was beyond the limits of the field of his telescope.

Uranus can, of course, be occulted by the Moon. The first such record seems to be due to Captain (later Rear-Admiral) Sir John Ross, on 6 August 1924, with a power

Table 11.3. Planetary conjunctions involving Uranus. Close conjunctions, 1900–2100.

	Date	UT	Separation (″)	Elongation (°)
Mars–Uranus	1947 Aug 6	01.49	+43	48W
Jupiter–Uranus	1955 May 10	20.39	−56	65E
Mars–Uranus	1988 Feb 22	20.48	+40	63W
Venus–Uranus	2077 Jan 20	20.01	−43	11W

There are conjunctions with Venus on 2000 Mar 4 (234″), 2003 Mar 28 (157″), 2015 Mar 4 (317″) and 2018 Mar 29 (243″); with Mercury on 2006 Feb 14 (83″) and with Mars on 2013 Mar 22 (39″)

of ×500 on a reflector of focal length 25 ft (7.26 m). On 4 October 1832 Thomas Henderson, from the Cape of Good Hope, observed an occultation.

EARLY OBSERVATIONS

Uranus was seen on a number of occasions before its identification in 1781. It was recorded on 23 December 1690 by the Astronomer Royal, John Flamsteed, when it was in Taurus; Flamsteed even gave it a stellar number – 34 Tauri. Altogether 22 pre-discovery observations have been listed, as follows:

Flamsteed, 1690, 1712, four times in 1715;

J. Bradley, 1748 and 1750;

P. Le Monnier, twice in 1750, 1764, twice in 1768, six times in 1769, 1771;

T. Mayer, 1756.

It is interesting that Le Monnier failed to identify Uranus from its movement. He observed it eight times in four weeks (27 December 1768 to 23 January 1769) without realizing that it was anything other than a star. He has often been ridiculed for this, but when he made his observations Uranus was near its stationary point, so it is hardly surprising that Le Monnier failed to identify it.

DISCOVERY AND NAMING

Uranus was discovered on 13 March 1781 by William Herschel, using a 6.2 inch (15.7 cm) reflector of 7 ft focal length and a magnification of ×227. Herschel realized that the object – in Gemini – was not a star, but he believed it to be a comet, and indeed his communication to the Royal Society was headed *An Account of a Comet*.

The object was first recognized as a planet, independently but about the same time, by the French amateur astronomer de Saron – later, in 1794, guillotined during the Revolution – and by the Finnish mathematician Anders Lexell. Lexell calculated an orbit, finding that the distance of the planet from the Sun was 19 a.u. – only slightly too small. He gave an orbital period of between 82 and 83 years, and stated that the apparent diameter was between 3 and 5 arcsec. In this case it was clear that Uranus was indeed a giant world, larger than any other planet apart from Jupiter and Saturn.

There was prolonged discussion over naming. J. E. Bode, in 1781, suggested Uranus, after the first ruler of Olympus (Uranus or Ouranos, Saturn's father). Other names were proposed – for example Hypercronius (J. Bernoulli, 1781) and 'the Georgian Planet', by Herschel himself in honour of his patron, King George III. Others called it simply 'Herschel'. Until 1850 the Nautical Almanac continued to call it the Georgian Planet, but in that year the famous mathematician and astronomer John Couch Adams suggested changing over to 'Uranus'. This was done, and the name became universally accepted.

DIAMETER AND ROTATION

The first attempt to measure the apparent diameter was made by Herschel in 1781. His value (4″.18) was rather too great. In 1788 he gave the diameter as 34 217 miles (55 067 km) with a mass 17.7 times that of the Earth; these values also were slightly too high. In 1792–4 Herschel also made an attempt to measure the polar flattening, and from his results rightly concluded that the rotation period must be short.

In 1856 J. Houzeau, in France, gave a rotation period of between $7\frac{1}{4}$ and $12\frac{1}{2}$ h; later, before the results from space-craft became available, the favoured period was 10 h 48 min, which is in fact decidedly too short. Uranus is unique in one respect; its axial inclination is more than a right angle, so that the rotation is technically retrograde, although not usually classed as such. From Earth, the equator of Uranus is regularly presented, as in 1923 and 1966; at other times a pole is presented – the south pole in 1901 and 1985, the north pole in 1946 and 2030. All this leads to a very peculiar Uranian calendar. Each pole has a 'night' lasting for 21 Earth years, with corresponding daylight at the opposite pole. For the rest of the orbital period conditions are less extreme.

The reason for this unique tilt is unclear. There have been suggestions that in its early career Uranus was struck by a massive impactor and literally knocked sideways. This does not sound very plausible, but it is not easy to think of anything better.

(There is, incidentally, some confusion about Uranus' poles. The International Astronomical Union has decreed that all poles above the ecliptic (i.e. the plane of the Earth's orbit) are north poles, while all poles below the ecliptic are south poles. In this case it was Uranus' south pole which was in sunlight during the pass of the Voyager 2 space probe in 1986. However, the Voyager team members reversed this and referred to the sunlit pole as the *north* pole. Take your pick!)

SURFACE MARKINGS FROM EARTH

In ordinary telescopes Uranus appears as a bland, rather greenish disk. Two bright spots were reported on 25 January 1870 by J. Buffham, using a power of ×320 on a 9 inch (23 cm) refractor; on 19 March he described a bright streak. It seems improbable that these were genuine features, although of course one cannot be sure. From Earth, even really large telescopes show virtually no surface details on Uranus.

Spectroscopic observations were made from 1869, when Angelo Secchi, from Italy, recorded dark lines in the spectrum; the lines were photographed in 1869 by W. Huggins and in 1902 H. Deslandres, from France, obtained spectroscopic confirmation of the retrograde rotation. Final confirmation of this was provided in 1911 by P. Lowell and V. M. Slipher, from the Lowell Observatory in Arizona, although their derived rotation period was several hours too short.

Meanwhile, very precise tables of the movements of Uranus had been compiled in 1875 by Simon Newcomb in America; earlier, in 1846, slight irregularities in the movements of Uranus had been used to identify the outer giant, Neptune, by J. Galle and H. D'Arrest, from Berlin.

Methane was identified in Uranus' atmosphere in 1933, by R. Mecke from Heidelberg; it had been suggested, on theoretical grounds, by R. Wildt in 1932. Confirmation was obtained by V. M. Slipher and A. Adel, from Flagstaff, in 1934. By then it had become clear that Uranus, like Jupiter and Saturn, was not a miniature sun; the outer layers at least were very cold indeed, and the visible surface was purely gaseous.

The first widely-accepted model of Uranus was that of R. Wildt, who in 1934 proposed that the planet must have a rocky core, overlaid with a thick layer of ice which was in turn overlaid by a hydrogen-rich atmosphere. In 1951 W. R. Ramsey, of Manchester University, proposed an alternative model, according to which Uranus was made up largely of methane, ammonia and water. However, reliable information was delayed until the mission of Voyager 2, the only probe so far to have by-passed the planet.

VOYAGER 2

It is not too much to say that most of our detailed knowledge of Uranus comes from Voyager 2, although in recent years good data have also been obtained from the Hubble Space Telescope. Voyager 2 was launched on 20 August 1977 (actually before Voyager 1). It by-passed Jupiter on 9 July 1979 and Saturn on 25 August 1981. It then went on to Uranus (24 January 1986) and finally Neptune (25 August 1989). Excellent images were obtained of all four giants, together with a mass of data. Voyager 2 is now leaving the Solar System, but was still in contact in 2000.

Because of Uranus' unusual tilt, Voyager 2 approached the planet more or less 'pole-on'. New satellites were discovered, and known satellites surveyed; studies were made of the Uranian magnetosphere, and radio waves were detected. Ultra-violet observation showed strong emissions on the day side of the planet, producing what was termed an 'electroglow'. On 'encounter day', 24 January, Voyager passed 107 100 km from the centre of the planet – that is to say, must over 80 000 km from the cloud tops. Closest approach occurred at 17h 59m UT. The Uranian equator was then in twilight, and it was found that the temperatures at the poles and over the rest of the planet were much the same.

CONSTITUTION OF URANUS

One important fact is that Uranus, unlike the other giant planets, seems to have little internal heat. Jupiter radiates 1.7 times as much energy as it would do if it depended entirely upon what it receives from the Sun; Saturn radiates 1.8 times as much and Neptune over 2. With Uranus, the upper limit is only 1.06, but with a possible uncertainty of 1 – so that there may be no excess energy at all. Moreover, the temperatures of Uranus and Neptune as measured from Earth are almost equal, even though Neptune is so much further away from the Sun.

Uranus is made largely of 'ices', but it is to be noted that these may not be in solid form. For planetary scientists, 'gas' is taken to mean helium and helium; 'ices' a solar mixture of water (H_2O), methane (CH_4) and ammonia (NH_4), with traces of other substances; water is the most abundant of the ices. 'Rock' is a mix of silicon dioxide (SiO_2), magnesium oxide (MgO) and either metallic iron (Fe) and nickel (Ni), or compounds of iron such as FeS and FeO. Inside planets such as Uranus and Neptune, the pressure and temperature conditions make all these materials behave as liquids.

The outer atmosphere is made up chiefly of hydrogen (probably 83% by number of molecules) and helium (15%); methane accounts for 2%, so that there are only traces of other substances. Methane freezes out at a very low temperature, and forms a thick cloud layer, above which comes the predominantly hydrogen atmosphere. Methane absorbs red light, which is why Uranus appears bluish-green. Minor constituents include acetylene (C_2H_2) and ethane (C_2H_6), which play a rôle in forming 'hazes'.

Below the atmosphere come the 'ices', and then a relatively small rocky core at a temperature of perhaps between 6000 and 7000 °C. However, we have to admit that our knowledge of the inner structure of Uranus is very fragmentary, and there is no definite proof that a rocky core exists, although it very probably does.

Voyager 2 detected few clouds, but in recent years cloud observations have been made with the Hubble Space Telescope. Good images were obtained in July to August 1997; these showed clouds in the northern hemisphere, which is now starting its 'spring' season – when Voyager 2 flew over the north pole it had been winter there, with the pole in total darkness. It has been found that features at different latitudes have rotation periods of between 14 and 17 h, so that there are winds blowing in an east–west direction; at high latitudes (around 60°) these are prograde, although there is a retrograde jet stream close to the equator. Observations made in 1998 with the Hubble Space Telescope recorded waves of massive storms on Uranus, with windspeeds in excess of 500 kilometres per hour; clearly the planet is much more dynamic than was originally thought. Obviously, the axial tilt means that wind conditions on Uranus are different from those on any other planet.

Radiation belts round Uranus exist; their intensity is similar to those round Saturn, but they differ in composition. They seem to be dominated by hydrogen ions; heavier ions are lacking.

MAGNETIC FIELD

Uranus has a magnetic field; the equatorial field strength at the equator is 0.25 G, as against 4.28 G for Jupiter (the value for Earth is 0.305 G). However, the magnetic axis is displaced from the rotational axis by 58°.6; neither does the magnetic axis pass through the centre of the globe – it is offset by 8000 km. The polarity is opposite to that of the Earth. The fact that the magnetic and the rotational poles are nowhere near each other means that auroræ, which were detected from Voyager 2, are a long way from the rotational pole. The magnetosphere of Uranus is relatively 'empty'; it extends to 590 000 km on the day side and around 6000 000 km on the night side.

The reason for the tilt of the magnetic is unknown. It was initially thought that Uranus might be experiencing a 'magnetic reversal', but subsequently it was found that the magnetic axis of Neptune was also displaced – and to assume that two reversals were occurring simultaneously would be too much of a coincidence.

THE RINGS

The discovery of the ring system of Uranus was accidental. It had been predicted that on 10 March 1977 the planet would occult the star SAO 158687, magnitude 8.9, and the occultation would provide a good opportunity to measure the diameter of Uranus. Calculations made by Gordon Taylor of the Royal Greenwich Observatory indicated that the occultation would be seen only from a restricted area in the southern hemisphere, and observations were made by J. Elliott, T. Dunham and D. Mink, flying at 12.5 km above the southern Indian Ocean in the Kuiper Airborne Observatory (KAO), which was in fact a modified C-141 aircraft carrying a 36 inch (91 cm) reflecting telescope. Close watches were also being kept from ground-based observatories, notably in South Africa. 35 min before occultation the star was seen by the KAO observers to 'wink' five times, so that apparently it was being temporarily obscured by material in the vicinity of Uranus. The occultation by Uranus began at 20.52 UT, and lasted for 25 min. After emersion there were more winks, and these were later found to be symmetrical with the first set, indicating a system of rings. The post-emersion winks were also recorded by J. Churms from South Africa.

Subsequent observations provided full confirmation. In 1978 G. Neugebauer and his colleagues imaged the rings with the Hale reflector at Palomar and in 1984 D. A. Allen, at Siding Spring, imaged them in infra-red, using the Anglo–Australian Telescope. They were surveyed in detail by Voyager 2 and also studied from the Hubble Space Telescope. Details of the system are given in Table 11.4.

The outermost ring – the ϵ ring – is not symmetrical, and is narrowest when closest to Uranus; the satellites Cordelia and Ophelia, discovered by Voyager 2, act as 'shepherds' to it. There is little obvious dust in the main rings, but Voyager 2 took a final picture, on its outward journey from Uranus, when the planet hid the Sun, showing 200 very diffuse, nearly transparent bands of microscopic dust surrounding the system. The rings are made up of particles a few metres in diameter, with not many centimetre- and millimetre-sized particles; they are as dark as coal, and it has been suggested, although without proof, that they may be relatively young and perhaps not even

Table 11.4. Rings of Uranus.

Ring	Distance from Uranus (km)	Eccentricity	Width (km)	Period (h)
6	41 837	1.01	~1.5	6.1988
5	42 235	1.90	~2	6.2875
4	42 571	1.06	~2.5	6.3628
α	44 718	0.76	4–10	6.8508
β	45 661	0.44	5–11	7.0688
η	47 176	0.004	1.6	7.4239
γ	47 626	0.0	1–4	7.5307
δ	48 303	0.0	3–7	7.6911
λ	50 024	0.0	~2	8.1069
ϵ	51 149	0.79	20–96	8.3823

permanent features of the Uranian system. Their thickness is from about 0.1 to 1 km.

The rings are not alike. 6, 5 and 4 show significant internal structure. Rings α and β lack sharp edges. Ring η does not show sensible inclination to Uranus' orbital plane, and is made up of two components – a sharp inner feature, and a much fainter one, which extends to about 55 km from the sharp feature. Both edges of the γ ring are sharp, while with the δ ring there is a faint component inside the main ring. Ring λ was discovered between the δ and ϵ rings; it is very thin and apparently circular. In addition to the 10 rings, there is a broad sheet of material closer-in than Ring 6, extending from 37 000 to 39 500 km from Uranus.

It is worth recalling that in 1787, a few years after his discovery of Uranus, William Herschel reported the existence of a ring; he was using his 20 ft focus reflector on 4 March, and reported the ring again on 22 February 1789. He described the ring as 'short, not like that of Saturn'. In fact he was being temporarily misled by optical effects; no telescope of that period could possibly show the true rings, and by the end of 1793 Herschel himself had realized that his 'ring' did not exist.

Certainly the ring system of Uranus is interesting, but there can be no comparison with the glorious, icy rings of Saturn.

Table 11.5. Satellites of Uranus. The main satellites have synchronous rotation, and this is probably true also for the minor satellites.

Name	Discoverer	Mean distance from Uranus (km)	Orbital period (d)	(h)	(m)	Mean synodic period (d)	(h)	(m)	(s)	Mean angular distance from Uranus, opposition (")	Orbital inclination (°)	Orbital eccentricity	Orbital velocity (km s^{-1})
Cordelia	Voyager 2, 1986	49 471	0.330	7	55						0.14	0.0005	10.5
Ophelia	Voyager 2, 1986	53 796	0.372	8	55						0.09	0.0101	10.4
Bianca	Voyager 2, 1986	59 173	0.433	10	23						0.16	0.0009	9.9
Cressida	Voyager 2, 1986	51 777	0.463	11	07						0.04	0.0001	9.7
Desdemona	Voyager 2, 1986	62 676	0.475	11	24						0.16	0.0002	9.6
Juliet	Voyager 2, 1986	64 352	0.493	11	50						0.04	0.0002	9.5
Portia	Voyager 2, 1986	66 085	0.513	12	19						0.09	0.0002	9.4
Rosalind	Voyager 2, 1986	69 941	0.558	13	24						0.08	0.0006	9.1
Belinda	Voyager 2, 1986	75 258	0.662	14	55						0.03	0.0001	8.8
1986 U10	Karkoschka, 1999	75,258	0.62	5	18						Low	Low	
Puck	Voyager 2, 1986	86 000	0.762	18	17						0.31	0.0001	8.2
Miranda	Kuiper, 1948	129 400	1.414	9	50	1	9	55	31	9.9	4.22	0.0027	6.7
Ariel	Lassell, 1851	191 000	2.520	12	29	2	12	29	39.0	14.5	0.31	0.0034	5.5
Umbriel	Lassell, 1851	256 300	4.144	3	28	4	3	28	25.8	20.3	0.36	0.0050	4.7
Titania	W Herschel, 1787	435 000	8.706	16	56	8	17	00	1.2	33.2	0.014	0.0022	3.6
Oberon	W Herschel, 1787	583 500	13.463	11	07	13	11	15	36.5	44.5	0.10	0.0008	3.2
Caliban	Nicholson *et al*, 1998	7 775 000	654		5	4						146	0.2	
Sycorax	Nicholson *et al*, 1998	8 845 000	795		9	5						154	0.34	
1999 U1	J Kavelaars	10 000 000	950										
1999 U2	J Kavelaars	25 000 000	3773										

Name	Rotation Period (days)*	Diameter (km)	Mass (kg)	Reciprocal mass, Uranus = 1	Density, water = 1	Escape velocity (km s^{-1})	Magnitude	Albedo	Apparent diameter, seen from Uranus (")
Cordelia	0.330	26	Low	Low	?	Very low	24.2	0.07?	?
Ophelia	0.372	32	Low	Low	?	Very low	23.9	0.07?	?
Bianca	0.433	42	Low	Low	?	Very low	23.1	0.07?	?
Cressida	0.463	62	Low	Low	?	Very low	22.3	0.07?	?
Desdemona	0.475	54	Low	Low	?	Very low	22.5	0.07?	?
Juliet	0.493	84	Low	Low	?	Very low	21.7	0.07?	?
Portia	0.513	105	Low	Low	?	Very low	21.1	0.07?	?
Rosalind	0.558	54	Low	Low	?	Very low	22.5	0.07?	?
Belinda	0.622	66	Low	Low	?	Very low	22.1	0.07?	?
1986 U10	0.62	40			?	Very low	23		
Puck	0.762	154	Low	Low	?	Very low	20.4	0.07?	?
Miranda	1.413	481 × 466 × 466	6.33 × 10^{19}	1 000 000	1.3	0.5	16.3	0.27	17.54
Ariel	2.250	1158	1.27 × 10^{21}	67 000	1.6	1.2	14.2	0.40	30.54
Umbriel	4.144	1169	1.27 × 10^{21}	67 000	1.4	1.2	14.8	0.18	14.12
Titania	8.706	1578	3.49 × 10^{21}	20 000	1.6	1.6	13.7	0.27	15.00
Oberon	13.463	1523	3.03 × 10^{21}	30 000	1.5	1.5	13.9	0.24	9.48
Caliban	654	60	Low	Low	?	Very low	22.3	?	?
Sycorax	795	120	Low	Low	?	Very low	20.7	?	?
1999 U1	950	40			?	Very low	24		
1999 U2	3773	40			?	Very low	24		

* The main satellites have synchronous rotation, and this is probably true also for the Ursa Minor satellites.

SATELLITES

Uranus has an extensive satellite system. Five were known before the Voyager 2 mission; Voyager added 10 more and another two with the Canada–France–Hawaii telescope on Mauna Kea. Since then two outer satellites have been detected from Palomar and another two have been suspected. Data are given in Table 11.5.

The first to be discovered were Oberon and Titania, by William Herschel in 1787. Both were seen on 11 January, although Herschel delayed making any announcement until he was certain of their nature. Between 1790 and 1802 Herschel claimed to have found four more satellites, but three of these are certainly non-existent; the fourth may have been Umbriel, but there is considerable uncertainty. Ariel and Umbriel were discovered on 24 October 1851 by the English amateur William Lassell (previous observations by Lassell in 1847 had been inconclusive). In 1894, 1897 and 1899 W. H. Pickering unsuccessfully searched for new satellites. The next satellite to be detected photographically from Earth was Miranda, on 16 February 1948, with the 82 inch reflector at the McDonald Observatory in Texas. Next, on 31 October 1997, P. Nicholson, J. Burns, B. Gladman and J. J. Kavelaars, using a CCD on the 5 m Hale Telescope at Palomar, found the outermost satellites, Caliban and Sycorax. In 1999 J. J. Kavelaars and his colleagues, using the 3.58 m Canada–France–Hawaii telescope on Mauna Kea, announced the discovery of two more small outer satellites, bringing the grand total to 20 – more than for any other planet. No doubt all these small outer satellites are asteroidal.

The names of the first four satellites were suggested by Sir John Herschel: two Shakesperean, one (Ariel) from both Shakespeare and Pope's *Rape of the Lock*, and Umbriel from *Rape of the Lock*. The name 'Miranda' was proposed by Kuiper and the names of the 10 new inner and the two new outer satellites were adopted by the International Astronomical Union. Many critics feel that this departure from conventional mythology is undesirable, and should not be regarded as a precedent.

Cordelia. The innermost and smallest of the Voyager 2 discoveries. It is named after the daughter of King Lear, in the play *King Lear*. It acts as the inner 'shepherd' for the ϵ ring. Like all the Voyager 2 satellites, it is presumably icy, but nothing definite is known about its composition.

Ophelia. Named after the daughter of Polonius, in the play *Hamlet*. It is the outer 'shepherd' for the ϵ ring and is slightly brighter and larger than the inner shepherd, Cordelia.

Next, beyond the ring system but of course within the magnetosphere, come eight icy satellites:

Bianca (named after the sister of Katherine in the play *The Taming of the Shrew*);
Cressida (Calchas' daughter, in *Troilus and Cressida*);
Desdemona (Othello's wife, in *Othello*);
Juliet (the heroine in *Romeo and Juliet*);
Portia (wife of Brutus, in *Julius Cæsar*);
Rosalind (the Duke's daughter, in *As You Like It*);
Belinda (the heroine in *The Rape of the Lock*);
Puck (from *A Midsummer Night's Dream*).

Of these, Puck was the first to be detected, on 30 December 1985, as Voyager 2 drew in toward Uranus. On 24 January 1986 – 'encounter day' – a single image of it was obtained from a range of 500 000 km. The resolution is of the order of 10 km. Puck proved to be roughly spherical, with a low albedo. Three craters were recorded and were named Bogle, Lob and Butz after various mischievous spirits of the Puck variety.

Miranda. Voyager 2 passed Miranda at only 3000 km, so that the surface features were imaged down to a resolution of 600 m. This was fortunate, because Miranda is indeed a remarkable world. Surface feature names are again Shakespearean. The measured temperature is given as $-187\,°C$, and the surface is icy. Naturally, only one hemisphere could be studied; the other was in darkness.

The landscape is amazingly varied, and there are several distinct types of terrain: old, cratered plains, brighter areas with cliffs and scarps, and 'ovoids' or coronæ, large, trapezoidal-shaped regions; one of these, Inverness Corona, was nicknamed 'the chevron', while another, Arden Corona, was 'the race-track'. The cliffs may tower to 20 km; large craters are lacking, but there are fault valleys, parallel ridges and graben up to 15 km across. A selected list of surface features is given in Table 11.6.

Miranda presents real problems of interpretation. It has been suggested that it has been shattered and re-formed several times, because the various types of terrain seem to have been formed at different periods, but this would involve considerable heating, which in view of Miranda's small size and icy nature does not sound probable.

Table 11.6. Features on Miranda. (Bold numbers indicate map references.)

Name	Lat. S	Long. E	Diameter/ length (km)
Craters			
Alonso **1**	44.0	352.6	25
Ferdinand **6**	34.8	202.1	17
Francisco **7**	73.2	236.0	14
Gonzalo **8**	11.4	77.0	11
Prospero **11**	32.9	329.9	21
Stephano **13**	41.1	234.1	16
Trinculo **14**	63.7	163.4	11
Coronæ			
Arden Corona **2**	10–60	30–120	318
Elsinore Corona **5**	10–42	215–305	323
Inverness Corona **9**	38–90	0–360	234
Regiones			
Dunsinane Regio **4**	20–75	345–65	244
Mantua Regio **10**	10–90	75–300	399
Silicia Regio **12**	10–50	295–340	174
Ropes			
Argier Rupes **3**	40–50	310–340	141
Verona Rupes **15**	10–40	340–350	116

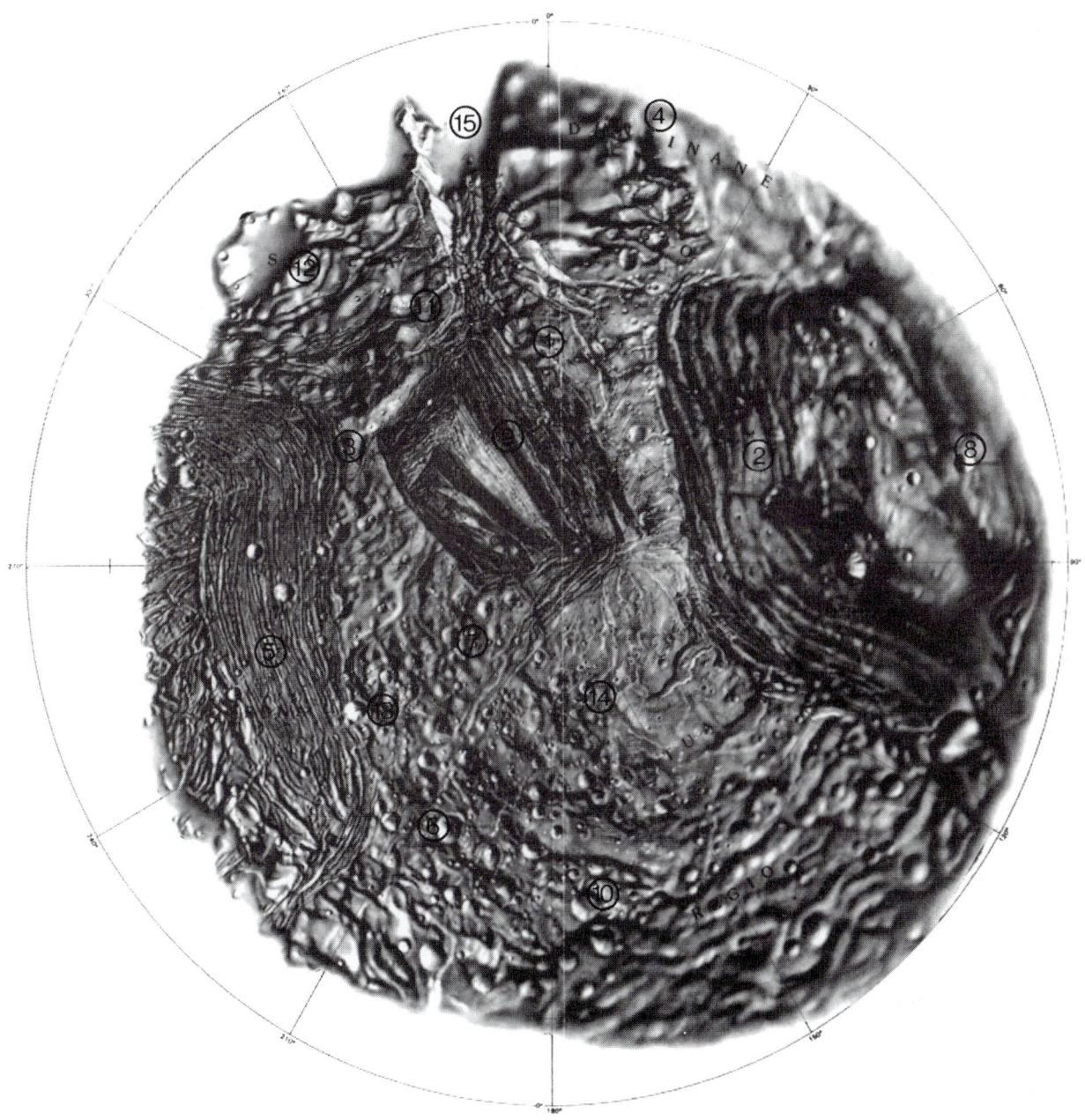

Figure 11.1. Miranda.

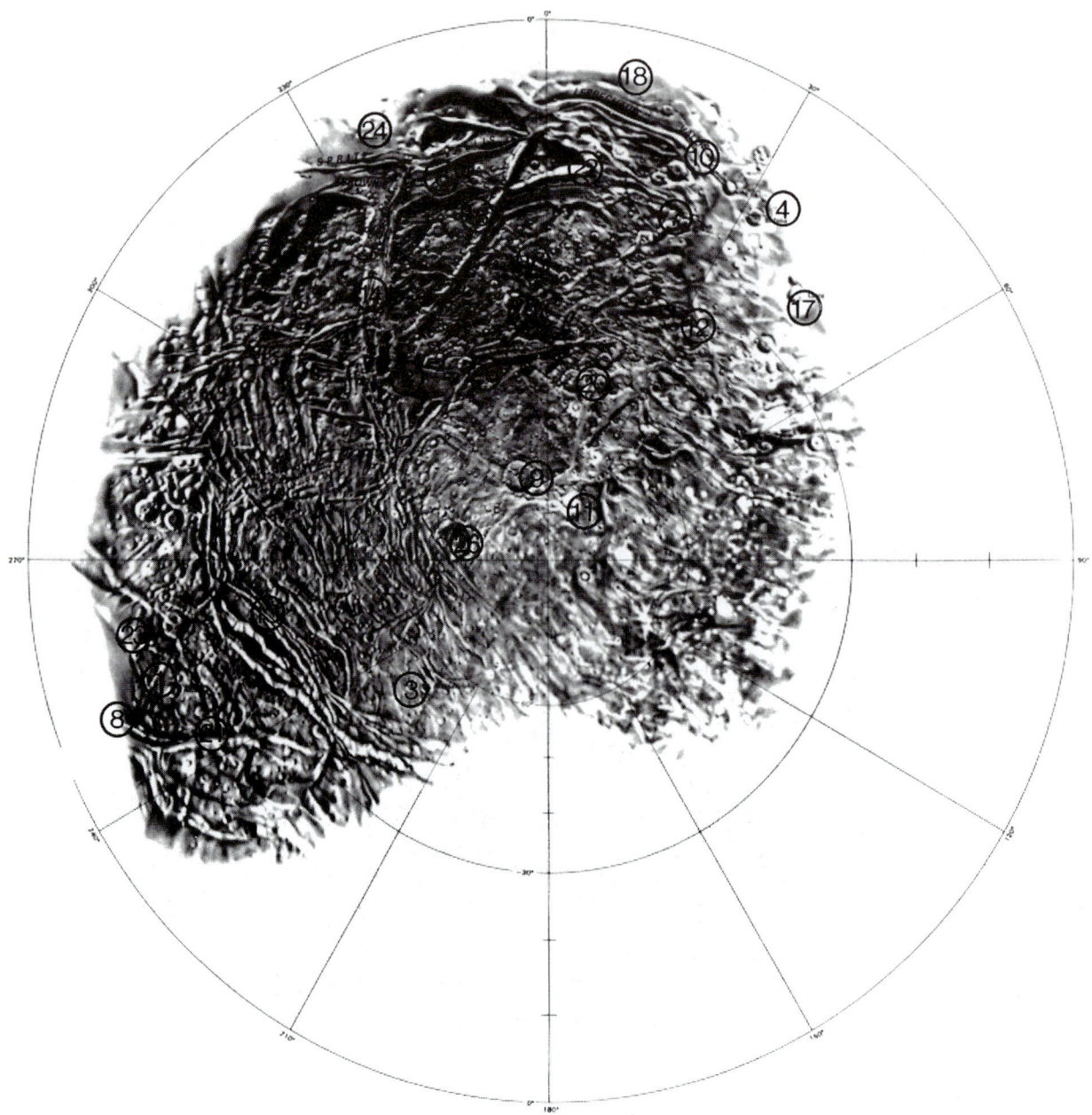

Figure 11.2. Ariel.

Ariel. Imaged by Voyager 2 from 130 000 km, to a resolution of 2.4 km. A selected list of surface features is given in Table 11.7. The name comes from *The Tempest*; Ariel is a friendly sprite.

There are plenty of craters, and the terrain is transected by fault scraps and graben, but the dominant features are the broad, branching, smooth-floored valleys such as Korrigan Chasma and Kewpie Chasma. These canyons look as though they have been smoothed by fluid, but water is not a likely candidate, because of Ariel's small size and low temperature. Ariel today is inert, but clearly it was once the site of great tectonic activity and 'icy vulcanism'. Certainly its surface appears to be younger than those of Umbriel, Titania or Oberon.

Table 11.7. Features on Ariel. (Bold numbers indicate map references.)

Name	Lat. S	Long. E	Diameter/ length (km)
Craters			
Abans **1**	15.5	251.3	20
Agape **2**	46.9	336.5	34
Ataksak **3**	53.1	224.3	22
Befanak **4**	17.0	31.9	21
Berylune **5**	22.5	327.9	29
Deive **7**	22.3	23.0	20
Djadek **8**	12.0	251.1	22
Domovoy **9**	71.5	339.7	71
Finvara **10**	15.8	19.0	31
Gwyn **11**	77.5	22.5	34
Huon **12**	37.8	33.7	40
Laica **17**	21.3	44.4	30
Mab **19**	38.8	352.2	34
Melusine **20**	52.9	8.9	50
Oonagh **21**	21.9	244.4	39
Rima **23**	18.3	260.8	41
Yangoor **26**	68.7	279.7	78
Chasmata			
Brownie **6**	16.0	337.6	343
Kachina **13**	33.7	246.0	622
Kewpie **14**	28.3	326.9	467
Korrigan **15**	27.6	347.5	365
Kra **16**	32.1	354.2	142
Pixie **22**	20.4	5.1	278
Sylph **25**	48.6	353.0	349
Valles			
Leprechaun **18**	10.4	10.2	328
Sprite **24**	14.9	340.0	305

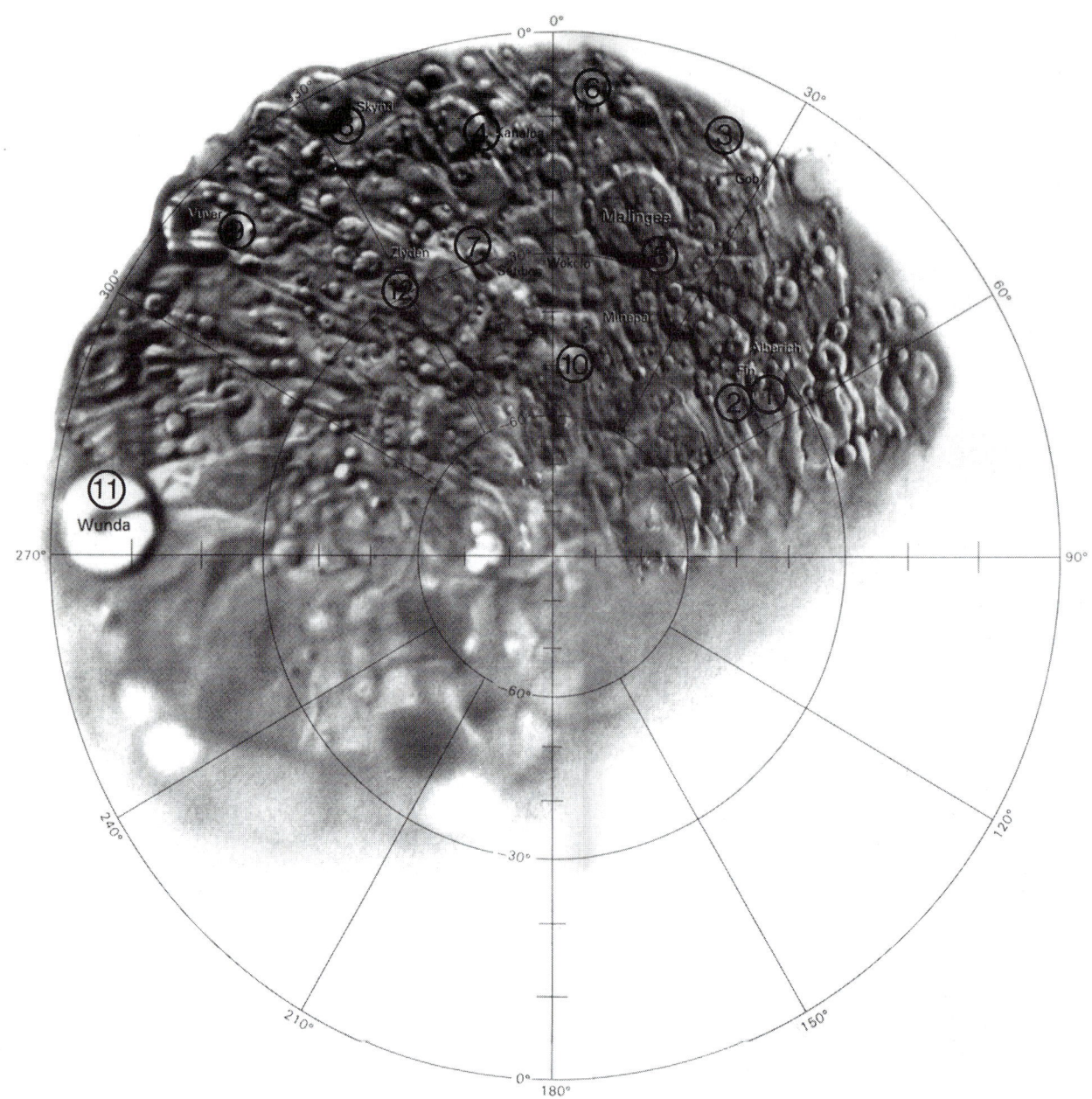

Figure 11.3. Umbriel.

Umbriel. Named for the dark sprite in *The Rape of the Lock*. While Ariel's surface features are named after benevolent spirits, those on Umbriel take their names from malevolent sprites. The surface is darker than those of the other main satellites; there are no ray-craters. Unfortunately the best image was taken when Voyager was still 557 000 km from Umbriel, so that the resolution is no better than 10 km. (See Table 11.8.)

The brightest crater, Skynd, has a central peak; other craters have darker floors, and since Umbriel is presumably icy there has been a surface coating of some kind. Wunda, at the edge of the image, is of uncertain nature; it is close to the equator (remember that Umbriel, like Uranus itself, was being imaged almost pole-on) and appears to be a ring, but it it so badly foreshortened that its form cannot be made out, and we cannot be sure that it is a crater at all. Nothing comparable has been found anywhere else on Umbriel.

Table 11.8. Features on Umbriel. (Bold numbers indicate map references.)

Name	Lat. S	Long. E	Diameter (km)
Craters			
Alberich **1**	33.6	42.2	52
Fin **2**	37.4	44.3	43
Gob **3**	12.7	27.8	88
Kanaloa **4**	10.8	345.7	86
Malingee **5**	22.9	13.9	164
Minepa	42.7	8.2	58
Peri **6**	9.2	4.3	61
Setibos **7**	30.8	346.3	50
Skynd **8**	1.8	331.7	72
Vuver **9**	4.7	311.6	98
Wokolo **10**	30.0	1.8	208
Wunda **11**	7.9	273.6	131
Zlyden **12**	23.3	326.2	44

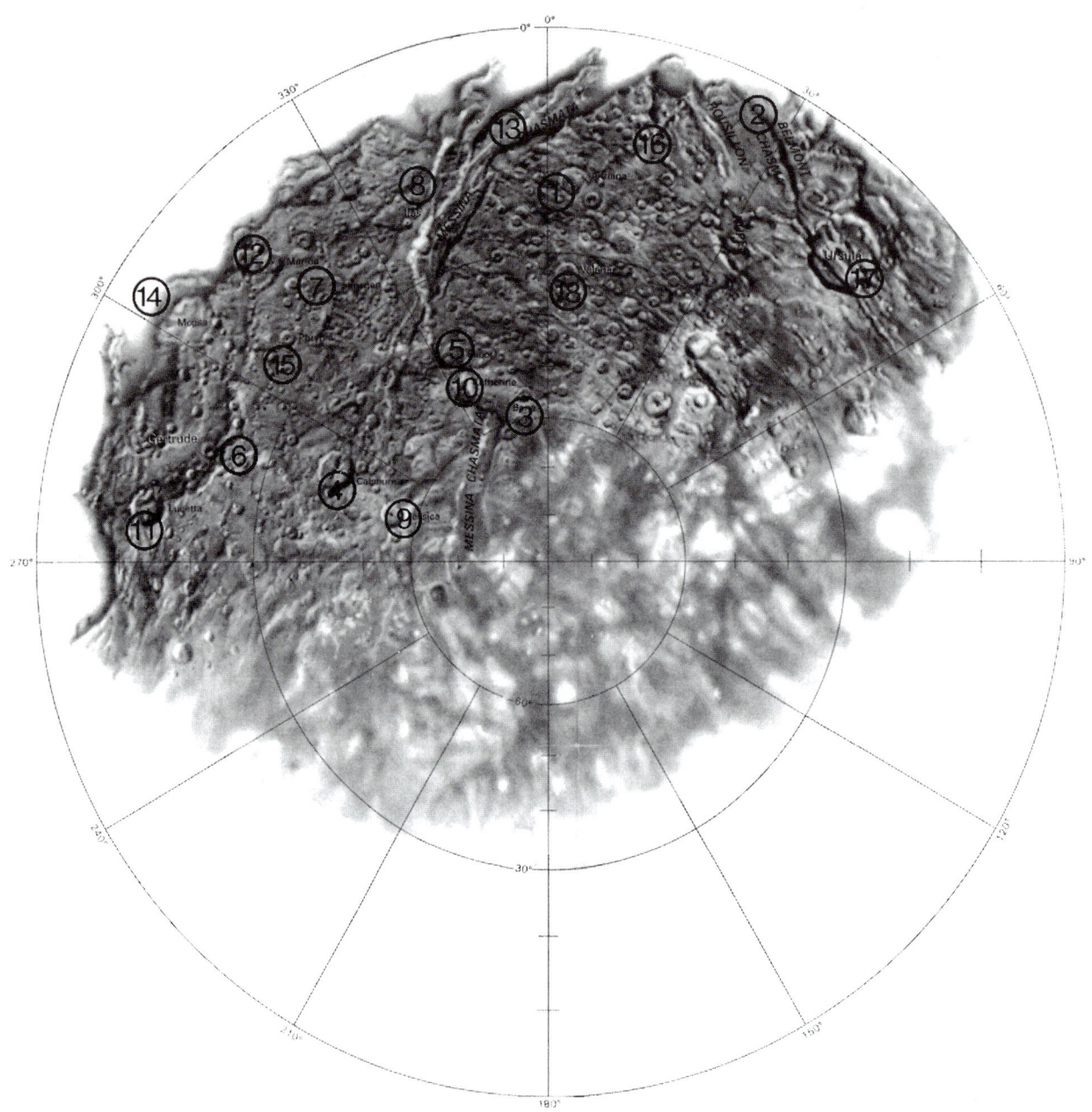

Figure 11.4. Titania.

Titania. The largest of Uranus' satellites; it is very slightly larger than Rhea and Iapetus in Saturn's system. It is named after the fairy queen in *A Midsummer Night's Dream*. Surface features are listed in Table 11.9. Voyager 2 imaged it from a range of 369 000 km.

Like Ariel, although to a lesser extent, Titania has clearly seen considerable tectonic activity in the past. There are many crates and ice cliffs, and trench-like features, notably Messina Chasmata, which extends for over 1490 km and crosses what was the boundary between the sunlit and dark sides of Titania at the time of the Voyager 2 pass. One crater, Gertrude, is over 320 km across. (Gertrude was King Claudius' wife in *Hamlet*; features on Titania are named after female Shakespearean characters.) The large crater Ursula is cut by a younger fault valley over 100 km wide. In size and mass Titania and Oberon are virtual twins, but Oberon does not show so much evidence of past activity.

Table 11.9. Features on Titania. (Bold numbers indicate map references.)

Name	Lat. S	Long. E	Diameter/length (km)
Craters			
Adriana **1**	20.1	3.9	50
Bona **3**	55.8	351.2	51
Calpurnia **4**	42.4	291.4	100
Elinor **5**	44.8	333.6	74
Gertrude **6**	15.8	287.1	326
Imogen **7**	23.8	321.2	28
Iras **8**	19.2	338.8	33
Jessica **9**	55.3	285.9	64
Katherine **10**	51.2	331.9	75
Lucetta **11**	14.7	277.1	58
Marina **12**	15.5	316.0	40
Mopsa **14**	11.9	302.2	101
Phrynia **15**	24.3	309.2	35
Ursula **17**	12.4	45.2	135
Valeria **18**	34.5	4.2	59
Chasmata			
Belmont Chasma **2**	4–25	25–35	258
Messina Chasmata **13**	8–28	325–5	1492
Rupes			
Rousillon Rupes **16**	7–25	17–38	402

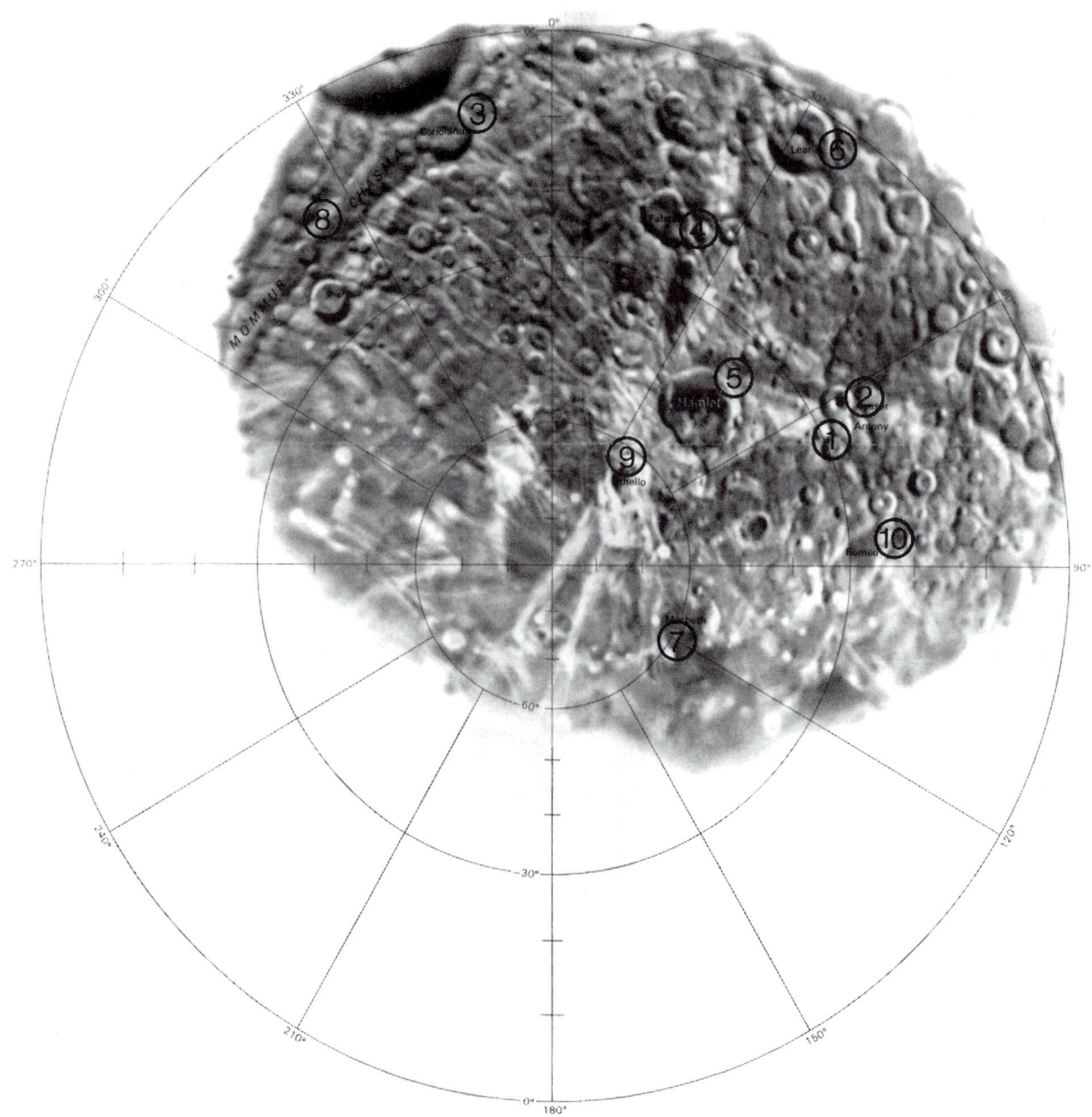

Figure 11.5. Oberon.

Oberon. Named for the fairy king in *A Midsummer Night's Dream*. It has a brownish surface, pitted with craters; although the average albedo is low, some of the larger craters, such as Othello, are the centres of bright ray-systems. Othello and other craters, such as Falstaff and Hamlet, have dark material inside them; this may be a mixture of ice and carbonaceous material erupted from the interior. Oberon was imaged by Voyager 2 from 660 000 km, to a resolution of 12 km. Selected features are listed in Table 11.10.

One interesting feature is what appears to be a lofty mountain, some 6 km high, shown on the best Voyager picture exactly at the edge of the disk, so that it protrudes

Table 11.10. Features on Oberon. (Bold numbers indicate map references.)

Name	Lat. S	Long. E	Diameter/ length (km)
Craters			
Antony **1**	27.5	65.4	47
Cæsar **2**	26.6	61.1	76
Coriolanus **3**	11.4	345.2	120
Falstaff **4**	22.1	19.0	124
Hamlet **5**	46.1	44.4	206
Lear **6**	5.4	31.5	126
Macbeth **7**	58.4	112.5	203
Othello **9**	66.0	42.9	113
Romeo **10**	28.7	89.4	159
Chasma			
Mommur Chasma **8**	16–20	240–343	537

from the limb (otherwise it might not be identifiable). Whether it is exceptional is something else about which we have as yet no definite information.

Caliban and Sycorax. They were discovered in October by Philip Nicholson and his colleagues from Palomar and named by them after characters in *The Tempest*. (It is reported that Nicholson wanted to name one of them Squeaker, after his cat, but Shakespeare prevailed!) Both are very faint and unusually red; they move far beyond the main satellite system and have non-circular orbits. Nothing definite is known about their nature; they may be captured Kuiper Belt objects. As yet nothing definite is known about the two outer satellites discovered by Kavelaars in 1999.

As the apparent diameter of the Sun as seen from Uranus is below 2 arcmin, all the satellites known before the Voyager 2 pass could produce total eclipses. From Uranus, Ariel would have much the largest diameter.

To an observer on Uranus – if he could get there! – sunlight would be relatively strong, ranging between 1068 and 1334 times that of full moonlight on Earth. Saturn would be fairly bright when well placed (every $45\frac{1}{2}$ years); Jupiter would have an apparent magnitude of 1.7, but would remain inconveniently close to the Sun in the Uranian sky. Neptune would be visible with the naked eye when near opposition.

Bland though it may appear, Uranus has proved to be of intense interest. It is very different from any other world known to us.

12 NEPTUNE

Neptune is the third most massive planet in the Solar System and, like Uranus, may be described as an 'ice giant'. It is too faint to be seen with the naked eye, but binoculars show it easily, and a small telescope will reveal its pale bluish disk. Data are given in Table 12.1.

Table 12.1. Data.

Distance from the Sun:
    max 4347 million km (30.316 a.u.)
    mean 4496.7 million km (30.058 a.u.)
    min 4456 million km (29.800 a.u.)
Sidereal period: 164.8 years (60 190.3 days)
Synodic period: 367.5 days
Rotation period: 16h 7m (16.1 h)
Mean orbital velocity: 5.43 km s^{-1}
Axial inclination: 28°48′
Orbital inclination: 1°45′19″.8
Orbital eccentricity: 0.009
Diameter (km): equatorial 50 538
                polar 49.600
Apparent diameter from Earth: max 2″.2
                                 min 2″.0
Reciprocal mass, Sun = 1: 19 300
Density, water = 1: 1.77
Mass, Earth = 1: 17.2
Volume, Earth = 1: 57
Escape velocity: 23.9 km s^{-1}
Surface gravity, Earth = 1: 1.2
Mean surface temperature: −220 °C
Oblateness: 0.02
Albedo: 0.35
Maximum magnitude: +7.7
Mean diameter of Sun, seen from Neptune: 1′04″

MOVEMENTS

Neptune is a slow mover; it takes almost 165 years to complete one journey round the Sun, so that it was discovered less than one 'Neptunian year' ago. Opposition dates are given in Table 12.2. Some close planetary conjunctions involving Neptune are listed in Table 12.3; occultations by the Moon can of course occur.

Table 12.2. Oppositions of Neptune, 1999–2005.

Neptune remains in Capricornus; opposition magnitude +7.7.

1999	July 26
2000	July 27
2001	July 30
2002	Aug 2
2003	Aug 4
2004	Aug 6
2005	Aug 8

Neptune was at perihelion on 28 August 1876, and will be again on 5 September 2042. Aphelion was reached on 13 July 1959.

EARLY OBSERVATIONS

Neptune was observed on several occasions before being identified as a planet. The first observation seems to have been made by Galileo on 27 December 1612. While drawing Jupiter and its four satellites, he recorded a 'star' which was certainly Neptune. He again saw it twice in January 1613, and noted its movement, but, not surprisingly, took it for a star. His telescope had a magnification of ×18 and a resolving power of 190 arcsec, with a field of view 17 arcmin in diameter, Neptune's magnitude was 7.7, and Galileo often plotted stars fainter than that. Jupiter actually occulted Neptune in 1613.

The next telescopic observation was made in May 1795 by J. J. de Lalande, but again Neptune was mistaken for a star. Further observations were made in 1845–6 by John Lamont (often referred to as Johann von Lamont; he was Director of the Munich Observatory).

DISCOVERY

Uranus was discovered in 1781 by William Herschel. Before long it became clear that it was not moving exactly as it had been expected to do; this was recognized by a mathematician, the Reverend Placidus Fixlmillner of Kremsmünster. It was reasonable to suppose that the old

Table 12.3. Close planetary conjunctions involving Neptune, 1900–2100.

	Date	UT	Separation (″)	Elongation (°)
Mercury–Neptune	1914 Aug 10	08.11	−30	18W
Venus–Neptune	2022 Apr 27	19.21	−26	43W
Venus–Neptune	2023 Feb 15	12.35	−42	28E
Mercury–Neptune	2039 May 5	09.42	−39	13W
Mercury–Neptune	2050 June 4	06.46	−46	17W
Mercury–Neptune	2067 July 15	12.04	+13	18W

There was a triple conjunction of Neptune and Uranus in September 1993, but the separation was never less than $1°08'$. The previous triple conjunction of Neptune and Uranus was in 1821; the next will be in 2164.

pre-discovery observations, used to calculate the orbit, were inaccurate, so Fixlmillner discarded them and re-worked the orbit. Within a few years new discrepancies became evident. In 1821 A. Bouvard produced new tables of Uranus' motion, but by 1832 G. Airy, then at Cambridge, found that these tables were wrong by half a minute of arc, which was unacceptable. It was around this time that there were the first definite suggestions than an unknown planet might be pulling Uranus out of position. The idea occurred to J. E. B. Valz, Director of the Marseilles Observatory, and to F. G. B. Nicolai, Director of the Mannheim Observatory. On 17 November 1834 an English amateur, the Reverend T. J. Hussey (Rector of Hayes in Kent) wrote a letter to Airy suggesting that it might be possible to work out a position for the perturbing planet. Airy's reply was not encouraging, and Hussey took the matter no further.

In 1840 F. W. Bessel, of Königsberg, returned to the possibility of a new planet and told Sir John Herschel – son of Sir William – that he intended to search for it, in collaboration with his pupil, F. W. Flemming. This never happened. Flemming died suddenly; Bessel became ill and he too died in 1846. Another interested astronomer was J. H. Mädler, co-author (with Beer) of the first really good map of the Moon; Mädler discussed the problem in 1841, but by then had left Germany to become Director of the new Dorpat Observatory in Estonia, and he never followed the Uranus problem through.

In 1841 John Couch Adams, then an undergraduate at Cambridge, 'formed a design of investigating as soon as possible, after taking my degree, the irregularities in the motion of Uranus', with the intention of tracking down the new planet. He began work in 1843, and by mid-1845 had calculated a position for the planet. He was in communication with James Challis, Professor of Astronomy at Cambridge, and also with Airy, now Astronomer Royal at Greenwich, but following a series of delays and misunderstandings no search was instigated. Meanwhile, U. J. J. Le Verrier, in France, had been working along similar lines, and by 1846 he had worked out a position very close to that given by Adams – about which, of course, Le Verrier was entirely ignorant.

When Airy saw Le Verrier's memoir, he realized that a search would have to be undertaken. There was no suitable telescope at Greenwich, but at Cambridge there was the 29.8 cm Northumberland refractor, and Airy instructed Challis to begin hunting. Unfortunately Challis had no up-to-date charts of the area of the sky concerned, and he was not enthusiastic; he adopted a cumbersome method of star-checking, and was in no hurry to compare his observations. Le Verrier had failed to persuade astronomers in Paris to begin a search, but the outcome was different. Patience was never Le Verrier's strong point, and instead of waiting he contacted Johann Galle, at the Berlin Observatory. Galle was interested, and asked permission from the Observatory director, J. F. Encke, to use the 23 cm Berlin refractor for the purpose. Encke agreed: 'Let us oblige the gentleman from Paris!' Galle, together with a young astronomer, H. D'Arrest, lost no time, and on 23 September 1846, the first night of their search, they identified the planet. Galle used the telescope, while D'Arrest checked the positions of the stars which came into view. Within minutes Galle described an eighth-magnitude star at RA 22h 53m 25s.84. D'Arrest called out. 'That star is not on the map!'.

Encke joined them in the dome, and they followed the object until it set. Next night they found that it had moved by the expected amount, and Encke wrote to Le Verrier; 'The planet whose position you have pointed out actually exists.' Le Verrier's predicted position was in error by only 55 arcmin.

It was also notable that the planet showed a definite disk; Le Verrier had predicted an apparent diameter of 3″.3, and Encke's first measures made it 3″.2, which is close to the correct value.

Subsequently Challis found that he had observed the planet twice soon after beginning his search, on 30 July and 4 August, and again on 12 August. Had Challis checked his records, he could not have failed to identify the planet, and on 12 August he even suspected a 'star' which showed a disk – yet he did not examine the suspect object with a higher magnification[1]. Following the announcement from Berlin the new planet was soon seen by J. R. Hind from London, using a 17.7 cm refractor, and by the well-known amateur William Lassell, who had set up a 61 cm reflector in Liverpool.

The first announcement of Adams' independent work (almost as accurate as Le Verrier's) was made on 3 October 1846 by Sir John Herschel, in the *Athenæum*. The announcement caused deep resentment in France, and led to acrimonious disputes, but in these Adams and Le Verrier took no part, and when they met they struck up an immediate and lasting friendship – even though Adams could not speak French and Le Verrier was equally unversed in English!

It is often said that Le Verrier and Adams should be recognized as co-discoverers of the planet, but this is not strictly correct; the true discoverers of Neptune were Johann Galle and Heinrich D'Arrest.

Adams made no personal search for the planet; he was not equipped to do so. The story that he wrote to

---

[1] In September 1988 I decided to check Challis' observations, so from the Royal Greenwich Observatory, than at Herstmonceux in Sussex, I looked at Neptune through the telescope used by Challis and with the same magnification (×117). There was no doubt that Neptune showed a disk, and with a power of ×250 the difference between the planet and a star was glaringly obvious.

Lassell, asking for assistance, and that Lassell failed to respond because he was hors de combât with a sprained ankle, is certainly untrue. Lassell did, however, discover Neptune's main satellite, Triton, on 10 October a few weeks after Neptune itself had been found. On 14 October 1846 Lassell also reported a ring, but later observations showed that this was an optical effect, as Lassell himself later realized. His spurious ring has no connection with the real ring system, which could not possibly have been detected with any telescope available at the time (in fact, even today it has never been seen with a ground-based instrument).

NAMING

A name had to be found, and the mythological tradition was followed. Challis suggested 'Oceanus'; Galle, in a letter to Le Verrier, preferred 'Janus', while Le Verrier's own choice was 'Neptune'. Le Verrier then changed his mind, and decided that the planet should be named after himself, but this was not well received, and before long 'Neptune' was universally accepted.

PHYSICAL CHARACTERISTICS

Neptune is a twin of Uranus – but it is a non-identical twin. It is very slightly smaller, but appreciably denser and more massive. Unlike Uranus, it has an internal heat source; it sends out 2.6 times more energy than it would do if it depended entirely upon what it receives from the Sun. It does not share in Uranus' unusual axial tilt; at the time of the Voyager 2 pass it was Neptune's south pole which was in sunlight. The rotation period of just over 16 h is slightly shorter than that of Uranus.

VOYAGER 2 AT NEPTUNE

Our first detailed information about Neptune was obtained from Voyager 2, in 1989; little can be seen on the planet's disk with Earth-based telescopes, and of course the Hubble Space Telescope was not launched until after the Voyager pass.

Voyager 2 had already encountered Jupiter, Saturn and Uranus. After the Saturn encounter, scientists at the Jet Propulsion Laboratory estimated that the chances of success at Uranus were about 60%, but no more than 40% at Neptune; in the event, both encounters were wellnigh faultless, and a tremendous amount of information was obtained.

At 06.50 GMT on 25 August 1989 Voyager 2 passed over the darkened north pole of Neptune, at a minimum relative velocity of 17.1 km s^{-1}. At that time the spacecraft was 29 240 km from the centre of Neptune, no more than 5000 km above the cloud tops, and 4 425 000 000 km from Earth, so that the 'light-time' was 4 h 6 min. The encounter was unlike anything previously experienced, because everything happened so quickly. The main picture-taking period was compressed into less than 12 h.

Before the closest approach over the north pole, many of the main discoveries had been made. The ring system was surveyed, and the magnetic field studied; lightning was detected from the charged particle (discharges recorded by the plasma wave equipment); auroræ were confirmed; surface features had been imaged, and six new satellites had been found. Neptune proved to be a dynamic world, very different from the bland Uranus.

After the polar pass, Voyager 2 passed through the plane of the rings, at a relative velocity of over 76 000 km h^{-1}. Impacts by ring particles were recorded 40 min before the crossing, and reached a peak of 300 per second for 10–15 min either side of the actual crossing; the Voyager team members were very relieved when the spacecraft emerged unscathed. Just over 5 h later, Voyager 2 passed Triton at a range of 38 000 km, and sent back images which were as fascinating as they were unexpected. One last picture was taken, showing Neptune and Triton together as crescents, and then Voyager began a never-ending journey out of the Solar System. Contact was still maintained in 2000, and probably signals will be received for several years before Voyager finally passes out of range. Its fate will never be known; it carries a plaque and recordings in case any alien civilization finds it, but the chances do not seem to be very high. Calculations indicate that it will by-pass several stars during the next 300 000 years (Table 12.4), but obviously all this is uncertain, and Voyager 2 may eventually be destroyed by a collision with some wandering body.

Since 1994, regular observations have been made with the Hubble Space Telescope, and these have been of great value, but it is true to say that the bulk of our present knowledge of Neptune has come from that single encounter with Voyager 2.

Table 12.4. Stars to be by-passed by Voyager 2 during the next 300 000 years.

Year (AD)	Star	Distance from Voyager to star (light-years)
8 600	Barnard's Star	4.0
20 300	Proxima Centauri	3.2
20 600	Alpha Centauri	3.5
23 200	Lalande 21185	4.7
40 100	Ross 248	1.7
44 500	DM-36°13940	5.6
46 300	AC +79°3888	2.8
129 000	Ross 154	5.8
129 700	DM +15°3364	3.5
296 000	Sirius	4.3

INTERIOR OF NEPTUNE

The four major planets are often called 'gas giants', but this is strictly true only for Jupiter and Saturn; Uranus and Neptune are better called 'ice giants', since their total mass of hydrogen and helium is no more than about two Earth masses, as against 300 Earth masses for Jupiter. The interiors of Uranus and Neptune are dominated by 'ices', primarily water; the main difference between the two is that Neptune has a marked internal heat source, which Uranus apparently lacks. Neptune may have a silicate core extending out to one-fifth of the planet's radius, but this is by no means certain, and in any case the core does not seem to be strongly differentiated from the ice components. The oblateness of the globe is 0.017, as against 0.029 for Uranus.

ATMOSPHERE

Predictably, Neptune's atmosphere consists mainly of hydrogen; it seems that the upper atmosphere is made up of 84% molecular hydrogen (H_2), 14% helium (He) and 2% methane (CH_4). Methane absorbs light of relatively long wavelength (orange and red), which is why Neptune looks blue. There are trace amounts of carbon monoxide (CO) and hydrogen cyanide (HCN) with even smaller amounts of acetylene (C_2H_2) and ethane (C_2H_6).

There are various cloud layers. At a level where the pressure is 3.3 bars there is a layer which seems to be made up of hydrogen sulphide, above which come layers of hydrocarbons, with a methane layer and upper methane haze. Above the hydrogen sulphide layer there are discrete clouds with diameters of the order of 100 km; these cast shadows on the cloud deck 50–75 km below. We are, in fact, dealing with clouds which may be described as methane cirrus.

Apparently there is a definite cycle of events. First, solar ultra-violet destroys methane high in Neptune's atmosphere by converting it to other hydrocarbons such as acetylene and ethane. These hydrocarbons sink to the lower stratosphere, where they evaporate and then condense. The hydrocarbon ice particles fall into the warmer troposphere, where they evaporate and are converted back to methane. Buoyant, convective methane clouds then rise up to the base of the stratosphere or higher, thereby returning methane vapour to the stratosphere and preventing any net methane loss. In the troposphere there are variable amounts of hydrogen sulphide, methane and ammonia, all of which are involved in the creation of the cloud layers and associated photochemical processes.

Temperature measurements show that there is a cold mid-latitude region, with a warmer equator and pole (we know very little about the north pole, which was in darkness during the Voyager 2 pass). Note, *en passant*, that the general temperature is much the same as that of Uranus; at the equator it is −226 °C, as against −214 °C, even though Neptune is so much further from the Sun. The planet's internal heat source more or less compensates for the increased distance.

There are violent winds on Neptune; most of them blow in a westerly direction, or retrograde (that is to say, opposite to the planet's rotation). The winds differ from those of Jupiter and Saturn, and are distinctively zonal. At the equator they blow westward at up to 450 m s^{-1}. Further south they slacken, and beyond latitude −50° they become eastward (prograde) at up to 300 m s^{-1}, decreasing once more near the south pole. In fact, a broad equatorial retrograde jet extends from approximately latitude +45° to −50°, with a relatively narrow prograde jet at around latitude −70°. Neptune is the 'windiest' planet in the Solar System; as the heat budget is only 1/20 of that at Jupiter, it seems that the winds are so strong because of the relative lack of turbulence.

SPOTS AND CLOUDS

At the time of the Voyager 2 encounter the most conspicuous feature on Neptune was a huge oval, the Great Dark Spot, with a longer axis of about 10 000 km; it lay at latitude −22° and its size, relative to Neptune, was about the same as that of the Great Red Spot relative to Jupiter. It had a rotation period of 18.296 h, and drifted westward at about 325 m s^{-1} relative to the adjacent clouds; it was about 10% darker than its surroundings, while the nearby material was about 10% brighter – indicative of the altitude difference between the two regions; it was a high-pressure area, rotating counter-clockwise and showing all the characteristics of an atmospheric vortex. Hanging above it were bright cirrus-type clouds, made up of methane ice; between the cirrus and the main cloud deck there was a clear region about 50 km deep. The cirrus changed shape quite quickly, and in some cases there were shadows cast on the cloud deck beneath – a phenomenon not observed on Jupiter or Saturn, and certainly not on Uranus.

The Great Dark Spot seemed to be migrating poleward at a rate of 15° per year, and was tacitly assumed to be a feature at least as semi-permanent as Jupiter's Red Spot. However, this proved not to be the case. By 1994, observations with the Hubble Space Telescope showed that it had disappeared, and there is no reason to expect that it will return, although we cannot be sure (after all, Jupiter's Red Spot vanishes sometimes, but only temporarily).

Voyager 2 also showed a smaller, very variable feature at latitude −42°, with a rotation period of 15.97 h; it was nicknamed the 'Scooter', and had a bright centre. Every few revolutions it caught up with the Great Dark Spot and passed it. Still further south, at latitude −54°, there was another dark spot, D2, which also had a bright core and showed small-scale internal details which changed markedly over periods of a few hours. These too have vanished, but in June 1994 Hubble images showed a bright cloud band near latitude +30° which apparently did not exist in 1989; subsequently it faded, while two bright irregular patches appeared in the southern hemisphere. Evidently Neptune's surface features are much more variable than had been expected.

MAGNETIC FIELD

It had been assumed that Neptune must have a magnetic field. As Voyager 2 drew inward, radio emissions were detected, but there was a delay before Voyager passed through the bow shock (that is to say, the region where the solar wind is heated and deflected by interaction with Neptune's magnetosphere). The bow shock was finally recorded at 879 000 km from the planet. The magnetic field itself was weaker than those of the other giants; the field strength at the surface is 1.2 G in the southern hemisphere but only 0.06 G in the northern. The real surprise was that the inclination of the magnetic axis relative to the axis of rotation is 47°, so that in this respect Neptune is not unlike Uranus; moreover the magnetic axis does not pass through the centre of the globe, but is displaced by 10 000 km or 0.4 Neptune radii. This indicates that the dynamo electric currents must be closer to the surface than to the centre of the globe. The magnetosphere has been described as comparatively 'empty'; it gyrates dramatically as the planet rotates, and the satellites are involved. Auroræ were found, but instead of being near the rotational poles they were closer to the equator – because the rotational and the magnetic poles are so far apart. At this time, of course, the northern hemisphere was experiencing its long winter. Neptunian auroræ are considerably weaker than those of the other giants.

NEPTUNE'S RINGS

Neptune has an obscure ring system. Indications of ring material were obtained in pre-Voyager times by the occultation method, which had been used to detect the rings of Uranus; for example, on 24 May 1981 observers from Arizona observed the close approach of Neptune to a star, and found that there was a very brief drop in the star's brightness. Other similar observations led to the suggestion that there might be incomplete rings – that is to say 'ring arcs', although nobody was quite sure how or why such arcs should develop. The mystery was solved by observations from Voyager 2. There are complete rings, but the main ring is 'clumpy'. Data are given in Table 12.5; the rings have been named after astronomers involved in the discovery of Neptune (although up to now, D'Arrest has been quite unfairly ignored). François Arago is included, as it was he

Table 12.5. The rings of Neptune.

Name	Distance from centre of Neptune (km)	Width (km)
Galle	41 900–49 000	1700
Le Verrier	53 200	50
'Plateau' (Lassell)	53 200–59 100	4000
Arago	57 600	30
Adams	62 900	50

who first drew Le Verrier's attention to the problem of the movements of Uranus.

The outer Adams ring is the most pronounced, and contains the three 'clumps' which led to the idea of ring arcs; these 'clumps' have been named Liberté, Egalité and Fraternité, for reasons which are, at best, obscure. The 'clumps' are not evenly distributed, but are grouped together over about 1/10 of the ring circumference; their lengths range from 5000 to 10 000 km. They may be due to the effects of the small satellite Galatea, which orbits very close to the rings. It has also been suggested that the cause may be small satellites embedded in the Adams ring, but so far these have not been seen; they may or may not exist. During the Voyager 2 pass, the Adams ring occulted the second-magnitude star Sigma Sagittarii, and proved to have a core only 17 km wide; the ring material is reddish, with a very low albedo. The Arago ring is dim and narrow. The Le Verrier ring is narrow and tenuous; it is not far from the orbit of the small satellite Despina, but shows no arcs. The Plateau is a diffuse band of material, containing a high percentage of very small particles. The inner Galle ring is much broader than the Adams or Le Verrier rings; there may be 'dust' extending down almost to the cloud tops. Unlike the Uranian rings, those of Neptune are brighter in forward-scattered light than in back-scattered light, so that they contain a larger fraction of very small particles than do the much narrower rings of Uranus. Even from Voyager, Neptune's rings were not easy to identify; they are blacker than soot! Possibly they are young, formed from the débris of small satellites.

Table 12.6. Satellites of Neptune.

Name	Discoverer	Mean distance from Neptune (km)	Sidereal period (days)
Naiad	R Terrile, 1989[a]	48 230	0.295 496
Thalassa	R Terrile, 1989[a]	50 074	0.311 485
Despina	S Synnott, 1989[a]	52 526	0.334 655
Galatea	S Synnott, 1989[a]	61 953	0.428 745
Larissa	H Reitsema, W Hubbard, L Lebo, 1981	73 548	0.554 654
Proteus	S Synnott, 1989[a]	117 647	1.122 315
Triton	W Lassell, 1846	354 760	5.876 854
Nereid	G Kuiper, 1949	5 513 400	360.136 19

Name	Orbital inclination (°)	Orbital eccentricity	Diameter (km)	Albedo	Magnitude
Naiad	4.74	0.000	58	0.06	25
Thalassa	0.21	0.000	80	0.06	24
Despina	0.07	0.000	148	0.06	23
Galatea	0.05	0.05	158	0.05	23
Larissa	0.20	0.001	208 × 178	0.06	21
Proteus	0.55	0.0004	436 × 416 × 402	0.05	20
Triton	157.345	0.000 02	2705	0.6–0.8	13.6
Nereid	7.23	0.7512	340	0.16	18.7

[a] From Voyager 2 images.

SATELLITES

Neptune has eight known satellites. The largest, Triton, is an easy telescopic object – easier than any of the satellites of Uranus – and was discovered soon after the identification of Neptune itself. The second satellite known before the Space Age, Nereid, was found by G. Kuiper in 1949. The remaining six were all discovered on images taken by Voyager 2, although one of them, Larissa, had been fortuitously noted on 24 May 1981, when it occulted a star; the observation was made by a team led by J. Reitsema, and they are surely entitled to be regarded as discoverers even though they could not, at the time, prove that a satellite was responsible. None of these small inner satellites can be seen with ground-based telescopes; all move more or less in the plane of the rings, with very low orbital eccentricity. Triton has retrograde motion and is almost certainly a captured body rather than a bona-fide satellite, while Nereid has a highly eccentric path round Neptune and could possibly be an escapee from the Kuiper Belt. All the

satellites are named after marine deities. Data are given in Table 12.6.

Naiad. The last satellite to be discovered, and the closest to the planet; it moves at only about 23 200 km above the Neptunian cloud tops. The name comes from a group of Greek water nymphs who were guardians of lakes, fountains, springs and rivers. Naiad seems to be rather irregular in shape, but no surface details were seen from Voyager.

Thalassa. Named for a Greek sea goddess, sometimes said to be the mother of Aphrodite. Like Naiad, Thalassa seems to be irregular in shape, but little else is known about it. It moves about 25 200 km above Neptune's cloud tops.

Despina. Named for the daughter of Neptune and Demeter. This is yet another irregularly-shaped satellite, moving in an orbit just closer to Neptune than the Le Verrier ring; it may have some effect upon this ring. In itself it seems to be similar to Naiad and Thalassa, but is rather larger.

Galatea. Named for one of the Nereids. Its orbit is very close to the obscure Arago ring. No close-range views were obtained from Voyager, and so no surface details have been recorded.

Larissa. Named after one of Neptune's numerous lovers. The satellite was almost certainly responsible for the occultation event of 1981, although its existence was not established until the Voyager 2 mission. Voyager 2 obtained an image of it on 24 August 1989 and showed a number of craters on its surface. With a longest diameter of 208 km, Larissa is decidedly larger than the four innermost satellites.

Proteus. In mythology Proteus was a marine deity, son of Oceanus and Tethys. This was the first satellite to be discovered from Voyager 2, and is larger than Nereid, but it is so close to Neptune – less than 93 000 km above the cloud tops – that it is unobservable with ground-based telescopes. Proteus has a very low albedo, and has been said to be 'as dark as soot'. It is rather irregular in form, and is triaxial. On 25 August 1989 Voyager 2 obtained an image of it. The main surface feature is a depression, Pharos (originally known as the Southern Hemisphere Depression), which dominates the southern part of the Neptune-facing hemisphere; like all the Neptunian satellites apart from Nereid, Proteus has synchronous rotation, so that its rotation period is the same as its orbital period (1d 2.9h). Pharos is basically circular, with a raised rim and a flat, rugged floor; it is 225 km across and 10–15 km deep. Proteus also shows linear streaks, which seem to be troughs 25–35 km wide and several kilometres deep; they form a global network,

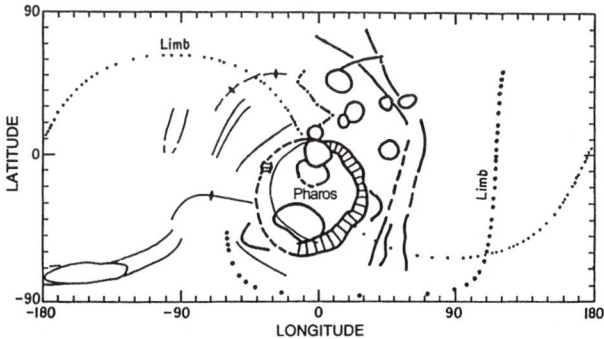

Figure 12.1. Proteus.

and may be classed as graben of extensional origin. There are no cryovolcanic structures or coronæ.

Triton. Here we have one of the most remarkable bodies in the entire Solar System. Before the Voyager pass it was believed to be large – possibly larger than Mercury – and to have chemical oceans on its surface, but this was found to be completely wrong. Triton is much smaller than had been expected, so that in size it ranks below all four of Jupiter's Galilean satellites, Titan in Saturn's system and our own Moon; it is also more reflective, with an albedo which in places rises to 0.8. It is very cold, with a temperature of $-235\,°C$ (a mere 38 K above absolute zero). The mass is 2.4×10^{22} kg, and the density fairly high (2.07 times that of water), so that the globe may be made up of a mixture of rock (2/3) and ice (1/3). The escape velocity is 1.44 km s^{-1}, lower than that of our Moon, but the intense cold means that Triton can retain an extensive though tenuous atmosphere; the ground atmospheric pressure is of the order of 14 microbars, about 1/70 000 the pressure of the Earth's air at sea-level. The main constituent is nitrogen, in the form of N_2, which accounts for 99%; most of the rest is methane, with a trace of carbon monoxide. There is considerable haze, seen by Voyager above the surface, and this is probably composed of microscopic ice crystals of methane or nitrogen. Winds in the atmosphere average around 5 m s^{-1} westward, although naturally they have very little force. There is a pronounced temperature inversion, since the temperature in the atmosphere rises to about $-173\,°C$ at a height of 600 km. This inversion occurs at a surprisingly high altitude, for reasons which are unclear.

In 1998 measurements with the Hubble Space Telescope found that the surface temperature had increased by about 2° since the Voyager pass nine years earlier, and that the atmospheric density had shown a slight but definite increase. This is certainly a seasonal effect, caused by changes in the polar cap.

The surface of Triton is very varied, and there is very marked evidence of past cryovulcanism (that is to say, icy vulcanism). There is a general coating of ice, presumably water ice overlaid by nitrogen and methane ices; water ice has not been detected spectroscopically, but it must exist, because nitrogen and methane ices are not hard enough to maintain surface relief over long periods. Not that there

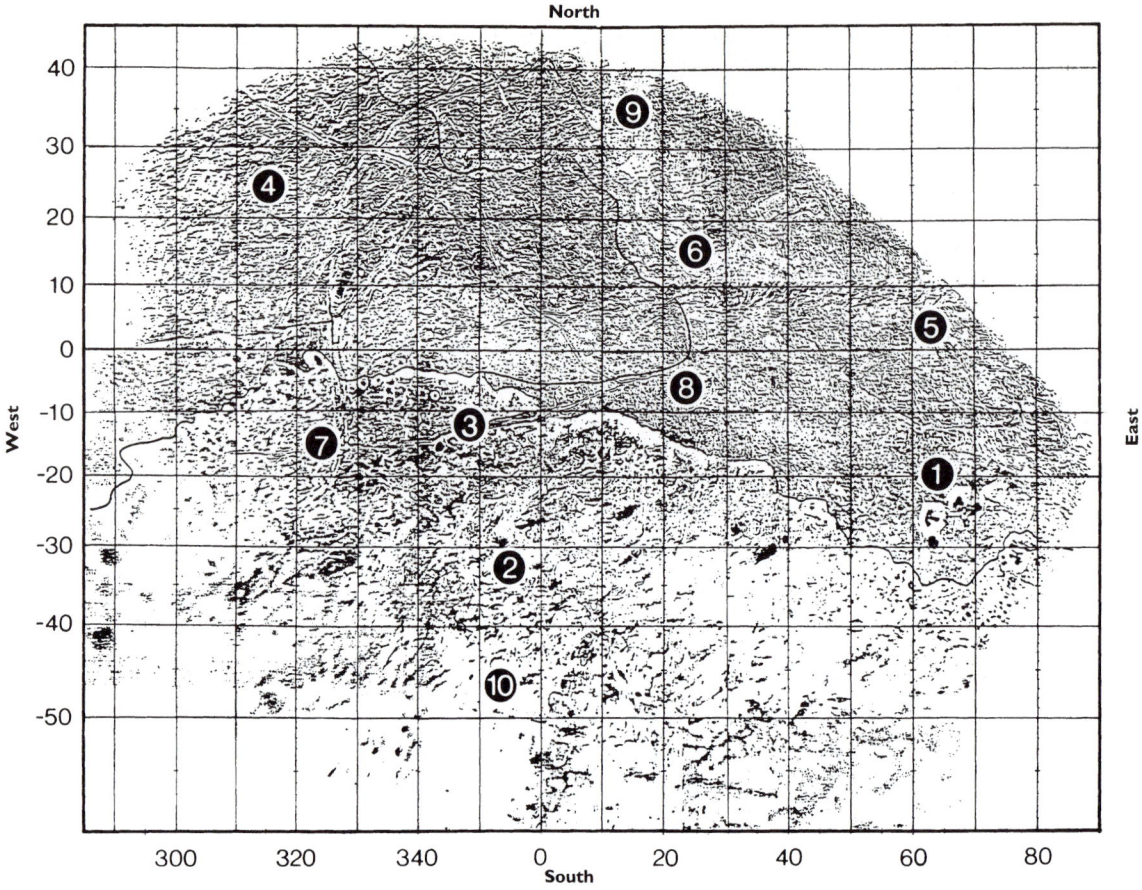

Figure 12.2. Triton.

is much surface relief on Triton; there are no mountains or deep valleys, and the total surface relief cannot amount to more than a few hundred metres. Normal craters are scarce, and even the largest of them is no more than 27 km in diameter. Extensive flows are seen, some of them up to 80 km wide, due probably to ammonia–water fluids.

The most striking feature is the southern polar cap, which is covered with pink nitrogen snow and ice. The areas surveyed by Voyager 2 have been divided into three main regions: Bubembe Region (western equatorial), Monad Regio (eastern equatorial) and Uhlanga Regio (polar). These names, and those of other features, have been allotted by the Nomenclature Committee of the International Astronomical Union. All the features have aquatic names, excluding Greek and Roman deities. Thus

Bubembe is the island location of the temple of Mukasa (Uganda), Monad is a Chinese symbol of duality in nature and Uhlanga is a Zulu reed from which mankind sprang!

Uhlanga is covered with the pink cap, with some of the underlying geological units showing through in places. It is here that we find the nitrogen ice geysers which were so totally unexpected. Apparently there is a layer of liquid nitrogen 20–30 m below the surface; here the pressure is high enough for nitrogen to remain liquid, but if for any reason it migrates upward it will come to a region where the pressure is about 1/10 that of the Earth's air at sea level. The nitrogen will then explode in a shower of ice and vapour (about 80% ice, 20% vapour) and will travel up the nozzle of the geyser-like vent at a rate of up to $150 \, \text{m s}^{-1}$, which is fast enough to make it rise to several kilometres before falling

Table 12.7. Surface features on Triton. Latitude and longitude are central values. (Bold numbers indicate map references.)

Name	Lat.	Long.	Name	Lat.	Long.
Crater			*Patera*		
Amarum	26.0N	24.5E	Dilolo Patera	26.0N	24.5E
Andvari	20.5N	34.0E	Gandvik Patera	28.0N	5.5E
Cay	12.0S	44.0E	Kasu Patera	39.0N	14.0E
Ilomba	14.5S	57.0E	Kibu Patera	10.5N	43.0E
Kurma	16.5S	61.0E	Leviathan Patera	17.0N	28.5E
Mazomba	18.5S	63.5E	*Planitia*		
Ravgga	3.0S	71.5E	Ruach Planitia	28.0N	24.0E
Tangaroa	25.0S	65.5E	Ryugu Planitia **8**	5.0S	27.0E
Vodyanoy	17.0S	28.5E	Sipapu Planitia	4.0S	36.0E
Catena			Tuonela Planitia **9**	34.0N	14.5E
Kraken Catena	14.0N	35.5E	*Planum*		
Set Catena	22.0N	33.5E	Abatos Planum **1**	21.5S	58.0E
Cavus			Cipango Planum	11.5N	34.0E
Apep Cavus	20.0N	301.5E	Medamothi Planum **5**	3.5N	69.0E
Bheki Cavus	16.0N	308.0E	*Plume*		
Dagon Cavus	29.0N	345.0E	Hili	57.0S	35.0E
Hekt Cavus	26.0N	342.0E	Mahilani	50.5S	359.5E
Hirugo Cavus	14.5N	345.0E	*Regio*		
Kasyapa Cavus	7.5N	358.0E	Bubembe Regio **4**	18.0N	335.0E
Kulilu Cavus	41.0N	4.0E	Monad Regio **6**	20.0N	37.0E
Mah Cavus	38.0N	6.0E	Uhlanga Regio **10**	37.0S	357.0E
Mangwe Cavus	7.0S	343.0E	*Sulcus*		
Ukupanio Cavus	35.0N	23.0E	Bia Sulci **2**	38.0S	3.0E
Dorsum			Boynne Sulci **3**	13.0S	350.0E
Awib Dorsa	7.0S	80.0E	Ho Sulci	2.0N	305.0E
Fossa			Kormet Sulci	23.0N	335.5E
Jumna Fossae	13.5S	44.0E	Leipter Sulci	7.0N	9.0E
Raz Fossae	8.0N	21.5E	Lo Sulci	3.8N	321.0E
Yenisey Fossa	3.0N	56.2E	Ob Sulci **7**	6.0S	328.0E
Macula			Ormet Sulci	17.0N	337.0E
Akupara Maculae	27.5S	63.0E	Slidr Sulci	23.5N	350.0E
Doro Macula	27.5S	31.7E	Tano Sulci	33.5N	337.0E
Kikimora Maculae	31.0S	78.0E	Vimur Sulci	11.0S	59.0E
Namazu Macula	25.5S	14.0E	Yaso Sulci	2.0N	347.0E
Rem Maculae	13.0N	349.5E			
Viviane Macula	31.0S	36.5E			
Zin Maculae	24.5S	68.0E			

back. The outrush sweeps dark débris along it, producing plumes such as Mahilani and Hili. The Mahilani plume is a narrow, straight cloud 90–150 km long, while Hili is a cluster of several plumes with a length of about 100 km. The ways in which the plume clouds move indicates the force and direction of the winds in the thin Tritonian atmosphere. The wind speeds here can reach 20 m s^{-1}; the direction is north-east near the surface, east at intermediate altitude and westward at the top of the troposphere, 8 km in altitude.

The edge of the cap, separating Uhlanga from Monad and Bubembe, is sharp and convoluted. The long seasons mean that the southern pole has been in constant sunlight for over a century now, and along the cap borders there are signs of evaporation; eventually the cap may appear to shift northward, so that the entire aspect of Triton may change according to the seasons – and as we have noted, this also affects the surface temperature and the atmospheric density. In reality it is not the cap shifting northward but evaporation/sublimation products moving northward in the atmosphere, where at the colder northern regions they are redeposited to form the raw materials for the process here to be repeated during the northern summer. North of the cap there is a darker, redder region; the colour may be due to the action of solar ultra-violet upon the methane. Running across this region, more or less parallel to the edge of the cap, is a slightly bluish layer, due possibly to tiny crystals of methane ice scattering the incoming sunlight.

Monad Regio is part smooth, with knobbly or hummocky terrain; there are rimless pits (paterae), with graben-like troughs (fossae) and strange, mushroom-like features (guttae) such as Zin and Akupara, whose origin is unclear. There are four walled plains or 'lakes' (Ruach, Ryugu, Sipapu, Tuonela) which have flat floors and are bounded by rougher plains units; they have been likened to calderae. They are edged with terraces, as though the original level has been changed several times by repeated melting and re-freezing. Undulating smooth plains cover most of the higher part of Monad.

Bubembe Regio is characterized by the so-called cantaloupe terrain – a nickname given to it because of its superficial resemblance to a melon-skin! Fissures cross it, meeting in elevated X or Y junctions. Liquid material, presumably a mixture of water and ammonia, seems to have forced its way up some of these fissures, so that there are central ridges; material has even flowed out on to the plain before freezing there. The overall pattern is a network of closely-spaced depressions (cavi), 25–35 km across; they do not overlap. The cantaloupe areas are probably the oldest parts of Triton's surface.

Quite apart from the nitrogen geysers, Triton's surface must be variable on a much larger scale. Southern midsummer will not fall until the year 2006, and there will no doubt be marked changes in the cap – and also in the north polar region, which is at the moment (2000) plunged into its long winter night.

Everything indicates that Triton was not originally a satellite of Neptune, but was captured well after its formation. Initially its orbit round Neptune would have been eccentric; over a period of perhaps 1000 million years it was forced into the present circular form, and during this time there was great internal heating, coupled with surface activity. There may also have been a dense atmosphere. A selected list of surface features is given in Table 12.7.

Mythologically, Triton was the sea-god son of Neptune and Amphitrite. Apparently the name was first suggested by the French astronomer Camille Flammarion. We have to agree that Neptune has certainly produced a remarkable child.

Nereid. Following the discovery of Triton, searches were made for the other satellites of Neptune. One was suspected by Lassell in 1852 and another by J. Schaeberle in 1892, using the 91 cm Lick refractor, but neither was confirmed, and a search by W. H. M. Christie, using the 152 cm reflector at Mount Wilson, was negative. Success then finally came in 1949, when G. P. Kuiper, using the 208 cm reflector at the McDonald Observatory in Texas, found a new faint satellite. It was named Nereid – and this was appropriate; the Nereids were the fifty daughters of Nereus, the sea-god, and Doris.

Nereid is exceptional inasmuch as its orbit is highly eccentric, and more like that of a comet than an asteroid. Its distance from Neptune ranges between 1 353 60 km out to 9 623 700 km, and it takes nearly a year to complete one orbit. Unfortunately it was not well imaged by Voyager 2, and the only reasonable image was obtained from a range of 4 500 000 km, so that little surface detail could be made out. It seems to be fairly regular in shape; it is more reflective than the small inner satellites, and is grey in colour, so that the surface may well be covered with rock or 'dirty ice'. No doubt there are craters, but for further information we must await the launch of a new space-craft to the Neptunian system – and this may well be delayed for a considerable time. Whether Nereid is a true satellite or whether it was captured is not known. It is unlikely that the rotation period is synchronous – one estimate, derived from variations in magnitude, gives 13.6 h, but it is quite probably that the period, like that of Hyperion in Saturn's system, is chaotic. To an observer on Neptune, the average apparent diameter of Nereid would be only about 19 arcsec.

There may well be other satellites of Neptune. If so, they will no doubt be discovered at some time in the future.

13 PLUTO

Pluto, the outermost known planet of the Solar System, was discovered in 1930 by Clyde Tombaugh, from the Lowell Observatory at Flagstaff in Arizona. It is the only planet not yet encountered by a space probe, and as a result our knowledge of it is very far from complete. It is accompanied by a smaller body, Charon, whose diameter is more than half that of Pluto itself.

MOVEMENTS

Pluto has a curiously eccentric orbit, which can bring it closer to the Sun than Neptune can ever be; the aphelion distance of Neptune is 4537 million km, the perihelion distance of Pluto 4446 million km – a difference of over 90 million km. Pluto was last at perihelion in 1989; it came within Neptune's orbit on 21 January 1979, and regained its status as 'the outermost planet' at 11.22 h GMT on 11 February 1999. It will retain this status until 5 April 2231. However, there is no fear of collision with Neptune, as Pluto's orbit is inclined at an angle of 17° and at the present epoch the distance between the two planets cannot be less than 348 million km. Pluto can depart from the official Zodiac, and for the 1999–2005 period it lies in Ophiuchus. Data for Pluto are given in Table 13.1, and opposition dates in Table 13.2.

STATUS

Pluto is an enigma, inasmuch as it does not fit easily into any category. It is smaller than the Moon, and also smaller than several planetary satellites, including Triton in Neptune's system. On the other hand it is too large to be classed as an asteroid or as a normal Kuiper Belt object. Suggestions to re-classify it as asteroidal and give it an asteroidal number (1000) were rejected in 1999 by the Nomenclature Committee of the International Astronomical Union, and Pluto is still officially ranked as a planet, although certainly its status must be regarded as dubious.

Table 13.1. Data.

Distance from the Sun:
   max 7381 200 000 km (49.28 a.u.)
   mean 5906 400 000 km (39.5 a.u.)
   min 4445 800 000 km (29.65 a.u.)
Sidereal period: 90 465 days (247.7 years)
Synodic period: 366.7 days
Rotation period: 6d 9h 17m (6.387 25 days)
Mean orbital velocity: 4.75 km s^{-1}
Axial inclination: 122.46°
Orbital inclination: 17.14°
Orbital eccentricity: 0.2488
Diameter: 2324 km
Density, water = 1: 2.05
Volume, Earth = 1: 0.006
Mass, Earth = 1: 0.0022
Reciprocal mass, Sun = 1: 135 500 000
Maximum surface temperature: about −233 °C
Escape velocity: 1.18 km s^{-1}
Surface gravity, Earth = 1: 0.06
Albedo: 0.55
Opposition magnitude at perihelion: 13.9
Distance from Earth: max 7533 300 000 km
               min 4293 700 000 km
Apparent diameter from Earth: max 0″.11
                      min 0″.06

DISCOVERY

As early as 1846 Le Verrier suggested that there might well be a planet moving beyond the orbit of the recently discovered Neptune. The first systematic search was made in 1877 by David Peck Todd, from the US Naval Observatory. From perturbations of Uranus, he predicted a planet at a distance of 52 a.u. from the Sun, with a diameter of 80 000 km. He conducted a visual search, using the 66 cm USNO reflector with powers of ×400 and ×600, hoping to detect an object showing a definite disk. He continued the hunt for 30 clear, moonless nights between

Table 13.2. Oppositions of Pluto.

	Declination
2000 June 1	−10°57′
2001 June 4	−11 49
2002 June 7	−12 39
2003 June 9	−13 27
2004 June 11	−14 14
2005 June 14	−14 59

Throughout this period Pluto remains in Ophiuchus. On 1 January 1999 Pluto passed 12″ south of the star ζ Ophiuchi

3 November 1877 and 5 March 1878, but with negative results. A second investigation was conducted in 1879 by the French astronomer Camille Flammarion, who based his suggestion upon the fact that several comets appeared to have their aphelia at approximately the same distance, well beyond the orbit of Neptune. In 1900 G. Forbes predicted two planets, at distances of 100 and 500 a.u. respectively, with periods of 1000 and 5000 years; both were assumed to be larger than Jupiter. In 1902 T. Grigull of Münster proposed a planet the size of Uranus, moving at 50 a.u. in a period of 360 years; Grigull even gave it a name – Hades. H. E. Lau, from Denmark (1900), suggested two planets, at 46.6 and 70.6 a.u., with masses 9 and 47 times that of the Earth; G. Gaillot (1901) also believed in two planets, and T. J. J. See in America increased the number to three, naming the innermost planet 'Oceanus'. A. Garnowsky, in Russia, went further and postulated the existence of four planets.

The first systematic photographic and visual search was made by Percival Lowell, from Flagstaff, from 1905 to 1907. Lowell had worked out a position for his 'Planet X', from perturbations of Neptune and (particularly) Uranus; his planet was believed to have a mass almost seven times that of the Earth, with a period of 282 years and a rather eccentric orbit (0.202). The date of perihelion was given as 1991. A second search at Flagstaff, carried out by C. O. Lampland in 1914, was equally fruitless, and was given up on Lowell's death in 1916, a year after Lowell had published his final orbit for the planet.

Also in the hunt was W. H. Pickering, who in 1898 had discovered Saturn's eighth satellite, Phœbe. His Planet O had a mass twice that of the Earth and a period of 248 years, again with an eccentric orbit. His method – unlike Lowell's – was essentially graphical, but his conclusions were much the same. On the basis of these results, Milton Humason at Mount Wilson Observatory undertook a photographic search in 1919, but again the results were negative.

In 1929 V. M. Slipher, by then Director of the Lowell Observatory, decided to make a fresh attempt. The search was entrusted to Clyde Tombaugh, using a 13 inch refractor specially acquired for the purpose. (Tombaugh was then a young, unqualified amateur; later he became one of America's most senior and respected astronomers.) Tombaugh used photographic methods, and before long success came; Pluto was detected upon plates taken on 23 and 29 January 1930, although the announcement was delayed until 13 March – 149 years after the discovery of Uranus and 78 years after Lowell's birth.

Examination of the earlier plates showed that Pluto had been recorded twice in 1915, but had been missed because it was unexpectedly faint. However, Humason's failure in 1919 was due to sheer bad luck. When his plates were re-examined, it was found that Pluto had been recorded twice – but once the image fell upon a flaw in the plate, and on the second occasion Pluto was masked by an inconvenient bright star.

It is interesting to compare predictions with fact:

Pluto	Actual	Planet X (Lowell)	Planet O (Pickering)
Mean distance from Sun (a.u.)	39.5	43.0	55.1
Period (years)	248	282	409
Eccentricity	0.249	0.202	0.248
Inclination (°)	17	10	15
Perihelion date	1989	1991	2129
Mass, Earth = 1	0.002	6.6	2.0

The first orbit for Pluto issued from Flagstaff gave an eccentricity of 0.909 and a period of 3000 years, but a few months later new observations yielded a better orbit. Pluto was detected near the star δ Geminorum. Slipher's initial announcement stated that 'the object was 7 seconds of time west from δ Geminorum, agreeing with Lowell's predicted longitude'.

Various names for the new planet were suggested, including Odin, Persephone, Chaos, Atlas, Tempus, Lowell, Minerva, Hercules, Daisy (!), Pax, Newton, Freya, Constance and Tantalus. The final choice was Pluto, suggested by an 11-year-old Oxford schoolgirl, Venetia Burney (now Mrs. Phair).

SIZE AND MASS

Lowell's prediction had been very accurate; but was it sheer luck? Before long, doubts began to creep in, because Pluto seemed to be far too small and lightweight to exert measurable perturbations upon giants such as Uranus and Neptune. (Uranus had been used for the main investigations, because its orbit was very well known, whereas Neptune had not completed a full revolution since its identification in 1846; in fact it has not done so even yet.) In 1936 A. C. D. Crommelin of Greenwich suggested the theory of specular reflection. If Pluto were highly reflective, the bright image of the Sun might falsify the diameter estimates, so that Pluto could be much larger – and hence more massive – than it seemed. This proved to be incorrect. In 1949 G. P. Kuiper, using the 82 in (208 cm) reflector at the McDonald Observatory in Texas, gave the diameter as 10 200 km, with a mass eight-tenths of that of the Earth, but even this proved to be far too great. In 1950 Kuiper and Humason made new measurements with the Hale reflector at Palomar, reducing the diameter to 5800 km – smaller than Mars. In 1965 a partial occultation of a star by Pluto showed that the diameter could not be more than 5800 km, and the current value is only 2324 km. In fact, Pluto could not have been predicted by any perturbations upon Uranus or Neptune. Either the accuracy was purely fortuitous, or else the planet for which Lowell was searching really exists, at a still greater distance from the Sun.

Incidentally, R. A. Lyttleton of Cambridge, in 1936, suggested that Pluto might be an escaped satellite of Neptune, an idea supported in 1956 by Kuiper, but the discovery of Pluto's companion, Charon, indicates that this cannot be the case.

CHARON

Pluto is not a solitary wanderer in space. It has a companion, Charon, discovered in 1978. Data are given in Table 13.3.

Table 13.3. Charon.

Distance from Pluto: 19 640 km
Orbital period: 6d 9h 17m (6.387 25 days)
Orbital eccentricity: 0.0076
Orbital inclination: 98°.80
Mean orbital velocity: 0.23 km s^{-1}
Diameter (km): 1270
Mean density; water = 1: 1.3
Escape velocity: 0.16 km s^{-1}
Albedo: 0.36
Magnitude at perihelic opposition: 16.8
Apparent diameter, seen from Pluto: 4″
Apparent magnitude, seen from Pluto: −9
Elongation from Pluto, as seen from Earth: 0″.6

The discovery was made on 22 June 1978 by James W. Christy, from the US Naval Observatory at Flagstaff. Pluto had been repeatedly photographed with the 1.55 m reflector with a view to predicting occultations of stars by the planet, which could be used to give an improved value for Pluto's diameter. Images taken in April and May showed that Pluto's image seemed to be elongated. Plates taken earlier (1965, 1970, 1971) were then checked, and the same effects were noted; further confirmation came on 6 July 1978 by J. Graham, using the 401 cm reflector at the Cerro Tololo Observatory in Chile. Either Pluto was curiously irregular in shape, or else it was attended by a satellite. The existence of the satellite as a separate body was established by D. Bonneau and P. Foy, using the 3.6 m Canada–France–Hawaii telescope on Mauna Kea; using the technique of speckle interferometry, they recorded the two bodies separately. The attendant was named Charon, after the gloomy boatman who ferried departed souls across the River Styx into the Underworld.

The Pluto–Charon system is unique, and better regarded as a binary planet (or a binary Kuiper Belt object) rather than as a planet and a satellite. The barycentre of the system (that is to say, the 'balancing point') lies between the two, whereas in all other cases the barycentre lies within the globe of the parent planet. Charon's orbit round Pluto is almost circular, and the surface-to-surface distance is only about 18 000 km. The main point is that the orbital period of Charon is exactly the same as the rotation period of Pluto: 6.4 days, so that to a Plutonian observer Charon would remain fixed in the sky.

The rotation period of Pluto had originally been measured in 1955, by M. Walker and R. Hardie, from variations in magnitude. It was also found that the axial inclination amounts to 122.5°, even more than that of Uranus. Charon moves in the plane of Pluto's equator, so that the phase effects will be very strange – particularly as there are about 14 000 Plutonian days in every Plutonian year. From one hemisphere of Pluto, of course, Charon will never be seen at all.

Charon has only about one-seventh the mass of Pluto, and Pluto contributes 80% of the total light we receive from the system, partly because it is larger and also because it has a higher albedo. Charon is 'grey', whereas Pluto is somewhat reddish. Very probably the two were once combined, and were separated following a massive impact in the early history of the Solar System. In 1999 A. Stern, R. Canup and D. Durda, of the Southwest Research Institute (Boulder, USA) suggested that some Kuiper Belt objects in neighbouring orbits to Pluto may in fact be débris produced by the impact.

INTERNAL STRUCTURE

We do not yet have much positive information about the internal structure of either Pluto or Charon. Pluto's density, over twice that of water, implies that there is more rock than in the icy satellites of the giant planets (perhaps around 70%). Below the frosty crust there may be a mantle of water ice, going down to between 200 and 300 km, and then a region of partially hydrated rock; the upper crust is not likely to be more than a few kilometres of a few tens of kilometres deep. It is unclear whether or not Pluto is differentiated, but it is probable that the gravitational pressure is inadequate to increase the rock density deep inside the globe to a marked degree. Charon is much less dense than Pluto, so that the percentage of rock is presumably lower.

SURFACE FEATURES

The apparent diameter of Pluto is so small that surface features are beyond the range of ordinary telescopes; only the

Hubble Space Telescope has been able to show anything at all definite. The first attempts to study the surface were made by L. Andersson and B. Fix, from 1973. They found that the amplitude of the light variations due to Pluto's rotation were increasing, but that the mean magnitude was becoming fainter. From this they inferred that there are bright poles and that one of these poles, previously presented to the Earth, was turning slowly away, so that more of the mid-latitude and equatorial zones were presented – where the ice would be 'dirtier' and older. Results from the IRAS satellite, in 1983, indicated the presence of a band near the equator which was bright at infra-red wavelengths but dark in visible wavelengths, so that it could be relatively frost-free. However, much better results came from a series of phenomena lasting from 1985 to 1990.

MUTUAL PHENOMENA

In 1978 Leif Andersson, of Sweden, pointed out that for a period occurring twice every Plutonian year – that is to say, every 124 Earth years – a situation arises when the orbits are positioned in a way which allows for mutual transits and occultations. The first observation, when the edge of Charon passed in front of Pluto, was made by R. Binzel on 17 February 1985. Total events began in 1987 and ended in 1988, while the whole series of phenomena ended in September 1990.

When Charon passed behind Pluto it was completely hidden, and Pluto's spectrum could be seen alone; when Charon passed in front of Pluto the two spectra were seen together, but that of Pluto could be subtracted. It was found that the surface of Pluto is covered with methane ice, together with large crystals of frozen nitrogen (N_2) and some of water ice and carbon monoxide. No methane ice was found on Charon, where the surface layer was apparently due to water ice.

These mutual phenomena were immensely informative. It is fortunate that they occurred when they did; the situation will not recur for well over a century.

In 1999 observations made with the 8.3 m Suburu Telescope on Mauna Kea, in Hawaii, detected solid ethane (C_2H_6), nitrogen (N_2), methane (CH_4) and carbon monoxide (CO) on Pluto; the surface temperature was found to be −233 °C. The water ice layer on Charon was confirmed.

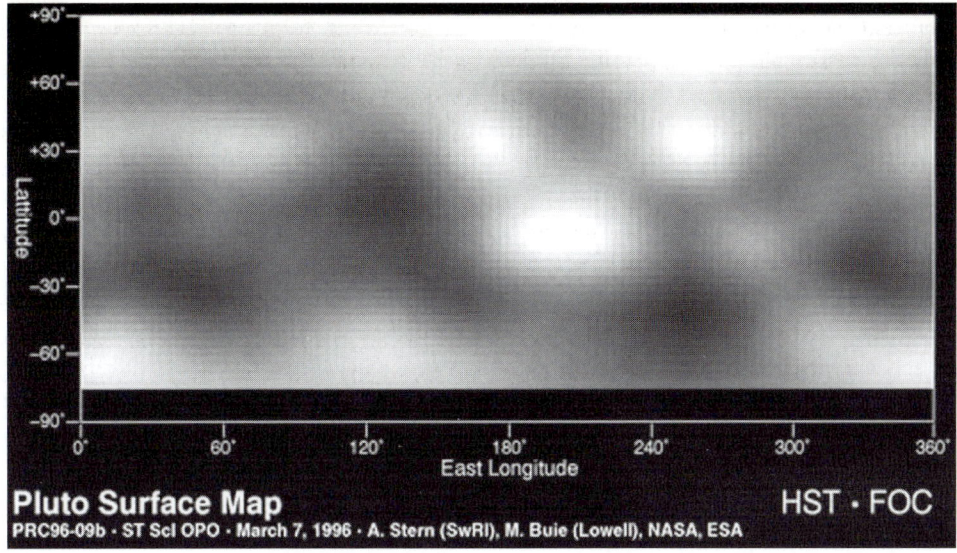

Figure 13.1. Pluto. (Courtesy: NASA.)

MAPS OF PLUTO

From July 1994 images of Pluto have been obtained with the Hubble Space Telescope, and definite features have been seen. Pluto is in fact an unusually complex object. As expected, there is a dark equatorial band and bright polar caps. Some of the features may be due to basins or impact craters, together with ridges, but others, including the prominent north polar cap, may be produced by the distribution of the frosts which migrate across Pluto's surface because of the orbital and seasonal cycles. No names have been given to the features (a system based upon Underworld deities, following my own suggestion, is now under consideration by the International Astronomical Union).

No well-marked features have been seen on Charon. There are vague indications of a darkish band in one hemisphere and a brighter band in the other, but nothing at all definite.

ATMOSPHERE

Pluto does have an atmosphere, albeit a very tenuous one, with a ground density of a few microbars – or a few tens of microbars at most. Preliminary spectroscopic searches by Kuiper in 1943–4 were unsuccessful, but occultations of stars by Pluto have given definite proof that the atmosphere not only exists, but is surprisingly extensive. The first occultation measures were made in 1980. On 9 June 1988 Pluto passed in front of a 12th-magnitude star; the star began to fade at a distance of 1500 km from the centre of Pluto and it seems that there is an upper transparent layer about 300 km deep, with haze below – still not opaque enough to hide the surface. The main constituent is nitrogen (N_2), together with very small amounts of methane and carbon monoxide. This means that the atmosphere is not very dissimilar to that of Triton, although the surface coating of Pluto is different.

Charon showed no trace of atmosphere, which in view of its much lower escape velocity is not surprising. It has been suggested that Charon's original atmosphere escaped and was captured by Pluto; alternatively, that an excessively tenuous atmosphere envelops both bodies, but as yet our information is very incomplete.

There is also the strong possibility that as Pluto moves out toward aphelion its atmosphere will freeze out, so that for part of each orbit there is no gaseous atmosphere at all. This is one reason why it is important to send a space-craft there before the atmosphere collapses, and the Kuiper Pluto Express mission is already being planned. If funded, it may be launched in late 2004; there will be a gravity assist from Jupiter in April–June 2006, and the probe will fly by Pluto and Charon in December 2012 at a minimum distance of 15 000 km. The fly-by velocity will be 17–18 km s^{-1}, and data will be transmitted back to Earth for a year following the encounter. The probe will then continue on to the Kuiper Belt, searching for new members and imaging any which are suitably placed.

Pluto is certainly a remote, lonely world, but it is not shrouded in permanent darkness; indeed, sunlight there will be at least 1500 times more powerful than full moonlight on Earth. Whether it will remain the outpost of the known planetary system remains to be seen.

PLANET X

There is still controversy as to whether or not a tenth planet exists. It is generally referred to as Planet X – not to be confused with Lowell's Planet X, which led to the identification of Pluto.

Comets have again been called in as evidence. In 1950 studies of eight cometary orbits led K. Schütte to assume the existence of a planet at 77 a.u., and his work was extended by H. H. Kritzinger, whose planet X moved at 65 a.u. Later, by 'pairing' data for two of Schütte's eight comets, he amended this distance to 75.1 a.u. and the period to 650 years, with an inclination of 40° and a magnitude of 10. A photographic search was undertaken in the indicated position, but with no result.

Less convincing was a theory by M. E. Savin, who believed in a planet moving at 78 a.u. His method was to divide the known planets into two groups, inner and outer, and 'pair' them, but for some obscure reason he included the tiny asteroid 944 Hidalgo. Pairing 'Planet X' with Mercury, he produced a planet with a period of 685.5 years, an eccentricity of 0.3 and a mass 11.6 times that of the Earth. Needless to say, no confirmation was forthcoming.

The 'comet family' idea has also been discussed by J. J. Matese and D. P. Whitmire (1986), V. P. Tomanov (1986) and A. S. Guliev (1987). Guliev claimed that a new cometary family could be identified, consisting of comets Halley, di Vico, Westphal, Pons–Gambart, Brorsen–Metcalf and Väisälä 2. Projections of their aphelia of their orbits on to the celestial sphere concentrate near a large circle, indicating the existence of Planet X moving within the corresponding plane; the distance was given by Guliev as 36.2 a.u. Tomanov came to much the same conclusion as Guliev. Matese and Whitmire believed that there are 'showers' or periodical comets associated with

the passage of Planet X through the Oort cloud, and they linked this with cratering and fossil records showing periods of major impacts on Earth, modulated with a period of about 30 000 000 years. Their Planet X moved between 50 and 100 a.u. In 1975 G. A. Chebotarev of what was then Leningrad used the aphelia of periodical comets to predict two outer planets, one at 53.7 a.u. and the other at 100 a.u. Much more recently (1999) J. Murray of the Armagh Observatory in Northern Ireland has used the orbits of very long-period comers to indicate the presence of a very massive planet orbiting the Sun at a distance of around one light-year. This is of course highly speculative, and the detection of a planet at so great a distance seems to be out of the question at the present time.

Halley's Comet was also involved. In 1952 R. S. Richardson made an attempt to measure the mass of Pluto by its perturbing effects on the comet. He decided that Pluto had no detectable influence, but that there might be an Earth-sized planet moving at 36.2 a.u, or 1 a.u. beyond the aphelion point of the comet; this would delay the return of the comet to perihelion by one day, while a similar planet at 35.3 a.u. would produce a delay of six days. A somewhat desultory search was put in hand, but with no result. In 1972 J. A. Brady, of the University of California, used the movements of Halley's Comet to indicate the presence of a Saturn-sized planet moving in a retrograde orbit at a distance of 59.9 a.u.; he believed the magnitude to be about 14 and indicated that the planet was situated in Cassiopiæ. Searches were made, but he planet refused to show itself, and it was generally agreed that Brady's calculations were fatally flawed.

In September 1988 new predictions, based on the movements of Uranus, were made by R. Harrington, from the US Naval Observatory in Washington. His planet had a period of 600 years and a mass two to five times that of the Earth, with a present distance of around 9.6 thousand million km; he gave a position in the Scorpius–Sagittarius area, but again there was no result.

A different line of approach as adopted by J. D. Anderson of the JPL (Jet Propulsion Laboratory at Pasadena, California). He claimed that there were genuine unexplained perturbations in the motions of Uranus and Neptune between 1810 and 1910, but not since. This would indicate a Planet X with a very eccentric orbit, now near its aphelion and therefore unable to produce measurable effects. Anderson gave it an inclination of about 90° and a period of between 700 and 1000 years, and a mass five times that of the Earth. Anderson suggested that the perturbations due to this planet would again become measurable about the year 2060. We must wait and see.

Uranus was again used by C. Powell, of JPL, in 1987; his planet would be located in Gemini and would move at about 39.8 a.u. with a period of 251 years – an orbit not unlike Pluto's. A brief search was made from the Lowell Observatory, but with the usual lack of success.

Four space-craft – Pioneers 10 and 11, and Voyagers 1 and 2 – are now leaving the planetary system in various directions, and it is possible that unexpected deviations might lead to the tracking down of Planet X, but this would require an enormous slice of luck.

(On a less serious note, I made some calculations in 1981, assuming that the real Planet X was in the same area of the sky as Pluto in 1930, that it moved about as far beyond the path of Neptune as Neptune is from Uranus, that its diameter was about equal to that of Neptune and that the eccentricity and inclination were low. From this, I worked out a 1981 position near the star χ Leonis. I was however hardly surprised when a search, carried out with the modest 39 cm reflector in my observatory, failed to show any new member of the Solar System.)

Recently, it has been claimed that improved values for the masses of Uranus and Neptune, obtained from the Voyager 2 data, show that no unexplained perturbations occur, and that Planet X does not exist. This may or may not be the case. If Planet X is real, it will no doubt be found eventually, but for the moment there is little more to be said.

One final comment may, however, be worth nothing. Assuming that Planet X comes to light, a name will have to be found for it and one favourite is 'Minerva'. In fact this was one of the names suggested for Tombaugh's planet in 1930, but the suggestion came from T. J. J. See, who was – to put it mildly – unpopular with his contemporaries. This is why Minerva is now called Pluto!

14 COMETS

Comets are the most erratic members of the Solar System. They may sometimes look spectacular, but they are not nearly so important as they then seem, and by planetary standards their masses are very low indeed. In most cases, though not all, their orbits round the Sun are highly eccentric. A comet has been aptly described as 'a dirty ice-ball'.

COMET PANICS

In earlier times comets were not classed as being celestial bodies, and were put down as atmospheric phenomena, although it is true that around 500 BC the Greek philosopher Anaxagoras regarded them as being due to clusters of faint stars. They were always regarded as unlucky. Recall the lines in Shakespeare's *Julius Cæsar*:

> When beggars die, there are no comets seen:
> The heavens themselves blaze forth the death of princes.

In 1578 the Lutheran bishop Andreas Calichus went further, and described comets as being 'the thick smoke of human sins, rising every day, every moment, full of stench and horror before the fact of God'. However, his Hungarian contemporary, Andreas Dudith, sagely pointed out that in this case the sky would never be comet-free! The first proof that comets were extraterrestrial came from the Danish astronomer Tycho Brahe, who found that the comet of 1577 showed no diurnal parallax, and must therefore be at least six times as far away as the Moon (actually, of course, it was much more remote than that).

Comets were viewed with alarm partly for astrological reasons and partly because it was thought that a direct collision between the Earth and a comet might mean the end of the world. In 1696 a book by William Whiston, who succeeded Isaac Newton as Lucasian Professor of Mathematics at Cambridge, predicted that Doomsday would come on 16 October 1736, when a comet would strike the Earth. In France, in 1773, a mathematical paper by the well-known astronomer J. J. de Lalande was misinterpreted, and led to the popular belief that a comet would strike the Earth on 20 or 21 May. (Seats in Paradise were sold by members of the Clergy at inflated prices.) Another alarm occurred in 1832, when it was suggested – wrongly – that there would be a very near encounter with Biela's periodical comet. In 1843, at the time of a particularly brilliant comet, there was a widespread end-of-the-world panic in America, due to the dire prophecies of one William Miller. And in 1910, when Halley's Comet was on view, a manufacturer in the United States made a large sum of money by selling what he called anti-comet pills, and many people sealed up their windows to keep out poisonous gases. Bennett's Comet of 1970 was mistaken by some Arabs for an Israeli war weapon, and Kohoutek's Comet of 1973 was also regarded as a threat; it was expected to become brilliant, although in the event it failed to do so. In 1994 a curious prophet named Sofia Richmond ('Sister Gabriel') achieved a degree of notoriety by predicting that a collision between Halley's Comet and the planet Jupiter (!) would result in the destruction of mankind. Finally, there are the books by an eccentric Russian-born psychoanalyst, Immanuel Velikovsky, who confused planets with comets, and believed that Venus had been a comet only a few thousands of years ago; he also maintained that Biblical events, notably the Flood, were due to comets. However, Velikovsky's ignorance of astronomy was so complete that trying to argue with him was a decidedly pointless exercise.

THE NATURE OF COMETS

In 1948 the Cambridge astronomer R. A. Lyttleton popularized the 'flying sandbank' theory of comets. He believed that dust particles were collected by the Sun during its passage through an interstellar cloud, and that these particles collected into 'clouds' which he identified as comets. There were fatal weaknesses in this theory, and in 1950 it was abandoned in favour of the model proposed by F. L. Whipple. The nucleus of a comet – the only reasonably 'solid' part – is made up of rocky fragments held together by frozen ices such as H_2O, methane, carbon dioxide and ammonia. When a comet is warmed as it approaches perihelion, the rise in temperature leads to evaporation, so that the comet develops a head or coma, often together with a tail or tails. Cometary tails always point more or less away from the Sun, and are of two types. There is a gas or ion tail; the molecules are repelled by the 'solar wind'. With a dust tail, the particles are driven out by the pressure of sunlight. This all means that when a comet is receding from the Sun, it travels tail-first; in general, ion tails are straight, while dust tails are curved. Many comets have tails of both types, although smaller, fainter comets never develop tails of any kind.

COMET NOMENCLATURE

Traditionally, comets are named after their discoverer or discoverers; thus the bright comet of 1996 was discovered by the Japanese amateur Y. Hyakutake and is named after him, while the even brighter comet of 1997 was detected independently from America by Alan Hale and Thomas Bopp, and is known as Hale–Bopp. No more than three names are now allowed. Sometimes the discoverers of different returns of the same comet are used; thus in 1881 W. F. Denning discovered a comet with a period of between 8 and 9 years, and it was not seen again until recovered in 1978 by S. Fujikawa, so that it is listed as Denning–Fujikawa. Occasionally the name used is that of the first computer of the orbit (as with comets Halley, Encke and Crommelin; this applies only to periodical comets).

Up to 1994 a comet was also assigned a letter in order of discovery during the year, and then a permanent designation using Roman numerals, in order of perihelion passage. Thus Halley's Comet was the ninth comet to be found in 1982 and became 1982i; it was the third comet to pass perihelion in 1986 and became 1986 III.

A new system was introduced in 1995, similar to that used for asteroids. Each year is divided into 24 sections, with its own letter (I and Z being omitted). Each comet is given a designation depending on the year of discovery, with a capital letter to indicate its half-month and a number to show the order of discovery in that half-month. Thus Comet Hale–Bopp, found on 23 July 1995 became 1995 O1,

Table 14.1. Letter designations for comets.

A	Jan	1–15
B	Jan	16–31
C	Feb	1–15
D	Feb	16–29
E	Mar	1–15
F	Mar	16–31
G	Apr	1–15
H	Apr	16–30
J	May	1–15
K	May	16–31
L	Jun	1–15
M	Jun	16–30
N	Jul	1–15
O	Jul	16–31
P	Aug	1–15
Q	Aug	16–31
R	Sep	1–15
S	Sep	16–30
T	Oct	1–15
U	Oct	16–31
V	Nov	1–15
W	Nov	16–30
X	Dec	1–15
Y	Dec	16–31

as it was the first comet to be discovered during the period between 16 and 30 July. A list of the letter designations is given in Table 14.1.

There are also prefixes. Basically, comets are divided into two classes; periodical, with orbital periods of less than 250 years, and non-periodical, with periods so long that they cannot be predicted with any accuracy. P/ indicates a periodical comet; D/ a periodical comet which has either disintegrated or been lost; C/ a non-periodical comet. (Strictly speaking, this is incorrect; all comets will return eventually unless they have been perturbed by planets and thrown into parabolic or hyperbolic paths.) Thus Halley's Comet, with a period of 76 years, is P/Halley; Westphal's Comet, which was seen to 'fade away' in 1913, is D/Westphal; Hale–Bopp, with a period of over 2000 years, is C/Hale–Bopp. A few comets which are observable all round their orbits are not given letter designations; such is P/Encke, with a period of 3.3 years.

Two comets have been given both cometary and asteroidal designations; Comet P/Wilson–Harrington is Asteroid 4015 and Comet P/Elst–Pizarro is Asteroid 7968. The distinction between comets and small asteroids has become decidedly blurred.

STRUCTURE OF COMETS

A comet is made up essentially of three parts; a nucleus, a head or coma, and a tail or tails.

The nucleus of an average comet is surprisingly small. Halley's Comet, which is large by cometary standards, had a nucleus measuring 15 km × 8 km × 8 km. Hale–Bopp, the bright comet of 1997, was one of the largest on record; even so, its nucleus was no more than about 40 km across, while Hyakutake, which was brilliant briefly in 1996, was very small indeed; the diameter of its nucleus can have been no more than 3 km at most.

Only one cometary nucleus has been really well seen; that of Halley's Comet in 1986, from the Giotto spacecraft. Otherwise, we are handicapped. When a nucleus is unshrouded, the comet is far away; as it draws inward, the nucleus is surrounded by gas and dust, which masks it effectively. Judging from the Halley results, a nucleus is covered with a layer of blackish organic material. Under this is an icy body; the ices are of various kinds, notably water, carbon monoxide and carbon dioxide. In general these ices are shielded from sunlight, but in isolated areas jets are able to spring out to produce material for the coma and tails. Some comets are active in this respect; others relatively inert.

A coma does not usually form until the comet is within about 3 a.u. of the Sun, when there is marked sublimation of the water ice. Gas flows outward and it is this which wrenches dust particles away from the main body of the comet. A coma may be very large (the coma of the Great Comet of 1811 was larger than the Sun) but is highly rarefied. With comets which are poor in dust, comæ are usually round; with dusty comets the comæ are fan-shaped or parabolic, and there is no hard, sharp boundary between the coma and the tail.

The rate of gas outflow is very high. Comet Hale–Bopp was probably losing 1000 tons of dust and 1200 tons of water per second when it was close to perihelion.

Some comets, such as P/Tempel 2, are rich in dust grains; observations made with IRAS (the Infra-Red Astronomical Satellite) in 1983 indicated that in this comet the particles ranged from tiny grains up to 'pebbles' as much as 6 cm in diameter.

Tails are of two main types. Ion tails (otherwise known as plasma tails or gas tails) are repelled by the solar wind, and are generally straight. Our information comes mainly from spectroscopic research (the first good cometary spectrum was obtained by the Italian astronomer G. Donati as long ago as 1864), and magnetic effects are all important, particularly as the solar wind carries a magnetic current. These tails consist largely of ionized carbon monoxide (CO^+) which tends to fluoresce under the influence of sunlight; this is why plasma tails have a bluish tinge. Tails can be very extended. That of the Great Comet of 1843 was 330 000 000 km long, which is greater than the distance between the Sun and the orbit of Mars.

In April 2000 the space-probe Ulysses, which had been launched to survey the poles of the Sun, passed fortuitously through the tail of Comet C/1996 B2 (Hyakutake) and found that the length of the tail was over 500 million km. This is the longest tail ever recorded.

Plasma tails show rapid changes, due largely to variations in the solar wind; these changes are much more evident in some comets than in others. Shock waves caused by solar flares may produce 'kinks' or even spiral effects. There are also marked 'disconnection' effects, caused when the comet crosses a region where the polarity of the solar wind changes; magnetic field lines inside the tail then cross and re-connect, severing the link with the region close to the nucleus on the sunward side. The tail breaks away and a new one is formed. This happened spectacularly with Halley's Comet at the 1986 return, and also with Comet Hyakutake in 1996.

Dust tails are generally curved; the particles in them are around the size of smoke particles and the tails are yellowish, since they shine only by reflected sunlight. Occasionally a comet may appear to have an 'anti-tail', pointing sunward – as with Comet Arend–Roland of 1957, nicknamed 'the spiked comet'. In fact, what is seen is not a tail, but material in the comet's orbit catching the sunlight at a suitable angle. Arend–Roland showed it particularly well, since it was exceptionally rich in dust. (*En passant,* this comet will never return. Planetary perturbations threw it into a hyperbolic orbit, and it has now made its permanent departure from the Solar System.)

In 1969 observations made from the OAO 2 (Orbiting Astronomical Observatory 2) led to the detection of a vast hydrogen cloud around a comet, Tago–Sago–Kosaka; the cloud was 1 600 000 km in diameter. Similar clouds were also found with other comets, notably Bennett (1970) and Kohoutek (1973), and there is no reason to doubt that they are quite common features of large comets.

The material forming the coma and tails is permanently lost to the comet. This indicates that comets are short-lived by cosmical standards; 0.1–1% of the total mass will be lost each time the comet passes through perihelion. Obviously, a comet of short period will 'waste away' much more quickly than a comet of longer period, and this explains why all the comets of really short period are faint. Very few of them ever achieve naked-eye visibility.

SHORT-PERIOD COMETS

Faint short-period comets are common, and more are discovered every year. Comets which have been observed at more than one return are given numbers, roughly in order of identification; thus Halley, the first to have its period recognized, is P/1, while Encke, which was next identified, is P/2. The first 140 periodical comets are listed in Table 14.12, and data for selected comets are given in Table 14.13. Some periodical comets seen at only one return are listed in Table 14.2; some of these may be recovered eventually.

In these tables, A indicates the absolute magnitude of the comet – that is to say, the magnitude it would have if seen at a distance of 1 a.u. from the Sun and 1 a.u. from the Earth.

A glance at Table 14.13 shows that many short-period comets have their aphelia at about the distance of Jupiter from the Sun (just over 5 a.u.). These are said to make up Jupiter's 'comet family', and as Jupiter is much the most massive planet in the Solar System it is bound to exert great influence. Comet families of the other giant planets are ill-defined, if they exist at all.

The comet with the shortest known period is Encke's (P/2), named in honour of the mathematician who first computed its orbit. The comet was first seen on 17 January 1786 by P. Méchain, from France. It was again recorded on 7 November 1795 by Caroline Herschel; the next returns to be seen were those of 1805 (discovered by Thulis at Marseilles, 19 October) and 1818 (Jean Pons, 26 November). J. F. Encke, at Berlin, decided that these comets must be identical, and predicted a return for 1822. He was correct, and subsequently the comet has been seen at every return except that of 1944, when it was badly placed and most astronomers were otherwise engaged. It can now be followed all round its orbit. Usually it is much too faint to be seen without optical aid, but occasionally it becomes a naked-eye object – as in 1947, when it rose to the fourth magnitude. It can also develop a short tail.

The orbital period has shortened slightly since the comet was first seen; Encke himself explained this by assuming the existence or a resisting medium near the Sun, but it is now known that the real cause is a sort of 'rocket' effect. Jets emitted from active areas on the nucleus will accelerate or retard the comet's motion according to the direction in which they leave the nucleus. All comets are rotating, and so the direction of the spin axis changes. For example, Halley's Comet returns to perihelion an average of 4.1 days late at every return, so that the nucleus must be rotating in the same direction as the comet's motion round the Sun; the thrust of the escaping gases reaches a peak in the 'afternoon' of the comet's 'day'. The resulting jet force pushes the comet forward in its orbit, the nucleus drifts outward from the Sun, the orbital period increases and perihelion occurs later than predicted.

Some comets have orbits of low eccentricity. The first discovered of these was Schwassmann–Wachmann 1 (1925); its path lies entirely between those of Jupiter and Saturn. Normally it is a very faint object, but it can show sudden, unpredictable outbursts which bring it within the range of small telescopes; thus in 1976 the magnitude rose to above 12. Other comets which remain in view throughout their orbits are Gunn and Smirnova–Chernykh. Oterma's Comet used to have low eccentricity and a period of 7.9 years, but a close approach to Jupiter in 1963 altered the period to over 19 years. Perihelion distance is now over 800 000 000 km, so that the comet is a very faint object.

Table 14.2. Periodical comets seen at only one return. It is unlikely that the comets in (a) will be recovered. There are, however, other short-period comets which will certainly be recovered, plus a few with longer periods. (b) All these are, of course, very uncertain.

(a)

Comet	Year	Period (years)	Perihelion distance, q (a.u.)	Aphelion distance, Q (a.u.)	Eccentricity	Inclination (°)
Helfenzrieder	1766	4.35	0.406	4.92	0.848	7.9
Blanpain	1819	5.10	0.892	5.03	0.699	9.1
Barnard 1	1884	5.38	1.279	4.86	0.583	5.5
Brooks 1	1886	5.44	1.325	4.86	0.571	12.7
Lexell	1770	5.60	0.674	5.63	0.786	1.6
Pigott	1783	5.89	1.459	5.06	0.552	45.1
Harrington–Wilson	1951	6.36	1.664	5.10	0.515	16.4
Barnard 3	1891	6.52	1.432	5.55	0.590	31.3
Giacobini	1896	6.65	1.455	5.62	0.588	11.4
Schorr	1918	6.67	1.884	5.21	0.469	5.6
Swift	1895	7.20	1.298	6.16	0.652	3.0
Denning	1894	7.42	1.147	6.01	0.698	5.5
Metcalf	1906	7.78	1.631	6.22	0.584	14.6
Linear	1999	12.53	1.872	5.395	0.653	20.4
Van Houten	1961	15.6	3.957	8.54	0.367	6.7
Pons–Gambart	1827	57.5	0.807	29.0	0.946	136.5
Dubiago	1921	62.3	0.929	30.3	0.929	22.3
de Vico	1846	76.3	0.664	35.3	0.963	85.1
Väisälä	1942	85.4	1.287	37.5	0.934	38.0
Barnard 2	1889	145	1.105	54.2	0.960	31.2
Mellish	1917	145	0.198	55.1	0.993	32.7
Wilk	1937	187	0.619	64.9	0.981	26.0

(b)

Comet	Period (years)	Last perihelion	Next due
D/1889 M1 Barnard 2	145	1889	2034
D/1984 A1 Bradfield	151	1983	2134
D/1989 A3 Bradfield 2	81.9	1988	2070
P/1983 V1 Hartley–IRAS	21.5	1984	2005
P/1997 B1 Kobayashi	24.5	1997	2021
P/1997 G1 Montani	21.8	1997	2019
D/1942 EA Väisälä 2	85	1942	2027

On 19 November 1949 A. G. Wilson and R. G. Harrington, using the Schmidt telescope at Palomar, discovered a 16th-magnitude comet with a short tail. The period was calculated to be 2.31 years, but the comet was not seen again until 1979, when it was recovered by E. Helin; it looked like an asteroid, and was given an asteroid number (4015) but the identity with the 1949 comet is not in doubt. It seems that the object is a largely inactive comet which produces occasional outbursts. On 7 August 1996 E. W. Elst discovered a comet on a plate exposed in July of that year by

Table 14.3. Lost periodical comets. Comets seen at more than one return.

Comet	Period (years)	Perihelion distance, q (a.u.)	Aphelion distance, Q (a.u.)	Eccentricity	Inclination (°)	Returns	Last seen	
11D Tempel–Swift	5.7	1.15	5.22	0.64	5.4	4	1908	Lost
25D Neujmin 2	5.4	1.34	5.43	0.57	10.6	2	1927	Probably disintegrated
3D Biela	6.6	0.86	6.19	0.76	12.6	6	1852	Broke up
5D Brorsen	5.5	0.59	5.61	0.81	29.4	5	1879	Lost; certainly disintegrated
34P Gale	11.3	1.21	5.02	0.76	10.7	2	1927	Lost
20D Westphal	61.9	1.25	30.8	0.76	12.6	2	1913	Faded out; no longer exists
— Shoemaker–Levy 9	—	—	—	—	—	(1)	1994	Impacted Jupiter

107P Wilson–Harrington (1949) was recovered in 1979 as an asteroid and given an asteroid number, 4015.
133P Elst-Pizarro (1996) has been given an asteroid number, 7968. It moves wholly within the main asteroid belt.

G. Pizarro; the magnitude was 18, and there was a definite though narrow tail. The period is 5.6 years and the orbit lies wholly within the main asteroid belt, so that the object has 'dual nationality'; it is Comet P/133 Elst–Pizarro and also Asteroid 7968. These cases add support to the suggestion that some Earth-grazing asteroids may well be ex-comets which have lost all their volatiles.

Some periodical comets have been 'lost', either because they have disintegrated or because we have failed to keep track of them (Table 14.3). The classic case is that of Biela's Comet. It was discovered in 1772 by Montaigne, from France; it returned in 1806 and again in 1826, when it was discovered by an Austrian amateur, W. von Biela. Biela recognized its identity with the 1772 and 1806 comets, and his name was justifiably attached to it. It returned in 1832, was missed in 1839 because it was badly placed, and recovered once more in 1846, when it astounded astronomers by dividing in two. The twins came back on schedule in 1852, but this was their last appearance. They were unfavourably placed in 1859, but should have been well seen in 1866; however, they failed to appear, and have never been seen since. Undoubtedly they have disintegrated, and their remnants were later seen in the form of meteors, although by now the meteor shower seems to have died out. Westphal's Comet was seen in 1852 and again in 1913, but failed to

survive perihelion and did not return on schedule in 1976; Brorsen's Comet was seen at five returns between 1846 and 1879, but has not appeared since, and has evidently broken up. However, one must beware of jumping to conclusions. Comet Di Vico–Swift was lost for 38 years after 1897, but was recovered in 1965; Holmes' Comet 'went missing' for 58 years prior to its recovery in 1964, following calculations by B. G. Marsden. It is important to note that flimsy objects such as comets are easily perturbed by planets, and no two cycles are exactly alike.

One comet which will certainly never be seen is Shoemaker–Levy 9. In July 1994 it impacted Jupiter, and is described on page 152.

It is worth noting that in 1886 Comet P/16 Brooks 2 had a close encounter with Jupiter, and passed within the orbit of Io. The encounter was not actually seen, but at the return of 1889 the comet was seen to be accompanied by four minor companions, which were classified as 'splinters' and did not last for long. Also, Comet P/82 Gehrels 3 was in orbit round Jupiter for some time during the 1970s, but escaped unharmed in 1973, and returned to solar orbit.

Comets can also make close approaches to the Earth (Table 14.4). Excluding the comet of 1491, whose orbit is highly uncertain, the approach record is held by D/1770

Table 14.4. Close-approach comets.

Comet	Name	Date	Distance (a.u.)	Magnitude
C/1491 B1	—	1491 Feb 20	0.094[a]	1?
D/1770 L1	Lexell	1770 July 1.7	0.0151	2
55P/1366 U1	Tempel–Tuttle	1366 Oct 26.4	0.0229	3
C/1983 H1	IRAS–Araki–Alcock	1983 May 11.5	0.0312	2
1/P 837 F1	Halley	837 Apr 10.5	0.0334	−3.5
3D/1805 V1	Biela	1805 Dec 9.9	0.0366	3
C/1743 C1	—	1743 Feb 8.9	0.0390	3
7/P	Pons Winnecke	1927 June 26.8	0.0394	3.4
C/1702 H1	—	1702 Apr 20.2	0.0437	3.5
73/P/1930 J1	Schwassmann–Wachmann 3	1930 May 31.7	0.0617	10
C/1983 J1	Sugano–Saigusa–Fujikawa	1983 June 12.8	0.0628	2
C/1760 A1	—	1760 Jan 8.2	0.0682	4
C/1853 G1	Schweizer	1853 Apr 29.1	0.0839	0
C/1797 P1	Bouvard–Herschel	1797 Aug 16.5	0.0879	3
1/P 374 E1	Halley	374 Apr 1.9	0.0884	0?
1/P 607 H1	Halley	607 Apr 19.2	0.0898	0?
C/1763 S1	Messier	1763 Sept 23.7	0.0934	6
C/1864 N1	Tempel	1865 Aug 8.4	0.0964	2.5
C/1862 N1	Schmidt	1862 July 4.6	0.0982	4.5
C/1996 B2	Hyakutake	1996 Mar 25.3	0.1018	0
C/1961 T1	Seki	1961 Nov 15.2	0.1019	4

[a] Very uncertain.

Table 14.5. Predicted returns of comets with periods of over 25 years.

Designation name	Discovered	Last perihelion	Period (years)	Next return
1P Halley	240 BC	1986	76.0	2061
12P Pons–Brooke	1812	1954	70.92	2024
13P Olbers	1815	1956	69.56	2024
23P Brorsen–Metcalf	1847	1989	70.54	2059
27P Crommelin	1818	1984	27.41	2011
35P Herschel–Rigollet	1788	1939	155	2092
38P Stephan–Oterma	1867	1980	37.70	2018
109P Swift–Tuttle	1862	1992	135.01	2126
122P di Vico	1846	1995	74.36	2069

L1 Lexell, discovered in 1770 by C. Messier; A. Lexell of St Petersburg computed the orbit. The minimum distance from Earth was 2 200 000 km, and the comet was visible with the naked eye. The period was then 5.6 years, but a subsequent encounter with Jupiter, in July 1779, changed the orbit completely; the current period is thought to be around 250 years, and from our point of view the comet is hopelessly lost.

All these comets have direct motion, but with longer periods we begin to encounter retrograde motions. Comet P/109 Swift–Tuttle will next return in 2127; for a time there were fears that it might be on a collision course, but this does not now seem to be the case. Halley's Comet also has retrograde motion. For comets seen at more than one return, the longest period is that of P/35 Herschel–Rigollet, seen in 1788 and 1939 (Table 14.5).

HALLEY'S COMET

Much the most famous of all comets is P/1 Halley. It may have been recorded by the Chinese as early as 1059 BC; since 240 BC it has been seen at every return. The mean period is 76 years. A list of known returns is given in Table 14.6.

There are many historical references to Halley's Comet. In 684 Ma-tuan-lin, the Chinese historian, refers to a comet seen in the western sky during September and October; this was certainly Halley's, and the first known drawing of it relates to this return. The drawing was published in the *Nürnberg Chronicle*; this was printed in 1493, and shows woodcuts by the German artist M. Wolgemuth. In 837 the comet was at its very best; on 11 April it was a mere 0.03 a.u. from the Earth (4 500 000 km) and the tail extended over 93°, while the brightness of the coma rivalled Venus. The return of 1066 was shown in the Bayeux Tapestry; King Harold is tottering on his throne, while his courtiers gaze up in horror.

The return of 1301 was favourable; one man who saw it was the Florentine painter Giotto di Bondone, who later used it as a model for the Star of Bethlehem in his *Adoration of the Magi*. (In fact there is no chance that the comet can be identified with the Star of Bethlehem; it returned years too early.) At the return of 1456 the comet was again bright, and, as usual, was regarded as an evil omen. At that time the Turkish forces were laying siege to Belgrade, and on the night of 8 June it was said that 'a fearsome apparition appeared in the sky, with a long tail like a dragon'. The current Pope, Calixtus III, went so far as to preach against the comet as an agent of the Devil, although it is unlikely that he excommunicated it, as has sometimes been claimed!

In 1672 the comet was seen by Edmond Halley (the actual discovery was made on 15 August of that year by G. Dorffel) and subsequently Halley decided that it must be identical with comets previously seen in 1607 and in 1531. He predicted a return of 1758. On Christmas Night of that year the comet was duly found, by the German amateur J. Palitzsch, and passed through perihelion in March 1759. This was the first predicted cometary return. Since then the comet has been back in 1835, 1910 and 1986.

In 1835 the comet was recovered on 6 August by Dumouchel and di Vico, from Rome, close to the predicted position near the star ζ Tauri. It remained prominent for weeks later in the year, and was followed until 20 May 1836; the last observation of it was made by Sir John Herschel from the Cape. For 1910 very accurate predictions were made by P. Cowell and A. C. D. Crommelin, from Greenwich; the discovery was made on 12 September 1909 by Max Wolf, from Germany, and the comet was followed until 15 June 1911, by which time its distance from the Sun was over 800 000 000 km. It was brilliant enough to cause general interest – although it was not so bright as the non-periodical 'Daylight Comet', which had been seen earlier in 1910, several weeks before Halley's Comet reached its brightest magnitude. On 18–19 May 1910 the comet passed in transit across the face of the Sun. The American astronomer F. Ellerman went to Hawaii to observe under the best possible conditions, but could seen no trace of the comet.

The 1910 return was the first occasion when the comet could be studied with photographic and spectroscopic equipment, and it was fortunate that the comet was well placed. The Earth was closest to the comet on 20 May, at a range of around 21 000 000 km; the closest encounter between the Earth and the comet's tail was about 400 000 km, and there was some public unease because it had become known that comet tails contain unpleasant substances such as cyanogen. This is true enough, but the density of a tail is so low that there can be no possible ill-effects on this score. At its best the tail was at least 140° long. The comet was indeed a magnificent sight, even if it could not equal the Daylight Comet of the preceding January.

The last perihelion occurred on 9 February 1986. The comet was recovered on 16 October 1982 by a team of astronomers at Palomar (Jewitt, Danielson and Dressler) who used the Hale reflector to detect the comet as a tiny blur of magnitude 24.3; it was a mere 8 arcsec away from its predicted position. The discovery was confirmed shortly afterwards from Kitt Peak. At the time of its recovery, the comet was still moving between the orbits of Saturn and Uranus.

Unfortunately, this was the most unfavourable return for many centuries, and although the comet became an

Table 14.6. Observed returns of Halley's Comet.

Year		Perihelion	First observed	Last observed	
BC	1059	Dec 3	?	?	Earliest probable recorded observation. Chinese annals.
	240	May 25	?	?	Mentioned in Chinese annals.
	164	Nov 12	Sept	Oct	Mentioned only by the Babylonians.
	87	Aug 6	?	?	Well established.
	12	Oct 10	Aug 26	Oct 20	Well established. Observed from Rome as well as China.
AD	26	Jan 25	Jan 31	Apr 11	'Like a sword hanging in the sky'.
	141	Mar 22	Mar 26	May ?	Fairly close approach to Earth on Apr 22 (0.17 a.u.).
	218	May 17	Apr	May	'A very fearful star' (Dion Cassius).
	295	Apr 20	May	May	Chinese records. Nothing said about the brightness.
	374	Feb 16	Mar 3	May	Close approach on Apr 2 (0.09 a.u.).
	451	June 28	June 10	Aug 16	Prominent; observed from Europe as well as China.
	530	Sept 27	Aug 28	Sept 27	Little information about this return.
	607	Mar 15	Apr 18	July	Close approach on Apr 19 (0.09 a.u.).
	684	Oct 2	Sept 6	Oct 24	Earliest recorded drawing (Nürnberg Chronicles, published 1493).
	760	May 20	May 16	July	Chinese report: 'like a great beam'.
	837	Feb 28	Mar 22	Apr 28	Most spectacular return; magnitude −3.5, close approach on Apr 10.5 (0.03 a.u.).
	912	July 18	July 19	July 28	Much less brilliant than in 837.
	989	Sept 5	Aug 11	Sept 11	Seen by the Chinese and by the Saxon historian Elmacin.
	1066	Mar 20	Apr 1	June 7	As bright as Venus. Shown in the Bayeux Tapestry.
	1145	Apr 18	Apr 26	July 9	Chinese described a long tail and a blue colour.
	1222	Sept 28	Sept 3	Oct 23	No special characteristics.
	1301	Oct 25	Sept 15	Oct 31	Seen by Giotto di Bondone, who used it in a famous painting.
	1378	Nov 10	Sept 26	Nov 10	Not favourable, but followed from Europe and China.
	1456	June 9	May 26	July 8	Condemned by Pope Calixtus III as an agent of the Devil.
	1531	Aug 26	Aug 1	Sept 8	Seen by Apian: 'reddish' or 'yellowish'.
	1607	Oct 27	Sept 21	Oct 26	Seen by Kepler. In size and brightness, compared with Jupiter.
	1682	Sept 15	Aug 24	Sept 22	Observed by Halley.
	1759	Mar 13	1758 Dec 2	1749 June 22	First predicted return.
	1835	Nov 16	1835 Aug 5	1836 May 19	Close approach on Oct 10 (0.05 a.u.).
	1910	Apr 20	1909 Aug 25	1911 June 15	Transited the Sun, May 18. Approach to Earth, shortly afterwards (0.14 a.u.).
	1986	Feb 9	1982 Oct 16	1994 Jan 11	First probes to the comet.

easy naked-eye object it was never brilliant. It rose to the sixth magnitude by early December 1985, and was at its best in mid-march 1986; the nucleus was then brighter than magnitude 2, and there was a very respectable tail, showing a great deal of structure. The comet was well south of the celestial equator when at its brightest; at one time it was close to the globular cluster ω Centauri, and with the naked eye the comet and the cluster looked very similar. On 24 April 1986 there was a total eclipse of the Moon and for many people (including myself) this was the last chance to see the comet without optical aid; the magnitude had by then fallen to 4.5, slightly brighter than the adjacent star α Crateris. The fan-shaped tail was still much in evidence.

By 1986 space probes had been developed, and five missions were dispatched; two Russian, two Japanese and one European (Table 14.7). (The Americans withdrew on the grounds of expense.) All the Halley probes were successful. The European mission, named Giotto in honour of the painter, was programmed to pass into the comet's inner coma and image the nucleus, but prior information sent back by the Japanese and Russian missions was invaluable. Giotto passed within 605 km of the comet's nucleus on the night of 13–14 March 1986. It carried a camera, the HMC (Halley Multicolour Camera) and this functioned until 14 seconds before closest approach to the nucleus, when it was made to gyrate by the impact of a dust particle probably about the size of a grain of rice and communications were temporarily interrupted; in fact the camera never worked again, and the closest image was obtained at 1675 km from the nucleus. The nucleus itself measured 15 km $\times$ 8 km $\times$ 8 km, and was shaped rather like a peanut; it had a total volume of over 500 km^3, and a mass of from 50 000 million to 100 000 million tons. The mean density was 0.1–0.2 g cm^{-3}; it would take 60 000 million comets of this mass to equal the mass of the Earth.

The nucleus was dark-coated, with an albedo of 2–4%. Water ice appeared to be the main constituent of the nucleus (84%) followed by formaldehyde and carbon dioxide (each around 3%) and smaller amounts of other volatiles, including nitrogen and carbon monoxide. The shape of the terminator showed that the central region was smoother than the ends; a bright patch 1.5 km in diameter was assumed to be a hill, and there were features which appeared to be craters, around 1 km across. Dust-jets were active, although from only a small area of the nucleus on the sunward side. The sunward side was found to have a temperature of 47 °C, far higher than expected, and from this it was inferred that the icy nucleus was coated with a layer of warmer, dark dust. The icy nucleus was eroded at around 1 cm per day near perihelion, and at each return the comet must lose around 300 000 000 tons of material. The rotation period was found to be 53 h with respect to the long axis of the nucleus, with a 7.3-day rotational period around the axis;

the nucleus was in fact 'precessing' rather in the manner of a toppling gyroscope.

As the comet drew away from the Sun, activity naturally died down. Observations made with large telescopes – notably by R. West with the 1.54 m Danish telescope at La Silla – showed that in April to May 1988 and January 1989 the images were still diffuse, indicating some residual activity or possibly a cloud of dust, but by February 1990, when the distance from the Sun was 12.5 a.u. and the magnitude had fallen to 24.3, the image appeared stellar. Then, on 12 February 1991, C. Hainaut and A. Smette, with the Danish telescope, recorded a major outburst; the magnitude rose to 18.9, even though the distance from the Sun had increased to 14.5 a.u.

On 22 February, Smette used the New Technology Telescope at La Silla to obtain a spectrum. The coma showed a solar-type spectrum, with no emission features, which indicated a dust composition. It was subsequently found that structures within the coma varied with time, while the central region faded by about 1 magnitude per month. It seems that a fan-like structure, in the approximate direction of the Sun, reached a radius of 61 000 km on 13 February, expanding to 142 000 km by 12 April. If the expansion of the coma material were about 14.5 m s^{-1}, the actual outburst would have occurred on 17 December 1990, lasting for three months or so. A short explosive event is ruled out – it would have involved higher velocities for the dust than were observed.

The cause of the outburst is uncertain. A collision with a wandering body is possible, but seems unlikely; possibly a pocket of volatile carbon monoxide ice was exposed to sunlight, and the vaporizing gases carried the dust particles away from the nucleus, but this also seems improbable in view of the comet's distance from the Sun. We may have to await the 2061 return before solving the problem.

The last image of the comet was obtained on 11 January 1994.

By June 1994 the comet had reached the halfway point between perihelion and aphelion. It will next reach perihelion in 2024. Unfortunately the return of 2061 will be as poor as that of 1986: for another really good view we must wait for the return of 2137.

Table 14.7. Cometary probes, 1978–2000.

Spacecraft	Launch date	Comet	Nearest to comet (km)	Closest approach to comet (km)
ISEE/ICE	12 Aug 1978	P/Giacobini–Zinner	11 Sept 1986	7800
Vega 1	15 Dec 1984	P/Halley	6 Mar 1986	8890
Vega 2	21 Dec 1984	P/Halley	9 Mar 1986	8030
Sakigake	8 Jan 1985	P/Halley	11 Mar 1986	7000 000
Giotto	2 July 1986	P/Halley,	14 Mar 1986	596
		P/Grigg–Skjellerup	10 July 1992	200
Suisei	18 Aug 1985	P/Halley	8 Mar 1986	150 000
Stardust	7 Feb 1999	P/Wild 2	Jan 2004	~145

MISSIONS TO OTHER COMETS

Although the Americans did not contribute to the Halley's Comet programme, they did at least send a probe to the periodical comet P/Giacobini–Zinner. They used an older probe, ISEE (the International Sun–Earth Explorer) which had been launched in 1978 for a completely different purpose, and had been orbiting the Earth monitoring the effects of the solar wind on the Earth's outer atmosphere. It carried a full complement of instruments, and had a large fuel reserve. On 10 June 1982 it was re-named ICE (the International Cometary Explorer) and began a series of manœuvres and orbital changes, involving a sequence of 'swing-by' passages around the Moon; at the pass of 22 December 1983 ICE was a mere 196 km from the lunar surface. The closest approach to the comet's nucleus occurred on 11 September 1985 (before the Halley armada reached its target); the range was 7800 km and the relative velocity was 20.5 km s^{-1}. The probe took 20 min to cross the ion tail, and collisions with dust grains were recorded as well as magnetic effects. The distance from Earth was then 70 000 000 km.

The Giotto probe was put into 'hibernation' in April 1986, and was re-activated on 19 February 1990. On 2 July it flew past Earth at 22 730 km, and used the gravity-assist technique to put it into a path to rendezvous with comet P/Grigg–Skjellerup. After a further hibernation period, Giotto was again reactivated on 4 May 1992, when it was 219 000 000 km from Earth, and on 10 July 1982 it encountered Grigg–Skjellerup, passing only 200 km from the nucleus. Most of the instruments on the space-craft were still working, apart from the camera, and valuable data

were secured. Grigg–Skjellerup is a much older comet than Halley, and seldom produces a tail; however, the density of the gas near the nucleus was greater than expected, and there was a good deal of fine 'dust'. It was found that the gas coma extended to at least 50 000 km beyond the visible boundary. Giotto suffered no damage, although it was hit by a particle about 3 mm across. Giotto is still in solar orbit, although it does not retain sufficient gas to send it on to another comet, as had originally been hoped.

On February 1999 NASA launched a new mission, Stardust, to rendezvous with comet P/Wild 2 in 2004; it is hoped that samples can be collected and returned to Earth in a capsule. Other missions are being planned, although, as usual, funding is always a problem.

BRILLIANT COMETS

Brilliant comets have been seen now and then all through the historical period, although early reports, most of them Chinese, are bound to be rather vague. A selected list of bright comets between the years 1500 and 1900 is given in Table 14.8. (In fact Sarabat's Comet of 1729 may have been the largest ever observed, but it was never less than 4.05 a.u. from the Sun and so did not become bright in our skies.)

The Great Comet of 1744 attained magnitude −7, and was visible in broad daylight when only 12° from the Sun. At perihelion it was only 33 000 000 km from the Sun, well inside the orbit of Mercury, and it had at least six bright, broad tails. The Great Comet of 1811, discovered by Honoré Flaugergues on 25 March, was also a daylight

Table 14.8. Selected list of brilliant comets, 1500–1900.

Comet	Name	Discovery	Perihelion	Maximum magnitude	Naked-eye visibility	
1577	Tycho Brahe	1577 Nov 1	1577 Oct 27	−4	1577 Nov–1578 Jan	Possibly brighter than −4.
1585	—	1585 Oct 13	1585 Oct 8	−4	1585 Oct–Nov	Discovered by Chinese.
1665	—	1665 Mar 27	1665 Apr 24	−4?	1665 Mar–Apr	Observed by Hevelius.
1677	Hevelius	1677 Apr 27	1677 May 6	−4?	1677 Apr–May	Long, thin tail.
1695	Jacob	1695 Oct 28	1695 Oct 23	−3?	1695 Oct–Nov	40° tail; probably a Sun-grazer.
1702	—	1702 Feb 20	1702 Feb 15?	?	1702 Feb–Mar	42° tail. Discovered at Cape.
1744	de Chéseaux	1743 Nov 29	1744 Mar 1	−7	1743 Dec–1744 Mar	Multi-tailed comet. Discovered by Klinkenberg, independently by de Chéseaux.
C/1811 F1 (1811 I)	Flaugergues	1811 Mar 25	1811 Sept 12	0	1811 Mar–1812 Jan	Great Comet; 20' coma, 24° ion tail.
C/1819 N1 (1819 II)	Tralles	1819 July 2	1819 June 28	1	1819 July	Transited Sun (unobserved), 26 June.
C/1843 D1 (1843 I)	Great Comet	1843 Feb 8	1843 Feb 27	−6	1843 Feb–Apr	Brighter than Comet of 1811. Sun-grazer.
C/1858 L1 (1858 VI)	Donati	1858 June 2	1858 Sept 30	−1	1858 June–Nov	Most beautiful of all comets; ion and dust tails, up to 60°.
C/1861 N1 (1861 II)	Tebbutt	1861 May 13	1861 June 12	−2	1861 May–Aug	Earth passed through the 100° tail on June 30.
C/1874 H1 (1874 III)	Coggia	1874 Apr 17	1874 July 9	−1	1874 Apr–Aug	63° tail.
C/1880 C1 (1880 I)	Great Comet	1880 Feb 1	1880 Jan 28	3	1880 Feb	Southern hemisphere comet. Sun-grazer.
C/1881 K1 (1881 III)	Tebbutt	1881 May 22	1881 June 16	1	1881 May–July	20° tail.
C/1882 F1 (1882 I)	Wells	1882 Mar 18	1882 June 11	0	1882 May–June	Yellow colour pronounced.
C/1882 R1 (1882 II)	Great Comet (Cruls)	1882 Sept 1	1882 Sept 14	−4	1882 Sept–1883 Feb	Transited Sun; perhaps as bright as magnitude −10 (transit unobserved). Sun-grazer.
C/1887 B1 (1887 I)	Great Comet	1887 Jan 18	1887 Jan 11	2	1887 Jan	'Headless' comet. Long, narrow tail.

object; it had a coma about 2 000 000 km in diameter, and a tail which extended for 160 000 000 km. *En passant*, the wine crop in Portugal was particularly good, and for years afterwards 'Comet Wine' appeared in the price lists of wine merchants. A bottle was sold at Sotheby's, in London, in 1984. (It would be interesting to know what it must have tasted like.)

The brightest comet of modern times was probaby that of 1843. According to the famous astronomer Sir Thomas Maclear, it surpassed the comet of 1811, and Maclear saw both. Donati's Comet of 1858 was said to be the most beautiful of all; it was discovered by G. Donati, from Florence, on 2 June 1858 and was finally lost on 4 March 1859. It had a wonderfully curved main tail and two smaller ones; the tail length was around 80 000 000 km. The period is unknown, but may be of the order of 2000 years.

Tebbutt's Comet of 1861 was brilliant, and it seems that the Earth passed through its tail on 30 June. Despite some unconfirmed reports of an unusual daytime darkness and a yellowish sky, no unusual phenomena were seen.

The first photograph of a comet (Donati's) was taken on 27 September 1858 by an English portrait artist, Usherwood, with a $f/2.4$ focal ratio portrait lens; but the first really good picture was taken in 1882 of Cruls' Comet, at the instigation of Sir David Gill. Many stars were also

shown, and it was this picture which made David Gill, Director of the Cape Observatory, appreciate the endless potentialities of stellar photography. Earlier in 1882, on 17 May, a comet was found on an image of the total eclipse of the Sun, seen from Egypt. The comet had never been seen before, and it was never seen again, so that this is the only record of it; it is generally referred to as Tewfik's Comet, in honour of the Khedive, ruler of Egypt at the time, who had made the astronomers very welcome.

THE TWENTIETH CENTURY

The last century has been relatively poor in brilliant comets; only those of 1910 and 1965 have come anywhere near to matching the splendour of the shadow-casting comets of the Victorian era. A list of selected bright 20th-century comets is given in Table 14.9.

The Daylight Comet of 1910 was first seen on 13 January by some diamond miners in South Africa. It passed perihelion on 17 January, and earlier had been seen with the naked eye when only 4.5° from the Sun. It was much brighter than Halley's – and people who claim to have seen Halley's Comet in 1910 usually saw the Daylight Comet instead. Its orbit is elliptical, but the period seems to be of the order of 4 000 000 years. The bright comet of 1948

Table 14.9. Bright naked-eye comets, 1900–2000.

Designation New	Designation Old	Name	Naked-eye visibility	Maximum magnitude
C/1901 G1	1901 I	Viscara	1901 Apr–May	−1.5
C/1910 A1	1910 I	Daylight Comet	1910 Jan–Feb	−4
1P	1910 II	Halley	1910 Feb–July	0
C/1911 S3	1911 IV	Beljawsky	1911 Sept–Oct	1
C/1911 O1	1911 V	Brooks	1911 Aug–Nov	2
C/1927 X1	1927 IX	Skjellerup–Maristany	1927 Dec–1928 Jan	−6
C/1941 B2	1941 IV	de Kock–Paraskevopoulos	1941 Jan–Feb	2
C/1947 X1	1947 XII	Southern Comet	1947 Dec	−1
C/1948 VI	1948 XI	Eclipse Comet	1948 Nov–Dec	−2
C/1956 R1	1957 III	Arend–Roland	1957 Mar–May	1
C/1957 P1	1957 V	Mrkós	1957 July–Sept	1
C/1961 O1	1961 V	Wilson–Hubbard	1961 July–Aug	3
C/1962 C1	1962 III	Seki–Lines	1962 Feb–Apr	−2.5
C/1965 S1	1965 VIII	Ikeya–Seki	1965 Oct–Nov	−10
C/1969 Y1	1970 II	Bennett	1970 Feb–May	0.5
C/1970 K1	1970 VI	White–Ortiz–Bolelli	1970 May–June	0.5
C/1973 E1	1973 XII	Kohoutek	1973 Nov–1974 Jan	0
C/1975 VI	1976 VI	West	1976 Feb–Apr	−2
P/1	—	Halley	1986 Jan–Dec	1
C/1996 B2	—	Hyakutake	1996 Mar–May	−0.2
C/1995 O1	—	Hale–Bopp	1996 July–1997 Oct	−1

was, like Tewfik's, discovered fortuitously during a total solar eclipse, but was subsequently followed and remained under observation until April 1949, when it had faded to the 17th magnitude. It will return in around 95 000 years. Of course, all periods of this order are very uncertain; some estimated values are given in Table 14.10.

The brightest 20th century comet was that of 1965, Ikeya-Seki, discovered on 18 September by two Japanese observers. It was a daylight object, and could be seen when only 2° from the Sun, but it faded quickly, and was never really well seen from Britain. The period has been given as 880 years.

Other fairly conspicuous comets were Arend–Roland (1957), Bennett (1970) and West (1975). Kohoutek's Comet of 1973 was a disappointment. It was found on 7 March by L. Kohoutek, from Hamburg, when it was still 700 000 000 km from the Sun. Few comets are detectable as far away as this, and the comet was expected to become a magnificent object in the winter of 1973–4, but it failed

to come up to expectations even though it was visible with the naked eye. It was, however, scientifically important, and was carefully studied by the astronauts then aboard the US space-station Skylab (Carr, Gibson and Pogue). Perhaps it will do better when it next returns to the Sun, in approximately 75 000 years' time.

Two splendid comets were seen near the close of the millennium. The first, C/1996 B2, was discovered on 30 January by the Japanese amateur Yuji Hyakutake. It passed perihelion on 1 May, and was then striking in the far north of the sky; it had a long tail – the length was subsequently found to be over 500 million km as found by the Ulysses space-probe which passed through it in 2000. Its beauty was enhanced by its greenish colour. It was in fact a very small comet, and owed its brilliance to its closeness to the Earth. Its original period seems to have been about 8000 years, but its orbit was altered during its journey through the inner Solar System, and the next return is likely to be postponed for 14 000 years.

Table 14.10. Comets of very long period. Obviously, the periods are very uncertain!

Comet	Year	Period (years)	Perihelion distance, q (a.u.)	Eccentricity	Inclination (°)
Great Comet	1861 II	409	0.822	0.985	85
Great Comet	1843 I	517	0.0055	0.999 91	14
Great Comet	1882 II	759	0.0077	0.999 91	14
Ikeya–Seki	1965 VIII	880	0.008	0.9999	14
Pereyra	1963 V	903	0.0051	0.999 95	14
Bennett	1970 II	1678	0.538	0.996	90
Donati	1858 VI	1951	0.578	0.996	11
Flaugergues	1811 I	3096	1.035	0.995	10
Hale–Bopp	1997	2360	0.913	0.9951	89
1680 Comet	1680	8917	0.006	0.9999	6
Hyakutake	1996	14 000	0.230	0.999	12.5

Comets now in hyperbolic orbits include Morehouse (1908), Arend–Roland (1957) and Kohoutek (1973).

Table 14.11. Selected list of Kreutz sun-grazing comets.

Comet	Name	Perihelion date	Perihelion distance (a.u.)	Magnitude
1106[a]	—	1106 Feb 2[a]	?	−5[a]
1668[a]	—	1668 Mar 1[a]	?	0[a]
1689[a]	—	1689 Sept 2[a]	?	3[a]
1695[a]	—	1695 Oct 23[a]	?	?
1702[a]	—	1702 Feb 15[a]	?	?
c/1843 D1	Great Comet	1843 Feb 27.9	0.0055	−6[a]
c/1880 C1	—	1880 Jan 28.1	0.0055	3
X/1882 K1	Tewfik	1882 May 17.5	?	−1?
C/1882 R1	—	1882 Sept 17.7	0.0077	−4
C/1887 B1	—	1887 Jan 11.9	0.0048	2
C/1945 X1	du Toit	1945 Dec 28.0	0.0075	7
C/1963 R1	Pereyra	1963 Aug 24.0	0.0051	2
c/1965 S1	Ikeya–Seki	1965 Oct 21.2	0.0078	−10
C/1970 K1	White–Ortiz–Bolelli	1970 May 14.5	0.0089	0.5
C/1979 Q1	Howard–Kooman–Michels (SOLWIND 1)	1979 Aug 30.9	Impacted	−4

Between 1979 and 1999 SOLWIND discovered six Sun-grazers, SMM (Solar Maximum Mission satellite) discovered 10, and SOHO (the Solar and Heliospheric Observatory satellite) discovered 46. By March 2000 SOHO had discovered over 100 comets, many of them very close to the sun.

[a] Very uncertain.

On 23 July 1995 two American astronomers, Alan Hale and Thomas Bopp, independently discovered the comet which was destined to become the most celebrated of recent years. Had it come as close to us as Hyakutake had done, it would have cast shadows; it was an exceptionally large comet – the nucleus was at least 40 km in diameter – and there were both plasma and dust tails, plus a third inconspicuous tail made up of sodium. By ill-fortune it never came near us; its perihelion distance from the Sun was 0.914 a.u., and its minimum distance from Earth, on 22 March 1997, was 1.3 a.u. – nearly 200 million km. However, it remained a naked-eye object for well over a year, and was truly beautiful – it must be the most photographed comet in history. There were marked changes in the tails, and a spiral structure in the coma. Apparently it was last at perihelion 4200 years ago, and will be back in 2360 years' time. Its orbital inclination is over 89°, so that its path lies almost at right angles to that of the Earth. Perihelion was passed on 1 April 1997. The axial rotation period was given as 11.4 h.

SUN-GRAZING COMETS

Some comets pass very close to the Sun; such were the comets of 1843 and 1965. During the last century H. Kreutz suggested that these 'sun-grazers' might be the remnants of a single giant comet which broke up near its perihelion, and the sun-grazers are often referred to as Kreutz comets. Some of them may hit the Sun, and are quickly destroyed; such was C/1979 Q1 (Howard–Kooman–Michels), on 31 August of that year. Its last moments were recorded by the SOLWIND satellite. Others have been recorded since, and it now seems that 'kamikaze' comets are relatively common. A selected list of Kreutz Comets is given in Table 14.11.

On 1 and 2 June 1998 the SOHO satellite recorded two comets plunging into the Sun, one after the other. By now the SOHO instruments have discovered over 100 comets, including a number of Sun-grazers.

THE ORIGIN OF COMETS

Comets are very ancient objects – as old as the Solar System itself. Since they lose material at every return to perihelion, it follows that the comets we now see cannot have remained in their present orbits for thousands of millions of years. They must have come from afar. They are almost certainly bona-fide members of the Solar System. If they came from interstellar space, they would move at greater velocities than are actually found.

In 1950 the Dutch astronomer J. H. Oort suggested that comets come from a cloud of bodies moving round the sun at between 30 000 and 50 000 a.u. from the Sun – that is to say, around 1 light-year (a light-year is equal to 63 240 a.u.). He proposed that the total cloud population could be as much as 200 000 million, with a total mass up to 100 times that of the Earth. If one of these primordial ice-rich bodies – never incorporated into a planet – were perturbed for any reason, perhaps by the pull of a passing star, it would start to fall in toward the Sun. It might then swing round the Sun and return to the Oort Cloud; it might be expelled from the Solar System altogether; it might be destroyed by collision with the Sun or a planet, or it might be forced into a short-period orbit. Today the existence of the Oort Cloud is generally accepted (it has also been referred to as the Öpik–Oort cloud, since a much less definite suggestion had been made by the Estonian astronomer E. J. Öpik), but it is now thought that though the long-period comets do come from the Oort Cloud, shorter-period comets – such as those of the Jupiter family – come from the Kuiper Belt, a disk-shaped region beyond Neptune, between about 30 and 100 a.u. from the Sun. Asteroidal-sized bodies have indeed been found in this region of the Solar System, and it is true that the distinction between comets and what are generally termed asteroids is much less clear-cut than was previously believed. The existence of this belt was proposed by G. P. Kuiper, although a much less definite suggestion had been made in 1943 by K. Edgeworth.

It may be that the Oort Cloud objects were formed closer to the Sun than the Kuiper Belt objects. Low-mass objects formed near the giant planets would have been ejected by gravitational encounters and sent to great distances, whereas Kuiper Belt objects, formed further out, were not so affected.

LIFE IN COMETS?

The 'panspermia' theory was due to the Swedish scientist Svante Arrhenius, whose work was good enough to win him the Nobel Prize for Chemistry in 1903. Arrhenius believed that life was brought to the Earth by way of a meteorite, but the theory never became popular, because it seemed to raise more problems than it solved. The same sort of theme has been followed up recently by Sir Fred Hoyle and C. Wickramasinghe, who believe that comets can actually deposit harmful bacteria in the Earth's upper air, thereby causing epidemics. Again there has been little support.

Table 14.12. Periodical Comets, Numbers 1–140. Of these comets, 18P Perrine–Mrkos, 34P Gale, 39P Oterma and 54P di Vico–Swift have been lost or at least mislaid. The missing numbers were filled by comets now given a D designation as being permanently lost or destroyed; 3 Biela, 5 Brorsen, 11 Tempel–Swift, 20 Westphal and 25 Neujmin 2.

Designation: P	Name	Discovery	Period (years)	Designation: P	Name	Discovery	Period (years)
1	Halley	240 BC	76.00	74	Smirnova–Chernykh	1975	8.57
2	Encke	1786	3.28	75	Kohoutek	1975	6.65
4	Faye	1843	7.34	76	West–Kohoutek–Ikemura	1975	6.41
6	D'Arrest	1851	6.30	77	Longmore	1975	6.98
7	Pons–Winnecke	1819	6.38	78	Gehrels 2	1973	7.94
8	Tuttle	1790	13.51	79	du Toit–Hartley	1945	5.21
9	Tempel 1	1867	5.50	80	Peters–Hartley	1846	8.13
10	Tempel 2	1873	5.48	81	Wild 2	1978	6.37
12	Pons–Brooks	1812	70.92	82	Gehrels 3	1975	8.11
13	Olbers	1815	69.56	83	Russell 1	1979	6.10
14	Wolf	1884	8.25	84	Giclas	1978	6.96
15	Finlay	1886	6.95	85	Boethin	1975	11.2
16	Brooks 2	1889	6.89	86	Wild 3	1980	6.91
17	Holmes	1892	7.09	87	Bus	1981	6.52
18	Perrine–Mrkos	1896	6.72	88	Howell	1981	5.58
19	Borrelly	1904	6.88	89	Russell 2	1980	7.38
21	Giacobini–Zinner	1900	6.61	90	Gehrels 1	1972	15.1
22	Kopff	1906	6.45	91	Russell 3	1983	7.50
23	Brorsen–Metcalf	1847	70.54	92	Sanguin	1977	12.50
24	Schaumasse	1911	8.22	93	Lovas 1	1980	9.09
26	Grigg–Skjellerup	1902	5.10	94	Russell 4	1984	6.57
27	Crommelin	1818	27.41	95	Chiron (Asteroid 2060)	1977	50.78
28	Neujmin 1	1913	18.21	96	Machholz 1	1986	5.24
29	Schwassmann–Wachmann 1	1927	14.85	97	Metcalf–Brewington	1906	7.76
30	Reinmuth 1	1928	7.31	98	Takamizawa	1984	7.22
31	Schwassmann–Wachmann 2	1929	6.39	99	Kowal 1	1977	15.02
32	Comas Solà	1926	8.83	100	Hartley 1	1985	6.02
33	Daniel	1909	7.06	101	Chernykh	1977	14.0
34	Gale	1927	11.0	102	Shoemaker 1	1984	7.26
35	Herschel–Rigollet	1788	155	103	Hartley 2	1986	6.26
36	Whipple	1933	8.53	104	Kowal 2	1979	6.39
37	Forbes	1929	6.13	105	Singer–Brewster	1986	6.43
38	Stephen–Oterma	1867	37.70	106	Schuster	1977	7.26
39	Oterma	1942	7.88	107	Wilson–Harrington	1949	4.29
40	Väisälä 1	1939	10.8		(Asteroid 4015)		
41	Tuttle–Giacobini–Kresák	1858	5.46	108	Ciffreo	1985	7.23
42	Neujmin 3	1929	10.63	109	Swift–Tuttle	1862	135.01
43	Wolf–Harrington	1924	6.51	110	Hartley 3	1988	6.84
44	Reinmuth 2	1947	6.64	111	Helin–Roman–Crockett	1989	8.16
45	Honda–Mrkos–Pajdusaková	1948	5.30	112	Urata–Niijima	1986	6.64
46	Wirtanen	1948	5.50	113	Spitaler	1890	7.10
47	Ashbrook–Jackson	1948	7.49	114	Wiseman–Skiff	1986	6.53
48	Johnson	1949	6.97	115	Maury	1985	8.74
49	Arend–Rigaux	1951	6.82	116	Wild 4	1990	6.16
50	Arend	1951	7.99	117	Helin–Roman–Alu 1	1989	9.50
51	Harrington	1953	8.78	118	Shoemaker–Levy 4	1991	6.51
52	Harrington–Abell	1955	7.59	119	Parker–Hartley	1989	8.89
53	van Biesbroeck	1954	12.43	120	Mueller 1	1987	8.41
54	di Vico–Swift	1844	6.31	121	Shoemaker–Holt 2	1989	8.05
55	Tempel–Tuttle	1865	32.9	122	di Vico	1846	74.36
56	Slaughter–Burnham	1958	11.59	123	West–Hartley	1989	7.57
57	du Toit–Neujmin–Delporte	1941	6.39	124	Mrkós	1991	5.64
58	Jackson–Neujmin	1936	8.24	125	Spacewatch	1991	5.57
59	Kwerns–Kwee	1963	8.96	126	IRAS	1983	13.29
60	Tsuchinshan 2	1965	6.82	127	Holt–Olmstead	1990	6.16
61	Shajn–Schaldach	1949	7.49	128	Shoemaker–Holt 1	1987	9.55
62	Tsuchinshan 1	1965	6.65	129	Shoemaker–Levy 3	1991	7.25
63	Wild 1	1960	13.3	130	McNaught–Hughes	1991	6.71
64	Swift–Gehrels	1889	9.21	131	Mueller 2	1990	7.05
65	Gunn	1970	6.83	132	Helin–Roman–Alu 2	1989	8.24
66	du Toit	1944	15.0	133	Elst–Pizarro (Asteroid 7968)	1996	5.61
67	Churyumov–Gerasimenko	1969	6.59	134	Kowal–Vavrova	1983	15.58
68	Klemola	1965	10.95	135	Shoemaker–Levy 8	1992	7.50
69	Taylor	1915	6.97	136	Mueller 3	1990	8.71
70	Kojima	1970	7.85	137	Shoemaker–Levy 2	1990	9.38
71	Clark	1973	5.50	138	Shoemaker–Levy 7	1991	6.73
72	Denning–Fujikawa	1881	9.01	139	Väisälä–Oterma	1979	9.55
73	Schwassmann–Wachmann 3	1930	5.34	140	Bowell–Skiff	1983	16.18

Table 14.13. Selected list of periodical comets.

Comet	Period (years)	Perihelion distance, q (a.u.)	Aphelion distance, Q (a.u.)	Eccentricity	Inclination (°)	Absolute magnitude
2 Encke	3.28	0.33	2.21	0.850	11.9	11
107 Wilson–Harrington (Asteroid 4015)	4.29	1.00	2.64	0.622	2.8	16
26 Grigg–Skjellerup	5.10	0.995	2.96	0.664	6.6	12
79 du Toit–Hartley	5.21	1.20	3.01	0.602	2.9	
96 Machholz 1	5.24	0.12	3.02	0.959	60.1	
10 Tempel 2	5.47	1.48	3.10	0.552	12.0	10
45 Honda–Mrkós–Pajdusaková	5.30	0.54	5.54	0.922	4.2	11
73 Schwassmann–Wachmann 3	5.35	0.93	3.06	0.695	11.4	11
41 Tuttle–Giacobini–Kresák	5.46	1.07	3.10	0.656	9.2	11
46 Wirtanen	5.46	1.07	3.10	0.657	11.7	16
9 Tempel 1	5.51	1.50	3.12	0.52	10.5	9
71 Clark	5.51	1.55	3.12	0.502	9.5	12
125 Spacewatch	5.56	1.54	3.14	0.36	10.4	
88 Howell	5.57	1.41	3.14	0.55	4.3	
133 Elst–Pizarro (Asteroid 7968)	5.61	2.62	3.67	0.166	1.4	14
100 Hartley 1	6.02	1.82	3.31	0.450	25.7	
116 Wild 4	6.16	1.99	2.36	0.407	3.7	
37 Forbes	6.13	1.44	3.34	0.578	7.2	10
104 Kowal 2	6.18	1.40	3.37	0.585	15.5	
103 Hartley 2	6.28	0.95	3.40	0.720	9.3	
127 Holt–Olmstead	6.33	2.15	3.42	0.370	17.7	
81 Wild 2	6.33	1.57	3.42	0.540	3.2	6
7 Pons–Winnecke	6.37	1.26	3.44	0.634	22.3	14
57 du Toit–Neujmin–Delporte	6.39	1.72	3.44	0.501	2.9	14
31 Schwassmann–Wachmann 2	6.39	2.07	3.44	0.399	3.8	11
105 Singer–Brewster	6.44	2.03	3.46	0.413	9.2	
76 West–Kohoutek–Ikemura	6.46	1.58	3.47	0.540	30.5	10
118 Shoemaker–Levy 4	6.51	2.02	3.49	0.420	8.5	
43 Wolf–Harrington	6.51	1.61	3.49	0.539	9.3	
6 D'Arrest	6.51	1.35	3.49	0.614	19.5	6
87 Bus	6.52	2.18	3.49	0.375	2.6	
94 Russell 4	6.58	2.23	3.51	0.365	6.2	
83 Russell 1	6.10	1.61	5.06	0.517	22.7	15
67 Churyumov–Gerasimenko	6.59	1.30	3.51	0.630	7.1	10
21 Giacobini–Zinner	6.61	1.03	3.52	0.706	31.86	10
49 Arend–Rigaux	6.61	1.37	3.52	0.611	18.2	9
62 Tsuschinshan 1	6.64	1.50	3.53	0.571	10.5	14
44 Reinmuth 2	6.64	1.89	3.53	0.454	7.0	10
75 Kohoutek	6.67	1.78	3.54	0.496	5.9	
130 McNaught–Hughes	6.69	2.12	3.55	0.40	18.29	
51 Harrington	6.78	1.57	3.58	0.561	8.7	15
19 Borrelly	6.80	1.37	3.59	0.623	30.2	13
60 Tsuchinshan 2	6.82	1.78	3.60	0.504	3.6	14
65 Gunn	6.83	2.46	3.59	0.306	5.5	13
110 Hartley 3	6.88	2.48	3.62	0.314	11.7	
16 Brooks 2	6.89	1.84	3.62	0.49	5.5	13
138 Shoemaker–Levy 7	6.89	1.76	3.62	0.531	10.1	
86 Wild 3	6.91	2.30	3.63	0.366	15.5	
15 Finlay	6.95	1.09	3.64	0.699	3.7	13
84 Giclas	6.96	1.85	3.65	0.494	7.3	
48 Johnson	6.97	2.30	4.98	0.367	13.7	10
69 Taylor	6.97	1.95	3.65	0.466	20.6	12
77 Longmore	6.98	2.40	3.65	0.343	26.4	
131 Mueller 2	7.05	2.41	3.68	0.344	14.1	
33 Daniel	7.06	1.65	3.68	0.551	20.1	11
17 Holmes	7.09	2.16	3.68	0.412	19.2	13
113 Spitaler	7.10	1.82	5.06	0.471	12.8	16
98 Takamizawa	7.21	1.57	3.73	0.575	0.49	
102 Shoemaker 1	7.25	1.98	3.75	0.471	26.3	
108 Ciffreo	7.25	1.71	3.74	0.542	13.1	
129 Shoemaker–Levy 3	7.25	2.82	3.75	0.248	5.01	
106 Schuster	7.29	1.55	3.76	0.688	20.1	
30 Reinmuth 1	7.31	1.87	3.77	0.502	8.1	14
54 di Vico–Swift	7.32	2.15	3.77	0.431	6.1	

Table 14.13. (Continued)

Comet	Period (years)	Perihelion distance, q (a.u.)	Aphelion distance, Q (a.u.)	Eccentricity	Inclination (°)	Absolute magnitude
4 Faye	7.34	1.59	3.78	0.578	9.1	8
89 Russell 2	7.38	2.28	3.79	0.40	12.0	
61 Shajn–Schaldach	7.46	2.32	5.31	0.390	6.1	12
47 Ashbrook–Jackson	7.46	2.30	3.81	0.396	12.5	7
52 Harrington–Abell	7.53	1.75	3.84	0.542	10.2	16
123 West–Hartley	7.59	2.13	3.86	0.447	15.3	
83 Russell 1	7.64	2.18	3.88	0.437	17.7	9
78 Gehrels 2	7.94	2.37	5.62	0.409	0.9	
70 Kojima	7.85	2.41		0.39		
121 Shoemaker–Holt 2	8.05	2.66	4.02	0.337	17.7	
82 Gehrels 3	8.45	3.62	4.15	0.125	1.1	9
80 Peters–Hartley	8.12	1.62	4.02	0.598	19.9	8
111 Helin–Roman–Crockett	8.16	3.49	4.04	0.139	4.2	
50 Arend	8.24	1.91	4.08	0.530	19.2	14
58 Jackson–Neujmin	8.24	1.38	4.08	0.661	13.7	17
14 Wolf	8.21	2.41	4.07	0.407	27.5	13
24 Schaumasse	8.22	1.20	4.07	0.705	11.9	11
120 Mueller 1	8.41	2.74	4.14	0.337	8.8	
36 Whipple	8.53	3.09	4.17	0.239	9.9	
74 Smirnova–Cbernykh	8.57	3.57	4.19	0.147	6.6	8
136 Mueller 3	8.71	3.01	4.23	0.289	9.4	
31 Schwassmann–Wachmann 2	8.72	3.42	4.23	0.195	4.5	
32 Comas Solà	8.83	1.85	4.27	0.568	12.9	8
119 Parker–Hartley	8.89	3.05	4.29	0.290	5.2	
72 Denning–Fujikawa	9.03	0.79	4.34	0.818	9.1	11
93 Lovas 1	9.14	1.69	4.37	0.613	12.2	
64 Swift–Gehrels	9.21	1.36	4.39	0.691	9.3	15
104 Kowal–Mrkós	9.24	2.67	4.40	0.394	5.3	
137 Shoemaker–Levy 2	9.38	1.87	4.45	0.580	4.7	
128 Shoemaker–Holt 1	0.51	3.05	4.49	0.321	4.4	
139 Väisälä–Oterma	9.54	3.38	4.58	0.048	2.4	
117 Helin–Roman–Alu 1	9.57	3.71	4.51	0.176	9.7	
59 Kwerns–Kwee	9.45	2.34	4.45	0.58	9.4	11
42 Neujmin 3	10.63	2.00	4.83	0.586	4.0	14
68 Klemola	10.82	1.75	4.89	0.641	11.1	
40 Väisälä 1	10.90	1.80	8.02	0.633	11.6	13
56 Slaughter–Burnham	11.6	2.54	7.71	0.504	8.2	14
85 Boethin	11.63	1.16	5.13	0.774	4.9	10
53 Van Biesbroeck	12.43	2.40	5.37	0.552	6.6	7
92 Sanguin	12.50	1.81	5.39	0.663	18.7	
126 IRAS	13.30	1.70	5.61	0.697	46.0	
63 Wild 1	13.3	1.98	9.24	0.647	9.2	14
8 Tuttle	13.51	0.997	5.67	0.824	54.7	8
101 Chernykh	13.97	2.35	5.80	0.593	5.1	
99 Kowal 1	15.08	4.67	6.20	0.234	4.4	
29 Schwassmann–Wachmann 1	14.85	5.77	6.04	0.045	9.4	16
66 du Toit	15.0	1.294	10.9	0.787	18.7	16
134 Kowal–Vavrova	15.57	2.58	6.23	0.587	4.34	
140 Bowell–Skiff	16.2	1.97	6.34	0.691	3.8	
28 Neujmin 1	18.2	1.55	12.3	0.776	12.3	10
39 Oterma	19.5	5.47	7.24	0.245	1.9	9

Comet	Period (years)	Perihelion distance, q (a.u.)	Aphelion distance, Q (a.u.)	Eccentricity	Inclination (°)	Next return	Absolute magnitude
27 Crommelin	27.4	0.74	17.4	0.919	19.1	2011	11
55 Tempel–Tuttle	33.2	0.98	10.3	0.905	162.5	2031	13
38 Stephan–Oterma	37.7	1.57	20.9	0.860	18.0	2018	5
13 Olbers	69.6	1.18	32.6	0.930	44.6	2024	5
23 Brorsen–Metcalf	70.6	0.48	17.8	0.972	19.3	2059	9
12 Pons–Brooks	70.9	0.77	33.5	0.955	74.2	2024	6
1 Halley	76.0	0.587	35.3	0.967	162.2	2061	4
109 Swift–Tuttle	135.0	0.96	51.7	0.964	113.4	2127	4
35 Herschel–Rigollet	155	0.75	56.9	0.974	64.2	2092	8

15 METEORS

Meteors are cometary débris. They are small and friable – usually no more than of centimetre size – and so never reach the Earth's surface intact. There are many well-defined showers, associated with comets which can often be identified; other meteors are sporadic, not associated with any known comet, and so may appear from any direction at any moment. Meteors can, of course, occur in daylight, as was pointed out by the Roman philosopher Seneca about 20 AD, and may be tracked by radio and radar.

Meteors are not associated with meteorites, which come from the asteroid belt. The link with comets was first proposed in 1861 by D. Kirkwood; he believed that meteors were the remnants of comets which have disintegrated – and in some cases this is true enough. In 1862 G. V. Schiaparelli demonstrated the link between the Perseid meteor shower and the periodical comet Swift–Tuttle, and other associations were soon established.

Some well-known periodical comets are the parents of meteor showers. Halley's Comet produces two, the η Aquarids of April and the Orionids of October; Comet P/Giacobini–Zinner can occasionally yield rich displays, as in 1933. Biela's Comet, which broke up and was last seen in 1852, produced 'meteor storms' in 1872 and in 1885; in recent years this shower (the Andromedids) has been almost undetectable, but it has been calculated that it may return around 2120, when the orbit of the stream will be suitably placed. The Lyrids, first recorded in 687, are linked with Thatcher's Comet of 1862, which has a period of over 400 years. The rich Geminid shower of December has an orbit very like that of asteroid 3100 Phæthon, and it is widely believed that Phæthon may be the parent of the stream, adding credibility to the suggestion that some near-Earth asteroids may be extinct comets.

EARLY THEORIES

Meteors were once regarded as atmospheric phenomena. Aristotle believed them to be due to vapours from Earth created by the warmth of the Sun; when they rose to great altitudes they caught fire, either by friction or because the column of air around them cooled, so squeezing out the hot vapours rather as toothpaste can be squeezed out of a tube. Even Newton believed that meteors were volatile gases which, when mixed with others, ignited to cause 'Lightning and Thunder and fiery Meteors'. Edmond Halley correctly maintained that they came from space and burned away in the upper air (although, curiously, he seems to have changed his mind later and reverted to the Aristotelian picture). Myths abounded. The Mesopotamians regarded meteors as evil portents, and to the Moslems they represented artillery in a war between devils and angels. In Sparta, around 1200 BC, the priests surveyed the sky on one special night once in eight years; if a meteor were seen, it indicated that the king had sinned and ought to be deposed. In mediæval Brunswick a meteor was a fiery dragon which could cause damage; however, if the observer sheltered and cried out 'Fiery Dragon, come to me', the dragon might relent, and even drop down a ham or a side of bacon!

NATURE OF METEORS

The status of meteors was solved in 1798 by two German students, H. W. Brandes and J. F. Benzenberg, of the University of Göttingen. Between 11 September and 4 November they observed meteors from sites 15.2 km apart, giving them a useful 'baseline', and made 402 measurements; in 22 cases they found that the same meteor had been seen from each site, and its track plotted. This made it possible to determine the height of the meteor by the method of triangulation. The heights at which the meteors disappeared ranged between 15 km and 226 km; the mean burnout altitude was found to be 89 km – now known to be very near the truth.

The total number of meteors entering the atmosphere daily has been given as 75 000 000 for meteors of magnitude 5 or brighter. An observer under ideal conditions would expect to see between about 5 and 15 naked-eye meteors per hour (except during a shower, when the number would

Table 15.1. Principal meteor showers.

Name	Begins	Max	Ends	ZNR	RA	Dec	Comet	
Quadrantids	1 Jan	3 Jan	6 Jan	100	15h28m	+50	—	Sharp maximum. Can be spectacular.
Virginids	7 Apr	10 Apr	18 Apr	5	13h36m	−11	—	Slow, long paths. Several radiants in Virgo, Mar–Apr.
Lyrids	19 Apr	22 Apr	25 Apr	10	18h08m	+32	Thatcher	Occasionally very rich, as in 1803, 1922, 1982.
η Aquarids	24 Apr	4 May	20 May	40	22h20m	−01	P/Halley	Multiple radiant, broad maximum.
α Scorpiids	20 Apr	27 Apr	19 May	5	16h32m	−24	—	Several weak radiants. One max on 12 May.
Ophiuchids	19 May	9 June	July	5	17h56m	−23	D/Lexell?	Weak activity from several radiants.
α Cygnids	July	21 July, 12 Aug	Aug	5	21h0m	+48	—	Weak but prolonged activity. Less rich than formerly.
Capricornids	July	8 July, 15, 26 July	Aug	5	20h44m	−15	P/Honda–Mrkós Pajdusakurá	Bright meteors. Three maxima, multiple radiant.
δ Aquarids	15 July	29 July, 6 Aug	20 Aug	20 10	22h36m	−17	—	Double radiant. Rich, but faint meteors.
Piscis Australids	13 July	31 July	20 Aug	5	22h40m	−30	—	Probably double maximum.
α Capricornids	15 July	2 Aug	20 Aug	5	20h36m	−10	—	Slow, yellow fireballs. Triple maximum.
ι Aquarids	July	6 Aug	Aug	8	22h10m	−15	—	Rich in faint meteors. Double radiant.
Perseids	23 July	13 Aug	20 Aug	80	03h04m	+58	P/Swift–Tuttle	Most reliable annual shower. Consistent.
Piscids	Sept	8, 21 Sept 13 Oct	Sept	10	00h36m	+07	—	Weak; multiple radiant.
Orionids	16 Oct	21 Oct	27 Oct	25	06h24m	+15	P/Halley	Swift, with fine trains. Flat maximum.
Draconids	10 Oct	10 Oct	10 Oct	var	18h00m	+54	P/Giacobini–Zinner	Usually weak, but occasional storms. Also known as the Giacobinids.
Taurids	20 Oct	3 Nov	30 Nov	10	03h44m	+14	P/Encke	Fine display in 1988. Slow meteors.
Puppids-Velids	27 Nov	9, 26 Dec	Jan	15	09h00m	−48	— Nov–Jan	Two of several radiants in Puppis, Vela and Carina.
Leonids	15 Nov	18 Nov	20 Nov	var	10h08m	+22	P/Tempel–Tuttle	Occasional storms (1799, 1833, 1866, 1966).
Andromedids	15 Nov	20 Nov	6 Dec	v low	00h50m	+55	D/Biela	Now virtually extinct.
Geminids	7 Dec	14 Dec	17 Dec	75	07h28m	+32	Phaethon? (asteroid)	Many bright meteors. Consistent. Can be even richer than the Perseids.
Ursids	17 Dec	21 Dec	25 Dec	10	14h28m	+78	P/Tuttle	Usually weak, but good displays 1945, 1982, 1986.

Permanent daytime showers include the Arietids (29 Mar–17 June), the ξ Perseids (1–15 June) and the β Taurids (23 June, 7 July). The β Taurids seem to be associated with Encke's Comet.

be higher). Meteors of magnitude −5 or brighter – that is to say, appreciably more brilliant than Venus – are conventionally termed fireballs. Very occasional fireballs, such as those of 20 November 1758 and 18 August 1783, may far outshine the Moon. The 1758 fireball was seen from England, and a contemporary eye-witness report is worth quoting:

'This night a surprising large meteor was seen at Newcastle, about 9 o'clock, which passed a little westward of the town, directly north, and illuminated the atmosphere to that degree, for a minute, that, though it was dark before, a pin might have been picked up in the streets. Its velocity was inconceivably great, and it seemed near the size of a man's head. It had a tail of between two and three yards long, and as it passed, some said that they saw sparks of fire fall from it.'

A meteor may enter the atmosphere at a velocity anywhere between 11 km s^{-1} and 72 km s^{-1}; it will be violently heated as it enters the upper atmosphere at an altitude of 150 km above the ground. It is vaporized; atoms from its outer surface are ablated and collide with molecules in the atmosphere, exciting and ionizing them, producing a trail which may extend for many kilometres. There is little deceleration before the meteor is destroyed. What we see is, therefore, not the particle itself, but the effects which it produces in the atmosphere during the final moments of its existence.

Particles below about 0.1 mm in diameter are termed micrometeorites, and do not produce luminous effects. Some are cometary, while others must be classed as Zodiacal 'dust'.

Meteors are easy to photograph – the earliest really good picture, of an Andromedid, was taken by L. Weinek, from Prague, as long ago as 27 November 1885 – but meteor spectra are much more difficult, because one never knows just when or where a meteor will appear. Many spectra have been obtained (largely by amateurs) and it seems that meteors are made up of material of the type only to be expected in view of their cometary origin.

Radar studies of meteor trails are now of great importance; the first systematic work was carried out in 1945 by J. S. Hey and his team, with the δ Aquarids. However, amateur observations are still very useful indeed.

Table 15.2. Selected list of minor annual meteor showers[a].

Name	Begins	Max	Ends	
ζ Aurigids	11 Dec	31 Dec	21 Jan	Slow meteors.
Boötids	9 Jan	15 Jan	18 Jan	
δ Cancrids	14 Dec	15 Jan	14 Feb	Weak shower.
η Carinids	14 Jan	21 Jan	27 Jan	
η Craterids	11 Jan	16 Jan	22 Jan	Rapid meteors.
ρ Geminids	1 Jan	10 Jan	15 Jan	Ill-defined.
α Hydrids	15 Jan	20 Jan	30 Jan	
Aurigids	31 Jan	7 Feb	23 Feb	Some bright fireballs.
α Centaurids	2 Feb	8 Feb	25 Feb	
δ Leonids	5 Feb	22 Feb	19 Mar	ZHR ∼3.
η Draconids	22 Mar	30 Mar	8 Apr	
β Leonids	14 Feb	20 Mar	25 Apr	ZHR 3–4.
δ Mensids	14 Mar	18 Mar	21 Mar	Weak shower.
γ Normids	11 Mar	16 Mar	21 Mar	
η Virginids	24 Feb	18 Mar	27 Mar	Diffuse.
π Virginids	13 Feb	6 Mar	8 Apr	ZHR 2–5.
θ Virginids	10 Mar	20 Mar	21 Apr	ZHR ∼2.
τ Draconids	13 Mar	31 Mar	17 Apr	
π Puppids	8 Apr	23 Apr	25 Apr	Weak; Comet P/Grigg-Skjellerup.
April Ursids	18 Mar	19 Apr	9 May	
α Virginids	10 Mar	13 Apr	6 May	Diffuse; complex radiants.
γ Virginids	5 Apr	14 Apr	21 Apr	
May Librids	1 Apr	6 May	9 May	
Pons–Winneckeids (June Boötids)	27 June	28 June	5 July	Comet P/Pons–Winnecke. Faint; good in 1921, 1927.
τ Herculids	19 May	9 June	19 June	Comet P/Schwassmann–Wachmann 3?
June Lyrids	10 June	15 June	21 June	(magnitude 3).
θ Ophiuchids	21 May	10 June	16 June	Flat maximum; ∼5 days.
φ Sagittariids	1 June	18 June	15 July	Weak shower.
χ and ω Scorpiids	6 May	4 June	11 July	Weak; diffuse radiants.
Scutids	2 June	27 June	29 July	ZHR ∼3.
α Lyrids	9 July	14 July	20 July	Fast; mainly telescopic.
Phœnicids	9 July	14 July	17 July	Diffuse radiant; ZHR 2.
κ Cygnids	26 July	19 Aug	1 Sept	Complex max.; ZHR 6?.
υ Pegasids	25 July	8 Aug	19 Aug	ZHR 2–5; swift, yellow.
α Ursæ Majorids	9 Aug	13 Aug	30 Aug	Weak shower.
γ Aurigids	1 Sept	7 Sept	14 Sept	ZHR can reach 4.
α Aurigids	25 Aug	1 Sept	6 Sept	ZHR may be 9; good in 1935 and 1986.
October Arietids	7 Sept	8 Oct	27 Oct	One of several radiants.
δ Aurigids	22 Sept	10 Oct	23 Oct	C/Bradfield, 1972 III?.
ε Geminids	10 Oct	18 Oct	27 Oct	ZHR 1–2.
November Monocerotids	13 Nov	21 Nov	2 Dec	
Coma Berenicids	8 Dec	25 Dec?	23 Jan	Weak; diffuse.
December Monocerotids	9 Nov	11 Dec	18 Dec	ZHR no more than 2.
χ Orionids	16 Nov	10 Dec	16 Dec	Bright meteors. ZHR 3.
December Phœnicids	29 Nov	3 Dec	9 Dec	Comet D/Blanpain? Rich in 1956.

[a] All data are rather uncertain. This table is derived from several sources, but mainly from the work of Gary W. Kronk.

METEOR RADIANTS

Because the meteors in any particular shower are moving through space in parallel paths (or virtually so), they seem to come from one set point in the sky, known as the radiant. (The effect may be likened to the view from a bridge overlooking a motorway; the parallel lanes of the motorway will seem to converge at a point near the horizon.) The shower is named after the constellation in which the radiant lies. One exception refers to the January meteors, the Quadrantids; they are named after Quadrans Muralis, a constellation added to the sky in Bode's maps of 1775 but later rejected – its stars are now included in Boötes, but the old name has been retained.

A list of the principal annual showers is given in Table 15.1. A selected list of minor showers is given in Table 15.2, although the low hourly rate of these showers means that the data are decidedly uncertain.

The ZHR, or Zenithal Hourly Rate, is given by the number of naked-eye meteors which would be expected to be seen by an observer under ideal conditions, with the radiant at the zenith. In practice these conditions are never met, so that the observed hourly rate is bound to be rather lower than the theoretical ZHR.

METEOR SHOWERS

On the night of 12–13 November 1833 there was a brilliant meteor shower; the meteors came from the constellation of Leo. It was observed from Yale, in the United States, by Denison Olmsted. H. A. Newton postulated the existence of definite showers; finding that the Leonids had appeared periodically, at intervals of 33 years, so that there should be another major meteor storm in 1866. It duly appeared, although unfortunately Olmsted did not see it (he died in 1859). Subsequently other showers were identified, initially the Perseids in 1834 (by J. Locke and A. Quetelét), the Lyrids in 1835 (by F. Arago), the Quadrantids in 1839 (by Quetelét and E. Herrick), the Orionids in 1839 (by Quetelét, Herrick and J. Benzenberg) and the Andromedids in 1838 (also by Quetelét, Herrick and Benzenberg).

Some of the recognized showers are consistent, notably the Perseids, while others, such as the Leonids, are very variable in richness. Some radiants are ill-defined, and while some showers are of brief duration

– such as the Quadrantids – others spread over weeks.

Material may leave a comet either in front of or behind the nucleus. Dust particles ejected from the nucleus may return to perihelion earlier than the comet itself, or may return later; gradually the material is distributed all around the comet's orbit, forming a loop. With older showers, such as the Perseids, this has had sufficient time to happen; with younger showers it has not, so that good displays are seen only when the Earth passes through the thickest part of the swarm. We must also consider what is termed the Poynting–Robertson effect. In re-radiating energy received from the Sun, a particle will lose orbital velocity and will spiral inward towards the Sun; therefore old streams are depleted in small particles (although even smaller particles are ejected altogether, by radiation pressure).

The Perseid shower of early August is consistent, and any observer who looks up into a dark, clear sky at any time during the first part of the month will be very unlucky not to see a few Perseids. The October Draconids, associated with Comet P/Giacobini–Zinner, are usually sparse, but produced a major storm in 1933, when for a while the ZHR reached an estimated 6000; a weaker but still rich storm occurred in 1946 (this was the first occasion on which meteors were systematically tracked by radar). Nothing comparable from the Draconids has been seen since. It must be remembered that meteor streams are easily perturbed by planets, and no two orbits are exactly alike.

The Leonids can produce the most spectacular storms of all; a selected list is given in Table 15.3 (drawn from the researches carried out by John Mason). In 1833 and 1866 it was said that meteors 'rained down like snowflakes'. No major displays were seen in 1899 and 1933, because the main swarm did not intersect the Earth's orbit at the critical time, but there was another storm in 1966 – unfortunately not seen from Europe, because it occurred during European daylight, but spectacular from parts of North America, such as Arizona.

Comet Tempel-Tuttle returned to perihelion in 1998, and was expected to produce another meteor storm. The predicted date was 17 November, 258 days after the comet had passed through perihelion, but in fact the richest display was seen on 16 November – not a 'storm', but certainly

Table 15.3. Leonid meteor storms.

902	Oct 12–13	South Europe, N Africa
934	Oct 13–14	Europe, N Africa, China
1002	Oct 14–15	China, Japan
1202	Oct 18–19	Japan
1238	Oct 18–19	Japan
1366	Oct 21–22	Europe, China
1533	Oct 25–27	Europe, China, Japan
1566	Oct 26–27	China, Korea
1601	Nov 5–6	China
1666	Nov 6–7	China
1698	Nov 8–9	Europe, Japan
1766	Nov 11–12	South America
1799	Nov 11–12	America
1833	Nov 12–13	North America
1866	Nov 13–14	Europe
1966	Nov 17	North America
1999	Nov 18	Europe, Middle East

Julian calendar dates before Oct 1582, Gregorian dates thereafter. It is of course possible that some storms were not recorded.

striking. It was calculated that the dust stream left behind by the comet does not have uniform cylindrical structure, but consists of a number of discrete, separate arcs of dust, each released at a different return of the comet. If the Earth passes through a thin filament, the meteor shower is brief but intense. If it passes through a broader filament, the shower is less intense, but lasts longer. If the Earth passes through a gap between filaments, the display is weak. If it passes through a broad filament first, and then through the edge of a narrower filament, there will be two peaks of activity.

The great storm of 1833 was caused by a dust trail generated in 1800, 33 years earlier; the 1966 storm was due to dust released from the comet in 1899. The displays of 1998 and 1999 were due to an arc-shaped cloud of dust shed by the comet in 1366. In 1999 there was indeed a meteor storm, peaking at 02 hours GMT on 18 November; if not as splendid as the storms of 1833 and 1866, it was very spectacular, with a peak ZHR of well over 2000. It was of brief duration, but was well seen from cloud-free areas of Europe. From Oban (Scotland) Iain Nicolson found the peak activity to be from 0200 to 0215 GNT, and had declined markedly by 0240. Many of the meteors were very bright, with long, sometimes persistent trains.

During the shower, there were two telescopic reports of flashes on the surface of the Moon, and it was suggested that these might be due to impacting Leonids, but this seems most improbable; a meteor could not produce a visible lunar flash – a meteorite-sized object would be needed, and meteorites are not associated with comets or with meteor showers.

Meteor Sounds?

During the Leonid shower of November 1998, the Croatian astronomer Dejan Vinkovic reported that he had recorded sounds from meteors. These have been reported before, and the deep, thunder-like or hissing noise seems to coincide with the visual appearance of the meteor. This would indicate that the sound-waves could travel at the speed of light, which seems impossible. However, the phenomenon – termed electrophonic sound – can be explained by radio waves, which do travel at the speed of light, interacting with objects at ground level to produce audible noise. Further research into this interesting problem is needed.

Danger From Meteors?

From ground level, meteors – unlike meteorites – are quite harmless. It was suggested that the expected 1998–9 Leonid shower might affect space-craft, such as the Hubble Telescope, but no damage was reported, and all in all it seems that the danger from meteors is not very great.

16 METEORITES

Meteorites reach ground level without being destroyed. They are not simply large meteors; they do not belong to showers, and have no definite association with comets, but seem to come mainly from the asteroid belt. It may well be that there is no difference between a large meteoroid and a small asteroid. The term 'meteorite' is used only for a meteoroid which has landed on Earth.

Normal meteorites are ancient; their ages are given as 4.6 thousand million years – the same as that of the Earth itself. Ages are measured chiefly by the method of radioactive decay. Meteorites contain radioactive isotopes which decay at a known rate; for instance, the half-life of uranium, U-235, is 704 million years. (Half-life indicates the time taken for half of the original material to decay.) U-235 ends up as lead, Pb-206. Half-life periods for materials found in meteorites are given in Table 16.1.

Table 16.1. Major isotopes used to date meteorites.

Parent isotope	Daughter	Half-life (years)
Carbon, C-14	Nitrogen, N-14	5730
Aluminium, Al-26	Magnesium, Mg-26	740 000
Iodine, I-129	Xenon, Xe-129	17 000 000
Uranium, U-235	Lead, Pb-207	704 000 000
Potassium, K-40	Argon, A-40	1300 000 000
Uranium, U-238	Lead, Pb-206	4500 000 000
Thorium, Th-232	Lead, Pb-208	14 000 000 000
Rubidium, Rb-87	Strontium, Sr-87	49 000 000 000

The first three parents in this table are extinct; all the material present when the Earth was formed has decayed. Thorium and uranium isotopes produce helium as well as lead.

It is widely believed that some meteorites come from the Moon (chapter 3) and others, the SNC meteorites, from Mars (chapter 7). It has also been suggested that some meteorites, known as achondrites, come from the asteroid Vesta. However, definite proof is lacking.

The earliest reports of meteoritic phenomena are recorded on Egyptian papyrus, around 2000 BC. Early meteorites falls are, naturally, poorly documented, but it seems that a meteorite fell in Crete in 1478 BC, stones near Orchomenos in Boetia in 1200 BC and an iron meteorite on Mount Ida in Crete in 1168 BC. According to Livy, 'stones' fell on Alban Hill in 634 BC, and there is evidence that in 416 BC a meteorite fell at Ægospotamos in Greece. A meteorite which fell at Nogara, in Japan, in 861 AD was placed in a Shinto shrine, and the Sacred Stone at Mecca is almost certainly a meteorite. The oldest meteorite which can be positively dated fell at Ensisheim, in Switzerland, on 16 November 1492 and is now on show at Ensisheim Church.

In India, it is said that the Emperor Jahangir ordered two sword blades, a dagger and a knife to be made from the Jalandhar meteorite of 10 April 1621. A sword was made form the meteorite which fell in Mongolia in 1670, and in the 19th century part of a South African meteorite was used to make a sword for the Emperor Alexander of Russia. Nowadays there is a flourishing trade in meteorites; for example, in 1999 a chip of the Dar al Gani meteorite, allegedly lunar, was sold at Sothebys in London for £9200. It was 1.75 cm in diameter.

Well over 10 000 meteorites have been identified (it is estimated that in each year the Earth sweeps up about 78 000 tons of extraterrestrial material). Relatively few have been seen to fall. Among famous falls which have resulted in meteorite discovery are those of the Přibram fireball (Czech, which was recorded as being of magnitude −19) on 7 April 1959, the Lost City meteorite (Oklahoma) in 1970, the Barwell meteorite (Leicestershire) on 24 December 1965 and the Sikhote–Alin fall in Siberia on 12 February 1947. Rather surprisingly, there are no authenticated records of any human death due to a meteorite. Reports that a monk was killed at Cremona in 1511, and another monk in Milan in 1650 are unsubstantiated. There have, however, been narrow escapes. In 1954 a woman in Alabama, USA, was disturbed by a meteorite which fell through the roof of her house, and she suffered a minor arm injury. On 21 June 1994 José Martin was driving his car from Madrid to Marbella, in Spain, when a 1.4 kg meteorite crashed through his windscreen, ricocheted off the dashboard and injured the driver's finger, fortunately not seriously. More than 50 fragments were later found within 200 m of the impact. Two boys, Brodie Spaulding and Brian Kinzie, were outdoors on 31 August 1991 in Noblesville, Indiana, when a meteorite landed 3.5 m away from them, making a crater 9 cm wide and 4 cm deep; the boys found a small black stone which was still warm – it proved to be an unusual sort of chondrite. On 15 August 1992 a piece of the Mbale meteorite (Uganda) struck a banana tree and then hit the head of a boy, again without real damage, and on 9 October of the same year a 12 kg meteorite landed on the bonnet of an unoccupied car at Peeksville, New York, belonging to Michelle Knapp. On 10 December, again in 1992, a house in Japan, belonging to Masaru and Maiko Matsumoto, was struck by a 6.5 kg meteorite.

The only definite fatality seems to have been an Egyptian dog, which was in the wrong place at the wrong time when the Nakhla meteorite fell on 28 June 1911.

Meteorites were recognized as extraterrestrial only a few centuries ago. The original suggestion was made by E. F. Chiadni in 1794, but met with considerable scepticism, and as recently as 1807 Thomas Jefferson, President of the United States, was quoted as saying 'I could more easily believe that two Yankee professors would lie than that stones would fall from heaven'. By then, however, proof had been obtained by the French astronomer J. B. Biot, following his investigation of the meteorite shower at L'Aigle on 26 April 1803.

On average, meteorites enter the Earth's atmosphere at a speed of 15 km s^{-1}, although the extreme range is probably between 11 km s^{-1} and 70 km s^{-1}. During entry the leading edge melts, and the ablation of molten material produces a smooth face. Melt droplets streaming along the sides of the meteorite collect at the opposite face and solidify, producing an oriented meteorite. Quenching of the molten coating leads to a dark, glassy fusion crust. For stone meteorites this crust is seldom more than 0.1 cm, thick and the contrast in colour with the underlying material, which is whitish grey with specks of iron, makes it easier to establish that the object really is meteoritic.

CLASSIFICATION OF METEORITES

Meteorites are of three main types: irons (siderites), stony-irons (siderolites) and stones (aerolites). Early systems of classification were due to G. Rose (1863), G. Tshcermak (1883) and A. Brezina (1904); these were extended by G. Prior (1920) and by G. J. H. McCall (1973). Stones are more commonly found than irons in the ratio of 96% to 4%, but this is misleading, as irons are much more durable and are more likely to survive. Antarctica is a particulary good area for meteorite collection, and many have been found there, initially by Japanese researchers in 1969.

Siderites (irons) are made up largely of metallic iron minerals. Kamacite is essentially metallic iron with up to 7.5% nickel in solid solution; taenite is iron with more than 25% nickel in solid solution. There is also plessite, which is a mixture of taenite and finite-grained kamacite.

Siderites are divided into three main groups: hexahedrites, octahedrites and ataxites. Hexahedrites are mainly of kamacite, with between 4 and 6% of nickel. Octahedrites contain between 7 and 12% of nickel. When etched with acid and polished, these types show what are termed Widmanstätten patterns, composed of parallel bands or plates of kamacite bordered by taenite, and intersecting one another in two, three or four directions. Widmanstätten patterns are unique to these meteorites. They do not appear in ataxites, which contain more than 16% of nickel.

Aerolites (stones) are made up chiefly of silicate minerals, and are again divided into two groups: chondrites (containing chondrules) and achondrites (without chondrules). Chondrules are small spherical particles; they are fragments of minerals, and show radiating structure; their average diameter is about 1 mm. They are formed from previously melted minerals which have combined with other mineral matter to form solid rock. Chondrites account for 86% of known specimens, and are believed to be among the oldest rocks in the Solar System, with ages of around 4.5 to 4.6 thousand million years. They contain pyroxenes, which are darkish minerals also common on Earth. A stone which was seen to fall at Monahans, Texas, on 22 March 1998, was found to contain salt crystals, inside which were tiny droplets of water.

Ordinary chondrites, much the commonest form, are divided into three groups. Those with 12–21% of metallic iron are known as bronzites (bronzite is usually green or brown; its chemical formula is $(MgFe)SiO_3$). Chondrites with 5–10% metallic iron are termed hypersthenes; darker than bronzite (chemical formula $(MgFe)SiO_3$). With about 2% metallic iron, the principal minerals are bronzite and olivine $(MgFe)_2SiO_4$; olivine is abundant in the mantle of the Earth. Much less common are the enstatites, containing 13–25% of low nickel–iron content metal; the formula for enstatite is $Mg_2Si_2O_6$, colour brown or yellowish.

Of special interest are the carbonaceous chondrites; on average the percentage of material by weight is 2.0 carbon, 1.8 metals, 0.2 nitrogen, 83.0 silicates and 11.0 water. There is almost no nickel–iron. Carbonaceous chondrites make up 50% of the asteroids at the inner edge of the main belt, and 95% at the outer edge.

Achondrites, accounting for about 7% of known specimens, contain no chondrules. Of special interest are eight meteorites known as the SNC meteorites after the regions in which they were found (Shergotty in India, Nakhla in Egypt and Chassigny in France). They seem to have crystallized only 1.3 thousand million years ago, and their composition and texture indicates that they formed on or in a planet which had a strong gravitational field. They have a concentration of volatile elements, and glassy incursions which were permanently formed in the extreme heat of whichever process ejected them from a parent body. These glassy incursions have trapped gases such as Ar, Kr, Xe and N. It has been suggested that they are of Martian origin, though this is of course highly speculative.

Siderolites (stony irons) are made up of a mixture of nickel–iron alloy and non-metallic mineral matter. Pallasites consist of a network of nickel–iron enclosing crystals of olivine; mesosiderites are heterogeneous aggregates of silicate minerals and nickel–iron alloy. There are two other groups, lodranites (iron, pyroxene, olivine) and siderophyres (iron, orthopyroxene), but these are excessively rare. Siderolites account for no more than 1.5% of known falls.

CHICXULUB IMPACT

One major problem in Earth history concerns the disappearance of the dinosaurs, around 65 000 000 years ago, at the end of the Cretaceous period. Not only the dinosaurs vanished; so did many other species of living things, and there was unquestionably a great 'extinction', although there have also been others (notably toward the end of the Permian Period). In 1980 Luiz and Walter Alvarez proposed that the K–T extinction, separating the Cretaceous (K) and Tertiary (T) eras, was due to the impact of a huge asteroid, meteoroid or comet, which threw up so much material that the world climate changed abruptly. Near the mediæval town of Gubbio, in Italy, they found limestone deposits laid down at this particular time which showed an abrupt change in fossil specimens, and moreover the centimetre thick layer was unusually rich in iridium, which is characteristic of certain meteorites. Subsequently, it was claimed that the point of impact had been found, near the village of Chicxulub on the Yucatán peninsula in Mexico. The crater is buried under a thick layer of sedimentary rock, and studies of the gravitational and magnetic fields indicate that the hidden crater is round 180 km in diameter; there are three major ring structures round its rim, and the whole multi-ring structure may have a diameter of at least 300 km.

The Alvarez theory is now widely accepted, and is certainly plausible, but final proof is lacking, and there remain some sceptics. However, there can be no serious doubt that a massive impactor did strike the Chicxulub area at about the time that the dinosaurs died out.

LARGE METEORITES

A selected list of large meteorites is given in Table 16.2. Pride of place must go to the Hoba West meteorite, near

Table 16.2. Selected list of large meteorites.

	Weight (tons)
Hoba West, Grootfontein, SW Africa	60
Ahnunghito (The Tent) Cape York, W Greenland	30.4
Bacuberito, Mexico	27
Mbosi, Zimbabwe	26
Agpalik, Cape York, W Greenland	20.1
Armanty, Outer Mongolia	20 (est)
Willamette, Oregon, USA	14
Chupaderos, Mexico	14
Campo del Cielo, Argentina	13
Mundrabilla, Western Australia	12
Morito, Mexico	11

Table 16.3. The largest meteorites found in different regions.

		Weight (tons)
Africa	Hoba West, Grootfontein	60
USA	Willamette, Oregon	14
Asia	Armanty, Outer Mongolia	20
South America	Campo del Oielo, Argentina	13
Australia	Mundrabilla	12
Europe	Magura, Czech Republic	1.5
Ireland	Limerick	48 kg
England	Barwell, Leicestershire	46 kg (total)
Scotland	Strathmore, Tayside, Perthshire	10.1 kg
Wales	Beddgelert, Gwynedd	723 g

Table 16.4. British Isles meteorites.

1623 Jan 10	Stretchleigh, Devon	12 kg
1628 Apr 9	Hatford, Berkshire	3 stones; about 33 kg
1719	Pettiswood, West Meath	?
1795 Dec 13	Wold Cottage, Yorkshire	25.4 kg
1804 Apr 5	High Possil, Strathclyde, Lanarkshire	4.5 kg
1810 Aug ?	Mooresfort, Tipperary	3.2 kg
1813 Sept 10	Limerick	48 kg (shower)
1830 Feb 15	Launton, Oxfordshire	0.9 kg
1830 May 17	Perth	about 11 kg
1835 Aug 4	Aldsworth, Gloucestershire	Small shower, over 0.5 kg
1844 Apr 29	Killeter, Tyrone	Small shower
1865 Aug 12	Dundrum, Tipperary	1.8 kg
1876 Apr 20	Rowton, Shropshire	3.2 kg (iron)
1881 Mar 14	Middlesbrough	1.4 kg
1902 Sept 13	Crumlin, County Antrim	4.1 kg
1914 Oct 13	Appley Bridge, Lancashire	33 kg
1917 Dec 3	Strathmore, Tayside, Perthshire	4 stones; 13 kg
1923 Mar 9	Ashdon, Essex	0.9 kg
1931 Apr 14	Pontlyfni, Gwynedd	723 g
1949 Sept 21	Beddgelert, Gwynedd	723 g
1965 Dec 24	Barwell, Leicestershire	46 kg (total)
1969 Apr 25	Bovedy, N Ireland	Main mass presumably fell in the sea
1991 May 5	Glatton, Cambridgeshire	767 g
1999 Nov 28	Leighlinbridge, County Carlow	220 g

Grootfontein in Namibia (South-West Africa), which is still lying where it fell in prehistoric times; the total weight is over 60 tons. It is now protected, since at one stage it was being vandalized by United Nations troops who were meant to be guarding it. All known meteorites weighing more than 10 tons are irons; the largest known aerolite fell in Kirin Province, Manchuria, on 8 March 1976. It weighs 1766 kg.

The largest meteorite on display in a museum is the Ahnighito (Tent), found by Robert Peary in Greenland in 1897; it is now in the Hayden Planetarium, New York, along with two other meteorites found at the same time and on the same site – known, appropriately, as The Woman and The Dog. Apparently the local Eskimos were rather reluctant to let them go. The Willamette meteorite is also in the Hayden Planetarium. (This meteorite was the subject of a lawsuit. It was found in 1902 on property belonging to the Oregon Iron and Steel Company. The discoverer moved it to his own property and exhibited it; the Company sued him for possession, but the Court ruled in favour of the discoverer.) Table 16.3 lists the largest meteorites found in different regions of the Earth.

BRITISH METEORITE FALLS

No really large meteorites have fallen in the British Isles in historic times, but there have been a number of small meteorites, listed in Table 16.4.

The Barwell fall was well observed. Many fragments of the meteorite were found; one was detected some time later nestling coyly in a vase of artificial flowers on the windowsill of a house in Barwell village. Although it broke up during descent, the stone is the largest known to have fallen over Britain.

The Bovedy meteorite was also well observed during its descent, but the main mass was not recovered, and almost certainly fell in the sea.

No British meteorite casualties have ever been reported. The Beddgelert meteorite – a small iron – scored a direct hit on the Prince Llewellyn Hotel, but caused no damage.

The 1991 meteorite fell at Glatton, in Cambridgeshire. It was found by A. Pettifor, who heard a loud whining noise and the crash of the stone into a conifer hedge some 20 metres away from him. He found the meteorite, which had made a shallow depression 2 cm deep. The meteorite was warm, not hot, when he picked it up. It has a granular structure, indicating that soon after its formation as part of an asteroid it had been hot, but did not melt, so that the mineral grains grew and interlocked. It is an ordinary chondrite of the low-iron Lodranite group, with 23% by weight of iron, about 5% of which is nickel–iron metal, with 18% of stony materials – mainly pyroxene and olivine, which are common components of terrestrial basaltic lavas.

BRILLIANT FIREBALLS

The brightest fireball ever seen may have been that of 1 February 1994. It passed over the Western Pacific at 22.38 GMT; the magnitude was about -25. Presumably this was a rocky object; if moving at 15 km s^{-1}, it would have been about 7 m across, weighing 400 tons. A remarkable phenomenon was seen on 9 February, 1913, from the North American continent, from Toronto (Canada) through to Bermuda. C. A. Chant, astronomer at Toronto University, recorded: 'At about 9.05 in the evening there suddenly appeared in the N.W. sky a fiery red body . . . it moved forward on a perfectly horizontal path with a peculiar, majestic, dignified deliberation . . . Before the astonishment caused by this first meteor had subsided other bodies were seen coming from the N.W., emerging from precisely the same point as the first one. Onward they moved at the same deliberate pace, in twos, threes or fours, with tails streaming behind . . . They all traversed the same path and were headed for the same point in the S.E. sky.' Because this was St. Cyril's Day, the objects are remembered as the Cyrillids. The whole display lasted for perhaps 3 minutes.

Were the Cyrillids meteoroids, which entered the Earth's upper air and then returned to space? Unfortunately, we do not have a definite explanation. Nothing similar had ever been seen before, and nothing similar has been seen since.

On 10 August 1972 a meteoroid was seen to enter the Earth's atmosphere and then leave it again. It seems to have approached the Earth from 'behind' at a relative velocity of 10 km s^{-1}, which increased to 15 km s^{-1} as it was accelerated by the Earth's gravity. The object entered the atmosphere at a slight angle, becoming detectable at a height of 76 km above Utah, and reaching its closest point to the ground at 58 km above Montana. It then began to move outward and became undetectable at just over 100 km above Alberta after a period of visibility of 1 min 41 s; the magnitude was estimated by eye-witnesses to be at least -15 and the diameter of the object may have been as much as 80 m. After emerging from the Earth's atmosphere it re-entered solar orbit, admittedly somewhat modified by its encounter, and presumably it is still orbiting the Sun.

THE TUNGUSKA FALL

The most famous fall of recent times was that of 30 June 1908, in the Tunguska region of Siberia. As seen from Kansk, 600 km away, the descending object was said to outshine the Sun, and detonations were heard 1000 km away; reindeer were killed, and pine-trees blown flat over a wide area. The first expedition to the site was led by L. Kulik, but did not arrive before 1927. No fragments were found, and it has been suggested that the impactor was the nucleus of a small comet or even a fragment of Encke's Comet – which, if icy in nature, would presumably evaporate during the descent and landing. (Inevitably, flying saucer enthusiasts have claimed that it must have been an alien space-ship in trouble!)

A second major fall occurred on 12 February 1947, in the Sikhote–Alin area of Siberia. The fall was observed, and many craters located. It is fortunate that both these Siberian falls struck uninhabited territory. If a meteorite of this size had hit a city, the death-roll would have been very high.

IMPACT CRATERS

Meteorite craters occur on the Earth, just as they do on the Moon, but lunar craters remain identifiable for much longer, because on the Moon there is no erosion.

Table 16.5. Terrestrial meteorite craters.

Name	Lat. (° ′)	Long. (° ′)	Diameter (km)	Age (years)	Discovered	
Acraman, Australia	32 01S	135 27E	160	570M	1986	Extensive ejecta; central seasonal lake.
Amguid, Algeria	26 05N	4 23E	0.45	100 000	1980	Circular; rim rises 30 m above floor.
Aouelloul, Mauritania	20 15N	12 41W	0.4	3.1M	1973	Impact glass found; associated with Tenoumer?.
Boxhole, Australia	22 37S	135 12E	0.17	30 000	1937	Rim raised 3–5 m. Many fragments found.
Brent, Canada	46 05N	78 29W	3.8	450M	1951	Partly filled with Lakes Gilmour, Tecumseh.
Campo del Cielo, Argentina	27 38S	61 42W	0.05	<40 000	1933	At least 11 craters, 1000 km NW of Buenos Aires.
Chicxulub, Mexico	21 24N	89 31W	54	65M	1990	Buried under almost 1 km sediments; NW corner, Yucatan; contains Merida town.
Clearwater Lake East, Quebec	56 05N	74 07W	20	290M	1965	Two lakes, emptying into Gulf of Richmond.
Clearwater Lake West, Quebec	56 13N	74 30W	32	290M	1965	Probable double impact.
Dalgaranga, Australia	27 43S	117 05E	0.021	30 000	1928	Fresh; fragments of mesodiertite collected.
Gosses' Bluff, Australia	23 50S	132 19E	22	142M	1972	Very eroded. Many shatter-cones. West of Alice Springs.
Henbury, Australia	24 35S	133 09E	0.157	10 000	1931	13 craters; largest measured 180 km × 140 m.
Holleford, Ontario, Canada	44 28N	76 38W	2.35	550M	1956	Barely recognizable at surface.
Kaalijarvi, Finland	58 24N	22 40E	0.10	40 000	1827	7 craters; main crater is lake-filled.
Lappajarvi, Finland	63 12N	23 42E	17	77M	1967	Eroded, partly exposed. Contains lake. 100 km E of Vaasa.
Lawn Hill, Queensland, Australia	18 40S	138 39E	20	540M	1987	Eroded. Near border with N Territories.
Lonar, India	19 59N	76 31E	1.83	52 000	1970	150 m deep. Rim rises 20 m above outer land. Possibly volcanic?.
Manicouagan, Quebec, Canada	51 23N	68 42W	100	212M	1964	Huge circle indicated by two narrow, semi-circular lakes.
Manson, Iowa	42 35N	94 31W	35	66M	1940	Buried; covered by recent sediments.
Meteor Crater, Arizona	35 02N	111 01W	1.186	49 000	1891	(Barringer Crater). Tourist attraction!
Morasko, Poland	52 29N	16 54E	0.1	10 000	1957	8 craters, 6 contain lakes. 9 km from Poznan. Contains village of Morasko.
New Quebec (Quebec)	61 17N	73 40W	3.44	1.4M	1943	Chubb Crater. Bowl-shaped, filled by lake 250 m deep at centre. Rim rises to 110 m. Somewhat eroded.
Odessa, Texas, USA	31 45N	102 29W	0.17	50 000	1921	Largest crater has raised rim. Many meteorites found.
Popigai, Russia (Siberia)	71 30N	111 00E	100	35M	1976	300 m deep; well preserved.
Ries, Germany	48 53N	10 37E	24	15M	1978	Large, flat, circular valley surrounded by low hills. Contains town of Nordlingen.
Sikhote-Alin, Russia	46 07N	134 40E	27 m	53	1947	Formed 12 Feb 1947; fall observed. At least 122 craters of 0.5 m diameter or greater.
Sobolev, Russia	46 18N	138 52E	0.053	200	1981	Asymmetrical; E rim lower than W rim.
Steinheim, Germany	48 02N	10 04E	3.4	15M	1905	Basin; there is a central peak, 50 m high. Basin contains villages Steinheim, Sontheim.
Sudbury, Ontario, Canada	46 36N	81 11W	200	1850M	1968	Contains extensive nickel–copper sulphide deposits. Possibly volcanic?.
Tenoumer, Mauritania	22 55N	10 24W	1.9	2.5M	1970	Rim 100 m high. Possibly volcanic?.
Tswaing, Pretoria, S Africa	23 24S	28 05E	1.1	0.2M	1991	'Saltpan'; depth 120 m. Partially lake-filled.
Vredefort, Pretoria, S Africa	27 00S	27 30E	140	1970M	1961	Contains Vredefort, Parys. Very possibly volcanic.
Waqar, Arabia	21 30N	50 28E	0.097	10 000	1932	Two craters, 500 km SE of Riyadh. Meteorites found.
Wolf Creek, Australia	19 18S	127 46E	0.875	300 000	1957	Alternatively, Wolfe Creek. Regular, well preserved.
Zhamanshin, Khazakstan	48 24N	60 58E	13.5	0.9M	1981	Basin partly filled with lake deposits. Tektites found nearby. 100 km N of the Aral Sea.

The most famous impact crater on Earth is Meteor Crater, in Arizona (it really should be Meteo*rite* Crater), also known as the Barringer Crater. It is 1186 m in diameter, and 175 m deep and is well preserved; it is a well-known tourist attraction. Its age is of the order of 50 000 years (older than was formerly believed), so that at the time of its fall the area was completely uninhabited. Another well-preserved impact crater is Wolf (or Wolfe) Creek, in Western Australia; it was found by the Aborigines long ago and called by them Kandimalal, but it was only identified as an impact crater in 1947, by aerial survey. Its diameter is between 870 and 950 m; the floor has a diameter of 675 km and lies 25 m below the level of the surrounding sand plain. The Tswaing or 'Saltpan' Crater near Pretoria, in South Africa, was identified much more recently; it is partially lake-filled.

Dozens of meteorite craters have been listed, but caution is needed. For example, all lists include the Vredefort Ring, again near Pretoria, but geologists who have made long-term studies of it are almost unanimous in classifying it as volcanic.

A selected list of terrestrial impact craters is given in Table 16.5.

MICROMETEORITES

Very small particles entering from the Earth's atmosphere from space are termed micrometeorites; if they are below about 1 mm in diameter, they cannot cause luminous effects, but end their journey to the ground in identifiable form. Many thousands of tons of micrometeoritic material reach the Earth's surface each year.

TEKTITES

These are glassy objects, found only in a few definite areas (Table 16.6). They are small; the largest, found in 1932 at Muong Nong in Laos, weighs 3.2 kg. They seem to have been heated twice, and are aerodynamically shaped. They were once believed to be meteoritic, and in 1897 R. O. M. Verbeek even suggested that they came from the Moon, but it now seems definite that they are of terrestrial origin, being blasted away from the surface by impact and, subsequently, re-entering the atmosphere.

Table 16.6. Main groups of tektites.

Region	Name	Geological age
Australasia	Australites	Middle/Late Pleistocene
Ivory Coast	Ivory Coast tektites	Lower Pleistocene
Czech Republic	Moldavites	Miocene
USA	Bediasites (Texas) Georgiates (Georgia)	Oligocene

17 GLOWS AND ATMOSPHERIC EFFECTS

AURORÆ

Auroræ or polar lights (Aurora Borealis in the northern hemisphere, Aurora Australis in the southern) must have been observed from early times, since they are often strikingly brilliant. The Northern and Southern Lights are of the same type, but obviously the northern displays are better known, because the Southern Lights are not well seen from inhabited countries. They are sometimes visible from the southern parts of South America and New Zealand, and occasionally from South Africa, but most of the displays over the ages must have been enjoyed only by penguins.

Legends and folklore

There are many legends about auroræ. In Scotland, the Lights were called the 'Merry Dancers', supernatural beings dancing in the heavens. In Norse mythology, the auroræ represented reflections from the shields of the Valkyries; in Denmark and Sweden it was said that auroræ came from an active volcano in the far north, put there by the gods to provide humanity with light and warmth. When the Lights flickered, the Greenlanders believed that the dead were trying to signal to their living kith and kin. The Inuit of Hudson's Bay pictured the sky as a solid dome, pierced by holes which allowed light from beyond to shine through; also, the spirits of the dead could pass through the holes into the heavenly regions. The Faroe Islanders kept their children indoors during brilliant displays, because of the fear that the Lights would swoop down and singe people's hair off. To some Eskimos, an aurora represented a game of football played by spirits using a walrus head as a ball, but in Siberia the tables were turned; here the walruses were playing, using a human skull as a ball!

Old ideas

Of course the ancients had no idea of the real nature of auroræ. The Greek philosopher Anaxagoras (c 500–428 BC) believed that they were due to fiery vapour poured down into the clouds from above, while the Roman writer Seneca (5 BC–65 AD) attributed them to currents of hot, boiling air in the highest parts of the sky; because they moved so quickly, the stars could generate enough heat to set them alight. The term 'aurora' was coined by the French astronomer P. Gassendi, following the brilliant display of 12 November 1621.

Proper scientific accounts had appeared earlier than this. A good description of a display of Aurora Borealis was given by K. Gesner of Zürich for the aurora of 27 December 1560, and in 1731 J. J. de Mairan wrote an excellent book, *Traité physique et historique de l'aurore boréale*. The first surviving description of the Southern Lights was written by Captain Cook for the display of 20 February 1773.

Auroræ and sunspots

The connection between auroræ and electrical discharges goes back to a suggestion made by Edmond Halley in 1716; he linked auroral displays with discharges associated with the Earth's magnetic field. (It must be remembered that Halley saw his first aurora in this year, following the end of the solar Maunder Minimum; between 1715 and 1745 there are no reports of auroræ from Britain.) De Mairan, in his book, saw a relationship between the frequency of sunspots and displays of auroræ, while Anders Celsius (1701–44) and Olof Hiorter (1696–1750) recognized the link between auroræ and disturbances of the magnetic needle, and in 1751 the Danish bishop Erik Pontoppidan came to the conclusion that the aurora was itself an electrical phenomenon.

The sunspot – aurora connection was refined in 1870 by E. Loomis of Yale, and in 1872 by the Italian astronomer G. Donati, following a major display on 4–5 February of that year. The foundations of modern theory were really laid down in 1895 by the Norwegian scientist Kristian Birkeland, who found that auroræ were due to charged particles moving under the influence of the Earth's magnetic field; he also reproduced 'miniature auroræ' by using a spherical electromagnet and a beam of cathode rays. In 1929 S. Chapman and V. Ferraro suggested that the auroræ were due to solar plasma, and it is indeed true that auroræ are produced by electrified particles from space, originating in the Sun, colliding with atoms and molecules in the Earth's upper atmosphere.

The streams of solar plasma sent out, particularly from coronal holes, take from 2 to 5 days to reach the Earth. When they meet the boundary of the magnetosphere they are deflected, and many of the particles are forced round the Earth's globe, following the magnetic field lines; the Earth's magnetic field is compressed on the day side, while on the night side a magnetic tail is produced, extending out to as much as at least 6 000 000 km. It may be said that the Earth's magnetic field creates a 'tunnel' in the plasma stream; the northern and southern sections are separated by a plasma sheet, and the particles in the opposite sides of the tunnel rotate in opposite directions, producing a dynamo effect. When the magnetic tail becomes unstable, due to an increase in the solar wind, the charged particles move inward to the centre of the tunnel, where they meet and cause a magnetic short-circuit. This circuit closes when the particles reach the top of the atmosphere, where the tenuous gases are made up of ionized particles and are therefore electrically conducting. The energy of the dynamo is converted into light, and auroræ appear.

Auroral ovals

The particles from the magnetic tail stream down toward the magnetic poles, but it is wrong to assume that the poles are the best sites from which to observe auroræ; they are not. Auroral activity is more or less permanent at high latitudes round the so-called auroral ovals, which are 'rings' asymmetrically displaced around the geometrical poles (at present the north magnetic pole lies in Nares Strait, at latitude 70°N, longitude 71°W, between Greenland and Ellesmere Island: the south magnetic pole lies at 79°S, 109°E, in Antarctica). Basically, the ovals remain more or less fixed in direction, and are wider on the night side of the Earth than on the day side. They are centred on the magnetic poles, while the Earth rotates around the geographical poles. When there are major disturbances on the Sun, producing energetic solar wind particles, the ovals broaden and expand, bringing auroral displays further north and further south of the main regions. The maximum activity of auroræ is seen along geomagnetic latitude 68° north or south. Thus from Tromsø in Norway (latitude

69°N) auroræ are far commoner than they are at the pole itself.

On average, auroræ of one kind or another are seen on 240 nights per year in North Alaska, North Canada, Iceland, North Norway and Novaya Zemlya; 25 nights per year along the Canada/United States border and in Central Scotland; only one night per year in Central France. Obviously, however, this varies according to the state of the solar cycle, and in general auroræ are most common about two years after a sunspot maximum. From Mediterranean countries there are only one or two good displays per century, and close to the equator they are very rare indeed, although on one occasion, in 1909, an aurora was seen from Singapore, latitude 1°25'N.

From South England, there were brilliant displays on 25–6 January 1938, 13 March 1989 and 8–9 November 1991. The 1938 display coloured the entire sky red, while those of 1989 and 1991 were bright enough to cast shadows. On these occasions the Sun was very active – but even so, auroræ cannot be reliably predicted.

A fairly bright aurora occurred on 6 April 2000, seen from much of England; from Selsey in Sussex I saw it as very conspicuous and decidedly red.

Altitudes

The heights of auroræ vary. In general, the sharp lower boundary lies at about 98 km above ground level, and the maximum region of auroral activity at about 110 km; the normal upper boundary is at 300 km, although in extreme cases this may reach up to 700 km or even 1000 km. Extremely low auroræ have been reported now and then, but not with any certainty. Auroræ always lie far above normal clouds, so that when the sky is overcast there is no chance of seeing an auroral display of any kind.

Form and brilliance

The form and brilliance of displays also varies and a definite scale of brightness has been drawn up, although it is not easy to be really precise. The accepted scale is given in Table 17.1.

Auroræ may be seen in many forms. There are glows; arcs, with or without rays; bands, more diffuse and irregular; draperies, or curtains made up of very long rays; individual

Table 17.1. Brilliance classes of auroral displays.

1   Rather faint, about equal in brightness to the Milky Way as seen on a clear night.
2   Brighter display, about ten times brighter than Class 1; about equal to thin cirrus clouds lit by moonlight.
3   Much brighter – about 100 times as bright as Class 1; comparable with cumulus clouds lit by moonlight.
4   Brilliant auroræ, 100 to 2000 times as bright as those of Class 1, sometimes casting shadows and even giving as much light as the full moon. From England, the last Class 4 auroræ were those of 1989 and 1991.

rays or streamers; coronæ, or radiating systems of rays converging at the zenith; 'surfaces', which are diffuse patches, sometimes pulsating; flaming auroræ, made up of quickly moving sheets of light; and ghost arcs, which may persist long after the main display has ended. There are also 'flickering auroræ', whose behaviour has been compared with that of a candle flame in a light breeze.

The colours of auroræ were first studied in detail by the Norwegian scientist Lars Vegard (1880–1963). The differences in hue are due to differences in composition of the atoms and molecules struck by the energetic particles which rain down from above, and each atmospheric gas produces its own characteristic colour. Oxygen atoms at the lower border of the auroral zone produce brilliant greens; the higher-altitude oxygen atoms yield the rare but spectacular all-red auroræ, and ionized nitrogen atoms are responsible for blue colours, while complete nitrogen atoms give red or purple edges to the lower borders and rippled edges of the display. Sometimes all these colours can be seen over a few minutes; at other times only one colour is seen. Over Norway, for example, by far the commonest auroræ are green.

Auroral noise

One interesting and frankly puzzling problem associated with auroræ is that of 'auroral noise'. There have been many reports of sounds: faint whistles, rustling, swishing, crackling and soft hissing. Many reported were listed by S. Tromholt, from Norway, in 1885 (reported in *Nature* **32** 499). A typical report is that of Dr. H. D. Curtis, who was in charge of the Labrador Station of the Lick Observatory in 1905. He wrote:

'Auroral displays were frequent and bright during July and August. On several nights I heard faint swishing and crackling sounds, which I could only attribute to the aurora.

There were times when large, faintly luminous patches or 'curtains' passed rapidly over our camp . . . . The faint hissing and crackling sounds were much more in evidence as such luminous patches swept past us. I tried in vain to assign the sounds to some reasonable cause other than the aurora, but was forced to exclude them as possible sources. In short, I feel certain that the sounds I heard were caused by the aurora and nothing else. There was, moreover, a certain synchronism between the maxima of these sounds and the sweeping of aurora curtains over the sky.'

It is extremely difficult to see how auroræ could produce sounds. Moreover, the speed of sound in the high atmosphere is such that any noise due to aurora should take minutes to reach ground level, making it hard to reconcile the noise with visual activity; in any cause auroræ are so high up that they are restricted to conditions of near-vacuum, so that there is no real chance of sounds being propagated downward. Electrical phenomena have been proposed; it has also been claimed that the reports are purely psychological, but we do not really know the answer, and so far I believe that no serious attempt has been made to tape-record the noise.

Slight odours have also been reported, but with no certainty, and it must be said that smelly auroræ seem only slightly less improbable than noisy auroræ!

Daytime auroræ occur, but are naturally difficult to detect because of the brightness of the sky. Moreover, the auroral belt is much narrower by day than by night. To see daytime auroræ from the Earth's surface, the observer must be within 10° to 15° of the magnetic pole, and the Sun must be at least 10° below the horizon. One favourable site for making these observations is Spitzbergen, but this is not a place which is particularly easy to reach.

A vivid description of an aurora was given in 1904 by the great Norwegian polar explorer Fridtjof Nansen, as he stood on the deck of his ship, the *Fram*, drifting over the Arctic pack-ice. He wrote:

'There is the supernatural for you – the Northern Lights flashing in matchless power and beauty over the sky in all the colours of the rainbow. Seldom or never have I seen the colours so brilliant. The prevailing one at first was yellow, but that gradually flickered over to green, and then a sparkling ruby-red began to show at the bottom of the rays on the underside of the arc. And now from the far-away western horizon a fiery serpent writhed up over the sky, growing brighter and brigher as it came . . . .'

There can be no auroræ on the Moon, which lacks atmosphere and has no detectable magnetic field; neither can they be expected on Mars, where the magnetic field is very weak. On the other hand, strong auroræ are associated with all four giant planets – Jupiter, Saturn, Uranus and Neptune, although with Uranus and Neptune the magnetic poles are nowhere near the rotational poles, and auroræ will be equatorial rather than polar.

THE AIRGLOW

The airglow (dayglow and nightglow) was named by Otto Struve in 1950, who suggested the name in correspondence with C. T. Elvey.

Nightglow prevents the sky from being completely dark at any time, quite apart from the diffusion of starlight. There are various reasons for it; chemical reactions of the neutral constituents of the upper atmosphere, with light emissions; reactions from ionized constituents, again producing light emission; the influx of incoming particles from space, and so on. There also seems to be a sort of 'geo corona' in the exosphere (1000 to 10 000 km altitude), due to emissions from hydrogen and helium atoms in the highly rarefied region of the atmosphere.

Diffused starlight is quite appreciable, coming as it does from the thousand million stars above magnitude 20, of which approximately one million are above magnitude 21.2, 4850 above magnitude 6, and 1620 above magnitude 5. Over a full visible hemisphere of the sky (20 626 square degrees) the total starlight is approximately equal to 52 stars the brilliance of Sirius or 4 planets the brilliance of Venus at maximum. It is, however, only a tiny fraction of the light sent to us by the Moon.

THE GREEN FLASH

The Green Flash (or Green Ray) is an atmospheric effect. As the Sun sinks below the horizon, the last segment of it may flash brilliant green for an instant. There are vague references to this in Egyptian and Celtic folklore, but the first scientific account of it was due to W. Swan, who saw it on 13 September 1865. However, Swan's account was not published until 1883. The first published reference to the phenomenon was due to J. P. Joule, in 1869. Much more recently, superb photographs of it have been taken by D. J. O'Connell at the Vatican Observatory; these were published in book form in 1958.

The Green Flash has even been seen with Venus; the first reference to it seems to be due to Admiral Murray, who saw it from HMS *Cornwall*, off Colombo, on 28 November 1939. Venus was setting over a sea horizon; Admiral Murray was using binoculars, and described the flash as 'emerald'. A green flash with Jupiter was seen by G. Verschuur on 8 June 1978.

THE ZODIACAL LIGHT

The Zodiacal Light may be seen as a faint cone of light rising from the horizon either after sunset or before sunrise. It extends away from the Sun, and is generally observable for only a fairly short period after the Sun has set or before it rises. On a clear, moonless night, under ideal conditions, it contributes about one-third of the total sky light, and may be brighter than the average Milky Way region. It is due to particles scattered in the Solar System along and near the main plane of the system; cometary débris is a major contributory factor. The diameters of the particles are of the order of 0.1 to 0.2 μm (one μm is equal to one-millionth of a metre). Since the Zodiacal Light extends along the ecliptic, it is best seen when the ecliptic is nearly vertical to the horizon – i.e. February to March and again in September to October.

The first scientific account of the Zodiacal Light was given by G. D. Cassini in 1683; he correctly suggested that it was due to sunlight reflected from tiny particles orbiting the Sun near the plane of the ecliptic. However, it had been observed much earlier by the Persians and the Arabs, and had been clearly described by Joshua Childrey in the *Britannia Baconica* published in London in 1661.

THE GEGENSCHEIN

The Gegenschein or Counterglow is seen as a faint patch of radiance in the position in the sky exactly opposite to the Sun. It is extremely elusive, and is generally visible only under near-perfect conditions. (From England I have seen it only once, in March 1942, when the whole country was blacked out as a precaution against German air-raids.) The best opportunities occur when the anti-Sun position is well away from the Milky Way (i.e. in February to April and September to November) and when the anti-Sun position is high, at local midnight. Generally it is oval in shape, measuring $10°$ by $22°$, so that its maximum diameter is roughly 40 times that of the full Moon.

The discovery of the Gegenschein seems to have been due to Esprit Pénézas, who reported it to the Paris Academy in 1731. It was named by Humboldt, who saw it on 16 March 1803. It was described in more detail by the Danish astronomer Theodor Brorsen, who saw it in 1854 and wrote about it in 1863. Contrary to statements made in many books. Brorsen did not claim the discovery for himself, as he was familiar with Humboldt's description.

THE ZODIACAL BAND

The Zodiacal Band is a very faint, parallel-sided band of radiance, which may extend to either side of the Gegenschein, or be prolonged from the apex of the Zodiacal Light cone to join the Zodiacal Light with the Gegenschein. It may be from $5°$ to $10°$ wide and is extremely faint. Like the Zodiacal Light and the Gegenschein, it is due to sunlight being reflected from interplanetary particles near the main plane of the Solar System.

18 THE STARS

Look at the sky on a dark, clear night and it may seem that millions of stars are visible. This is not so. Only about 5780 stars are visible with the naked eye, and this means that it is seldom possible to see more than 2500 naked-eye stars at any one time, but much depends upon the visual acuity of the observer. People with average sight can see stars down to magnitude 6, but very keen-eyed observers can reach at least 6.5. On the magnitude scale, a star of magnitude 1 is exactly 100 times as bright as a star of magnitude 6.

The proper names of stars are usually Arabic, although a few (such as Sirius) are Greek. In general, proper names are used only for the stars conventionally classed as being of the first magnitude (down to Regulus in Leo, magnitude 1.36), plus a few special stars, such as Mizar in Ursa Major and Mira in Cetus. The system of using Greek letters was introduced by J. Bayer in 1603; also in wide use are the numbers given in Flamsteed's catalogue.

DISTANCES OF THE STARS

It had long been known that the stars are suns, and are very remote, but early efforts to measure their distances ended in failure. William Herschel tried the method of parallax; he reasoned – quite correctly – that if a relatively nearby star is observed at an interval of six months, it will seem to shift slightly against the background of more remote stars, because in the interim the Earth will have moved from one side of its orbit to the other. At that time it was believed that double stars resulted from line of sight effects, so that the nearer component would show a parallax shift against its companion. Herschel's measurements were not sufficiently precise to show parallaxes, but in 1801 he did show that many doubles, such as Castor in Gemini, are binary systems, with the components moving around their common centre of gravity.

In 1838 F. W. Bessel, from Königsberg, successfully measured the parallax of the star 61 Cygni, finding it to be about 11 light-years away. From the Cape, T. Henderson measured the parallax of the brilliant star α Centauri, although his results were not announced until after those of Bessel; and about the same time F. G. W. Struve, from

Dorpat, made a rather less accurate measurement of the parallax of Vega.

The nearest stars within 13 light-years of the Sun are listed in Table 18.1. Most are very faint dwarfs; only α Centauri, Sirius, ε Eridani, ε Indi, τ Ceti and Procyon are visible with the naked eye, and only α Centauri, Sirius and Procyon are more luminous than the Sun.

The nearest star beyond the Sun is Proxima, a member of the α Centauri system. (In 1976 O. J. Eggen reported the discovery of a red star, magnitude 10.8, in the constellation of Sculptor near τ Sculptoris which was regarded as being very close, but later work showed that it was well beyond the 13 light-year limit of the table.) It is most unlikely that any stars exist closer to the Sun than Proxima.

All parallax shifts are very small. A star showing a parallax of 1 arcsec would lie at a distance of 3.26 light-years, but even Proxima is well beyond this limit.

Absolute magnitude is the apparent magnitude that a star would have if seen from a standard distance of 10 parsecs (32.6 light-years). From this range only α Centauri, Sirius, Procyon and τ Ceti would be naked-eye objects. It is interesting to note that if our Sun could be observed from Sirius, it would be a second-magnitude star at right ascension 18h44m, declination $-16°4'$.

Most of the stars in Table 18.1 are red dwarfs; the companions of Sirius and Procyon are white dwarfs. There are four flare stars; UV Ceti B (Luyten L 726-8B), Proxima, Krüger 60B and Ross 154. Lalande 21185 and Groombridge 34A are excessively close binaries (too close to be separated visually) and there is a third body, of low mass, in the 61 Cygni system. There are strong suspicions that ε Eridani is accompanied by a body of planetary mass.

STELLAR MOTIONS

All proper motions are very slight, so that to all intents and purposes the constellations we see today are the same as those visible in ancient times – although in 1718 Edmond Halley was able to show that Sirius, Arcturus, Aldebaran and Betelgeux had shifted perceptibly

Table 18.1. The nearest stars.

Star	Constellation	Right ascension h	m	s	Declination °	′	Proper motion (″)	Parallax (″)	Distance (light-years)	Radial velocity (km s^{-1})
Proxima Centauri	Cen	14	29	43	−62	40.8	3.816	0.772	4.249	−22
α Centauri A	Cen	14	39	36	−60	50.0	3.698	0.749	4.35	−25
α Centauri B	Cen	14	39	35	−60	50.2	3.698	0.749	4.35	−21
Barnard's Star	Oph	17	57	49	+4	41.6	10.374	0.545	5.98	−111
Wolf 359	Leo	10	56	29	+7	00.9	4.689	0.418	7.80	+13
Lalande 21185	UMa	11	03	20	+35	58.2	4.819	0.395	8.23	−84
UV Ceti A	Cet	01	39	01	−17	57.0	3.366	0.381	8.57	+29
UV Ceti B	Cet	01	39	01	−17	57.0	3.366	0.381	8.57	+32
Sirius A	CMa	06	45	09	−16	43.0	1.326	0.380	8.57	−9
Sirius B	CMa	06	45	09	−16	43.0	1.326	0.380	8.57	−9
Ross 154	Sgr	18	49	50	−23	50.2	0.721	0.341	9.56	−12
Ross 248	And	23	41	55	+44	10.5	1.626	0.316	10.33	−78
ε Eridani	Eri	03	32	56	−09	27.5	0.976	0.306	10.67	+17
Ross 128	Vir	11	47	45	+00	48.3	1.347	0.301	10.83	−31
Luytens 789-6	Aqr	22	38	33	−15	18.1	3.259	0.294	11.08	−60
Groombridge 34 A	And	00	18	23	+44	01.4	2.912	0.290	11.27	+12
Groombridge 34 B	And	00	18	26	+44	01.7	2.912	0.290	11.27	+11
ε Indi	Ind	22	03	22	−56	47.2	4.707	0.289	11.29	−40
61 Cygni A	Cyg	21	06	54	+38	45.0	5.231	0.289	11.30	−65
61 Cygni B	Cyg	21	06	55	+38	44.5	5.231	0.289	11.30	−64
BD +59°1915 A	Dra	18	42	45	+59	37.9	2.269	0.286	11.40	−1
BD +59°1915 B	Dra	18	42	46	+59	37.6	2.268	0.286	11.40	+1
τ Ceti	Cet	01	44	04	−15	56.2	1.921	0.286	11.40	−17
Procyon A	CMi	07	39	18	+05	13.5	1.244	0.286	11.41	−4
Procyon B	CMi	07	39	18	+05	13.5	1.244	0.286	11.41	−4
Lacaille 9352	PsA	23	05	52	−35	51.2	6.895	0.284	11.47	+10
GJ 1111	Cnc	08	29	49	+26	46.6	1.288	0.276	11.83	−5
GJ 1061	Hor	03	36	00	−44	30.8	0.836	0.270	12.06	−20
Luytens 723-32	Cet	01	12	31	−17	00.0	1.345	0.267	12.20	−20
BD +05°1668	CMi	07	27	24	+05	13.5	3.758	0.264	12.34	+18
Lacaille 8760	Mic	21	17	15	−38	52.1	3.452	0.259	12.61	+28
Kapteyn's Star	Cep	05	11	40	−45	01.1	8.654	0.258	12.63	+246
Krüger 60 A	Cep	22	28	00	+57	41.8	0.943	0.252	12.95	−33
Krüger 60 B	Cep	22	28	00	+57	41.8	0.943	0.252	12.95	−32

since Ptolemy had drawn them. The star with the greatest proper motion is Barnard's Star, discovered in June 1916 by E. E. Barnard at the Yerkes Observatory. The annual proper motion is over 10 arcsec, so that in 170 years it crosses the sky by a distance equal to the diameter of the full moon.

Radial velocities (the towards or away movements of the stars, relative to the Sun) are given as negative for a velocity of approach, positive for a velocity of recession. Barnard's Star is approaching us at 111 km s^{-1}, so that its distance is decreasing at the rate of 0.036 light-years per century. The proper motion is increasing by about 0.0013

arcsec per year, and will reach 25 arcsec by 11 800 AD, when the star will be at its closest to us – 3.85 light-years, which is nearer than Proxima; the parallax will then be 0″.87 and the apparent magnitude will have risen to 8.5. Subsequently, the star will begin to recede.

Over sufficient periods of time, of course, the skies will alter. For example, Arcturus must first have become visible with the naked eye half a million years ago, and has brightened steadily; it is now about at its nearest and will subsequently recede, dropping below naked-eye visibility half a million years from now.

Table 18.1. (Continued)

Star	Spectrum	Apparent magnitude	Absolute magnitude	Luminosity, Sun = 1	Alternative name
Proxima Centauri	M5.5V	11.09	+15.5	0.00006	α Centauri C
α Centauri A	G2	−0.01	+4.4	1.6	
α Centauri B	K0	1.34	+5.7	0.45	
Barnard's Star	M4V	9.55	+13.2	0.00045	Munich 15040
Wolf 359	M6V	13.45	+16.6	0.00002	
Lalande 21185	M2V	7.47	+10.5	0.0055	BD +36ᵖ2147
UV Ceti A	M5.5V	12.41	15.3	0.00006	
UV Ceti B	M6V	13.2 var	+16.1	0.00004	Luyten's Flare Star
Sirius A	A1	1.43	+1.5	26	α (9) Canis Majoris
Sirius B	DA2	8.44	+11.3	0.003	
Ross 154	M3.5V	10.47	+13.1	0.00048	
Ross 248	M5.5V	12.29	+14.8	0.00011	
ε Eridani	K2	3.73	+6.2	0.303	18 Eridani
Ross 128	M4V	11.12	+13.5	0.00036	
Luytens 789-6	M5V	12.33	+14.7	0.00014	
Groombridge 34 A	M1.5V	8.08	+10.4	0.0061	GX Andromedæ
Groombridge 34 B	M3.5V	11.07	+13.4	0.00039	GQ Andromedæ
ε Indi	K5	4.68	+7.0	0.14	
61 Cygni A	K5	5.22	+7.5	0.082	
61 Cygni B	K7	6.03	+8.3	0.039	
BD +59°1915 A	M3V	8.90	+11.2	0.003	
BD +59°1915 B	M3.5V	9.68	+12.0	0.0015	
τ Ceti	G8	3.50	+5.8	0.45	52 Ceti
Procyon A	F5	0.40	+2.6	7	α (10) Canis Minoris
Procyon B	DA	10.7	+13.0	0.00055	
Lacaille 9352	M1.5V	7.34	+9.6	0.013	
GJ 1111	M6.5V	14.79	+17.0	0.00001	
GJ 1061	M5.5V	13.03	+15.2	0.0055	
Luytens 723-32	M4.5V	12.05	+14.2	0.0002	
BD +05°1668	M4.5V	9.86	+12.0	0.0015	
Lacaille 8760	M0	6.67	+8.7	0.028	
Kapteyn's Star	M0V	8.84	+10.9	0.0039	
Krüger 60 A	M3V	9.85	+11.9	0.0016	
Krüger 60 B	M4V	11.3 var	+13.3	0.0004	DO Cephei

According to calculations by Jocelyn Tomkin, the brightest star in the sky a million years ago was ζ Leporis, which now shines modestly at magnitude 3.55. A list of stars which achieve prime position is given in Table 18.2, but there are bound to be uncertainties; for instance, estimates of the luminosity of Canopus vary widely.

Inevitably there are stars whose proper motions will carry them from one constellation into another constellation. The only modern case of a 'migrating' star of naked-eye visibility is ρ Aquilæ; in 1992 its proper motion carried it across the boundary of Delphinus. Table 18.3 gives a list of other stars which will move from one constellation to another before the year 7000; data are according to R. W. Sinnott.

STARS OF THE FIRST MAGNITUDE

Stars down to magnitude 1.4 are usually said to be of the first magnitude. Obviously the distances and luminosities of the stars, apart from the very close ones, are not known precisely, and in Table 18.4 the values have been 'rounded off'. The distances are given in light-years, although in the main catalogue which follows it seems better to use parsecs (1 parsec is equal to 3.2633 light-years).

Table 18.2. Brightest stars in the past and future.

Years past/future	Star	Magnitude	Closest distance (light-years)	Present magnitude	Present distance (light-years)
−4 700 000	ε Canis Majoris	−4.0	34	1.50	490
4 400 000	β Canis Majoris	−3.7	37	1.98	710
1 200 000	ζ Sagittarii	−2.7	8	2.60	78
1 000 000	ζ Leporis	−2.1	5.3	3.55	78
300 000	Aldebaran	−1.5	21.5	0.87	65
240 000	Capella	−0.8	28	0.08	42
60 000	Sirius	−1.6	7.8	1.44	8.6
+ 300 000	Vega	−0.8	17	0.03	25
1 190 000	β Aurigæ	−0.4	28	1.90	82
1 250 000	δ Scuti	−1.8	9.2	4.70	160
1 500 000	γ Draconis	−1.4	28	2.24	101
2 290 000	υ Libræ	−0.5	30	3.60	127
2 900 000	HR 2853	−0.9	14	5.60	280
3 500 000	γ Herculis	−0.6	44	3.75	137
4 600 000	β Cygni	−0.5	80	3.08	390

Canopus is not included because of the uncertainty about its real luminosity, but it will certainly be among the brightest stars between 3 700 000 years ago and 1 000 000 years in the future. Data are according mainly to J. Tomkin; modern distances from the Cambridge catalogue.

Some variable stars can reach the first magnitude, although in general they remain fainter; at one time in the 19th century the unique η Carinæ outshone every star in the sky apart from Sirius. Mira Ceti reached 1.2 at the maximum of 1772, according to the Swedish astronomer Per Wargentin, although most maxima are below 2; the official maximum magnitude is usually given as 1.7. γ Cassiopeiæ is an irregular variable; in 1936 it reached magnitude 1.6, but at present (2000) and for some years past it has been around 2.2. Different catalogues give different values; the data given in this book are from the Cambridge catalogue (Sky Catalogue 2000), which is as authoritative as any.

Some stars have been suspected of permanent or secular variation. Thus Ptolemy ranked both β Leonis and θ Eridani as being of the first magnitude, although in the latter case there may well have been an error in identification. Ptolemy ranked Castor and Pollux as equal, although Pollux is now much the brighter of the two. Stars which are now decidedly brighter than as listed by Ptolemy

Table 18.3. Stars which will change constellations.

Star	Magnitude	Enters	Year
ρ Aquilæ	5.0	Delphinus	1992
γ Cæli	4.6	Columba	2400
ε Indi	4.7	Tucana	2640
ε Sculptoris	5.3	Fornax	2920
λ Hydri	5.1	Tucana	3200
μ Cygni	4.8	Pegasus	4500
χ Pegasi	4.8	Pisces	5200
μ Cassiopeiæ	5.1	Perseus	5200
η Sagittarii	3.1	Corona Australis	6300
ζ Doradûs	4.7	Pictor	6400

include β Canis Majoris, β Canis Minoris, γ Cygni, δ Draconis, β Eridani, γ Geminorum, ε Sagittarii and Polaris; those which are decidedly fainter include ζ Eridani, α Sagittarii and α Microscopii. However, it is very unwise to place much reliance on those old records, and there is no proven case of a star of naked-eye brightness which has shown permanent change since Ptolemy's time.

Table 18.4. The 50 brightest stars. Many of these stars (inc. Polaris) are slightly variable, but only Betelgeux has an appreciable range.

Star	Name	Apparent magnitude	Luminosity, Sun = 1	Spectrum	Distance (light-years)
1. α Canis Majoris	Sirius	−1.44	26	A1	8.6
2. α Carinæ	Canopus	−0.72	200 000	F0	1200
3. α Centauri		−0.27	1.7 + 0.45	K1 + G2	4.395
4. α Boötis	Arcturus	−0.05	115	K2	36.7
5. α Lyræ	Vega	0.03	52	A0	25.3
6. α Aurigæ	Capella	0.08	90 + 70	G8 + G1	42
7. β Orionis	Rigel	0.12	60 000	B8	900
8. α Canis Minoris	Procyon	0.40	7	F5	11.4
9. α Eridani	Achernar	0.45	400	B5	144
10. α Orionis	Betelgeux	0.5 var	15 000	M2	310
11. β Centauri	Agena	0.61	10 500	B1	460
12. α Aquilæ	Altair	0.76	10	A7	16.8
13. α Crucis	Acrux	0.83	3200 + 2000	B1 + B3	360
14. α Tauri	Aldebaran	0.87	100	K5	65
15. α Scorpii	Antares	0.96	7500	M1	330
16. α Virginis	Spica	0.98	2100	B1	260
17. β Geminorum	Pollux	1.16	60	K0	33.7
18. α Piscis Australis	Fomalhaut	1.17	13	A3	25.1
19. α Cygni	Deneb	1.25	70 000	A2	1800
β Crucis		1.25	8200	A2	425
21. α Leonis	Regulus	1.36	87	B7	78
22. ε Canis Majoris	Adhara	1.50	5000	B2	490
23. α Geminorum	Castor	1.58	45	A0	52
24. γ Crucis		1.59	160	M3	88
25. λ Scorpii	Shaula	1.63	1300	B2	275
26. γ Orionis	Bellatrix	1.64	2200	B2	360
27. β Tauri	Alnath	1.65	400	B7	131
28. β Carinæ	Miaplacidus	1.67	130	A0	111
29. ε Orionis	Alnilam	1.70	23 000	B2	1200
30. α Gruis	Alnair	1.73	230	B5	101
31. ε Ursæ Majoris	Alioth	1.76	60	A0	81
32. ζ Orionis	Alnitak	1.77	19 000	09.5	1100
33. γ Velorum	Regor	1.78	3800	WC7	520
34. ε Sagittarii	Kaus Australis	1.79	110	B9	145
35. α Persei	Mirphak	1.80	6000	F5	620
36. α Ursæ Majoris	Dubhe	1.81	60	K0	75
37. η Ursæ Majoris	Alkaid	1.85	450	B3	108
38. δ Canis Majoris	Wezea	1.86	139 000	F8	3000
39. θ Scorpii	Sargas	1.87	14 000	G0	900
40. α Trianguli Aust.	Atria	1.92	96	K2	55
41. δ Velorum	Koo She	1.93	50	A0	80
β Aurigæ	Menkarlina	1.93	50	A2	82
γ Geminorum	Alhena	1.93	82	A0	105
44. α Pavonis		1.94	700	B3	183
45. β Canis Majoris	Mirzam	1.98	7200	B1	710
46. γ Leonis	Algieba	1.99	60	K0 + G7	90
α Hydræ	Alphard	1.99	105		177
α Ursæ Minoris	Polaris	1.99	6000	F8	680
49. α Arietis	Hamal	2.00	96	K2	85
50. σ Sagittarii	Nunki	2.02	525	B3	209

Excluding Mira Ceti and η Carinæ, the other stars above magnitude 2.5 are as follows:

β Ceti	Diphda	2.04	γ Draconis	Eltamin	2.25
β Andromedæ	Mirach	2.06	ι Carinæ	Tureis	2.25
α Andromedæ	Alpheratz	2.06	ε Scorpii	Wei	2.29
θ Centauri	Haratan	2.06	α Lupi	Men	2.30
κ Orionis	Saiph	2.06	ε Centauri		2.30
β Ursæ Minoris	Kocab	2.08	η Centauri		2.31
α Ophiuchi	Rasalhague	2.08	δ Scorpii	Dschubba	2.32
ζ Ursæ Majoris	Mizar	2.09	β Ursæ Majoris	Merak	2.37
β Gruis	Al Dhanab	2.11	ε Boötis	Izar	2.38
β Persei	Algol	2.12 max	ε Pegasi	Enif	2.38
γ Andromedæ	Algieba	2.14	α Phœnicis	Ankaa	2.30
β Leonis	Denebola	2.14	β Pegasi	Schaet	2.4 var
γ Cassiopeiæ		2.2 var	κ Scorpii	Girtab	2.41
γ Cygni	Sadr	2.20	γ Ursæ Majoris	Phad	2.44
λ Velorum	Al Suhail al Wazn	2.21	η Canis Majoris	Aludra	2.44
α Coronæ Borealis	Alphekka	2.23	α Cephei	Alderamin	2.44
α Cassiopeiæ	Shedir	2.23 var?	ε Cygni	Gïenah	2.46
ζ Puppis	Suhai; Hadar	2.25	α Pegasi	Markab	2.49
δ Orionis	Mintaka	2.23 var	κ Velorum	Markeb	2.50

19 STELLAR SPECTRA AND EVOLUTION

The stars show a tremendous range in luminosity, though much less in mass. Some known stars are millions of times more luminous than the Sun, while others are remarkably feeble. At its peak, in the 1840s, the erratic variable η Carinæ was estimated to be 6 000 000 times as powerful as the Sun; S Doradûs, in the Large Magellanic Cloud, has an absolute magnitude of −8.9, so that it is at least a million times as luminous as the Sun – yet because of its great distance (170 000 light-years) it cannot be seen with the naked eye. At the other end of the scale is MH18, discovered in 1990 by M. Hawkins at what was then the Royal Observatory, Edinburgh, from plates taken with the UKS telescope in Australia. It has 1/20 000 the luminosity of the Sun, and is presumably a brown dwarf (see below). Its mass is 5% that of the Sun, and its distance is 68 light-years.

The first attempt to classify the stars according to their spectra was made by the Italian Jesuit astronomer, Angelo Secchi, in 1863–7. He divided the stars into four main types:

(1) White or bluish stars, with broad, dark lines of hydrogen but obscure metallic[1] lines. Example: Sirius.
(2) Yellow stars; hydrogen lines less prominent, metallic lines more so. Examples: Capella, the Sun.
(3) Orange stars; complicated, banded spectra. Examples: Betelgeux, Mira. The class included many long-period variables.
(4) Red stars, with prominent carbon lines; all below magnitude 5. Example: R Cygni. This class also included many variables.

Secchi's work was followed up enthusiastically by the English amateur W. Huggins, but the modern system of classification was developed in America, at the Harvard College Observatory. It was introduced by E. C. Pickering in 1890, and was extended by two famous women astronomers, Annie Jump Cannon and Wilhelmina Fleming, who produced the Draper Catalogue – so named because money for its development was provided by the widow of Henry Draper, who in 1872 had been the first to photograph the spectrum of a star (Vega). Secchi's catalogue had contained over 500 stars; the Draper Catalogue contained spectra of 225 000 stars down to the ninth magnitude.

The spectral types were allotted letters in order of decreasing temperature, A, B, C . . ., but before long it became clear that some of the original types were unnecessary, and others out of order. The final sequence was alphabetically chaotic: O, B, A, F, G, K, M, R, N, S. A famous mnemonic runs 'O Be a Fine Girl Kiss Me Right Now Sweetie' (or, if you prefer it, Smack)[2].

The original scheme depended mainly on surface temperature. Conventionally, the first in the sequence were called 'early' types and the K, M and other red stars as 'late', but this is now known to have nothing to do with evolutionary sequences.

In the Hertzsprung–Russell or HR diagram, drawn up in the early part of the twentieth century by E. J. Hertzsprung in Denmark and H. N. Russell in America, stars are plotted according to their luminosities and their spectral types. Most of the stars lie on a line extending from the upper left of the diagram to the lower right; this is termed the Main Sequence. Main Sequence stars (such as the Sun) are officially regarded as dwarfs. Characteristics of the main spectra types are given in Table 19.1. Stars of very 'early' type (W and O) and very 'late' type (R, N and S) are relatively rare. Types R and N are now often combined as type C.

In 1998 astronomers in the United States added a new spectral type, L, to accommodate very cool dwarfs of low mass, no more than 1/20 that of the Sun. These L dwarfs

---

[1] Conventionally but confusingly, all elements except hydrogen and helium are classed as 'metals' from the spectroscopist's point of view.

[2] Naturally, this mnemonic has been met with howls of anguish from the Politically Correct fanatics, I wonder how long it will be before these curious people begin agitating about red giants, brown dwarfs and, above all, black holes?

Table 19.1. Stellar spectra.

Type	Surface temperature (°C)	Spectrum	Examples	Notes
W	Up to 80 000	Many bright lines; few absorption lines. Broad emission lines of hydrogen, ionized helium, carbon, nitrogen and oxygen.	γ Vel, WC7	Wolf–Rayet stars. Expanding shells, moving outward at up to 3000 km s^{-1}. All very remote. They may exist in binary systems where the companion star has stripped away the Wolf–Rayet's outer layers.
O	40 000–35 000	Both bright and dark lines; singly ionized helium lines in either emission or absorption. Strong ultra-violet continuum. Very massive and luminous.	ζ Pup, 05.8 ξ Per, 07 τ CMa, 09 ζ Ori, 09.5 10 Lac, 09	Represent a transition between W and B stars, although this does not imply any evolutionary sequence.
B	Over 26 000 for B0 to 12 000 for B9	No emission lines, but dominant absorption lines of neutral helium; hydrogen lines also prominent.	β Cru, B0 ϵ CMa, B2 α Eri, B5 α Leo, B7 β Ori, B8	Bluish (B0) to white (B9). Some, such as Rigel (β Ori) are exceptionally luminous.
A	11 000–7500	Spectra dominated by hydrogen lines.	α Lyr, A0 α CMa, A1 α Cyg, A2 α Aql, A7	White stars, although some such as α Lyræ (Vega) are bluish.
F	7500–6000	Hydrogen lines less prominent. Metallic lines become noticeable; calcium very conspicuous.	α Car, F0 θ Sco, F0 α CMi, F5 δ CMa, F8 α UMi, F8	Yellowish, although the hue is so subdued that most F-type stars look white, Canopus being a good example.
G	Giants: 4200 Dwarfs: 5500	Solar-type spectra. Conspicuous metallic lines, hydrogen lines weaker.	η Boö, G0 δ Boö, G8 τ Cet, G8	Beginning of division between giants and Main Sequence stars (dwarfs). Yellow stars.
K	Giants: 4000–3000 Dwarfs: 5000–4000	Metallic lines dominant; hydrogen weak; weak blue continuum.	β Gem, K0 α Boö, K2 α Tau, K5 ϵ Eri, K2 ϵ Ind, K5	Orange stars. K-stars are more numerous than any other type. Division between giants and dwarfs now very marked.
M	Giants: 3000 Dwarfs: 3400	Very complicated spectra, with many bands due to molecules. Molecular bands of titanium oxide noticeable.	α Sco, M1 α Ori, M2 δ Oph, M1 μ Cep, M2 α Her, M5 τ Cet, M7	Orange-red stars, many of them variable. Many M-type stars are highly luminous giants, such as Antares, or supergiants, such as Betelgeux; many red dwarfs are very feeble, such as Proxima Centauri.

Table 19.1. (Continued)

Type	Surface temperature (°C)	Spectrum	Examples	Notes
R	2600	Carbon lines very prominent	V Ari, R	Red stars, most of them variable.
N	2500	Similar to R stars, but rather cooler.	R Lep, N TX Psc, N	Types R and N are now often combined as type C.
S	2600	Prominent bands of heavier elements such as zirconium, yttrium and barium.	χ Cyg, X Aqr R Cyg	Red stars; many are long-period variables.

Table 19.2. Stellar luminosity classes.

Ia	Very luminous supergiants
Ib	Less luminous supergiants
II	Luminous giants
III	Normal giants
IV	Subgiants
V	Dwarfs (Main Sequence stars)

Table 19.3. Additional spectral classification.

comp	Composite spectrum; two stars are involved, indicating a very close binary system.
e	Emission lines are present (generally hydrogen).
m	Abnormally strong metallic lines; usually applied to A stars.
n	Broad (nebulous) lines, indicating fast rotation.
nn	Very broad lines, indicating very fast rotation.
neb	The spectrum of the star is mixed with that of a nebula.
p	Unspecified peculiarity, except with type A, where it denotes abnormally strong metallic lines.
s	Very narrow (sharp) lines.
sh	Shell star (a B to F Main Sequence star with emission lines from a shell of gas).
var	Varying spectral type.
wl	Weak lines, indicating an ancient, metal-poor star.

are no doubt very numerous, and it has even been suggested that they outnumber all other classes of stars combined.

The original Harvard scheme has been modified into what is known as the Yerkes or MKK classification, based on the work of W. W. Morgan, P. V. Keenan and E. Kellman. It takes into account the fact that the gravitational acceleration on the surface of a giant star is much lower than for a dwarf, so that the gas pressures and densities are much lower in giants than in dwarfs. There are six luminosity classes (Table 19.2), so that the Sun would be specified as a G2V star. There are other characteristics of importance, denoted by lower case letters (Table 19.3); thus Alioth (Epsilon Ursæ Majoris) in the Great Bear is specified as of type A0pIV(CrEu), indicating that the spectrum shows strong lines of the elements chromium (Cr) and europium (Eu). The famous red variable Mira Ceti is classified as M7IIIe; the e indicates that emission lines are present. The very luminous, unstable stars of type W have spectra dominated by bright lines; they are known as Wolf–Rayet stars, after the two astronomers who first drew attention to them.

STELLAR EVOLUTION

The stars show a tremendous range in luminosity. In Table 19.4, absolute magnitude is given in terms of luminosity compared with that of the Sun. Most stars have absolute magnitudes of between −10 and +16; the extremes are supernovæ (brighter than −16) and brown dwarfs, which may be said to be links between stars and planets even though this analogy should not be carried too far.

It was originally believed that a star began its career as a large, red body, condensing out of nebular material. Its spectral type would then be M. It would contract and heat up, to become a Main Sequence star (type B or A), and then cool down while continuing to shrink, ending up as a red dwarf (type M once more). In this case the HR

Table 19.4. Conversion of absolute magnitude (*A*) to luminosity (*L*) in terms of the Sun. This table is no more than approximate, but serves as a general guide.

A	L	A	L
−16	200 000 000	+0.5	52
−15	80 000 000	+1	33
−14	30 000 000	+1.5	21
−13	13 000 000	+2	13
−12	5 000 000	+2.5	8.3
−11	2 000 000	+3	5.2
−10	800 000	+3.5	3.3
−9	330 000	+4	2.1
−8.5	200 000	+4.5	1.3
−8	132 000	+4.83	1 (Sun)
−7.5	83 000	+5	0.8
−7	52 500	+5.5	0.5
−6.5	33 000	+6	0.3
−6	21 000	+6.5	0.2
−5.5	13 200	+7	0.1
−5	8 300	+7.5	0.08
−4.5	5 200	+8	0.05
−4	3 300	+8.5	0.03
−3.5	2 000	+9	0.02
−3	1 300	+10	0.008
−2.5	800	+11	0.003
−2	520	+12	0.001
−1.5	330	+13	0.0005
−1	200	+14	0.0022
−0.5	130	+15	0.000 08
0	83	+16	0.000 03
		+17	0.000 01
		+18	0.000 005

diagram would represent a definite evolutionary sequence, and certainly the Main Sequence is very marked, as are the giant and supergiant areas to the upper right. White dwarfs, to the lower left, were unrecognized in 1913, when the HR diagram was drawn up in its present form.

At that time the source of stellar energy was generally assumed to be gravitational, so that the star would begin at the top right of the HR diagram, cross to the Main Sequence at the top left, and then pass down the Main Sequence as it cooled. This would mean that red giants and supergiants would be very young. In fact this is not true; they are very advanced in their evolution.

Neither can gravitational contraction account for the radiation of a normal star. A star such as the Sun is several thousands of millions of years old, and simple gravitational contraction could not sustain it for anything like this period. Russell himself proposed that the energy source could be due to the annihilation of matter, so that certain types of particles were wiping each other out and releasing energy in the process. However, this would lead to a life-cycle of millions of millions of years, which was as obviously too long as previous estimates had been too short.

The key to the problem was found in 1939 by H. Bethe and, at about the same time, by G. Gamow. (Bethe actually worked it out during a train journey from Washington to Cornell University!) Normal stars shine by means of nuclear reactions. Thus, deep inside the Sun, hydrogen is being converted into helium. It takes four hydrogen nuclei to make one helium nucleus, and each time this happens a little mass is lost and a little energy is released. Each second, the Sun converts 600 million tons of hydrogen into helium – and loses 4 million tons in mass. This may sound a great deal, but the Sun is very massive, and has a great deal of hydrogen 'fuel', which lasts for a long time. The age of the Earth is 4.6 thousand million years; the Sun is certainly older than that, but even so it is no more than half-way through its main career.

What are termed 'stellar populations' were first described by W. Baade in the early 1950s. Population I stars are metal-rich, with about 2% of their mass being made up to elements heavier than helium. The most brilliant Population I stars are of types W, O and B. The disk and arms of the Galaxy (and other spirals) are mainly Population I.

Population II stars are metal-poor, and are clearly older, so that their most brilliant members are red giants and supergiants which have already left the Main Sequence. They are dominant in the halo and nucleus of the Galaxy (and other galaxies) and in globular clusters. However, there is no hard and fast boundary between Population I and Population II regions.

There are numerous cases in which many stars of similar type, and presumably of similar age, are concentrated in a limited area. These are known as *stellar associations*.

Star birth

A star begins its career by condensing out of the gas and dust in the interstellar medium – that is to say, the gas and dust lying between existing stars. Vast amounts of material are concentrated in what are termed giant molecular clouds (GMCs); for example, a GMC covers almost the whole of the constellation of Orion, and the famous nebula M.42 is only a small feature of it. Nebulæ are, in fact, stellar birthplaces.

Star formation in a given area may well be triggered off by a nearby supernova explosion – the death of a very massive star, which literally blows itself to pieces, or the destruction of the white dwarf component of a binary pair. After the outburst, a shell of gas passes through the interstellar medium; if it encounters a GMC, then the GMC is compressed, and this stimulates star birth.

In nebulæ we can observe small dark 'globules', known as Bok globules in honour of the Dutch astronomer Bart Bok, who first drew attention to them in the 1940s. A typical Bok globule measures 1 to 2 light-years across, through some are larger; they are intensely cold – less than 15° above absolute zero. The denser regions inside them contract gravitationally and heat up, so that eventually they form masses of material known as protostars; these draw in further material by the process of accretion.

The subsequent career of the protostar depends entirely upon its initial mass, so let us consider the different categories.

1. *Stars with mass below 0.08 that of the Sun*, or 80 times that of Jupiter. The core temperature never becomes high enough to trigger off nuclear reactions such as the conversion of hydrogen into helium; for this, the temperature must rise to around 10 000 000 °C. Very low-mass stars which cannot achieve this are known as brown dwarfs, a term coined in 1975 by Jill Tarter; in fact it is rather misleading, since visually a brown dwarf would look dull red. The pressure due to emitted radiation counteracts gravitation in a young star of this kind, so that the size remains more or less constant for several thousands of millions of years; initially the dwarf shines feebly, although because no nuclear reactions are going on it gradually cools down and eventually ceases to shine at all. It becomes a black dwarf (though whether the universe in its current

form is old enough for this to have happened is still problematical).

It has been said that a brown dwarf is a cross between a star and a planet, but this analogy must not be taken too far; a planet is formed by accretion from the material surrounding a young star, and its internal structure is different. Moreover a planet has no light of its own, and depends upon reflecting the light of a nearby star, whereas a young brown dwarf may have an absolute magnitude of around +17 and a surface temperature of the order of 2000 °C. The luminosity may then be around 1/10 000 that of the Sun.

Because they shine so feebly (if at all), brown dwarfs are not easy to identify, and for many years searches for them proved to be fruitless. Then, in 1995, a brown dwarf was detected in the Pleiades star-cluster; it was catalogued as PPl 15. The Pleiades cluster is only about 125 000 000 years old, so that PPl 15 was young; the surface temperature was about 2000 °C. Spectroscopic examination revealed the presence of the element lithium. Newly-born stars contain small amounts of lithium (about 1 atom per thousand million), but an increased temperature will destroy the lithium in about 100 million years – so that a cool object containing lithium must be a brown dwarf. Other brown dwarfs in the Pleiades, such as Teide 1, were also reported.

However, even stronger evidence comes from the star Gleise 229, a red dwarf in the constellation of Lepus; it is 19 light-years away. It has a companion, Gleise 229B, found on 27 October 1993 with the 60 inch (152 cm) and 200 inch (5 m) telescopes at Palomar, and quickly confirmed by the Hubble Space Telescope. The distance between Gleise A and B is around 40 a.u. (much the same as the distance of Pluto from our Sun), and the mass seems to be between 20 and 50 times that of Jupiter, putting it firmly in the brown dwarf category. The surface temperature is around 700 °C; since it is between one and five thousand million years old, it has had plenty of time to cool down. In 1997 a spectrum taken with UKIRT, the United Kingdom Infra-Red Telescope on Mauna Kea in Hawaii, revealed the presence of methane and water vapour. Methane is destroyed at a temperature of 2500 °C, and so Gleise 229B cannot be as hot as this; it can only be a brown dwarf.

No doubt brown dwarfs are very common. We have even found one brown dwarf binary pair, again in the Pleiades cluster; the separation between the two components is only 0.03 a.u., and the revolution period is seven days.

An isolated brown dwarf of special interest was detected in 1997 by Maria Teresa Ruiz, using the 3.6 m telescope at the La Silla Observatory in Chile. It was tracked down because of its unusually large proper motion. It seems to be 33 light-years away, and is of magnitude 22. The surface temperature is of the order of 1700 °C, and the mass is 75 times that of Jupiter, which is equivalent to 6% of the mass of the Sun; its spectrum shows unmistakable traces of lithium. It is three million times fainter than the dimmest object which can be seen with the naked eye; Ruiz named it Kelu-1, since that word means 'red' in the language of the Mapuche people, the ancient inhabitants of that part of Chile.

The coolest known methane brown dwarf, NNTDF J1205-0744, was detected in 1999 in Virgo, following a deep-field exposure with the Hubble Space Telescope. Its temperature is around 700 °C, suggesting an age of 500 to 1000 million years, and a mass 20 to 50 times that of Jupiter. It is 300 light-years away. Lacking a stable energy source at its core, it is becoming fainter and cooler, and will continue to do so for thousands of millions of years.

The coolest brown dwarf of any kind is Gliese 570D, identified in 2000 by A. Burgasser and D. Kirkpatrick with the Two Micron All Sky Survey at infra-red wavelengths. A spectrum taken with the Cerro Tololo reflector (aperture 4 metres) shows lithium, which cannot survive in hotter stars. The temperature of Gliese 570D is about 750 K (480 °C). The dwarf orbits the triple star system of Gliese 570 at a distance of 1500 astronomical units, in a period of around 40 000 years. It is about the same size as Jupiter, but 50 times more massive. The distance from Earth is 19 light-years; the system lies in the constellation of Libra.

It has been said that a brown dwarf is a star which has failed its entrance examination, and this certainly seems appropriate!

2. *Low-mass stars*; too massive to remain as brown dwarfs, but below 1.4 solar masses. The process of increasing mass in the central region of a dense cloud (accretion) continues, and eventually a protostar develops; nuclear processes have not begun, but the protostar glows because of the heat generated by the compression of the gas. If the dense core is rotating it collapses into a disk, which may then condense into several stars; if the core is not rotating rapidly, it will form a single star (such as the Sun).

At first the protostar will be large and red, although by no means the same as red giants such as Arcturus or Aldebaran. As the contraction goes on, the surface temperature remains the same, so that the luminosity decreases. On the HR diagram, the star passes along what are known as the Hayashi and Henyey tracks (named after the astronomers who first described them) and finally joins the Main Sequence. T Tauri stars are still contracting toward the Main Sequence, and are irregularly variable; they are also sources of strong 'stellar winds', which blow away the original surrounding cocoon of dust. The removal of the cocoon makes the star brighten, and there is strong emission of infra-red radiation.

When the cocoon of gas and dust is finally blown away, and mass accretion stops, the protostar joins the Main Sequence at a point known as the ZAMS (Zero Age Main Sequence). Just where it joins depends upon its mass, and this also regulates the subsequent course of events. Once nuclear reactions begin – the conversion of hydrogen into helium – the star settles down to a period of stability; a star with a mass similar to that of the Sun will spend about 10 000 million years on the Main Sequence, so that the Sun is now about half-way through this stage in its evolution. At first it was only about 70% as luminous as it is now, but as the hydrogen-into-helium process continued the core temperature rose to its present value of about 15 000 000 °C. The conversion process – known, rather misleadingly, as hydrogen burning – is accomplished by a rather roundabout process known as the proton–proton reaction. With hotter stars, what is termed the carbon–nitrogen cycle is dominant, so that these two elements are used as catalysts; however, the end result is much the same – hydrogen is converted into helium, with release of energy and loss of mass. Gas radiation pressure (tending to make the star expand) and gravitation (tending to make it contract) balance each other out. In fact, the star adjusts its size to make this happen, so that the star remains stable.

This stage of evolution lasts for a long time, but not for ever. Helium is built up in the core, and eventually the supply of available hydrogen runs low. This means that less energy is generated, and finally the core can no longer support the weight of the star's outer layers pressing down on it from all directions. The inner temperature rises, with hydrogen burning continuing round the now helium-rich core. As the central temperature continues to rise, the outer layers expand and cool, so that the star becomes a red giant – as Arcturus and Aldebaran are now. It is then said to be on the Asymptotic Giant Branch (AGB) of the HR diagram. The luminosity has increased a hundredfold, and the bloated globe is now so large that in the case of the Sun it will engulf the inner planets. Certainly the Earth cannot hope to survive as a habitable world.

When the core temperature reaches $100\,000\,000\,°C$, helium suddenly reacts to form carbon and oxygen. This is known as the 'helium flash', and is very sudden indeed; subsequently the energy output declines, and so the outer layers again contract. The star is left smaller, hotter and dimmer, and cannot return to the Main Sequence. Very often the star becomes variable; stars of about the same mass as the Sun become what are termed W Virginis variables, while higher-mass stars become Cepheid variables. The variations are due to the fact that the stars are pulsating, changing both in luminosity and in radiation output. Following this period of instability, the outer layers of the star are puffed off; subsequently the expelled material becomes visible, excited by radiation from the star, to produce what is called a planetary nebula – a misleading term, because a planetary nebula has nothing whatsoever to do with a planet. This may dispose of 20% of the star's mass. The expelled material will not fall back, but will expand and dissipate in space; planetary nebulæ, as such, are therefore relatively short-lived, and are unlikely to last for as long as $100\,000$ years. Some are symmetrical, while others are irregular in form. Undeniably they are beautiful objects.

What, then, of the core, in which nuclear reactions have ceased? The temperature is not high enough to make carbon and oxygen react, and what remains of the old star (that is to say, the burnt-out core) settles into a very small, super-dense object known as a white dwarf. The dim companion of Sirius is of this type, and was in fact the first white dwarf to be recognized as such. Its escape velocity is $3400\,\mathrm{km\,s^{-1}}$, and its surface gravity is 900 times that of the Sun.

In a white dwarf, the atoms are broken up, and their component parts are packed tightly together with little waste of space. This leads to amazing density values, so that a teaspoonful of white dwarf material would weigh several tons. They are of two main types, DA (with hydrogen-rich atmospheres) and DB (with more complex spectra). They are very small; we know white dwarfs which are smaller than the Moon but are as massive as the Sun. They have been described as 'bankrupt stars', although in some cases their surface temperatures are still high, and may exceed $100\,000\,°C$.

White dwarf material is officially termed 'degenerate'. The star cannot contract further, and it will simply cool down until it turns into a cold, dead globe – a black dwarf, sending out no energy at all. It is significant that no white dwarfs are known with surface temperatures much below $3000\,°C$, so that evidently the Galaxy is not yet old enough for even the most ancient white dwarfs to have cooled down below this temperature.

One very hot white dwarf, RE 1738 + 665, was discovered in 1994 by M. Barstow of Leicester University, from data supplied by the ROSAT artificial satellite. The surface temperature is of the order of $90\,000\,°C$, and the atmosphere is pure hydrogen. The star seems to form a link between the cooler, hydrogen-rich white dwarfs and the very hot hydrogen-rich stars at the centres of planetary nebulæ.

The Indian astronomer S. Chandrasekhar has shown that if a star is more than 1.4 times as massive as the Sun, it cannot become a white dwarf unless it sheds some of its mass during its evolutionary career – which happens, of course, during the unstable and the planetary nebula stages. If the mass remains above the Chandrasekhar limit, the star will meet with a very different fate; it will not subside quietly as a white dwarf, but will implode to become a neutron star via a supernova explosion.

3. *Stars more than 1.4 times as massive as the Sun*. Here, everything happens at an accelerated rate; for example, if the proto-star is 1.5 times as massive as the Sun it may reach the Main Sequence in a few thousand years, and will remain on the Main Sequence for a mere 10 million years before it has exhausted all its available hydrogen 'fuel'. The core heats up to 100 million degrees, and helium is

converted to carbon, although the helium 'burns' steadily after the helium flash. By the time that helium burning has come to an end, the outer layers of the star have extended further still, and the star has become much brighter than a low-mass red giant such as Arcturus; it has turned into a supergiant, as Betelgeux in Orion is now. The absolute magnitude of a supergiant may attain −10, of the order of a million times the luminosity of the Sun (in the case of Betelgeux, the absolute magnitude is between −5 and −6). Massive though they are, supergiants are very rarefied, with average densities of less than one-millionth that of the Sun. Many are variable; these include the Cepheid variables, to be discussed below.

As the core temperature continues to rise, carbon reacts in its turn; next, oxygen and silicon are produced, and then iron. This is where the whole situation changes. Iron will not react, and so energy production stops. Disaster follows. The process is out of control; in a matter of seconds the core collapses, and the electrons and protons are fused into neutrons. There is an 'implosion', followed by an explosion; shock waves race through the star, and literally blow it to pieces in what is termed a type II supernova outburst. At its peak the luminosity may be 1000 million times that of the Sun. Heavy elements, produced in the supernova, are hurled away into space, later to form new stars; the remnant of the supernova remains as a neutron star, although it will eventually end up as a cold, dead globe.

A rough timetable for a star 25 times as massive as the Sun is given in Table 19.5.

Table 19.5. Evolution stages for a star 25 times as massive as the sun.

Hydrogen burning	7 000 000 years
Helium burning	500 000 years
Carbon burning	600 years
Oxygen burning	6 months
Silicon burning	1 day
Core collapse	c 0.1 s
Core bounce	a few milliseconds
Explosive burning	c 10 s
Surface blows away	c 1 h

Neutron stars. The concept of neutron stars was first proposed in 1932 by the Russian physicist L. Landau, and again in 1934 by F. Zwicky and W. Baade at Caltech (USA), but the first neutron star was not detected until 1967 – not by its visible light, but by its radio emissions. A neutron star is indeed an amazing object. The diameter is of the order of 20 km or even less, and the star cannot contract further, because of neutron degeneracy. The density is perhaps a thousand million million tons per cubic metre, or 1000 million million times that of water. A teaspoonful of neutron star material would weigh a thousand million tons, and a pin's head of the material would balance the weight of an ocean liner.

According to current theory, the surface layer of a neutron star is crystalline, and around 100 metres deep; it is iron-rich, and composed of what we may call 'normal' matter. There are mountains on the surface – but these can be no more than around a centimetre high, and the 'atmosphere' above can extend for no more than a few centimetres. Cracks occur in the surface, producing 'starquakes' which affect the rotation period. Below the crust comes the neutron-rich liquid mantle; below again, a superfluid core composed mainly of neutrons, and at the centre material made up of 'hyperons', about which we can only speculate. Stand on a neutron star, and you will weigh 10 000 million times as much as you do on Earth. As the original star collapsed the magnetic field was concentrated, and became very strong, reaching perhaps a hundred million tesla, as against 30 millionths of a tesla for Earth.

A neutron star with an even stronger magnetic field is termed a magnetar; the field may be 5000 million times stronger than that of the Earth. The field is assumed to cause 'starquakes' in the solid crust, releasing energy in the form of gamma-rays, X-rays, and sub-atomic particles travelling at nearly the speed of light. The first confirmation as obtained in 1998, when astronomers using the VLA (Very Large Array) in New Mexico identified a short-lived 'afterglow' of particles emitted by a magnetar. The object, SGR 1900+14, was 15 000 light-years away, and in August 1998 had emitted a powerful burst of X-rays and gamma-rays, which is why attention had been drawn to it.

A typical neutron star has a mass around 1.5 times that of the Sun. When first created it is very hot, but has a tiny surface area and appears very faint; as it ages, it naturally cools down.

Pulsars. The first pulsar was discovered in 1967 by Jocelyn Bell (now Jocelyn Bell Burnell) from Cambridge. She was using a special 'radio telescope', designed by A. Hewish, which was not a 'dish', but looked remarkably like a collection of barbers' poles; the aim was to study the scintillation of distant radio sources. Stars twinkle, or scintillate, because of effects in the Earth's atmosphere; radio sources do so because they are affected by the clouds of electrons in the solar wind.

During her surveys, Jocelyn Bell came across a discrete radio source which 'pulsed' quickly, almost as though ticking. It was catalogued as CP 1919. Bell found the period to be 1.337 3011 s, which means that there are around 60 000 pulses per day; the period was absolutely regular, and the object was so extraordinary that Bell's colleagues were initially sceptical. However, following the initial announcement, made on 29 February 1965, other pulsars were soon found, and by now many hundreds are known.

It was some time before astronomers found out just what they were. There was even the short-lived LGM or Little Green Men theory – that the signals were artificial. Rotating white dwarfs were next suggested, but it became clear that a pulsar had to be even smaller than a white dwarf. It could be nothing other than a rotating neutron star.

A pulsar is spinning rapidly, and beams of radio radiation are sent out from its magnetic poles, which do not coincide with the poles of rotation. This leads to the 'lighthouse' effect. Each time the Earth passes through a beam, we receive a pulse of radiation. The 'normal' pulsar with the longest known period (4.3 s) is PSR 1845 − 19. The present holder of the short-period record (0.016 s) was identified in 1999 in the supernova remnant N157B, in the Large Cloud of Magellan, 170 000 light-years away; its age is around 5000 years. Next comes NP 0532, in the Crab Nebula, with a period of 0.339 s, so that it pulses 30 times in each second.

The Crab Nebula, in Taurus, is known to be the remnant of a supernova seen in the year 1054 (although since it is 6000 light-years away, the actual outburst took place in prehistoric times). It contains the first pulsar to be seen visually. This was achieved in January 1969 by a team at the Steward Observatory in Arizona, using a 36 inch reflector. They identified a faint, flashing object whose mean magnitude was about 17, and whose period was the same as that of the pulsar. Soon afterwards it was photographed from the Kitt Peak Observatory, also in Arizona. (Interestingly, R. Minkowski and W. Baade had observed it as far back as 1942, and had even suspected that it was the centre of activity in the Crab, but, not surprisingly, had failed to interpret it.) The second visual identification was that of the pulsar 0833 − 45 in the southern constellation of Vela; the pulsar lies in the Gum Nebula (named after its discoverer, C. S. Gum) which is unquestionably a supernova remnant. The pulsar itself was found in 1968 from the Molongo Radio Astronomy Observatory in Australia, and in 1977 a team working with the 3.9 m Anglo–Australian Telescope at Siding Spring in New South Wales identified the pulsar as a faint, flashing object with a mean magnitude of 24.2. The period is 0.089 s, the third shortest known for a 'normal' pulsar.

As a pulsar spins and emits energy, its rate of rotation slows down by measurable amounts. Thus the period of CP 1919, Bell's original pulsar, is lengthening by a thousand millionth of a second each month, so that in 3000 years' time the period will be 1.3374 s instead of the current 1.3373 s; this applies to all pulsars, including Bell's original, CP 1919. The period of the Crab pulsar is increasing by 3×10^{-8} s per day. Some pulsars show sudden, irregular changes in period, known as glitches; thus on 1 March 1969 the pulsar PSR 0833 − 45 speeded up by a full quarter of a millionth of a second. Glitches are due to disturbances in the pulsars – in other words, starquakes.

Obviously, there must be many pulsars whose beams do not sweep over the Earth. For example, there is the relatively close neutron star known as Geminga; it is a radio and gamma-ray source, lying 300 light-years away. Its optical counterpart is a 25th magnitude star which has moved by 1.8 arcsec in eight years. Geminga emits light rays, X-rays and gamma-rays in all directions, but there are no pulses, so that presumably its beams never sweep across our line of sight. The closest known pulsar is JO108 − 1431, in the constellation of Cetus; it is a thousand times weaker than any other known pulsar, and its age has been estimated as 160 000 000 years. The distance is 280 light-years. In the Guitar Nebula in Cepheus we find the fast-moving pulsar PSR 2224 + 65, 6000 light-years away; it spins 1.47 times per second, and is moving at around

900 km s^{-1}, presumably as the result of an asymmetrical supernova blast. A few pulsars, such as Hercules X1 and Centaurus X3, pulse only in the X-ray region of the electromagnetic spectrum. Binary pulsars are also known; the first-discovered of these, PSR 1913+16, consists of two neutron stars orbiting each other in a period of just under 8 h. It is possible that a binary pair of this type may eventually coalesce into a single object.

Millisecond pulsars. In October 1982 a team at Berkeley in California, led by D. Backer, found a pulsar in the constellation of Vulpecula – now catalogued as PSR 1937+215 – which flashed at a record 642 pulses per second, regularly spaced at intervals of 1.557 ms. This was clearly different from a 'normal' pulsar, and still has the shortest period known, although by now many other 'millisecond pulsars' have been found. Many of them, though not all, are members of binary systems; the other component can be any kind of highly evolved star.

Apparently a millisecond pulsar is an old neutron star which has only a weak magnetic field, and was originally a member of a binary system; the other member was a red giant star, whose outer layers were only weakly held by gravitation. The neutron star pulled material away from its companion, and the gases pulled toward it were heated sufficiently to produce X-rays. As the material struck the surface of the neutron star tangentially, it increased the rotation speed. In some cases the two components separated, leaving the millisecond pulsar isolated[3].

Millisecond pulsars are often found in globular clusters; the first of these was identified in the cluster M.28, in Sagittarius, by Andrew Lyne and his team at Jodrell Bank. In the famous globular cluster 47 Tucanæ there are more than a dozen millisecond pulsars, but not all are members of binary systems. In some cases the companion star will be literally evaporated by the neutron star over a sufficient period of time, and only the millisecond pulsar will be left. Proof of this theory was obtained in 1988 by J. Taylor and A. Fruchter (Princeton University), who found that the pulsar PSR 1957+20 in Sagitta – spin rate 622 per second – is evaporating its companion, which is now left with only 1/50 of the mass of the Sun. Not inappropriately, the pulsar is known as the Black Widow. A second case, PSR 1744−24a, has been found in the globular cluster Terzan 5, where the pulsar seems to be engulfed in the clouds of material which it is pulling away from its doomed companion.

4. *Stars too massive initially to form neutron stars.* Here we come to the real cosmic heavyweights, with masses of the order of 40 times that of the Sun or even greater. Their fate will be different again; they will produce black holes, a term coined by Archibald Wheeler 30 years ago.

The basic principle was suggested as long ago as 1783 by the English natural philosopher John Michell, and again by the great French mathematician Laplace in 1796. Like Newton, Laplace believed light to consist of a stream of particles, and wrote that if a body were sufficiently small and dense it would be invisible, since the light particles would not be able to travel fast enough to escape from it. In fact this is reasonable enough; if the escape velocity exceeds the speed of light, then light cannot break free – and if light cannot do so, then certainly nothing else can.

Relativity theory explains the situation rather differently. A massive body distorts the curvature of space (or, more precisely, spacetime), and the paths of rays of light or particles of matter are regulated by the curvature of the space in which they are travelling. At a critical radius from the star, the curvature of space will be so great that it prevents light from escaping; this is known as the Schwarzschild radius, after the German astronomer Karl Schwarzschild, who drew up the theory in 1916. The value of the Schwarzschild radius, in kilometres, is $3.0M$, where M is the mass of the body in terms of the Sun. The Schwarzschild radius of the Sun is therefore 3 km; for a star 60 times as massive as the Sun, it will be 60 km; for the Earth, only about 9 mm. The boundary of the Schwarzschild radius determines the size of the black hole, and is termed the event horizon, because we have no positive information about events taking place inside it; the black hole is to all intents and purposes cut off from the outer universe.

A black hole is the result of the collapse of a very massive star. If the mass of the collapsing core exceeds the maximum possible for a neutron star, the collapse continues to a central point of infinite density, termed a singularity

---

[3] There is an easy way to demonstrate this principle. Fasten a table-tennis ball to the end of a string, suspend it, and blow on it through a drinking straw. The ball will at once begin to rotate.

– a concept which is impossible to describe in everyday language. But before this, the collapsing star will have passed through its event horizon, and will have vanished as effectively as the hunter of the Snark.

Close to a black hole there are some very strange effects involving time dilation. A clock will run slow if it is within a strong gravity field (this has been experimentally proved). Picture an astronaut who carries a very accurate clock, and is falling towards a black hole, watched by an observer from a safe distance. The observer will conclude that the astronaut's clock is slowing down as it nears the event horizon; the interval between the ticks will increase. At the event horizon itself, the interval between the ticks will become infinitely long, so that the observer will assume that the traveller is left poised on the event horizon; but time will pass naturally insofar as the traveller is concerned, and he will simply fall through the event horizon and crash on to the singularity. By that time he will be in poor shape, because the tidal pull on his feet will be far stronger than that on his head (assuming that he is moving feet first), and he will be stretched out – a sort of cosmical Procrustes bed!

What is the true situation within a black hole, and does the original star's core crush itself out of existence altogether? We have to admit that we do not know, and since there can be no communication with the region beyond the event horizon it will be very difficult to find out.

If the black hole is rotating, the situation is rather different (a black hole of this kind is known as a Kerr black hole, after John Kerr, who first described it). A Kerr black hole is assumed to be surrounded by an ellipsoidal area or ergosphere, in which nothing can avoid being dragged round in the direction of the rotation; the singularity is a ring rather than a point of zero dimensions. There have been suggestions that an astronaut could enter a Kerr black hole, avoid the singularity, and emerge elsewhere, either in our universe or in a totally different universe, via a 'wormhole' linking one location in spacetime with another. It is an interesting theory, but one feels that there would be a definite shortage of volunteers willing to test it.

The famous British mathematician Stephen Hawking has suggested that black holes may finally evaporate. We know that pairs of particles and antiparticles form spontaneously in space, and almost immediately annihilate each other. If this happens close to the event horizon,

one particle may fall into the black hole and the other be allowed to escape; the net result is that the mass of the black hole is slightly reduced, and eventually this process might accelerate, in which case the black hole would evaporate completely. Obviously we are theorizing on the basis of very meagre data, and speculation is almost endless. In any case, the universe is not nearly old enough for any solar-mass black hole to have evaporated in this way, and though 'mini black holes' have been postulated there is no evidence that they actually exist.

Black holes are best detected by their gravitational effects upon objects which we can see. One good candidate is Cygnus X-1, so-called because it is an X-ray source. The system consists of a B0 type supergiant, HDE 226868, with about 30 times the mass of the Sun and a diameter 23 times that of the Sun (18 000 000 km), together with an invisible secondary with 14 times the mass of the Sun. The orbital period is 5.6 days, the distance from us is 6500 light-years, and the magnitude of the primary star is 9 (it lies at RA 19h 56m 295.3, declination $+35°3'55''$, near the star η Cygni). The secondary would certainly be visible if it were a normal star, and it is almost certainly a black hole which is pulling material away from the supergiant; before this material disappears over the event horizon, it is heated sufficiently to give off the X-rays which we receive. Another good candidate is V404 Cygni, which (having presumably been created by a much larger supernova explosion in the remote past) flared up as a nova in 1938, and rose from magnitude 19 to 12 before fading back to its normal brightness. In 1989 observations from the Japanese satellite Ginga showed that X-rays were being emitted, and studies by P. Charles, using the William Herschel telescope at La Palma, indicated that the system consists of a visible star 70% as massive as the Sun together with a black hole of 12 solar masses. The revolution period is 6.5 days.

In January 2000 it was announced that isolated black holes had been identified for the first time. Using the gravitational lensing technique, a team using the Australian and Chilean telescopes found evidence of two black holes, each about six times as massive as the Sun, lensing the light of stars in the background. If the lensing objects were ordinary stars they would have outshone the background star; they seem too massive to be white dwarfs or neutron stars, but more observations are needed before any certain conclusions can be drawn.

In January 2000 it was announced that the X-ray source V4641 Sagittarii, at a distance of 1600 light-years from Earth, is almost certainly attended by a black hole companion. R. Hjellming examined the system with the Very Large Array, and discovered twin jets shooting out from the system at over 90% of the speed of light. Only three other known systems eject material at such a velocity, and are termed 'microquasars'. Only the intense gravity of a black hole can generate so much power. This means that the companion to V4641 Sgr is the closest known black hole.

We do not yet have final proof that black holes really exist, but all the evidence points that way. If they are real, then they must surely rank as the most bizarre objects known to us.

Sakurai's Object. Before leaving the subject of stellar evolution, it is worth saying a little about a case in which a star seems to have been caught in the act of changing from a red giant into a white dwarf. This is V4334 in Sagittarius, known as Sakurai's Object because attention was first drawn to it in February 1996, when the Japanese amateur astronomer Yukio Sakurai observed that it was rapidly brightening. This was due to a flash of helium fusion taking place in a shell surrounding the giant's carbon/oxygen core; in six months the star swelled from being a hot dwarf, with a surface temperature of $50\,000°$, into a yellow supergiant with a surface temperature of $6000°$C. The inner core contracted, and generated enough heat to start helium burning. The outburst also revealed a previously-unknown planetary nebula associated with the star, and since then there have been interesting spectral changes. The nebula is expanding at the rate of 31 km s^{-1}, and now has an apparent diameter of 44 arcsec; its age has been given as between 3800 and 27 000 years. It has also been found that the rapidly evolving star showed a marked drop in temperature between 1996 and 1997. Its subsequent career will be followed with special interest.

Another star which seems to be evolving at breakneck speed is FU Sagittæ, which has been seen to change from a B4-type blue giant to an F5-type red giant over a period of less than a century (1900 to the present time). At one stage it passed through the Cepheid instability strip, and now shows variations more like those of an R Coronæ Borealis star. Its mass is 0.6 that of the Sun.

Colour of Sirius. En passant, some old reports, dating back to Classical times, describe Sirius as a red star, whereas it is now pure white (type A). There have been suggestions that several thousands of years ago the companion, now a white dwarf, was a red giant, but this explanation seems to be definitely out of court; the time-scale is wrong, and moreover a red giant shining together with the present Sirius would be so bright that it would have caused particular comment. There is little doubt that the description of Sirius as a red star was due to an error either in observation or (much more probably) in interpretation. Of course, as seen from Britain and other northern countries Sirius seems to flash various colours; it twinkles strongly, partly because it is so bright and partly because it is never very high up. From more southerly latitudes it naturally twinkles much less.

20   Extra-solar Planets

Our Sun is one of at least a hundred thousand million stars in our Galaxy alone, and many of these stars are of solar type. Therefore, there is every reason to assume that planetary systems are common. If our Solar System had been formed by the near-collision between the Sun and a passing star, it would certainly have been a rarity, or even unique; but this theory has long since been discarded, and there is nothing unusual about the Sun.

Brown dwarfs were elusive; extra-solar planets would be even more so. In any case, how does one differentiate between the two classes of objects? A brown dwarf does radiate, albeit very feebly. Like normal stars, brown dwarfs form by the gravitational shrinkage of a rotating mass of gas, whereas planets form by accretion in a disk of material round the parent star – and they have definite cores, whereas brown dwarfs are fully convective. Finally, there is the question of mass. If a low-mass companion of a star has a mass greater than around 13 times that of Jupiter, the most massive planet in our Solar System, it is probably a brown dwarf.

The first positive clues came in 1983, from IRAS, the Infra-Red Astronomical Satellite. While calibrating the on-board instruments, H. Aumann and F. Gillett, at the Rutherford Appleton Laboratory in England, found that the star Vega had 'a huge infra-red excess', due presumably to solid particles moving in an extended region round the star and stretching out to about 80 a.u. The temperature of the material was around $-185\,^{\circ}$C, and it was suggested that it might indicate a planetary system in the making.

Other cases followed; and by now there are several stars with known 'dust disks'. One of these is Fomalhaut, in Piscis Australis, one of the closest of the bright stars; it is 25 light-years away and 13 times as luminous as the Sun, with an A3-type spectrum. There seems to be a central hole in the dust disk, which may have been excavated by a planet. There is also ϵ Eridani, a mere 10.8 light-years away and one of the two closest stars to bear marked resemblance to the Sun (τ Ceti is the other). In 1998 observations made at submillimetre wavelengths with the James Clerk

Maxwell Telescope on Mauna Kea (Hawaii) showed that ϵ Eridani is surrounded by a ring of dust particles, 120 a.u. in diameter, remarkably like the inner comet zone in our Solar System. The star is thought to be no more than 1000 million years old, so that its planetary system – if it has one – is presumably less evolved than ours. A region close to the star seems to be more or less dust-free, so that any particles there may have been absorbed by a planet or planets.

However, the most significant case is that of the southern β Pictoris, 78 light-years away according to the Cambridge catalogue and 58 times as luminous as the Sun; it is of spectral type A5. As with Vega, IRAS showed a marked infra-red excess, and the material round it was subsequently detected optically by R. Terrile and B. Smith from the Las Campanas Observatory in Chile. The disk of material extends to nearly 80 000 000 000 km for the star; we see it almost edge-on, and it may be no more than a few hundred million years old. The cloud of material shows a depleted region in an area extending from 20 to 30 a.u., and this could be due to the presence of planets, which would sweep up much of the dust.

Observations were also made with the IUE (International Ultraviolet Explorer) satellite. Then, in May 1991, came results from the Goddard High Resolution Spectrograph on the Hubble Space Telescope, indicating that the solid particles in the disk discovered by IRAS are distributed over a radius of about 50 a.u.; the main gas-cloud is smaller, and extends to a few astronomical units from the star. The gas consists of several main components: first a diffuse disk circling β Pictoris in a stable manner; second, an inner gas disk which is slowly drifting inwards towards the star; and third, isolated clumps of gas which are spiralling in towards β Pictoris at velocities of up to 200 m s^{-1}. As the gas clumps pass in front of β Pictoris, the star's ultra-violet spectrum shows marked changes. All three gaseous components seem to be embedded in the more rarefied circumstellar gas-cloud which is diffusing outwards, possibly because of the radiation pressure from β Pictoris. The origin of the circumstellar gas-cloud is uncertain. It may be due to the slow decomposition of the solid particles in the disk discovered by IRAS, or it may have evaporated from cool objects in the suspected protoplanetary system. Later observations with the Hubble Space Telescope (1997–8) reveal significant 'warps' in the disk which may be caused by a planet orbiting the star in a slightly inclined orbit. β Pictoris is a young star, and though definite evidence is lacking it is certainly a promising candidate for a planetary system. In 1998 Sally Heap, one of the principal investigators at NASA, commented that 'the shape of the warp is a telltale indicator that favours the existence of a planet'. It must however by added that alternative explanations have been proposed; and it has even been suggested that the 'warp' may be caused by a brown dwarf circling β Pictoris at a much greater distance. In 2000, observations with the Hubble Space Telescope indicated that dynamically, the dust ring around β Pictoris was 'ringing like a bell'. This was attributed to the long-term gravitational effects of a star which by-passed β Pictoris about 100 000 years ago which is certainly possible, though there is no definite proof.

A massive planet orbiting a fairly low-mass star could conceivably produce measurable perturbations in the star's motion, and efforts to detect planets by this astrometric method go back to the mid-20th century. In 1943 perturbations were suspected in the proper motions of two red dwarfs, 70 Ophiuchi and 61 Cygni, but were found to be due to causes other than planets. At the Sproule Observatory in the United States, P. van de Kamp began in 1937 a long series of observations of Barnard's Star, a red dwarf only six light-years away (in fact, the nearest star apart from those of the α Centauri group). The luminosity of the star is 0.000 45 that of the Sun, and it has the greatest known proper motion, so that in 170 years it crawls across the sky by a distance equal to the apparent diameter of the full moon. In 1963 van de Kamp announced the detection of a planet with a mass similar to that of Jupiter, moving round the star at a distance of 4.4 a.u. He later claimed that there were at least two planets in the system. However, it was later found that the perturbations were spurious, and were due to faults in van de Kamp's telescopic equipment.

In 1983 R. Harrington, using the 155 cm reflector at the US Naval Observatory, announced that there was a low-mass companion to the faint red dwarf star Van Biesbroeck 8 B. It was said to have a mass from five to eight times that

of Jupiter, and to have a surface temperature of 1700 °C – not because it was self-luminous, but because it was being heated by its parent star. Further searches failed to locate it, and it is certainly non-existent. Next, in 1988, slight effects were noted with the naked-eye star γ Cephei (Alrai) which were attributed to the presence of a planet, but these were subsequently found to be due to changes in the star itself.

In 1991 came a startling announcement from Jodrell Bank. Using the Lovell radio telescope, A. Lyne and his colleagues announced the discovery of a planet orbiting a pulsar, PSR 1829-10. The method used was that of timing the pulsations, which would be affected by the pull of an orbiting body. The period was given as exactly six months. This too proved to be spurious; the researchers had not taken into account the fact that the Earth's orbit is an ellipse rather than a circle. As soon as they realized this, they promptly and generously admitted their mistake.

There was a major development in September 1995, when two Swiss astronomers, Michel Mayor and Didier Queloz, made some new observations with the 30 m reflector at the Haute Provence Observatory, in France, and announced the detection of a planet orbiting the star 51 Pegasi, a G4-type star 54 light-years away; it is fractionally less luminous than the Sun, and at an apparent magnitude of 5.5 is easily visible with the naked eye. Mayor and Queloz used an indirect method. As a planet orbits a star, the star itself swings round the common centre of gravity of the system. This affects its radial velocity as seen from Earth, and this can be measured by making use of the Doppler effect; the equipment used by Mayor and Queloz was sensitive enough to detect changes down to 12 m s^{-1}. The result seemed reliable, but the planet was very peculiar. The mass was about half that of Jupiter, but the distance from the parent star was a mere 7 000 000 km – about one-eighth the distance between the Sun and Mercury, the innermost planet of our Solar System. In this case, the surface should be baked to a temperature of around 1300 °C; the orbital period was 4.3 days. All in all, it seemed to be a most improbable object, but other stars were soon found to have planetary companions of the same kind. Among them were τ Boötis, ρ Coronæ Borealis and ρ' Canori. Another early report referred to a planet orbiting the star 70 Virginis, but this was much more massive, so that it

Table 20.1. Selected list of extra-solar planets.

Star	Mag.	Spectrum	Distance (light-years)	Period (days)	Mass, Jupiter = 1	Separation (a.u.)	Eccentricity
Gleise 876	10.17	M4	15	60.85	2.1	0.21	0.27
47 Ursæ Majoris	5.05	G1	42	1108	2.4	2.1	0.09
υ Andromedæ	4.63	F8	44	4.62	0.71	0.06	0.03
				241.2	2.11	0.83	0.18
				1266.5	4.61	2.50	0.41
ρ′ (55) Cancri	5.95	G8	45	14.65	0.84	0.11	0.05
51 Pegasi	5.49	G2	50	4.23	0.47	0.05	0.015
ι Horologii	5.42	G0	56	320	2.3	0.93	0.38
τ Boötis	4.50	F6	51	3.31	3.87	0.05	0.02
ρ Coronæ Bor.	5.43	G0	57	39.64	1.1	0.23	0.03
14 Herculis	6.67	K0	59	1619	3.3	2.6	0.35
70 Virginis[a]	5.16	G4	59	116.6	6.6	0.43	0.40
16 Cygni B	6.2	G1	70	804	1.5	1.70	0.57

[a] Possibly a brown dwarf.

Reported pulsar planets

Pulsar or neutron star	Period	Mass	Separation (a.u.)	Eccentricity
PSR 1257 + 12	25.3	0.015 (Earth = 1)	0.19?	0.00?
	66.54	3.4 (Earth = 1)	0.36	0.02
	98.22	2.8 (Earth = 1)	0.47	0.03
	170 years	95 (Jupiter = 1)	35	?
PSR B1620 − 26	100 years	10 (Jupiter = 1)	35	?
Geminga	5.1 years	1.7 (Earth = 1)?	3.3?	?

(The quoted mass figures are minimum values, because the orbital inclinations are not known. All that can be calculated is $M \sin i$ rather than M (M being the mass, i the orbital inclination). Therefore, the masses could well be greater than the values listed here.)

may well be a brown dwarf or at least something different from a true planet. One star, υ Andromedæ, has apparently three planets; one is rather less massive than Jupiter, but the outermost member of the family could match well over four Jupiters. In November 1999 a team of astronomers using the Keck I telescope in Hawaii (S. Vogt, P. Butler and K. Apps) announced the discovery of six more extra-solar planets orbiting solar-type stars. In most cases the masses of the planets were comparable with that of Jupiter. In March 2000 G. Marcy, P. Butler and S. Vogt announced the discovery of a planet orbiting the star 79 Ceti, 117 light-years away. The orbit is elliptical, with a mean distance of 0.35 a.u. from the star. The temperature must be of the order of 830 °C. The mass is probably about 70% that of Saturn.

Other planetary attendants have been suggested; it is even claimed that the brilliant Aldebaran has a companion, 11 times as massive as Jupiter, at a distance of 1.3–1.4 a.u. – although this is very tentative, and if the companion really exists it could be a brown dwarf. Another possible way of detecting extra-solar planets is to observe the effects produced when the companions pass in front of the parent star.

In December 1999 it was announced that a team of British astronomers, using the 4.2 m William Herschel Telescope on La Palma, had obtained visual confirmation of a planet orbiting the star τ Boötis. C. Cameron, K. Horne and D. James from the University of St. Andrews, together with A. Penney, developed a sensitive new computer program which was able to untangle the faint light of the planet from the glare of the parent star. The computer analysis revealed a faint copy of the star's spectrum, embedded in the starlight and wobbling back and forth in a sense exactly opposite to the star's own, smaller wobble. This is exactly how light reflected from a closely orbiting planet should behave; the planet's light will be blue-shifted when the planet is approaching Earth, red-shifted when it is receding (the overall motion of the system being taken into account). The wobble was only half the size expected for a planet in a nearly edge-on orbit. This indicates that the orbit was strongly tilted to the line of sight. Knowing the tilt enabled the team to find that the planet's mass is 8 times that of Jupiter. The tilt also explains why the planet is so faint. Even when the planet is on the far side of the star, we never see the face fully illuminated, so that the planet is always 30 000 times fainter than the star.

It has been suggested that the planet's atmosphere may be a chemical cauldron at a temperature of around 1700 °C. D. Sudarsky and his colleagues at the University of Arizona recently predicted an alien environment in which elements such as sodium and potassium, which take the form of metals on Earth, are present as trace gases in a hot hydrogen atmosphere above a cloud-deck of magnesium silicate droplets.

Certainly the sighting of the τ Boötis planet is a major advance. No doubt other visual discoveries will be made in the near future.

The European Space Agency is already planning what is known as the Darwin project. This involves sending several small infra-red telescopes into the outer part of the Solar System; these will be linked to form an interferometer. The aim will be to search for planets of stars within a range of 50 light-years. It is hoped that the first of the so-called 'Cornerstone' telescopes will be launched in 2009.

The first 'pulsar planet' proved to be spurious, but several others have been reported since, mainly by A. Wolszczan, working at the Penn State University together with D. Frail. In 1990 Wolszczan used the Arecibo radio telescope to study the pulsar 1257 + 12, which has a period of 6.2 ms and its between 1500 and 3000 light-years away.

He found evidence of two companions of planetary mass, both very close to the pulsar. Planets have also been reported with the pulsar PSR B1620 − 26 and the strange neutron star Geminga, which is probably a pulsar but is not detectable as such because its radio beams do not sweep over the Earth (though in the case of Geminga there are serious doubts). Pulsar planets are admittedly very difficult to explain, and if they really exist, as most researchers believe, they must indeed be bizarre worlds.

A selected list of extra-solar planets is given in Table 20.1.

THE DRAKE EQUATION

Assuming that extra-solar planets exist – and of this there no longer seems any reasonable doubt – we must speculate as to the chances of intelligent civilizations with which we might communicate. In 1961 Frank Drake worked out a method of estimating mathematically the numbers of worlds which might support civilizations able and willing to communicate with us. The celebrated Drake Equation is as follows:

$$N = R^* \, f_p \, n_e \, f_l \, f_i \, f_c \, L$$

where

N = the number of communicative civilizations,
R^* = the rate of formation of suitable solar-type stars,
f_p = the fraction of those stars which have planetary systems,
n_e = the number of Earth-like worlds per planetary system,
f_l = the fraction of those Earth-like worlds where life actually develops,
f_i = the fraction of life-sites where intelligence develops,
f_c = the fraction of planets supporting a civilization technologically capable of communicating with us, and
L = the fraction of a planet's life in which a communicative civilization endures.

It is all too obvious that most of these figures are highly uncertain; for example, consider L – does a civilization wipe itself out when it has the technical capability to do so, and lacks the will to refrain? (After all, we on Earth are in precisely this position at the present moment.) Drake came to the final conclusion that there ought to be about 10 000 communicative civilizations in our Galaxy. Whether or not he was correct remains to be seen!

21 DOUBLE STARS

Double stars are of two types; optical pairs (that is to say, line-of-sight effects) and binaries (physically-associated pairs). Binaries are much the more frequent. They range from 'contact pairs', where the components are almost or quite touching, to very distant pairs separated by at least a light-year. In a binary system the components move round their common centre of gravity. For visual binaries, the shortest period is that of Wolf 630 Ophiuchi (1.725 years), but shorter periods are known; the record-holder is X-1820-303, an X-ray star in the globular cluster NGC 6623, distance 30 000 light-years. Its period is 685 s or 11 min. It was discovered in 1987 by the aptly-named Luigi Stella and collaborators with the Exosat satellite. It is impossible to say which is the binary with the longest period, and all we can say is that very widely-separated components share a common motion through space.

Early observations

The term 'double star' was first used by Ptolemy, who wrote that η Sagittarii was '$\delta\iota\pi\lambda\upsilon\zeta$'. There are of course several doubles which can be separated with the naked eye, so that presumably they have been known since antiquity; of these the most celebrated is Mizar (ζ Ursæ Majoris), which makes a naked-eye pair with Alcor (80 Ursæ Majoris). The Arabs described it –although they regarded Alcor as a rather difficult object. This is not true today, but it is most unlikely that there has been any real change.

The first double star to be discovered telescopically was Mizar itself, which is made up of two rather unequal components separated by 14″.5. The discovery was made by Riccioli in 1651. Alcor is 700″ from the main pair, which is rather too wide to be classed as a recognized 'double' in the official catalogues. The duplicity of γ Arietis was discovered in 1665 by Robert Hooke, while he was searching telescopically for a comet.

The first southern double star to be discovered was α Crucis, by Father Guy Tachard, in 1685. Tachard was on his way to Siam, by sea, and stopped off at the Cape of Good Hope, where he was warmly welcomed by the Dutch settlers and set up a temporary observatory, mainly for navigational purposes. He recorded that 'the foot of the Crozier marked in Bayer's map is a Double Star, that is to say consisting of two bright stars distant from one another about their own Diameter, only much like the most northern of the Twins, not to speak of a third much less, which is also to be seen but further from these two'. The Crozier is the Southern Cross; 'Bayer' refers to J. Bayer's famous catalogue of 1603, and the 'northern of the Twins' is Castor, which was already known to be double. (Like most of his contemporaries, Tachard believed the stars to show definite apparent diameters rather than being virtual point sources.)

Other doubles discovered at an early stage were α Centauri (1689), γ Virginis and the 'Trapezium', θ Orionis, in the Orion Nebula, which is a multiple system.

Double Star catalogues

The first true list of doubles was published in 1771 by C. Mayer of Mannheim. This list included γ Andromedæ, ζ Cancri, α Herculis and β Cygni. Most of his observations were made with an 80 ft mural quadrant, with magnifications of 60 and 80. Many catalogues have appeared since. Among them are those by F. G. W. Struve (Dorpat 1822, with later additions); E. Dembowski (Naples 1852; over 20 000 measures), S. W. Burnham (1870 and again in 1906, listing 13 665 pairs – he was personally responsible for the discovery of 1340 of them); R. Aitken (1932, 17 180 pairs) and the Lick Index Catalogue or IDS (1963, 65 000 pairs, of which 40 000 are binaries). Work of the greatest importance was carried out at the Republic Observatory, Johannesburg (formerly the Union Observatory) between 1917 and 1965, under the successive directorships of R. T. A. Innes (1917–27), H. E. Wood (1927–41), W. H van den Bos (1941–56) and W. S. Finsen (1957–65); the telescope used was the 27 inch (69 cm) Innes refractor.

William Herschel, of course, discovered large numbers of double stars, and in our own time superbly accurate measurements have been made from the Hipparcos astrometric satellite; it was launched on 8 August 1989, and its main catalogue appeared seven years later. The catalogue includes details of over 12 000 double stars, of which 3000 were new discoveries. The accuracy was truly amazing. For example, images were obtained of the double star HIP 46706 in Hydra, which is 34 light-years away; the components are separated by only about 1/5000 of a degree. This is a binary system (period 18.3 years); the masses of the components are 0.42 and 0.41 that of the Sun.

The position angle of a visual double star (either an optical or a binary pair) is measured according to the angular direction of the secondary (B) from the primary (A), reckoned from 000 at north round by east (090), south (180) and west (270), back to north. With rapid binaries the position angles and separations alter quickly; a good example is the fine binary ζ Herculis, where the magnitudes are 3 and 5.6, and the period only 34 years. Some of the published catalogues are already out of date, and need revising. There is scope here for the skilful and well-equipped amateur as well as for the professional.

Binary systems

The first suggestion that some double stars might be physically-associated pairs was made by the Rev. John Michell in 1766, who wrote: 'It is highly probable, in particular, and next to a certainty, in general, that such double stars as appear to consist of two or more stars placed very near together, do really consist of stars under the influence of some general law.' Michell repeated this view in 1784. However, in 1782 William Herschel commented that it was 'much too soon to form any theories of small stars revolving around large ones'. The actual proof was given by Herschel himself in 1802. From 1779 he had been attempting to measure the parallaxes of stars, since if one member of the pair were more remote than the other it followed that the closer member should show an annual parallax relative to the more distant component. He failed, because his equipment was not sufficiently sensitive, but he made the fortuitous discovery that some of the pairs under study (such as Castor) showed orbital motion, and by 1802 he was confident enough to publish his findings. His classic paper actually appeared in the *Philosophical Transactions* on 9 June 1803.

The first reliable orbit for a binary pair (ξ Ursæ Majoris) was worked out by the French astronomer Felix Savary in 1830. (The period is 60 years.) Such calculations are of great importance, since they lead to a determination of the combined masses of the components – something which is much more difficult to calculate for a single star. In fact, our knowledge of stellar masses depends very largely upon the orbital movements of binaries. It has been calculated that more than 50% of all stars are members of binary systems, with a mean separation of 10–20 a.u., although this may be too high a ratio. It is important to note that both components of a binary move round their common centre of gravity, and the orbits are not in general so dissimilar as might be thought, since in mass there is a far lower spread among the stars than there is in luminosity and in size.

Incidentally, it was the measurements of a binary – 61 Cygni – which led F. W. Bessel, in 1838, to make the first successful determination of the distance of a star. 61 Cygni was selected because it had a large proper motion and was a wide binary, indicating that by stellar standards it must be comparatively close (the distance is 11.1 light-years).

The first *spectroscopic binary* (Mizar A) was discovered in 1889 by E. C. Pickering at Harvard; another identification (β Aurigæ) soon followed. Spectroscopic binaries have too small a separation for the components to be seen individually, but the binary nature of the system betrays itself because of the Doppler shifts in the spectra. If both spectra are visible, the absorption lines will be periodically doubled; if one spectrum is too faint to be seen, the lines due to the primary will oscillate around a mean position. There are some 'borderline' cases; thus Capella was long known to be a spectroscopic binary, but only the world's largest telescopes can just indicate that it is not a single star. The orbital period is 100 days.

The first *astrometric binaries* to be studied were Sirius and Procyon, by F. W. Bessel in 1844. In an astrometric binary, the presence of an invisible companion is inferred from slight displacements of the primary. (This, of course, is also one important method for detecting planetary or brown dwarf companions of visible stars.)

There is an interesting corollary with regard to the Companion of Sirius. It was first seen in 1862 by Clark, using the Washington refractor, almost exactly where Bessel had predicted. The orbital period is 50 years, and the

Table 21.1. Selected list of prominent double stars.

Star	RA h	min	Declination °	''	Magnitude	Separation ('')	P.A.	
β Tuc	00	31.5	−62	58	4.4, 4.8	27.1	170	Both components again double.
η Cas	00	49.1	+57	49	3.4, 7.5	12.7	315	Creamy, bluish. Binary, 480 y.
ζ Psc	01	13.7	+07	35	5.6, 6.5	23.1	063	
γ Ari	01	53.6	+19	18	4.8, 4.8	7.5	001	
α Psc	02	02.0	+02	46	4.2, 5.1	1.8	274	Binary, 933 y.
γ And	02	03.9	+42	20	2.3, 5.0	9.6	062	B is double; 5.5, 6.3; 0''.5', 61°.
ι Cas	02	29.2	+67	25	4.9, 6.9	3.0	227	Binary, 840 y. 3rd star at 7''.2, PA 118, mag 8.4.
ω For	02	33.8	−28	14	5.0, 7.7	10.8	245	Fixed. Common proper motion.
γ Cet	02	43.3	+03	14	3.5, 7.3	2.6	298	
θ Eri	02	58.3	−40	18	3.4, 4.5	8.2	088	Fine pair. Both white. Acamar.
ε Ari	02	59.2	+21	20	5.2, 5.5	1.5	208	
α For	03	12.1	−28	59	4.0, 7.0	4.8	301	Binary, 314 y.
ι Pic	04	50.9	−53	28	5.6, 6.4	12.3	058	Fixed.
κ Lep	05	13.2	−12	56	4.5, 7.4	2.2	357	
β Ori	05	14.5	−08	12	0.1, 6.8	9.5	202	Rigel. Fixed. Common proper motion.
η Ori	05	24.5	−02	24	3.8, 4.8	1.8	078	3rd star, mag 9.4, 115''.1, 051°.
λ Ori	05	35.1	+09	56	3.6, 5.5	4.3	043	Fixed.
θ Ori AB	05	35.3	−05	23	6.7, 7.9	8.8	032	Trapezium. In M.42.
CD	''		''		5.1, 6.7	13.4	061	
σ Ori AC	05	38.7	−02	36	4.0, 10.3	11.4	238	
ζ Ori	05	40.8	−01	57	1.9, 4.0	2.4	162	Alnitak. Binary, 1509 y
η Gem	06	14.9	+22	30	var, 6.5	1.6	257	Propus. A is orange. Binary, 474 y.
γ Vol	07	08.8	−70	30	4.0, 5.9	14.1	298	Combined mag 3.6.
δ Gem	07	20.1	+21	59	3.5, 8.2	5.8	225	Binary, 1200 y. Yellow, pale blue.
α Gem	07	34.6	+31	53	1.9, 2.9	3.7	067	Castor. Binary, 420 y. Widening.
k Pup	07	38.8	−26	48	4.5, 4.7	9.8	318	Combined mag 3.8.
ζ Cnc AB+C	08	12.2	+17	39	5.3, 6.0	5.9	074	Binary, 1150 y.
δ Vel	08	44.7	−54	43	2.1, 5.1	2.6	153	
ε Hya	08	46.8	+06	25	3.3, 6.8	3.3	298	A is a close binary, 890 y.
υ Car	09	47.1	−65	04	3.1, 6.1	5.0	127	Fixed.
γ Leo	10	20.0	+19	51	2.2, 3.5	4.6	125	Binary, 619 y. 2 distant companions.
μ Vel	10	46.8	−49	25	2.7, 6.4	2.6	057	Binary, 116 y. Closing.
ξ UMa	11	18.2	+31	32	4.3, 4.8	1.6	080	Binary, 59.8 y. Opening.
ι Leo	11	23.9	+10	32	4.0, 6.7	1.7	118	Binary, 192 y.
N Hya	11	32.3	−29	16	5.8, 5.9	9.5	210	Fixed.
D Cen	12	14.0	−45	43	5.6, 6.8	2.8	243	Orange, white. Closing.
α Cru	12	26.6	−63	06	1.4, 1.9	4.0	113	Acrux. Combined mag 0.8. C at 90''.1, 202°, mag 4.9.
γ Cru	12	31.2	−57	07	1.6, 6.7	110.6	031	C at 155''.2, 082°, mag 9.5.
γ Cen	12	41.5	−48	58	2.9, 2.9	1.0	347	Binary, 84.5 y. Closing.
γ Vir	12	41.7	−01	27	3.5, 3.5	1.6	264	Binary, 171.4 y. Closing.
β Mus	12	46.3	−68	06	3.7, 4.0	1.2	039	
μ Cru	12	54.6	−57	11	4.0, 5.2	34.9	017	
α CVn	12	56.0	+38	19	2.9, 5.5	19.1	299	Cor Caroli. Yellow, bluish.
ζ UMa	13	23.9	+54	56	2.3, 4.0	14.4	152	Mizar. Alcor at 708''.7, mag 4.0, 0.71''.
α Cen	14	39.6	−60	50	0.0, 1.2	14.8	221	Binary, 79.9 y. Closing.
ζ Boö	14	41.1	+13	44	4.5, 4.6	0.8	300	Closing. Binary, 123.3 y.
ε Boö	14	45.0	+27	04	2.5, 4.9	2.9	341	Yellow, blue.
ξ Boö	14	51.4	+19	06	4.7, 6.8	6.7	319	Binary, 150 y.
π Lup	15	18.5	−47	53	4.6, 4.7	1.7	067	Widening. 7.2 mag star at 23''.7, 130°.
γ Cir	15	23.4	−59	19	5.1, 5.5	0.8	011	Closing. Binary, 180 y.
δ Ser	15	34.8	+10	32	4.1, 5.2	4.0	175	Binary, 3168 y.
ζ CrB	15	39.4	+36	38	5.1, 6.0	6.3	305	
γ Lup	15	56.9	−33	58	5.3, 5.8	10.2	049	Fixed.
σ CrB	16	14.7	+33	52	5.6, 6.6	7.0	236	Binary, 1000 y.
α Sco	16	29.4	−26	26	1.2, 5.4	2.7	274	Antares. Red, green. Binary, 878 y.
ζ Her	16	41.3	+31	36	2.9, 5.5	0.9	029	Binary, 34.5 y. PA and separation change quickly.
μ Dra	17	05.3	+54	28	5.7, 5.7	2.0	017	Binary, 482 y. Closing.
α Her	17	14.6	+14	23	var, 5.4	4.6	105	Red. Green. Binary, 3600 y.
70 Oph	18	05.5	+02	30	4.2, 6.0	3.4	152	Binary, 88.1 y. Widening.
ε¹ Lyr	18	44.3	+39	40	5.0, 5.1	2.6	357	Separation 207''.7. Quadruple.
ε² Lyr	18	44.3	+39	40	5.2, 5.5	2.3	094	
θ Ser	18	56.2	+04	12	4.5, 4.5	22.4	104	Fixed.
γ CrA	19	06.4	−37	04	4.8, 5.1	1.3	061	Binary, 120.4 y.
β Cyg	19	30.7	+27	58	3.1, 5.1	34.4	054	Albireo. Yellow, blue.
δ Cyg	19	45.0	+45	07	2.9, 6.3	2.5	226	Widening.
ε Dra	19	48.2	+70	16	3.8, 7.4	3.1	019	Slow binary.
γ Del	20	46.7	+16	07	4.5, 5.5	9.2	206	Both yellowish.
61 Cyg	21	06.9	+38	45	5.2, 6.0	30.5	150	Binary, 722 y.
θ Ind	21	19.9	−53	27	4.5, 7.0	6.8	271	Yellow and red. Slow binary.
μ Cyg	21	44.1	+28	45	4.8, 6.1	2.0	206	Binary, 716 y.
ξ Cep	22	03.8	+64	38	4.4, 6.5	8.0	276	White and blue.
ζ Aqr	22	28.8	−00	01	4.3, 4.5	1.9	187	Binary, 856 y. Widening.
δ Aps	16	20.3	−78	42	4.7, 5.1	102.9	012.	
α Aql	19	50.8	+08	52	0.8, 9.5	165.2	301	Altair. Optical pair.
θ Aur	05	59.7	+37	13	2.6, 7.1	3.6	313	10.6 mag. star at 50'', 297°.

Table 21.1. (Continued)

Star	RA h	RA min	Declination °	Declination ″	Magnitude	Separation (″)	P.A.	
ε CMa	06	58.6	−28	58	1.5, 7.4	7.5	161.	
α Cap	20	18.1	−12	33	3.6, 4.2	377.7	291.	
β Cap	20	21.0	−14	47	3.1, 6.0	205	267	B is a close double.
β Cep	21	28.7	+70	34	3.2, 7.9	13.3	239.	
δ Cep	22	29.2	+58	25	var, 7.5	41.0	191.	
o Cet	02	19.3	−02	59	var, 9.5	0.6	085	Mira. Binary, 400 y. B is VZ Ceti.
κ CrA	18	33.4	−38	44	5.9, 5.9	21.6	359.	
δ Her	17	15.0	+24	50	3.7, 8.2	8.9	236	Optical pair.
β Hya	11	52.9	−33	54	4.7, 5.5	0.9	008	
γ Lep	05	44.5	−22	27	3.7, 6.3	96.3	350	
α Lib	14	50.9	−16	02	2.8, 5.2	231.0	314	
ζ Lyr	18	44.8	+37	36	4.3, 5.9	43.7	150	
ε Mon	06	23.8	+04	36	4.5, 6.5	13.4	027	
θ Mus	13	08.1	−65	18	5.7, 7.3	5.3	187	
ε Nor	16	27.7	−47	33	4.8, 7.5	22.8	335	
λ Oct	21	50.9	−82	43	5.4, 7.7	3.1	070	
ρ Oph	16	25.6	−23	27	5.3, 6.0	3.1	344	
β Phe	01	06.1	−46	43	4.0, 4.2	1.4	346	
β PsA	22	31.5	−32	21	4.4, 7.9	10.3	172	Optical pair.
ζ Ret	03	18.2	−62	30	4.7, 5.2	310.0	218	Common proper motion.
β Sco	16	05.4	−19	48	2.6, 4.9	13.6	021	A is a close double.
θ Tau	04	28.7	+15	32	3.4, 3.8	337.4	346	White, orange. Optical pair.
κ + 67 Tau	04	25.4	+22	18	4.2, 5.3	339	173	Optical pair.
α UMi	02	31.8	+89	16	2.0, 9.0	18.4	218	Polaris.
α + 8 Vul	19	28.7	+24	40	4.4, 5.8	413.7	028	Optical pair.

maximum separation is 11″.5 (as in 1975). For many years after its discovery the companion was assumed to be faint because it was cool and red, but in 1915 W. S. Adams, at Mount Wilson, studied its spectrum and found that it was white; the surface temperature was at least 8000 °C. Since the luminosity was only 1/10 000 of that of Sirius itself, the companion had to be small – no larger than a planet such as Uranus or Neptune. The companion was, in fact, the first known white dwarf, with a density 125 000 times that of water. The absolute magnitude is +11.4, and if we could bring a cubic inch of its material back to Earth the weight would be about two and a half tons. Since Sirius is commonly known as the Dog Star, the companion has predictably been nicknamed the Pup. The companion of Procyon, first seen in 1896 by J. M. Schaeberle with the aid of the Yerkes refractor, is also a white dwarf; the separation was greatest in 1990, and is now decreasing again to a minimum of no more than 2 arcsec. The real separation is 2 250 000 000 km, rather less than the distance between our Sun and the planet Uranus.

Binary systems are of many different kinds; sometimes the components are dissimilar (as with Sirius), sometimes they are identical twins – as with γ Virginis, which has a period of 171 years. Several decades ago γ Virginis was wide and easy to split with a very small telescope, but it is now closing, and by 2016 will be single except when seen with a very powerful telescope, after which it will start to open out again. This does not indicate any actual change in separation; everything depends upon the angle from which we see the pair. Another binary, formerly easy to split but now much less so, is Castor. Here, both components of the bright pair are spectroscopic binaries, and also associated with the system is Castor C or YY Geminorum, made up of two red dwarfs; it is an eclipsing binary. Castor therefore consists of six stars, four luminous and two very dim.

Multiple stars are not uncommon. Of special note is ε Lyræ, near Vega in the sky; it has two main components, making up a naked-eye pair, and each component is again double, making up a quadruple system. θ Orionis, in the Orion Nebula (M.42) has been nicknamed the Trapezium, for obvious reasons; it lies on the outskirts of the nebula and is responsible for making the nebulosity luminous.

Some double stars show beautiful contrasting colours. Thus β Cygni has a golden-yellow primary with a vivid blue companion; Antares and α Herculis have fainter green secondaries; with δ Geminorum the primary is yellow and the companion pale blue. A list of prominent double stars is given in Table 21.1.

Origin of binary systems

The old theory – that a binary was formed as a result of the fission of a single star – has been abandoned, and neither is it likely that binaries are due to the mutual capture of the components. It seems that the components were formed from the same cloud of interstellar material in the same region of space. When there is a marked difference in brightness between the two components, the spectra also differ. If both stars belong to the Main Sequence, the primary is usually of earlier type than the secondary, while if the primary is a giant the secondary is either a giant of earlier type or else a dwarf of similar spectral type. Novæ are binary systems (see below).

If the two components of a binary system are close together, the evolution of one component may profoundly affect the other. In such a system there is an hourglass-shaped region bounded by the points where the two stars will equally affect a small particle; each of the two segments of the 'hourglass' encloses a region termed a Roche lobe. The two Roche lobes may touch at what is termed the Inner Lagrangian Point. The giant component of a close binary will evolve more quickly than its lower-mass companion; as it expands it may fill its Roche lobe completely, and material will flow across the Inner Lagrangian Point on to the second star, so that there is actual transfer of mass from one component to the other, and the original secondary may in time become the more massive of the two. There are also cases in which the two components share a common envelope as the expansion continues, and if one component accumulates enough mass it may explode as a supernova. The evolutionary careers of the members of a close binary system are decidedly complicated.

White dwarf binaries are known; the first of these was found by W. Luyten and P. Higgins in 1973 (RA 9h 42m, declination $+23°41'$). The separation is 13 arcsec, the position angle 052 and the period 12 000 years; at present the components are 600 a.u. apart. X-ray binaries were reported following the launch of the first X-ray astronomical satellite, UHURU, in 1970. A system of this kind consists of a white dwarf, neutron star or black hole orbiting another star which may, in some cases, be a high-mass O- or B-type star, and in others a B- or K-type star of mass similar to that of the Sun.

Eclipsing binaries

As the two components of a binary move round their common centre of gravity, it may happen that one component will pass in front of the other as seen from Earth; this will cause a change in brightness. Stars of such a kind are usually called eclipsing variables, although 'eclipsing binary' is much more accurate. The first to be discovered was Algol (β Persei), by G. Montanari in 1669. It is not now thought that the variability was known in ancient times, even though Algol was always called 'the Demon Star'. With Algol, one component is considerably brighter than the other, so that there is a deep minimum when the primary star is eclipsed (even though the eclipse is not total). The drop in magnitude when the fainter member of the two is hidden is too slight to be noticed with the naked eye. With stars of the β Lyræ type the two components are much less unequal, and are almost or quite touching, so that changes in brightness are always going on. Since eclipsing binaries do show changes in brightness, they are dealt with in the next section.

Incidentally, even the term 'eclipsing binary' is technically wrong. It really ought to be 'occulting binary'.

22 VARIABLE STARS

Variable stars are of many types. Elaborate systems of classifying them have been proposed, and the data given here are not intended to be more than a general guide. Seven major categories are now recognized.

(1) Eclipsing stars (more properly eclipsing binaries, because they are not intrinsically variable).
(2) Pulsating variables; either radial or non-radial pulsations.
(3) Eruptive variables, where the changes are caused by flares or the ejection of shells of material.
(4) Cataclysmic variables, where the changes are due to explosions in the star or in an accretion disk round it. Novæ, dwarf novæ and supernovæ come into this category.
(5) Rotating variables, where the changes are caused by star-spots, non-spherical shape or magnetic effects.
(6) X-ray variables, usually inherent in the neutron star or black hole companion of a binary.
(7) Unclassifiable stars, which do not fit into any accepted category.

Mention should also be made of what are termed secular variables: stars which have permanently brightened or faded in historic times. Thus Ptolemy ranked β Leonis and θ Eridani as of the first magnitude, whereas today they are below magnitude 2 and 3 respectively; α Ophiuchi was ranked of magnitude 3, but is now 2.1. However, these changes must be regarded as highly suspect. It is unwise to trust the old observations too far.

Early identifications

The first variable to be positively identified as such was Mira (o Ceti) in 1638. It had been recorded by Fabricius (1596) and Bayer (1603), and Bayer had even allotted it a Greek letter, so that it is surprising that its fluctuations were not recognized until 1638 (by Phocylides Holwarda). In the latter part of the 17th century two more variables were identified, Algol and χ Cygni. Table 22.1 lists the variables identified between 1638 and 1850, excluding novæ.

Table 22.1. The first known variable stars.

Star	Discoverer of variability	Date
Mira (o Ceti)	Holwarda	1638
Algol (β Persei)	Montanari	1669
χ Cygni	Kirch	1686
R Hydræ	Maraldii	1704
Rasalgethi (α Herculis)	W Herschel	1759
μ Cephei	W Herschel	1782
R Leonis	Koch	1782
δ Cephei	Goodricke	1784
β Lyræ	Goodricke	1784
η Aquilæ	Pigott	1784
R Scuti	Pigott	1795
R Coronæ Borealis	Pigott	1795
R Virginis	Harding	1809
R Aquarii	Harding	1811
ε Aurigæ	Fritsch	1821
R Serpentis	Harding	1826
η Carinæ	Burchell	1827
S Serpentis	Harding	1828
U Virginis	Harding	1831
δ Orionis	J Herschel	1834
S Vulpeculæ	Rogerson	1837
Betelgeux (α Orionis)	J Herschel	1840
β Pegasi	Schmidt	1847
λ Tauri	Baxendell	1848
R Orionis	Hind	1848
R Pegasi	Hind	1848
R Capricorni	Hind	1848
S Hydræ	Hind	1848
S Cancri	Hind	1848
S Geminorum	Hind	1848
R Geminorum	Hind	1848
T Geminorum	Hind	1848
R Tauri	Hind	1849
T Virginis	Boguslawsky	1849
T Cancri	Hind	1850
R Piscium	Hind	1850

The recognized abbreviations for the different classes of variables are given in Table 22.2. A more detailed scheme is given in Table 22.3 (page 286).

Table 22.2. Variable star types.

E	Eclipsing binary
EA	Algol type
EB	Beta Lyræ type
EW	W Ursæ Majoris type
M	Mira type (long period)
SR	Semi-regular
SRa	Semi-regular; well-defined periodicity
SRb	Semi-regular; poorly-defined periodicity
SRc	Semi-regular; disk component stars
SRd	Semi-regular; types F, G or K
RR	RR Lyræ variable
RRa	RR Lyræ; sharp asymmetrical light curve
RRa, b	RR Lyræ; asymmetrical light curve
RRc	RR Lyræ; symmetrical sinusoidal light curve
RV	RV Tauri type
RVa	RV Tauri; constant mean brightness
RVb	RV Tauri; varying mean brightness
L	Irregular
Lb	Slow irregular variations
Lc	Irregular supergiants
Cep	Cepheid
δ Cep	Classical Cepheid
CW	W Virginis star (type II Cepheid)
DSct	δ Scuti type
SXPHE	SX Phœnicis type
ACYG	Deneb (α Cygni) type
β Cep	β Cephei type
ZZ	ZZ Ceti type
FU	Fuors (FU Orionis type)
GCas	γ Cassiopeiæ type
IN	Orion variables
IT	T Tauri variables
RCB	R Coronæ Borealis variables
RS	RS Canum Venaticorum variables
SDOR	S Doradûs variables
UV	Flare stars (UV Ceti)
W	Unstable Wolf–Rayet stars
AM	Polars (AM Herculis type)
UG	Dwarf novæ (U Geminorum or SS Cygni)
UGZ	Dwarf novæ Z Camelopardalis type
ZABD	Symbiotic stars (Z Andromedæ type)
N	Novæ
RN	Recurrent novæ
NL	Novalike variables
SN	Supernovæ
ACV	Magnetic variables (α^2 Canum Venaticorum)
BY	BY Draconis type
ELL	Ellipsoidal variables
FKCOM	FK Comæ variables
SXARI	SX Arietis type
Pec	Peculiar; not fitting into any class

Eclipsing binaries

The light changes are due entirely to mutual eclipses (or, to be accurate, occultations) of the two components.

Algol type (EA). The components are more or less spherical, and are unequal, so that an Algol variable remains at maximum for most of the time. The secondary minimum, when the faint component is hidden, is often very slight.

With Algol (β Persei), the prototype EA star, the primary eclipse it not total. The main component (Algol A) is of type B, 105 times as luminous as the Sun, with a diameter of about 4 000 000 km. Algol B, the companion, is not genuinely dark; it is of type G, and about three times as luminous as the Sun. It has a diameter of 5 500 000 km, so that it is larger than the primary and qualifies as a sub-giant. The secondary minimum is less than 0m.1 in amplitude. The distance from the Earth is 95 light-years.

EA stars are common enough, but few are naked-eye objects; apart from Algol, only λ Tauri, δ Libræ and ζ Phœnicis rise above the fifth magnitude. A list of bright eclipsing stars is included in the bright-variable catalogue (Table 22.4).

β Lyræ type (EB). Here the two components are much less unequal, so that there are alternate deep and shallow minima. The prototype star, β Lyræ itself, has a period of 13 days. The maximum magnitude is 3.4. The star fades to magnitude 3.8, recovers and then goes through its primary minimum, which takes it down to below 4. It then returns to maximum, and the cycle is repeated. The components are almost touching each other, and are tidally distorted into egg-like shapes; the more massive component has filled its Roche lobe, and material is streaming through the Inner Lagrangian Point to form an accretion disk round the less massive star. There is evidence that the components are connected by huge streamers of gas moving at over 300 km s^{-1}. The primary minimum as seen from Earth is caused by a total eclipse, the secondary minimum by a partial eclipse.

W Ursæ Majoris type (EW). Dwarf binaries of types F or G; the components are almost or quite in contact, and the two minima are more or less equal. The periods are less than one day. There are no naked-eye examples; the maximum magnitude of the prototype, W UrsæMajoris itself, is only 7.9.

Very long-period eclipsing binaries. A few eclipsing stars have periods of years. The most celebrated of these is ε Aurigæ, one of a triangle of stars close to Capella in the sky (they are often known as the Hædi, or Kids). ε Aurigæ is generally just above the third magnitude, but every 27 years it fades, taking over five months to drop down to magnitude 3.8. The minimum lasts for just over a year, followed by a gradual recovery. The eclipsing companion has never been seen, and neither has its spectrum; the primary is a very luminous supergiant of type F – according to the Cambridge catalogue, 200 000 times as powerful as the Sun, although this may be something of an over-estimate. It is generally believed that the secondary is a fairly normal, very hot blue star which is contained inside a huge shell of gas which is virtually opaque and which causes the eclipses. The last eclipse began on 22 July 1982; totality lasted from 11 January 1983 to 16 January 1984, and the partial phase ended on 25 June 1984, so that the next eclipse is not due until 2009.

Also in the triangle of the 'Kids' is the other celebrated long-period eclipsing binary, ζ Aurigæ (this is sheer coincidence; ζ is much the closer of the two). The period is 972 days, and both spectra are visible; the primary is a supergiant, while the companion is much smaller and hotter. As the supergiant begins to hide the secondary, at the start of the eclipse, the light of the secondary comes to us through the supergiant's outer layers, and there are complicated spectral changes which are highly informative. Totality lasts for 38 days, and the magnitude falls from 3.7 to 4.2.

Pulsating variables

Here the variations are intrinsic. The star expands and contracts, changing its surface temperature and its output as it does so.

Mira variables (M), often called long-period variables; Mira Ceti is the brightest member of the class. They are late-type red giants with emission lines in their spectra. The periods range from 80 days to over 1000 days, and the amplitudes are large – well over 10 magnitudes in some cases (as with χ Cygni, where the extreme range is from magnitude 3.3 to 14.2). Neither the periods nor the amplitudes are constant, and no two cycles are exactly alike. At some maxima Mira may rise to the second magnitude,

and it is reported that in 1779 it matched Aldebaran, but other maxima are no brighter than magnitude 4, and on average Mira is a naked-eye object for only a few weeks in every year. Minima are always about magnitude 10. The period is 332 days, but this may vary by a day or two either way.

Mira is a huge star. Measurements with the Hubble Space Telescope give the angular diameter as 60 milliarcsec. This corresponds to a true diameter 700 times that of the Sun. Yet it is not so massive or so luminous as might be expected; the absolute magnitude ranges between −2.5 and +4.7, so that even at its peak it is not much more than 1000 times as powerful as the Sun.

Mira has a binary companion, which is itself variable and has been given a variable star designation (VZ Ceti). The range is from magnitude 9.5 to 12. The companion is probably a white dwarf in interaction with Mira, surrounded by an accretion disk of material which it has captured from the primary. The orbital period is 400 years, and the separation 70 a.u. The Hubble Space Telescope has also detected a small, hook-like appendage extending from Mira in the direction of the companion, and Mira is itself rather football-shaped instead of spherical. No doubt the presence of the companion is responsible for this.

Several Mira variables reach naked-eye visibility – for instance, χ Cygni (which is a particularly strong infra-red source) and the southern R Carinæ. All are of spectral types M or later.

Semi-regular variables (SR). These also are of late spectral type. In some cases the periods are so rough that they are almost unrecognizable, and again no two cycles are alike. The amplitudes are much less than with Mira stars; the periods range from 20 days to over 1000 days. There are various sub-divisions:

SRa: late-type supergiants with relatively stable cycles;
SRb: less regular cycles;
SRc; red supergiants;
SRd; orange to yellow supergiants.

The brightest semi-regular variable is Betelgeux, in Orion. The official magnitude range is from 0.2 to 0.9, but there are indications that it may occasionally become as

Table 22.3. The classification of variable stars.

Eclipsing variables (or eclipsing binaries)

EA	Algol	Period 0.2 day to over 27 years. Almost spherical components. Maximum for most of the time.
EB	β Lyræ	Periods over 1 day; spectra B to A. Ellipsoidal components; magnitude continuously changing.
EW	W Ursæ Majoris	Dwarfs, periods usually less than 1 day. Almost or quite in contact. Primary and secondary minima almost equal.

Pulsating variables

M	Mira	Long-period late-type giants; spectra M–C–S, periods 80 to 1000 days. Amplitude may exceed 10 magnitudes. Periods and amplitudes vary from cycle to cycle.
SR	Semi-regular	Late-type giants (spectra M–C–S). Periods from 20 days to several years, SRA: persistent periodicity (Z Aquarii). SRB: rough periodicity (RR Coronæ Borealis). SRG: red supergiants (μ Cephei). SRD: F to K type giants and supergiants (SX Herculis).
PVTEL	PV Telescopii	Helium supergiants; periods from a few hours to a year; spectra Bp; small amplitudes (0.1 mag).
RR	RR Lyræ	Spectra A to F; periods 0.2 to 1.2 days; formerly called cluster-Cepheids. RRAV: steep rise to maximum. RRB: almost symmetrical light curves. RRC: sinusoidal light curves.
RV	RV Tauri	Supergiants, usually type F to K, Periods 30 to 150 days. RVA: stars with constant mean magnitude (AC Herculls). RVB: stars with variable mean magnitude.
L	Irregular	Types K, M, C, S. LB: slow giant variables (CO Cygni). LC: slow supergiant variables (TZ Cassiopeiæ).
CEP	Cepheids	Radial pulsating stars, Periods 1 to 135 days; spectra F to K.
CW	W Virginis	Population II Cepheids. CWA: longer period (W Virginis). CWB: shorter period (BL Herculis).
DSCT	δ Scuti	Types A to F; periods less than 1 day; small amplitude.
SXPHE	SX Phoenicis	Population II sub-dwarfs resembling δ Scuti stars. Types A to F, periods less than 1 day, amplitude up to 0.7 magnitude.
ACYG	α Cygni	Types 8 to A; supergiants, with low amplitudes. Most have short periods.
BCEP	β Cephei (or β Canis Majoris)	B-type subgiants. Low amplitude and short period.
ZZ	ZZ Ceti	Non-radially pulsating white dwarfs; amplitude up to 0.2 magnitude; periods 30 seconds to 1500 seconds. ZZA: hydrogen absorption lines only (spectrum DA). ZZB: helium absorption lines only (spectrum DV). ZZO: very hot stars (spectrum DO).

Eruptive variables

FU	Fuors	Prototype, FU Orionis. Types A to G; slow rise to maximum (many years) and slower decline.
GCAS	γ Cassiopeiae	Shell stars; type B; rapid rotaters. Amplitudes usually below 2 magnitudes.
I	Irregular	Types O to M. IA: early spectral type. 1B: intermediate and late spectral type.
IN	Orion	INA: early type (T Orionis). INB: later type (spectra F to M).
IT	T Tauri	Similar to Orion variables: types F to M (RW Aurigæ). INT: types F to M; contained in diffuse nebulæ (T Tauri itself). IS: rapidly-varying stars, amplitudes up to 1 magnitude; ISA (early type, spectra B to A). ISB (later type, spectra F to M).
RCB	R Coronæ Borealis	Types B to R. Occasional deep minima; large amplitude (over 9 magnitudes in some cases).
RS	RS Canum Venaticorurn	Small amplitude. Close binaries with active chromospheres.
SDOR	S Doradûs	Types Bp to Fp. Very luminous supergiants with expanding shells, often in diffuse nebulæ.
UV	UV Ceti	Dwarfs of types K to M. Flare stars; amplitudes may be as much as 6 magnitudes (UV Ceti itself).
W	Wolf-Rayet	Very low amplitudes; spectra of type W. Non-stable mass outflow.

Table 22.3. (Continued)

Rotating variables

ACV	α^2 Canum Venaticorum	Types B to A; strong magnetic fields; spectra rich in silicon, strontium, chromium and rare earth lines. Periods from 12 hours up to 160 days.
BY	BY Draconis	Types G to M; periods up to 20 days. Rotating dwarfs with starspots and active chromospheres.
ELL	Ellipsoidal	Close binaries with no eclipses, but changing visible area. Low amplitudes (up to 0.1 magnitude).
FKCOM	FK Com	Types G to K. Periods up to several days. Amplitudes up to half a magnitude. Rapidly rotating giants with non-uniform surface brightness.
SXARI	SX Arietis	Helium stars (type B); amplitude around 0.1 magnitude and periods of around 1 day. High-temperature versions of α^2 Canum Venaticorum stars.

Cataclysmic variables

AM	AM Herculis	Polars; close binaries with one compact component. Amplitude up to 5 magnitudes.
UG	U Geminorum	Dwarf novae. Periods from 10 to 1000 days, amplitudes from 2 to 9 magnitudes. SS Cygni stars have outbursts lasting for several days.
UGSU	SU Ursæ Majoris	Dwarf novae, with occasional supermaxima brighter and longer than normal maxima.
UGZ	Z Camelopardalis	Dwarf novae which have occasional 'standstills' when the normal cycle of variation is suspended.
ZAND	Z Andromedæ	Symbiotic stars; close binaries.
N	Novæ	Thermonuclear outburst on the white dwarf component of a binary system. NA: fast, fading by 3 magnitudes in 100 days or less (GK Persei 1901). NAB: fading at intermediate speed. NB: slow fading, no more than 3 magnitudes in 150 days (RR, Pictoris 1925). NC: very slow novæ with maxima which may last for years, as with RR Telescopii.
NL	Nova-like	NL: poorly-studied stars with nova-like outbursts (V Sagittæ). NR: recurrent novæ (such as T Coronæ Borealis) which flared in 1866 and again in 1946.
SN	Supernovæ	SNI: explosion and destruction of the white dwarf component of a binary system. SNII: collapse of a very massive star, often leaving a neutron star or pulsar.

A few variables, such as VY Canis Majoris, do not seem to fit into any class, and there are also very exceptional stars such as η Carinæ. Pulsars and X-ray binaries are often included in variable star classification lists.

bright as Rigel. The period is given as 2110 days, but this is very rough indeed. The distance given in the Cambridge catalogue (310 light-years) may be an underestimate; other catalogues increase it to 520 light-years. The angular diameter is $0''.048$, and this is large enough for surface details to be seen with the largest telescopes. There are apparently huge convection cells rising to the surface. The real diameter is greater than that of the orbit of the Earth. At its peak, Betelgeux is 15 000 times as luminous as the Sun, but its mass is only between 10 and 20 times that of the Sun; there is a vast, extended atmosphere. Betelgeux is well advanced in its evolutionary sequence, and eventually it will no doubt explode as a supernova. When this happens, the apparent magnitude as seen from Earth will be around -9.

μ Cephei – nicknamed the 'Garnet Star' by William Herschel because of its colour – is much larger even than Betelgeux, and may in fact be the largest star known, but is further away (well over 1000 light-years). The range is from magnitude 3.4 to just below 5, and the period is officially given as 730 days, but, as with Betelgeux, this is very rough indeed. Other bright semi-regulars are α Herculis, η Geminorum and β Pegasi. With β Pegasi, in the 'Square', the 38 day period is reasonably well defined.

Cepheids (CEP) take their name from δ Cephei, the first-discovered member of the class. They are radially pulsating yellow giants, with periods of from a few days to over 100 days; the amplitudes are from 0.1 to 2 magnitudes. They are of importance because their periods are linked

Table 22.4. A catalogue of bright variable stars. The following list includes variable stars with a maximum of magnitude 6 or brighter, and a range of at least 0m.4.

	Max.	Min.	Period (d)	Spectrum
Eclipsing binaries				
(Algol type)				
R Ara	6.0	6.9	4.4	B
WW Aur	5.8	6.5	2.5	A + A
R CMa	5.7	6.3	1.1	F
RS Cha	6.0	6.7	0.1	A + F
δ Lib	4.9	5.9	2.3	B
U Oph	5.9	6.6	1.7	B + B
β Per	2.2	3.4	2.9	B + G Algol
ζ Phe	3.9	4.4	1.7	B + B
RS Sgr	6.0	6.9	2.4	B + B
λ Tau	3.3	3.8	3.9	B + A
HU Tau	5.9	6.7	2.1	A
(Beta Lyræ type)				
UW CMa	4.0	5.3	4.3	07
u Her	4.6	5.3	2.0	B + B
GG Lup	5.4	6.0	2.1	B + A
β Lyr	3.3	4.3	12.9	B + A Sheliak
V Pup	4.7	5.2	1.4	B + B
(Long period)				
ε Aur	2.9	3.8	9892	F
ζ Aur	3.7	4.1	972	K + B
VV Cep	4.8	5.4	7430	M + B
Mira variables				
R And	5.8	14.9	409	S
R AqI	5.5	12.0	284	M
R Car	3.9	10.5	309	M
S Car	4.5	9.9	149	K–M
R Cas	4.7	13.5	430	M
R Cen	5.3	11.8	546	M
o Cet	1.7	10.1	332	M Mira
S CrB	5.8	14.1	360	M
X Cyg	3.3	14.2	407	S
U Cyg	5.9	12.1	462	N
R Gem	6.0	14.0	370	S
S Gru	6.0	15.0	401	M
R Hor	4.7	14.3	404	M
R Hya	4.0	10.0	390	M
R Leo	4.4	11.3	312	M
R Lep	5.5	11.7	432	N
V Mon	6.0	13.7	334	M

Table 22.4. (Continued)

	Max.	Min.	Period (d)	Spectrum
Mira variables				
X Oph	5.9	9.2	334	M + K
U Ori	4.8	12.6	372	M
RU Sgr	6.0	13.8	240	M
RT Sgr	6.0	14.1	305	M
RR Sco	5.0	12.4	279	M
S Scl	0.5	13.6	365	M
R Ser	5.1	14.4	356	M
R Tri	5.4	12.6	266	M
SS Vir	6.0	9.6	355	N
R Vir	6.0	12.1	146	M
Semi-regular variables				
UU Aur	5.1	6.8	234	N
W Boö	4.7	5.4	450	M
VZ Cam	4.7	5.2	24	M
X Cnc	5.6	7.5	195	N
TU CVn	5.6	6.6	50	M
S Cen	6.0	7.0	65	N
T Cen	5.5	9.0	90	K–M
T Cet	5.0	6.9	159	M
FS Com	5.3	6.1	58	M
W Cyg	5.0	7.6	126	M
U Del	5.7	7.6	110	M
EU Del	5.8	6.9	59	M
R Dor	4.8	6.6	338	M
μ Cep	3.4	5.1	750	M
UX Dra	5.9	7.1	168	N
RY Dra	5.6	8.0	173	N
η Gem	3.2	3.9	233	M Propus
π' Gru	5.4	6.7	150	S
g Her	5.7	7.2	70	M
α Her	3	4	±100	M Rasalgethi
R Lyr	3.9	5.0	46	M
ε Oct	4.9	5.4	55	M
α Ori	0.1	0.9	2110	M Betelgeux
W Ori	5.9	7.7	212	N
CK Ori	5.9	7.1	120	K
Y Pav	5.7	8.5	233	N
SX Pav	5.4	6.0	50	M
β Peg	2.3	2.8	38	M Scheat
ρ Per	3	4	33–55	M
TV Psc	4.6	5.4	70	M
L² Pup	2.6	6.2	140	M
R Scl	5.8	7.7	370	N
RR UMi	6.0	6.5	40?	M

Table 22.4. (Continued)

	Max.	Min.	Period (d)	Spectrum
Eruptive variables				
U Ant	5.7	6.8		N
η Car	−0.8	7.9		Pec
ρ Cas	4.1	6.2		F–K (Occasional fades)
α Cas	2. 1?	2.5?		K (Suspected variable)
γ Cas	1.6	3.3		B
μ Cen	2.9	3.5		B
θ Cir	5.0	5.4		B
P Cyg	3	4		Bp
T Cyg	5.0	5.5		K
BU Gem	5.7	7.5		M
RX Lep	5.0	7.0		M
S Mon	4	5		07
BO Mus	6.0	7.7		M
χ Oph	4.2	5.0		B
λ Pav	3.4	4.3		B
X Per	6.0	7.0		09.5 X-ray star
d Ser	4.9	5.9		G+A
VY UMa	5.9	6.5		N
BU Tau	4.8	5.5		Bp Pleione
Cepheids				
η Aql	3.5	4.4	7.2	F–G
RT Aur	5.0	5.8	3.7	F–G
ZZ Car	3.3	4.2	36.5	F–K
U Car	5.7	7.0	38.8	F–G
SU Cas	5.7	6.2	1.9	F
δ Cep	3.5	4.4	5.4	F–G

Table 22.4. (Continued)

	Max.	Min.	Period (d)	Spectrum
Cepheids				
AX Cir	5.6	6.1	5.3	F–G
X Cyg	5.9	6.9	16.4	F–G
β Dor	3.7	4.1	9.8	F–G
ζ Gem	3.7	4.1	10.1	F–G
T Mon	6.0	6.6	27.0	F–K
S Mus	5.9	6.4	9.7	F
R Mus	5.9	6.7	7.5	F
Y Oph	5.9	6.4	17.1	F–G
κ Pav	3.9	4.7	9.1	FSv (W Virginis type)
S Sge	5.3	6.0	8.4	F–G
X Sgr	4.2	4.8	7.0	F
W Sgr	4.3	5.1	7.6	F–G
Y Sgr	5.4	6.1	5.8	F
AH Vel	5.5	5.9	4.2	F
T Vul	5.4	6.1	4.4	F–G
RV Tauri variable				
R Sct	4.4	8.2	140	G–K
Symbiotic variables				
R Aqr	5.8	12.4	387	M+P
AG Peg	6.0	9.4	830	WN+M

	Max.	Min.	Spectrum	Outbursts
Recurrent novæ				
T CrB	2.0	10.8	M+Q	1866,1946 Blaze Star
RS Oph	5.3	12.3	O+M	1901,1933, 1958, 1967
R Coronæ Borealis variables				
R CrB	5.7	15	Fp	
RY Sgr	6.0	15	Gp	

with their real luminosities, and this means that once the period is known the distance can be found; Cepheids act as 'standard candles' in space, and because they are highly luminous they can be seen over vast distances. The Cepheid period – luminosity law states that the longer the period, the more powerful the star. The period of pulsation is the time taken for a vibration to travel from the surface of the star towards the centre and back again, so that these periods are longer for larger and brighter stars; at its peak δ Cephei is at least 6000 times more luminous than the Sun. Other bright Cepheids are η Aquilæ, ζ Geminorum and β Doradûs.

Polaris is a Cepheid of very small amplitude. In 1899 it was found to have a range of magnitude from 1.92 to 2.07 and a period of 3.969 778 days. Subsequently the amplitude decreased, and by 1992 was down to 0.010 of a magnitude. It was thought that the pulsations might cease altogether, but since 1995 the amplitude has stabilized at 0m.03. It remains in the 'instability strip' of the H–R diagram, where every star ought to pulsate; this strip occupies a region between the Main Sequence and the red giant branch.

Unlike Mira stars, Cepheids are perfectly regular, and the cycles repeat each other, so that the magnitude at any particular moment can be predicted. Either the rise of maximum is sharper than the subsequent decline, or else the light-curve is virtually symmetrical.

W Virginis stars (CW) are associated with Cepheids, and may be termed Type 2 Cepheids or Population II Cepheids. They have lower masses than the classical Cepheids, and are about two magnitudes fainter; they are also metal-poor rather than metal-rich, and their periods are less precise. The only naked-eye example is κ Pavonis, in the southern sky.

RR Lyræ variables (RR). These were once called cluster-Cepheids, because they are common in globular clusters; however, many of them (including the prototype, RR Lyræ) are not cluster members. The spectra are of type A or F. They are old, with masses lower than that of the Sun, but radii four to five times greater, and all are of about the same luminosity, 95 times that of the Sun, so – like the Cepheids – they can be used as 'standard candles'. The amplitude is around 1 magnitude, and the periods are from 0.2 to 1.2 days,. There are various sub-divisions. RRAB stars (formerly divided into two sub-classes, RRa and RRb) have amplitudes of 0.3 to 1 magnitude, and periods of from 0.5 to 0.7 days. RRc stars have almost sinusoidal light-curves, with amplitudes around 0.5 magnitude and periods around 0.3 days. Some RR Lyræ stars, such as AR Herculis, have several pulsation periods, and this results in a continuous deformation of the light curve (Blazhko effect). Dwarf Cepheids were once classed with the RR Lyræ stars, as RRs; the prototype is AI Velorum. They are of types A to F, with absolute magnitudes of from +1 to +5, and periods from 0.05 to 0.25 day.

RV Tauri variables (RV). Radially pulsating yellow to red supergiants, usually of types F to K, but occasionally M. There are alternate deep and shallow minima; the interval between successive primary minima may be from 30 to 150 days, with amplitudes up to 4 magnitudes. There are several subdivisions:

RVa: the period is not constant, but is at least reasonably consistent for most of the time; the brightest member of the class, R Scuti, is of this type;

RRb: several superimposed cycles, with rough periods of from 30 to 100 days.

δ Scuti variables (DSCT). These are stars belonging to the galactic disk (Population I) and are young; types A to F. The amplitudes range from below 0.1 up to 0.9 magnitude, and many are spectroscopic binaries. Sub-dwarfs showing the same characteristics are termed *SX Phœnicis stars.*

In 1999 a new class was proposed. The prototype star is γ Doradûs, a 4th-magnitude star in the southern hemisphere. These stars have spectra around type F0; they are main sequence stars slightly hotter and more massive than the Sun, but cooler than δ Scuti stars. Stars such as δ Scuti are expected to pulsate, since they lie in the instability strip of the H–R Diagram, but γ Doradûs stars are outside the instability strip. It is suggested that they undergo non-radial pulsations; portions of their outer layers expand outward while other portions contract, as against stars such as Cepheids which expand and contract as a unit, thereby preserving spherical symmetry. The amplitudes of γ Doradûs stars are very small (several hundredths of a magnitude).

β Cephei variables (BCEP) (sometimes called β Canis Majoris variables). B0 to B3 giants or sub-giants, with periods of from 0.1 to 0.7 day, and amplitudes from 0.1 to 0.3 magnitude. They are relatively massive stars which have almost exhausted their core hydrogen. Ultra-short periods of a few hundredths of a day are suffixed s; a typical example is χ Centauri.

α Cygni variables (ACYG). Pulsating supergiants of types B or A; very small amplitudes. Deneb is the best-known example.

ZZ Ceti variables (ZZ). Pulsating white dwarfs, with periods which may be as short as 30 s and never as long as half an hour. The amplitudes are below 0.2 mag. Suffixes A, B or O indicate spectral features such as lines of hydrogen, helium or carbon.

Eruptive variables

Fuors (FU). Named after the prototype star, FU Orionis. In a star of this kind there is a very slow rise to a maximum which may last for years, followed by a slower decline. Emission lines develop in the spectrum. The types are usually A to G.

γ Cassiopeiæ variables (GCAS). Rapidly rotating blue giants, usually of type B, which occasionally throw off shells of material. *γ* Cassiopeiæ itself is the prototype. Its usual magnitude is just below 2, but in the late 1930s it rose to 1.6 before declining to 3.2; for decades now the magnitude has hovered around 2.2.

T Tauri variables. These are very young stars, still contracting towards the Main Sequence and varying irregularly; they are strong infra-red emitters, and send out pronounced stellar winds. T Tauri itself lies in a dark dust cloud; many are found inside nebulæ. The Sun certainly passed through a T Tauri stage early in its evolution.

R Coronæ Borealis variables (RCB). These are hydrogen-poor, but rich in carbon. They are highly luminous, and remain at a maximum for most of the time; they then undergo sudden, unpredictable falls to a minimum, taking from several weeks to many months to recover. The amplitudes are large – at least 10 magnitudes in the case of R Coronæ itself, which is usually on the fringe of naked-eye visibility (magnitude 6) but at some minima falls below magnitude 15. The fadings are due to clouds of soot accumulating in the star's atmosphere. R Coronæ stars are rare.

RS Canum Venaticorum variables (RS). Close binaries with chromospheric activity, causing slight light variations.

S Doradûs variables (SDOR). Massive, very luminous blue supergiants, usually surrounded by expanding envelopes. They are often found in diffuse nebulæ; generally the amplitude is small, but there may be occasional outbursts up to 7 magnitudes, lasting for many weeks. These are due to the ejection of shells of material. S Doradûs itself lies in the Large Cloud of Magellan, at a distance of 169 000 light-years, and has a mass about 60 times that of the Sun; although it is almost a million times as luminous as the Sun, it is too faint to be seen with the naked eye (magnitude 11).

Flare stars: UV Ceti type (UV). Red dwarfs, of type K or (usually) M. Flare activity produces sudden outbursts of from 1 to 6 magnitudes, lasting for several minutes. All flare stars show emission lines in their spectra. UV Ceti itself is normally of magnitude 13.4, but on one occasion brightened abruptly to 6.8. Many red dwarfs show flare activity of a less spectacular kind; among these is Proxima Centauri, the nearest star to the Sun. Flash variables (UVa) are of earlier spectral type, and are more luminous; they are associated with nebulosity.

Wolf–Rayet stars (W). These are highly luminous, and are unstable, with expanding envelopes; this causes random, low-amplitude changes in brightness. Their spectra show emission lines of nitrogen and carbon, and there is evidence for unstable mass outflow as a stellar wind.

The exceptional star *ρ* Cassiopeiæ, near the W of Cassiopeia, is officially classed as an eruptive variable, but it is very much of a puzzle. Normally it is of magnitude 5.1, with slight fluctuations, but there are occasional minima, and in 1946 it dropped briefly to magnitude 6.2. It is immensely luminous, with an absolute magnitude of at least −8, and may be one of the most powerful stars in the Galaxy. Its spectral type is F8.

Rotating variables

α² Canum Venaticorum stars (ACV). These are white stars, usually of type B or A, which have intense magnetic fields. It seems that there are huge 'starspots', produced by the magnetic fields; there are variable spectral lines due to silicon, strontium, chromium and rare-earth elements. Periods range between 12 h and 160 days. The spectra are very variable, but the light fluctuations are very small.

BY Draconis stars (BY). Red dwarfs, with large starspots and active chromospheres. The periods may reach 20 days.

Ellipsoidal variables (ELL). Tidally distorted close binaries. The changes in light are due to the varying areas presented, but there are no eclipses, as with *β* Lyræ stars. As the fluctuations are not intrinsic, it would be more accurate to refer to these stars as ellipsoidal binaries.

FK Comæ variables (FKCOM). Rapidly rotating yellow or orange giants, types G to K, with periods up to several days and amplitudes of no more than half a magnitude. The surfaces are of non-uniform brightness, and it is this which causes the fluctuations, so that again it cannot really be said that these stars are intrinsically variable.

SX Arietis variables (SXARI). These are similar to the α^2 Canum Venaticorum stars, except that they are richer in helium and have higher temperatures. The amplitudes never exceed 0.1 magnitude.

Finally there are *reflection binaries (R)*, where a large, cool component is illuminated by its hotter companion, leading to variations of up to a magnitude as the system rotates.

η Carinæ. Unquestionably the most erratic of all variables is η Carinæ (before the dismemberment of the old constellation of Argo Navis, it was known as η Argûs). It has a strange history. It was recorded by Halley, in 1677, as being of the fourth magnitude. It remained between magnitudes 4 and 2 until 1827, when it blazed up to the first magnitude. In 1837 John Herschel, from the Cape, made it as bright as α Centauri, and following a slight fade it reached its greatest brilliance in April 1843, when it outranked Canopus and almost matched Sirius. Then a decline set in. By 1870 η was only of the sixth magnitude, and it has remained at around this level ever since. At its peak, the luminosity was of the order of 6000 000 times that of the Sun, and it has not really declined a great deal, since most of its emission is in the infra-red. The distance is of the order of 8000 light-years, and the mass is at least 100 times that of the Sun. It is associated with nebulosity, and superimposed on this is a dark, dusty region known as the Keyhole Nebula from its shape.

What apparently happened in the years following 1834 is that there was a massive explosion, which threw off a shell of material from the surface of the star. As the shell expanded, the star seemed to brighten; after 1843 the shell itself cooled, dimmed and finally became opaque, hiding the light of the star beneath. After a century and a half of expansion, at a rate of 700 km s^{-1}, we now see the shell as a tiny nebula, nicknamed the Homunculus Nebula from its shape; it is orange in colour, and telescopically η Carinæ looks quite unlike an ordinary star. Images taken with the Hubble Space Telescope show the billowing clouds of expanding material.

η Carinæ is unstable, and there is no doubt that it will explode as a supernova. This will probably happen within the next few hundreds of thousands of years.

P Cygni. Another very remote, very luminous star is P Cygni. It was first recorded in 1600, when it was of the third magnitude. By 1626 it had dropped below naked-eye visibility, but brightened again in 1655. After further fluctuations is settled down in 1715 at magnitude 5, and since then there has been little change. It was once classed as a nova, but is now officially listed as a variable of the S Doradûs type. However, its luminosity may be comparable with that of η Carinæ, and it too is very unstable. Certainly it ranks as one of the most powerful stars in the Galaxy. Its distance is not known accurately; one estimate is just under 6000 light-years. A few other stars of the same general type are known.

Cataclysmic variables

Novæ. Novæ can be spectacular in the extreme. The name is misleading, because a nova is not a new star. What happens is that a formerly faint star suffers a tremendous outburst, and flares up to many times its normal brightness, remaining bright for a few days, weeks or months before fading back to obscurity.

Some novæ remain bright only briefly. The rise may take only a few hours, as with the bright nova V1500 Cygni of 1975, which reached an absolute magnitude of −10 and an apparent magnitude of 1.8; within a week it had fallen below naked-eye visibility, whereas slow novæ, such as HR Delphini, of 1967, are much more gradual in their decline. A suffix 'a' indicates a fall of three magnitudes in less than 100 days; 'b', three magnitudes in 100 to 150 days; 'c', three magnitudes in over 150 days.

A list of naked-eye novæ seen since 1600 is given in Table 22.5. The brightest nova, V603 Aquilæ of 1918, briefly outshone every star in the sky apart from Sirius.

A nova is a binary system, made up of a low-density red star with a white dwarf companion. The white dwarf pulls material away from the giant star, and this material builds up into an accretion disk round the white dwarf. Over long periods of time more and more material collects; it is hydrogen-rich and at a high temperature, and the very strong surface gravity of the white dwarf generates tremendous pressures. Eventually the situation becomes unstable, and a runaway nuclear reaction begins, hurling material into space at speeds up to to 1500 km s^{-1}. The brilliancy may

Table 22.5. Naked-eye Novæ. The following list includes all novæ since 1600 to have attained magnitude 6.0 or brighter.

		Max. mag.	Discoverer
CK Vulpeculæ	1670	3	Anthelm
WY Sagittæ	1783	6	D'Agelet
V. 841 Ophiuchi	1848	4	Hind
Q Cygni	1876	3	Schmidt
T Aurigæ	1891	4.2	Anderson
V. 1059 Sagittarii	1898	4.9	Fleming
GK Persei	1901	0.0	Anderson
DM Geminorum	1903	5.0	Turner
OY Aræ	1910	6.0	Fleming
DI Lacertæ	1910	4.6	Espin
DN Geminorum	1912	3.3	Enebo
V. 603 Aquilæ	1918	−1.1	Bower
GI Monocerotis	1918	5.7	Wolf
V. 476 Cygni	1920	2.0	Denning
RR Pictoris	1925	1.1	Watson
XX Tauri	1927	6.0	Schwassmann and Wachmann
DQ Herculis	1934	1.2	Prentice
V. 368 Aquilæ	1936	5.0	Tamm
CP Lacertæ	1936	1.9	Gomi
V. 630 Sagittarii	1936	4.5	Okabayasi
BT Monocerotis	1939	4.3	Whipple and Wachmann
CP Puppis	1942	0.4	Dawson
DK Lacertæ	1950	6.0	Bertaud
RW Ursae Minoris	1956	6.0	Satyvaldiev
V. 446 Herculis	1960	5.0	Hassell
V. 533 Herculis	1963	3.2	Dahlgren and Peltier
HR Delphini	1967	3.7	Alcock
LV Vulpeculæ	1968	4.9	Alcock
FH Serpentis	1970	4.4	Honda
V. 1500 Cygni	1975	1.8	Honda
NQ Vulpeculæ	1976	6.0	Alcock
V. 1370 Aquilæ	1982	6.0	Honda
QU Vulpeculæ	1984	5.6	Collins
V. 842 Centauri	1986	4.6	McNaught
V. 838 Herculis	1991	5.0	Alcock
V 1974 Cygni	1992	4.3	Collins
V. 705 Cassiopeiæ	1993	5.4	Kanatsu
V. 382 Velorum	1999	2.5	Williams and Gilmore
V. 1494 Aquilæ	1999	3.6	Pereira

increase by a factor of at least 1000, but the outburst is brief; after a few days, weeks or months the nova returns to its old state. The mass of the ejected material is no more than 1/1000 of a solar mass. The spectra of novæ show absorption lines which are blue-shifted, indicating that gas thrown off during the outburst is moving towards us.

At the peak of the outburst, a nova may be highly luminous. Nova Puppis 1942 sent out as much radiation as 1 600 000 Suns. Compared with this, the fluctuations of variable stars such as Mira seem very minor.

In some cases an old nova may be seen to be surrounded by a gas cloud. GK Persei 1901 is a case in point. Some months after maximum it was found that nebulosity was appearing to one side of the star, and to be expanding at the speed of light. This was clearly unacceptable; in fact the nova lay in a dark nebula, and the radiation spreading out from the outburst illuminated more and more of the nebula each year. The even brighter nova V.603 Aquilæ developed a tiny surrounding disk, which grew steadily in size and became fainter; by 1941 it had become too dim to be followed further. Many old novæ are now seen as eclipsing binaries; such is DQ Herculis 1934, which has a period of 4h 39m. The least luminous 'old nova' is CK Vulpeculæ of 1670, which is 2000 light-years away and has only 1/100 the luminosity of the Sun. (See Table 22.5.)

Some stars have been known to undergo more than one outburst; these are the *recurrent novæ* (see Table 22.6). The best known of these is the 'Blaze Star', T Coronæ Borealis. Usually it is of around magnitude 10, but in 1866 it flared up briefly to magnitude 2; it was then regarded as a normal nova, but in 1946 it flared up again, this time to magnitude 3. Several recurrent novæ are now under observation.

Faint novæ are by no means uncommon, and many of these are discovered by amateurs. Ten naked-eye novæ have appeared since 1970; one of these, V.382 Velorum of 1999, reached magnitude 2.5. It was discovered on 22 May by Alan Gilmore in New Zealand, and independently by Peter Williams in Australia; it was too far south in the sky to be seen from Europe. At its peak it shone 75 000 times more brightly than the Sun; the distance is believed to be 2600 light-years, and before the outburst its magnitude was below 16.

Table 22.6. Selected list of recurrent novæ.

Star	Outbursts	Max. mag.
T Coronæ Borealis	1866, 1946	2.0
RS Ophiuchi	1901, 1933, 1958, 1967	5.1
T Pyxidis	1890, 1902, 1920, 1925, 1945	7.0
WZ Sagittæ	1913, 1946, 1979	7.0

WZ Sagittæ is an exceptional system, consisting of a white dwarf and a less massive 'normal' star. Over the ages, the white dwarf had drained material from its companion, so that the companion is now small and exceptionally cool – surface temperature about 1450 °C. It may eventually end up as an unique type of stellar end-product.

Dwarf novæ. These are usually termed *U Geminorum* stars, although much the brightest member of the class is SS Cygni. They show minor outbursts at roughly regular intervals. Of the two components, one is a K or M type dwarf and the other is a white dwarf. The amplitude is of the order of from 2 to 9 magnitudes (usually less), and the interval between successive outbursts may be from 10 days to several years – 103 days for U Geminorum (magnitude 14.9 to 8.2), 50 days for SS Cygni (12.4 to 8.2). The basic cause of the outbursts is the same as for true novæ, but on a much reduced scale. *SU Ursæ Majoris* stars have both normal maxima and occasional 'supermaxima' of greater amplitude, while *Z Camelopardalis* stars show outbursts of from 2 to 5 magnitudes every 10 to 40 days, but with unpredictable 'standstills' when variations are temporarily suspended. *Z Andromedæ* stars or symbiotic variables are close binaries, where the hot companion actually orbits within the envelope of its cool red giant companion; the variations are caused by pulsations in the red star together with interactions between the two, so that the light curves are decidedly complicated. *RR Telescopii* stars show slow increases which may be of very long duration. *Polars*, or AM Herculis stars, show sudden outbursts of up to three magnitudes, caused by the accretion of material on to the magnetic poles of a compact star; the light is strongly polarized – hence the name.

X-ray novæ. An X-ray nova is a binary system in which there are sudden outbursts at X-ray wavelengths. Around 200 X-ray binaries are known in our Galaxy, and of these about one-third have been seen to suffer outbursts. They are of two main classes:

(a) High mass; a hot blue star (type O or B0) together with a neutron star. There are more or less regular periodic outbursts.

(b) Low mass; a cool red star (M or K type) with a compact object which may be either a neutron star or a black hole. The outbursts are unpredictable.

The strong gravitational pull of the compact member of the pair collects material from its companion to form an accretion disk; material is then sucked down on to the surface of the compact member, and an outburst occurs when there is a sudden increase in the amount of material striking the compact object. X-rays are emitted, but optical and radio outbursts can also be detected. The term 'X-ray nova' is rather misleading; 'X-ray variable' would have been better.

Supernovæ

Supernovæ are among the most colossal outbursts known in nature. Many have been seen in external galaxies – over 700 since 1885 – but the last supernova to be seen in our Galaxy was observed as long ago as 1604, before the invention of the telescope. It has been estimated that a typical galaxy should produce one supernova every 25 to 100 years, but some of these will be concealed by interstellar dust.

Types. Supernovæ are of two definite types. With type Ia, we are dealing with a binary system, in which one component (A) is initially more massive than its companion (B), and therefore evolves more quickly into the red giant stage. Material from it is pulled across to B, so that B grows in mass while A declines; eventually B becomes the more massive of the two. Meanwhile, A is reduced to a very small, dense core, made up chiefly of carbon. The situation is then reversed; B evolves in its turn, swells out, and starts to lose material back to the shrunken A, which is now a white dwarf. The white dwarf builds up a gaseous layer. When the mass exceeds the Chandrasekhar limit (about 1.4 times the mass of the Sun) the carbon detonates, and in a matter of a few seconds the white dwarf is completely destroyed. The resulting outburst takes some time to die down. When the carbon white dwarf explodes, it creates other elements, including neon, oxygen and silicon, ending up as nickel. Nickel decays to cobalt and then to iron, so that these elements can be detected spectroscopically; there are no lines due to hydrogen. At its peak, a type Ia supernova can reach an absolute magnitude of -19, or around 4000 million times the luminosity of the Sun.

Some type I supernovæ may be due to collapsing single stars. These are classed as of type Ib and Ic, to distinguish them from the type Ia supernovæ described above.

A type II supernova is due to the collapse of a very massive star, at least five times as massive as the Sun, which has used up its nuclear fuel and has produced a nickel–iron core with a mass about 1.5 times that of the Sun. The structure of the star just prior to the outburst has been likened to that of an onion. Outside the core there is a zone of silicon and sulphur; next comes a layer of neon and magnesium; then a layer of carbon, neon and oxygen; then a layer of helium; and finally an outer layer of hydrogen. Eventually the core can no longer support the weight of the outer layers, and collapses to form a neutron star of very great density (you could pack over 2500 million tons of neutron star material into a matchbox). Vast numbers of neutrinos are produced, most of which pass straight through the star into space. When the collapsing layers hit the unyielding neutron star, they 'bounce', and a shock-wave moves outwards, colliding with the material which is still falling inwards. The result is a catastrophic explosion, and most of the star is blown away into space, leaving only the core – which may be a neutron star or, in extreme cases, a black hole. The peak luminosity is not so great as with a type Ia supernova, but the absolute magnitude may reach -17, which is more than 500 million times as luminous as the Sun.

Type Ia supernovæ may occur anywhere, but type II seem to be confined to the spiral arms of galaxies. The light curves differ markedly. A type I shows a steep rise, early fading and then a long, slow decline. With a type II there is also a steep rise and an early decline, after which the brightness falls off more sharply until settling down to a gradual decline. Supernova remnants (SNR) may be found as patches of gas, sometimes containing pulsars; they emit radiation over a wide range, from radio waves to gamma-rays; and many of them are very strong radio sources.

Galactic supernovæ. Observable supernova remnants show that there have been a number of outbursts in our Galaxy over the past few tens of thousands of years. One of these is the Cygnus Loop; the lovely Veil Nebula forms one section of the circular structure. The distance is thought to be about 1400 light-years, and the age 5000 years. The Vela SNR, in the southern Gum Nebula, is about 11 000 years old, and contains one of the few pulsars to have been optically identified with a very faint object, flashing at the same rate as the pulsar and certainly identical with it. In historic times, eight galactic supernovæ have been reported. These are listed in Table 22.7.

185 December. Discovered 7 December, near α and β Centauri; Chinese sources. The optical object RCW86 (G.315-4-2.3) is an X-ray emitter. However, it has recently been suggested that there may have been an error, so that the position was near α and ξ Centauri and the object was a comet.

386. Chinese sources. Several radio sources lie in this area, near λ Sagittarii. Probably a supernova.

1006 3 April (discovery). This was certainly the brightest known galactic supernova, and may have been as brilliant as the quarter-moon, but it is not well documented.

1054 July. Chinese, Japanese and Korean sources. It was seen for 23 days in daylight over 22 months in all. This is the Crab Nebula progenitor, and is what is termed a *plerion*, filling the whole area which it occupies rather than producing a shell. It radiates over the whole of the electromagnetic spectrum. The pulsar, NP 0532, has a period of 0.333 s, and is identified with a dim, flashing object which can reach magnitude 15. There are suggestions that the outburst may have been shown in some American Indian cave paintings, but the evidence for this is not conclusive. The nebula is officially catalogued as M.1; the 'Crab' nickname was bestowed on it by the Earl of Rosse, who drew it with the great Birr Castle reflector.

1181 August. Between ϵ and ι Cassiopeiæ. Chinese and Japanese sources. Almost certainly a supernova.

1572, Tycho's Star, B Cassiopeiæ. Seen by W. Schüler of Wittenberg on 6 November. Tycho saw it on 11 November; it may have been seen as early as 3 November. Fortunately, Tycho left a very full account of it. The remnant, near ξ Cassiopeiæ, is an X-ray source.

Table 22.7. Galactic supernovæ.

Date	Constellation	Maximum magnitude	Time visible with naked-eye (months)	Remnant name	Remnant diameter (light-years)	Estimated distance (light-years)	Maximum absolute magnitude
185	Centaurus	−8	20	RCW 86?	115	10 000	−19 ± 2
386	Sagittarius	+1.5	3	—	20?	15 000	
393	Scorpius	0	8	CTB 37A-B?	80	35 000	
1006	Lupus	−9.5	24	PKS 1459-41	30	3000	−19.8 ± 1
1054	Taurus	−5	22	Crab, 3C 144	9	7000	
1181	Cassiopeia	0	6	3C 58	17	9000	
1572	Cassiopeia	−4	16	3C 10	18	8000	−17.6 ± 0.5
1604	Ophiuchus	−3	12	3C 358	12	15 000	−19.6 ± 0.5

1604 October. Kepler's Star, near λ Ophiuchi. Discovered on 9 October; Kepler first saw it on 11 October. Korean astronomers followed it for several months. Like Tycho's Star, this was probably a type I, so that the remnant is a dim gas cloud with no pulsar.

There must have been a supernova in Cassiopeia around 1667, because we have here the radio source Cassiopeia A – actually the brightest X-ray emitter in the sky apart from the Sun. It was not observed, as it was too heavily obscured by interstellar dust near the main plane of the Galaxy.

Extragalactic supernovæ. Supernovæ in outer galaxies are discovered regularly, often by amateurs (one Australian amateur, the Rev. Robert Evans, now has more than two dozen discoveries to his credit). Since their peak luminosities are probably about equal, they are invaluable as 'standard candles', and they can of course be seen over immense distances. On one occasion, in April 1991, two supernovæ flared up in the same galaxy, NGC 10-24-007 in Draco, within a month of each other.

In 1885 what we now know to be a supernova was seen in M.31, the Andromeda Spiral, over 2 000 000 light-years away. It reached the fringe of naked-eye visibility, but at the time its nature was not recognized. It is now catalogued as S Andromedæ, and on 4 November 1988 R. A. Feisen, using the 4 m telescope at Kitt Peak in Arizona, identified its remnant. The iron-rich remains of the supernova showed up as a dark patch against the background.

S Andromedæ was discovered on 20 August 1885 by Hartwig, at the Dorpat Observatory in Estonia, but there is an interesting aside here. On 22 August, the Hungarian Baroness de Podmaniczky was holding a house party, and for the entertainment of her guests had set up a small telescope on the lawn. Using this, she saw a 'small star' in M.31. Her observation was quite independent of Hartwig's, though she certainly did not appreciate its importance!

The most recent naked-eye supernova was 1987A, in the Large Cloud of Magellan, at a distance of 169 000 light-years. It was discovered on 24 February by Ian Shelton, at the Las Campanas Observatory in Chile, on a routine photograph he had taken, and almost at the same time by O. Duhalde, also at Las Campanas, with the naked eye. (It had been photographed earlier by R. McNaught, from Australia, but McNaught had not checked his observations.)

When discovered it was of magnitude 4.5, but brightened to magnitude 2.3 before starting to fade. One surprise was that the progenitor star, Sanduleak −69°202, was identifiable, and was a blue giant rather than a red star; it was thought to be about 20 000 000 years old, with a mass from five to seven times that of the Sun. The peak luminosity was low by supernova standards: only about 250 000 000 times that of the Sun – because the progenitor star was smaller, though hotter, than a red supergiant. It seems that the progenitor was originally very hot and massive; as it used up its available hydrogen it went through the red supergiant stage, before shedding its outer layers and contracting. It was then, with the star a blue supergiant, that the outburst happened.

Maximum brightness was reached in mid-May 1987, and then came the expected decline. Various phenomena were seen, including 'light rings' due to interstellar dust which was illuminated by the outburst. The supernova was observed at many wavelengths, including X-rays (by the Japanese satellite Ginga) and ultraviolet (by the IUE satellite). Neutrinos were also recorded. So far, no pulsar has been detected. In fact, a pulsar was reported on 18 January 1989 by astronomers at Cerro Tololo in Chile, led by J. Middleditch; it was claimed that the pulsar was rotating at a rate of 1968.63 turns per second. Various explanations were offered, but it was then found that the observed effects were due to the mechanism of the telescope. Whether a pulsar will eventually form remains to be seen.

Paul Murdin has given[1] a very interesting time-scale, linking the evolution of the progenitor star with geological periods on Earth, It is as follows:

Formation of the star. 20 000 000 years ago, during the Miocene period. Hydrogen burning continued for 15 000 000 years.

Helium burning. End of the Pliocene period.

Carbon burning. Pleistocene period; first men on Earth.

Neon and oxygen burning. 1980; Mrs. Thatcher as British Prime Minister.

Silicon burning. 1987 February 20.

Explosion of the star. 1987 February 23, 17h 35m.

Of course, all this actually happened 169 000 years ago!...

Hypernovæ

Can there be outbursts which will dwarf even supernovæ? Very recently it has been suggested that this may be the case.

Violent outbursts of gamma radiation have been traced; these bursts (GRBs) are the most energetic events known, and they have never been satisfactorily explained. In 1997 Bohdan Paczynski, of Princeton University in the United States, proposed that they might be due to 'hypernovæ', which would outshine even the most energetic supernova by a factor of 100. Paczynski described a model in which a massive rotating star collapses into a black hole, leaving behind a disk of material which releases an incredible amount of energy. Another explanation involved a collision between two neutron stars. In 1999 astronomers at Northwestern University and the University of Illinois detected the first observational evidence of hypernova remnants; Daniel Wang, of Northwestern University, identified two hypernova remnants in the Pinwheel Galaxy, M.101, at a distance of 25 000 000 light-years. The remnants had previously been classed as the results of ordinary supernovæ, but Wang's X-ray analyses indicated a far more energetic process. One remnant, MF83, has a diameter of 860 light-years; the other, NGC 5471B, is expanding at a rate of at least 60 km s^{-1}.

The actual existence of hypernovæ remains to be proved. We must await the results of further research.

---

[1] 1990 *End in Fire* (Cambridge University Press). I have adapted it slightly!

23 STELLAR CLUSTERS

Clusters and nebulæ are among the most striking of stellar objects. Several are easily visible with the naked eye. Few people can fail to recognize the lovely star cluster of the Pleiades or Seven Sisters, which has been known since prehistoric times and about which there are many old legends. The nebula in the Sword of Orion, the Sword-Handle in Perseus, Præsepe in Cancer and the Jewel Box cluster in Crux are other objects easily visible without optical aid. Keen-sighted people have little difficulty in locating the great Andromeda Spiral and the globular cluster in Hercules, while in the far south there are the two Clouds of Magallan, which cannot possibly be overlooked, as well as the bright globular clusters ω Centauri and 47 Tucanæ.

The most famous of all catalogues of nebulous objects was compiled by the French astronomer Charles Messier, and published in 1781. Ironically, Messier was not interested in the objects he listed; he was a comet-hunter, and merely wanted a quick means of identifying misty patches which were non-cometary in nature. In 1888 J. L. E. Dreyer, Danish by birth (although he spent much of his life in Ireland, and finally in England) published his New General Catalogue (NGC), augmented in 1898 and again in 1908 by his Index Catalogue (I or IC).

Messier's original catalogue included 103 objects; later it was extended to 110, although not by Messier himself. However, Messier excluded many bright clusters and nebulæ, either because they could not be confused with comets, and were therefore of no interest to him, or because they were too far south in the sky to be seen from France. In 1995 I compiled the Caldwell Catalogue[1], of 109 objects arranged in order of declination and omitted by Messier. This now seems to be widely used.

The first-known clusters and nebulæ are listed in Table 23.1. There are many nicknames for astronomical objects; some of these are given in Table 23.2.

[1] Obviously I could not use the letter M. However, my surname is actually a hyphenated one (Caldwell-Moore), so I used C.

Table 23.1. The first known clusters and nebulæ. It seems that a few nebulous objects have been known since prehistoric times. This certainly applies to the Pleiades, which can hardly be overlooked. In his great work the Almagest, Ptolemy (circa 120–180 AD) records the Pleiades, and also the Sword-Handle in Perseus (NGC 869 and 884), M.44 (Præsepe) and, with almost certain identification, the open cluster M.7 in Scorpius; probably Ptolemy himself discovered the last of these. We may also assume that ancient men of the southern hemisphere knew the Magellanic Clouds, though they did not come to the notice of European astronomers until about 1520.

From the year 1745 more and more objects were found, mainly by astronomers such as De Chéseaux, Legentil, Lacaille and, of course, the two great Frenchmen, Messier and Méchain. By 1781, when Messier published his Catalogue, 138 nebulous objects were known. The list in Table 23.1 includes all the objects found before 1745.

This makes in all 21 objects. Many others had been previously listed, but subsequently found to be mere groups of stars rather than true clusters or nebulæ. One remarkable fact relates to the Great Spiral in Andromeda, which was recorded by the Persian astronomer Al-Sûfi. It was not noted again until 1612, when Simon Marius described it. Amazingly, it was completely overlooked by the greatest observer of pre-telescopic times, Tycho Brahe.

Number		Discoverer
M.45	Pleiades	(Prehistoric)
NGC 869/884, C.14	Sword-Handle in Perseus	(Listed by Ptolemy)
M.44	Præsepe	(Listed by Ptolemy)
M.7	Open cluster in Scorpius	Ptolemy, c 140 AD
M.31	Andromeda Spiral	Al-Sûfi, c 964
IC 2391, C.85	o Velorum cluster	Al-Sûfi, c 964
Large Cloud of Magellan		1519
Small Cloud of Magellan		1519
M.42	Great Nebula in Orion	N. Peiresc, 1610
M.22	Globular in Sagittarius	A. Ihle, 1665
NGC 5139, C.80	ω Centauri	Halley, 1677
M.8	Lagoon Nebula	Flamsteed, 1680
M.11	Wild Duck Cluster	G. Kirch, 1681
NGC 2244	12 Monocerotis	Flamsteed, 1690
M.41	Open cluster in Canis Major	Flamsteed, 1702
M.5	Globular in Serpens	G. Kirch, 1702
M.50	Open cluster in Monoceros	G. D. Cassini, c 1711
M.13	Hercules Globular	Halley, 1714
M.43	Part of the Orion Nebula	de Mairan, c 1731
M.1	Crab Nebula	J. Bevis, 1731

Table 23.2. Some astronomical nicknames. There are many unofficial names for stellar objects in common use. The following list is far from complete, but does include some familiar nicknames.

Ant Nebula	Bipolar nebula PK 331-1.1, at RA 16h 17m.2, dec. −51°59′
Antennæ	Colliding galaxies NGC 4038 and 4039; in Corvus (C.60 and 61)
Barnard's Galaxy	NGC 6822; in the Local Group. RA 19h 45, dec. −14 deg 48 min, C.57
Barnard's Loop	Extensive ring of nebulosity in Orion
Barnard's Star	Munich 15040; the nearest star apart from the α Centauri group
Bear Paw Galaxy	NGC 2537; galaxy in Lynx
Becklin-Neugebauer Object	Infra-red source in M.42 (Orion Nebula)
Beehive Cluster	M.44, Præsepe
Black-eye Galaxy	M.64; in Coma (spiral galaxy)
Blaze Star	The recurrent nova T Coronæ Borealis
Blinking Nebula	NGC 6826; planetary nebula in Cygnus, C.15
Blue Planetary	NGC 3918; planetary nebula in Centaurus
Bode's Nebula	M.81; spiral galaxy in Ursa Major
Boomerang Nebula	Bipolar nebula at RA 12h 44m.8, dec. −54°31′
Box Nebula	NGC 4169
Bubble Nebula	NGC 7635; nebula in Cassiopeia
Bug Nebula	NGC 6302; planetary nebula in Scorpius, C.69
Burnham's Nebula	T Tauri nebula
Butterfly Cluster	M.6; open cluster in Scorpius
California Nebula	NGC 1499; nebula in Perseus
Cartwheel Galaxy	Ring galaxy at RA 1h 37m.4, dec. −33°45′
Centaurus A	Radio galaxy NGC 5128, C.77
Christmas Tree Cluster	NGC 2264
Coal Sack	Dark nebula in Crux, C.99
Coat-Hanger	Cluster in Vulpecula (also known as Brocchi's Cluster)
Cocoon Nebula	IC 5146; nebula in Cygnus, C.19
Cone Nebula	IC 2264 (S Monocerotis); nebula in Monoceros
Crab Nebula	M.I; supernova remnant in Taurus; NGC 1952
Crescent Nebula	NGC 6888; nebula in Cygnus, C.27
Crimson Star	The Mira variable R Leporis
Demon Star	Algol (β Persei)
Dog Star	Sirius (α Canis Majoris)
Dumbbell Nebula	M.27; planetary nebula in Vulpecula
Eagle Nebula	M.16; nebula in Serpens Cauda
Egg Nebula	CRL 2688, at RA 21h 02m.3, dec. 36°42′
Eskimo Nebula	NGC 2392; planetary nebula in Gemini (also known as the Clown Face Nebula), C.38
Flaming Star Nebula	IC 405; nebula associated with AE Aurigæ
Flying Star	61 Cygni (because of its large proper motion)
Garnet Star	μ Cephei
Geminga	Gamma-ray source in Gemini
Ghost of Jupiter	NGC 3242; planetary nebula in Hydra, C.59
Gum Nebula	Nebula in Vela (supernova remnant)
Helix Nebula	NGC 7293; planetary nebula in Aquarius
Hind's Variable Nebula	NGC 1555; nebula near T Tauri
Homunculus Nebula	Core of η Carinæ Nebula
Hourglass Nebula	The brightest part of M.8
Horse's Head Nebula	Barnard 33; dark nebula in Orion
Hubble's Variable Nebula	NGC 2261; nebula round R Monocerotis, C.46
Hyades	Mel. 25, open cluster in Taurus, C.41
Innes' Star	Proxima Centauri
Intergalactic Tramp	NGC 2419; globular cluster in Lynx, C.25
Jewel Box	NGC 4755; open cluster (κ Crucis), C.94
Kepler's Star	The supernova of 1604 (in Ophiuchus)
Keyhole Nebula	Dark nebula; in the η Carinæ nebulosity
La Superba	Y Canum Venaticorum
Lacework Nebula	NGC 6960
Lagoon Nebula	M.8; nebula in Sagittarius
Little Dumbbell	M.76; planetary nebula in Perseus
Markarian's Chain	M.85 to M.88, in the Virgo Cluster
Network Nebula	NGC 6992-5, in the Veil Nebula
North America Nebula	NGC 7000; nebula in Cygnus, C.10
Omega Nebula	M.17; nebula in Sagittarius (also known as the Horseshoe Nebula)
Owl Nebula	M.97; planetary nebula in Ursa Major
Pelican Nebula	IC 5067/70; nebula in Cygnus
Pinwheel Galaxy	Triangulum Galaxy, M.33
Pinwheel Nebula	M.99 (NGC 4254)

Table 23.2. (Continued)

Pipe Nebula	Barnard 59, 65-7, 76; dark nebulosity in Ophiuchus
Plaskett's Star	Massive O-type binary star; HD 47129
Pleiades	M.45; open cluster in Taurus
Polaris Australis	σ Octantis
Pup	Sirius B (companion of Sirius, the Dog Star)
Red Rectangle	Planetary nebula at RA 6h 20m, dec. $-10°39'$
Ring Nebula	M.57; planetary nebula in Lyra
Ring-tail Galaxies	Another name for the Antennæ
Rosette Nebula	NGC 2237-9 C.49; nebula in Monoceros
Rotten Egg Nebula	OH231-8 +4.2
Runaway Star	Another nickname for Barnard's Star
Sagittarius A-*	Centre of the Galaxy
Saturn Nebula	NGC 7009; planetary nebula in Aquarius, C.55
Seven Sisters	The Pleiades, M.45
Sidus Ludovicianum	Star near Mizar and Alcor
Snake Nebula	Barnard 72; dark nebula near θ Ophiuchi
Snickers	Nearby dwarf galaxy (RA 6h 28m, dec. +15 deg.)
Sombrero Hat Galaxy	M.104; galaxy in Virgo
Southern Pleiades	The θ Carinæ cluster, C.102
Spindle Galaxy	NGC 3115; galaxy in Sextans
Star Queen Nebula	Another name for M.16 (Eagle Nebula)
Stephan's Quintet	NGC 7317-20; galaxies in Pegasus
Sunflower Galaxy	M.63; spiral galaxy in Canes Venatici
Swan Nebula	Alternative name for the Omega Nebula
Sword of Orion	Region of M.42; nebula in Orion
Sword-Handle	NGC 869-884; double cluster in Perseus, C.14
Tarantula Nebula	NGC 2070; 30 Doradûs; nebula in Large Magellanic Cloud, C.103
Toby Jug Nebula	IC 2220, at RA 7h 56m.9, dec. $-59°07'$
Trapezium	The multiple star θ Orionis
Trifid nebula	M.20; nebula in Sagittarius
Tycho's Star	B Cassiopaiæ; the 1572 supernova
Veil Nebula	NGC 6960-92-95; nebulosity in Cygnus, C.33.34
Whirlpool Galaxy	M.51; spiral galaxy in Canes Venatici
Wild Duck Cluster	M.11; cluster in Scutum
Witch Head Nebula	IC 2118, at RA 5h 06m.9, dec. $-7°13'$

All catalogues include objects of different kinds: open or loose clusters, globular clusters, supernova remnants, diffuse of galactic nebulæ, planetary nebulæ and galaxies. The Messier Catalogue is given in Table 23.3 (page 303), and the Caldwell Catalogue in Table 23.4 (page 305).

Open clusters

Open or loose clusters are aggregations of stars arranged in no particular shape; the Pleiades, Hyades and Præsepe are good examples. Some clusters are sparse, containing no more than 10 stars, while others may include at least 3000 stars. The diameter of an average cluster is of the order of 65 light-years, and in its richest part the distances between individual stars will be no more than a light-year. They are of various ages. The Pleiades are about 76 000 000 years old, but the lovely Jewel Box in the Southern Cross has an age of little more than 7 000 000 years, and others are even younger, at around 1 000 000 years. On the other hand, M.67, an open cluster in Cancer, has an age of 4000 million years.

Interstellar space is never completely empty, but the density is very low indeed. Here and there, denser clouds of dust and gas form, and are very cold. When a cloud reaches a mean density a thousand times greater than that of 'normal' space, atoms combine into molecules, and we have a molecular cloud, which may be from 1 to 300 light-years across, containing enough material to produce at least 10 000 stars. If the mass of the cloud exceeds 100 000 times that of the Sun, it is classed as a giant molecular cloud (GMC). Clouds of this type are bitterly cold, at around $-263\,°C$ (only $10°$ above absolute zero), and as they are made up chiefly of hydrogen molecules (H_2) they are hard to detect optically; radio telescopes have to be used. However, in recent years the Hubble Space Telescope has been used to observe colour changes of background stars seen through the clouds, which leads on to information about the distribution of matter in the clouds themselves.

Still denser clumps of material accumulate in the GMCs, and star formation follows; many stars may be produced from the same cloud, resulting in a stellar cluster.

It is logical to assume that the stars in any particular cluster are of the same age, and of the same initial chemical composition, so that present differences between them are due to initial differences in mass. Drawing up an HR diagram of a cluster shows where there is a 'turn-off' from the Main Sequence to the giant branch, and this gives the age of the cluster. For example, the brightest stars in the Pleiades are blue and very hot, while the much older Hyades contain stars which have already evolved into red giants. In young open clusters there is still visible material which shows that star formation may still be going on. The nebulosity in the Pleiades is easy to detect, but nebulosity is totally absent from the Hyades, where all the star-forming material has been used up.

Open clusters are not permanent structures; they are disrupted by passing 'field' stars, and will eventually disperse. The oldest open clusters, such as M.67, survive because they are well away from the main plane of the Galaxy and are not likely to encounter many non-cluster stars, while most of the open clusters are close to the galactic plane, so that encounters are frequent.

The nearest cluster is that of Ursa Major, but the average distance of the cluster stars from us is only about 80 light-years, so that they are scattered round the sky; members include five of the stars of the famous Plough pattern in the Great Bear (α and η being the exceptions); Sirius is also a member. The Hyades cluster comes next; new measurements by the Hubble Space Telescope give its distance as 151 light-years. The Coma cluster is 270 light-years away and a newly-discovered, much fainter cluster near η Chamæleontis lies at 315 light-years. The η Chamæleontis cluster is only 2 light-years across, and as seen from Earth its component stars are of between magnitudes 10 and 14, so that it is not surprising that the cluster was overlooked until recently.

A selected list of bright open clusters is given in Table 23.5 (page 306). Many of these are naked-eye objects, and all in the list are easily seen with binoculars. Fainter clusters are plentiful, and there may be some 20 000 of them in our Galaxy. Of course, open clusters are also seen in external galaxies.

The most famous of all open clusters is that of the Pleiades (M.45), in Taurus. The brightest star is η Tauri, Alcyone (magnitude 2.9); then follow Atlas (3.6), Electra (3.7), Maia (3.9), Merope (4.2), Taygete (4.3), Pleione (5.1, but variable), Celæno (5.4) and Asterope (5.6). The cluster is always known as the Seven Sisters, and certainly people with average eyesight can see seven individual stars on a clear night; Asterope is on the fringe of naked-eye visibility, while Pleione and Atlas are so close together that binoculars are needed to separate them easily. It is said that the 19th century German astronomer E. Heis could see 19 stars in the cluster without optical aid; many people can manage a dozen. The total membership of the cluster is around 500 stars, contained in an area 50 light-years across, and embedded in nebulosity which shines by reflected light from the stars within. The distance is 375 light-years, rather less than was believed before new measurements were made with the Hubble Space Telescope.

Several brown dwarfs have been identified in the Pleiades. The 19th century German astronomer J. H. Mädler, renowned for his studies of the Moon, believed Alcyone to be the central star of the Galaxy; it is not clear how he arrived at this conclusion, which is quite erroneous.

The Hyades, also in Taurus, are different; they were not listed either by Messier or Dreyer. They are numbered 41 in the Caldwell catalogue and 25 in the catalogue by Melotte. The area is dominated by Aldebaran, but in fact Aldebaran is not a cluster member; at 68 light-years it lies roughly midway between the Hyades and ourselves. The main stars of the Hyades are arranged in a V-shape. ϵ (3.5) and γ (3.6) are orange stars of type K, an there are some interesting pairs; δ (3.8) makes a pair with 64 Tauri (4.8), and θ is a naked-eye double, made up of a white star of magnitude 3.4 and a K-type orange companion of magnitude 3.8. The two are not particularly close, as the distance between them is 15 light-years, although of course both condensed out of the same cloud which produced the rest of the Hyades. The proper motions of the Hyades stars show that they are converging to a point east of Betelgeux, at RA 6h declination 9°N, so that in the far future the cluster will look more condensed than it does now.

Præsepe in Cancer (M.44) is also prominent with the naked eye; it was well known to the ancient Greeks, and Hipparchus called it a 'cloudy star'. Bayer even gave it a

Greek letter (ϵ Cancri). It is over 500 light-years away, and there have been suggestions that it may once have had close associations with the Hyades. In Perseus we find the Sword-Handle (NGC 869 and 884) (C.14); two rich open clusters in the same telescopic field, with a red star between them – a magnificent sight. In the far south there is the Jewel Box, NGC 5139 (C.80) round κ Crucis in the Southern Cross; the brightest stars are blue, but there is one prominent red giant, providing a beautiful colour contrast.

Globular clusters

Globular clusters are quite different from open clusters. They are vast symmetrical systems, containing from 10 000 to at least a million stars, and so strongly condensed that near their centres the individual stars may be only light-months apart. An average globular cluster has a diameter of 65 light-years, but there is a wide range in both size and mass. About 200 globular clusters are known in our Galaxy, but other galaxies have their own systems of clusters.

Our Galaxy is a flattened system, and is rotating round its centre; the Sun takes about 225 000 000 years to complete one orbit, and lies not far from the main plane. The globular clusters, on the other hand, are members of what is termed the galactic halo, which is roughly spherical and extends out to several hundreds of thousands of light-years. The globulars move in highly eccentric orbits which take them far outside the main Galaxy, and they do not share in the disk rotation; this means that their velocities relative to the Sun may be very high (up to 100 km s^{-1}). One globular, NGC 2419 (C.25), in Lynx, seems to be escaping from the Galaxy altogether, and has been called an Intergalactic Tramp!

Globular clusters are very ancient, and are indeed as old as the Galaxy itself. They are very poor in heavy elements, since they were formed from primitive material, whereas the open clusters were formed from 'recycled' material. The globulars are much more stable than the open clusters, and though they are subjected to disruptive forces (such as tidal effects from the parent Galaxy) they are very long-lived.

Huge though they are, most globulars are so remote that they appear faint, and not many are visible with the naked-eye. They contain short-period variables (RR Lyræ stars) which enables their distances to be measured. This was first done in 1917 by Harlow Shapley, who also realized that most of the globulars lie in the southern sky; there is a heavy concentration in the area of Scorpius, Ophiuchus and Sagittarius. From this Shapley deduced, quite correctly, that we are having a lop-sided view, because the Sun is well away from the centre of the Galaxy.

The two brightest globulars, ω Centauri and 47 Tucanæ, lie in the far south of the sky. ω Centauri (C.80) was recorded by Ptolemy as a star; Bayer gave it a Greek letter, but it was Edmond Halley who first recognized its true nature. Even at a range of 16 000 light-years it is a very prominent naked-eye object, and a small telescope will resolve its outer parts into stars, although near its centre the star density is very high. It is much the largest and most massive of the Galaxy's globular clusters, and its mass, 5000 000 times that of the Sun, is as great as that of a dwarf galaxy. Also in the far south is NGC 104 (C.107), 47 Tucanæ, which is almost silhouetted against the Small Cloud of Magellan, although the two are in no way associated; the Cloud is an independent galaxy, over 170 000 light-years away, while the distance of 47 Tucanæ is a mere 15 000 light-years. It contains about 2000 000 stars, in a system around 200 light-years across.

In the northern hemisphere, the brightest globular cluster is M.13 in Hercules, discovered by Halley in 1714. It is just visible with the naked eye, between the stars η and ζ Herculis. The distance is 22 500 light-years, and the diameter has been given as about 100 light-years.

Because globular clusters are so old, their leading stars are highly-evolved red giants. There are, however, some stars which are hot and blue. A 'blue straggler' of this kind seems to be a rejuvenated star – either the result of a stellar collision, or the merger of the two components of a binary pair. 47 Tucanæ is particularly rich in blue stragglers.

The view from a planet orbiting a star near the centre of a globular cluster would indeed be fascinating. There would be many stars bright enough to cast shadows, and many of these would be red; there would be no darkness at night. Against this, an observer on such a world would have a very restricted view of the outer universe.

Table 23.3. The Messier Catalogue. Messier's original catalogue ends with M.103. The later numbers are objects which, apart from the last, were discovered by Méchain. (M.110 may have been discovered by Messier himself, but it is still always known as NGC 205, one of the companions to the Great Spiral in Andromeda.) M.104 was added to the Messier catalogue by Camille Flammarion in 1921, on the basis of finding a handwritten note about it in Messier's own copy of his 1781 catalogue. M.105 to 107 were listed by H. S. Hogg in the 1947 edition of the catalogue; M.108 and 109 by Owen Gingerich in 1960. The naming of NGC 205 as M.110 was proposed by K. G. Jones in 1968, but has never been accepted.

M	NGC	Constellation	Type	RA h	RA min	Dec. °	Dec. ′	Magnitude	Dimensions (′)	Distance (thousands of light-years)*	
1	1952	Taurus	Supernova remnant	05	34.5	+22	01	8.2	6.4	6.3	Crab Nebula
2	7089	Aquarius	Globular	21	33.5	−00	49	6.5	12.9	36.2	
3	5272	Canes Venatici	Globular	13	42.2	+28	23	6.2	16.2	30.6	
4	6121	Scorpius	Globular	16	23.6	−26	32	5.6	26.3	6.8	
5	5904	Serpens	Globular	15	18.6	+02	05	5.6	17.4	22.8	
6	6405	Scorpius	Open cluster	17	40.1	−32	13	5.3	15.0	2.0	Butterfly cluster
7	6475	Scorpius	Open cluster	17	53.9	−34	49	4.1	80.0	1.0	
8	6523	Sagittarius	Nebula	18	03.8	−24	23	6.0	60 × 35	6.5	Lagoon Nebula
9	6333	Ophiuchus	Globular	17	19.2	−18	31	7.7	9.3	26.4	
10	6154	Ophiuchus	Globular	16	57.1	−04	06	6.6	15.1	13.4	
11	6705	Scutum	Open cluster	18	51.1	−06	16	6.3	14.0	6.0	Wild Duck cluster
12	6218	Ophiuchus	Globular	16	47.2	−01	57	6.7	14.5	17.6	
13	6205	Hercules	Globular	16	41.7	+36	28	5.8	16.6	22.2	Hercules cluster
14	6402	Ophiuchus	Globular	17	37.6	−03	15	7.6	11.7	27.4	
15	7078	Pegasus	Globular	21	30.0	+12	10	6.2	12.3	32.6	
16	6611	Serpens	Nebula and embedded cluster	18	18.8	−13	47	6.4	7.0	7.	Eagle Nebula
17	6618	Sagittarius	Nebula	18	20.8	−16	11	7.0	11.0	5.0	Omega Nebula
18	6613	Sagittarius	Open cluster	18	19.9	−17	08	7.5	9.0	6.0	
19	6273	Ophiuchus	Globular	17	02.6	−26	16	6.8	13.5	27.1	
20	6514	Sagittarius	Nebula	18	02.6	−23	02	9.0	28.0	2.2	Trifid Nebula
21	6531	Sagittarius	Open cluster	18	04.6	−22	30	6.5	13.0	4.3	
22	6656	Sagittarius	Globular	18	36.4	−23	54	5.1	24.0	10.1	
23	6494	Sagittarius	Open cluster	17	56.8	−19	01	6.9	27.0	4.5	
24	6603	Sagittarius	Star-cloud	18	16.9	−18	29	4.6	90.0	10.0	Not a cluster
25	IC 4725	Sagittarius	Open cluster	18	31.6	−19	15	6.5	40.0	2.0	
26	6694	Scutum	Open cluster	18	45.2	−09	24	9.3	15.0	5.0	
27	6853	Vulpecula	Planetary nebula	19	59.6	+22	43	7.4	8.0 × 5.7	1.3	Dumbbell Nebula
28	6626	Sagittarius	Globular	18	24.5	−24	52	6.8	11.2	17.9	
29	6913	Cygnus	Open cluster	20	23.9	+38	32	7.1	7.0	7.2	
30	7099	Capricornus	Globular	21	40.4	−23	11	7.2	11.0	24.8	
31	224	Andromeda	Spiral galaxy	00	42.7	+41	16	4.8	178	2200	Great Spiral
32	221	Andromeda	Elliptical galaxy	00	42.7	+40	52	8.7	8 × 6	2200	Companion to M.31
33	598	Triangulum	Spiral galaxy	01	33.9	+30	39	6.7	73 × 45	2300	Triangulum spiral
34	1039	Perseus	Open cluster	02	42.0	+42	47	5.5	35.0	1.4	
35	2168	Gemini	Open cluster	06	08.9	+24	20	5.3	28.0	2.8	
36	1960	Auriga	Open cluster	05	36.1	+34	08	6.3	12.0	4.1	
37	2099	Auriga	Open cluster	05	52.4	+32	33	6.2	24.0	4.6	
38	1912	Auriga	Open cluster	05	28.4	+35	50	7.4	21.0	4.2	
39	7092	Cygnus	Open cluster	21	32.2	+48	26	5.2	32.0	0.8	
40	—	Missing.	Possibly a comet?								
41	2287	Canis Major	Open cluster	06	47.0	−20	44	4.6	38.0	2.4	
42	1976	Orion	Nebula	05	35.4	−05	27	4.0	85 × 60	1.6	Great Nebula in Orion
43	1982	Orion	Nebula	05	35.6	−05	27	9.0	20 × 15	1.6	Part of Orion Nebula
44	2632	Cancer	Open cluster	08	40.1	+19	59	3.7	95.0	0.5	Præsepe
45	—	Taurus	Open cluster	03	47.0	+24	07	1.6	110	0.4	Pleiades
46	2437	Puppis	Open cluster	07	41.8	−14	49	6.0	27.0	5.4	
47	2422	Puppis	Open cluster	07	36.6	−14	30	4.5	30.0	1.6	
48	2548	Hydra	Open cluster	08	13.8	−05	48	5.5	54.0	1.8	
49	4472	Virgo	Elliptical galaxy	12	29.8	+08	00	8.5	9 × 7.5	60 000	
50	2323	Monoceros	Open cluster	07	03.2	−08	20	6.3	16.0	3.0	
51	5194	Canes Venatici	Spiral galaxy	13	29.9	+47	12	8.1	11 × 7	37 000	Whirlpool Galaxy
52	7654	Cassiopeia	Open cluster	23	24.2	+61	35	7.3	13.0	7.0	

* Distances of galaxies in the Virgo Cluster have been 'rounded off' to 60 million light-years (60,000,000).

Table 23.3. (Continued)

M	NGC	Constellation	Type	RA h	RA min	Dec. °	Dec. ′	Magnitude	Dimensions (′)	Distance (thousands of light years)	
53	5024	Coma	Globular	13	12.9	+18	10	7.6	12.6	56.4	
54	6715	Sagittarius	Globular	18	55.1	−30	29	7.6	9.1	82.8	
55	6809	Sagittarius	Globular	19	40.0	−30	58	6.3	19.0	16.6	
56	6779	Lyra	Globular	19	16.6	+30	11	8.3	7.1	31.6	
57	6720	Lyra	Planetary nebula	18	53.6	+33	02	8.8	1.4 × 1.0	4.1	Ring Nebula
58	4579	Virgo	Spiral galaxy	12	37.7	+11	49	9.2	5.5 × 4.5	60 000	
59	4621	Virgo	Elliptical galaxy	12	42.0	+11	39	9.6	5 × 3.5	60 000	
60	4649	Virgo	Elliptical galaxy	12	43.7	+11	33	6.9	7 × 6	60 000	
61	4303	Virgo	Spiral galaxy	12	21.9	+04	28	9.6	6 × 5.5	60 000	
62	6266	Ophiuchus	Globular	17	01.2	−30	07	6.5	14.1	21.5	
63	5055	Canes Venatici	Spiral galaxy	13	15.8	+42	02	9.5	10 × 6	37 000	Sunflower Galaxy
64	4826	Coma	Spiral galaxy	12	56.7	+21	41	8.8	9.3 × 5.4	12 000	Black-eye Galaxy
65	2612	Leo	Spiral galaxy	11	18.9	+13	05	9.3	8 × 1.5	35 000	
66	3627	Leo	Spiral galaxy	11	20.2	+12	59	8.2	8 × 2.5	35 000	
67	2682	Cancer	Open cluster	08	50.4	+11	49	6.1	30.0	2.3	Famous old cluster
68	4590	Hydra	Globular	12	39.5	−26	45	7.8	12.0	32.3	
69	6637	Sagittarius	Globular	18	31.4	−32	21	7.6	7.1	25.4	
70	6681	Sagittarius	Globular	18	43.2	−32	18	7.9	7.8	28.0	
71	6838	Sagitta	Globular	19	53.8	+18	47	8.2	7.2	11.7	
72	6981	Aquarius	Globular	20	53.5	−12	32	9.3	5.9	52.8	
73	6994	Aquarius	Four faint stars	20	58.9	−12	38	9	3	—	Not a cluster
74	628	Pisces	Spiral galaxy	01	36.7	+15	47	9.2	10.2 × 9.5	35 000	
75	6864	Sagittarius	Globular	20	06.1	−21	55	8.5	6.0	57.7	
76	650	Perseus	Planetary nebula	01	42.4	+51	34	10.1	2.7 × 1.8	3.4	Little Dumbbell
77	1068	Cetus	Spiral galaxy	02	42.7	−00	01	8.9	7 × 6	60 000	
78	2068	Orion	Nebula	05	46.7	+00	03	8.3	8 × 6	1.6	
79	1904	Lepus	Globular	05	24.5	−24	33	7.7	8.7	39.8	
80	6093	Scorpius	Globular	16	17.0	−22	59	7.3	8.9	27.4	
81	3031	Ursa Major	Spiral galaxy	09	55.6	+69	04	6.8	12 × 10	11 000	Bode's Nebula
82	3034	Ursa Major	Irregular galaxy	09	55.8	+69	41	8.4	9 × 4	11 000	
83	5236	Hydra	Spiral galaxy	13	37.0	−29	52	7.6	11 × 10	10 000	
84	4374	Virgo	Spiral galaxy	12	25.1	+12	53	9.3	5.0	60 000	
85	4382	Coma	Spiral galaxy	12	25.4	+18	11	9.3	7.1 × 5.2	60 000	
86	4406	Virgo	Elliptical galaxy	12	26.2	+12	57	9.7	7.5 × 5.5	60 000	
87	4486	Virgo	Elliptical galaxy	12	30.8	+12	24	8.6	7.0	60 000	Giant Seyfert galaxy
88	4501	Coma	Spiral galaxy	12	32.0	+14	25	10.2	7.0 × 4	60 000	
89	4552	Virgo	Elliptical galaxy	12	35.7	+12	33	9.5	4.0	60 000	
90	4569	Virgo	Spiral galaxy	12	36.8	+13	10	10.0	9.5 × 4.5	60 000	
91	—	Not identified	Possibly a comet?								
92	6341	Hercules	Globular	17	17.1	+43	08	6.4	11.2	26.1	
93	2447	Puppis	Open cluster	07	44.6	−23	52	6.9	22.0	4.5	
94	4736	Canes Venatici	Spiral galaxy	12	50.9	+41	07	7.9	7 × 3	14 500	
95	3351	Leo	Barred spiral galaxy	10	44.0	+11	42	10.4	4.4 × 3.3	38 000	
96	3368	Leo	Spiral galaxy	10	46.8	+11	49	9.1	6 × 4	38 000	
97	3587	Ursa Major	Planetary nebula	11	14.8	+55	01	9.9	3.4 × 3.3	2.6	Owl Nebula
98	4192	Coma	Spiral galaxy	12	13.8	+14	54	1.7	9.5 × 3.2	60 000	
99	4524	Coma	Spiral galaxy	12	18.8	+14	25	10.1	5.4 × 4.8	60 000	
100	4321	Coma	Spiral galaxy	12	22.9	+15	49	10.6	7 × 6	60 000	
101	5457	Ursa Major	Spiral galaxy	14	03.2	+54	21	9.6	22.0	24 000	
102	—	Missing	Possibly a faint spiral in Draco, or else identical with M.101								
103	581	Cassiopeia	Open cluster	01	33.2	+60	42	7.4	6.0	8.0	
104	4594	Virgo	Spiral galaxy	12	40.0	−11	37	8.7	9 × 4	50 000	Sombrero Galaxy
105	3379	Leo	Elliptical galaxy	10	47.8	+12	35	9.2	2.0	38 000	
106	4528	Ursa Major	Spiral galaxy	12	19.0	+47	18	8.6	19 × 8	25 000	
107	6171	Ophiuchus	Globular	16	32.5	−13	03	9.2	10.0	19.5	
108	3556	Ursa Major	Spiral galaxy	11	11.5	+55	40	10.7	8 × 1	45 000	
109	3992	Ursa Major	Spiral galaxy	11	57.6	+53	23	10.8	7 × 4	55 000	

Table 23.4. The Caldwell Catalogue.

C	NGC/IC	Constellation	Type	RA (2000.0) h	min	Dec. °	′	Magnitude	Size (′)	
1	188	Cepheus	Open cluster	00	44.4	+85	20	8.1	14	Very old cluster
2	40	Cepheus	Planetary nebula	00	13.0	+72	32	10.7	1.0 × 0.7	
3	4236	Draco	Sb galaxy	12	16.7	+69	27	9.6	23 × 8	
4	7023	Cepheus	Reflection nebula	21	01.8	+68	10	—	18 × 18	
5	IC 342	Camelopardalis	SBc galaxy	03	46.8	+68	06	9.2	18 × 17	
6	6543	Draco	Planetary nebula	17	58.6	+66	38	8.1	22 × 16	Cat's Eye Nebula
7	2403	Camelopardalis	Sc galaxy	07	36.9	+65	36	8.4	18 × 10	
8	559	Cassiopeia	Open cluster	01	29.5	+63	18	9.5	4	
9	Sh2-155	Cepheus	Bright nebula	22	56.8	+62	37	—	50 × 10	Cave Nebula
10	663	Cassiopeia	Open cluster	01	46.0	+61	15	7.1	16	
11	7635	Cassiopeia	Bright nebula	23	20.7	+61	12	8	15 × 8	Bubble Nebula
12	6946	Cepheus	SAB galaxy	20	34.8	+60	09	8.9	11 × 10	
13	457	Cassiopeia	Open cluster	01	19.1	+58	20	6.4	13	φ Cas Cluster
14	869/884	Perseus	Double Cluster	02	20.0	+57	08	4.3	30 and 30	Sword-Handle
15	6826	Cygnus	Planetary nebula	19	44.8	+50	31	8.8	27 × 24	Blinking Nebula
16	7243	Lacerta	Open cluster	22	15.3	+49	53	6.4	21	
17	147	Cassiopeia	dE4 galaxy	00	33.2	+48	30	9.3	13 × 8	
18	185	Cassiopeia	dE0 galaxy	00	39.0	+48	20	9.2	12 × 10	
19	IC 5146	Cygnus	Bright nebula	21	53.5	+47	14	—	12 × 12	Cocoon Nebula
20	7000	Cygnus	Bright nebula	20	58.8	+44	33	—	175 × 110	North America Nebula
21	4449	Canes Venatici	Irregular galaxy	12	28.2	+44	06	9.4	6 × 5	
22	7662	Andromeda	Planetary nebula	23	25.9	+42	33	8.3	17 × 14	Blue Snowball Nebula
23	891	Andromeda	Sb galaxy	02	22.6	+42	21	9.9	14 × 3	
24	1275	Perseus	Seyfert galaxy	03	19.8	+41	31	11.6	3.5 × 2.5	Perseus A
25	2419	Lynx	Globular cluster	07	38.1	+38	53	10.4	4.1	
26	4244	Canes Venatici	Scd galaxy	12	17.5	+37	49	10.2	18 × 2	
27	6888	Cygnus	Bright nebula	20	12.0	+38	20	—	20 × 10	Crescent Nebula
28	752	Andromeda	Open cluster	01	57.8	+37	41	5.7	50	
29	5005	Canes Venatici	Sb galaxy	13	10.9	+37	03	9.5	6 × 3	
30	7331	Pegasus	Sb galaxy	22	37.1	+34	25	9.5	11 × 4	
31	IC 405	Auriga	Bright nebula	05	16.2	+34	16	—	37 × 19	Flaming Star Nebula
32	4631	Canes Venatici	Sc galaxy	12	42.1	+32	32	9.3	17 × 3	Whale Galaxy
33	6992/5	Cygnus	SN remnant	20	56.4	+31	43	—	60 × 8	Eastern Veil Nebula
34	6960	Cygnus	SN remnant	20	45.7	+30	43	—	70 × 6	Western Veil Nebula
35	4889	Coma Berenices	E4 galaxy	13	00.1	+27	59	11.4	3 × 2	Brightest in Coma Cluster
36	4559	Coma Berenices	Sc galaxy	12	36.0	+27	58	9.8	13 × 5	
37	6885	Vulpecula	Open cluster	20	12.0	+26	29	5.9	7	
38	4565	Coma Berenices	Sb galaxy	12	36.3	+25	59	9.6	16 × 2	Needle Galaxy
39	2392	Gemini	Planetary nebula	07	29.2	+20	55	8.6	47 × 43	Eskimo Nebula
40	3626	Leo	Sb galaxy	11	20.1	+18	21	10.9	3 × 2	
41	Melotte 25	Taurus	Open cluster	04	27	+16		0.5	330	Hyades
42	7006	Delphinus	Globular cluster	21	01.5	+16	11	10.6	2.8	
43	7814	Pegasus	Sb galaxy	00	03.3	+16	09	10.3	6 × 3	
44	7479	Pegasus	SBb galaxy	23	04.9	+12	19	10.9	4.4 × 3.4	
45	5248	Boötes	Sc galaxy	13	37.5	+08	53	10.2	7 × 5	
46	2261	Monoceros	Bright nebula	06	39.2	+08	44	—	2 × 1	Hubble's Variable Nebula
47	6934	Delphinus	Globular cluster	20	34.2	+07	24	2	5.9	
48	2775	Cancer	Sa galaxy	09	10.3	+07	02	10.1	5 × 4	
49	2237-9	Monoceros	Bright nebula	06	32.3	+04	59	—	80 × 70	Rosette Nebula
50	2244	Monoceros	Open cluster	06	32.4	+04	52	4.8	24	
51	IC 1613	Cetus	Irregular galaxy	01	04.8	+02	07	9.2	12 × 11	
52	4697	Virgo	E4 galaxy	12	48.6	−05	48	9.3	6 × 4	
53	3115	Sextans	S0 galaxy	10	05.2	−07	43	8.9	8 × 3	Spindle Galaxy
54	2506	Monoceros	Open cluster	08	00.2	−10	47	7.6	7	
55	7009	Aquarius	Planetary nebula	21	04.2	−11	22	8.0	28 × 23	Saturn Nebula
56	246	Cetus	Planetary nebula	00	47.0	−11	53	8.6	4 × 3	
57	6822	Sagittarius	Irregular galaxy	19	44.9	−14	48	8.8	20 × 10	Barnard's Galaxy
58	2360	Canis Major	Open cluster	07	17.8	−15	37	7.2	13	
59	3242	Hydra	Planetary nebula	10	24.8	−18	38	8.6	40 × 35	Ghost of Jupiter
60	4038	Corvus	Sc galaxy	12	01.9	−18	52	10.7	2.6 × 2	Antennæ
61	4039	Corvus	Sp galaxy	12	01.9	−18	53	10.7	2.6 × 2	Antennæ
62	247	Cetus	SAB galaxy	00	47.1	−20	46	9.1	20 × 7	
63	7293	Aquarius	Planetary nebula	22	29.6	−20	48	6.3	16 × 12	Helix Nebula
64	2362	Canis Major	Open cluster	07	18.8	−24	57	4.1	8	τ CMa Cluster
65	253	Sculptor	Scp galaxy	00	47.6	−25	17	7.1	25 × 7	Sculptor Galaxy
66	5694	Hydra	Globular cluster	14	39.6	−26	32	10.2	3.6	
67	1097	Fornax	SBb galaxy	02	46.3	−30	16	9.2	9 × 7	
68	6729	Corona Australis	Bright nebula	19	01.9	−36	58	—	1.0	R CrA Nebula
69	6302	Scorpius	Planetary nebula	17	13.7	−37	06	9.6	2 × 1	Bug Nebula
70	300	Sculptor	Sd galaxy	00	54.9	−37	41	8.1	20 × 15	
71	2477	Puppis	Open cluster	07	52.3	−38	33	5.8	27	
72	55	Sculptor	SB galaxy	00	15.1	−39	13	7.9	25 × 4	Brightest in Sculptor Group
73	1851	Columba	Globular cluster	05	14.1	−40	03	7.3	11	
74	3132	Vela	Planetary nebula	10	07.7	−40	26	8.2	1.4 × 0.9	Southern Ring Nebula
75	6124	Scorpius	Open cluster	16	25.6	−40	40	5.8	29	

Table 23.4. (Continued)

C	NGC/IC	Constellation	Type	RA (2000.0) h	min	Dec. °	′	Magnitude	Size (′)	
76	6231	Scorpius	Open cluster	16	54.0	−41	48	2.6	15	
77	5128	Centaurus	Radio galaxy	13	25.5	−43	01	6.8	18 × 14	Centaurus A
78	6541	Corona Australis	Globular cluster	18	08.0	−43	42	6.6	13	
79	3201	Vela	Globular cluster	10	17.6	−46	25	6.7	18	
80	5139	Centaurus	Globular cluster	13	26.8	−47	29	3.6	36	ω Centauri
81	6352	Ara	Globular cluster	17	25.5	−48	25	8.1	7	
82	6193	Ara	Open cluster	16	41.3	−48	46	5.2	15	
83	4945	Centaurus	SBc galaxy	13	05.4	−49	28	8.7	20 × 4	
84	5286	Centaurus	Globular cluster	13	46.4	−51	22	7.6	9	
85	IC 2391	Vela	Open cluster	08	40.2	−53	04	2.5	50	O Velorum Cluster
86	6397	Ara	Globular cluster	17	40.7	−53	40	5.7	26	
87	1261	Horologium	Globular cluster	03	12.3	−55	13	8.4	7	
88	5823	Circinus	Open cluster	15	05.7	−55	36	7.9	10	
89	6087	Norma	Open cluster	16	18.9	−57	54	5.4	12	S Noræ Cluster
90	2867	Carina	Planetary nebula	09	21.4	−58	19	9.7	12	
91	3532	Carina	Open cluster	11	06.4	−58	40	3.0	55	
92	3372	Carina	Bright nebula	10	43.8	−59	52	—	120 × 120	η Carinæ Nebula
93	6752	Pavo	Globular cluster	19	10.9	−59	59	5.4	20	
94	4755	Crux	Open cluster	12	53.6	−60	20	4.2	10	Jewel Box, κ Crucis
95	6025	Triangulum Aus.	Open cluster	16	03.7	−60	30	5.1	12	
96	2516	Carina	Open cluster	07	58.3	−60	52	3.8	30	
97	3766	Centaurus	Open cluster	11	36.1	−61	37	5.3	12	
98	4609	Crux	Open cluster	12	42.3	−62	58	6.9	5	
99	—	Crux	Dark nebula	12	53	−63		—	420 × 300	Coal Sack
100	IC 2944	Centaurus	Cluster and nebulosity	11	36.6	−63	02	4.5	60 × 40	λ Centauri Cluster
101	6744	Pavo	SBb galaxy	19	09.8	−63	51	8.3	16 × 10	
102	IC 2602	Carina	Open cluster	10	43.2	−64	24	1.9	50	θ Carinæ Cluster 'Southern Pleiades'
103	2070	Dorado	Bright nebula	05	38.7	−69	06	—	30 × 20	Tarantula Nebula
104	362	Tucana	Globular cluster	01	03.2	−70	51	6.6	13	
105	4833	Musca	Globular cluster	12	59.6	−70	53	7.3	14	
106	104	Tucana	Globular cluster	00	24.1	−72	05	4.0	31	47 Tucanæ
107	6101	Apus	Globular cluster	16	25.8	−72	12	9.3	11	
108	4372	Musca	Globular cluster	12	25.8	−72	40	7.8	19	
109	3195	Chamæleon	Planetary nebula	10	09.5	−80	52	8.4	40 × 30	

Table 23.5. Some bright star clusters.

M	NGC	C	Constellation	Magnitude	
—	752	28	Andromeda	5.7	Open cluster
2	7089	—	Aquarius	6.5	Globular cluster
—	6193	82	Ara	5.2	Open cluster
36	1960	—	Auriga	6.0	Open cluster
37	2099	—	Auriga	5.6	Open cluster
38	1912	—	Auriga	6.4	Open cluster
44	2632	—	Cancer	3.1	Open cluster: Præsepe
67	2682	—	Cancer	6.9	Open cluster
41	2287	—	Canis Major	4.5	Open cluster
—	2362	64	Canis Major	4.1	Open cluster: τ Canis Majoris
—	2516	96	Carina	3.8	Open cluster
—	IC 2602	102	Carina	1.9	Open cluster: θ Carinæ
—	457	13	Cassiopeia	6.4	Open cluster: φ Cassiopeiæ
—	5139	80	Centaurus	3.6	Great globular: ω Centauri
—	4755	94	Crux	4.2	Open cluster: Jewel Box (κ Crucis)
—	6394	47	Delphinus	5.9	Globular cluster
35	2168	—	Gemini	5.0	Open cluster
13	6205	—	Hercules	5.9	Globular cluster
92	6341	—	Hercules	6.5	Globular cluster
—	7243	16	Lacerta	6.4	Open cluster
—	2244	50	Monoceros	4.8	Open cluster in Rosette Nebula
50	2323	—	Monoceros	5.9	Open cluster
—	6752	93	Pavo	5.4	Globular cluster
—	869/884	14	Perseus	4.3, 4.4	Sword-handle; two open clusters
47	2422	—	Puppis	4.4	Open cluster
25	IC 4725	—	Sagittarius	4.6	Open cluster; υ Sagittarii
23	6494	—	Sagittariuis	5.5	Open cluster
6	6405	—	Scorpius	4.2	Open cluster: Butterfly cluster
7	6475	—	Scorpius	3.3	Open cluster
11	6705	—	Scutum	5.8	Open cluster: Wild Duck cluster
45	1432/5	—	Taurus	1.2	Open cluster: Pleiades
—	—	41	Taurus	1	Open cluster: Hyades (Melotte 25)
—	6025	95	Triangulum Aust.	5.1	Open cluster
—	104	106	Tucana	4.0	Globular cluster: 47 Tucanæ
—	IC 2391	85	Vela	2.5	Open cluster: o Velorum

24 NEBULÆ

Planetary nebulæ

Planetary nebulæ are very inappropriately named. They were so-called by Sir William Herschel, because their pale, often greenish disks made them look superficially like the planets Uranus and Neptune when seen through a telescope, but they are not true nebulæ and have nothing whatsoever to do with planets.

A planetary nebula is in fact a very late stage in the evolution of a star up to eight times the mass of the Sun (stars which are more massive than this explode as supernovæ). At the end of the Main Sequence stage, the core contracts and the outer layers expand and cool; the star becomes a red giant. Eventually the outer layers are ejected, and move away into space at a speed of around 10 to 30 km s^{-1} from what is left of the star – which is now very hot, with a surface temperature of from 20 000 to as much as 100 000 °C; it is in the process of turning into a white dwarf. Ultraviolet radiation from the star ionizes the ejected material and makes it glow; the result is a planetary nebula. The common greenish tint is due to emissions from doubly ionized oxygen.

Planetary nebulæ are transient. They cannot persist, as such, for more than about 50 000 years before the material dissipates and becomes too dim to be seen, while the 'pre-planetary' stage, before the nebula itself is produced, is even briefer – perhaps no more than 1000 years. This is why planetary nebulæ seem to be uncommon. About a thousand are known in our Galaxy; doubtless there are many more which have not been observed, but they do not last for long. One case of development into a planetary nebula is Henize 1357, in Ara; it is 17 000 light-years away, and has now assumed the characteristic form of a planetary. Although the central star may have started to expel gas a few thousands of years ago, only very recently has there been enough radiation from the star to make the gas glow.

The best-known planetary is the Ring Nebula in Lyra, M.57, which is easy to locate because it lies midway between the stars β and γ Lyræ; a small telescope shows it as a tiny, shining ring, and a larger instrument will show

the central star. The ejected material takes the form of a sphere, but we see from long range (about 1400 light-years) and the eye observes more material at the periphery than at the centre – hence the appearance of a luminous ring; at the centre we can see 'straight through'. The material is indeed tenuous. If you take the amount of gas in an average village hall and spread it out over an area equivalent to the volume of the Earth, the resulting density will be about the same as that of the gas in a planetary nebula.

The Ring Nebula looks fairly symmetrical, but other planetaries assume bizarre forms; the Helix, the Hourglass, the Egg Nebula and the complex Cat's Eye Nebula are good examples. Often, too, a shell of material ejected by a fast stellar wind will catch up with, and collide with, a shell ejected much earlier at a slower rate.

Most planetaries are fairly remote; the Helix Nebula in Aquarius (NGC 7293, C.63) is the closest, at 450 light-years, and with a magnitude of 6.5 it is also the brightest. The Hourglass Nebula, MyCn18, is 8000 light-years from us, and has been beautifully imaged by the Hubble Space Telescope. The gas coming from the poles is less dense than that coming from the equator, and moves faster, which results in the characteristic hourglass shape. Another remarkable object, also imaged by Hubble, is the protoplanetary Egg Nebula (CRL 2688), 3000 light-years away and 0.6 of a light-year across; the central star is shrouded in a cocoon of dust. The star was a red giant only a few centuries ago, and is now in the process of becoming a white dwarf.

One very interesting object, imaged by the Hubble Space Telescope in 1999, is OH231.8 + 4.2 which is a remarkable example of a star which is in the process of changing from a normal red giant star into a planetary nebula. Infra-red pictures with the HST show dust and gas being blown out in opposite directions at up to 700 000 km h^{-1}. Radio observations indicate many unusual molecules of gas surrounding the star, including sulphur compounds such as hydrogen sulphide and sulphur dioxide. This is why the object has been nicknamed the Rotten Egg Nebula!

Planetary nebulæ are not unique to our Galaxy. In 1991 a planetary, N.66, was detected in the Large Cloud of Magellan; it is almost two light-years in diameter.

Diffuse nebulæ

Diffuse nebulæ, otherwise termed galactic or gaseous nebulæ, are among the most beautiful objects in the sky. Photographs taken with large ground-based telescopes, as well as with the Hubble Space Telescope, bring out their vivid colours and their bizarre forms, and even when observed visually with a small telescope they are fascinating.

Nebulæ are of three main types:

(1) Emission nebulæ, where the material is illuminated by short-wave radiation from very hot stars of types O and B and emits a certain amount of light on its own account. The ultra-violet radiation ionizes the hydrogen in the nebulosity and makes it glow, so that these nebulæ are also termed H.II regions.

(2) Reflection nebulæ, which shine only by light reflected from stars either in or very close to the nebulosity. The light is reflected from particles of dust.

(3) Absorption or dark nebulæ, where there are no suitable stars in the vicinity, and the nebula remains dark, showing up only because the dust that it contains blots out the light of stars beyond. The most famous dark nebula is the Coal Sack in the Southern Cross (C.90). There is no basic difference between a bright and a dark nebula; for all we know there may be stars illuminating the far side of the Coal Sack, so that if we could observe it from a different vantage point in space it would appear bright.

Catalogues also include other types of objects which are of a different nature. For example, there are supernova remnants such as the Crab Nebula and the Cygnus Loop. There are planetary nebulæ, which, as we have noted, are highly-evolved stars which have cast off their outer layers. Finally, there are the objects once termed spiral nebulæ, which we now know to be external galaxies; the term 'spiral nebula' is misleading, and has now gone out of use, to be replaced by 'spiral galaxy'. (Of course, by no means all galaxies are spirals.)

True nebulæ are stellar birthplaces. When formed they are made up chiefly of hydrogen, with some helium and only trace amounts of other elements, which is why first generation stars are 'metal-poor' (remember that 'metal', confusingly, is used to embrace all elements heavier than helium). The young clouds also contain small quantities of the light element lithium. When a star begins to shine by nuclear reactions the lithium is destroyed – so that if a star shows traces of lithium in its spectrum, we know that it is not yet old enough to have started radiating by nuclear fusion.

The most famous of all diffuse nebulæ is M.42, in the Sword of Orion, which is easily visible with the naked eye. In fact, it is only the most prominent part of a Giant Molecular Cloud which covers almost the whole of Orion; it is 1600 light-years away, and has been intensively studied. The material in it is almost incredibly rarefied, and it has been said that if you could take a 3 cm core sample right through it, the total weight of material collected would just about balance a pound coin – even though the nebula is 16 light-years in diameter (twice the distance between our Sun and Sirius).

Inside a nebula of this sort, star formation begins with the development of a rather denser area – perhaps because of the shock-wave from a supernova or by a 'compressional wave' sweeping through the material. Protostars are formed, usually shrouded by an envelope of opaque material and are detectable only at infra-red and millimetre wavelengths. We find Bok globules and also what are called Herbig–Haro objects (from their co-discoverers, G. Herbig of the United States and G. Haro of Mexico), formed when jets sent out from very young stars strike material which is already present and compress it. When a protostar blows away its opaque cocoon and becomes visible at optical wavelengths, it is termed a T Tauri star, after the first example to be identified. Classical T Tauri stars are from 1 to 10 million years old, and vary irregularly, because they are still unstable and have yet to join the Main Sequence. They send out violent 'stellar winds' and are usually accompanied by thick disks of material (proplyds: *pro*to-*pl*anetary *d*isks), many of which have been found inside the Orion Nebula. Most stars seem to go through a T Tauri stage, when the stellar winds blow much of the remaining débris out of the system. Contraction time to the

Table 24.1. Selected list of prominent diffuse nebulæ.

Name		Constellation	Distance (light-years)	Diameter (light-years)
M.42	Great Nebula	Orion	1600	16
M.20	Trifid Nebula	Sagittarius	2200	12
M.17	Omega Nebula	Sagittarius	5000	30
M.8	Lagoon Nebula	Sagittarius	6500	30
M.16	Eagle Nebula	Serpens	7000	20

Main Sequence depends upon the mass of the star – perhaps 100 million years for a star of low mass, 40 million years for a star of solar mass and as little as one million years for a star of much greater mass. As we have seen, a star which has a mass of less than 8% that of the Sun (equivalent to 80 times the mass of Jupiter) will never reach the Main Sequence at all, and will shine feebly as a brown dwarf.

T Tauri stars are characterized by strong emission lines in their spectra, produced by interactions between the star and its disk. Once the disk dissipates, these lines vanish, but 'weak-lined' T Tauri stars produce X-rays in the hot plasma trapped in the magnetic fields above the surface of the star, and these X-rays are very evident in observations from satellites such as the recent Chandra and Newton. It may well be that in many cases the disk material collects into planetesimals, which in turn collide and produce true planets.

The Orion Nebula shines because of the very hot, massive stars making up the Trapezium, θ Orionis – which can be seen well with a small telescope. However, the dust means that we cannot see into the heart of the Nebula, where there are stars which will never become visible to us at optical wavelengths because their lifetimes will be too short for them to 'burn' a passage into outer space. Such is the Becklin–Neugebauer (BN) Object, deep inside the Nebula, which is a very young, immensely powerful star which we can detect only because of its infra-red radiation.

A selected list of other well known diffuse nebulæ is given in Table 24.1. Proplyds have been detected inside the Lagoon Nebula, which is much larger than M.42 but is also much further away. The Eagle Nebula, M.16 in Serpens, is truly magnificent; a photograph taken from the Hubble Space Telescope shows three columns of gas and dust, each 1 to 3 light-years long, inside which stars are forming. Near the tops of the pillars, the radiation from the fledgling stars is evaporating the gas, so that it will eventually be dissipated; these structures are known as evaporating gaseous globules (EGGs). Then the star-forming material in a nebula is exhausted, all that is left is an open star cluster.

Another interesting nebula is the North America Nebula NGC 7000 (C.10). Its shape really does recall the outline of the North American continent, and there is darker material in the position of the 'Gulf of Mexico'! Many dark rifts are seen in the Milky Way, notably in the region of Cygnus, and also in other nebulæ, notably the Trifid Nebula in Sagittarius (M.20).

25 THE GALAXY

The Sun is one of around 100 000 million stars making up our Galaxy. The system is often referred to as the Milky Way galaxy, but nowadays the term 'Milky Way' is usually restricted to the luminous band stretching across the sky. It was Galileo, in 1610, who first realized that the Milky Way is made up of stars; it must have been known since antiquity, and almost all early civilizations have legends about it.

The Galaxy is flattened. There is a main disk, around 100 000 light-years in diameter (some authorities regard this as a slight overestimate); the disk is 2000 light-years thick. There is a nearly spherical central bulge with a diameter of about 15 000 light-years. Surrounding the main system is the galactic halo, which may have a diameter of at least 100 kpc[1], and contains isolated stars as well as the globular clusters. The disk is rotating round the centre of the Galaxy, in the bulge; the Sun takes 225 000 000 years to complete one circuit. The halo objects move in very inclined orbits, and do not share in the disk rotation.

The disk is made up chiefly of young objects (Population I), and star formation is still going on; the halo consists of older Population II objects, and the star-forming material has been used up. Within 1 parsec of the centre of the Galaxy the average distance between stars may be no more than 1000 a.u., so that there could be ten million stars per cubic parsec. Near the Sun, the density is only 0.2 of a star per cubic parsec, and in the halo the density is much lower still.

As well as stars, the Galaxy contains a vast amount of thinly-spread interstellar matter. The first proof of this was obtained by W. Hartmann in 1904, when he was studying the spectrum of the star δ Orionis. δ Orionis is a spectroscopic binary, so that its lines show Doppler shifts corresponding to the orbital motions of the components. However, Hartmann found that some of the lines remained stationary, so that

clearly they were associated not with the star itself, but with material lying between δ Orionis and ourselves. Further proof was obtained by J. Trümpler in 1930; some of the Milky Way clusters were reddened and appeared fainter than logically they ought to have done, so that they were being dimmed by intervening material. The gas between the stars is made up of hydrogen and helium, with smaller amounts of carbon, nitrogen, oxygen and neon. In the 1930s astronomers at Mount Wilson found the first indications of interstellar molecules, but firm proof was postponed until 1963, when the hydroxyl radical, OH, was identified.

In 1968 C. Townes and his colleagues at Berkeley detected interstellar ammonia (NH_3) at a wavelength of 1.26 cm. The first organic molecules (that is to say, molecules containing carbon) were identified in 1969 by L. Snyder, B. Buhl, B. Zuckerman and P. Palmer, who detected formaldehyde (H_2CO). Many others have since been found, including carbon monoxide, hydrogen cyanide and methanoic acid. Ethyl alcohol (CH_3CH_2OH) has also been detected. One 'cloud' contains enough of it to fill the entire globe of the Earth with alcohol, or make 10^{28} bottles of whisky. More molecules are being detected yearly.

However, the average density of interstellar gas in our Galaxy is very low – about 500 000 hydrogen atoms per cubic metre, which corresponds to a very good laboratory vacuum. There are also dust grains, about 0.1 μm (10^{-7} m) in radius, which are chiefly responsible for the reddening and dimming of distant objects and prevent us from seeing through to the galactic centre.

Originally it was believed that the Sun lay in the centre of the Galaxy; this was assumed by Herschel, and also by the Dutch astronomer J. C. Kapteyn, who carried out a detailed investigation in the early part of the 20th century. However, in 1917 H. Shapley proved that this is not so. He realized that the globular clusters are not spread symmetrically around the sky, but are concentrated in the southern hemisphere; using short-period variable stars he measured the distances of the clusters, and established that the Sun lies well away from the centre, although very

---

[1] When dealing with vast distances, parsecs are often used in preference to light-years. One parsec is the distance from which a star would show a *par*allax of 1 *sec*ond of arc (in fact there is no star within 1 parsec of the Sun). One parsec is equal to 3.26 light-years. One kiloparsec is equal to 1000 parsecs, and one megaparsec is equal to 1000 000 parsecs.

close to the main plane. This, of course, explains the Milky Way band; we are looking along the main plane of the Galaxy, and seeing many stars in much the same line of sight. The centre of the galaxy lies at RA 17h 45m, declination $-28°66'$, in Sagittarius; the galactic pole is at RA 12h 51m.4, declination $-27°07'$, in the star-poor region of Sculptor. Shapley gave the diameter of the Galaxy as 30 000 light-years, although this proved to be very much of an under-estimate.

The Galaxy is a spiral system, classed officially as of type Sbc. This is no surprise; many galaxies are spiral in form, and are rotating round their nuclei, as was established by Shapley and confirmed in more detail by the Swedish astronomer B. Lindblad. The Sun lies in a segment of what is termed the Orion arm, which also includes the Orion Nebula; inward, towards the galactic centre, is the Sagittarius arm, while on the opposite side is the Perseus arm; also recognized are the Centaurus and Cygnus arms.

It is now believed that individual spiral arms are not permanent features. 'Density waves' sweep around the Galaxy, and produce what may be called cosmic traffic jams, compressing the material and triggering off star formation. Very luminous O and B stars develop, but do not shine as such for long enough to leave the spiral arms, so that these O–B associations characterize the arms. In the future our Galaxy will still have spiral arms – but they will not be the same as those of the present time.

We cannot see through to the centre of the Galaxy, because there is too much obscuring material in the way, but we know that it lies at a distance of about 27 000 light-years, beyond the star clouds in Sagittarius. Our knowledge of it is drawn from infra-red and radio observations, since the dust cannot block radiations at these wavelengths. For example, in 1997 two huge star clusters less than 100 light-years from the centre were identified by the NICMOS[2] camera on the Hubble Space Telescope. These clusters are known as the Arches and Quintuplet clusters, and are 10 times the size of the typical open clusters spread through the Galaxy. The Arches cluster, thought to be around two million years old, is so packed that a region of space equivalent to a sphere bounded by the Sun and α Centauri would contain 100 000

[2] NICMOS: *N*ear *I*nfrared *C*amera and *M*ulti*p*robe *S*pectrometer.

stars; the Quintuplet cluster, rather more dispersed, seems to be around four million years old. It contains what has become known as the Pistol Star, which may be the most luminous star in the Galaxy – at least 10 000 000 times more powerful than the Sun, and emitting as much energy in six seconds as the Sun does in a year. Its nickname comes from the bright nebula associated with it, which does look a little like the shape of a pistol, and may have been formed from the star no more than 6000 years ago. Within a million years or so the Pistol Star is likely to explode as a supernova. It must, however, be remembered that we can never have a direct optical view of it, and it is always possible that we are dealing with a closely-packed group of stars rather than a single giant sun.

Round the galactic centre is what is termed the circum-nuclear disk (CND), a torus of warm atomic molecular gas and dust. It is rotating, and extends from about 1.5 to 8 parsecs from the centre. It may well be that the actual centre is marked by a strong radio source, Sagittarius A, which has two components: Sgr A East (probably an expanding gas-bubble due to a former supernova) and Sgr A West, which contains a compact source, Sgr A^{-*} (pronounced Sagittarius A-star) which could consist of a massive black hole at least 2 000 000 times as massive as the Sun, surrounded by an accretion disk. Whether or not Sgr A^{-*} marks the true centre, it is certain that the whole region is the site of violent activity.

We also know that the Galaxy contains a great deal of material which we cannot detect in any way. This is shown by the manner in which the disk system rotates. Almost all the visible matter in the Galaxy is contained within a radius of no more than 15 kiloparsecs from the centre, and it would be expected that the rotational speeds of objects would lessen with increased distance from the centre, as happens in the Solar System (Kepler's Laws). Thus a star, say 40 000 light-years from the centre would be expected to have a rotational speed less than that of the Sun, at 27 000 light-years. Yet this does not happen; the rotational velocities do not become less, but may even increase. This proves that the main mass of the Galaxy is not contained in the innermost region (in the Solar System, over 99% of the total mass is concentrated in the Sun). The same is true of other galaxies, and we are faced with what was once called the 'missing mass' problem, although it is now thought that there is a vast amount of undetectable 'dark matter'.

The Galaxy must be regarded as a large system, although it is in no way exceptional (it is not nearly so large or so massive as the Andromeda Spiral). Whether it can be classed as a 'barred spiral' is uncertain; at any rate the bar – if it exists at all – is not pronounced. It is one of a number of galaxies making up what we call the Local Group, of which the other senior members are the Andromeda and Triangulum spirals.

26 GALAXIES

Galaxies are external systems; many of them are much smaller than our Galaxy, while others are far larger and more populous. Many are included in Messier's Catalogue, but their true nature was not then recognized. A major breakthrough came in 1845, when the third Earl of Rosse, using his remarkable home-made 72 inch reflector at Birr Castle in Ireland, found the first spiral – M.51 in the constellation of Canes Venatici, now generally called the Whirlpool. Other spirals were soon found, although for some years only the Birr telescope was capable of showing them as such; other 'starry nebulæ' were spherical in form, so that superficially they looked very like globular clusters, while others were elliptical, and some irregular in outline.

The first proof that the resolvable 'nebulæ' are external systems came in 1923, when Edwin Hubble detected Cepheid variables in some of the spirals (including M.31 in Andromeda). He could therefore find their distances, which also gave the distances of the systems in which the Cepheids lay; at once it became clear that they were so far away that they could not possibly belong to the Milky Way. This idea was not new. It had been suggested as early as 1755 by Immanuel Kant, and had at one time been supported by William Herschel, but it had remained very much of an open question, and was the subject of a great debate in 1920 between two leading American astronomers, Harlow Shapley and Heber D. Curtis. By studying short-period variables in globular clusters, Shapley had given the first reasonable estimate of the size of the Milky Way system, but he regarded the spirals as being contained in our Galaxy, while Curtis believed them to be external. Hubble's work, carried out with the Hooker reflector at Mount Wilson (then the only telescope in the world of sufficient power) proved that Curtis had been right. The distance of the Andromeda Spiral was given as 900 000 light-years, which was subsequently reduced to 750 000 light-years.

This was later found to be an under-estimate. In 1952 W. Baade, using the new Hale reflector at Palomar, realized that there had been a major error in the Cepheid scale. There are two types of Cepheids, one more luminous than the other, and those in the spirals had been wrongly identified; they were more powerful than had been believed, and hence more remote. The distance of the Andromeda Spiral is now known to be 2 200 000 light-years[1], and is the most remote object which is clearly visible with the naked-eye (very keen-sighted observers can make out the Triangulum Spiral, M.33, but most people will need binoculars to identify it). The only two external systems brighter than these two spirals are the far-southern and much nearer Magellanic Clouds.

In 1925 Hubble drew up a simple system of classifying galaxies by their shapes, and his 'tuning fork' diagram is still used, even though more complex classifications have been proposed. Hubble's classes are as follows:

Spirals, resembling Catherine wheels. Sa: conspicuous, often tightly-wound arms issuing from a well-defined nucleus. Sb: arms looser, nucleus less condensed. Sc: inconspicuous nucleus, loose arms.

Barred spirals, in which the spiral arms issue from the ends of a kind of 'bar' through the nucleus. They are divided into types SBa, SBb and SBc in the same way as for the normal spirals.

Elliptical systems, with no sign of spirality; they range from E7 (highly flattened) down to E0 (virtually spherical, looking rather like globular clusters). Many ellipticals are dwarf systems (dE), but there are also giant elliptical galaxies, such as the immense M.87.

Irregular systems, with no definite shape.

It is thought that about 30% of galaxies are spiral, 60% elliptical and 10% irregular, though these values are very approximate.

Active galaxies are galaxies with unusual characteristics. Most active galaxies contain a compact, brilliant and often variable central core, known as an active galactic

---

[1] Very recent results indicate that even this may be an understatement; if so, the distances of other Local Group galaxies may also have to be revised.

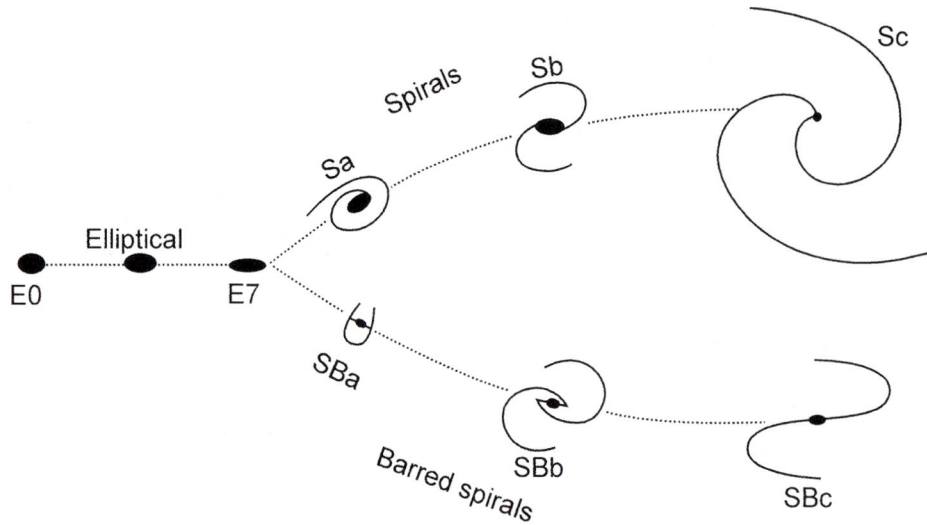

Figure 26.1. Hubble's 'tuning fork' diagram.

nucleus (AGN). Whereas most of the energy radiated by an ordinary galaxy is starlight (the combined output of its constituent stars), an active galaxy radiates strongly over a much wider range of wavelengths; much of this radiation is emitted by energetic charged particles moving in magnetic fields. There are many types of active galaxies. For example, *radio galaxies* may send out a thousand to a million times more energy at radio wavelengths than is the case for a normal galaxy. Many radio galaxies are giant elliptical systems; there are vast clouds of radio-emitting material to either side of the visible system. *Starburst galaxies* are highly disturbed systems in which star formation is proceeding at a furious rate. *Seyfert galaxies*, first identified by the American astronomer Carl Seyfert in 1943, have bright, variable nuclei and weak spiral arms; they are very energetic, particularly at infra-red and radio wavelengths; the luminosity of the core may equal the total luminosity of a galaxy such as our own. It may well be that the centre of a Seyfert galaxy may be occupied by a supermassive black hole. *Quasars* and *BL Lacertæ* objects are immensely luminous, and are now known to be the cores of very active galaxies, probably powered by central black holes. The word 'quasar' is an abbreviation of 'quasi-stellar radio source', since it was originally thought that all members of the class were strong radio emitters. In fact this

is not the case, and the term QSO or quasi-stellar object is now widely used. Violently active systems are often termed 'blazars'.

The first quasar to be identified was 3C-273, in Virgo. (The prefix 3C indicates the third Cambridge catalogue of radio sources, published in 1962.) 3C-273 was known to be a strong radio emitter, but identifying it with a visual object proved to be difficult. However, on 5 August 1962 radio astronomers in Australia, working with the Parkes telescope, followed an occultation of the radio source by the Moon, and were able to pinpoint its position very accurately. From this, the source was identified with what seemed to be a faint bluish star of magnitude of 12.8. In 1963 M. Schmidt, at Palomar, obtained an optical spectrum, and found that the object was not a star at all; its spectrum was quite different, and showed lines which could not at first be identified. They proved to be due to hydrogen, but were tremendously red-shifted, indicating a high recessional velocity and therefore immense distance and luminosity. In fact, it is now known that the recessional velocity is 47 400 km s^{-1}, and the distance is 2.2 thousand million light-years – yet even so, 3C-273 is one of the closest of the quasars, and the only one which is within the range of small telescopes. It is not hard to locate, but visually it looks exactly like an ordinary dim star.

Other quasars were soon found. Their luminosities range from 10 to 10 000 times that of a normal galaxy such as our own, and they vary in luminosity over short periods, so that most of their radiation must come from a small region, only about one light-week (or 1200 a.u.) in diameter. Again the best explanation is that the power comes from a central, very massive black hole. All quasars are very remote, so that we see them as they used to be when the universe was comparatively young; it may well be that a quasar is simply one stage in the evolution of a massive galaxy.

The distances of quasars are, of course, measured by the red shifts of their spectral lines. For example, consider the quasar PC 1247+3406, where $z = 4.897$ (z is a measure of the red shift: $z = (L - L_0)/L_0$, where L_0 is the change in wavelength and L is the laboratory wavelength). The recessional velocity here is 94% of the velocity of light.

One interesting development has been the detection of what are termed 'gravitational lenses'. If a galaxy lies on or near the line of sight of a more remote quasar, the result will be that the quasar will show multiple images, simply because the intervening galaxy acts as a lens. The first instance of this, with the galaxy Q0956+561, was found in 1979, but many examples are now known – such as the 'Clover Leaf', H 1413 + 117, discovered from La Silla in 1988, where there are four images of the distant quasar. There is also the 'Einstein ring' found in 1985 by astronomers using the Canada–France–Hawaii telescope on Mauna Kea. They detected a giant luminous arc around the cluster of galaxies Abell 370; the arc is 500 000 light-years long and 25 000 light-years wide. The effect is due to the light from a background galaxy being bent in the gravitational field of the Abell 370 cluster; this sort of effect had been predicted by Albert Einstein – hence the nickname.

However, one must be wary of jumping to conclusions. In 1987 a binary quasar was detected by S. Bjorgevski, G. Meylan, R. Perley and P. McCarthy. The binary quasar (QQ 1145-071) is at least 10 000 million light-years away. The spectra of the two images are not identical, so that we are dealing with two separate objects rather than a gravitational lens effect. There are, too, cases when a very remote background object has been detected only because of the 'magnifying' effect of a massive system lying in almost the same line of site[2]. The bending of light gathers most of the light of the distant object, causing it to appear brighter than it would do if the gravitational lens were not present.

We also have to consider BL Lacertæ objects, which are related to quasars. The first to be studied was BL Lacertæ itself, which was noted in 1941 and taken for an ordinary variable star. In 1968 it was identified as a radio source (by Maarten Schmidt, at Palomar) and found to have a spectrum which was virtually featureless. Other similar objects, such as AP Libræ and W Comæ, were then found; they too had been taken for variable stars. In 1973 astronomers at Palomar found fuzzy 'surrounds' of the objects which did show spectral lines, and in 1976 J. Wampler, at the Lick Observatory, found lines in the spectrum of the BL Lac object 0548-322, so that he could measure the red shift and find the distance.

Very probably BL Lacertæ objects and quasars are of the same nature, but with a quasar we are observing at a substantial angle to the jet, whereas with a BL Lacertæ object we are looking 'straight down' a jet, so that the jet appears as a bright spot overpowering the dim galaxy surrounding it.

It was once believed that Hubble's 'tuning fork' diagram had an evolutionary significance, so that spirals could develop into ellipticals or *vice versa*, but it now seems that the situation is much less straightforward than this, and it has to be admitted that our knowledge of the evolution of galaxies is still very incomplete; one fact about which we may be confident is that the quasar and BL Lacertæ stages are temporary and comparatively brief. Elliptical systems consist mainly of Population II objects, so that the most brilliant stars are old red giants, while Population I objects are dominant in the spirals, although the populations are always to some extent mixed, and no hard and fast boundaries can be drawn. Certainly the ellipticals contain

---

[2] *Microlensing* makes use of the same principle; when an image is gravitationally lensed by a star rather than a galaxy, it is termed microlensing. Studies were carried out from 1993 by astronomers using the Melbourne telescope at Mount Stromlo in Australia. In 1986 B. Paczynski had proposed that if what is termed a MACHO (Massive Astronomical Compact Halo Object) passes in front of a star, it will act in the manner of a lens and cause a temporary increase in the star's apparent brightness. Effects of this type have been reported, although the precise nature of MACHOs is very uncertain; brown dwarfs, black holes, neutron stars and normal stars have all been suggested as possible candidates.

Table 26.1. Selected list of galaxies in the Local Group.

Name	RA h	RA m	Dec. °	Dec. ′	Type	Distance (thousands of light-years)	diameter (thousands of light-years)	Absolute magnitude
Our Galaxy	—	—	—	—	Sbc	—	100	−20.6
SagDEG	18	51.9	−30	30	dE7	80	10	−14.0
LMC	05	24.0	−69	48	SBc	170	30	−18.1
SMC	00	51.0	−73	06	Irr	190	16	−16.2
Ursa Minor dwarf	15	08.2	+67	23	dE5	250	2	−8.9
Draco dwarf	17	19.2	+57	58	dE3	260	2	−8.6
Carina dwarf	06	40.4	−50	55	dE4	550	—	−9.2
Sculptor dwarf	00	57.6	−33	58	dE	280	5	−10.7
Fornax dwarf	02	37.8	−34	44	dE3	420	7	−13.0
Leo II	11	10.8	+22	26	dE4	750	3	−10.2
Leo I	10	05.8	+12	33	dE3	880	2	−12.0
NGC 6822 (Barnard's Galaxy)	19	42.1	−14	56	Irr	1700	5	−16.4
M.31	00	40.0	+40	59	Sb	2200	130	−21.1
M.32	00	40.0	+40	36	E2	2200	6	−10.1
NGC 205	00	37.6	+41	25	E5	2200	12	−16.3
NGC 185	00	36.2	+48	04	E5	2200	8	−16.3
NGC 147	00	30.5	+48	14	E4	2200	2	−15.7
Andromeda I	00	43.0	+37	44	dE0	2200	2	−11.7
Andromeda II	01	13.5	+33	09	dE3	2200	2	−11.7
Andromeda III	00	32.6	+36	12	dE6	2200	2	−10.2
IC 1613	01	02.2	+01	51	Irr	2400		−14.9
M.33	01	31.1	+30	24	Sc	2900	52	−18.0

very little remaining interstellar material, whereas there is plenty of it in the spirals, where star formation is still going on.

Galaxies congregate in clusters, and our Galaxy is a member of what we call the Local Group. It contains over two dozen galaxies, but most of them are small; the senior members are the Andromeda Spiral, our Galaxy and the Triangulum Spiral. The two Magellanic Clouds are regarded as satellites of our Galaxy, while M.32 and NGC 205 rank as satellites of the Andromeda Spiral. A selected list of the Local Group systems is given in Table 26.1.

The Magellanic Clouds are bright naked-eye objects, and contain objects of all kinds; in 1987 the Large Cloud even produced a supernova. They were formerly classified as irregular, but it now seems that the Large Cloud is a rather poorly-defined barred spiral, while the Small Cloud may be double. They are linked with our Galaxy by the Magellanic Stream, which is made up of neutral hydrogen,

stretching in a continuous arc beyond the Clouds and tracing out their orbits round our Galaxy. The origin of the Stream is probably tidal. The Large Cloud is about one-quarter the size of our Galaxy; for the Small Cloud the figure is one-sixth.

The Andromeda Spiral, M.31, is one and a half times the size of our Galaxy, and contains over 300 globular clusters, as well as variable stars of all types, doubles, diffuse nebulæ, planetary nebulæ and novæ. There has also been one recorded supernova, S Andromedæ of 1885, which reached the fringe of naked-eye visibility. Although its nature was not realized at the time (M.31 was still generally believed to be a minor feature of our Galaxy), its remnant has been identified since. In 1993 results from the Hubble Space Telescope showed that M.31 has a double nucleus. It is a well-formed spiral system of class Sb, and it is unfortunate that it lies at a narrow angle to us, so that the full beauty of the spiral is lost.

At present M.31 is approaching our Galaxy at a rate of almost 500 km h^{-1}. Whether there will be an eventual collision is not clear; at any rate, no such collision can occur for at least 5000 million years in the future, by which time the Sun will long since have left the Main Sequence and the Earth will no longer exist. If a collision occurs, the spiral forms of both systems will be destroyed.

The Triangulum Spiral, M.33, is half the size of our Galaxy, and contains around 10 000 million stars. The other members of the Local Group are dwarfs, either elliptical or irregular.

The closest of all galaxies is the Sagittarius Dwarf Elliptical Galaxy, SagDEG (not to be confused with another dwarf system in Sagittarius, the Sagittarius Dwarf Spheroidal Galaxy or SagDIG). SagDEG was discovered in 1994 by R. Ibata, M. Irwin and G. Gilmore. It is 80 000 light-years from the Sun, and 50 000 light-years from the centre of the Galaxy. In the sky it extends over an area of 5 × 10 degrees, but it is far from prominent. It orbits the Galaxy in a period of less than 1000 million years, and will eventually be disrupted. It seems to contain the globular cluster M.54; probably it includes about a million stars, many of which are old red giants. Its current length is around 10 000 light-years, and is elongated in the direction pointing to the centre of our Galaxy.

The dwarf members of the Local Group are mainly less than 5000 light-years in diameter. Some of them, such as the Ursa Minor, Draco, Sculptor and Fornax galaxies, are within a million light-years of us, but are very sparse. No doubt many more dwarfs exist, hidden from us by interstellar matter. Indeed, few galaxies can be seen near the dust-rich main plane of our Galaxy; this region is known as the Zone of Avoidance.

Adjoining the Local Group are other groups, such as the Maffei group; this contains a giant elliptical system (Maffei 1), discovered in 1968 by the Italian astronomer of that name. It is heavily obscured by material in our Galaxy, and was at first thought to be a true member of the Local Group; not far from it is a second galaxy, Maffei 2, and there are other smaller systems. Also comparatively close is the Sculptor group, dominated by NGC 253 (C.65), a beautiful edgewise-on spiral.

The nearest really large group is the Virgo Cluster, which contains about 2000 members and is the physical centre of what is termed the Local Supercluster; the average distance from us of the galaxies in the Virgo cluster is 60 000 000 light-years. The largest member is the giant elliptical M.87, which is a powerful radio source, and which has sent out a curious jet of material. The Virgo cluster has a pronounced effect upon our Local Group, and we are also affected by the 'Great Attractor', which may be a concentration of massive systems heavily obscured by intervening dust.

From 1912 V. M. Slipher, at the Lowell Observatory in Arizona, examined the spectra of galaxies, using the Lowell refractor. The galaxies are composed of millions of stars, and the spectra are bound to be something of a jumble, but the absorption lines can be measured easily enough. Slipher found that apart from a few of the very nearest galaxies – those now known to belong to the Local Group – all the Doppler shifts were to the red, indicating velocities of recession. At the time the significance of this was not fully appreciated, and it was still widely believed that the 'starry nebulæ' were contained in our Galaxy.

Edwin Hubble found the answer. First he proved, by observations of Cepheids, that the galaxies are external systems. Two years later, in 1925, he established that there is a definite link between distance and recessional velocity; the further away a galaxy lies, the faster it is receding. In fact, the entire universe is expanding, and every group of galaxies is moving away from every other group. There is no 'centre', and by now our telescopes can reach out to systems which are over 10 000 million light-years away, receding at well over 90% of the velocity of light.

It is of vital importance to define the value of the 'Hubble constant', which defines the rate at which the universe if expanding. Many estimates have been made, ranging from 30 up to 100 km s^{-1} Mpc^{-1}. In 1999 new measurements, made with the Space Telescope, gave a value of 70 km s^{-1} Mpc^{-1}, with a possible error of only 10%. The Hubble team observed galaxies out to 64 000 000 light-years, and identified 800 Cepheid variables, so that the result seems to be much more reliable than any previous estimate.

Assuming that the expansion has gone on ever since the universe, in its present form, came into existence –

however that may have happened! – the new data indicate that the universe cannot be much more than about 12 000 million years old, which is considerably younger than has been previ- ously believed[3]. Moreover, if the rule of 'the further, the faster' holds good, we will eventually peer out to a distance at which a system would be racing away at the full velocity of light. We would then be unable to see it, and we would have reached the boundary of the observable universe, although not necessarily the boundary of the universe itself.

Inside clusters, collisions between galaxies can and do occur. For example there are the Antennæ galaxies in Corvus, NGC 4033 and 4039 (C.61 and 62), where the two systems are meeting and triggering off intense star formation; the nickname comes from the fact that the long streamers of gas issuing from the systems resemble an insect's antennæ. The Cartwheel Galaxy in Sculptor, 500 million light-years away, was once a normal spiral, but a smaller system plunged into it, careering through the core and sending ripples of energy outward; these compressed the dust and gas, and the result was an expanding ring made up of several thousands of millions of new stars. The ring seems to have resulted from a collision which dates back 200 000 000 years. The diameter of the Cartwheel is 15 0000 light-years.

In 1999 a superb picture of two interacting spiral galaxies in Canis Major, NGC 2207 and IC 2163, was taken by the Hubble Space Telescope's Wide Field Planetary Camera 2. Strong tidal forces from the larger system, NGC 2207, have distorted the shape of IC 2163, flinging out stars and gas into long streamers stretching out for at least 100 000 light-years. It seems that IC 2163 is swinging past NGC 2207 in a counter-clockwise direction, having made its closest approach 40 000 000 years ago. However, IC 2163 does not have sufficient energy to escape from the gravitational pull of NGC 2207, and in the future it will be pulled back, again swinging past the larger system. Trapped in their mutual orbits round each other, the two galaxies will continue to distort and disrupt each other until finally, perhaps several thousands of millions of years from now, they will merge into a single, more massive galaxy.

Some galaxies are known as ULIRG systems (Ultra-Luminous Infra-red Galaxies). In 1999 observations of some of these, made with the Hubble Space Telescope, showed that there are 'nests' of these systems, apparently engaged in multiple collisions which involve three, four or even five galaxies. Previously it had been thought that only pairs of galaxies took part in these collisions.

Yet what does all this tell us about the origin and evolution of the universe itself?

---

[3] Some astronomers still maintain that 15 000 million years is nearer the truth. We cannot yet be confident of the precise value.

27 THE EVOLUTION OF THE UNIVERSE

Most astronomers believe that the universe was created at one set moment, in what is usually called the 'Big Bang'. Matter did not simply erupt into pre-existing space; space, time and matter came into existence simultaneously. It is impossible to discuss what happened before that, because there was no 'before'. Neither can we say just where the Big Bang happened, because if it included the whole universe it happened 'everywhere'. Expansion began at once, and has been continuing ever since. To be accurate, it is space which is expanding, carrying all matter – and, of course, the galaxies –with it. The concept was originally described in 1927 by the Belgian abbé Georges Lemaître.

Theory can take us back to 10^{-43} second after the Big Bang. The temperature at that time was of the order of 10^{32} degrees C, and the universe was dominated by radiation. Energetic particles were moving around, and some of this radiation turned into particles of matter and anti-matter – including what are termed quarks, the 'building blocks' of protons and neutrons. If a particle and an anti-particle meet, both vanish, and if the numbers had been equal there would have been nothing left of the fledgling universe. However, there were rather more particles than anti-particles, so that most of the anti-particles were rapidly annihilated. About one-millionth of a second after the beginning of time, quarks clumped together to form protons and neutrons.

There were more protons than neutrons. After about 100 seconds, nuclear reactions began, and protons and neutrons combined to form the first elements, hydrogen and helium. The universe was opaque, because photons of light could not travel far before being blocked by collisions.

About 300 000 years later, when the universe had cooled to around 3000 degrees C, electrons were captured by nuclei to make complete atoms. Light could now travel for vast distances without being blocked, and the universe became transparent to radiation; this is known as the decoupling stage. It followed that the radiation content of the universe was free to spread out all over the expanding volume of space. This expansion diluted the radiation, and the wavelengths of the light were increased. Today the radiation is still there, and is detectable, although very diluted and also shifted into the millimetre and centimetre range of the electromagnetic spectrum. It remains a faint glow pervading all space – the last remnant of the Big Bang.

All this may sound highly speculative, but we do have the evidence of the microwave radiation, indicating an overall temperature of 3 K – that is to say, 3 degrees above absolute zero, the coldest temperature there can possibly be ($-273\,°C$). The microwave radiation was detected in 1965 by A. Penzias and R. Wilson, using a special radio antenna which had been built for a completely different investigation. However, the background radiation had been predicted earlier by R. Dicke, who calculated that the temperature of the universe should by now have fallen to precisely this value: 3 K.

One initial problem was that the background radiation seemed to be quite uniform, and it was not easy to see how a 'lumpy' universe could have been formed from a very smooth expansion; how could galaxies begin to condense? To the relief of theorists, measurements carried out in 1992 from an artificial satellite, COBE (the Cosmic Background Explorer), showed that there were tiny irregularities in the background radiation, and these were confirmed in 1993 by S. S. Meyer and his team using a balloon-borne radiometer; independent confirmation came from a Jodrell Bank team using a radiometer on Mount Teide in Tenerife.

Many authorities believe that about 10^{-35} second after the Big Bang there was a brief period of rapid inflation, when the size of the universe increased enormously. Subsequently matter clumped together to form galaxies and clusters of galaxies, and the rate of expansion settled down. But will expansion continue indefinitely? This depends upon the overall density of matter of the universe. If it is above a certain critical value, expansion will stop and be followed by a period of contraction, ending in what may be called a Big Crunch. The critical value is around 3 atoms of hydrogen per cubic metre. If the overall density is less than this, the universe will expand for ever.

The ratio of the actual mean density to the critical density is usually denoted by the Greek letter omega, Ω. If Ω is greater than 1, the universe is closed. If it is less than 1, the universe is open.

The amount of luminous matter which we can actually see – galaxies, stars and so on – is not nearly enough to close the universe; Ω works out at around 0.01. However, there seems to be a vast amount of 'dark matter' which we cannot see, and this could change the whole situation. We have seen indications of this in the way in which our Galaxy rotates, and the same applies to other spirals, which may well contain five to ten times as much dark matter as luminous matter. Moreover, individual member galaxies of a cluster are moving around, and unless something were holding the cluster together it would disperse. Typical clusters of galaxies need 30 to 50 times as much dark matter as luminous matter to stop them from flying apart. We have to admit that we have no real idea of the nature of this dark matter. It has been proposed that neutrinos may have a little rest mass; it could even be that dark matter is of a completely different nature to ordinary matter, so that our present-day equipment is unable to detect it.

In an *open universe*, expansion will continue indefinitely, with the clusters of galaxies moving further and further apart. Eventually, in perhaps 10^{14} years' time, all the nuclear reactions of all the stars in all the galaxies will cease, and the galaxies will become cold, dead places.

In a *closed universe*, with Ω greater than 1, the clusters of galaxies will eventually start to draw together again; red shifts will be succeeded by blue shifts, and the temperature of the background radiation will rise. The 'Big Bang' will be repeated in what has been called the 'Big Crunch'. By around 10 000 million years before the Big Crunch, the temperature will have climbed back to its present value (3 K). A hundred million years before the Big Crunch, galaxies will merge and lose their separate identity. A million years before the end, the whole of space will be warmer than the present-day temperature of the surface of the Earth. About 100 000 years before the Crunch, the temperature everywhere will be around 10 000 K, hotter than the present surface of the Sun; stars will explode, and the whole universe will become opaque, consisting of a mass of plasma together with radiation. A hundred seconds before the Crunch, atomic nuclei will disintegrate

into protons and neutrons. When the Crunch comes, it may be the end of everything, because time itself may cease.

On the other hand, it may be that the Crunch will be succeeded by another Big Bang, and the entire cycle will begin again. We cannot tell; our current theories of the forces of nature are hopelessly inadequate to deal with problems of this kind.

We must also consider a situation in which Ω is exactly 1. The galaxies will then have just enough energy to continue moving apart for ever; their velocities will fall closer and closer to zero, but will not actually become zero until an infinite time in the future. This is usually termed *'flat' universe*, because space would have no curvature.

Obviously there are many details about which we are unsure; we cannot even give a definite 'age' for the universe it its present form. It has been estimated that the Big Bang happened 12 000 million to 15 000 million years ago. Very recently, there have been suggestions that the rate of expansion may be increasing, and this would certainly pose a major problem for theorists. It might even be necessary to bring back the 'cosmological constant', a force acting against that of gravitation – once introduced by Einstein, but later abandoned by him. However, as yet the evidence for an accelerating expansion of the universe is very uncertain.

In 1947 H. Bondi and T. Gold proposed a different theory – that the universe was in a steady state, which had no beginning and will have no end. The discovery of the 3 K background radiation effectively disproved this concept, but it is not so easy to dispose of the ideas of Halton Arp, who has produced images of galaxies and quasars which are clearly joined by luminous 'bridges', and are therefore presumably associated, but which have completely different red shifts. Arp believes that the shifts are not pure Doppler effects, but that there is a marked non-velocity component, so that all our distance measures beyond the Local Group are unreliable; he even thinks it possible that quasars are minor features shot out of galaxies. This is a most unpopular view among cosmologists, and it may be significant that Arp was refused further observing time with the large American telescopes because he was obtaining results which were embarrassing to official theory (subsequently he went to Germany and is now researching at the Max Planck Institute in Munich). If Arp is right, then many of our cherished theories will have to be drastically modified, or even abandoned. Among Arp's strong supporters is Sir Fred Hoyle.

New data concerning the very early universe were obtained in 1999, from the project termed Balloon Observations of Millimetric Extragalactic Radiation and Geophysics ('BOOMERANG'). The main telescope had a 1.2 m primary mirror, and weighed 2 tons. The equipment was carried in a giant helium-filled balloon which flew around Antarctica from 29 December 1998 to 9 January 1999, covering over 8000 km at a maximum altitude of 37 km; the launch took place from McMurdo Base, and the landing was within 50 km of this. Over 1800 square degrees of the sky were covered. Antarctica was chosen because of the stable prevailing winds at high altitude and, of course, the constant sunshine. The scientists came from Britain, Canada, Italy and the United States.

The resolution of the BOOMERANG images was 35 times better than with COBE, and were the first to bring the cosmic microwave background into sharp focus. The images reveal hundreds of complex regions visible as tiny variations – of the order of 0.0001 °C – in the temperature of the microwave background. The results give the best measurements to date of the geometry of space-time. They support the inflation theory – that the entire universe grew from a tiny subatomic region during a period of violent expansion a fraction of a second after the Big Bang. This expansion 'stretched' the geometry of space and made it to all intents and purposes 'flat'. The complex patterns visible in the images confirm predictions of the patterns which would result from sound-waves racing through the young universe, creating structures which subsequently evolved into giant clusters and superclusters of galaxies.

All in all, it is clear that our present knowledge of the evolution of the universe is very incomplete. Moreover, no theory can explain how matter came into existence in the first place, so that we are not genuinely discussing the origin of the universe at all; we are dealing with its development, which is a very different thing.

28 THE CONSTELLATIONS

From BC 4000 constellation patterns were drawn up. All these were different; the Chinese and Egyptian constellations, for example, are quite unlike ours (for example, our Draco seems to correspond with the Egyptian hippopotamus). Our system is derived from that of Ptolemy (Table 28.1) (it may originally have been Cretan, though opinions differ). In all, 88 separate constellations are now in use. A list of these is given in Table 28.2.

Ptolemy gave a list of 48 constellations: 21 northern, 12 Zodiacal and 15 southern (Table 28.2). All these are to be found on modern maps, although in many cases their boundaries have been altered – and the huge Argo Navis, the Ship Argo, has been chopped up into a Keel (Carina), sails (Vela) and a poop (Puppis).

Surviving post-Ptolemaic constellations are given in Table 28.3. Many of the original names have been shortened; this Piscis Volans, the Flying Fish, has become simply Volans, while Mons Mensæ, the Table Mountain, has become Mensa. There was some confusion over two of Bayer's constellations, Apis (the Bee) and Avis Indica (the Bird of Paradise); modern maps give it as Apus. There were also two Muscas, one formed by Lacaille to replace Bayer's Apis, and the other (rejected) formed by Bode out of stars near Aries. One discarded constellation, Quadrans, has at least given its name to the Quadrantid meteor shower of early January; the stars of Quadrans are now included in Boötes (near β Boötis).

The list of rejected constellations is very long (Table 28.4). Few will be regretted, though there are some people who are sad to see the demise of Noctua and Felis (the Owl and the Pussycat!) There have been occasional attempts to revive the whole constellation system, but none has met with much support.

However, the constellations are very unequal in size and importance. Apart from the dismembered Argo, the largest constellation, Hydra, covers 1303 square degrees, while the smallest, Crux, covers a mere 68 square degrees. Orion has five stars above the second magnitude and 42 above the fifth magnitude; against this, Mensa has no star

Table 28.1. Ptolemy's original 48 constellations.

Northern	Zodiacal	Southern
Ursa Minor	Aries	Cetus
Ursa Major	Taurus	Orion
Draco	Gemini	Eridanus
Cepheus	Cancer	Lepus
Boötes	Leo	Canis Major
Corona Borealis	Virgo	Canis Minor
Hercules	Libra	Argo Navis
Lyra	Scorpio (Scorpius)	Hydra
Cygnus	Sagittarius	Crater
Cassiopeia	Capricornus	Corvus
Perseus	Aquarius	Centaurus
Auriga	Pisces	Lupus
Ophiuchus		Ara
Serpens		Corona Australis
Sagitta		Piscis Australis
Aquila		
Delphinus		
Equuleus		
Pegasus		
Andromeda		
Triangulum		

above magnitude 5, while Cælum, Horologium and Sextans each have only two stars above magnitude 5.

'Star density' is also very uneven. For stars above magnitude 5, the greatest 'density' is for Crux (19.12 stars per 100 square degrees), at the other end of the scale comes Mensa (0.0), and then Sextans (0.63).

There have been a few cases of stars which have been arbitrarily transferred from one constellation to another; thus δ Pegasi has become α Andromedæ, γ Aurigæ has become β Tauri and γ Scorpii has become σ Libræ.

The current star-map is so well established that it will certainly not be altered, but one cannot but agree with the 19th-century Sir John Herschel, who commented that the constellation patterns seemed to have been designed so as to cause as much confusion and inconvenience as possible.

Table 28.2. The Constellations.

		1st mag	Stars to mag: 2.00	4.00	5.00	Area sq. deg.	Number of stars above mag 5 per 100 square degrees (star density)
Andromeda	Andromeda	—	0	7	25	722	9.19
Antlia	The Airpump	—	0	0	4	239	1.67
Apus	The Bird of Paradise	—	0	2	6	206	2.91
Aquarius	The Water-bearer	—	0	7	31	980	3.16
Aquila	The Eagle	Altair	1	8	16	652	2.45
Ara	The Altar	—	0	7	10	237	4.21
Aries	The Ram	—	1	4	11	441	2.49
Auriga	The Charioteer	Capella	2	7	21	657	3.20
Boötes	The Herdsman	Arcturus	1	8	24	907	2.65
Cælum	The Graving Tool	—	0	0	2	125	1.60
Camelopardus	The Giraffe	—	0	0	11	757	1.45
Cancer	The Crab	—	0	1	6	506	1.19
Canes Venatici	The Hunting Dogs	—	0	1	7	465	1.51
Canis Major	The Great Dog	Sirius	4	10	26	380	6.84
Canis Minor	The Little Dog	Procyon	1	2	4	183	2.19
Capricornus	The Sea-Goat	—	0	5	16	414	3.86
Carina	The Keel	Canopus	3	14	40	494	8.10
Cassiopeia	Cassiopeia	—	0	7	23	598	3.85
Centaurus	The Centaur	α Cen., Agena	2	14	49	1060	4.62
Cepheus	Cepheus	—	0	8	20	588	3.40
Cetus	The Whale	—	0	8	24	1232	1.95
Chamæleon	The Chameleon	—	0	0	6	132	4.55
Circinus	The Compasses	—	0	1	4	93	4.30
Columba	The Dove	—	0	4	9	270	3.33
Coma Berenices	Berenice's Hair	—	0	0	8	386	2.07
Corona Australis	The Southern Crown	—	0	0	7	128	5.47
Corona Borealis	The Northern Crown	—	0	3	10	179	5.59
Corvus	The Crow	—	0	5	6	184	3.26
Crater	The Cup	—	0	1	6	282	2.13
Crux Australis	The Southern Cross	Acrux, β Crucis	3	5	13	68	19.12
Cygnus	The Swan	Deneb	1	11	43	804	5.35
Delphinus	The Dolphin	—	0	4	6	189	3.17
Dorado	The Swordfish	—	0	1	8	179	4.47
Draco	The Dragon	—	0	11	26	1083	2.40
Equuleus	The Foal	—	0	0	3	72	4.17
Eridanus	The River	Achernar	1	12	43	1138	3.78
Fornax	The Furnace	—	0	1	5	398	1.26
Gemini	The Twins	Pollux	3	13	23	514	4.47
Grus	The Crane	—	1	4	13	366	3.55
Hercules	Hercules	—	0	15	37	1225	3.02
Horologium	The Clock	—	0	1	2	249	0.80
Hydra	The Watersnake	—	1	9	32	1303	2.46
Hydrus	The Little Snake	—	0	3	9	243	3.70
Indus	The Indian	—	0	2	7	294	2.38

Table 28.2. (Continued)

		1st mag	Stars to mag: 2.00	4.00	5.00	Area sq. deg.	Number of stars above mag 5 per 100 square degrees (star density)
Lacerta	The Lizard	—	0	1	11	201	5.47
Leo	The Lion	Regulus	2	10	26	947	2.75
Leo Minor	The Little Lion	—	0	1	6	232	2.59
Lepus	The Hare	—	0	7	14	290	4.83
Libra	The Balance	—	0	5	13	538	2.42
Lupus	The Wolf	—	0	8	32	334	9.58
Lynx	The Lynx	—	0	2	12	545	2.20
Lyra	The Lyre	Vega	1	4	11	286	3.85
Mensa	The Table	—	0	0	0	153	0.00
Microscopium	The Microscope	—	0	0	4	210	1.90
Monoceros	The Unicorn	—	0	2	13	482	2.70
Musca Australis	The Southern Fly	—	0	4	11	138	7.97
Norma	The Rule	—	0	0	6	165	3.64
Octans	The Octant	—	0	1	4	291	1.37
Ophiuchus	The Serpent-bearer	—	0	12	36	948	3.80
Orion	Orion	Rigel, Betelgeux	5	15	42	594	7.07
Pavo	The Peacock	—	1	4	14	378	3.70
Pegasus	The Flying Horse	—	0	9	29	1121	2.59
Perseus	Perseus	—	1	10	34	615	5.52
Phœnix	The Phœnix	—	0	7	17	469	3.62
Pictor	The Painter	—	0	2	5	247	2.02
Pisces	The Fishes	—	0	3	24	889	2.70
Piscis Australis	The Southern Fish	Fomalhaut	1	1	7	245	2.86
Puppis	The Poop	—	0	11	42	673	6.24
Pyxis	The Compass	—	0	1	7	221	3.17
Reticulum	The Net	—	0	2	7	114	6.14
Sagitta	The Arrow	—	0	2	5	80	6.25
Sagittarius	The Archer	—	1	14	33	867	3.80
Scorpius	The Scorpion	Antares	3	17	38	497	7.64
Sculptor	The Sculptor	—	0	0	6	475	1.26
Scutum	The Shield	—	0	0	6	109	5.50
Serpens	The Serpent	—	0	9	17	637	2.67
Sextans	The Sextant	—	0	0	2	314	0.63
Taurus	The Bull	Aldebaran	2	14	44	797	5.52
Telescopium	The Telescope	—	0	1	4	252	1.59
Triangulum	The Triangle	—	0	2	3	132	2.27
Triangulum Australe	The Southern Triangle	—	1	3	6	110	5.45
Tucana	The Toucan	—	0	2	7	295	2.37
Ursa Major	The Great Bear	—	3	19	35	1280	2.73
Ursa Minor	The Little Bear	—	1	3	9	256	3.51
Vela	The Sails	—	2	10	30	500	6.00
Virgo	The Virgin	Spica	1	8	26	1294	2.01
Volans	The Flying Fish	—	0	3	7	141	4.96
Vulpecula	The Fox	—	0	0	10	268	3.73
	Totals	21	50	455	1417		(Average 3.8)

These counts do not include variable stars which can rise above the fifth magnitude, but whose average magnitude is below this limit. For instance, Mira Ceti is excluded even though its brightest maxima exceed magnitude 2.

Table 28.3. Surviving post-Ptolemaic constellations.

Added: *By Tycho Brahe, c. 1590:*
Coma Berenices
By Bayer, 1603:
Pavo
Tucana
Grus
Phœnix
Dorado
Volans (originally Piscis Volans)
Hydrus
Chamæleon
Apus (originally Avis Indica)
Triangulum Australe
Indus

By Royer, 1679:
Columba (originally Columba Noachi, Noah's Dove)
Crux Australis

By Hevelius, 1690:
Camelopardus
Canes Venatici
Vulpecula (originally Vulpecula et Anser, the Fox and Goose)
Lacerta
Leo Minor
Lynx
Scutum Sobieskii
Monoceros
Sextans (originally Sextans Uraniæ, Urania's Sextant)

By La Caille, 1752:
Sculptor (originally Apparatus Sculptoris, the Sculptor's Apparatus)
Fornax (originally Fornax Chemica, the Chemical Furnace)
Horologium
Reticulum (originally Reticulus Rhomboidalis, the Rhomboidal Net)
Cælum (originally Cæla Sculptoris, the Sculptor's Tools)
Vulpecula (originally Vulpecula et Anser, the Fox and Goose)
Pictor (originally Equuleus Pictoris, the Painter's Easel)
Pyxis (originally Pyxis Nautica, the Mariner's Compass)
Antila (originally Antila Pneumatica, the Airpump)
Octans
Circinus
Norma (or Quadra Euclidis, Euclid's Square)
Telescopium
Microscopium
Mensa (originally Mons Mensæ, the Table Mountain)

Table 28.4. Rejected constellations. The list of rejected constellations is very long. One of these, Quadrans, has at least given its name to the Quadrantid meteor shower of early January.

Name		Introduced by:	Position	Now in:
Anser	The Goose	Hevelius 1690	Cygnus/Vulpecula	Vulpecula
Antinoüs	Antinoüs	Tycho 1559	Near λ Aquilæ	Aquila
Cancer Minor	The Little Crab	Lubinietzki 1650	Cancer/Gemini	Cancer
Cor Caroli	Charles' Heart	Flamsteed 1700	α Canum Venaticorum	Canes Venatici
Cerberus	Cerberus	Hevelius 1687	Hercules	Hercules
Custos Messium	Messier's Equipment	Lalande 1776	Camelopardalis	Cepheus/Cassiopeia
Felis	The Cat	Bode 1775	Hydra/Antlia	Hydra
Gallus	The Cock	Plancius 1613	Argo Navis	Argo
Globus Ærostaticus	The Air Balloon	Lalande 1798	Capricornus	Piscis Australis/Microscopium
Honores Frederici	The Honours of Frederick	Bode 1787	Andromeda/Pegasus	Andromeda
Jordanus Fluvius	The River Jordan	Plancius 1613	Position uncertain	
Lochium Funis	The Log Line	Bode 1787	Argo Navis	Argo
Lilium	The Lily	Royer 1679	Aries	Aries
Machina Electrica	The Electrical Machine	Bode 1787	Cetus	Fornax/Sculptor
Mons Mænalus	Mount Mænalus	Flamsteed 1700	Boötes	Boötes
Musca Borealis	The Northern Fly	Hevelius 1687	Aries	Aries
Noctua	The Night Owl	Burritt 1833	Hydra	Hydra/Libra/Virgo
Norma Nilotica	The Nilometer	Burritt 1833	Capricornus	Aquarius
Nubes Major	The Great Cloud	Royer 1679	Nubecula Major	Dorado
Nubes Minor	The Small Cloud	Royer 1679	Nubecula Minor	Tucana
Officina Typographica	The Printing Press	Bode 1787	Monoceros	Puppis
Psalterium Georgii	George's Lute	Hell 1780	Taurus	Eridanus
Quadrans Muralis	The Mural Quadrant	Bode 1775	Boötes	Boötes
Quadratum	The Square	Allard 1700	Near Nubecula Major	Dorado
Robur Carolinum	Charles' Oak	Halley 1680	Argo Navis	Argo
Sagitta Australis	The Southern Arrow	Hevelius 1645	Scorpius/Sagittarius	Scorpius/Sagittarius
Sceptrum Brandenburgicum	The Sceptre of Brandenburg	Kirch 1688	Eridanus/Lepus	Lepus
Solarium	The Sundial	Burritt 1833	Hydrus/Dorado	Hydrus/Dorado
Solitarius	The Solitaire	Le Monnier 1776	Hydra	Hydra
Taurus Poniatowski	The Bull of Poniatowski	Poczobut 1777	Ophiuchus	Ophiuchus
Tarandus	The Reindeer	Le Monnier 1776	Camelopardalis/Cassiopeia	Cepheus/Cassiopeia
Telescopium Herschelii Major	Herschel's Large Telescope	Hell 1781	Lynx/Gemini	Auriga
Telescopium Herschelii Minor	Herschel's Small Telescope	Hell 1781	Taurus	Taurus
Tigris	The Tigris	Bartschius 1624	Ophiuchus/Pegasus	Ophiuchus
Triangulum Minor	The Little Triangle	Hevelius 1677	Triangulum/Aries	Aries

29 THE STAR CATALOGUE

This catalogue has been compiled from various sources; positions of the brightest stars (down to magnitude 4.75) are given for epoch 2000.0. The published values for absolute magnitudes and distances differ somewhat, and are subject to uncertainty; therefore the values given here have been in general 'rounded off' except for those stars for which the data are known with true precision. I have followed the data given in the authoritative *Cambridge Sky Catalogues* (1987).

The list of double stars, variable stars, and nebular objects does not pretend to be complete, and could be extended almost *ad infinitum*, but I have included most variable stars whose maximum magnitudes are 8.0 or brighter and where the range is at least half a magnitude. For double stars, I have given pairs which are within the range of modest telescopes together with some which may be regarded as test objects. For binary stars of reasonably short period, the values of position angle and distance are for approximate date 2000.

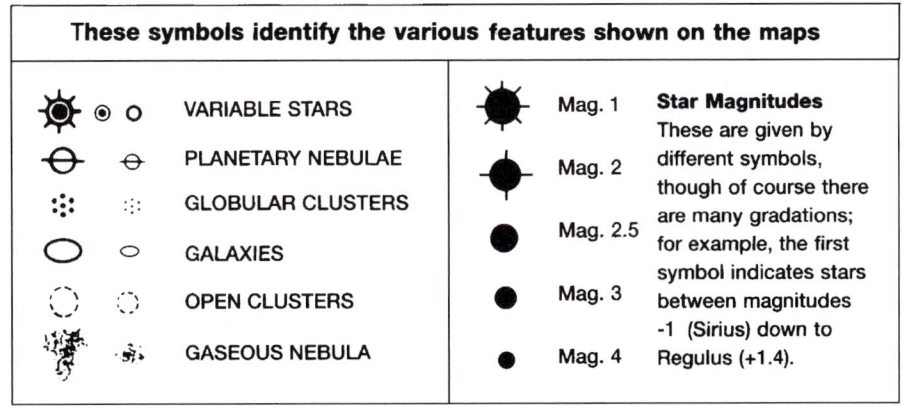

Data given are:

VARIABLE STARS. Range, type, period in days, and spectral type.
DOUBLE STARS. Position angle in degrees, separation in seconds of arc, and magnitudes of the components.
OPEN CLUSTERS. Diameter in minutes of arc, magnitude, and approximate number of stars (though in many cases this is subject to great uncertainty).
GLOBULAR CLUSTERS. Diameter in minutes of arc, and approximate total magnitude.
PLANETARY NEBULÆ. Dimensions in seconds of arc, total magnitude, and magnitude of central star.
NEBULÆ. Dimensions in minutes of arc, and, where appropriate, the magnitude of the illuminating star.
GALAXIES. Magnitude, dimensions in minutes of arc, and type.

For nebular objects I have given NGC numbers, together with Messier or Caldwell numbers where appropriate. Right ascensions and declinations for all objects are for epoch 2000.

ANDROMEDA

(Abbreviation: And).

A large and important northern constellation; one of Ptolemy's 'originals'. It contains the Great Spiral, M.31. One of the leading stars in the constellation – Alpheratz or α Andromedæ– is actually a member of the Square of Pegasus, and was formerly known, more logically, as δ Pegasi.

In mythology, Andromeda was the beautiful daughter of King Cepheus and Queen Cassiopeia. Cassiopeia offended the sea god Neptune by her boasting about Andromeda's beauty, which, she claimed, was greater than that of any sea-nymph. Neptune thereupon sent a sea-monster to ravage the kingdom, and the Oracle stated that the only solution was to chain Andromeda to a rock by the shore where she would be devoured by the monster.

This was duly done, but the situation was redeemed by the hero Perseus, who was on his way home after killing the Gorgon, Medusa. Mounted upon his winged sandals, Perseus arrived in the nick of time, turned the monster to stone by the simple expedient of showing it Medusa's head, and then, in the best story-book tradition, married Andromeda. Cepheus, Cassiopeia and of course Perseus are to be found in the sky; the Gorgon's head is marked by the 'Demon Star' Algol, and the sea-monster has sometimes been identified with the constellation of Cetus.

Andromeda has no first-magnitude star, but there are eight above magnitude 4.

	Mag.	Luminosity Sun = 1	Dist lt-yrs
β	2.06	115	88
α	2.06	96	72
γ	2.14	95	121
δ	3.27	105	160
51	3.57	105	186
o	3.6v	26	114
λ	3.82	16	78
μ	3.87	11	81

The solar-type star υ is of special interest, since it is believed to be attended by three planets.

BRIGHTEST STARS

Star	R.A. h	m	s	Dec. deg	min	sec	Mag.	Abs. mag.	Spec.	Dist pc	
1 o	23	01	55.1	+42	19	34	3.6v	1.4	B6+A2	35	
7	23	12	32.9	+49	24	33	4.52	2.6	F0	23	
16 λ	23	37	33.7	+46	27	30	3.82v	1.8	G8	24	
17 ι	23	38	08.0	+43	16	05	4.29	−0.2	B8	79	
19 κ	23	40	24.4	+44	20	02	4.14	−0.2	B8	60	
21 α	00	08	23.2	+29	05	26	2.06	−0.1	A0p	22	Alpheratz
24 θ	00	17	05.4	+38	40	54	4.61	1.4	A2	44	
25 σ	00	18	19.6	+36	47	07	4.52	1.4	A2	42	
29 π	00	36	52.8	+33	43	09	4.36	−1.1	B5	120	
30 ε	00	38	33.3	+29	18	42	4.37	0.3	G8	41	
31 δ	00	39	19.6	+30	51	40	3.27	−0.2	K3	49	
34 ξ	00	47	20.3	+24	16	02	4.06v	−2.2	K1	48	Al Dhail
35 υ	00	49	48.8	+41	04	44	4.53	−1.1	B5	130	
37 μ	00	56	45.1	+38	29	58	3.87	2.1	A5	25	
38 η	00	57	12.4	+23	25	04	4.42	1.8	G8	35	
42 ϕ	01	09	30.1	+47	14	31	4.25	−1.2	B8	120	
43 β	01	09	43.8	+35	37	14	2.06	−0.4	M0	27	Mirach
50 υ	01	36	47.8	+41	24	20	4.09	4.0	F8	13	
51	01	37	59.5	+48	37	42	3.57	−0.2	K3	57	
57 γ	02	03	53.9	+42	19	47	2.18	−0.1	K2 ⎫	37	Almaak
	02	03	54.7	+42	19	51	5.03		A0 ⎭		
65	02	25	37.3	+50	16	43	4.71	−0.3	K4	82	

o is a β-Lyræ variable with a small range (3.5–3.7), period 1.6 days.
ζ is an ellipsoidal variable with a small range, 4.06–4.20.
λ is slightly variable (3.7–4.1) and is probably a spectroscopic binary.

ANDROMEDA (continued)

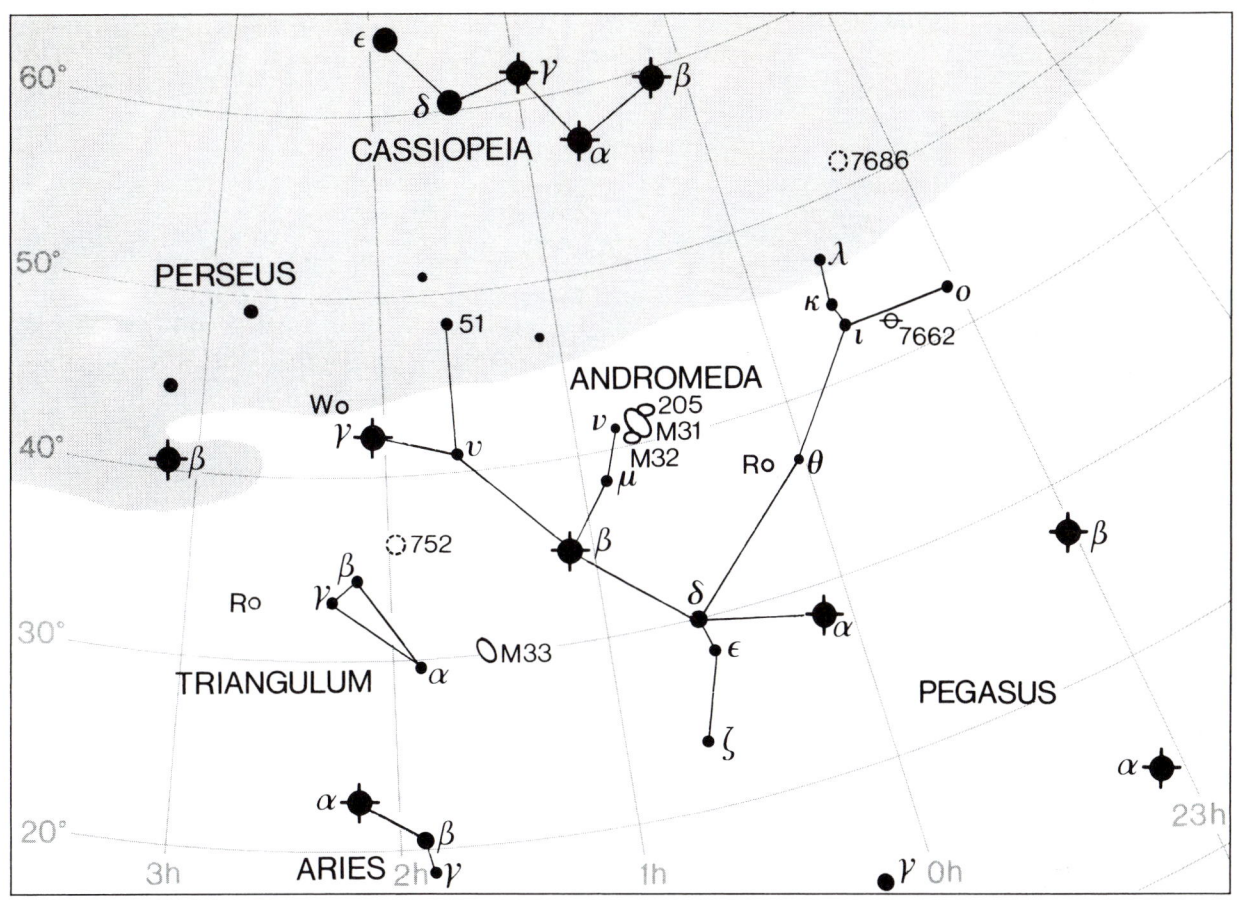

Also above mag. 5:

	Mag.	Abs. mag.	Spectrum	Dist	
3	4.65	0.2	K0	71	
5	5.70	3.4	F5	29	
8	4.85	−0.5	M2	100	
46 ξ	4.88	1.7	K0	35	Adhil
48 ω	4.83	2.0	F4	36	
52 χ	4.98	0.3	G8	86	
53 τ	4.94	−0.6	B8	130	
58	4.80	1.9	A4	38	
60	4.83	−0.3	K4	97	

VARIABLE STARS

	R.A. h	m	Dec. deg	min	Range	Type	Period, d.	Spec.
Z	23	33.7	+48	49	8.0–12.4	Z And	—	M
ST	23	38.8	+35	46	7.7–11.8	Semi-reg.	328	R
SV	00	04.3	+40	07	7.7–14.3	Mira	325.2	M
KU	00	06.9	+43	05	6.5–10.5	Mira	750	M
VX	00	19.9	+44	43	7.8–9.3	Semi-reg.	369	N
T	00	22.4	+27	00	7.7–14.5	Mira	280.8	M
R	00	24.0	+38	35	5.8–14.9	Mira	409.3	S
TU	00	32.4	+26	02	7.8–13.1	Mira	316.8	M
RW	00	47.3	+32	41	7.9–15.7	Mira	430.3	M
W	02	17.6	+44	18	6.7–14.6	Mira	395.9	S

ANDROMEDA (continued)

DOUBLE STARS

	R.A. h	m	Dec. deg	min	P.A. deg	Sep. sec	Mags.	
π	00	36.9	+33	43	173	35.9	4.4, 8.6	
γ²	02	03.9	+42	20	063	9.8	2.3, 4.8	
γ²					106	0.5	5.5, 6.3	Binary, period 61 years

OPEN CLUSTERS

M	C	NGC	R.A. h	m	Dec. deg	min	Diameter min	Mag.	No. of stars
		7686	23	30.2	+49	08	15	5.6	20
	28	752	01	57.8	+37	41	50	5.7	60

PLANETARY NEBULA

M	C	NGC	R.A. h	m	Dec. deg	min	Diameter sec	Mag.	Mag. of central star
	22	7662	23	25.9	+42	33	20 × 130	9.2	13.2

GALAXIES

M	C	NGC	R.A. h	m	Dec. deg	min	Mag.	Dimensions min	Type	
		205	00	40.4	+41	41	8.0	17.4 × 9.8	E6	Companion to M.31
31		224	00	42.7	+41	16	3.5	17.8 × 6.3	Sb	
32		221	00	42.7	+40	52	8.2	7.6 × 5.8	E2	Companion to M.31
	23	891	02	22.6	+42	21	9.9	13.5 × 2.8	Sb	

ANTLIA

(Abbreviation: Ant).

A small southern constellation, covering 239 square degrees, added to the sky by Lacaille in 1752; its original name was Antlia Pneumatica. It contains no star brighter than magnitude 4.4, and no mythological legends are associated with it.
See chart for Carina.

BRIGHTEST STARS

Star	R.A. h	m	s	Dec. deg	min	sec	Mag.	Abs. mag.	Spec.	Dist pc
ε	09	29	14.7	−35	57	05	4.51	−0.4	M0	96
α	10	27	09.1	−31	04	04	4.25	−0.4	M0	85
ι	10	56	43.0	−37	08	16	4.60	0.3	F2	79

Also above mag. 5:

	Mag.	Abs. mag.	Spectrum	Dist
θ	4.79	3.8	F7	14

VARIABLE STARS

	R.A. h	m	Dec. deg	min	Range	Type	Period, d.	Spectrum
S	09	32.3	−28	38	6.4–6.9	W Uma	0.65	A
U	10	35.2	−39	34	5.7–6.8	Irreg.	—	N

DOUBLE STARS

	R.A. h	m	Dec. deg	min	P.A. deg	Sep. sec	Mags.
θ	09	44.2	−27	46	005	0.1	5.4, 5.6
δ	10	29.6	−30	36	226	11.0	5.6, 9.6

GALAXIES

M	NGC	R.A. h	m	Dec. deg	min	Mag.	Dimensions min	Type
	2997	09	45.6	−31	11	10.6	8.1 × 6.5	Sc
	3223	10	21.6	−34	16	11.8	4.1 × 2.6	Sb
	3347	10	42.8	−36	22	12.5	4.4 × 2.6	SBb

APUS

(Abbreviation: Aps).

Originally Avis Indica; it was introduced by Bayer in 1603. It lies in the far south, and has only two stars brighter than magnitude 4. See charts for Musca.

	Mag.	Luminosity Sun = 1	Dist lt-yrs
α	3.83	110	218
γ	3.89	4.5	46

BRIGHTEST STARS

Star	R.A. h	m	s	Dec. deg	min	sec	Mag.	Abs. mag.	Spec.	Dist pc
α	14	47	51.6	−79	02	41	3.83	−0.3	K5	67
δ	16	20	20.7	−78	41	44	4.68	−0.5	M4 + K5	110
γ	16	33	27.1	−78	53	49	3.89	3.2	K0	14
β	16	43	04.5	−77	31	02	4.24	0.2	K0	42
ζ	17	21	59.3	−67	46	13	4.78	−0.3	gK5	100

The magnitude of δ is given as the combination of δ^1 and δ^2, which are separated by 103 sec and make up a very wide physically connected pair.

Also above mag. 5:

	Mag.	Abs. mag.	Spectrum	Dist
η	4.91		A2p	

VARIABLE STARS

	R.A. h	m	Dec. deg	min	Range	Type	Period, d.	Spec.
θ	14	05.3	−76	48	6.4–8.6	Semi-reg.	119	M
S	15	04.3	−71	53	9.5–1.5	R CrB	—	R
DW	17	23.5	−67	56	7.9–9.1	Algol	2.31	B

(R Apodis, magnitude 5.3, has been suspected of variability.)

DOUBLE STAR

	R.A. h	m	Dec. deg	min	P.A. deg	Sep. sec	Mags.
δ	16	20.3	−78	42	012	102.9	4.7, 5.1

GLOBULAR CLUSTERS

M	NGC	R.A. h	m	Dec. deg	min	Diameter min	Mag.
	IC 4499	15	00.3	−82	13	7.6	10.6
	6101	16	25.8	−72	12	10.7	9.3

GALAXY

M	NGC	R.A. h	m	Dec. deg	min	Mag.	Dimensions min	Type
	5967	15	48.1	−75	40	12.5	2.9 × 1.8	SBc

AQUARIUS

(Abbreviation: Aqr).

A Zodiacal constellation, and of course one of Ptolemy's 'originals'. Oddly enough there are no well-defined legends attached to it, though it has been associated with Ganymede, cup-bearer of the Olympian gods. There are nine stars brighter than magnitude 4. Aquarius is distinguished by the presence of two splendid planetary nebulæ, C55 (the Saturn Nebula) and C63 (the Helix Nebula).

	Mag.	Luminosity Sun = 1	Dist lt-yrs
β	2.91	5250	980
α	2.96	5250	945
δ	3.27	105	98
ζ	3.6	50	98
c^2	3.66	60	107
λ	3.74	120	256
ε	3.77	28	107
γ	3.84	50	91
b^1	3.97	60	160

AQUARIUS (continued)

BRIGHTEST STARS

Star	R.A. h	m	s	Dec. deg	min	sec	Mag.	Abs. mag.	Spec.	Dist pc	
2 ε	20	47	40.3	−09	29	45	3.77	1.2	A1	33	Albali
3 k	20	47	44.0	−05	01	40	4.42	−0.5	M3	91	
6 μ	20	52	39.0	−08	59	00	4.73	1.0	A8	30	
13 ν	21	09	35.4	−11	22	18	4.51	0.3	G8	70	
22 β	21	31	33.3	−05	34	16	2.91	−4.5	G0	300	Sadalsuud
23 ξ	21	37	44.9	−07	51	15	4.69	2.4	A7	36	Bunda
31 o	22	03	18.7	−02	09	19	4.69	−0.2	B8	95	
34 α	22	05	46.8	−00	19	11	2.96	−4.5	G2	290	Sadalmelik
33 ι	22	06	26.1	−13	52	11	4.27	−0.2	B8	78	
43 θ	22	16	49.9	−07	47	00	4.16	1.8	G8	26	Ancha
48 γ	22	21	39.2	−01	23	14	3.84	0.6	A0	28	Sadachiba
52 π	22	25	16.4	+01	22	39	4.66	−4.1	B0	450	Seat
55 ζ^1	22	28	49.5	−00	01	13	4.53	0.6	F2	30	
ζ^2	22	28	49.9	−00	01	12	4.31		F2		
62 η	22	35	21.2	−00	07	03	4.02	−0.2	B8	46	
71 τ	22	49	35.3	−13	35	33	4.01	−0.4	M0	74	
73 λ	22	52	36.6	−07	34	47	3.74	−0.5	M2	71	
76 δ	22	54	38.8	−15	49	15	3.27	−0.2	A2	30	Scheat
88 c^2	23	09	26.6	−21	10	21	3.66	0.2	K0	33	
90 ϕ	23	14	19.2	−06	02	56	4.22	−0.5	M2	75	
91 ψ^1	23	15	53.4	−09	05	16	4.21	1.8	K0	30	
93 ψ^2	23	27	54.1	−09	10	57	4.39	−1.1	B5	130	
98 b^1	23	22	58.0	−20	06	02	3.97	0.2	K0	49	
99 b^2	23	26	02.5	−20	38	31	4.39	−0.3	K5	87	
101 b^3	23	33	16.4	−20	54	32	4.71	—	A0	25	
105 ω^2	23	42	43.2	−14	32	42	4.49	0.4	B9.5	43	

Also above mag. 5:

	Mag.	Abs. mag.	Spectrum	Dist
57 σ	4.82	0.0	A0	92
95 ψ^3	4.98	0.6	A0	75
102 ω^1	5.00	—	A5	18
104 A^2	4.82	−2.0	G0	220

VARIABLE STARS

	R.A. h	m	Dec. deg	min	Range	Type	Period, d.	Spec.
W	20	46.4	−04	05	8.4–14.9	Mira	381.1	M
V	20	46.8	+02	26	7.6–9.4	Semi-reg.	244	M
T	20	49.9	−05	09	7.2–14.2	Mira	202.1	M
X	22	18.7	−20	54	7.5–14.8	Mira	311.6	S
S	22	57.1	−20	21	7.6–15.0	Mira	297.3	M
R	23	43.8	−15	17	5.8–12.4	Symb.	387	M+Pec

DOUBLE STARS

	R.A. h	m	Dec. deg	min	P.A. deg	Sep. sec	Mags.	
β	21	31.6	−05	34	AB 321	35.4	2.9, 10.8	
					AC 186	57.2	11.4	
41	22	14.3	−21	04	AB 114	5.0	5.6, 7.1	
					AC 043	212.1	9.0	
51	22	24.1	−04	50	AB 324	0.5	6.5, 6.5	
					AB+D 191	116.0	10.1	
					AC 342	54.4	10.2	
					AE 133	132.4	8.6	
ζ	22	28.8	−00	01	187	1.9	4.3, 4.5	Binary, p. 856 y
89	23	09.9	−22	27	007	0.4	5.1, 5.9	
107	23	46.0	−18	41	136	6.6	5.7, 6.7	

ASTERISM

M	C	NGC	R.A. h	m	Dec. deg	min	
73		6994	20	58.9	−12	38	Four stars; not a true cluster.

AQUARIUS (continued)

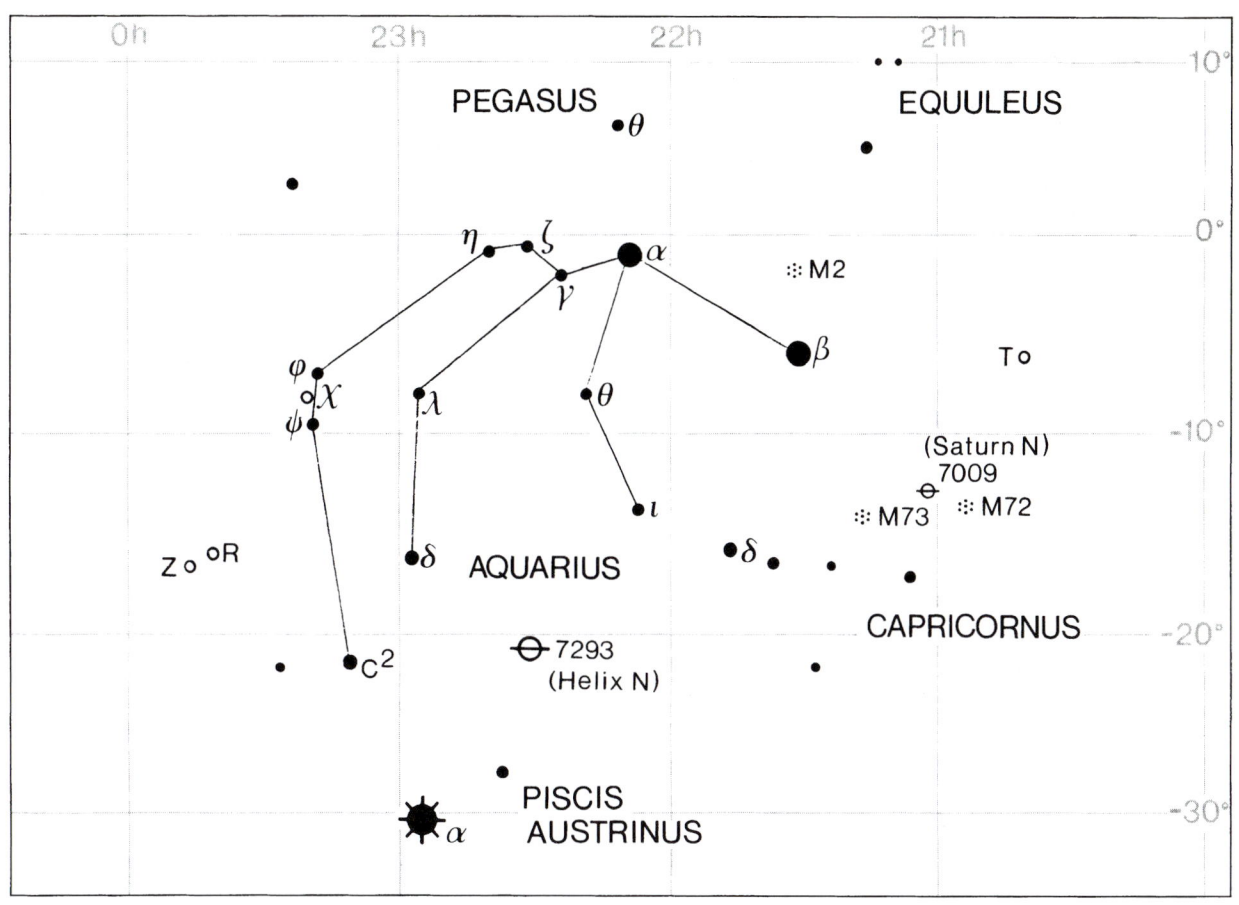

GLOBULAR CLUSTERS

M	C	NGC	R.A.		Dec.		Diameter	Mag.
			h	m	deg	min	min	
72		6981	20	53.5	−12	32	5.9	9.3
2		7089	21	33.5	−00	49	12.9	6.5

PLANETARY NEBULÆ

M	C	NGC	R.A.		Dec.		Dimensions	Mag.	Mag. of	
			h	m	deg	min	sec		central star	
	55	7009	21	04.2	−11	22	2.5 × 100	8.3	11.5	Saturn Nebula
	63	7293	22	29.6	−20	48	770	6.5	13.5	Helix Nebula

GALAXIES

M	C	NGC	R.A.		Dec.		Mag.	Dimensions	Type
			h	m	deg	min		min	
		7184	22	02.7	−20	49	12.0	5.8 × 1.8	Sb
		7606	23	19.1	−08	29	10.8	5.8 × 2.6	Sb
		7723	23	38.9	−12	58	11.1	3.6 × 2.6	Sb
		7727	23	39.9	−12	18	10.7	4.2 × 3.4	SBap

AQUILA

(Abbreviation: Aql).

One of the most distinctive of all the northern constellations, and, of course, an 'original'. Mythologically it represents an eagle which was sent by Jupiter to collect a Phrygian shepherd-boy, Ganymede, who was destined to become cup-bearer of the Gods – following an unfortunate episode in which the former holder of the office, Hebe, tripped and fell during a particularly solemn ceremony.

The leading star is Altair. Altogether there are eight stars above magnitude 4.

	Mag.	Luminosity Sun = 1	Dist lt-yrs
α	0.77	10	16.6
γ	2.72	700	186
ζ	2.99	60	104
θ	3.23	180	199
δ	3.36	11	52
λ	3.44	82	98
η	3.5 (max)	5000v	1400
β	3.71	4.5	36

Altair is distorted by its very rapid rotation; the equatorial diameter is 2 260 000 km. It is an optical double, not a binary.

η Aquilæ and δ Cephei are the brightest Cepheids in the sky.

ρ Aquilæ is now in Delphinus, because of its proper motion. Its Flamsteed number is 67 Aquilæ. The annual proper motion is $+0.004''$ in R.A., $+0.06''$ in dec. 2000 position, R.A. 20h 14m 16$''$.4, dec. $+15°$ 11$'$ 51$''$. Its apparent magnitude is 4.95; spectrum B2; absolute magnitude $+1.4$.

On 1 December 1999 A. Pereida discovered a nova at R.A. 19h 23m 5s.3, dec. $+4°$ 57$'$ 20$''$.1. It reached magnitude 3.5, and was the brightest northern-hemisphere nova since 1975.

BRIGHTEST STARS

Star	R.A. h	m	s	Dec. deg	min	sec	Mag.	Abs. mag.	Spec.	Dist pc	
13 ε	18	59	37.2	+15	04	06	4.02	−0.1	K2	65	
12 i	19	01	40.7	−05	44	20	4.02	0.0	K1	64	
17 ζ	19	05	24.4	+13	51	48	2.99	0.2	B9	32	Dheneb
16 λ	19	06	14.7	−04	52	57	3.44	0.0	B8.5	30	Althalimain
30 δ	19	25	29.7	+03	06	53	3.36	2.1	F0	16	
32 ν	19	26	30.9	+00	22	44	4.66	−4.6	F2	530	
38 μ	19	34	05.2	+07	22	44	4.45	−0.2	K3	39	
41 ι	19	36	43.1	−01	17	11	4.36	−2.2	B5	180	
50 γ	19	46	15.4	+10	36	48	2.72	−2.3	K3	87	Tarazed
53 α	19	50	46.8	+08	52	06	0.77	2.2	A7	5.1	Tarazed
55 η	19	52	28.1	+01	00	20	var.	−4.5v	G0v	440	
59 ξ	19	54	14.7	+08	27	41	4.71	0.2	K0	73	
60 β	19	55	18.5	+06	24	24	3.71	3.2	G8	11	Alshain
65 θ	20	11	18.1	−00	49	17	3.23	−0.8	B9	61	
71	20	38	20.1	−01	06	19	4.32	0.3	G8	64	

Also above mag. 5:

	Mag.	Abs. mag.	Spectrum	Dist	
39 κ	4.95		B0.5		
67 ρ	4.95	1.4	A2	51	(Now in Delphinus)
69	4.91	−0.1	K2	100	
70	4.89	−2.3	K5	240	

VARIABLE STARS

	R.A. h	m	Dec. deg	min	Range	Type	Period, d.	Spec.
V	19	04.4	−05	41	6.6–8.4	Semi-reg.	353	N
R	19	06.4	+08	14	5.5–12.0	Mira	284.2	M
TT	19	08.2	+01	18	6.4–7.7	Cepheid	13.75	F–G
W	19	15.4	−07	03	7.3–14.3	Mira	490.4	S
U	19	29.4	−07	03	6.1–6.9	Cepheid	7.02	F–G
X	19	51.5	+04	28	8.3–15.5	Mira	347.0	M
η	19	52.5	+01	00	3.5–4.4	Cepheid	7.18	F–G
RR	19	57.6	−01	53	7.8–14.5	Mira	394.8	M

AQUILA (continued)

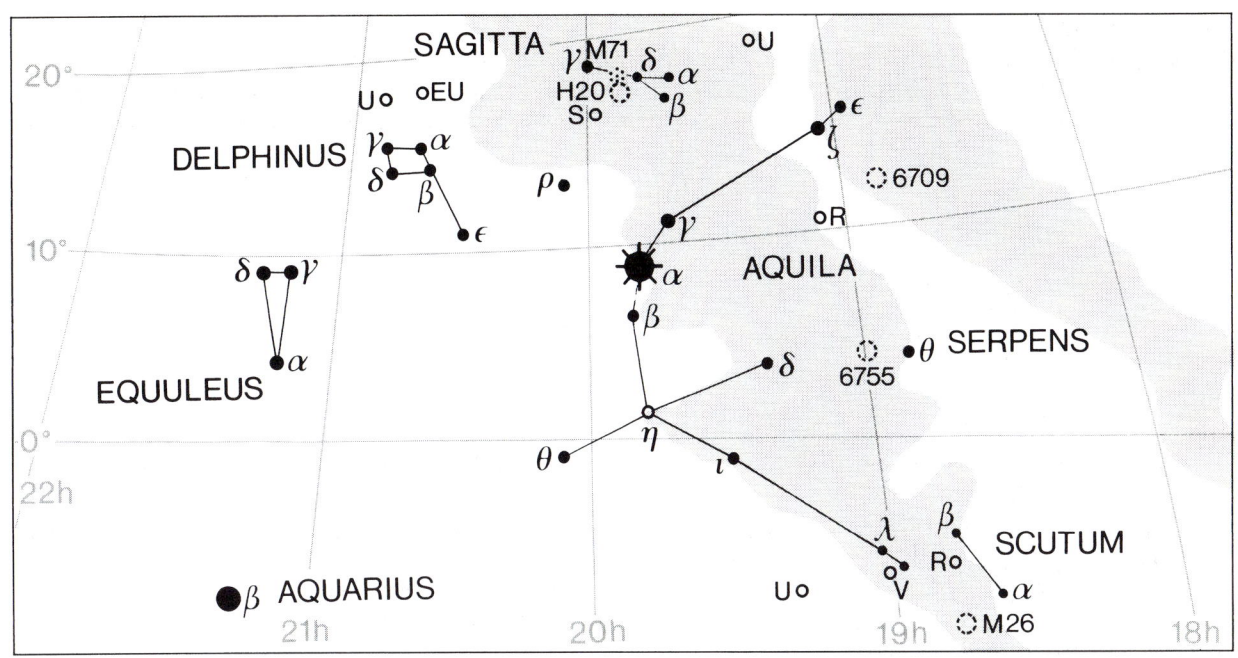

DOUBLE STARS

	R.A.		Dec.		P.A.	Sep.	Mags.	
	h	m	deg	min	deg	sec		
ε	18	59.6	+15	04	187	131.1	4.0, 9.9	
23	19	18.5	+01	05	005	3.1	5.3, 9.3	
31	19	25.0	+11	57	343	105.6	5.2, 8.7	
δ	19	25.5	+03	07	271	108.9	3.4, 10.9	
ν	19	26.5	+00	20	288	201.0	4.7, 8.9	
U	19	29.4	−07	03	228	1.5	var., 11.7	
χ	19	42.6	+11	50	077	0.5	5.6, 6.8	
γ	19	46.3	+10	37	258	132.6	2.7, 10.7	
π	19	48.7	+11	49	110	1.4	6.1, 6.9	
α	19	50.8	+08	52	301	165.2	0.8, 9.5	optical
57	19	54.6	−08	14	170	35.7	5.8, 6.5	

OPEN CLUSTERS

M	C	NGC	R.A.		Dec.		Diameter	Mag.	No. of stars
			h	m	deg	min	min		
		6709	18	51.5	+10	21	13	6.7	40
		6755	19	07.8	+04	14	15	7.5	100

PLANETARY NEBULÆ

M	C	NGC	R.A.		Dec.		Diameter	Mag.	Mag. of central star
			h	m	deg	min	sec		
		6741	19	02.6	−00	27	6	10.8	14.7
		6751	19	05.9	−06	00	20	12.5	13.9
		6790	19	23.2	+01	31	7	10.2	13.5
		6803	19	31.3	+10	03	6	11.3	15.2

ARA

(Abbreviation: Ara).

An original constellation, though apparently without any definite legends attached to it. There are seven stars above magnitude 4.00.

	Mag.	Luminosity Sun = 1	Dist lt-yrs
β	2.85	5000	780
α	2.95	450	190
ζ	3.13	110	137
γ	3.34	5000	1075
δ	3.62	105	95
θ	3.66	9000	1570
η	3.76	110	190

BRIGHTEST STARS

Star	R.A. h	m	s	Dec. deg	min	sec	Mag.	Abs. mag.	Spec.	Dist pc	
η	16	49	47.0	−59	02	29	3.76	−0.3	K5	58	
ζ	16	58	37.1	−55	59	24	3.13	−0.3	K5	42	
ε¹	16	59	34.9	−53	09	38	4.06	−0.5	M1	82	
β	17	25	17.9	−55	31	47	2.85	−4.4	K3	240	
γ	17	25	23.5	−56	22	39	3.34	−4.4	B1	330	
δ	17	31	05.8	−60	41	01	3.62	−0.2	B8	29	
α	17	31	50.3	−49	52	34	2.95	−1.7	B3	58	Choo
σ	17	35	39.4	−46	30	20	4.59	—	A0	47	
θ	18	06	37.6	−50	05	30	3.66	−5.1	B1	480	

Also above mag. 5:

	Mag.	Abs. mag.	Spectrum	Dist
λ	4.77	3.4	dF5	19

VARIABLE STARS

	R.A. h	m	Dec. deg	min	Range	Type	Period, d.	Spectrum
X	16	36.4	−55	24	8.0–13.5	Mira	175.8	M
R	16	39.7	−57	00	6.0–6.9	Algol	4.42	B
U	17	53.6	−51	41	7.7–14.1	Mira	225.2	M

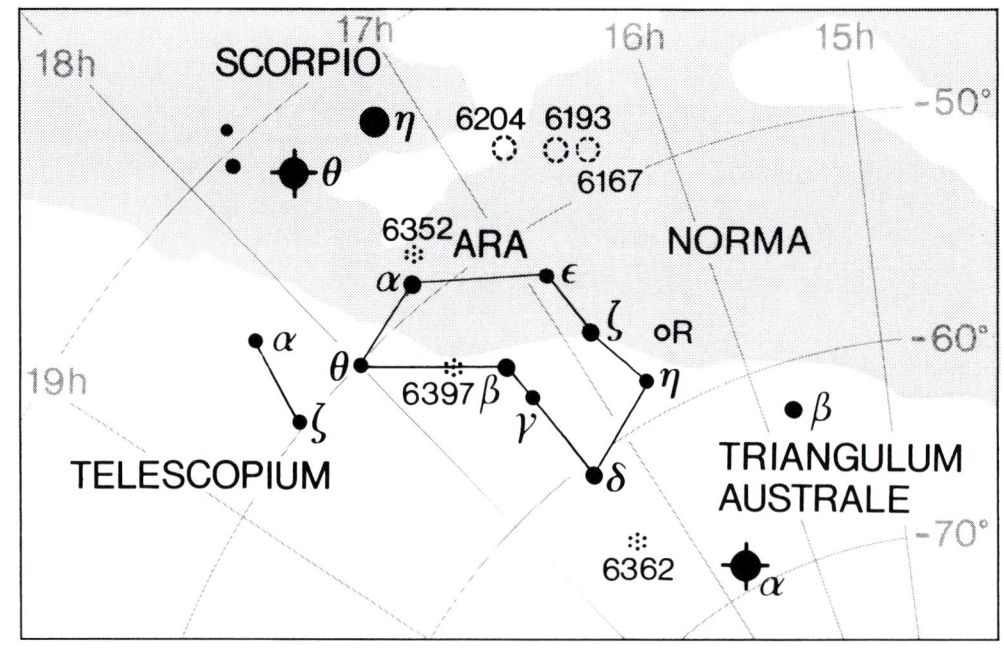

ARA (continued)

DOUBLE STARS

	R.A.		Dec.		P.A.	Sep.	Mags.
	h	m	deg	min	deg	sec	
γ	17	25.4	−56	23	AB 328	17.9	3.3, 10.3
					AC 066	41.6	11.8

OPEN CLUSTERS

M	C	NGC	R.A.		Dec.		Diameter	Mag.	No. of stars
			h	m	deg	min	min		
	82	6193	16	41.3	−48	46	15	5.2	—
		6204	16	46.5	−47	01	5	8.2	45
		6208	16	49.5	−53	49	16	7.2	60
		6250	16	58.0	−45	48	8	5.9	60
		6253	16	59.1	−52	43	5	10.2	30
		H.13	17	05.4	−48	11	15	—	15
		IC 4651	17	24.7	−49	57	12	6.9	80

GLOBULAR CLUSTERS

M	C	NGC	R.A.		Dec.		Diameter	Mag.
			h	m	deg	min	min	
	81	6352	17	25.5	−48	25	7.1	8.1
		6362	17	31.9	−67	03	10.7	8.3
	86	6397	17	40.7	−53	40	25.7	5.6

GALAXIES

M	C	NGC	R.A.		Dec.		Mag.	Dimensions	Type
			h	m	deg	min		min	
		6215	16	51.1	−58	59	11.8	2.0 × 1.6	Sc
		6221	16	52.8	−59	13	11.5	3.2 × 2.3	SBc

ARIES

(Abbreviation: Ari).

The first constellation of the Zodiac – though since the vernal equinox has shifted into Pisces, Aries should logically be classed as second! In mythology it represents a ram with a golden fleece, sent by the god Mercury (Hermes) to rescue the children of the king of Thebes from an assassination plan by their stepmother. The ram, which had the remarkable ability to fly, carried out its mission; unfortunately, the girl (Helle) lost her balance and fell to her death in that part of the sea now called the Hellespont, but the boy (Phryxus) arrived safely. After the ram's death, its golden fleece was hung in a sacred grove, from which it was later removed by the Argonauts commanded by Jason.

Aries is moderately conspicuous. There are four stars above the fourth magnitude.

	Mag.	Luminosity Sun = 1	Dist lt-yrs
α	2.00	96	85
β	2.64	11	46
c	3.63	105	117
γ	3.9	60 + 56	117

Of these, γ is an excellent double with two equal components.

See chart for Andromeda.

BRIGHTEST STARS

Star	R.A.			Dec.			Mag.	Abs. mag.	Spec.	Dist pc	
	h	m	s	deg	min	sec					
5 α	01	53	31.7	+19	17	45	4.68	0.4	A	36	Mesartim
	01	53	31.8	+19	17	37	4.59	0.2	B9	36	
6 β	01	54	38.3	+20	48	29	2.64	2.1	A5	14	Sheratan
13 α	02	07	10.3	+23	27	45	2.00	−0.1	K2	26	Hamal
35	02	43	27.0	+27	42	26	4.66	−1.7	B3	180	
39	02	47	54.5	+29	14	50	4.51	0.0	K1	58	
41 c	02	49	58.9	+27	15	38	3.63	−0.2	B8	38	Nair al Butain
48 ε	02	59	12.6	+21	20	25	4.63	1.4	A2	48	
57 δ	03	11	37.7	+19	43	36	4.35	−0.1	K2	78	Boteïn

ARIES (continued)

Also above mag. 5:

	Mag.	Abs. mag.	Spectrum	Dist
9 λ	4.79	1.7	F0	42
14	4.98	0.6	F2	75
58 ζ	4.89	0.0	A0	91

VARIABLE STARS

	R.A.		Dec.			Range	Type	Period, d.	Spectrum
	h	m	deg	min					
R	02	16.1	+25	03		7.4–13.7	Mira	186.8	M
T	02	48.3	+17	31		7.5–11.3	Semi-reg.	317 (var.)	M
U	03	11.0	+14	48		7.2–15.2	Mira	371.1	M

DOUBLE STARS

	R.A.		Dec.		P.A.	Sep.	Mags.
	h	m	deg	min	deg	sec	
γ	01	53.5	+19	18	007	7.8	4.8, 4.8
π	02	49.3	+17	28	AB 120	3.2	5.2, 8.7
					AC 110	25.2	10.8
ε	02	59.2	+21	20	208	1.5	5.2, 5.5

GALAXIES

M	C	NGC	R.A.		Dec.		Mag.	Dimensions	Type	
			h	m	deg	min		min		
		772	01	59.3	+19	01	10.3	7.1 × 4.5	Sb	Arp 78
		976	02	34.0	+20	59	12.4	1.7 × 1.5	Sb	

AURIGA

(Abbreviation: Aur).

One of the most brilliant northern constellations, with Capella outstanding. Mythologically it honours Erechthonius, son of Vulcan, the blacksmith of the gods, who became King of Athens and also invented the four-horse chariot. There are eight stars brighter than the fourth magnitude. ε, ζ and η are often nicknamed 'the Kids'.

	Mag.	Luminosity Sun = 1	Dist lt-yrs
α	0.08	90 + 70	43
β	1.90	50	46
θ	2.62	75	82
ι	2.69	700	267
ε	2.99v	200 000	4600
η	3.17	450	200
δ	3.72	60	163
ζ	3.75v	700	520
ν	3.97	60	162

In addition, Al Nath used to be called γ Aurigæ, but has now been given a free transfer, and is included in Taurus as β Tauri.

Capella is an extremely close binary. Both are yellow; the brighter component is of type G, the fainter either late G or early F. The diameters are respectively 18 000 000 km and 9 600 000 km; masses in terms of the Sun, 3.09 and 2.95; separation 11 300 000 km; period 104 days. The apparent separation ranges between 0.04 and 0.05 of a second of arc.

ε and ζ are the two very long-period eclipsing binaries, described on page 282.

M 36, 37 and 38 are very easy binocular objects.

AURIGA (continued)

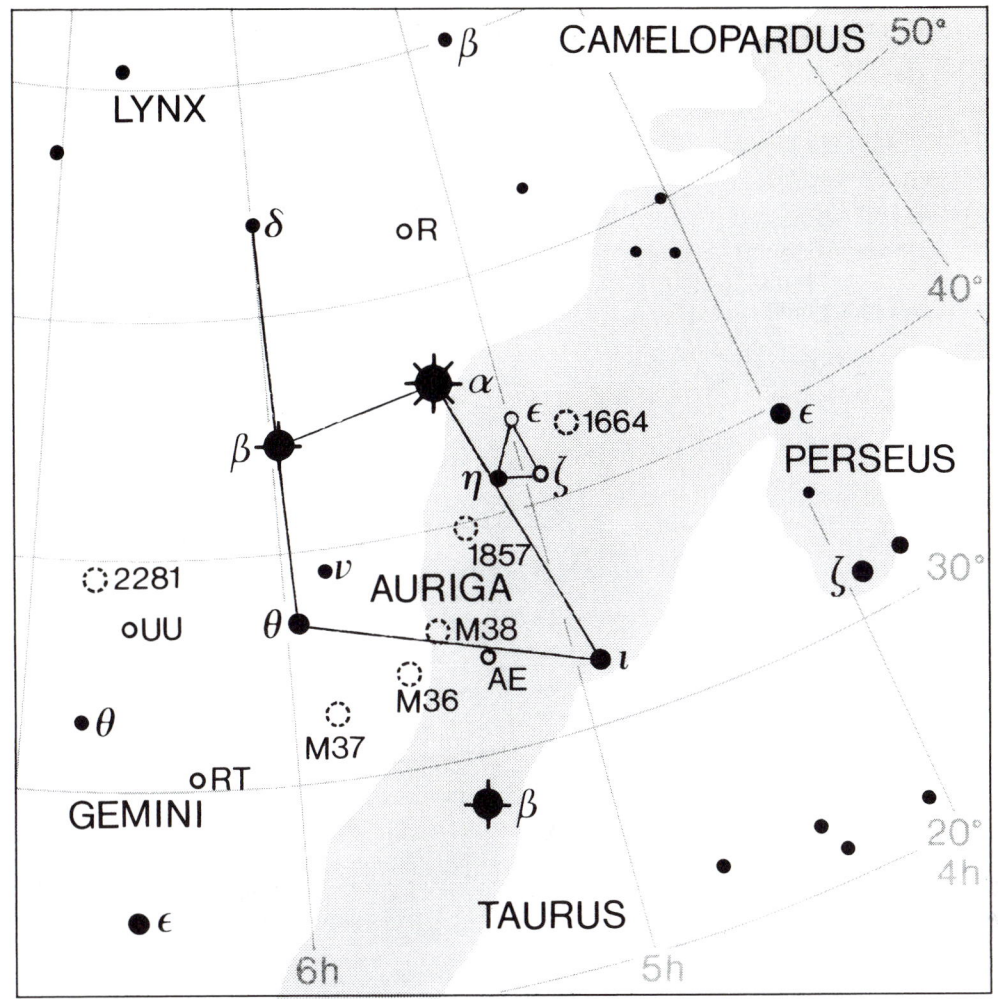

BRIGHTEST STARS

Star	R.A. h	m	s	Dec. deg	min	sec	Mag.	Abs. mag.	Spec.	Dist pc	
3 ι	04	56	59.5	+33	09	58	2.69	−2.3	K3	82	Hasseleh
7 ε	05	01	58.1	+43	49	24	2.99v	−8.5	F0	1400	Almaaz
8 ζ	05	02	28.6	+41	04	33	3.75v	−2.3	K4	160	Sadatoni
10 η	05	06	30.8	+41	14	04	3.17	−1.7	B3	61	
13 α	05	16	41.3	+45	59	53	0.08	0.3	G8	13	Very close double Capella
15 λ	05	19	08.4	+40	05	57	4.71	4.4	G0	13	
29 τ	05	49	10.4	+39	10	52	4.52	0.3	G8	70	
31 υ	05	51	02.4	+37	18	20	4.74	−0.5	GM1	110	
32 ν	05	51	29.3	+39	08	55	3.97	0.2	K0	45	
33 δ	05	59	31.6	+54	17	05	3.72	0.2	K0	50	
34 β	05	59	31.7	+44	56	51	1.90	0.6	A2	22	Menkarlina
37 θ	05	59	43.2	+37	12	45	2.62	0.1	A0p	25	
35 π	05	59	56.1	+45	56	12	4.26	−2.4	M3	200	
44 κ	06	15	22.6	+29	29	53	4.35	0.3	G8	46	

AURIGA (continued)

Also above mag. 5:

	Mag.	Abs. mag.	Spectrum	Dist	
4 ω	4.94	0.6	A0	69	
2	4.78	−0.2	K3	190	
9	5.00	2.6	F0	30	
11 μ	4.86	1.8	A3	22	
21 σ	4.89	−0.3	K4	110	
25 χ	4.76	−6.3	B5	930	
30 ξ	4.99	0.8	A2		
46 ψ^1	4.91	−5.7	M0	600	Dolones

VARIABLE STARS

	R.A. h	m	Dec. deg	min	Range	Type	Period, d.	Spec.
RX	05	01.4	+39	58	7.3–8.0	Cepheid	11.62	F–G
ε	05	02.0	+43	49	2.9–3.8	Eclipsing	9892	F
ζ	05	02.5	+41	05	3.7–4.1	Eclipsing	972.1	K + B
R	05	17.3	+53	35	6.7–13.9	Mira	457.5	M
UV	05	21.8	+32	31	7.4–10.6	Mira	394.4	M
U	05	42.1	+32	02	7.5–15.5	Mira	408.1	M
X	06	12.2	+50	14	8.0–13.6	Mira	163.8	M
RT	06	28.6	+30	30	5.0–5.8	Cepheid	3.73	F–G
WW	06	32.5	+38	27	5.8–6.5	Algol	2.53	A + A
UU	06	36.5	+38	27	5.1–6.8	Semi-reg.	234	N

DOUBLE STARS

	R.A. h	m	Dec. deg	min	P.A. deg	Sep. sec	Mags.
ω	04	59.3	+37	53	359	5.4	5.0, 8.0
R	05	17.3	+53	35	339	47.5	var., 8.6
ν	05	51.5	+39	09	206	54.6	4.9, 9.3
δ	05	59.5	+54	17	AB 271	115.4	3.7, 9.5
					AC 067	197.1	9.5
θ	05	59.7	+37	13	AB 313	3.6	2.6, 7.1
					AC 297	50.0	10.6

OPEN CLUSTERS

M	C	NGC	R.A. h	m	Dec. deg	min	Diameter min	Mag.	No. of stars
		1664	04	51.1	+43	42	18	7.6	—
		1778	05	08.1	+37	03	7	7.7	25
		1857	05	20.2	+39	21	6	7.0	40
		1893	05	22.7	+33	24	11	7.5	60
38		1912	05	28.7	+35	50	21	6.4	100
36		1960	05	36.1	+34	08	12	6.0	60
37		2099	05	52.4	+32	33	24	5.6	150
		2126	06	03.0	+49	54	6	10.2	40
		2281	06	49.3	+41	04	15	5.4	30

NEBULA

M	C	NGC	R.A. h	m	Dec. deg	min	Dimensions min	Mag. of illuminating star	
	31	IC 405	05	16.2	+34	16	30 × 19	6v	AE Aurigæ; Flaming Star Nebula

BOÖTES

(Abbreviation: Boö).

An original constellation, dominated by Arcturus. There are various myths attached to it, but none is very definite. According to one version, Boötes was a herdsman who invented the plough drawn by two oxen, for which service he was transferred to the sky. There are eight stars above magnitude 4.

	Mag.	Luminosity Sun = 1	Dist lt-yrs
α	−0.04	115	36
ε	2.37	200	150
η	2.68	6.5	32
γ	3.03	53	104
δ	3.47	58	140
β	3.50	58	137
ρ	3.58	105	183
ζ	3.78	105	205

Arcturus is the brightest star in the northern hemisphere of the sky (marginally superior to Vega and Capella). The discarded constellation Quadrans was made up of dim stars near β Boötis.

Arcturus is a giant, with a probably diameter of about 32 000 000 km. The heat from it has been given as the same as that received from a candle at a range of 8 km.

τ has a large planetary companion – the first to be seen optically (1999).

BRIGHTEST STARS

Star	R.A. h	m	s	Dec. deg	min	sec	Mag.	Abs. mag.	Spec.	Dist pc	
4 τ	13	47	15.6	+17	27	24	4.50	3.8	F7	16	
5 υ	13	49	28.6	+15	47	52	4.06	−0.3	K5	72	
8 η	13	54	41.0	+18	23	51	2.68	2.7	G0	9.8	
17 κ	14	13	27.6	+51	47	15	4.40	1.5	A7	38	
16 α	14	15	39.6	+19	10	57	−0.04	−0.2	K2	11	Arcturus
21 ι	14	16	09.8	+51	22	02	4.75	2.4	A7	28	
19 λ	14	16	22.9	+46	05	18	4.18	1.8	A0p	29	
23 θ	14	25	11.7	+51	51	02	4.05	3.8	F7	13	
25 ρ	14	31	49.7	+30	22	17	3.58	−0.2	K3	56	
27 γ	14	32	04.6	+38	13	30	3.03	0.5	A7	32	Seginus
28 σ	14	34	40.7	+29	44	42	4.46	3.0	F2	17	
30 ζ	14	41	08.8	+13	43	42	3.78	−0.2	A2	63	
36 ε	14	44	59.1	+27	04	27	2.37	−0.9	K0	46	Izar
35 o	14	45	14.4	+16	57	51	4.60	0.2	K0	29	
37 ξ	14	51	23.2	+19	06	04	4.55	5.5	G8	6.8	
42 β	15	01	56.6	+40	23	26	3.50	0.3	G8	42	Nekkar
43 ψ	15	04	26.6	+26	56	51	4.54	−0.1	K2	75	
49 δ	15	15	30.1	+33	18	53	3.47	0.3	G8	43	
51 μ¹	15	24	29.3	+37	22	38	4.31	2.6	F0	18	Alkalurops

Also above mag. 5:

	Mag.	Abs. mag.	Spectrum	Dist	
6	4.91	−0.3	K4	110	
12	4.83	2.4	F8	29	
20	4.86	−0.2	K3	100	
29 π¹	4.93	−0.3	B9	40	Alazal
31	4.86	0.3	G8	75	
34	4.81		gM0	30	
41 ω	4.81	−0.3	K4	91	
44	4.76	4.4	G0	12	
45	4.93	3.4	F5	19	
A	4.83	0.7	K1		

VARIABLE STARS

	R.A. h	m	Dec. deg	min	Range	Type	Period, d.	Spec.
ZZ	13	56.2	+25	55	5.8–6.4	W UMa	5.0	G2 + G2
S	14	22.9	+53	49	7.8–13.8	Mira	270.7	M
V	14	29.8	+38	52	7.0–12.0	Semi-reg.	258	M
R	14	37.2	+26	44	6.2–13.1	Mira	223.4	M
W	14	43.4	+26	32	4.7–5.4	Semi-reg.	450	M
i (44)	15	03.8	+47	39	6.5–7.1	W UMa	0.27	G + G

BOÖTES (continued)

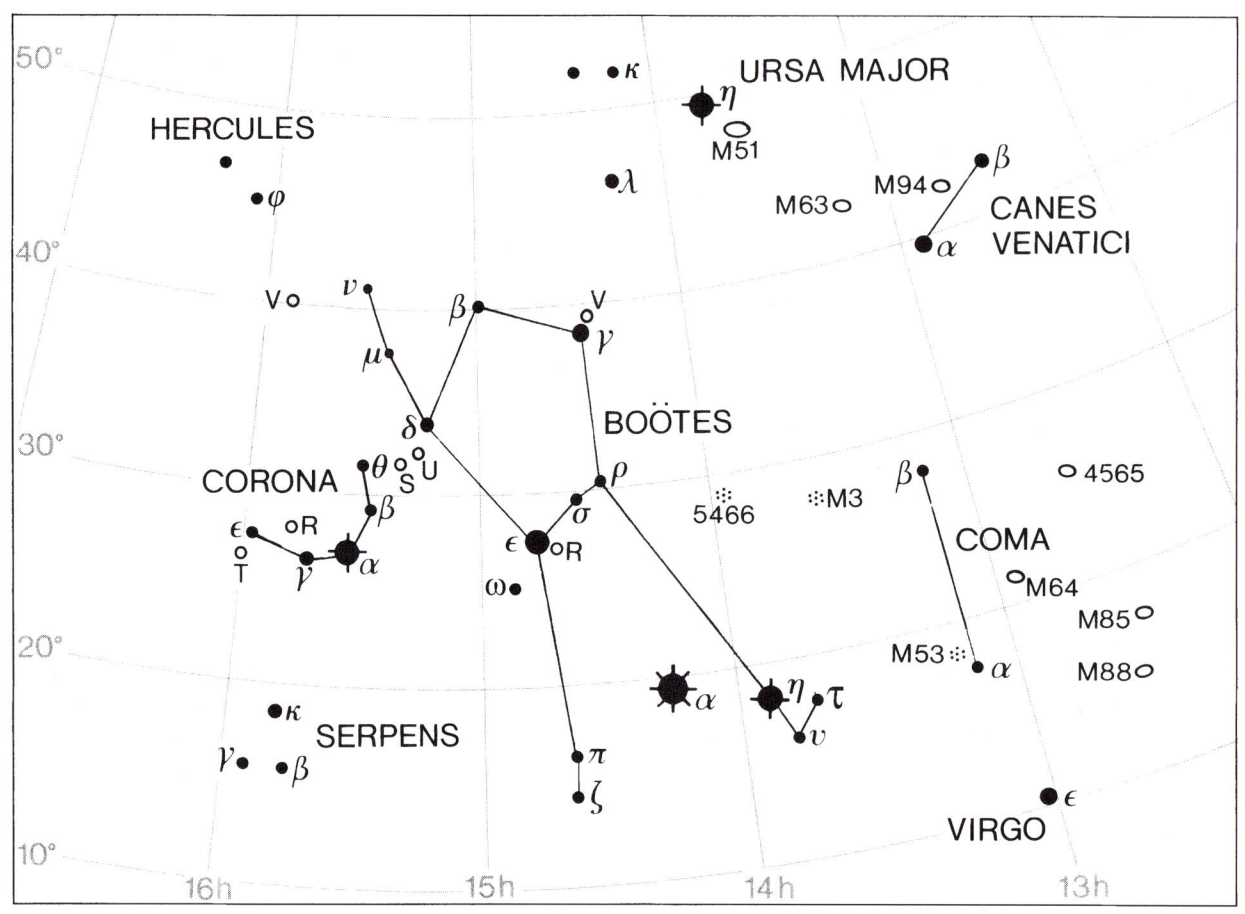

DOUBLE STARS

	R.A.		Dec.		P.A.	Sep.	Mags.	
	h	m	deg	min	deg	sec		
κ	14	13.5	+51	47	236	13.4	4.6, 6.6	
ι	14	16.2	+51	22	033	38.5	4.9, 7.5	
π	14	40.7	+16	25	108	5.6	4.9, 5.8	
ζ	14	41.1	+13	44	AB 303	1.0	4.5, 4.6	Binary, 123.3 y
					AC 259	99.3	10.9	
i (44)	15	03.8	+47	39	040	1.0	5.3v, 6.2	Binary, 225 y
μ	15	24.5	+37	23	171	108.3	4.3, 7.0	
ε	14	45.0	+27	04	341	2.9	2.5, 4.9	
ξ	14	51.4	+19	06	319	6.7	4.7, 6.8	Binary, 150 y

GLOBULAR CLUSTER

M	C	NGC	R.A.		Dec.		Diameter	Mag.
			h	m	deg	min	sec	
		5466	14	05.5	+28	32	11	9.1

GALAXIES

M	C	NGC	R.A.		Dec.		Mag.	Dimensions	Type
			h	m	deg	min		min	
	45	5248	13	37.5	+08	53	10.2	6.5 × 4.9	Sc
		5676	14	32.8	+49	28	10.9	3.9 × 2.0	Sc

CÆLUM

(Abbreviation: Cae).

This entirely unremarkable constellation was introduced in 1752 by Lacaille, under the name of Cæla Sculptoris. It has no star brighter than the fourth magnitude, and only two which are brighter than the fifth.

See chart for Columba.

BRIGHTEST STARS

Star	R.A.			Dec.			Mag.	Abs.	Spec.	Dist
	h	m	s	deg	min	sec		mag.		pc
α	04	40	33.6	−41	51	50	4.45	3.0	F2	20
γ	05	04	24.3	−35	29	00	4.55	0.2	gK0	52

VARIABLE STARS

	R.A.		Dec.		Range	Type	Period, d.	Spec.
	h	m	deg	min				
R	04	40.5	−38	14	6.7–13.7	Mira	390.9	M
T	04	47.3	−36	13	7.0–9.8	Semi-reg.	156	N

DOUBLE STARS

	R.A.		Dec.		P.A.	Sep.	Mags.
	h	m	deg	min	deg	sec	
α	04	40.6	−41	52	121	6.6	4.5, 12.5
γ	05	04.4	−35	29	308	2.9	4.6, 8.1

CAMELOPARDUS

(Abbreviation: Cam).

A very barren northern constellation. It was introduced to the sky by Hevelius in 1690, and some historians have maintained that it represents the camel which carried Rebecca to Isaac! It is interesting to note that several of the apparently faint stars are in fact highly luminous and remote; for instance α Cam, which is below the fourth magnitude, is well over 20 000 times more luminous than the Sun.

There are no stars in Camelopardus above the fourth magnitude.

See chart for Cassiopeia.

BRIGHTEST STARS

Star	R.A.			Dec.			Mag.	Abs.	Spec.	Dist
	h	m	s	deg	min	sec		mag.		pc
2 H	03	29	04.1	+59	56	25	4.21	−7.1	B9	1100
γ	03	50	21.5	+71	19	57	4.63	0.9	A3	56
9 α	04	54	03.0	+66	20	34	4.29	−6.2	09.5	860
7	04	57	17.1	+53	45	08	4.47	1.2	A1	45
10 β	05	03	25.1	+60	26	32	4.03	−4.5	G0	460

Also above mag. 5:

	Mag.	Abs. mag.	Spectrum	Dist
22	4.80	0.6	A0	66

VARIABLE STARS

	R.A.		Dec.		Range	Type	Period, d.	Spec.
	h	m	deg	min				
U	03	37.5	+62	29	7.7–8.9	Semi-reg.	400	N
RV	04	30.7	+57	25	7.1–8.2	Semi-reg.	101	M
T	04	40.1	+66	09	7.3–14.4	Mira	373.2	M
X	04	45.7	+75	06	7.4–14.2	Mira	143.6	K–M
S	05	41.0	+68	48	7.7–11.6	Semi-reg.	327	R
V	06	02.5	+74	30	7.7–16.0	Mira	522.4	M
VZ	07	31.1	+82	25	4.8–5.2	Semi-reg.	24	M
R	14	17.8	+83	50	7.0–14.4	Mira	270.2	S

CAMELOPARDUS (continued)

OPEN CLUSTER

M	C	NGC	R.A.		Dec.		Diameter	Mag.	No. of stars
			h	m	deg	min	min		
		1502	04	07.7	+62	20	8	5.7	45

PLANETARY NEBULA

M	C	NGC	R.A.		Dec.		Diameter	Mag.	Mag. of
			h	m	deg	min	sec		central star
		IC 3568	12	32.9	+82	33	6	11.6	12.3

GALAXIES

M	C	NGC	R.A.		Dec.		Mag.	Dimensions	Type
			h	m	deg	min		min	
		IC 342	03	46.8	+68	06	9.2	17.8 × 17.4	SBc
		1961	05	42.1	+69	23	11.1	4.3 × 3.0	Sb
		2146	06	18.7	+78	21	10.5	6.0 × 3.8	SBb
		2366	07	28.9	+69	13	10.9	7.6 × 3.5	Irreg.
	7	2403	07	36.9	+65	36	8.4	17.8 × 11.0	Sc
		2460	07	56.9	+60	21	11.7	2.9 × 2.2	Sb
		2655	08	55.6	+78	13	10.1	5.1 × 4.4	SBa
		2715	09	08.1	+78	05	11.4	5.0 × 1.9	Sc

CANCER

(Abbreviation: Cnc).

Cancer is an obscure constellation, redeemed only by the presence of two famous star-clusters, Præsepe and M.67. However, it lies in the Zodiac, and it has a legend attached to it. It represents a sea-crab which Juno, queen of Olympus, sent to the rescue of the multi-headed hydra which was doing battle with Hercules. Not surprisingly, Hercules trod on the crab, but as a reward for its efforts Juno placed it in the sky!

There are only two stars above the fourth magnitude.

	Mag.	Luminosity Sun = 1	Dist lt-yrs
β	3.52	110	170
δ	3.94	60	153

Præsepe (M.44) was actually given a Greek letter (ε) but this is never used.

BRIGHTEST STARS

Star	R.A.			Dec.			Mag.	Abs. mag.	Spec.	Dist pc	
	h	m	s	deg	min	sec					
16 ζ	08	12	12.6	+17	38	52	4.67	3.2, 4.1	F7, G2	16	Tegmine
17 β	08	16	30.9	+09	11	08	3.52	−0.3	K4	52	Altarf
43 γ	08	43	17.1	+21	28	06	4.66	1.2	A1	48	Asellus Borealis
47 δ	08	44	41.0	+18	09	15	3.94	0.2	K0	47	Asellus Australis
48 ι¹	08	46	41.8	+28	45	36	4.02	−2.1	G8	130	
65 α	08	58	29.2	+11	51	28	4.25	1.9	A3	23	Acubens

VARIABLE STARS

	R.A.		Dec.		Range	Type	Period, d.	Spec.
	h	m	deg	min				
R	08	16.6	+11	44	6.1–11.8	Mira	361.6	M
V	08	21.7	+17	17	7.5–13.9	Mira	272.1	S
X	08	55.4	+17	14	5.6–7.5	Semi-reg.	195	N
T	08	56.7	+19	51	7.6–10.5	Semi-reg.	482	R–N
W	09	09.9	+25	15	7.4–14.4	Mira	393.2	M
RS	09	10.6	+30	58	6.2–7.7	Semi-reg.	120	M

DOUBLE STAR

	R.A.		Dec.		P.A.	Sep.	Mags.	
	h	m	deg	min	deg	sec		
ζ	08	12.2	+17	39	AB + C 094	5.9	5.0, 6.2	Binary, 1150 y
					AB 182	0.6	5.3, 6.0	Binary, 59.7 y
					AB + D 108	287.9	9.7	

CANCER (continued)

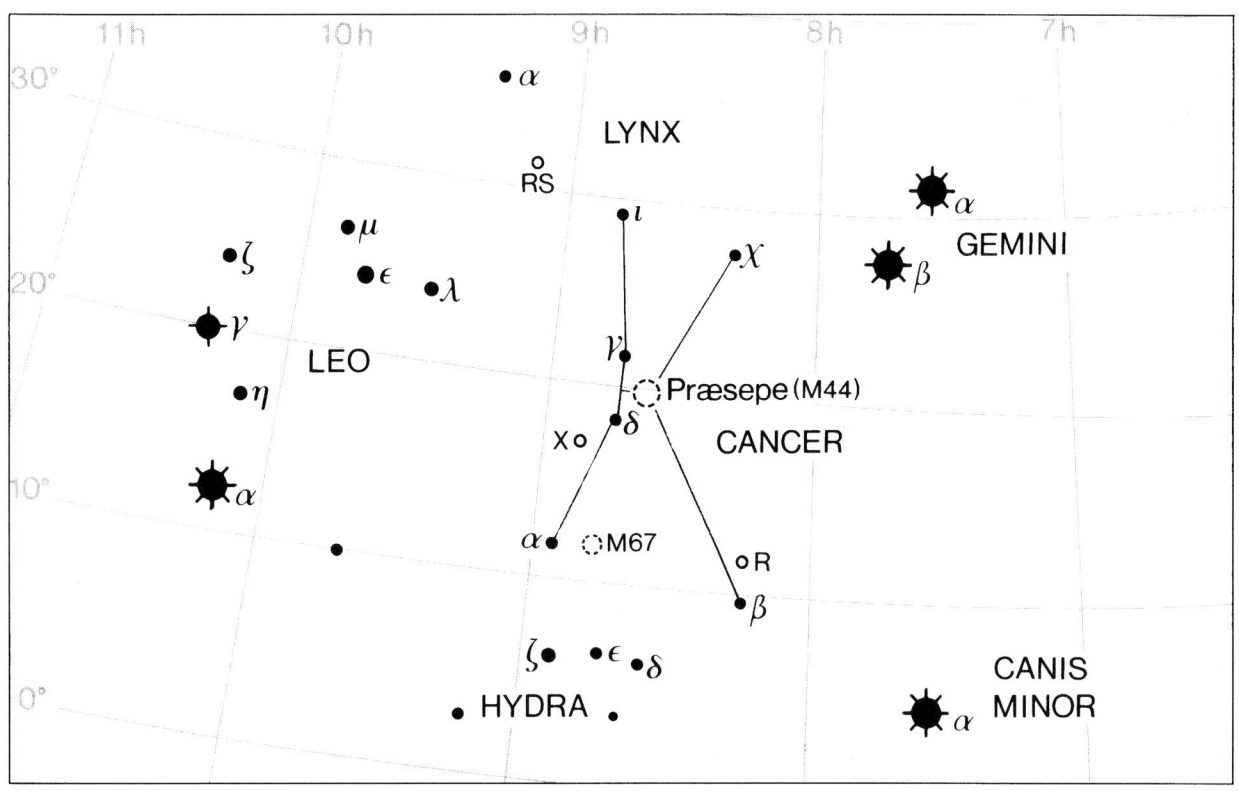

OPEN CLUSTERS

	M	C	NGC	R.A. h	m	Dec. deg	min	Diameter min	Mag.	No. of stars	
(ϵ)	44		2632	08	40.1	+19	59	95	3.1	50	Præsepe
	67		2682	08	50.4	+11	49	30	6.9	200	

GALAXY

M	C	NGC	R.A. h	m	Dec. deg	min	Mag.	Dimensions min	Type
		2775	09	10.3	+07	02	10.3	4.5 × 3.5	Sa

CANES VENATICI

(Abbreviation: CVn).

One of Hevelius' constellations, dating only from his maps of 1690, and evidently representing two hunting dogs (Asterion and Chara) which are being held by the herdsman Boötes – possibly to stop them from chasing the two Bears round and round the celestial pole. The name of Cor Caroli was given to α^2 CVn by Edmond Halley in honour of King Charles I. Cor Caroli (mag. 2.90) is the only star above the fourth magnitude. It is also the prototype 'magnetic variable'. α^1 and α^2 share common proper motion in space, and must be associated, but they are a long way apart.

See chart for Ursa Major.

	Mag.	Luminosity Sun = 1	Dist lt-yrs
α^2	2.90	75	65

CANES VENATICI (continued)

BRIGHTEST STARS

Star	R.A.			Dec.			Mag.	Abs.	Spec.	Dist	
	h	m	s	deg	min	sec		mag.		pc	
8 β	12	33	44.4	+41	21	26	4.26	4.5	G0	9.2	Chara
12 α²	12	56	01.6	+38	19	06	2.90	0.1	A0p	20	Cor Caroli
20	13	17	32.5	+40	34	21	4.73	−0.7	F0	52	
24	13	34	27.2	+49	00	57	4.70	1.9	A4	35	

Also above mag. 5:

	Mag.	Abs. mag.	Spectrum	Dist
5	4.80	0.3	G7	38
25	4.82	0.5	A7	73

VARIABLE STARS

	R.A.		Dec.		Range	Type	Period, d.	Spec.	
	h	m	deg	min					
T	12	30.2	+31	30	7.6–12.6	Mira	290.1	M	
Y	12	45.1	+45	26	7.4–10.0	Semi-reg.	157	N	La Superba
U	12	47.3	+38	23	7.2–11.0	Mira	345.6	M	
TU	12	54.9	+47	12	5.6–6.6	Semi-reg.	50	M	
V	13	19.5	+45	32	6.5–8.6	Semi-reg.	192	M	
R	13	49.0	+39	33	6.5–12.9	Mira	328.5	M	

DOUBLE STAR

	R.A.		Dec.		P.A.	Sep.	Mags.
	h	m	deg	min	deg	sec	
α²	12	56.0	+38	19	229	19.1	2.9, 5.5 (α¹)

GLOBULAR CLUSTER

M	C	NGC	R.A.		Dec.		Diameter	Mag.
			h	m	deg	min	min	
3		5272	13	42.2	+28	23	16.2	6.4

GALAXIES

M	C	NGC	R.A.		Dec.		Mag.	Dimensions	Type	
			h	m	deg	min		min		
		4111	12	07.1	+43	04	10.8	4.8 × 1.1	S0	
		4138	12	09.5	+43	41	12.3	2.9 × 1.9	E4	
		4145	12	10.0	+39	53	11.0	5.8 × 4.4	Sc	
		4151	12	10.5	+39	24	10.4	5.9 × 4.4	Pec.	
		4618	12	41.5	+41	09	10.8	4.4 × 3.8	Sc	
		4214	12	15.6	+36	20	9.8	7.9 × 6.3	Irreg.	
		4217	12	15.8	+47	06	11.9	5.5 × 1.8	Sb	
		4242	12	17.5	+45	37	11.0	4.8 × 3.8	S	
	26	4244	12	17.5	+37	49	10.7	13.0 × 10	Sb	
106		4258	12	19.0	+47	18	8.3	18.2 × 7.9	Sb	
		4395	12	25.8	+33	33	10.1	12.9 × 11.0	S	
	21	4449	12	28.2	+44	06	9.4	5.1 × 3.7	Irreg.	
		4490	12	30.6	+41	38	9.8	5.9 × 3.1	Sc	
	32	4631	12	42.1	+32	32	9.3	15.1 × 3.3	Sc	
		4656–7	12	44.0	+32	10	10.4	13.8 × 3.3	Sc	
94		4736	12	50.9	+41	07	8.2	11.0 × 9.1	Sb	
	29	5005	13	10.9	+37	03	9.8	5.4 × 2.7	Sb	
		5033	13	13.4	+36	36	10.1	10.5 × 5.6	Sb	
63		5055	13	15.8	+42	02	8.6	12.3 × 7.6	Sb	
		5112	13	21.9	+38	44	11.9	3.9 × 2.9	Sc	
51		5194	13	29.9	+47	12	8.4	11.0 × 7.8	Sc	Whirlpool
		5195	13	30.0	+47	16	9.6	5.4 × 4.3	Pec.	Companion to M.51
		5371	13	55.7	+40	28	10.7	4.4 × 3.6	Sb	

CANIS MAJOR

(Abbreviation: CMa).

An original constellation, representing one of Orion's hunting dogs. Though dominated by Sirius, it contains several other bright stars, of which one (Adhara) is only just below the first magnitude. Altogether there are 11 stars brighter than the fourth magnitude.

	Mag.	Luminosity Sun = 1	Dist lt-yrs
α	−1.46	26	8.6
ε	1.50	5000	490
δ	1.86	132 000	3060
β	1.98v	7200	720
η	2.44	52 500	2500
ζ	3.02	450	290
o^2	3.03	43 000	2800
o^1	3.86	21 000	1700
ω	3.86	700	550
ν^2	3.95	82	235
κ	3.96	800	650

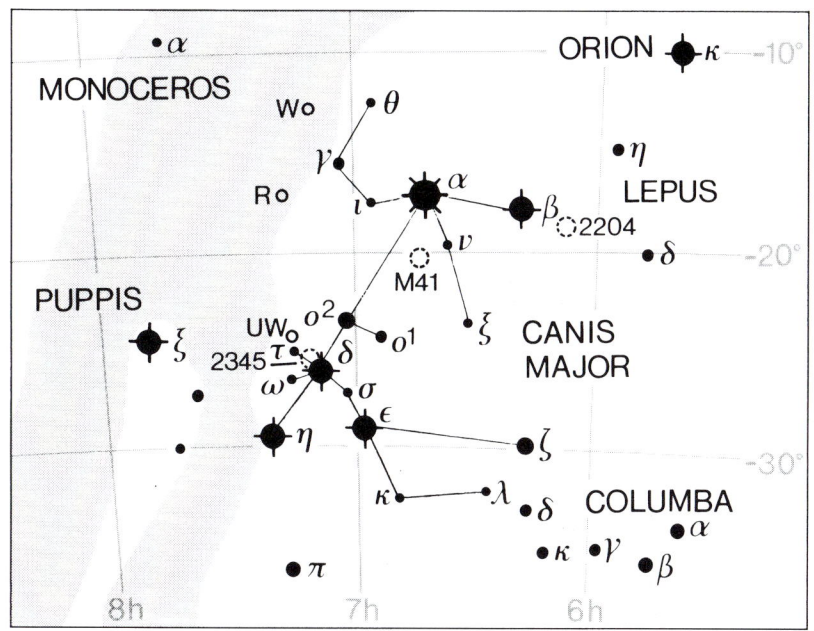

It is interesting to note that three of these stars – Wezea, Aludra and o^2 – are extremely luminous. Compared with them, Sirius is extremely feeble, as it has only 26 times the Sun's luminosity.

Sirius has a white dwarf companion, mag. 8; binary, period 50 years. Though the companion does not appear faint, it is difficult to see because of the overpowering glare of the primary. The separation ranges between 3 sec to 11 sec .5. It was widest in 1975, so that it is now closing up. β (Mirzam) is variable over a very small range, and is also a spectrum variable. ξ^1 is also a spectrum variable, and optically variable over a very small range of less than 0.1 magnitude.

At EUV (extreme ultra-violet) wavelengths, Adhara is the brightest object in the entire sky.

CANIS MAJOR (continued)

BRIGHTEST STARS

Star	R.A.			Dec.			Mag.	Abs. mag.	Spec.	Dist pc	
	h	m	s	deg	min	sec					
1 ζ	06	20	18.7	−30	03	48	3.02	−1.7	B3	88	Phurad
2 β	06	22	41.9	−17	57	22	1.98v	−4.8	B1	220	Mirzam
λ	06	28	10.1	−32	34	48	4.48	−0.5	B5	17	
4 ξ¹	06	31	51.2	−23	25	06	4.34	−3.9	B1	440	
5 ξ²	06	35	03.3	−22	57	53	4.54	0.6	A0	61	
7 ν²	06	36	41.0	−19	15	22	3.95	0.0	K1	72	
8 ν³	06	37	53.3	−18	14	15	4.43	0.0	K1	72	
9 α	06	45	08.9	−16	42	58	−1.46	1.4	A1	2.7	Sirius
13 κ	06	49	50.4	−32	30	31	3.96	−2.5	B2	200	
16 o¹	06	54	07.8	−24	11	02	3.86	−6.0	K3	520	
14 θ	06	54	11.3	−12	02	19	4.07	−0.3	K4	73	
19 π	06	55	37.3	−20	08	11	4.68	0.6	gF2	65	
20 ι	06	56	08.1	−17	03	14	4.38	−3.9	B3	410	
21 ε	06	58	37.5	−28	58	20	1.50	−4.4	B2	150	Adhara
22 σ	07	01	43.1	−27	56	06	3.46	−5.7	M0	460	
24 o²	07	03	01.4	−23	50	00	3.03	−6.8	B3	860	
23 γ	07	03	45.4	−15	38	00	4.11	−3.4	B8	320	Muliphen
25 δ	07	08	23.4	−26	23	36	1.86	−8.0	F8	940	Wezea
27	07	14	15.1	−26	21	09	4.66	−2.9	B3	320	
28 ω	07	14	48.6	−26	46	22	3.86	−2.3	B3	170	
29 UW	07	18	40.3	−24	33	32	4.98v	—	O7.8		
30 τ	07	18	41.4	−24	57	15	4.39	−7.0	O9	2900	
31 η	07	24	05.6	−29	18	11	2.44	−7.0	B5	760	Aludra

Also above mag. 5:

		Mag.	Abs. mag.	Spectrum	Dist
15	EY	4.82	−3.9	B1	560
29		4.98v		O7.8	

VARIABLE STARS

	R.A.		Dec.		Range	Type	Period, d.	Spec.
	h	m	deg	min				
W	07	08.1	−11	55	6.4–7.9	Irreg.	—	N
RY	07	16.6	−11	29	7-7–8.5	Cepheid	4.68	F–G
UW	07	18.4	−24	34	4.0–5.3	Beta Lyræ	4.39	O7
R	07	19.5	−16	24	5.7–6.3	Algol	1.14	F

DOUBLE STARS

	R.A.		Dec.		P.A.	Sep.	Mags.	
	h	m	deg	min	deg	sec		
ν¹	06	36.4	−18	40	262	17.5	5.8, 8.5	
α	06	45.1	−16	43	005	4.5	−1.5, 8.3	Binary, 50 y
17	06	55.0	−20	24	AB 147	44.4	5.8, 9.3	
					AC 184	50.5	9.0	
					AD 186	129.9	9.5	
π	06	55.6	−20	08	018	11.6	4.7, 9.7	
μ	06	56.1	−14	03	AB 340	3.0	5.3, 8.6	Isis
					AC 288	88.4	10.5	
					AD 061	101.3	10.7	
ε	06	58.6	−28	58	161	7.5	1.5, 7.4	

OPEN CLUSTERS

M	C	NGC	R.A.		Dec.		Diameter	Mag.	No. of stars	
			h	m	deg	min	min			
		2204	06	15.7	−18	39	13	8.6	80	
41		2287	06	47.0	−20	44	38	4.5	80	
		2345	07	08.3	−13	10	12	7.7	20	
	58	2360	07	17.8	−15	37	13	7.2	80	
	64	2362	07	17.8	−24	57	8	4.1	60	τ CMa

GALAXIES

M	C	NGC	R.A.		Dec.		Mag.	Dimensions	Type
			h	m	deg	min	min		
		2207	06	16.4	−21	22	10.7	4.3 × 2.9	Sc
		2217	06	21.7	−27	14	10.4	4.8 × 4.4	SBa
		2223	06	24.6	−22	50	11.4	3.3 × 3.0	SBb
		2280	06	44.8	−27	38	11.8	5.6 × 3.2	Sb

CANIS MINOR

(Abbreviation: CMi)

The second of Orion's two dogs. There are two stars above the fourth magnitude.

	Mag.	Luminosity Sun = 1	Dist lt-yrs
α	0.38	7	11.4
β	2.90	105	137

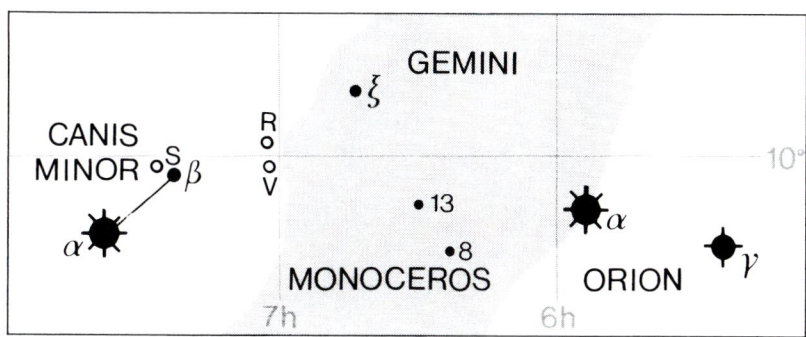

Procyon has a white dwarf companion; magnitude 13, period 40 years. It is very difficult to observe because of the glare from the primary.

The diameter of Procyon itself is thought to be about 15 700 000 km.

BRIGHTEST STARS

Star	R.A. h	m	s	Dec. deg	min	sec	Mag.	Abs. mag.	Spec.	Dist pc	
3 β	07	27	09.0	+08	17	21	2.90	−0.2	B8	42	Gomeisa
4 γ	07	28	09.7	+08	55	33	4.32	−0.2	K3	65	
6	07	29	47.7	+12	00	24	4.54	−0.1	K2	70	
10 α	07	39	18.1	+05	13	30	0.38	2.6	F5	3.5	Procyon
HD 66141	08	02	15.8	+02	20	04	4.39	−0.1	K2	70	

Also above mag. 5:

	Mag.	Abs. mag.	Spectrum	Dist
2 ε	4.99	0.3	G8	78

VARIABLE STARS

	R.A. h	m	Dec. deg	min	Range	Type	Period, d.	Spectrum
V	07	07.0	+08	53	7.4–15.1	Mira	366.1	M
R	07	08.7	+10	01	7.3–11.6	Mira	337.8	S
S	07	32.7	+08	19	6.6–13.2	Mira	332.9	M

DOUBLE STAR

	R.A. h	m	Dec. deg	min	P.A. deg	Sep. sec	Mags.	
α	07	39.3	+05	14	021	5.2	0.4, 12.9	Binary, 40.7 y

CAPRICORNUS

(Abbreviation: Cap).

A Zodiacal constellation, though by no means brilliant. It has been identified with the demigod Pan, but there seems to be no well-defined mythological associations. There are five stars above magnitude 4:

	Mag.	Luminosity Sun = 1	Dist lt-yrs
δ	2.87v	13	49
β	3.08	2	104
α^2	3.57	60	117
γ	3.68	28	65
ζ	3.74	5200	1470

α^1 and α^2 make up a naked-eye pair, separated by 376 sec (position angle 291 deg), but the two components are not genuinely associated, and each is itself double; the fainter component of α^2 is also double!

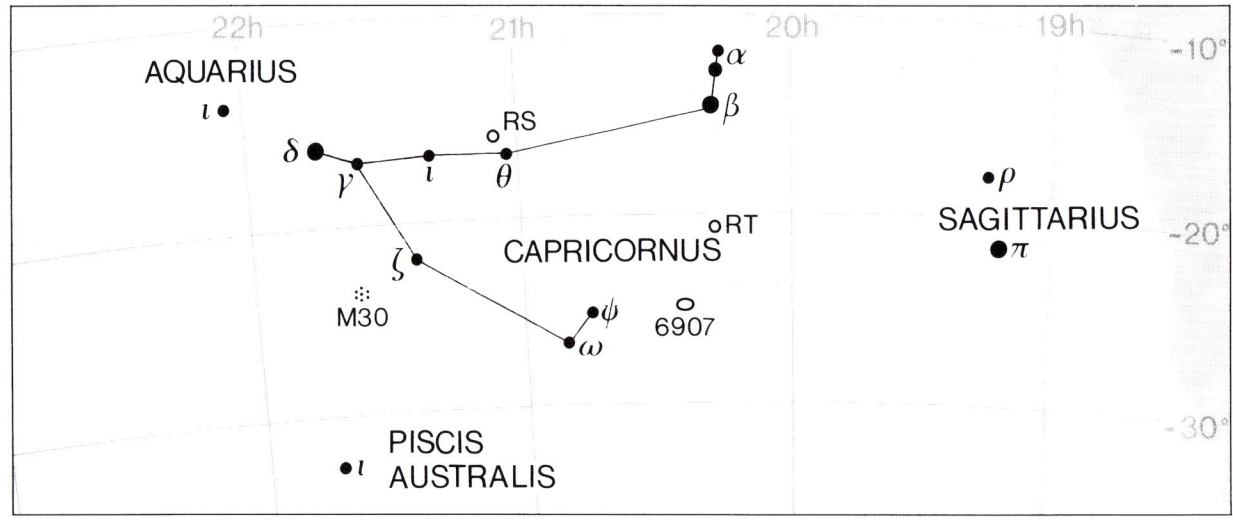

BRIGHTEST STARS

Star	R.A. h	m	s	Dec. deg	min	sec	Mag.	Abs. mag.	Spec.	Dist pc	
5 α^1	20	17	38.6	−12	30	30	4.24	−4.5	G3	490	Al Geidi
6 α^2	20	18	03.1	−12	32	42	3.57	0.2	G9	36	
9 β	20	21	00.5	−14	46	53	3.08	4.0	F8	32	Dabih
16 ψ	20	46	05.5	−25	16	16	4.14	3.4	F5	12	Wei
18 ω	20	51	49.1	−26	55	09	4.11	−0.3	K5	64	
23 θ	21	05	56.6	−17	13	58	4.07	0.6	A0	49	
24 A	21	07	07.5	−25	00	21	4.50	−0.5	M1	95	
32 ι	21	22	14.6	−16	50	05	4.28	0.3	G8	63	
34 ζ	21	26	39.9	−22	24	41	3.74	−4.5	G4	450	Yen
36 b	21	28	43.2	−21	48	26	4.51	0.3	gG5	67	
39 ε	21	37	04.7	−19	27	58	4.68	−2.3	B3	250	
40 γ	21	40	05.2	−16	39	45	3.68	1.2	F0p	18	Nashira
43 κ	21	42	39.3	−18	51	59	4.73	0.3	G8	77	
49 δ	21	47	02.3	−16	07	38	2.87v	2.0	A5	15	Deneb al Giedi

Also above mag. 5:

	Mag.	Abs. mag.	Spectrum	Dist	
8 ν	4.76	−0.3	B9	99	Alshat
11 ρ	4.78	0.6	F2	32	
22 η	4.84	2.6	A4	12	Chow

CAPRICORNUS (continued)

VARIABLE STARS

	R.A.		Dec.		Range	Type	Period, d.	Spec.
	h	m	deg	min				
RT	20	17.1	−21	19	6.5–8.1	Semi-reg.	393	N
RR	21	02.3	−27	05	7.8–15.5	Mira	277.5	M
RS	21	07.2	−16	25	7.0–9.0	Semi-reg.	340	M

DOUBLE STARS

	R.A.		Dec.		P.A.	Sep.	Mags.	
	h	m	deg	min	deg	sec		
α	20	18.1	−12	33	291	377.7	3.6, 4.2	(α^1–α^2)
α^1	20	17.6	−12	30	AB 182	44.3	4.2, 13.7	
					AC 221	45.4	9.2	
α^2	20	18.1	−12	33	AB 172	6.6	3.6, 11.0	
					AD 156	154.6	9.3	
					BC 240	1.2	11.3	
σ	20	19.6	−19	07	179	55.9	5.5, 9.0	
β	20	21.0	−14	47	267	205	3.1, 6.0	B is double
π	20	27.3	−18	13	148	3.2	5.3, 8.9	
ρ	20	28.9	−17	49	158	0.5	5.0, 10.0	
ε	21	37.1	−19	28	047	68.1	4.7, 9.5	
τ	20	39.3	−14	57	118	0.3	5.8, 6.3	Binary, 200 y

GLOBULAR CLUSTER

M	C	NGC	R.A.		Dec.		Diameter	Mag.
			h	m	deg	min	min	
30		7099	21	40.4	−23	11	11.0	7.5

GALAXY

M	C	NGC	R.A.		Dec.		Mag.	Dimensions	Type
			h	m	deg	min		min	
		6907	20	25.1	−24	49	11.3	3.4 × 3.0	SBb

CARINA

(Abbreviation: Car).

The keel of the ship Argo, in which Jason and his companions sailed upon their successful if somewhat unprincipled expedition to remove the golden fleece of the sacred ram (Aries) from its grove. Carina is the brightest part of Argo, and contains Canopus, which is second only to Sirius in brilliance. Canopus is a highly luminous star, but estimates of its power and distance vary considerably. A recent estimate makes it over 200 000 times as luminous as the Sun; other estimates are much lower. Also in this constellation is η Carinæ, which during part of the last century rivalled Sirius, but which is now well below naked-eye visibility. It is unique, and at its peak may have been the most luminous star in the Galaxy. Excluding η, there are 15 stars in Carina above the fourth magnitude.

	Mag.	Luminosity Sun = 1	Dist lt-yrs
α	−0.72	200 000	1200
β	1.68	130	85
ε	1.86	600	200
ι	2.25	7500	800
θ	2.76	3800	750
υ	2.97	520	320
ρ	3.32v	450	310
ω	3.32	200	230
q	3.40	5000	900
a	3.44	1300	620
χ	3.47	1300	590
u	3.78	17	85
R	3.8–10	var	800
c	3.84	200	300
x	3.91	132 000	4600

CARINA (continued)

ι and ε Carinæ are two of the stars of the False Cross (the other two stars are κ and δ Velorum). The False Cross is often mistaken for the Southern Cross, but is larger and less brilliant.

By 5000 AD ω Carinæ will be the South Pole Star.

BRIGHTEST STARS

Star	R.A. h	m	s	Dec. deg	min	sec	Mag.	Abs. mag.	Spec.	Dist pc	
α	06	23	57.1	−52	41	44	−0.72	−8.5	F0	360	Canopus
N	06	34	58.5	−52	58	32	4.39	−0.8	B9	100	
A	06	49	51.3	−53	37	20	4.40	0.4	gG3	52	
χ	07	56	46.7	−52	58	56	3.47	−3.0	B2	180	
d	08	40	37.0	−59	45	40	4.33	−3.6	B2	350	
ε	08	22	30.8	−59	30	34	1.86	−2.1	K0	62	Avior
f	08	46	42.6	−56	46	11	4.49	−2.5	B2	250	
c	08	55	02.8	−60	38	41	3.84	−1.0	B8	93	
G	09	05	38.4	−70	32	20	4.71	−2.5	B2	280	
a	09	10	57.9	−58	58	01	3.44	−3.0	B2	190	
i	09	11	16.7	−62	19	02	3.97	−2.3	B3	180	
β	09	13	12.2	−69	43	02	1.68	−0.6	A0	26	Miaplacidus
g	09	16	12.2	−57	32	28	4.34	−0.3	gK5	70	
ι	09	17	05.4	−59	16	31	2.25	−4.7	F0	250	Tureis (or Aspidiske)
R	09	32	14.7	−62	47	19	3.8–10	var.	M5	250	
h	09	34	26.6	−59	13	46	4.08	−3.7	B5	350	
m	09	39	20.9	−61	19	40	4.52	0.2	B9	73	
l(ZZ)	09	45	14.8	−62	30	28	4.1v	4.4	G0	16	
υ	09	47	06.1	−65	04	18	2.97	−2.0	A7	99	
ω	10	13	44.3	−70	02	16	3.32	−1.0	B7	70	
q	10	17	04.9	−61	19	56	3.40	−4.4	K5	280	
l	10	24	23.7	−74	01	54	4.00	0.6	F2	17	
s	10	27	24.4	−57	38	19	4.68	−8.5	F0	3600	
ρ	10	32	01.4	−61	41	07	3.32v	−1.7	B3	96	
r	10	35	35.2	−57	33	27	4.45	−0.4	gM0	92	
t^2	10	38	45.0	−59	10	58	4.66	−5.9	cK	1300	
θ	10	42	57.4	−64	23	39	2.76	−4.1	B0	230	
W	10	43	32.1	−60	33	59	4.57	−0.3	K5	75	
u	10	53	29.6	−58	51	12	3.78	1.7	K0	26	
x	11	08	35.3	−58	58	30	3.91	−8.0	G0	1400	
y	11	12	36.0	−60	19	03	4.60	−8.5	F0	3300	

Also above mag. 5:

	Mag.	Abs. mag.	Spectrum
Q	4.92	−0.8	M0
D^1	4.96	−2.4	B3
B	4.80	3.5	F5
e^2	4.80	0.2	G6
p	4.94	—	FI
k	4.87	1.5	G4
K	4.94	−0.5	A2
Z^1	4.76	1.1	G5

VARIABLE STARS

	R.A. h	m	Dec. deg	min	Range	Type	Period, d.	Spectrum	
V	08	28.7	−60	07	7.1–7.8	Cepheid	6.70	F–H	
X	08	31.3	−59	14	7.9–8.6	Beta Lyræ	1.08	A+A	
R	09	32.2	−62	47	3.9–10.5	Mira	308.7	M	
ZZ	09	45.2	−62	30	3.3–4.2	Cepheid	35.53	F–K	(I Carinæ)
S	10	09.4	−61	33	4.5–9.9	Mira	149.5	K–M	
η	10	45.1	−59	41	−0.8–7.9	Irreg.	—	Pec.	
BO	10	45.8	−59	29	7.2–8.5	Irreg.	—	M	
U	10	57.8	−59	44	5.7–7.0	Cepheid	38.77	F–G	

DOUBLE STARS

	R.A. h	m	Dec. deg	min	P.A. deg	Sep. sec	Mags.
η	10	45.1	−59	41	195	0.2	var., 8.6
u	09	47.1	−65	04	127	5.0	3.1, 6.1

CARINA (continued)

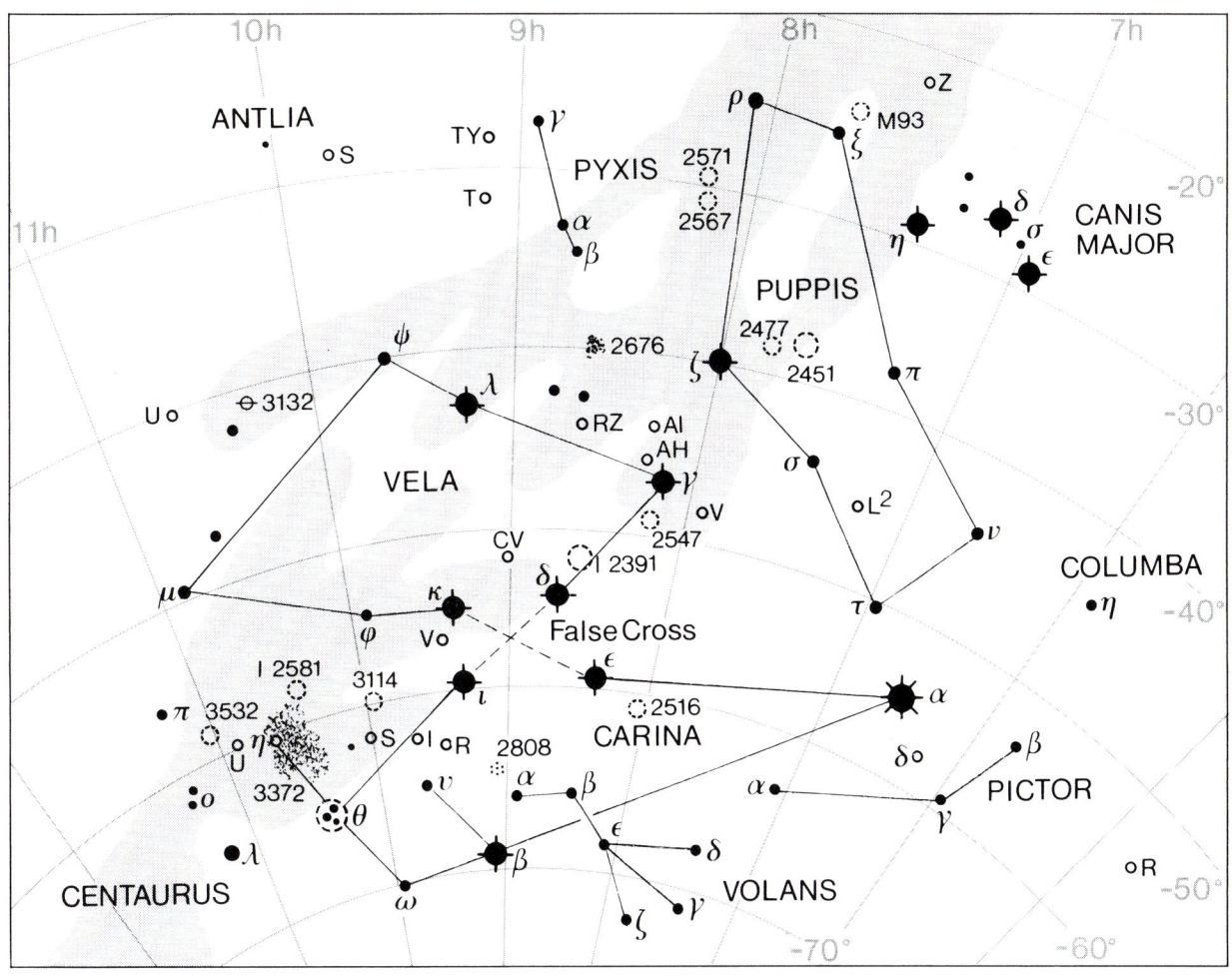

OPEN CLUSTERS

M	C	NGC	R.A. h	m	Dec. deg	min	Diameter min	Mag.	No. of stars	
	96	2516	07	58.3	−60	52	30	3.8	80	
		3114	10	02.7	−60	07	35	4.2	—	
		IC 2581	10	27.4	−57	38	8	4.3	25	
	102	IC 2602	10	43.2	−64	24	50	1.9	60	θ Carinæ cluster
	91	3532	11	06.4	−58	40	55	3.0	150	
		Mel 101	10	42.1	−65	06	14	8.0	50	
		3572	11	10.4	−60	14	7	6.6	35	
		3590	11	12.9	−60	47	4	8.2	25	
		Mel 105	11	19.5	−63	30	4	8.5	70	
		IC 2714	11	17.9	−62	42	12	8.2	100	
		3680	11	25.7	−43	15	12	7.6	30	

GLOBULAR CLUSTER

M	C	NGC	R.A. h	m	Dec. deg	min	Diameter min	Mag.
		2808	09	12.0	−64	52	13.8	6.3

CARINA (continued)

NEBULA

M	C	NGC	R.A.		Dec.		Mag.	Mag. of illuminating star
			h	m	deg	min		
	92	3372	10	43.8	−59	52	6.2	var.	η Carinæ nebula

PLANETARY NEBULÆ

M	C	NGC	R.A.		Dec.		Diam.	Mag.	Mag. of
			h	m	deg	min	sec		central star
		IC 2448	09	07.1	−69	57	8	11.5	12.9
	90	2867	09	21.4	−58	19	11	9.7	13.6
		IC 2501	09	38.8	−60	05	25	11.3	—
		3211	10	17.8	−62	40	12	11.8	—
		IC 2621	11	00.3	−65	15	5	—	13.6

CASSIOPEIA

(Abbreviation: Cas).

An original constellation – one of the most distinctive in the sky. Mythologically, Cassiopeia was Andromeda's mother and wife of Cepheus; it was her boasting which led to the unfortunate contretemps with Neptune's sea-monster.

There are seven stars above magnitude 4. Of these, γ (Cih) is variable, and it is also suspected that α (Shedir) is variable over a small range – possibly 2.1 to 2.4, though opinions differ. κ, which looks obscure, is extremely luminous.

There are eight stars brighter than magnitude 4.

	Mag.	Luminosity Sun = 1	Dist lt-yrs
γ	2.2v	6000v	780
α	2.23v?	200	120
β	2.27	14	42
δ	2.68	11	62
ε	3.38	1200	520
η	3.44	1.2	19
ζ	3.67	830	550
ι	3.98	28	117

ρ is extremely powerful; one estimate gives its absolute magnitude as −9.5 and its distance as 8000 light-years, in which it is even more luminous than Canopus. The spectrum is variable, but generally of around type F8.

6 Cassiopeiæ (magnitude 5.43) is also very luminous; estimated absolute magnitude −7.6, spectrum A4. Its position is R.A. 23 h 48 m 50.0 s, declination +62° 12′ 53″. The distance has been given as 6500 light-years, but this is decidedly uncertain.

BRIGHTEST STARS

Star	R.A.			Dec.			Mag.	Abs. mag.	Spec.	Dist pc	
	h	m	s	deg	min	sec					
7 ρ	23	54	22.9	+57	29	58	var.	−8.0	F8	1500	
11 β	00	09	10.6	+59	08	59	2.27	1.9	F2	13	Chaph
14 λ	00	31	46.3	+54	31	20	4.73	−0.9	B8	13	
15 κ	00	32	59.9	+62	55	55	4.16	−6.6	B1	930	
17 ζ	00	36	58.2	+53	53	49	3.67	−2.5	B2	170	
18 α	00	40	30.4	+56	32	15	2.23v?	−0.9	K0	37	Shedir
22 o	00	44	43.4	+48	17	04	4.54	−2.5	B2	210	
24 η	00	49	05.9	+57	48	58	3.44	4.6	G0	5.9	Achird
28 υ^2	00	56	39.7	+59	10	52	4.63	1.8	G8	29	
27 γ	00	56	42.4	+60	43	00	2.2v	−4.6	B0p	240	Cih
33 θ	01	11	06.1	+55	09	00	4.33	2.4	A7	37	Marfak
37 δ	01	25	48.9	+60	14	07	2.68	2.1	A5	19	Ruchbah
36 ψ	02	25	55.9	+68	07	48	4.74	0.2	K0	74	
39 χ	01	33	55.8	+59	13	56	4.71	0.2	K0	80	
45 ε	01	54	23.6	+63	40	13	3.38	−2.9	B3	160	Segin
48	02	01	57.3	+70	54	26	4.48	1.9	A4	33	
50 ι	02	03	26.0	+72	25	17	3.98	1.2	A1	36	

CASSIOPEIA (continued)

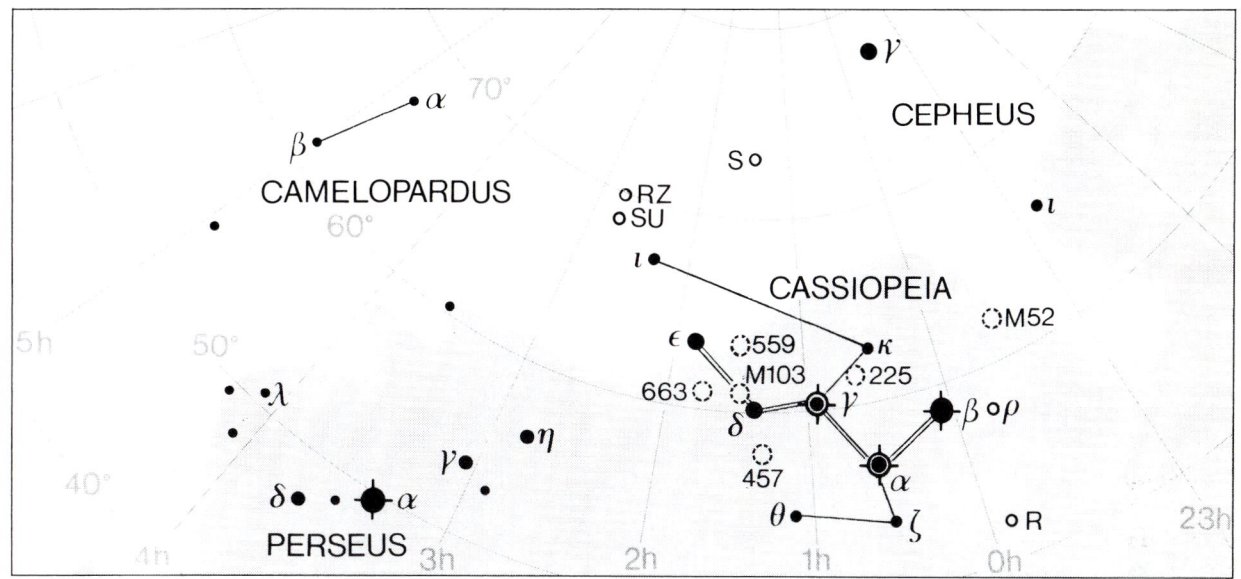

Also above mag. 5:

	Mag.	Abs. mag.	Spectrum	Dist	
1	4.85	−4.4	B1	570	
5 τ	4.87	0.0	K1	92	
4	4.97	−0.3	K5	86	
8 σ	4.88	−3.5	B1	400	
10 ξ	4.80	−2.5	B2	260	
25 υ	4.89		B9		
26 υ¹	4.83	−0.1	K2	91	Castula
46 ω	4.99		B8		

VARIABLE STARS

	R.A. h	m	Dec. deg	min	Range	Type	Period, d.	Spec.
V	23	11.7	+59	42	6.9–13.4	Mira	228.8	M
ρ	23	54.4	+58	30	4.1–6.2	?	—	F–K
R	23	58.4	+51	24	4.7–13.5	Mira	430.5	M
T	00	23.2	+55	48	6.9–13.0	Mira	444.8	M
TU	00	26.3	+51	17	6.9–8.1	Cepheid	2.14	F
α	00	40.5	+56	22	?2.1–2.5	Suspected	—	K
U	00	46.4	+48	15	8.0–15.7	Mira	277.2	S
RV	00	52.7	+47	25	7.3–16.1	Mira	331.7	M
W	00	54.9	+58	34	7.8–12.5	Mira	405.6	N
γ	00	56.7	+60	43	1.6–3.3	Irreg.	—	B
S	01	19.7	+72	37	7.9–16.1	Mira	612.4	S
SU	02	52.0	+68	53	5.7–6.2	Cepheid	1.95	F
RZ	02	48.9	+69	38	6.2–7.7	Algol	1.19	A

DOUBLE STARS

	R.A. h	m	Dec. deg	min	P.A. deg	Sep. sec	Mags.	
λ	00	31.8	+54	31	176	0.5	5.3, 5.6	
η	00	49.1	+57	49	293	12.2	3.4, 7.5	Binary, 480 y
ψ	01	25.9	+68	08	113	25.0	4.7, 9.6	
ι	02	29.1	+67	24	232	2.4	4.6, 6.9	Binary, 840 y
σ	23	59.0	+55	45	326	3.0	5.0, 7.1	

CASSIOPEIA (continued)

OPEN CLUSTERS

M	C	NGC	R.A. h	m	Dec. deg	min	Diameter min	Mag.	No. of stars	
52		7654	23	24.2	+61	35	13	6.9	100	
		7788	23	56.7	+61	24	9	9.4	20	
		7789	23	57.0	+56	44	16	6.7	300	
		H.21	23	54.1	+61	46	4	9.0	6	
		129	00	29.9	+60	14	21	6.5	35	Contains DL Cas
		133	00	31.2	+63	22	7	9.4	5	
		146	00	33.1	+63	18	7	9.1	20	
		225	00	43.4	+61	47	12	7.0	15	
		381	01	08.3	+61	35	6	9.3	50	
		436	01	15.6	+58	49	6	8.8	30	
	13	457	01	19.1	+58	20	13	6.4	80	φ Cas cluster
	8	559	01	29.5	+63	18	4.4	9.5	60	
		IC 1805	02	32.7	+61	27	22	6.5	40	
103		581	01	33.2	+60	42	6	7.4	25	
		637	01	42.9	+64	00	3.5	8.2	20	
		1027	02	42.7	+61	33	20	6.7	40	
		654	01	44.1	+61	53	5	6.5	60	
		659	01	44.2	+60	42	5	7.9	40	
	10	663	01	46.0	+61	15	16	7.1	80	

NEBULÆ

M	C	NGC	R.A. h	m	Dec. deg	min	Diameter	Mag. of illuminating star	
	11	7635	23	20.7	+61	12	15 × 8	7	Bubble Nebula
		281	00	52.8	+56	36	35 × 30	8	
		IC 1805	02	33.4	+61	26	60 × 60	—	
		IC 1848	02	51.3	+60	25	60 × 30	—	

GALAXIES

M	C	NGC	R.A. h	m	Dec. deg	min	Mag.	Dimensions min	Type
	17	147	00	33.2	+48	30	9.3	12.9 × 8.1	dE4
	18	185	00	39.0	+48	20	9.2	11.5 × 9.8	dE0

CENTAURUS

(Abbreviation: Cen).

A brilliant southern constellation (one of Ptolemy's originals), containing many important objects, including the nearest of all the bright stars (α) and the superb globular cluster ω. There are 19 stars brighter than the fourth magnitude.

	Mag.	Luminosity Sun = 1	Dist lt-yrs
α	−0.27	1.7+0.45	4.3
β	0.61	10 000	460
θ	2.06	17	46
γ	2.17	130	110
ε	2.30	2100	490
η	2.31v	1200	360
ζ	2.55	1300	360
δ	2.60v	830	325
ι	2.75	26	52

CENTAURUS (continued)

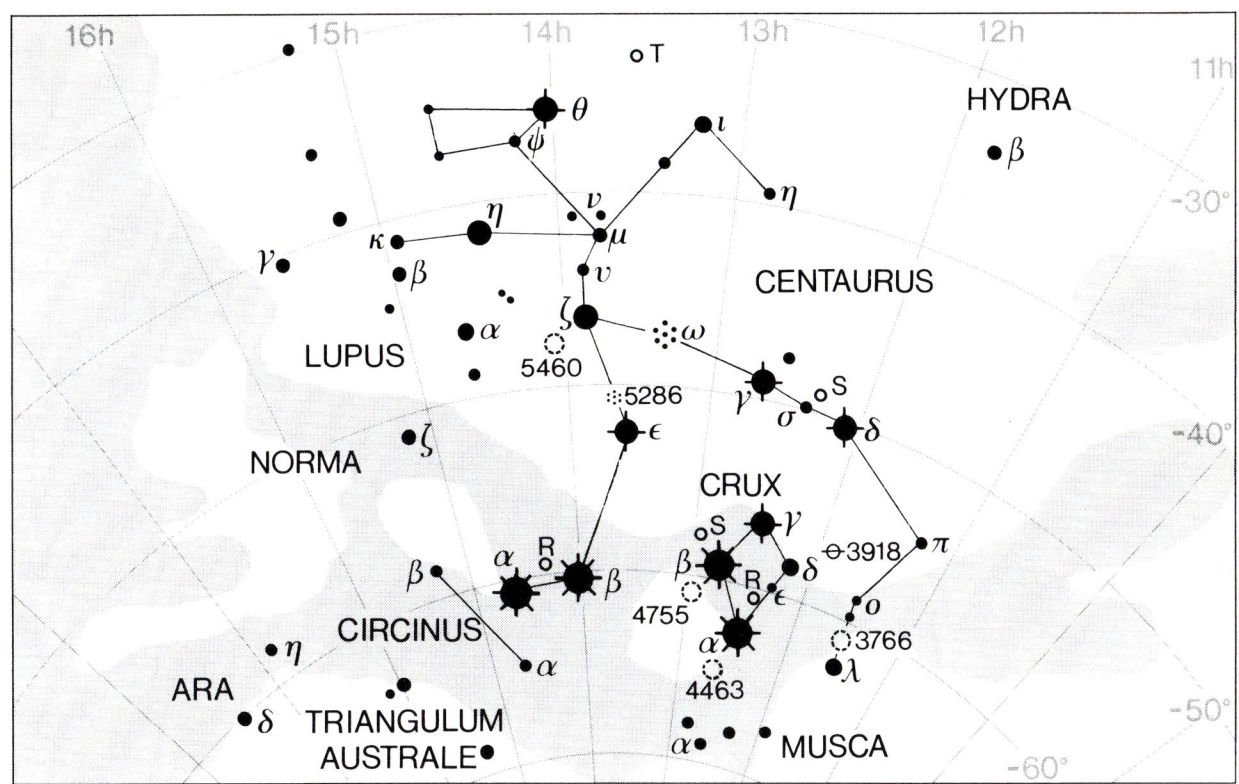

	Mag.	Luminosity Sun = 1	Dist lt-yrs
μ	3.04v	450	300
κ	3.13	830	420
λ	3.13	180	190
ν	3.41	2600	1000
ϕ	3.83	830	620
τ	3.86	26	98
υ^1	3.87	700	550
d	3.88	60	120
π	3.89	230	323
σ	3.91	450	425

Strangely, α has no official proper name. It has been called Al Rijil Rigel Kent, and also Toliman, but astronomers in general prefer to call it simply α Centauri.

α and β are known as the Pointers, because they show the way to the Southern Cross. The Australian Aborigines called them the Tho Brothers.

β and its faint companion have common proper motion, and must therefore be associated.

ω (listed by Ptolemy as a star) is the largest Milky Way globular, with a mass 5 000 000 times that of the Sun – equal to a small galaxy. It is about 10 times as massive as other big globulars. Distance, 16 000 000 light-years.

CENTAURUS (continued)

BRIGHTEST STARS

Star	R.A. h	m	s	Dec. deg	min	sec	Mag.	Abs. mag.	Spec.	Dist pc	
π	11	21	00.4	−54	29	27	3.89	−1.1	B5	99	
λ	11	35	46.8	−63	01	11	3.13	−0.8	B9	57	
65G	11	46	30.7	−61	10	12	4.11	−2.0	G0	140	
j	11	49	41.0	−45	10	25	4.32	−1.1	B5	120	
B	11	51	08.5	−45	43	05	4.46	−0.3	K4	90	
δ	12	08	21.5	−50	43	20	2.60v	−2.5	B2	100	
ρ	12	11	39.1	−52	22	07	3.96	−1.4	B4	120	
τ	12	37	42.1	−48	32	28	3.86	1.4	A2	30	
σ	12	28	02.4	−50	13	51	3.91	−1.7	B3	130	
γ	12	41	30.9	−48	57	34	2.17	−0.6	A0	34	Menkent
w	12	42	35.3	−48	48	47	4.66	0.0	gK1	86	
e	12	53	06.8	−48	56	35	4.33	−0.1	gK2	54	
n	12	53	26.1	−40	10	44	4.27	0.5	A7	57	
ξ²	13	06	54.5	−49	54	22	4.27	−3.0	B2	280	
ι	13	20	35.8	−36	42	44	2.75	1.4	A2	16	
J	13	22	37.8	−60	59	18	4.53	−1.1	B5	130	
m	13	24	0.05	−64	32	09	4.53	1.8	G5	35	
d	13	31	02.6	−39	24	27	3.88	0.3	G8	36	
ε	13	39	53.2	−53	27	58	2.30	−3.5	B1	150	
1	13	45	43.0	−33	02	30	4.23	0.8	F2	28	
M	13	46	39.3	−51	25	58	4.65	0.2	gK0	78	
2 g	13	49	26.6	−34	27	03	4.19	−0.5	m1	87	
ν	13	49	30.2	−41	41	16	3.41	−2.5	B2	150	
μ	13	49	36.9	−42	28	25	3.04	−1.7	B3	89	
3	13	51	50.0	−32	59	41	4.32	−1.6	B5	91	
4	13	53	12.4	−31	55	39	4.73	−1.6	B5	190	
ζ	13	55	32.3	−47	17	17	2.55	−3.0	B2	110	Al Nair al Kentaurus
294G	13	57	38.9	−63	41	11	4.71	−0.3	K4	100	
φ	13	58	16.2	−42	06	02	3.83	−2.5	B2	190	
υ¹	13	58	40.7	−44	48	13	3.87	−2.3	B3	170	
υ²	14	01	43.3	−45	36	12	4.34	−2.0	F5	160	
β	14	03	49.4	−60	22	22	0.61	−5.1	B1	140	Agena
χ	14	16	02.7	−41	10	47	4.36	−2.5	B2	240	
5 θ	14	06	40.9	−36	22	12	2.06	1.7	K0	14	Haratan
υ	14	20	19.4	−56	23	12	4.33	−3.7	B5	310	
ψ	14	20	33.3	−37	53	07	4.05	0.0	A0	65	
a	14	23	02.1	−39	09	34	4.44	−1.9	B6	180	
η	14	35	30.3	−42	09	28	2.31v	−2.9	B3	110	
α²	[14	39	35.4	−60	50	13	1.39	5.7	K1	1.3]	
α¹	[14	39	36.7	−60	50	02	0.00	4.4	G2	1.3]	
b	14	41	57.5	−37	47	37	4.00	−1.7	B3	140	
c¹	14	43	39.3	−35	10	25	4.05	−0.3	K5	71	
κ	14	59	09.6	−42	06	15	3.13	−2.5	B2	130	Ke Kwan

Also above mag. 5:

	Mag.	Abs. mag.	Spectrum	Dist
	4.85	0.6	A0	67
1	4.79	0.5	B8p	230
o¹	4.96	1.1	G4	
4 n	4.76	−1.3	B7	
f	4.96	−2.1	B3	

VARIABLE STARS

Star	R.A. h	m	Dec. deg	min	Range	Type	Period, d.	Spec.
RS	11	20.5	−61	52	7.7–14.1	Mira	164.4	M
X	11	49.2	−41	45	7.0–13.8	Mira	315.1	M
W	11	55.0	−59	15	7.6–13.7	Mira	201.6	M
S	12	24.6	−49	26	6.0–7.0	Semi-reg.	65	N
U	12	33.5	−54	40	7.0–14.0	Mira	220.3	M
RV	13	37.5	−56	29	7.0–10.8	Mira	446.0	N
XX	13	40.3	−57	37	7.3–8.3	Cepheid	10.95	F–G
T	13	41.8	−33	36	5.5–9.0	Semi-reg.	60	K–M
μ	13	49.6	−42	28	2.9–3.5	Irreg.	—	B
R	14	16.6	−59	55	5.3–11.8	Mira	546.2	M
V	14	32.5	−56	53	6.4–7.2	Cepheid	5.49	F–G

CENTAURUS (continued)

DOUBLE STARS

	R.A. h	m	Dec. deg	min	P.A. deg	Sep. sec	Mags.	
D	12	140	−45	43	243	2.8	5.6, 6.8	
γ	12	41.5	−48	58	347	1.0	2.9, 2.9	Binary, 84.5 y
ε	13	39.9	−53	28	158	36.0	2.3, 12.7	
3	13	51.8	−33	00	108	7.9	4.5, 6.0	
4	13	53.2	−31	56	185	14.9	4.8, 8.4	
β	14	03.8	−60	22	251	1.3	0.7, 31.9	
η	14	35.5	−42	09	270	5.0	2.6, 13.5	
α	14	39.6	−60	50	221	14.8	0.0, 1.2	Binary, 79.9 y

OPEN CLUSTERS

M	C	NGC	R.A. h	m	Dec. deg	min	Diam. min	Mag.	No. of stars	
	97	3766	11	36.1	−61	37	12	5.3	100	
	100	IC 2944	11	36.6	−63	02	15	4.5	30	λ Cen cluster, Collinder 249
		3960	11	50.9	−55	42	7	8.3	45	
		5138	13	27.3	−59	01	8	7.6	40	
		5281	13	46.6	−62	54	5	5.9	40	
		5316	13	53.9	−61	52	14	6.0	80	
		5460	14	07.6	−48	19	25	5.6	40	
		5617	14	29.8	−60	43	10	6.3	80	
		5662	14	35.2	−56	33	12	5.5	70	

GLOBULAR CLUSTERS

M	C	NGC	R.A. h	m	Dec. deg	min	Diameter min	Mag.	
	80	5139	13	26.8	−47	29	36.3	3.6	ω Centauri
	84	5286	13	46.4	−51	22	9.1	7.6	

PLANETARY NEBULA

M	C	NGC	R.A. h	m	Dec. deg	min	Diameter sec	Mag.	Mag. of central star	
		3918	11	50.3	−57	11	12	8.4	10.9	Blue Planetary

NEBULA

M	NGC	R.A. h	m	Dec. deg	min	Diameter min	
	5367	13	57.7	−39	59	4.3 Includes IC 4347. Double nucleus	

GALAXIES

M	C	NGC	R.A. h	m	Dec. deg	min	Mag.	Dimensions min	Type	
		4603	2	40.9	−40	59	12.0	3.8 × 2.5	Sc	
		4696	2	48.8	−41	19	10.7	3.5 × 3.2	Elp	
	83	4945	13	05.4	−49	28	9.5	20.0 × 4.4	SBc	
		4976	13	08.6	−49	30	10.2	4.3 × 2.6	E4p	
	77	5128	13	25.5	−43	01	7.0	18.2 × 14.3	S0p	Centaurus A
		5253	13	39.9	−31	39	0.6	4.0 × 1.7	E5	
		5483	14	10.4	−43	19	12.0	3.1 × 2.8	Sc	
		3557	11	10.0	−37	32	10.4	4.0 × 2.7	E3	

CEPHEUS

(Abbreviation: Cep).

A rather undistinguished constellation, though it contains δ Cephei – the prototype Cepheid. Mythologically, Cepheus was Andromeda's father and Cassiopeia's husband. There are eight stars above the fourth magnitude – though one of these, the 'Garnet Star' μ, is a variable which is generally rather below this limit.

	Mag.	Luminosity Sun = 1	Dist lt-yrs
α	2.44	14	46
γ	3.21	10	59
β	3.23v	2200	750
ζ	3.35	5000	700
η	3.43	4.5	46
ι	3.52	82	130
μ	3.6v	130 000?	1560
δ	3.7v	6000v	1340

CEPHEUS (continued)

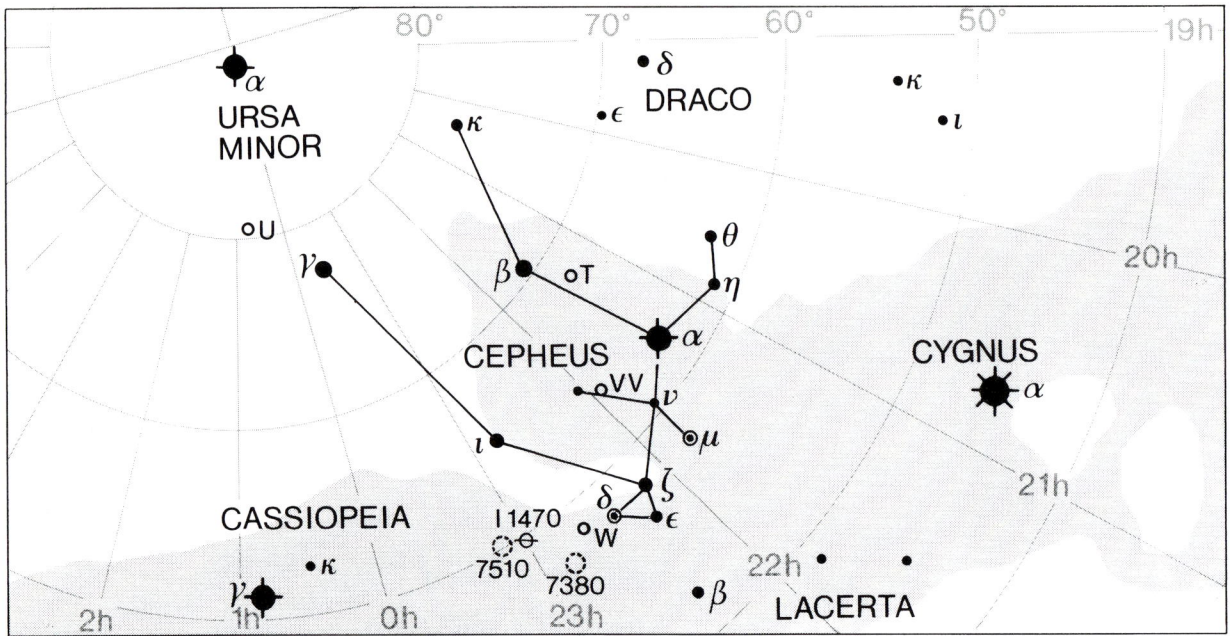

The relatively nearby double Krüger 60 lies near δ; R.A. 22 h 26 m.3, dec. +57 deg 27 min. The mean separation is between 2 and 3 seconds of arc, but the period is only 44.5 years, so that the separation and position angle alter quite quickly. The magnitudes are 9.8 and 11.4. The real separation is about the same as that between the Sun and Saturn. Krüger 60B, the fainter component, is a flare star, and in variable star catalogues is listed as DO Cephei.

μ Cephei–Herschel's 'Garnet Star' – may be one of the largest stars known. It is much more luminous and remote than Betelgeux, but with the naked eye its colour is less striking simply because it appears fainter. Its distance is not known accurately. It is now classed as a semi-regular variable, but the period given is at best extremely rough.

BRIGHTEST STARS

Star	R.A. h	m	s	Dec. deg	min	sec	Mag.	Abs. mag.	Spec.	Dist pc	
1 κ	20	08	53.1	+77	42	41	4.39	−0.8	B9	100	
2 θ	20	29	34.7	+62	59	39	4.22	1.8	A0	34	
3 η	20	45	17.2	+61	50	20	3.43	3.2	K0	14	
5 α	21	18	34.6	+62	35	08	2.44	1.9	A7	14	Alderamin
8 β	21	28	39.4	+70	33	39	3.23v	−3.6	B2	230	Alphirk
9	21	37	55.0	+62	04	55	4.73	−5.7	B2	640	
11	21	41	55.1	+72	18	42	4.56	0.2	K0	60	
μ	21	43	30.2	+58	46	48	var.	−8.0?	M2	1700	The 'Garnet Star'
10 ν	21	45	26.8	+61	07	15	4.29	−7.5	A2	1200	
17 ξ	22	03	45.7	+64	37	42	4.29	2.3	F7+G	37	Kurdah
21 ζ	22	10	51.1	+58	12	05	3.35	−4.4	K1	220	
23 ε	22	15	01.8	+57	02	37	4.19	1.7	F0	30	
27 δ	22	29	10.1	+58	24	55	var.	−4.6v	F8	410	Prototype Cepheid
32 ι	22	49	40.6	+66	12	02	3.52	0.0	K1	39	
33 π	23	07	53.7	+75	23	16	4.41	0.4	G2	59	
35 γ	23	39	20.7	+77	37	57	3.21	2.2	K1	16	Alrai
43 H	01	08	44.6	+86	15	26	4.25	−0.1	K2	68	

CEPHEUS (continued)

Also above mag. 5:

	Mag.	Abs. mag.	Spectrum	Dist
24	4.79	0.3	G8	360
34 o	4.90	1.2	G7	

VARIABLE STARS

	R.A. h	m	Dec. deg	min	Range	Type	Period, d.	Spec.
T	21	09.5	+68	29	5.2–11.3	Mira	388.1	M
VV	21	56.7	+63	38	4.8–5.4	Eclipsing	7430	M + B
S	21	35.2	+78	37	7.4–12.9	Mira	486.8	N
μ	21	43.5	+58	47	3.4–5.1	Semi-reg.	730	M
δ	22	29.2	+58	25	3.5–4.4	Cepheid	5.37	F–G
W	22	36.5	+58	26	7.0–9.2	Semi-reg.	Long	K–M
U	01	02.3	+81	53	6.7–9.2	Algol	2.49	B + G

DOUBLE STARS

	R.A. h	m	Dec. deg	min	P.A. deg	Sep. sec	Mags.	
κ	20	08.9	+72	43	122	7.4	4.4, 8.4	
β	21	28.7	+70	34	249	13.3	3.2, 7.9	
ξ	22	03.8	+64	38	277	7.7	4.4, 6.5	Binary, 3800 y
δ	22	29.2	+58	25	191	41.0	var., 7.5	
π	23	07.9	+75	23	346	1.2	4.6, 6.6	Slow binary
o	23	18.6	+68	07	220	2.9	4.9, 7.1	Binary, 796 y

OPEN CLUSTERS

M	C	NGC	R.A. h	m	Dec. deg	min	Diam. min	Mag.	No. of stars
		IC 1396	21	39.1	+57	30	50	3.5	50
		7160	21	53.7	+62	36	7	6.1	12
		7235	22	12.6	+57	17	4	7.7	30
		7261	22	20.4	+58	05	6	8.4	30
		7380	22	47.0	+58	6	12	7.2	40
		7510	23	11.5	+60	34	4	7.9	60
	1	188	00	44.4	+85	20	14	8.1	120

PLANETARY NEBULA

M	C	NGC	R.A. h	m	Dec. deg	min	Dimensions sec	Mag.	Mag. of central star
	2	40	00	13.0	+72	32	37	10.7	11.6

NEBULÆ

M	C	NGC	R.A. h	m	Dec. deg	min	Dimensions min	Mag.	
	4	7023	21	01.8	+68	12	18 × 18	6.8	
	9	Sh2-155	22	56.8	+62	37	50 × 30	7.7	Cave Nebula

GALAXIES

M	C	NGC	R.A. h	m	Dec. deg	min	Mag.	Dimensions min	Type
	12	6946	20	34.8	+60	09	9.7	11.0 × 9.8	Sc
		6951	20	37.2	+66	06	12.2	3.8 × 3.3	Sbp

CETUS

(Abbreviation: Cet).

One of the largest of all constellations; sometimes associated with the sea-monster of the Perseus legend, at others relegated to the status of a harmless whale. It contains the prototype long-period variable Mira, which can occasionally rise to magnitude 1.7, but which spends most of its period below naked-eye visibility. On average, Mira is visible with the naked eye for only a few weeks in every year. Mira is a huge star; its diameter varies, but is of the order of 650 000 000 km. The absolute magnitude varies between −2.5 and +4.7. Its companion is the flare star VZ Ceti. Cetus abounds in faint galaxies.

Excluding Mira, there are eight stars above the fourth magnitude.

	Mag.	Luminosity Sun = 1	Dist lt-yrs
β	2.04	60	68
α	2.53	120	130
η	3.45	96	117
γ	3.47	26	75
τ	3.50	0.45	11.8
ι	3.56	96	160
θ	3.60	60	114
ζ	3.73	96	190

BRIGHTEST STARS

Star	R.A. h	m	s	Dec. deg	min	sec	Mag.	Abs. mag.	Spec.	Dist pc	
2	00	03	44.3	−17	20	10	4.55	−0.3	B9	91	
7	00	14	38.4	−18	55	58	4.44	−0.5	gM1	86	
8 ι	00	19	25.6	−08	49	26	3.56	−0.1	K2	50	Baten Kaitos Shemali
16 β	00	43	35.3	−17	59	12	2.04	0.2	K0	21	Diphda
17 ϕ^1	00	44	11.3	−10	36	34	4.75	0.2	K0	81	Al Nitham
31 η	01	08	35.3	−10	10	56	3.45	−0.1	K2	36	
45 θ	01	24	01.3	−08	11	01	3.60	0.2	K0	35	
53 χ	01	49	35.0	−10	41	11	4.67	1.9	F2	30	
52 τ	01	44	04.0	−15	56	15	3.50	5.7	G8	3.6	
55 ζ	01	51	27.6	−10	20	06	3.73	−0.1	K2	58	Baten Kaitos
59 υ	02	00	00.2	−21	04	40	4.00	−0.5	M1	79	
65 ξ^1	02	12	59.9	+08	50	48	4.37	−2.1	G8	200	
68 o	02	19	20.6	−02	58	39	var.	1.2v	M	400	Mira
73 ξ^2	02	28	09.5	+08	27	36	4.28	−0.8	B9	98	
76 σ	02	32	05.1	−15	14	41	4.75	2.1	F5	33	
82 δ	02	39	28.9	+00	19	43	4.07	−3.0	B2	120	
86 γ	02	43	18.0	+03	14	09	3.47	1.4	A2	23	Alkaffaljidhina
89 π	02	44	07.3	−13	51	32	4.25	−0.6	B7	93	
87 μ	02	44	56.4	+10	06	51	4.27	1.7	F0	30	
91 λ	02	59	42.8	+08	54	27	4.70	−2.2	B5	220	
92 α	03	02	16.7	+04	05	23	2.53	−0.5	M2	40	Menkar

Also above mag. 5:

	Mag.	Abs. mag.	Spectrum	Dist
6	4.89	3.7	F6	18
20	4.77	−0.4	M0	110
46	4.90	−0.2	K3	110
72 ρ	4.89	0.2	B9	85
78 ν	4.86	0.3	G8	82
83 ε	4.84	2.8	F5	22
96 κ	4.83	5.0	G5	9.3

VARIABLE STARS

	R.A. h	m	Dec. deg	min	Range	Type	Period, d.	Spec.
W	00	02.1	−14	41	7.1–14.8	Mira	351.3	S
T	00	21.8	−20	03	5.0–6.9	Semi-reg.	159	M
S	00	24.1	−09	20	7.6–14.7	Mira	320.5	M
U	02	33.7	−13	09	6.8–13.4	Mira	234.8	M
UV	01	38.8	−17	58	6.8–13.0	Flare	–	dM
o	02	19.3	−02	59	1.7–10.1	Mira	332.0	M
R	02	26.0	−00	11	7.2–14.0	Mira	166.2	M

CETUS (continued)

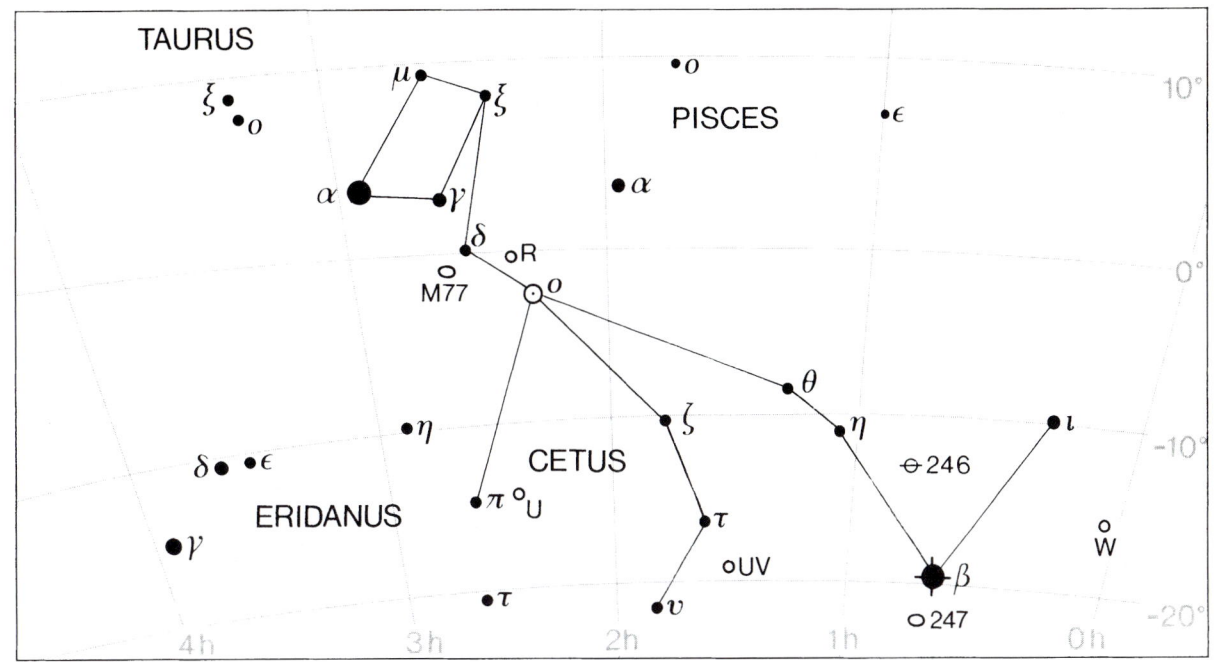

DOUBLE STARS

	R.A.		Dec.		P.A.	Sep.	Mags.	
	h	m	deg	min	deg	sec		
37	01	14.4	−07	55	331	49.7	5.2, 8.7	
χ	01	49.6	−10	41	250	183.8	4.9, 6.9	
66	02	12.8	−02	24	AB 234	16.5	5.7, 7.5	
					AC 061	172.7	11.4	
o	02	19.3	−02	59	085	0.6	var., 9.5	Binary (400y) (B is VZ Ceti)
ν	02	35.9	+05	36	081	8.1	4.9, 9.5	
ε	02	39.6	−11	52	039	0.1	5.8, 5.8	Binary, 2.7 y
γ	02	43.3	+03	14	294	2.8	3.5, 7.3	

PLANETARY NEBULA

M	C	NGC	R.A.		Dec.		Diameter	Mag.	Mag. of
			h	m	deg	min	sec		central star
		246	00	47.0	−11	53	225	8.0	11.9

GALAXIES

M	C	NGC	R.A.		Dec.		Mag.	Dimensions	Type
			h	m	deg	min		min	
		45	00	14.1	−23	11	10.4	8.1 × 5.8	S
	62	247	00	47.1	−20	46	8.9	20.0 × 7.4	S
		428	01	12.9	+00	59	11.3	4.1 × 3.2	Scp
		578	01	30.5	−22	40	10.9	4.8 × 3.2	Sc
		584	01	31.3	−06	52	10.3	3.8 × 2.4	E4
		720	01	53.0	−13	44	10.2	4.4 × 2.8	E3
		864	02	15.5	+06	00	11.0	4.6 × 3.5	Sc
		895	02	21.6	−05	31	11.8	3.6 × 2.8	Sb
		908	02	23.1	−21	14	10.2	5.5 × 2.8	Sc
		936	02	27.6	−01	09	10.1	5.2 × 4.4	SBa
	51	IC1613	01	04.8	+02	07	9.9	12.0 × 11.2	Irr
		1042	02	40.4	−08	26	10.9	4.7 × 3.9	Sc
		1055	02	41.8	+00	26	10.6	7.6 × 3.0	Sb
		1068	02	42.7	−00	01	8.8	6.9 × 5.9	SBp
		1073	02	43.7	+01	23	11.0	4.9 × 4.6	SBc
		1087	02	46.4	−00	30	11.0	3.5 × 2.3	Sc

CHAMÆLEON

(Abbreviation: Cha).

A small southern constellation of no particular note; there are no legends attached to it, and there are no stars above the fourth magnitude.

See chart for Musca.

BRIGHTEST STARS

Star	R.A.			Dec.			Mag.	Abs.	Spec.	Dist
	h	m	s	deg	min	sec		mag.		pc
α	08	18	31.7	−76	55	10	4.07	2.2	F6	24
θ	08	20	38.7	−77	29	04	4.35	1.7	K0	24
γ	10	35	28.1	−78	36	27	4.11	−0.4	M0	77
δ²	10	45	46.7	−80	32	24	4.45	−1.7	B3	170
β	12	18	20.7	−79	18	43	4.26	−0.9	B6	110

Also above mag. 5:

	Mag.	Abs. mag.	Spectrum	Dist
ε	4.91	0.2	B9	88

VARIABLE STARS

	R.A.		Dec.		Range	Type	Period, d.	Spec.
	h	m	deg	min				
R	08	21.8	−76	21	7.5–14.2	Mira	334.6	M
RS	08	43.2	−79	04	6.0–6.7	Algol + δ Scuti	1.67	A–F

DOUBLE STARS

	R.A.		Dec.		P.A.	Sep.	Mags.
	h	m	deg	min	deg	sec	
δ	10	45.3	−80	28	076	0.6	6.1, 6.4
ε	11	59.6	−78	13	188	0.9	5.4, 6.0

PLANETARY NEBULA

M	C	NGC	R.A.		Dec.		Diameter
			h	m	deg	min	sec
	109	3195	10	09.5	−80	52	38

CIRCINUS

(Abbreviation: Cir).

A very small southern constellation, in the area of α and β Centauri. It is associated with no legends. The only star above magnitude 4 is α.

See chart for Centaurus.

	Mag.	Luminosity Sun = 1	Dist lt-yrs
α	3.19	7	46

BRIGHTEST STARS

Star	R.A.			Dec.			Mag.	Abs.	Spec.	Dist
	h	m	s	deg	min	sec		mag.		pc
α	14	42	28.0	−64	58	43	3.19	2.6	F0	14
β	15	17	30.8	−58	48	04	4.07	1.7	A3	22
γ	15	23	22.6	−59	19	14	4.51	−1.0	B5	84

Also above mag. 5:

	Mag.	Abs. mag.	Spectrum	Dist
ε	4.86	0.3	K4	110

CIRCINUS (continued)

VARIABLE STARS

	R.A.		Dec.		Range	Type	Period, d.	Spec.
	h	m	deg	min				
AX	14	52.6	−63	49	5.6–6.1	Cepheid	5.27	F–G
θ	14	56.7	−62	47	5.0–5.4	Irreg. γ C	—	B

DOUBLE STARS

	R.A.		Dec.		P.A.	Sep.	Mags.	
	h	m	deg	min	deg	sec		
α	14	42.5	−64	59	232	5.7	3.2, 8.6	
δ	15	16.9	−60	57	270	50.0	5.1, 13.4	
γ	15	23.4	−59	19	011	0.8	5.1, 5.5	Binary, 180 y

OPEN CLUSTERS

M	C	NGC	R.A.		Dec.		Diam.	Mag.	No. of stars
			h	m	deg	min	min		
	88	5823	15	05.7	−55	36	10	7.9	100

COLUMBA

(Abbreviation: Col).

A 'modern' constellation dating from 1679, when it was introduced to the sky by Royer. Apparently it represents the dove which Noah released from the Ark; it was originally called Columba Noachi. There are five stars above magnitude 4.

	Mag.	Luminosity Sun = 1	Dist lt-yrs
α	2.64	105	120
β	3.12	96	143
δ	3.85	120	117
ε	3.87	60	137
η	3.96	60	143

δ Columbæ was formerly called 3 Canis Majoris.

μ (mag. 5. 16), position R.A. 05.44. 1 dec. −32 deg 10 min, is one of the three 'runaway stars' which seem to be moving away from the nebulous region of Orion (the others are 53 Arietis and AE Aurigæ). μ Columbæ has an annual proper motion of 0 sec .025. Its spectral type is O9.5.

BRIGHTEST STARS

Star	R.A.			Dec.			Mag.	Abs. mag.	Spec.	Dist pc	
	h	m	s	deg	min	sec					
ε	05	31	12.7	−35	28	15	3.87	0.2	gK0	42	
α	05	39	38.9	−34	04	27	2.64	−0.2	B8	37	Phakt
β	05	50	57.5	−35	46	06	3.12	−0.1	K2	44	Wazn
γ	05	57	32.2	−35	17	00	4.36	−2.3	B3	210	
η	05	59	08.8	−42	48	55	3.96	0.2	K0	44	
κ	06	16	33.0	−35	08	26	4.37	0.3	G8	58	
δ	06	22	06.7	−33	26	11	3.85	0.5	gG1	36	

Also above mag. 5:

	Mag.	Abs. mag.	Spectrum	Dist
ξ	4.97	0.0	K1	97
λ	4.87	−1.1	B5	160
o	4.83	3.2	K0	21

VARIABLE STARS

	R.A.		Dec.		Range	Type	Period, d.	Spec.
	h	m	deg	min				
T	05	19.3	−33	42	6.6–12.7	Mira	225.9	M
R	05	50.5	−29	12	7.8–15.0	Mira	327.6	M

COLUMBA (continued)

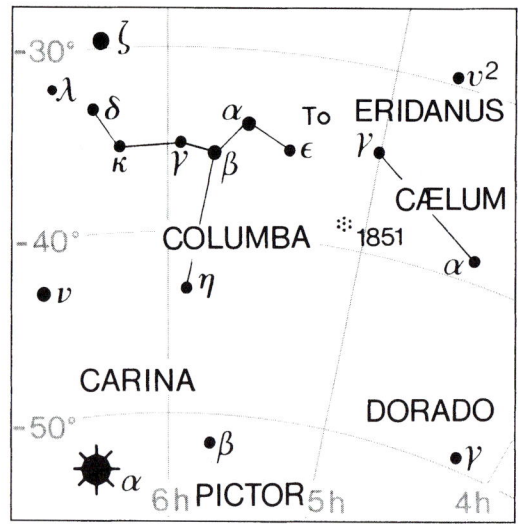

DOUBLE STARS

	R.A. h	m	Dec. deg	min	P.A. deg	Sep. sec	Mags.
α	05	39.6	−34	04	359	13.5	2.6, 12.3
γ	05	57.5	−35	17	110	33.8	4.4, 12.7
π^2	06	07.9	−42	09	150	0.1	6.2, 6.3

GLOBULAR CLUSTER

M	C	NGC	R.A. h	m	Dec. deg	min	Diameter min	Mag.	
	73	1851	05	14.1	−40	03	11.0	7.3	X-ray source

GALAXIES

M	NGC	R.A. h	m	Dec. deg	min	Mag.	Dimensions min	Type
	1792	05	05.2	−37	59	10.2	4.0 × 2.1	Sb
	1808	05	07.7	−37	32	9.9	7.2 × 4.1	SBa
	2090	05	47.0	−34	14	11.7	4.5 × 2.3	Sc

COMA BERENICES

(Abbreviation: Com).

At first glance this constellation gives the impression of being a vast, dim cluster. Coma has no star above magnitude 4.3, but it abounds in faint ones, and there are many telescopic galaxies.

Though the constellation is not 'original', there is a legend attached to it. When Ptolemy Euergetes, King of Egypt, set out in an expedition against the Assyrians, his wife Berenice vowed that if he returned safely she would cut off her lovely hair and place it in the temple of Venus. The King returned; the Queen kept her vow, and Jupiter placed the shining tresses in the sky.

See chart for Boötes.

BRIGHTEST STARS

Star	R.A. h	m	s	Dec. deg	min	sec	Mag.	Abs. mag.	Spec.	Dist pc	
15 γ	12	26	56.2	+28	16	06	4.35	−0.2	K1	31	
42 α	13	09	59.2	+17	31	45	4.32	3.4	F5	18	Diadem
43 β	13	11	52.3	+27	52	41	4.26	4.7	G0	8.3	

α is a very close binary, with a period of 26 years; the separation is never more than 0 sec .3. The components are approximately equal.

COMA BERENICES (continued)

Also above mag. 5:

	Mag.	Abs. mag.	Spectrum	Dist
12	4.79	0.7	F8	27
7	4.94	0.2	K0	89
11	4.74	0.3	G8	69
14	4.93		F0	24
16	4.99		A4	21
23	4.81	−0.6	A0	110
31	4.94	0.6	G0	74
36	4.78	−0.5	M1	110
37	4.90		K1	27
41	4.80	−0.3	K5	110

VARIABLE STARS

	R.A. h	m	Dec. deg	min	Range	Type	Period, d.	Spec.
R	12	04.0	+18	49	7.1–14.6	Mira	362.8	M
FS	13	06.4	+22	37	5.3–6.1	Semi-reg.	58	M

GALAXIES

M	C	NGC	R.A. h	m	Dec. deg	min	Mag.	Dimensions min	Type	
		4136	12	09.3	+29	56	11.7	4.1 × 3.9	Sc	
98		4192	12	13.8	+14	54	10.1	9.5 × 3.2	Sb	
		4251	12	18.1	+28	10	11.6	4.2 × 1.9	E7	
99		4254	12	18.8	+14	25	9.8	5.4 × 4.8	Sc	
		4278	12	20.1	+29	17	10.2	3.6 × 3.5	E1	
		4314	12	22.6	+29	53	10.5	4.8 × 4.3	SBa	
100		4321	12	22.9	+15	49	9.4	6.9 × 6.2	Sc	
		4448	12	28.2	+28	37	11.1	4.0 × 1.6	Sb	
		4450	12	28.5	+17	05	10.1	4.8 × 3.5	Sb	
		4459	12	29.0	+13	59	10.4	3.8 × 2.8	E2	
		4473	12	29.8	+13	26	10.2	4.5 × 2.6	E4	
		4477	12	30.0	+13	38	10.4	4.0 × 3.5	SBa	
88		4501	12	32.0	+14	25	9.5	6.9 × 3.9	SBb	
		4548	12	35.4	+14	30	10.2	5.4 × 4.4	SBb	
	36	4559	12	36.0	+27	58	9.8	10.5 × 4.9	Sc	
	38	4565	12	36.3	+25	59	9.6	16.2 × 2.8	Sb	
		4651	12	43.7	+16	24	10.7	3.8 × 2.7	Sop	
		4689	12	47.8	+13	46	10.9	4.0 × 3.5	Sb	
		4725	12	50.4	+25	30	9.2	11.0 × 7.9	SBb	
64		4826	12	56.7	+21	41	6.6	9.3 × 5.4	Sb	Black-Eye Galaxy
	35	4889	13	00.1	+27	58	11.4	3 × 2	E4	

OPEN CLUSTER

M	C	NGC	R.A. h	m	Dec. deg	min	Diam. min	Mag.	No. of stars	
		Mel 111	12	25	+26		275	4	80	Coma Berenices

GLOBULAR CLUSTER

M	C	NGC	R.A. h	m	Dec. deg	min	Diameter min	Mag.
53		5024	13	12.9	+18	10	12.6	7.7

CORONA AUSTRALIS

(Abbreviation: CrA).

An original constellation. It has no bright stars, but is easy to recognize because of its distinctive shape.
See chart for Sagittarius.

BRIGHTEST STARS

Star	R.A. h	m	s	Dec. deg	min	sec	Mag.	Abs. mag.	Spec.	Dist pc	
θ	18	33	29.9	−42	18	45	4.64	0.3	G5	60	
ζ	19	03	06.7	−42	05	43	4.75	0.6	A0	68	
γ	19	06	24.9	−37	03	48	4.21	4.0	F8	12	
δ	19	08	20.6	−40	29	48	4.59	0.2	gK0	63	
α	19	09	28.2	−37	54	16	4.11	1.6	A2	14	Meridiana
β	19	10	01.5	−39	20	27	4.11	0.3	gG5	34	

CORONA AUSTRALIS (continued)

Also above mag. 5:

	Mag.	Abs. mag.	Spectrum	Dist
ε	4.8v	2.6	F0	28

DOUBLE STARS

	R.A.		Dec.		P.A.	Sep.	Mags.	
	h	m	deg	min	deg	sec		
κ	18	33.4	−38	44	359	21.6	5.9, 5.9	
λ	18	43.8	−38	19	214	29.2	5.1, 9.7	
γ	19	06.4	−37	04	061	1.3	4.8, 5.1	Binary, 120.4 y

GLOBULAR CLUSTER

M	C	NGC	R.A.		Dec.		Diameter	Mag.
			h	m	deg	min	min	
	78	6541	18	08.0	−43	42	13.1	6.6

NEBULA

M	C	NGC	R.A.		Dec.		Diameter	Mag.	Mag. of illuminating star	
			h	m	deg	min	min			
	68	6729	19	01.9	−36	57	1(var.)	var.	9.7v	(R Coronæ Australis)

PLANETARY NEBULA

M	C	NGC	R.A.		Dec.		Diameter	Mag.	Mag. of central star	
			h	m	deg	min	sec			
		IC 1297	19	17.4	−39	37	7	—	12.9v	RU Coronæ Australis

CORONA BOREALIS

(Abbreviation: CrB).

A small but very distinctive constellation representing a crown given by Bacchus to Ariadne, the daughter of King Minos of Crete. There are three stars above magnitude 4; α is an eclipsing binary with a very small range.

	Mag.	Luminosity Sun = 1	Dist lt-yrs
α	2.23v	130	78
β	3.68	28	59
γ	3.84v	110	210

See chart for Boötes.

The solar-type star ρ Coronæ (mag. 5.4, absolute mag. 4.7, spectrum G2, distance 52 light-years) is believed to be attended by a planet 3.9 times as massive as Jupiter, at a separation of 0.23 astronomical units.

The 'Blaze Star', T Coronæ, rose to naked-eye visibility in 1866 and again in 1946.

BRIGHTEST STARS

Star	R.A.			Dec.			Mag.	Abs. mag.	Spec.	Dist pc	
	h	m	s	deg	min	sec					
3 β	15	27	49.7	+29	06	21	3.68	1.2	F0p	18	Nusakan
4 θ	15	32	55.7	+31	21	32	4.14	−1.1	B5	110	
5 α	15	34	41.2	+26	42	53	2.23v	0.6	A0	24	Alphekka
7 ζ²	15	39	22.6	+36	38	09	4.7	−0.6	B7	130†	
8 γ	15	42	44.5	+26	17	44	3.84v	−0.3	A0	64	
10 δ	15	49	35.6	+26	04	06	4.63	1.8	G5	38	
13 ε	15	57	35.2	+26	52	40	4.15	−0.2	K3	74	

† Wide binary with ζ¹, mag. 6.0. α is an eclipsing binary with a very small range (0m.1).
γ is a δ Scuti variable; range only 0.06 mag.
β is a magnetic spectrum variable.

Also above mag. 5:

	Mag.	Abs. mag.	Spectrum	Dist
2 η	4.98	4.4	G0	14
ζ	5.0+6.0	−0.6	B6+B7	130
17 σ	5.22	4.4	G0	21
14 ι	4.99	−0.6	A0	130
16 τ	4.76	0.2	K0	52
19 ξ	4.85	0.2	K0	85
11 κ	4.82	1.7	K0	33

CORONA BOREALIS (continued)

VARIABLE STARS

	R.A. h	m	Dec. deg	min	Range	Type	Period, d.	Spec.	
U	15	18.2	+31	39	7.7–8.8	Algol	3.45	B + F	
S	15	21.4	+31	22	5.8–14.1	Mira	360.3	M	
R	15	48.6	+28	09	5.7–15	R CrB	—	F8p	
V	15	49.5	+39	34	6.9–12.6	Mira	357.6	N	
T	15	59.5	+25	55	2.0–10.8	Recurrent nova		M+Q	(1866, 1946)
W	16	15.4	+37	48	7.8–14.3	Mira	238.4	M	

DOUBLE STARS

	R.A. h	m	Dec. deg	min	P.A. deg	Sep. sec	Mags.	
o	15	20.1	+29	37	337	147.3	5.5, 9.4	
η	15	23.2	+30	17	AB 030	1.0	5.6, 5.9	Binary, 41.6 y
					AC 012	57.7	12.5⁻	
					AB + D 047	215.0	10.0	
ζ	15	39.4	+36	38	305	6.3	5.1, 6.0	
γ	15	42.7	+26	18	118	0.6	4.1, 5.5	Binary, 91 y
ε	15	57.6	+26	53	003	1.8	4.2, 12.6	
ρ	16	01.0	+33	18	071	89.8	5.5, 8.7	
σ	16	14.7	+33	52	236	7.0	5.6, 6.6	Binary, 1000 y

CORVUS

(Abbreviation: Crv).

An original group. When the god Apollo became enamoured of Coronis, mother of the great doctor Æsculapius, he sent a crow to watch her and report on her behaviour. To be candid, the crow's report was decidedly adverse; but Apollo rewarded the bird with a place in the sky!

Corvus is distinctive, since its leading stars form a quadrilateral. There are four stars above magnitude 4.

	Mag.	Luminosity Sun = 1	Dist lt-yrs
γ	2.59	250	185
β	2.65	600	290
δ	2.95	60	117
ε	3.00	96	104

Curiously, the star lettered α is more than a magnitude fainter than any of these.

See chart for Hydra.

BRIGHTEST STARS

Star	R.A. h	m	s	Dec. deg	min	sec	Mag.	Abs. mag.	Spec.	Dist pc	
1 α	12	08	24.7	−24	43	44	4.02	1.9	F2	21	Alkhiba
2 ε	12	10	07.4	−22	37	11	3.00	−0.1	K2	32	
4 γ	12	15	48.3	−17	32	31	2.59	−1.2	B8	57	Minkar
7 δ	12	29	51.8	−16	30	55	2.95	0.2	B9	36	Algorel
8 η	12	32	04.1	−16	11	46	4.31	1.7	F0	29	
9 β	12	34	23.2	−23	23	48	2.65	−2.1	G5	89	Kraz

VARIABLE STARS

	R.A. h	m	Dec. deg	min	Range	Type	Period, d.	Spec.
R	12	19.6	−19	15	6.7–14.4	Mira	317.0	M
SV	12	49.8	−15	05	6.8–7.6	Semi-reg.	70	M

DOUBLE STAR

	R.A. h	m	Dec. deg	min	P.A. deg	Sep. sec	Mags.
δ	12	29.9	−16	31	214	24.2	3.0, 9.2

CORVUS (continued)

PLANETARY NEBULA

M	C	NGC	R.A. h	m	Dec. deg	min	Diameter sec	Mag.	Mag. of central star
		4361	12	24.5	−18	48	45 × 110	10.3	13.2

GALAXIES

M	C	NGC	R.A. h	m	Dec. deg	min	Mag.	Dimensions min	Type	
	60	4038	12	01.9	−18	52	11.3	2.6 × 1.8	Sc	Antennæ
	61	4039	12	01.9	−18	53	13	3.2 × 2.2	Smp	Antennæ

CRATER

(Abbreviation: Crt).

Like Corvus, a small constellation adjoining Hydra; it has been identified with the wine goblet of Bacchus. The only star above the fourth magnitude is δ.

	Mag.	Luminosity Sun = 1	Dist lt-yrs
δ	3.56	16	72

See chart for Hydra.

BRIGHTEST STARS

Star	R.A. h	m	s	Dec. deg	min	sec	Mag.	Abs. mag.	Spec.	Dist pc	
7 α	10	59	46.4	−18	17	56	4.08	0.2	K0	37	Alkes
11 β	11	11	39.4	−22	49	33	4.48	0.2	A2	72	Al Sharasif
12 δ	11	19	20.4	−14	46	43	3.56	1.8	G8	22	
15 γ	11	24	52.8	−17	41	02	4.08	2.1	A5	24	
21 θ	11	36	40.8	−09	48	08	4.70	0.2	B9	79	
27 ζ	11	44	45.7	−18	21	03	4.73	0.3	G8	74	

Also above mag. 5:

	Mag.	Abs. mag.	Spectrum	Dist
ε	4.83	−0.3	K5	96

DOUBLE STAR

	R.A. h	m	Dec. deg	min	P.A. deg	Sep. sec	Mags.
γ	11	24.9	−17	41	096	5.2	4.1, 9.6

GALAXIES

M	C	NGC	R.A. h	m	Dec. deg	min	Mag.	Dimensions min	Type
		3511	11	03.4	−23	05	11.6	5.4 × 2.2	Sc
		3513	11	03.8	−23	15	12.0	2.8 × 2.3	SBc
		3571	11	11.5	−18	17	12.8	3.3 × 1.3	Sa
		3672	11	25.0	−09	48	11.5	4.1 × 2.1	Sb
		3887	11	47.1	−16	51	11.0	3.3 × 2.7	Sc
		3981	11	56.1	−19	54	12.4	3.9 × 1.5	Sb

CRUX AUSTRALIS

(Abbreviation: Cru).

Though Crux is the smallest constellation in the entire sky, it is also one of the most famous. Before Royer introduced it, in 1679, it had been included in Centaurus. Strictly speaking it is more like a kite than a cross. As well as its brilliant stars it contains the glorious 'Jewel Box' cluster, and also the dark nebula known as the Coal Sack. There are five stars above the fourth magnitude.

	Mag.	Luminosity Sun = 1	Dist lt-yrs
α	0.83	3200 + 2000	360
β	1.25v	8200	425
γ	1.63	120	88
δ	2.80v	1320	260
ε	3.59	80	59

CRUX AUSTRALIS (continued)

Of the four main stars, any casual glance will show the difference between γ, a red giant, and the other three, which are hot and bluish-white.

See chart for Centaurus.

The lovely Jewel Box cluster (κ Crucis) is probably no more than 7 100 000 years old. The three brightest stars are of type B, while the fourth is a red supergiant, mag. 7.56. Close by is the Coal Sack (C.99), 60 to 70 light-years in diameter, and 500 to 600 light-years away.

BRIGHTEST STARS

Star	R.A. h	m	s	Dec. deg	min	sec	Mag.	Abs. mag.	Spec.	Dist pc	
θ^1	12	03	01.6	−63	18	46	4.33	−0.5	A0	17	
θ^2	12	04	19.2	−63	09	56	4.72	−3.0	B2	290	
η	12	06	52.8	−64	36	49	4.15	0.6	F0	33	
δ	12	15	08.6	−58	44	55	2.80	−3.0	B2	79	
ζ	12	18	26.1	−64	00	11	4.04	−2.3	B3	150	
ε	12	21	21.5	−60	24	04	3.59	0.0	K2	18	
α	12	26	35.9	−63	05	56	1.41	−3.9	B1	110	Acrux
	12	26	36.5	−63	05	58	1.88	−3.4	B3		
γ	12	31	09.9	−57	06	47	1.63	−0.5	M3	27	
ι	12	45	37.8	−60	58	52	4.69	0.0	K1	87	
β	12	47	43.2	−59	41	19	1.25v	−5.0	B0	130	(Mimosa)
μ^1	12	54	35.6	−57	10	40	4.03	−2.3	B3	190	

β is variable over a small range (below 0 m.1); βCanis Majoris type.

VARIABLE STARS

	R.A. h	m	Dec. deg	min	Range	Type	Period, d.	Spec.
BH	12	16.3	−56	17	7.2–10.0	Mira	421	S
T	12	21.4	−62	17	6.3–6.8	Cepheid	6.73	F
R	12	23.6	−61	38	6.4–7.2	Cepheid	5.83	F–G
S	12	54.4	−58	26	6.2–6.9	Cepheid	4.69	F–G

DOUBLE STARS

	R.A. h	m	Dec. deg	min	P.A. deg	Sep. sec	Mags.
θ^1	12	03.0	−63	19	325	4.5	4.3, 13.6
η	12	06.9	−64	37	299	44.0	4.2, 11.7
α	12	26.6	−63	06	AB 113	4.0	1.4, 1.9
					AC 202	90.1	1.0, 4.9
γ	12	31.2	−57	07	AB 031	110.6	1.6, 6.7
					AC 082	155.2	9.5
ι	12	45.6	−60	59	022	26.9	4.7, 9.5
μ^1	12	54.6	−57	11	017	34.9	4.0, 5.2

OPEN CLUSTERS

M	C	NGC	R.A. h	m	Dec. deg	min	Diam. min	Mag.	No. of stars	
		4052	12	01.9	−63	12	8	8.8	80	
		4103	12	06.7	−61	15	7	7.4	45	
		4337	12	23.9	−58	08	3.5	8.9	—	
		4349	12	24.5	−61	54	16	7.4	30	
		H.5	12	29.0	−60	46	5	9.0	40	
		4439	12	28.4	−60	06	4	8.4	—	
	98	4609	12	42.3	−62	58	5	6.9	40	
	94	4755	12	53.6	−60	20	10	4.2	50+	Jewel Box (κ Crucis)

DARK NEBULA

M	C	NGC	R.A. h	m	Dec. deg	Dimensions min	Area, sq. deg
	99		12	53	−63	400 × 300	26.2 Coal Sack

CYGNUS

(Abbreviation: Cyg).

Cygnus is one of the richest constellations in the sky; it is often nicknamed the Northern Cross – certainly it is much more nearly cruciform than is Crux. Various legends are associated with it. According to one, the group was placed in the sky to honour a swan into which Jupiter once transformed himself when on a visit to the wife of the King of Sparta!

There are 14 stars above the fourth magnitude. To these must be added the red variable χ, which can rise above magnitude 4 at times.

	Mag.	Luminosity Sun = 1	Dist lt-yrs
α	1.25	70 000	1800
γ	2.20	6000	750
ε	2.46	60	81
δ	2.87	130	160
β	3.08	700	390
ζ	3.20	600	390
ξ	3.72	5000	950
τ	3.72	17	68
κ	3.77	105	170
ι	3.79	11	134
o^1	3.79	650	520
η	3.89	60	170
υ	3.94	50	147
o^2	3.98	1900	910

The immense power of Deneb makes it act upon the nebulosity in its area, exciting it to luminosity. This may well apply to the famous North America Nebula (NGC 7000, C.20), which is about 70 light-years from Deneb.

P Cygni is exceptionally luminous, with an estimated absolute magnitude of −9. Its distance has been given as 5900 light-years (1800 parsecs). When it rose to magnitude 3, in 1600, it was regarded as a nova, but is now classed as a variable of the S Doradus type. For many years its magnitude has hovered between 4.8 and 5.2.

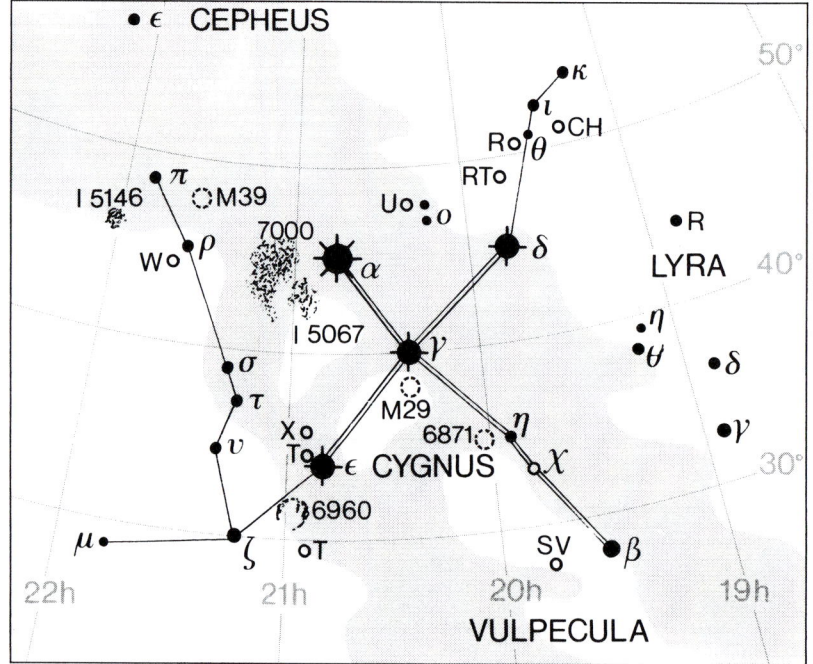

CYGNUS (continued)

BRIGHTEST STARS

Star	R.A. h	m	s	Dec. deg	min	sec	Mag.	Abs. mag.	Spec.	Dist pc	
1 κ	19	17	06.0	+53	22	07	3.77	0.2	K0	52	
10 ι	19	29	42.1	+51	43	47	3.79	2.1	A5	41	
6 β	19	30	43.1	+27	57	35	3.08	−2.3	K5	120	Albireo
8	19	31	46.1	+34	27	11	4.74	−2.3	B3	240	
13 θ	19	36	26.2	+50	13	16	4.48	2.1	F5	19	
12 φ	19	39	22.4	+30	09	12	4.69	1.8	G8	34	
18 δ	19	44	58.4	+45	07	51	2.87	−0.6	A0	49	
χ	19	50	33.7	+32	54	51	var.	0.2v	S	25	
21 η	19	56	18.2	+35	05	00	3.89	0.2	K0	52	
33	20	13	23.7	+56	34	04	4.30	1.3	A3	39	
31 o¹	20	13	37.7	+46	44	29	3.79	−2.2	K2	160	
32 o²	20	15	28.1	+47	42	51	3.98	−3.4	K3	280	
34 P	20	17	47.0	+38	01	59	var.	var.	B1p	6000	
37 γ	20	22	13.5	+40	15	24	2.20	−4.6	F8	230	Sadr
39	20	23	51.4	+32	11	25	4.43	−0.2	K3	78	
41	20	29	23.6	+30	22	07	4.01	−2.0	F5	160	
47	20	33	54.0	+35	15	03	4.61	−4.4	K2	570	
50 α	20	41	25.8	+45	16	49	1.25	−7.5	A2	560	Deneb
52	20	45	39.6	+30	43	11	4.22	0.2	K0	59	
53 ε	20	46	12.5	+33	58	13	2.46	0.2	K0	25	Gieneh
54 λ	20	47	24.3	+36	29	27	4.53	−1.1	B5	130	
58 υ	20	57	10.2	+41	10	02	3.94	0.6	A0	45	
59 f¹	20	59	49.3	+47	31	16	4.74v	−3.9	B1	480	(V.832 Cyg)
62 ξ	21	04	55.7	+43	55	40	3.72	−4.4	K5	290	
63	21	06	35.9	+47	38	54	4.55	−2.3	K4	220	
64 ζ	21	12	56.0	+30	13	37	3.20	−2.1	G8	120	
65 τ	21	14	47.3	+38	02	44	3.72	1.7	F0	21	
67 σ	21	17	24.7	+39	23	41	4.23	−7.1	B9	1600	
66 υ	21	17	54.9	+34	53	48	4.43	−2.5	B2	240	
73 ρ	21	33	58.7	+45	35	30	4.02	0.3	G8	56	
80 π¹	21	42	05.5	+51	11	23	4.67	−1.7	B3	180	Azelfafage
78 {μ²	21	44	08.2	+28	44	35	6.14	3.7	dF3 }	17	Combined mag. 4.4.
{μ¹	21	44	08.4	+28	44	34	4.78		F6 }		
81 π²	21	46	47.4	+49	18	35	4.23	−2.9	B3	250	

Also above mag. 5:

	Mag.	Abs. mag.	Spec.	Dist	
29 b³	4.97	1.4	A2	33	
30	4.83	0.0	A3	92	
28 b²	4.93	−1.7	B3	210	
22	4.94	−1.9	B6	220	
24 ψ	4.92	1.3	A3	52	
17	4.99		F5		
15	4.89	0.3	G8	83	
2	4.97	−2.3	B3	260	
45 ω¹	4.95	−2.5	B2	270	Ruchba
55	4.84	−6.8	B3	1000	
57	4.78	−1.1	B5	150	
68	5.00		O8		
72	4.90	0.0	K1	96	
34 P	4.8v	−9	B2p	1800	

DOUBLE STARS

	R.A. h	m	Dec. deg	min	P.A. deg	Sep. sec	Mags.	
β	19	30.7	+27	58	054	34.4	3.1, 5.1	Yellow, blue
δ	19	45.0	+45	07	226	2.5	2.9, 6.3	Binary, 828 y
ψ	19	55.6	+52	26	178	3.2	4.9, 7.4	
γ	20	22.2	+40	15	196	41.2	2.2, 9.9	B is a close dble
61	21	06.9	+38	45	150	30.3	5.2, 6.0	Binary 653 y
τ	21	14.8	+38	03	015	0.5	3.8, 6.4	Binary, 50 y
μ	21	44.1	+28	45	206	2.0	4.8, 6.1	Binary, 713 y

CYGNUS (continued)

VARIABLE STARS

	R.A.		Dec.		Range	Type	Period, d.	Spec.
	h	m	deg	min				
CH	19	24.5	+50	14	6.4–8.7	Z Andromedæ	±97	M + B
R	19	36.8	+50	12	6.1–14.2	Mira	426.4	M
RT	19	43.6	+48	47	6.4–12.7	Mira	190.2	M
SU	19	44.8	+29	16	6.5–7.2	Cepheid	3.84	F
χ	19	50.6	+32	55	3.3–14.2	Mira	406.9	S
Z	20	01.4	+50	03	7.4–14.7	Mira	263.7	M
RS	20	13.4	+38	44	6.5–9.3	Semi-reg.	417	N
P	20	17.8	+38	02	3–6	S Dor.	—	B2p
CN	20	17.9	+59	48	7.3–14.0	Mira	198.5	M
U	20	19.6	+47	54	5.9–12.1	Mira	462.4	N
V	20	41.3	+48	09	7.7–13.9	Mira	421.4	N
X	20	43.4	+35	35	5.9–6.9	Cepheid	16.39	F–G
T	20	47.2	+34	22	5.0–5.5	Lb?	—	K
W	21	36.0	+45	22	5.0–7.6	Semi-reg.	126	M
SS	21	42.7	+43	35	8.4–12.4	SS Cygni	±50	A–G
WY	21	48.7	+44	15	7.5–14·0	Mira	304.5	M

OPEN CLUSTERS

M	C	NGC	R.A.		Dec.		Diam.	Mag.	No. of stars	
			h	m	deg	min	min			
		6811	19	38.2	+46	34	13	6.8	70	
		6819	19	41.3	+40	11	5	7.3	—	
		6834	19	52.2	+29	25	5	7.8	50	
		6866	20	03.7	+44	00	7	7.6	80	
		6871	20	05.9	+35	47	20	5.2	15	27 Cygni
		6910	20	23.1	+40	47	8	7.4	50	
29		6913	20	23.9	+38	32	7	6.6	50	
		6939	20	31.4	+60	38	8	7.8	80	
		7067	21	24.2	+48	01	3	9.7	20	
39		7092	21	32.2	+48	26	32	4.6	30	

PLANETARY NEBULÆ

M	C	NGC	R.A.		Dec.		Dimensions	Mag.	Mag. of central star	
			h	m	deg	min	sec			
		6826	19	44.8	+50	31	30 × 140	9.8	10.4	Blinking Nebula
		7048	21	14.2	+46	16	61	11.3	18	

NEBULÆ

M	C	NGC	R.A.		Dec.		Dimensions	Mag. of illuminating star	
			h	m	deg	min	min		
	27	6888	20	12.0	+38	21	20 × 10	7.4	Crescent Nebula
	34	6960	20	45.7	+30	43	70 × 6	—	Filamentary Nebula, 52 Cygni
		IC 5067/70	20	50.8	+44	21	80 × 70	—	Pelican Nebula
	33	6992/5	20	56.4	+31	43	60 × 8	—	Veil Nebula: SNR
	20	7000	20	58.8	+44	20	120 × 100	6	North America Nebula
	19	IC 5146	21	53.5	+47	16	12 × 12	10	Cocoon Nebula, with sparse cluster

DELPHINUS

(Abbreviation: Del).

A small but compact constellation; one of Ptolemy's originals. It honours the dolphin which carried the great singer Arion to safety, after he had been thrown overboard by the crew of the ship carrying him home after winning all the prizes in a competition. The curious names of α and β were allotted by one Nicolaus Venator, for reasons which are obvious!

There are three stars above the fourth magnitude.

	Mag.	Luminosity Sun = 1	Dist lt-yrs
β	3.54	46	108
α	3.77	60	170
γ	3.9 (4.3 + 5.2)	4.5	75

See chart for Aquila.

The proper motion of ρ Aquilæ (+0.004″ in R.A., +0.06″ in dec) has now carried it into Delphinus. It was also listed as 67 Aquilæ.

DELPHINUS (continued)

BRIGHTEST STARS

Star	R.A. h	m	s	Dec. deg	min	sec	Mag.	Abs. mag.	Spec.	Dist pc	
2 ε	20	33	12.6	+11	18	12	4.03	−1.9	B6	150	Deneb Dulfine
4 ζ	20	35	18.4	+14	40	27	4.68	1.7	A3	38	
6 β	20	37	32.8	+14	35	43	3.54	0.7	F5	33	Rotanev
9 α	20	39	38.1	+15	54	43	3.77	0.2	B9	52	Svalocin
11 δ	20	43	27.3	+15	04	28	4.43	0.5	A7	56	
12 γ	20	46	39.3	+16	07	27	3.9	3.2	G5+F8	23	

VARIABLE STARS

	R.A. h	m	Dec. deg	min	Range	Type	Period, d.	Spec.
R	20	14.9	+09	05	7.6–13.8	Mira	284.9	M
EU	20	37.9	+18	16	5.8–6.9	Semi-reg.	59	M
HR	20	42.3	+19	10	3.7–12.7	Nova	—	Q
U	20	45.5	+18	05	5.7–7.6	Semi-reg.	110	M
S	20	43.1	+17	05	8.3–12.4	Mira	277.2	M
V	20	47.8	+19	20	8.1–16.0	Mira	533.5	M

DOUBLE STARS

	R.A. h	m	Dec. deg	min	P.A. deg	Sep. sec	Mags.	
1	20	30.3	+10	54	AB 346	0.9	6.1, 8.1	
					AC 349	16.8	14.1	
β	20	37.5	+14	36	167	0.3	4.0, 4.9	Binary, 26.7 y
α	20	39.6	+15	55	AB 224	29.5	3.8, 13.3	
					AC 272	43.4	11.8	
K	20	39.1	+10	05	286	28.8	5.1, 11.7	
γ	20	46.7	+16	07	206	9.2	4.5, 5.5	
13	20	47.8	+06	00	194	1.6	5.6, 9.2	

GLOBULAR CLUSTERS

M	C	NGC	R.A. h	m	Dec. deg	min	Diameter min	Mag.
	47	6394	20	34.2	+07	24	5.9	8.9
	42	7006	21	01.5	+16	11	2.8	10.6

PLANETARY NEBULA

M	C	NGC	R.A. h	m	Dec. deg	min	Dimensions sec	Mag.	Mag. of central star
		6891	20	15.2	+12	42	12 × 74	11.7	12.4

DORADO

(Abbreviation: Dor).

A 'modern' southern constellation. The only star above the fourth magnitude is α, but Dorado contains part of the Large Magellanic Cloud (LMC), in which is the magnificent Tarantula looped nebula round 30 Doradûs.

	Mag.	Luminosity Sun = 1	Dist lt-yrs
α	3.27	130	192

See chart for Reticulum.

γ is the prototype star of a class of variables of very low amplitude. Its fluctuations are much too small to be detected without photometric measurements.

BRIGHTEST STARS

Star	R.A. h	m	s	Dec. deg	min	sec	Mag.	Abs. mag.	Spec.	Dist pc
γ	04	16	01.6	−51	29	12	4.25	2.6	F0	45
α	04	33	59.8	−55	02	42	3.27	−0.6	A0	59
ζ	05	05	30.6	−57	28	22	4.72	4.0	F8	13
β	05	33	37.5	−62	29	24	var.	−8.0	F9v	2300
δ	05	44	46.5	−65	44	08	4.35	2.4	A7	450
HD 40409	05	54	06.1	−63	05	23	4.65	1.4	K0	13

DORADO (continued)

Also above mag. 5:

	Mag.	Abs. mag.	Spec.	Dist
θ	4.83	−0.1	K2	80

VARIABLE STARS

	R.A. h	m	Dec. deg	min	Range	Type	Period, d.	Spec.
R	04	36.8	−62	05	4.8–6.6	Semi-reg.	338	M
β	05	33.6	−62	29	3.7–4.1	Cepheid	9.84	F–G

DOUBLE STAR

	R.A. h	m	Dec. deg	min	P.A. deg	Sep. sec	Mags.
α	04	34.0	−55	03	AB 182	0.2	3.8, 4.3
					AB + C 101	77.7	9.8

GALAXIES

M	C	NGC	R.A. h	m	Dec. deg	min	Mag.	Dimensions min	Type
		1549	04	15.7	−55	36	9.9	3.7 × 3.2	E0
		1553	04	16.2	−55	47	9.5	4.1 × 2.8	S0
		1596	04	27.6	−55	02	11.0	3.9 × 1.2	S0
		1617	04	31.7	−54	36	10.4	4.7 × 2.4	SBa
		1672	04	45.7	−59	15	11.0	4.8 × 3.9	SBb
		LMC	05	23.6	−69	45	0.1	650 × 550	Large Cloud of Magellan. Contains 30 Doradûs and three planetary nelbulæ, NGC 1714, 1722 and 1743.
		1947	05	26.8	−63	46	10.8	3.0 × 1.6	SOp

NEBULA

M	C	NGC	R.A. h	m	Dec. deg	min	Dimensions min	
	103	2070	05	38.7	−69	06	40 × 25	30 Doradûs. In the LMC; Tarantula Nebula.

DRACO

(Abbreviation: Dra).

A long, sprawling northern group. In mythology it has been identified either with the dragon killed by Cadmus before the founding of the city of Bœotia, or with the dragon which guarded the golden apples in the Garden of the Hesperides. Thuban (α) was the Pole Star in ancient times.

There are 12 stars above the fourth magnitude.

	Mag.	Luminosity Sun = 1	Dist lt-yrs
γ	2.23	110	101
η	2.74	110	81
β	2.79	600	270
δ	3.07	60	117
ζ	3.17	500	316
ι	3.29	96	156
χ	3.57	2	25
α	3.65	130	230
χ	3.75	96	189
ε	3.83	58	166
λ	3.84	115	212
κ	3.87	60	72

DRACO (continued)

BRIGHTEST STARS

Star	R.A. h	m	s	Dec. deg	min	sec	Mag.	Abs. mag.	Spec.	Dist pc	
HD											
81817	09	37	05.2	+81	19	35	4.29	−0.2	K3	57	
1 λ	11	31	24.2	+69	19	52	3.84	−0.4	M0	65	Giansar
5 κ	12	33	28.9	+69	47	17	3.87	0.2	B7	22	
10 i	13	51	25.8	+64	43	33	4.66	−0.5	gM3	110	
11 α	14	04	23.2	+64	22	33	3.65	−0.6	A0	71	Thuban
12 ι	15	24	55.6	+58	57	58	3.29	−0.1	K2	48	Edasich
13 θ	16	01	53.2	+58	33	55	4.01	3.2	F8	16	
14 η	16	23	59.3	+61	30	50	2.74	0.3	G8	25	Aldhibain
22 ζ	17	08	47.0	+65	42	53	3.17	−1.9	B6	97	Aldhibain
23 β	17	30	25.8	+52	18	05	2.79	−2.1	G2	82	Alwaid
31 ψ	17	41	56.1	+72	08	56	4.58	2.8	F5	23 ⌉	Dziban
	17	41	57.7	+72	09	24	5.79	4.0	F0	23 ⌋	
32 ξ	17	53	31.5	+56	52	21	3.75	−0.1	K2	58	Juza
33 γ	17	56	36.2	+51	29	20	2.23	−0.3	K5	31	Eltamin
43 φ	18	20	45.2	+71	20	16	4.22	−0.6	A0p	33	
44 χ	18	21	03.0	+72	43	58	3.57	4.1	F7	7.8	
47 o	18	51	12.0	+59	23	18	4.66	−0.9	K0	67	
57 δ	19	12	33.1	+67	39	41	3.07	0.2	G9	36	Taïs
60 τ	19	15	32.8	+73	21	24	4.45	−0.2	K3	70	
58 π	19	20	39.9	+65	42	52	4.59	0.6	A2	63	
61 σ	19	32	21.5	+69	39	40	4.68	5.9	K0	5.7	Alrakis
63 ε	19	48	10.2	+70	16	04	3.83	0.3	G8	51	Tyl
67 ρ	20	02	48.9	+67	52	25	4.51	−0.2	K3	82	

Also above mag. 5:

	Mag.	Abs. mag.	Spec.	Dist	
42	4.82	−0.1	K2	96	
39	4.98	1.2	A1	53	
28 ω	4.80	3.4	F5	22	Al Dhih
24 ν¹	4.88		A8	62 ⌉	
25 ν²	4.87		A4	62 ⌋	Kuma
21 μ	4.92		F5		
4	4.95	−0.5	M4	120	
6	4.94	−0.1	K2	84	
18 g	4.83	−4.5	K1		
19	4.89	3.7	F6	17	
45	4.77	−4.6	F7	450	
52 υ	4.82	0.2	K0	64	
54	4.99	−0.1	K2	100	

VARIABLE STARS

	R.A. h	m	Dec. deg	min	Range	Type	Period, d.	Spec.
RY	12	56.4	+66	00	5.6–8.0	Semi-reg.	173	N
R	16	32.7	+66	45	6.7–13.0	Mira	245.5	M
T	17	56.4	+58	13	7.2–13.5	Mira	421.2	N
UW	17	57.5	+54	40	7.0–8.0	Irreg.	—	K
UX	19	21.6	+76	34	5.9–7.1	Semi-reg.	168	N

DOUBLE STARS

	R.A. h	m	Dec. deg	min	P.A. deg	Sep. sec	Mags.	
η	16	24.0	+61	31	142	5.2	2.7, 8.7	
μ	17	05.3	+54	28	017	2.0	5.7, 5.7	Binary, 482 y
ν	17	32.2	+55	11	312	61.9	4.9, 4.9	
ψ	17	41.9	+72	09	015	30.3	4.9, 6.1	
ε	19	48.2	+70	16	019	3.1	3.8, 7.4	Slow binary

DRACO (continued)

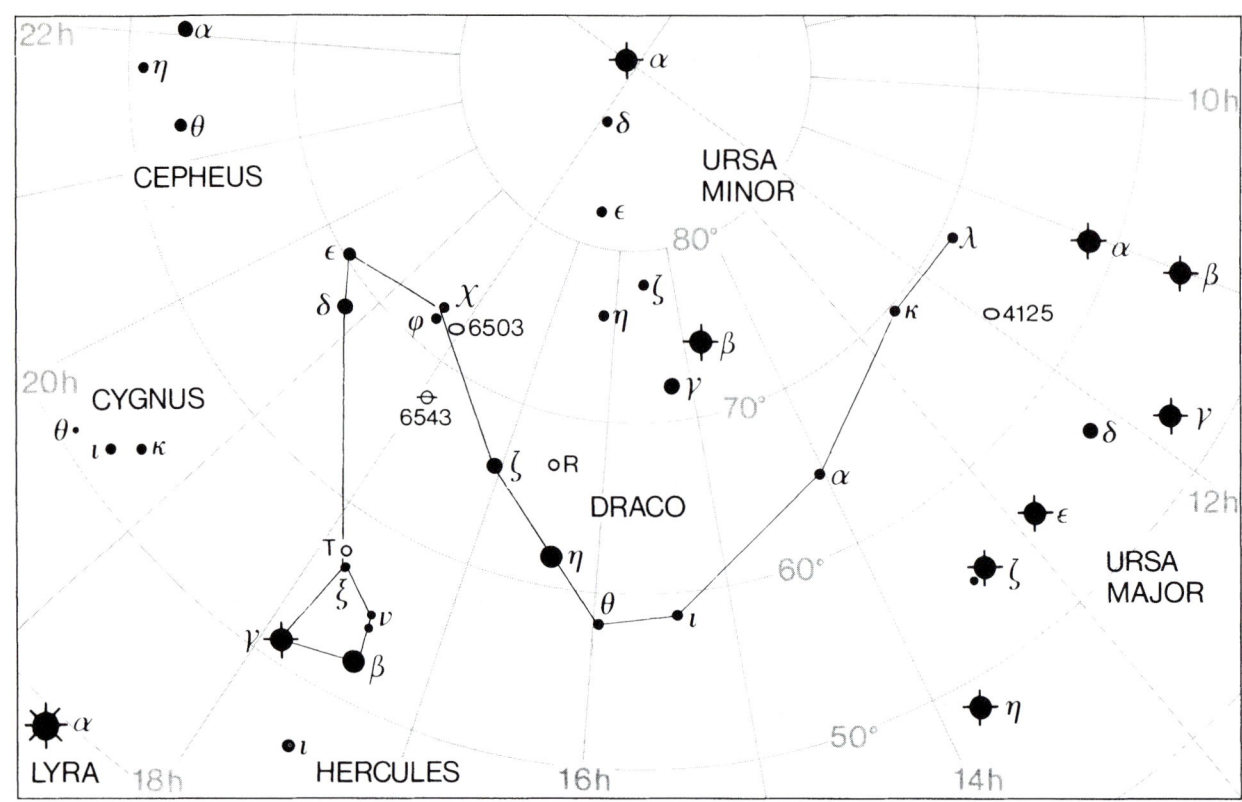

PLANETARY NEBULA

M	C	NGC	R.A.		Dec.		Dimensions	Mag.	Mag. of
			h	m	deg	min	sec		central star
	6	6543	17	58.6	+66	38	18 × 350	8.8	9.5

GALAXIES

M	C	NGC	R.A.		Dec.		Mag.	Dimensions	Type
			h	m	deg	min		min	
		3147	10	16.9	+73	24	10.6	4.0 × 3.5	Sb
		4125	12	08.1	+65	11	9.8	5.1 × 3.2	E5p
	3	4236	12	16.7	+69	28	9.7	18.6 × 6.9	Sb
		5866	15	06.5	+55	46	10.0	5.2 × 2.3	E6p
		5879	15	09.8	+57	00	11.5	4.4 × 1.7	Sb
		5907	15	15.9	+56	19	10.4	12.3 × 1.8	Sb
		5985	15	39.6	+59	20	11.0	5.5 × 3.2	Sb
		6015	15	51.4	+62	19	11.2	5.4 × 2.3	Sc
		5907	15	15.9	+56	19	10.4	12.3 × 1.8	Sb
		5985	15	39.6	+59	20	11.0	5.5 × 3.2	Sb
		6503	17	49.4	+70	09	10.2	6.2 × 2.3	Sb

EQUULEUS

(Abbreviation: Eql).

A very obscure and small constellation, but one of the 'originals'. It represents a foal given by Mercury to Castor, one of the Heavenly Twins. There is only one star above the fourth magnitude: α. The name often applied to α has a decidedly modern flavour!

	Mag.	Luminosity Sun = 1	Dist lt-yrs
α	3.92	50	150

See chart for Aquila.

BRIGHTEST STARS

Star	R.A. h	m	s	Dec. deg	min	sec	Mag.	Abs. mag.	Spec.	Dist pc	
5 γ	21	10	20.3	+10	07	53	4.69	1.0	F0p	64	
7 δ	21	14	28.7	+10	00	25	4.49	4.0	F8	15	
8 α	21	15	49.3	+05	14	52	3.92	0.6	G0	46	Kitalpha

VARIABLE STAR

	R.A. h	m	Dec. deg	min	Range	Type	Period, d.	Spec.
S	20	57.2	+05	05	8.0–10.1	Algol	3.44	B + F

DOUBLE STARS

	R.A. h	m	Dec. deg	min	P.A. deg	Sep. sec	Mags.	
ε	20	59.1	+04	18	AB 285	1.0	6.0, 6.3	Binary, 101.4 y
					AB + C 070	10.7	7.1	
					AD 280	74.8	2.4	
γ	21	10.4	+10	08	AB 268	1.9	4.7, 11.5	
					AC 005	47.7	12.5	
δ	21	14.5	+10	00	029	0.3	5.2, 5.3	Binary, 5.7 y
β	21	22.9	+06	49	257	34.4	5.2, 13.7	

ERIDANUS

(Abbreviation: Eri).

An immensely long constellation, extending from Achernar in the far south as far as Kursa, near Orion. Achernar is the only brilliant star. Mythologically, Eridanus is the river Po – and this was the river into which the youth Phæthon was plunged when he had obtained permission to drive the Sun-chariot for a day, and had lost control of it, so that Jupiter was forced to strike him down with a thunderbolt.

Achernar is the closest really brilliant star to the south celestial pole. The pole lies in a very barren area, around midway between Achernar and the Southern Cross.

There are 15 stars above magnitude 4.

	Mag.	Luminosity Sun = 1	Dist lt-yrs
α	0.46	400	85
β	2.79	82	100
θ	2.92	50+17	55
γ	2.95	110	144
δ	3.54	3	29
υ^4	3.56	82	130
ϕ	3.56	105	120
τ^4	3.69	120	225
χ	3.70	4.5	49
ε	3.73	0.3	10.7
υ^2	3.82	60	173
53	3.87	96	144
η	3.89	37	75
υ	3.93	2200	1140
υ^3	3.96	120	255

ERIDANUS (continued)

θ Eridani (Acamar) was rated of the 1st magnitude by Ptolemy, but there may be confusion with Achernar, and any real diminution seems unlikely. Acamar is a fine double; the two components certainly make up a binary system, but the period must be of the order of several thousands of years. it has been suggested that A may itself be a spectroscopic binary.

ε Eridani is one of the two closest solar-type stars (τ Ceti is the other), and is a favourite candidate as a centre of a planetary system, though proof is lacking.

BRIGHTEST STARS

Star	R.A. h	m	s	Dec. deg	min	sec	Mag.	Abs. mag.	Spec.	Dist pc	
α	01	37	42.9	−57	14	12	0.46	−1.6	B5	26	Achernar.
χ	01	55	57.5	−51	36	32	3.70	3.2	G5	15	
κ	02	26	59.1	−47	42	14	4.25	−2.2	B5	190	
ϕ	02	16	30.6	−51	30	44	3.56	−0.2	B8	37	
ι	02	40	40.0	−39	51	19	4.11	0.2	K0	61	
1 τ^1	02	45	06.1	−18	34	21	4.47	3.7	F6	15	
2 τ^2	02	51	02.2	−21	00	15	4.75	0.2	K0	81	
3 η	02	56	25.6	−08	53	54	3.89	0.9	K1	23	Azha
θ	02	58	15.6	−40	18	17	2.92	0.6 + 1.7	A3 + A2	17	Acamar
11 τ^3	03	02	23.4	−23	37	28	4.09	2.1	A5	23	
13 ζ	03	15	49.9	−08	49	11	4.80	1.8	A3	16	Zibal
16 τ^4	03	19	30.9	−21	45	28	3.69	−0.5	gM3	69	Angetenar
e	03	19	55.7	−43	04	10	4.27	5.3	G5	6.2	
17 v	03	30	37.0	−05	04	30	4.73	−0.2	B8	97	
18 ε	03	32	55.8	−09	27	30	3.73	6.1	K2	3.3	
19 τ^5	03	33	47.2	−21	37	58	4.27	−0.2	B8	78	
y	03	37	05.6	−40	16	29	4.58	0.2	K0	89	
h	03	42	50.0	−37	18	49	4.59	−0.3	gK5	95	
δ	03	43	14.8	−09	45	48	3.54	3.8	K0	9	Rana
26 π	03	46	08.4	−12	06	06	4.42	−1.1	gMa	50	
27 τ^6	03	46	50.8	−23	14	59	4.23	3.1	F3	17	
g	03	48	35.3	−37	37	20	4.27	0.6	A0	50	
33 τ^8	03	53	42.6	−24	36	45	4.65	−1.1	B5	140	
32 w	03	54	17.4	−02	57	17	4.46	0.3	G8	68	
34 γ	03	58	01.7	−13	30	31	2.95	−0.4	M0	44	Zaurak
36 τ^9	03	59	55 4	−24	00	59	4.66	−0.4	A0	110	
38 o^1	04	11	51.8	−06	50	15	4.04	−0.7	F2	85	Beid
40 o^2	04	51	16.2	−07	39	10	4.43	6.0	K1	4.9	Keid
41 v^4	04	17	53.6	−33	47	54	3.56	0.0	B8.5	40	
43 v^3	04	24	02.1	−34	01	01	3.96	−0.5	M1	78	
50 v^1	04	33	30.6	−29	46	00	4.51	0.3	gG6	63	
52 v^2	04	35	33.0	−30	33	45	3.82	0.2	K0	53	Theemini
48 v	04	36	19.1	−03	21	09	3.93	−3.6	B2	350	
53 I	04	38	10.7	−14	18	15	3.87	−0.1	K2	44	Sceptrum
54	04	40	26.4	−19	40	18	4.32	−0.5	gM4	92	
57 μ	04	45	30.1	−03	15	17	4.02	−1.6	B5	130	
61 ω	04	52	53.6	−05	27	10	4.39	1.6	A9	36	
67 β	05	07	50.9	−05	05	11	2.79	0.0	A3	28	Kursa
69 λ	05	09	08.7	−08	45	15	4.27	−3.0	B2	280	

Also above mag. 5:

	Mag.	Abs. mag.	Spec.	Dist
45	4.91	−1.3	K3	170
39	4.87	−0.2	K3	100
15	4.88	0.3	G6	82
64 S	4.79	1.7	F0	42
65 ψ	4.81	−2.3	B2	290

VARIABLE STARS

	R.A. h	m	Dec. deg	min	Range	Type	Period, d.	Spec.
Z	02	47.9	−12	28	7.0–8.6	Semi-reg.	80	M
RR	02	52.2	−08	16	7.4–8.6	Semi-reg.	97	M
T	03	55.2	−24	02	7.4–13.2	Mira	252.2	M
W	04	11.5	−25	08	7.5–14.5	Mira	376.7	M
RZ	04	43.8	−10	41	7.8–8.7	Algol	39.28	A + G

ERIDANUS (continued)

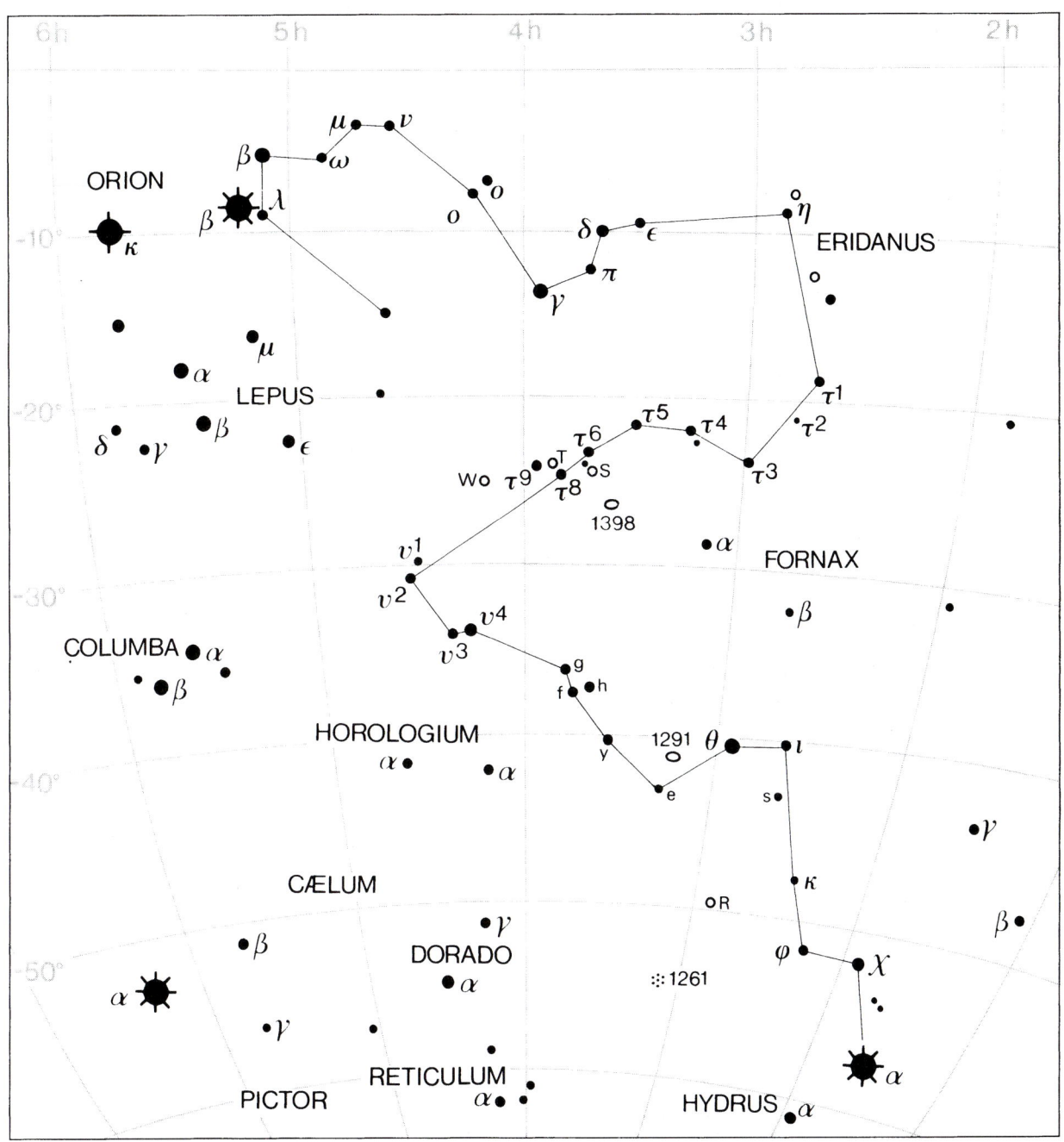

ERIDANUS (continued)

DOUBLE STARS

	R.A.		Dec.		P.A.	Sep.	Mags.	
	h	m	deg	min	deg	sec		
χ	01	56.0	−51	37	202	5.0	3.7, 10.7	
P	01	39.8	−56	12	194	11.2	5.5, 5.8	Binary, 484 y
θ	02	58.3	−40	18	088	8.2	3.4, 4.5	
τ^4	03	19.5	−21	45	AB 288	5.7	3.7. 9.2	
					AC 112	39.2	10.7	
υ^4	04	17.9	−33	48	013	49.2	3.6, 11.8	A is a close
ρ^2	03	02.7	−07	41	075	1.8	5.3, 9.5	
o^2	04	15.2	−07	39	AB 104	83.4	4.4, 9.5	Binary, 248 y
					BC 279	7.6	9.5, 11.8	

PLANETARY NEBULA

M	C	NGC	R.A.		Dec.		Dimensions	Mag.	Mag. of
			h	m	deg	min	sec		central star
		1535	04	14.2	−12	44	18 × 44	9.6	12.2

GALAXIES

M	C	NGC	R.A.		Dec.		Mag.	Dimensions	Type
			h	m	deg	min		min	
		1084	02	46.0	−07	35	10.6	2.9 × 1.5	Sc
		1179	03	02.6	−18	54	11.8	4.6 × 3.9	Sp
		1187	03	02.6	−22	52	10.9	5.0 × 4.1	SBc
		1291	03	17.3	−41	08	8.5	10.5 × 9.1	SBa
		1300	03	19.7	−19	25	10.4	6.5 × 4.3	SBb
		1332	03	26.3	−21	20	10.3	4.6 × 1.7	E7
		1337	03	28.1	−08	23	11.7	6.8 × 2.0	S
		1395	03	38.5	−23	02	11.3	3.2 × 2.5	E3
		1407	03	40.2	−18	35	9.8	2.5 × 2.5	E0
		1532	04	12.1	−32	52	11.1	5.6 × 1.8	Sb
		1637	04	41.5	−02	51	10.9	3.3 × 2.9	Sc

FORNAX

(Abbreviation: For).

A southern group, originally Fornax Chemica (the Chemical Furnace). It has no bright stars, but is notable for containing a large number of faint galaxies.

The only star above the fourth magnitude is α.

	Mag.	Luminosity Sun = 1	Dist lt-yrs
α	3.87	4	46

See chart for Eridanus.

BRIGHTEST STARS

Star	R.A.			Dec.			Mag.	Abs. mag.	Spec.	Dist pc
	h	m	s	deg	min	sec				
υ	02	04	29.4	−29	17	49	4.69	−0.6	A0	110
β	02	49	05.4	−32	24	22	4.46	0.3	G6	61
α	03	12	04.2	−28	59	13	3.87	3.3	F8	14

Also above mag. 5:

	Mag.	Abs. mag.	Spec.	Dist.
δ	5.00		B5	
ω	4.90	0.2	B9	

VARIABLE STARS

	R.A.		Dec.		Range	Type	Period, d.	Spec.
	h	m	deg	min				
R	02	29.3	−26	06	7.5–13.0	Mira	387.9	N
ST	02	44.4	−29	12	7.7–9.0	Semi-reg.	277	M
S	03	46.2	−24	24	?5.6–8.5?	Suspected	?	M

FORNAX (continued)

DOUBLE STARS

	R.A. h	m	Dec. deg	min	P.A. deg	Sep. sec	Mags.	
ω	02	33.8	−28	14	244	10.8	5.0, 7.7	
γ^1	02	49.8	−24	34	AB 145	12.0	6.1, 12.5	
					AC 143	40.9	10.5	
η^2	02	50.2	−35	51	014	5.0	5.9, 10.1	
α	03	12.1	−28	59			4.0, 7.0	Binary, 314 y
χ^3	03	28.2	−35	51	248	6.3	6.5, 10.5	

PLANETARY NEBULA

M	C	NGC	R.A. h	m	Dec. deg	min	Diameter sec	Mag.	Mag. of central star
		1360	03	33.3	−25	51	390	—	11.3

GALAXIES

M	C	NGC	R.A. h	m	Dec. deg	min	Mag.	Dimensions min	Type
		986	02	33.6	−39	02	11.0	3.7 × 2.8	SBb
	67	1097	02	46.3	−30	17	9.2	9.3 × 6.6	SBb
		1201	03	04.1	−26	04	10.6	4.4 × 2.8	Sa
		1255	03	13.5	−25	44	11.1	4.1 × 2.8	Sa
		1302	03	19.9	−26	04	11.5	4.4 × 4.2	SBa
		1316	03	22.7	−37	12	8.8	7.1 × 5.5	SB0p
		1326	03	23.9	−36	28	10.5	4.0 × 3.0	SB0
		1344	03	28.3	−31	04	10.3	3.9 × 2.3	E3
		1350	03	31.1	−33	38	10.5	4.3 × 2.4	SBb
		1365	03	33.6	−36	08	9.5	9.8 × 5.5	SBb
		1371	03	35.0	−24	56	11.5	5.4 × 4.0	SBa
		1380	03	36.5	−34	59	11.1	4.9 × 1.9	S0
		1385	03	37.5	−24	30	11.2	3.0 × 2.0	Sc
		1399	03	38.5	−35	27	9.9	3.2 × 3.1	Elp
		1398	03	38.9	−26	20	9.7	6.6 × 5.2	SBb
		1404	03	38.9	−35	35	10.2	2.5 × 2.3	E1
		1425	03	42.2	−29	54	11.7	5.4 × 2.7	Sb

GEMINI

(Abbreviation: Gem).

A brilliant Zodiacal constellation. In mythology, Castor and Pollux were the twin sons of the King and Queen of Sparta. Pollux was immortal, but Castor was not. When Castor was killed, Pollux pleaded with the gods to be allowed to share his immortality; so Castor was brought back to life, and both youths placed in the sky.

Today Pollux is the brighter of the two stars, but in ancient times Castor was recorded as being the more brilliant. If any change has occurred (and this is by no means certain) it is more likely to have been in the late-type Pollux than in Castor, which is a multiple system. There are 13 stars in Gemini above the fourth magnitude.

	Mag.	Luminosity Sun = 1	Dist lt-yrs
β	1.14	60	36
α	1.58	45	52
γ	1.93	82	85
μ	2.88v	120	230
ε	2.98	5200	680
η	3.1 (max)	120v	186
ξ	3.36	46	75
δ	3.53	14	59
κ	3.57	58	147
λ	3.58	17	81
θ	3.60	82	166
ζ	3.7 (max)	5200	1400
ι	3.79	60	163

GEMINI (continued)

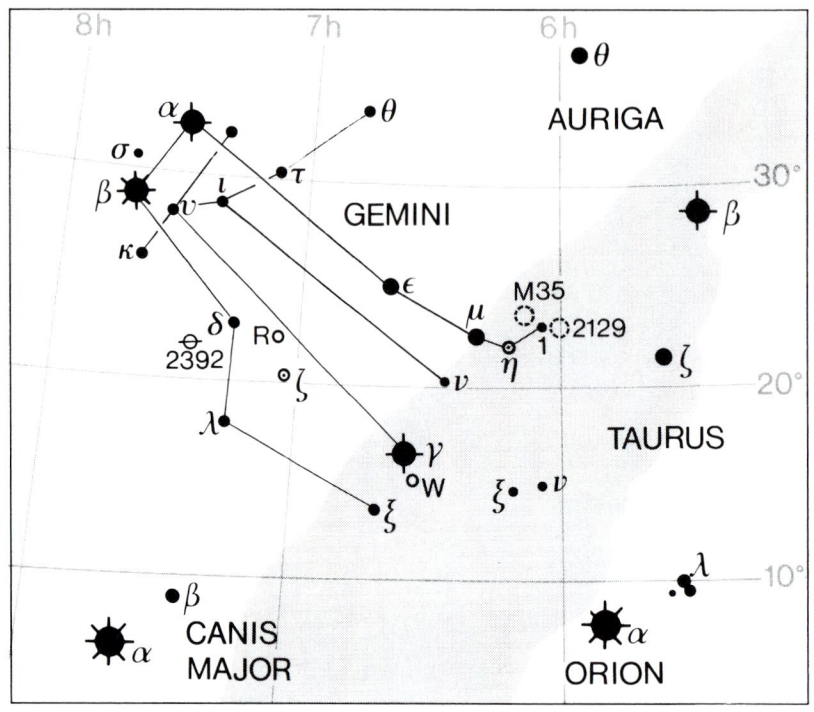

BRIGHTEST STARS

Star	R.A.			Dec.			Mag.	Abs. mag.	Spec.	Dist pc	
	h	m	s	deg	min	sec					
1	06	04	07.2	+23	15	48	4.16	0.3	gG5	59	
7 η	06	14	52.6	+22	30	24	var.	−0.5v	M3	57	Propus
13 μ	06	22	57.6	+22	30	49	2.88v	−0.5	M3	71	Tejat
18 ν	06	28	57.7	+20	12	43	4.15	−1.0	B7	110	
24 γ	06	37	42.7	+16	23	57	1.93	0.0	A0	26	Alhena
27 ε	06	43	55.9	+25	07	52	2.98	−4.5	G8	210	Mebsuta
30	06	43	59.2	+13	13	40	4.49	0.0	K1	70	
31 ξ	06	45	17.3	+12	53	44	3.36	0.7	F5	23	Alzirr
34 θ	06	52	47.3	+33	57	40	3.60	0.0	A3	51	
38 e	06	54	38.6	+13	10	40	4.65	2.6	F0	26	
43 ζ	07	04	06.5	+20	34	13	var.	−4.5	G0	430	Mekbuda
46 τ	07	11	08.3	+30	14	43	4.41	−0.1	K2	67	
54 λ	07	18	05.5	+16	32	25	3.58	1.7	A3	25	
55 δ	07	20	07.3	+21	58	56	3.53	1.9	F2	18	Wasat
60 ι	07	25	43.5	+27	47	53	3.79	0.2	K0	50	
62 ρ	07	29	06.6	+31	47	03	4.18	2.6	F0	19	
66 α	07	34	35.9	+31	53	18	1.58	1.2	A0	14	Castor
69 υ	07	35	55.3	+26	53	45	4.06	−0.4	M0	78	
75 σ	07	43	18.7	+28	53	01	4.28	0.0	K1	40	
77 κ	07	44	26.8	+24	23	52	3.57	0.3	G8	45	
78 β	07	45	18.9	+28	01	34	1.14	0.2	K0	11	Pollux

Also above mag. 5:

	Mag.	Abs. mag.	Spec.	Dist
51	5.00	−0.5	M4	120
71 o	4.90	0.6	F3	70
81	4.88	−0.3	K5	110
83 φ	4.97	1.7	A3	45

GEMINI (continued)

VARIABLE STARS

	R.A.		Dec.		Range	Type	Period, d.	Spec.
	h	m	deg	min				
BU	06	12.3	+22	54	5.7–7.5	Irreg.	—	M
η	06	14.9	+22	30	3.2–3.9	Semi-reg.	233	M
W	06	35.0	+15	20	6.5–7.4	Cepheid	7.91	F–G
X	06	47.1	+30	17	7.5–13.6	Mira	263.7	M
ζ	07	04.1	+20	34	3.7–4.1	Cepheid	10.15	F–G
R	07	07.4	+22	42	6.0–14.0	Mira	369.8	S
V	07	23.2	+13	06	7.8–14.9	Mira	275.1	M
T	07	49.3	+23	44	8.0–15.0	Mira	287.8	S
U	07	55.1	+22	00	8.2–14.9	SS Cygni	±103	M + WD

DOUBLE STARS

	R.A.		Dec.		P.A.	Sep.	Mags.	
	h	m	deg	min	deg	sec		
η	06	14.9	+22	30	267		var.	Binary, 474 y
μ	06	22.9	+22	31	077	72.7	3.0, 9.8	
ν	06	29.0	+20	13	329	112.5	4.2, 8.7	
ε	06	43.9	+25	08	094	110.3	3.0, 9.0	
38	06	54.6	+13	11	147	7.0	4.7, 7.7	Binary, 3190 y
ζ	07	04.1	+20	34	AB 084	87.0	var., 10.5	
					AC 350	96.5	8.0	
λ	07	18.1	+16	32	033	9.6	3.6, 10.7	
δ	07	20.1	+21	59	255	5.8	3.5, 8.2	Binary, 1200 y
α	07	34.6	+31	53	AB 067	3.7	1.9, 2.9	Binary, 420 y
					AC 164	72.5	1.6, 8.8	
κ	07	44.4	+24	24	240	7.1	3.6, 8.1	

OPEN CLUSTERS

M	C	NGC	R.A.		Dec.		Diameter	Mag.	No. of stars	
			h	m	deg	min	min			
		2129	06	01.0	+23	18	7	6.7	40	
		IC 2157	06	05.0	+24	00	7	8.4	20	
		2169	06	08.4	+13	57	7	5.9	30	
35		2168	06	08.9	+24	20	28	5.0	200	
		2266	06	43.2	+26	58	7	9.5	30	
		2355	07	16.9	+13	47	9	9.7	40	
		2395	07	27.1	+13	35	12	8.0	30	Asterism?

PLANETARY NEBULA

M	C	NGC	R.A.		Dec.		Dimensions	Mag.	Mag. of central star	
			h	m	deg	min	sec			
	39	2392	07	29.2	+20	55	13 × 44	10	10.5	Eskimo Nebula

GRUS

(Abbreviation: Gru).

The most distinctive of the 'Southern Birds'. The contrast in colour between α and β is very marked.

There are six stars above the fourth magnitude.

	Mag.	Luminosity Sun = 1	Dist lt-yrs
α	1.74	230	68
β	2.11v	800	173
γ	3.01	250	228
ε	3.49	26	81
ι	3.90	60	173
δ¹	3.97	58	140

β has a very small range (2.11 to 2.23), so that its fluctuations are not detectable with the naked eye.

δ and μ give the impression of being very wide doubles, but in neither case do the two components share a common motion in space. There is a good colour contrast between μ (type G) and the red δ (M4).

There is also a good colour contrast between Alnair (white) and Al Dhanab (orange).

GRUS (continued)

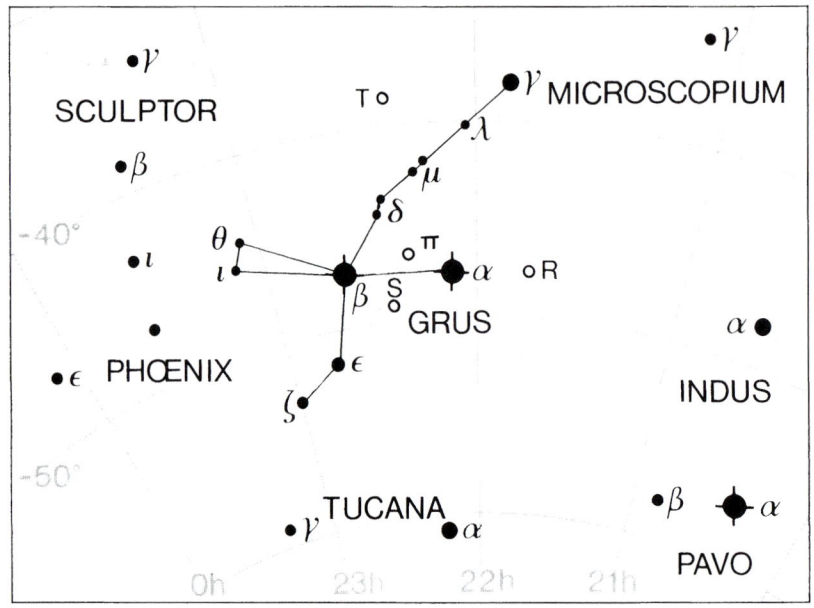

BRIGHTEST STARS

Star	R.A.			Dec.			Mag.	Abs.	Spec.	Dist	
	h	m	s	deg	min	sec		mag.		pc	
γ	21	53	55.6	−37	21	54	3.01	−1.2	B8	70	
λ	22	06	06.7	−39	32	36	4.46	−0.4	M0	94	
α	22	08	13.8	−46	57	40	1.74	−1.1	B5	21	Alnair
δ¹	22	29	15.9	−43	29	45	3.97	0.3	gG5	43	
δ²	22	29	45.3	−43	44	58	4.11		M4	27	
β	22	42	39.9v	−46	53	05	2.11v	−2.4	M3	53	Al Dhanab
ε	22	48	33.1	−51	19	01	3.49	1.4	A2	25	
ζ	23	00	52.6	−52	45	15	4.12	0.3	G5	50	
θ	23	06	52.6	−43	31	14	4.28	2.2	F6	26	
ι	23	10	21.4	−45	14	48	3.90	0.2	K0	53	

Also above mag. 5:

	Mag.	Abs. mag.	Spec.	Dist	
δ²	4.11		M4	727	
μ¹	4.79	0.3	G4	79	(μ² is magnitude 5.1)
ρ	4.85	0.2	K0	83	
η	4.85	0.2	K0	62	

VARIABLE STARS

	R.A.		Dec.		Range	Type	Period, d.	Spec.
	h	m	deg	min				
RS	21	43.1	−48	11	7.9–8.5	Delta Scuti	0.15	A–F
R	21	48.5	−46	55	7.4–14.9	Mira	331.9	M
π¹	22	22.7	−45	57	5.4–6.7	Semi-reg.	150	S
T	22	25.7	−37	34	7.8–12.3	Mira	136.5	M
S	22	26.1	−48	26	6.0–15.0	Mira	401.4	M

GRUS (continued)

DOUBLE STARS

	R.A. h	m	Dec. deg	min	P.A. deg	Sep. sec	Mags.
θ	23	06.9	−43	31	075	1.1	4.5, 7.0
υ	23	06.9	−38	54	211	1.1	5.7, 8.0

GALAXIES

M	NGC	R.A. h	m	Dec. deg	min	Mag.	Dimensions min	Type
	7144	22	52.7	−48	15	10.7	3.5 × 3.5	E0
	7213	22	09.3	−47	10	10.4	1.9 × 1.8	Sa
	7410	22	55.0	−39	40	10.4	5.5 × 2.0	SBa
	7412	22	55.8	−42	39	11.4	4.0 × 3.1	SBb
	7418	22	56.6	−37	02	11.4	3.3 × 2.8	SBc
	IC 1459	22	57.2	−36	28	10.0	—	E3
	IC 5267	22	57.2	−43	24	10.5	5.0 × 4.1	S0
	7424	22	57.3	−41	04	11.0	7.6 × 6.8	SBc
	IC 5273	22	59.5	−37	42	11.4	2.9 × 2.1	SBc
	7456	23	02.1	−39	35	11.9	5.9 × 1.8	Sc
	7496	23	09.8	−43	26	11.1	3.5 × 2.8	SBb
	7531	23	14.8	−43	36	11.3	3.5 × 1.5	Sb
	7552	23	16.2	−42	35	10.7	3.5 × 2.5	SBb
	7582	23	18.4	−42	22	10.6	4.6 × 2.2	SBG
	7599	23	19.3	−42	15	11.4	4.4 × 1.5	Sc

HERCULES

(Abbreviation: Her).

A large but by no means brilliant constellation, commemorating the great hero of mythology. The most interesting objects are the red supergiant α and the globular clusters M.13 and M.92; M.13 is the brightest globular in the northern hemisphere of the sky, and is surpassed only by the southern ω Centauri and 47 Tucanae.

Hercules contains fifteen stars above the fourth magnitude.

	Mag.	Luminosity Sun = 1	Dist lt-yrs
β	2.77	58	100
ζ	2.81	5.2	31
α	3.0 (max)	700v	220
δ	3.14	37	91
π	3.16	700	390
μ	3.42	2.5	26
η	3.53	16	68
ξ	3.70	60	163
γ	3.75	50	137
ι	3.80	450	420
o	3.83	60	170
109	3.84	96	108
θ	3.86	650	420
τ	3.89	400	425
ε	3.92	50	85

α is a huge red supergiant. For most of the time its magnitude is between 3.4 and 3.7, and the official period, 100 days, is very rough indeed. Its companion is green, though this is due largely to contrast with the red primary.

HERCULES (continued)

BRIGHTEST STARS

Star	R.A. h	m	s	Dec. deg	min	sec	Mag.	Abs. mag.	Spec.	Dist pc	
1 χ	15	52	40.4	+42	27	05	4.62	4.2	F9	15	
11 φ	16	08	46.0	+44	56	05	4.26	0.0	B9p	23	
22 τ	16	19	44.3	+46	18	48	3.89	−1.6	B5	130	
20 γ	16	21	55.1	+19	09	11	3.75	0.6	A9	42	
24 ω	16	25	24.8	+14	02	00	4.57	1.8	A0p	33	Cujam
27 β	16	30	13.1	+21	29	22	2.77	0.3	G8	31	Kornephoros
35 σ	16	34	06.0	+42	26	13	4.20	0.2	B9	61	
40 ζ	16	41	17.1	+31	36	10	2.81	3.0	G0	9.6	Rutilicus
44 η	16	42	53.7	+38	55	20	3.53	1.8	G8	21	
58 ε	17	00	17.2	+30	55	35	3.92	0.6	A0	26	
64 α	17	14	38.8	+14	23	25	var.	−2.3v	M5	67	Rasalgethi
65 δ	17	15	01.8	+24	50	21	3.14	0.9	A3	28	Sarin
67 π	17	15	02.6	+36	48	33	3.16	−2.3	K3	120	
68 u	17	17	19.4	+33	06	00	var.	−2.9v	B3	350	
69 e	17	17	40.1	+37	17	29	4.65	1.4	A2	45	
75 ρ	17	23	40.8	+37	08	45	4.17	0.6	A0	52	
76 λ	17	30	44.1	+26	06	39	4.41	−0.3	K4	85	Masym
85 ι	17	39	27.7	+46	00	23	3.80	−1.7	B3	130	
86 μ	17	46	27.3	+27	43	15	3.42	3.9	G5	8.1	
91 θ	17	56	15.1	+37	15	02	3.86	−2.2	K1	130	
92 ξ	17	57	45.7	+29	14	52	3.70	0.2	K0	50	
94 ν	17	58	30.0	+30	11	22	4.41	−2.0	F2	190	
93	18	00	03.2	+16	45	03	4.67	−0.9	K0	100	
95	18	01	30.3	+21	35	44	4.27	0.5	A7	48	
103 o	18	07	32.4	+28	45	45	3.83	0.2	B9	52	
102	18	08	45.4	+20	48	52	4.36	−2.5	B2	240	
109	18	23	41.7	+21	46	11	3.84	−0.1	K2	33	
110	18	45	39.6	+20	32	47	4.19	3.7	F6	15	
111	18	47	01.1	+18	10	53	4.36	1.7	A3	34	
113	18	54	44.7	+22	38	43	4.56	1.7	A3	36	

Also above mag. 5:

	Mag.	Abs. mag.	Spec.	Dist
6 υ	4.76		B9	
7 κ	5.00	−0.1	K2	87
29	4.84	−0.3	K5	70
g	5.0v	var.	M6	23
52	4.82	0.9	A2p	
60	4.91	0.9	A3	62
104	4.97	−0.5	M3	120
106	4.95	−0.4	M0	110

VARIABLE STARS

	R.A. h	m	Dec. deg	min	Range	Type	Period, d.	Spec.
X	16	02.7	+47	14	7.5–8.6	Semi-reg.	95	M
R	16	06.2	+18	22	7.13–15.0	Mira	318.4	M
RU	16	10.2	+25	04	6.8–14.3	Mira	485.5	M
U	16	25.8	+18	54	6.5–13.4	Mira	406.0	M
g (30)	16	28.6	+41	53	5.7–7.2	Semi-reg.	70	M
W	16	35.2	+37	21	7.6–14.4	Mira	280.4	M
S	16	51.9	+14	56	6.4–13.8	Mira	307.4	M
α	17	14.6	+14	23	3–4	Semi-reg.	±100?	M
u (68)	17	17.3	+35	06	4.6–5.3	Beta Lyrae	2.05	B + B
RS	17	21.7	+22	55	7.0–13.0	Mira	219.6	M
Z	17	58.1	+15	08	7.3–8.1	Algol	3.99	F + K
T	18	09.1	+31	01	6.8–13.9	Mira	165.0	M
AC	18	30.3	+21	52	7.4–9.7	RV Tauri	75.5	F + K
RX	18	30.7	+12	37	7.2–7.8	Algol	1.78	A + A

HERCULES (continued)

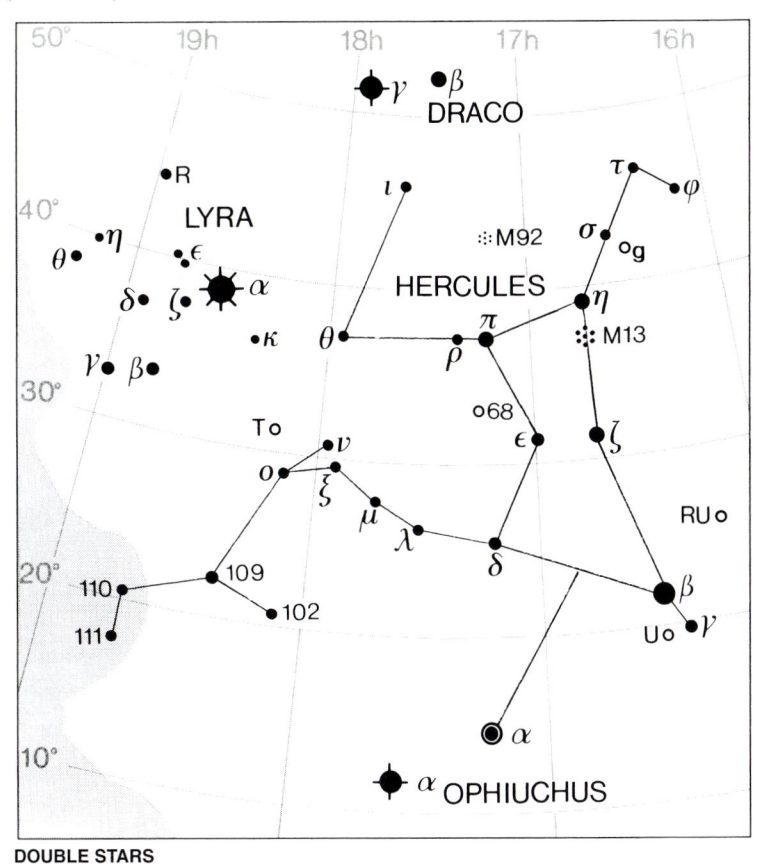

DOUBLE STARS

	R.A.		Dec.		P.A.	Sep.	Mags.	
	h	m	deg	min	deg	sec		
κ	16	08.1	+17	03	012	28.4	5.3, 6.5	
γ	16	21.9	+19	09	233	41.6	3.8, 9.8	
ω	16	25.4	+14	02	AB 223	1.0	4.6, 11.6	
					AC 096	28.4	11.1	
37	16	40.6	+04	13	230	69.8	5.8, 7.0	
ζ	16	41.3	+31	36	029	1.6	2.9, 5.5	Binary, 34.5 y
54	16	55.4	+18	26	183	2.5	5.4, 12.7	
α	17	14.6	+14	23	107	4.7	var., 5.4	Binary, 3600 y
δ	17	15.0	+24	50	236	8.9	3.7, 8.2	Optical pair
u (68)	17	17.3	+35	06	060	4.4	var., 10.2	
ρ	17	23.7	+37	09	316	4.1	4.6, 5.6	
μ	17	46.5	+27	43	247	33.8	3.4, 10.1	Binary. Very long period. B is itself a close binary.

GLOBULAR CLUSTERS

M	C	NGC	R.A.		Dec.		Diameter	Mag.
			h	m	deg	min	min	
92		6341	17	17.1	+43	08	11.2	6.5
13		6205	16	41.7	+36	28	16.6	5.9

PLANETARY NEBULÆ

M	C	NGC	R.A.		Dec.		Dimensions	Mag.	Mag. of
			h	m	deg	min	sec		central star
		6058	16	04.4	+40	41	23	13.3	13.8
		IC 4593	16	12.2	+12	04	12 × 120	10.9	11.3
		6210	16	44.5	+23	49	14	9.3	12.9

HOROLOGIUM

(Abbreviation: Hor).

A very obscure southern constellation. The only star above the fourth magnitude is α. The only other star above the fifth magnitude is δ; magnitude 4.93.

	Mag.	Luminosity Sun = 1	Dist lt-yrs
α	3.86	82	192

See chart for Eridanus.

ι (mag. 5.42, absolute mag. 3.0, type G3, R.A. 2h 42m 33s.4, dec. $-50°$ 48′ 01″, is believed to have a planetary companion 2.3 times as massive as Jupiter.

BRIGHTEST STAR

Star	R.A. h	m	s	Dec. deg	min	sec	Mag.	Abs. mag.	Spec.	Dist pc
α	04	14	00.0	-42	17	40	3.86	0.0	K1	59

Also above mag. 5:

	Mag.	Abs. mag.	Spec.	Dist
δ	4.93	2.6	F0	28

VARIABLE STARS

	R.A. h	m	Dec. deg	min	Range	Type	Period, d.	Spec.
R	02	53.9	-49	53	4.7–14.3	Mira	404.0	M
T	03	00.9	-50	39	7.2–13.7	Mira	217.7	M
V	03	03.5	-58	56	7.8–8.9	Semi-reg.	?	M
U	03	52.8	-45	50	7.8–15.1	Mira	348.4	M

GLOBULAR CLUSTER

M	C	NGC	R.A. h	m	Dec. deg	min	Diameter min	Mag.
	87	1261	03	12.3	-55	13	6.9	8.4

GALAXIES

M	C	NGC	R.A. h	m	Dec. deg	min	Mag.	Dimensions min	Type
		1249	03	10.0	-53	21	11.7	5.2×2.7	SBc
		1411	03	38.8	-44	05	11.9	2.8×2.3	S0
		1433	03	42.0	-47	13	10.0	6.8×6.0	SBa
		1448	03	44.5	-44	39	11.3	8.1×1.8	Sc
		1493	03	57.5	-46	12	11.8	2.6×2.3	SBc
		1512	04	03.9	-43	21	10.6	4.0×3.2	SBa

HYDRA

(Abbreviation: Hya).

The largest constellation in the sky, representing the hundred-headed monster which lived in the Lernæan marshes until it was killed by Hercules. It extends for over six hours of R.A. α (Alphard) is often known as 'the Solitary One', because there are no other bright stars anywhere in the neighbourhood.

Hydra contains ten stars above the fourth magnitude.

	Mag.	Luminosity Sun = 1	Dist lt-yrs
α	1.98	105	85
γ	3.00	58	104
ζ	3.11	60	124
ν	3.11	96	127
π	3.27	96	153
ε	3.38	50	110
ξ	3.54	58	144
λ	3.61	60	150
μ	3.81	110	192
θ	3.88	50	147

HYDRA (continued)

BRIGHTEST STARS

Star	R.A. h	m	s	Dec. deg	min	sec	Mag.	Abs. mag.	Spec.	Dist pc	
4 δ	08	37	39.3	+05	42	13	4.16	0.6	A0	43	
5 σ	08	38	45.4	+03	20	29	4.44	−0.1	K2	75	Al Minliar al Shuja
7 η	08	43	13.4	+03	23	55	4.30	−1.7	B3	160	
12 D	08	46	22.4	−13	32	52	4.32	0.3	G8	64	
11 ε	08	46	46.5	+06	25	07	3.38	0.6	G0	34	
13 ρ	08	48	25.9	+05	50	16	4.36	0.6	A0	57	
16 ζ	08	55	23.6	+05	56	44	3.11	0.2	K0	38	
22 θ	09	14	21.8	+02	18	51	3.88	0.6	A0	45	
30 α	09	27	35.2	−08	39	31	1.98	−0.2	K3	26	Alphard
31 τ¹	09	29	08.8	−02	46	08	4.60	3.7	F6	15	
32 τ²	09	31	58.9	−01	11	06	4.57	0.0	A3	82	
35 ι	09	39	51.3	−01	08	34	3.91	−0.2	K3	63	
39 υ¹	09	51	28.6	−14	50	48	4.12	0.3	G8	58	
40 υ²	10	05	07.4	−13	03	53	4.60	−1.2	B8	140	
41 λ	10	10	35.2	−12	21	15	3.61	0.2	K0	46	
42 μ	10	26	05.3	−16	50	11	3.81	−0.3	K4	59	
ν	10	49	37.4	−16	11	37	3.11	−0.1	K2	39	
ξ	11	33	00.1	−31	51	27	3.54	0.3	G7	44	
o	11	40	12.9	−34	44	40	4.70	0.3	B9	?	
β	11	52	54.5	−33	54	28	4.28	−0.3	B9	82	
46 γ	13	18	55.2	−23	10	17	3.00	0.3	G5	32	
R	13	29	42.7	−23	16	52	var.	var.	Md	100	(Very red)
49 π	14	06	22.2	−26	40	56	3.27	−0.1	K2	47	
58 E	14	50	17.2	−27	57	37	4.41	−0.3	gK4	88	

Also above mag. 5:

	Mag.	Abs. mag.	Spec.	Dist
6	4.98	−0.3	K4	110
9	4.88	0.0	K1	95
18 ω	4.97	−1.1	K2	170
26	4.79	0.3	G8	79
27 P	4.80	1.8	G8	39
F	4.70	−4.1	G4	(=31 Mon)
HD 83953	4.77	−1.1	B5	150
φ	4.91	0.2	K0	88
χ¹	4.94	3.3	F4	21
45 ψ	4.95	0.2	K0	81
51 κ	4.77	−0.3	K5	48
52	4.97	−0.6	B8	130
54	4.94	0.6	F0	51

VARIABLE STARS

	R.A. h	m	Dec. deg	min	Range	Type	Period, d.	Spec.
RT	08	29.7	−06	19	7.0–11.0	Semi-reg.	253	M
S	08	53.6	+03	04	7.4–13.3	Mira	256.4	M
T	08	55.7	−09	08	6.7–13.2	Mira	289.2	M
X	09	35.5	−14	42	8.0–13.6	Mira	301.4	M
U	10	37.6	−13	23	4.8–5.8	Semi-reg.	450	N
V	10	49.2	−20	59	6.0–12.5	Mira	533	N
R	13	29.7	−23	17	4.0–10.0	Mira	389.6	M
TT	11	13.2	−26	28	7.5–9.5	Algol	6.95	A+G
HZ	11	26.3	−25	45	7.6–8.2	Semi-reg.	95	M
W	13	49.0	−28	22	7.7–11.6	Semi-reg.	397	M
RU	14	11.6	−28	53	7.2–14.3	Mira	333.2	M

HYDRA (continued)

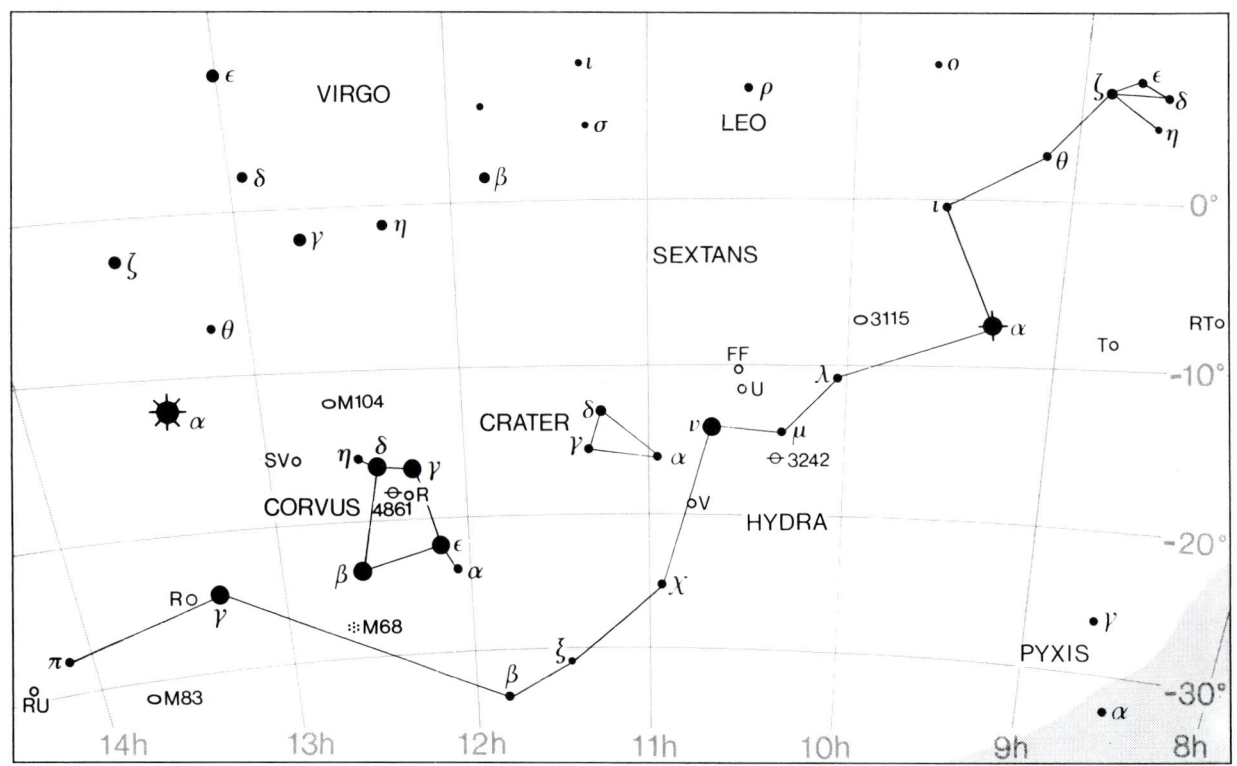

DOUBLE STARS

	R.A.		Dec.		P.A.	Sep.	Mags.	
	h	m	deg	min	deg	sec		
ε	08	46.8	+06	25	AB 295	0.2	3.8, 4.7	Binary, 890 y
					AB+C 298	3.3	6.8	
θ	09	14.4	+02	19	197	29.4	3.9, 9.9	
α	09	17.6	−08	40	153	283.1	2.0, 9.5	
N	11	32.3	−28	16	210	9.5	5.8, 5.9	
β	11	52.9	−33	54	008	0.9	4.7, 5.5	
R	13	29.7	−23	17	324	21.2	var., 12.0	
52	14	28.2	−29	30	AB 130	0.1	5.8, 5.8	
					AB+C 279	4.2	10.0	
					AB+D 282	140.8	12.0	
59	14	58.7	−27	39	335	0.8	6.3, 6.6	

OPEN CLUSTER

M	C	NGC	R.A.		Dec.		Diameter	Mag.	No. of stars
			h	m	deg	min	min		
48		2548	08	13.8	−05	48	54	5.8	80

GLOBULAR CLUSTER

M	C	NGC	R.A.		Dec.		Diameter	Mag.
			h	m	deg	min		
68	66	5694	14	39.6	−26	32	3.6	10.2

HYDRA (continued)

GALAXIES

M	C	NGC	R.A.		Dec.		Mag.	Dimensions	Type
			h	m	deg	min		min	
		2784	09	12.3	−24	10	10.1	5.1 × 2.3	S0
		2835	09	17.9	−22	21	11.1	6.3 × 4.4	Sp
		3109	10	03.1	−26	09	10.4	14.5 × 3.5	Irreg.
		3585	11	13.3	−26	45	10.0	2.9 × 1.6	E5
		3621	11	18.3	−32	49	9.9	10.0 × 6.5	Sc
		3923	11	51.0	−28	48	10.1	2.9 × 1.9	E3
		5078	13	19.8	−27	24	12.0	3.2 × 1.7	Sa
		5085	13	20.3	−24	26	11.9	3.4 × 3.0	Sb
		5061	13	18.1	−26	50	11.7	2.6 × 2.3	E2
		5101	13	21.8	−27	26	11.7	5.5 × 4.9	Sa
83		5236	13	37.0	−29	52	8.2	11.2 × 10.2	Sc

PLANETARY NEBULA

M	C	NGC	R.A.		Dec.		Dimensions	Mag.	
			h	m	deg	min	sec		
	59	3242	10	24.8	−18	38	16 × 1250	12.0	Ghost of Jupiter

HYDRUS

(Abbreviation: Hyi).

A constellation in the far south – remarkably lacking in interesting objects. There are three stars above the fourth magnitude.

	Mag.	Luminosity Sun = 1	Dist lt-yrs
β	2.80	3	20.5
α	2.86	7	36
γ	3.24	115	160

β is the closest star to the south celestial pole which is above the third magnitude.

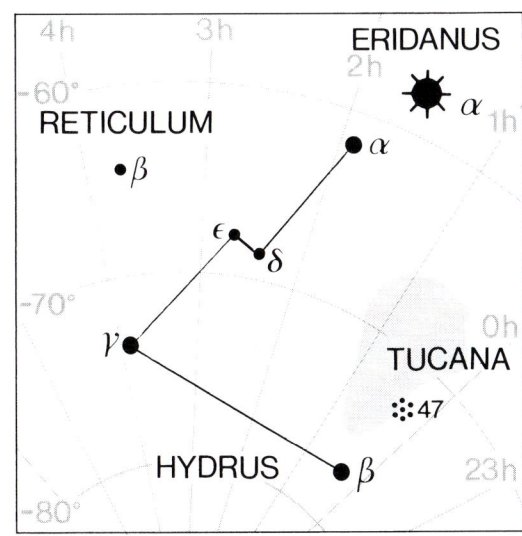

HYDRUS (continued)

BRIGHTEST STARS

Star	R.A. h	m	s	Dec. deg	min	sec	Mag.	Abs. mag.	Spec.	Dist pc
β	00	25	46.0	−77	15	15	2.80	3.8	G1	6.3
η^2	01	54	56.1	−67	38	50	4.69	0.3	G5	69
α	01	58	46.2	−61	34	12	2.86	2.6	F0	11
δ	02	21	45.0	−68	39	34	4.09	1.4	A2	35
ε	02	29	35.5	−68	16	00	4.11	−0.8	B9	91
ν	02	50	28.7	−75	04	00	4.75	−0.3	gK6	100
γ	03	47	14.5	−74	14	20	3.24	−0.4	M0	49

Also above mag. 5:

	Mag.	Abs. mag.	Spec.	Dist
ζ	4.84	−1.6	A2	

VARIABLE STAR

	R.A. h	m	Dec. deg	min	Range	Type	Period, d.	Spec.
VW	04	09.1	−71	18	8.4–14.4	SS Cygni	100	M

INDUS

(Abbreviation: Ind).

An undistinguished little constellation, but there are two stars above the fourth magnitude. ε is one of our closest stellar neighbours.

	Mag.	Luminosity Sun = 1	Dist lt-yrs
α	3.11	60	124
β	3.65	60	121

See chart for Grus.

BRIGHTEST STARS

Star	R.A. h	m	s	Dec. deg	min	sec	Mag.	Abs. mag.	Spec.	Dist pc	
α	20	37	33.9	−47	17	29	3.11	0.2	K0	38	Persian
η	20	44	02.2	−51	55	16	4.51	2.4	dA7	26	
β	20	54	48.5	−58	27	15	3.65	0.2	K0	37	
θ	21	19	51.1	−53	26	57	4.39	2.1	A5	28	
δ	21	57	55.0	−54	59	34	4.40	1.7	F0	35	
ε	22	03	21.5	−56	47	10	4.69	7.0	K5	3.4	

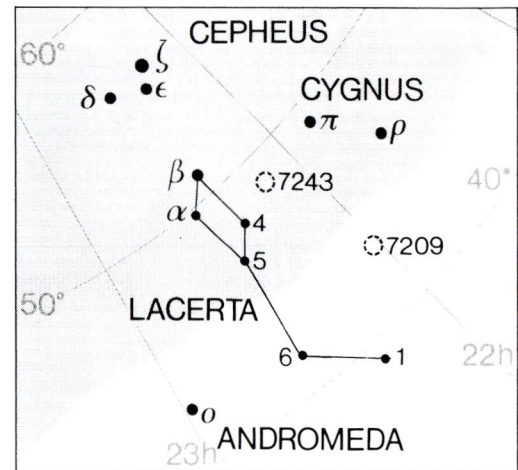

INDUS (continued)

Also above mag. 5:

	Mag.	Abs. mag.	Spec.	Dist
ζ	4.89	−0.5	M1	120

VARIABLE STARS

	R.A.		Dec.		Range	Type	Period, d.	Spec.
	h	m	deg	min				
S	20	56.4	−54	19	7.4–14.5	Mira	399.9	M
T	21	20.2	−45	01	7.7–9.4	Semi-reg.	320	N

DOUBLE STARS

	R.A.		Dec.		P.A.	Sep.	Mags.	
	h	m	deg	min	deg	sec		
θ	21	19.9	−53	27	271	6.8	4.5, 7.0	
δ	21	57.9	−55	00	323	0.1	5.3, 5.3	Binary, 12 y

GALAXIES

M	NGC	R.A.		Dec.		Mag.	Dimensions	Type
		h	m	deg	min		min	
	7049	21	19.0	−48	34	10.7	2.8 × 2.2	S0
	7083	21	35.7	−63	54	11.8	4.5 × 2.9	Sb
	7090	21	36.5	−54	33	11.1	7.1 × 1.4	SBc
	7168	22	02.1	−51	45	12.6	2.0 × 1.6	E3
	7205	22	08.5	−57	25	11.4	4.3 × 2.2	Sb

LACERTA

(Abbreviation: Lac)

An obscure constellation, with only one star above the fourth magnitude (α).

	Mag.	Luminosity Sun = 1	Dist lt-yrs
α	3.77	26	98

BRIGHTEST STARS

Star	R.A.			Dec.			Mag.	Abs. mag.	Spec.	Dist pc
	h	m	s	deg	min	sec				
HD 211073 IH	22	13	52.5	+39	42	54	4.49	−0.2	K3	77
1	22	15	58.1	+37	44	56	4.13	−1.3	K3	100
2	22	21	01.4	+46	32	12	4.57	−1.3	B6	150
3 β	22	23	33.4	+52	13	44	4.43	0.2	G9	66
4	22	24	30.8	+49	28	35	4.57	−6.5	B9	1500
5	22	29	31.7	+47	42	25	4.36	−2.4	M0	230
6	22	30	29.1	+43	07	25	4.51	−3.0	B2	280
7 α	22	31	17.3	+50	16	57	3.77	1.4	A2	30
9	22	37	22.3	+51	32	43	4.63	1.5	A7	41
11	22	40	30.7	+44	16	35	4.46	−0.2	K3	84

Also above mag. 5:

	Mag.	Abs. mag.	Spec.	Dist
15	4.94	−0.4	M0	120
10	4.88	−4.8	O9	780

LACERTA (continued)

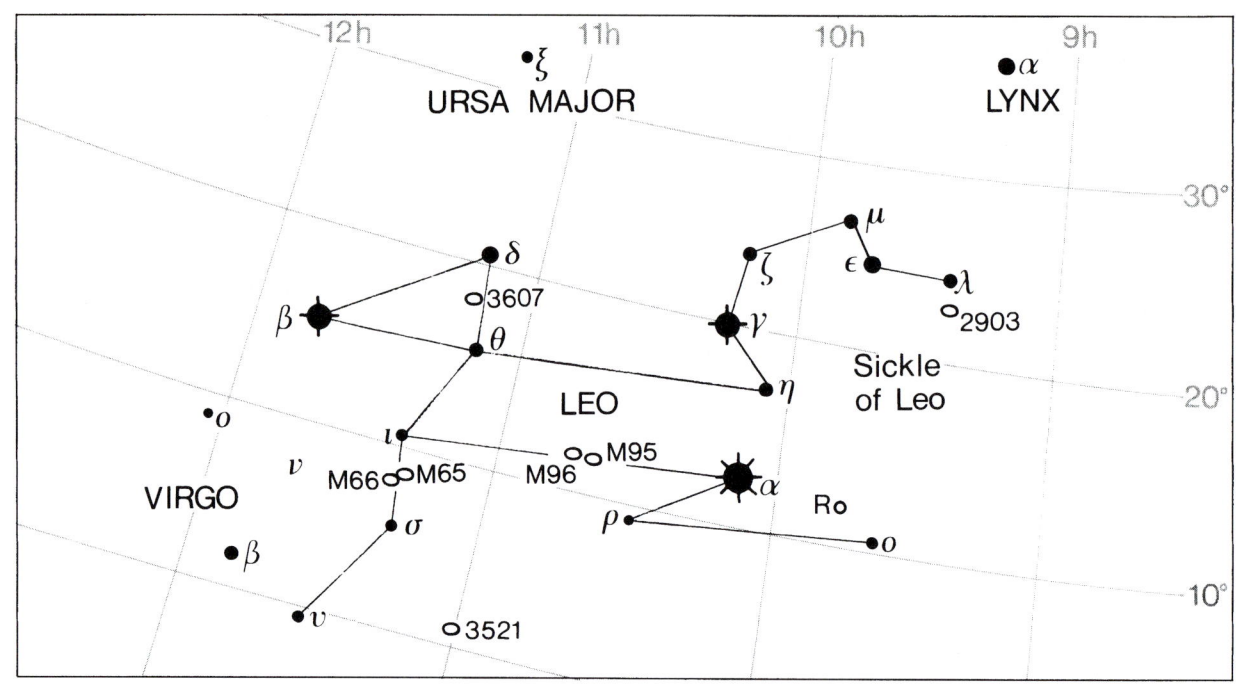

VARIABLE STARS

	R.A.		Dec.		Range	Type	Period, d.	Spec.
	h	m	deg	min				
S	22	29.0	+40	19	7.6–13.9	Mira	241.8	M
Z	22	40.9	+56	50	7.9–8.8	Cepheid	10.89	F–G
R	22	43.3	+42	22	8.5–14.8	Mira	299.9	M

OPEN CLUSTERS

M	C	NGC	R.A.		Dec.		Diameter	Mag.	No. of stars
			h	m	deg	min	min		
	16	7243	22	15.3	+49	53	21	6.4	40
		7296	22	28.2	+52	17	4	9.7	20

LEO

(Abbreviation: Leo).

An important Zodiacal constellation – mythologically, the Nemæan lion was killed by Hercules. There are 12 stars above the fourth magnitude.

	Mag.	Luminosity Sun = 1	Dist lt-yrs
α	1.38	130	85
γ	1.99	60	90
β	2.14	17	39
δ	2.56	14	52
ε	2.98	520	310
θ	3.34	26	78
ζ	3.44	50	117
η	3.52	9500	1800
o	3.52	11	55
ρ	3.85	16 000	2500
μ	3.88	96	200
ι	3.94	14	78

LEO (continued)

β (Denebola) was ranked as of the first magnitude by Ptolemy and others, and there is therefore a suspicion that it has faded, though the evidence is not conclusive.

Regulus has a distant companion which is a close and difficult binary. As Regulus and the companion have common motion in space, they are presumably associated.

BRIGHTEST STARS

Star	R.A. h	m	s	Dec. deg	min	sec	Mag.	Abs. mag.	Spec.	Dist pc	
1 κ	09	24	39.2	+26	10	56	4.46	−0.1	K2	73	Al Minliar al Asad
4 λ	09	31	43.1	+22	58	04	4.31	−0.3	K5	79	Alterf
14 o	09	41	09.0	+09	53	32	3.52	2.1	A5	17	Subra
17 ε	09	45	51.0	+23	46	27	2.98	−2.0	G0	95	Asad Australis
24 μ	09	52	45.8	+26	00	25	3.88	−0.1	K2	55	Rassalas
29 π	10	00	12.7	+08	02	39	4.70	−0.5	M2	110	
30 η	10	07	19.9	+16	45	45	3.52	−5.2	A0	560	
31 A	10	07	54.2	+09	59	51	4.37	−0.3	K4	81	
32 α	10	08	22.2	+11	58	02	1.35	−0.6	B7	26	Regulus
36 ζ	10	16	41.4	+23	25	02	3.44	0.6	F0	36	Adhafera
41 γ	10	19	58.3	+19	50	30	1.99	0.2	K0 + G7	28	Algieba
47 ρ	10	32	48.6	+09	18	24	3.85	−5.7	B1	770	
54	10	55	37.2	+24	44	55	4.32	1.2	A1	42	
61	11	01	49.6	−02	29	04	4.74	−0.3	K5	86	
60 b	11	02	19.7	+20	10	47	4.42	0.4	A0	22	
63 χ	11	05	01.0	+07	20	10	4.63	1.3	F2	47	
68 δ	11	14	06.4	+20	31	25	2.56	1.9	A4	16	Zosma
70 θ	11	14	14.3	+15	25	46	3.34	1.4	A2	24	Chort
72	11	15	12.2	+23	05	43	4.63	−0.5	gM2	100	
74 φ	11	16	39.6	−03	39	06	4.47	1.5	A7	39	
77 σ	11	21	08.1	+06	01	45	4.05	0.2	B9	59	
78 ι	11	23	55.4	+10	31	45	3.94	1.9	F2	24	
91 υ	11	36	56.9	−00	49	26	4.30	0.2	G9	64	
93	11	47	59.0	+20	13	08	4.53	2.2	A0	22	
94 β	11	49	03.5	+14	34	19	2.14	1.7	A3	12	Denebola

Also above mag. 5:

	Mag.	Abs. mag.	Spec.	Dist
5 ξ	4.97	0.2	K0	84
10	5.00	0.0	K1	100
40	4.79	2.2	F6	29
58	4.84	0.0	K1	83
84 τ	4.95	−0.9	G8	150

VARIABLE STAR

	R.A. h	m	Dec. deg	min	Range	Type	Period, d.	Spec.
R	09	47.6	+11	25	4.4–11.3	Mira	312.4	M

DOUBLE STARS

	R.A. h	m	Dec. deg	min	P.A. deg	Sep. sec	Mags.	
ω	09	28.5	+09	03	053	0.5	5.9, 6.5	Binary, 118 y
α	10	08.4	+11	58	307	176.9	1.4, 7.7	
γ	10	20.0	+19	51	AB 125	4.6	2.2, 3.5	Binary, 619 y
					AC 291	259.9	9.2	
					AD 302	333.0	9.6	
TX	10	35.0	+08	39	157	2.4	5.8, 8.5	
ι	11	23.9	+10	32	131	1.5	4.0, 6.7	Binary, 192 y
τ	11	27.9	+02	51	176	91.1	4.9, 8.0	

LEO (continued)

GALAXIES

M	C	NGC	R.A. h	m	Dec. deg	min	Mag.	Dimensions min	Type	
		3190	10	18.1	+21	50	11.0	4.6 × 1.8	SG	
95		3351	10	44.0	+11	42	9.7	7.4 × 5.1	SBb	
96		3368	10	46.8	+11	49	9.2	7.1 × 5.1	Sb	
		3377	10	47.7	+13	59	10.2	4.4 × 2.7	E5	
105		3379	10	47.8	+12	35	9.3	4.5 × 4.0	E1	
		3384	10	48.3	+12	38	10.0	5.9 × 2.6	E7	
		3412	10	50.9	+13	25	10.6	3.6 × 2.0	E5	
		3489	11	00.3	+13	54	10.3	3.7 × 2.1	E6	
		3521	11	05.8	−00	02	8.9	9.5 × 5.0	Sb	
		3593	11	14.6	+12	49	11.0	5.8 × 2.5	Sb	
		3596	11	15.1	+14	47	11.6	4.2 × 4.1	Sc	
		3607	11	16.9	+18	03	10.0	3.7 × 3.2	E1	
65		3623	11	18.9	+13	05	9.3	10.0 × 3.3	Sb	
	40	3626	11	20.1	+18	21	10.9	3.1 × 2.2	Sb	
66		3627	11	20.2	+12	59	9.0	8.7 × 4.4	Sb	
		3628	11	20.3	+13	36	9.5	14.8 × 3.6	Sb	Arp 317
		3630	11	20.3	+02	58	12.8	2.3 × 0.9	E7	
		3640	11	21.1	+03	14	10.3	4.1 × 3.4	EI	
		3646	11	21.7	+20	10	11.2	3.9 × 2.6	Sc	
		3686	11	27.7	+17	13	11.4	3.3 × 2.6	Sc	
		3810	11	41.0	+11	28	10.8	4.3 × 3.1	Sc	

LEO MINOR

(Abbreviation: LMi).

A small and obscure constellation. The only star above the fourth magnitude is 46.

	Mag.	Luminosity Sun = 1	Dist lt-yrs
46	3.83	17	75

See chart for Ursa Major.

BRIGHTEST STARS

Star	R.A. h	m	s	Dec. deg	min	sec	Mag.	Abs. mag.	Spec.	Dist pc	
10	09	34	13.3	+36	23	51	4.55	0.3	G8	71	
21	10	07	25.7	+35	14	41	4.48	2.4	A7	26	
30	10	25	54.8	+33	47	45	4.74	2.6	F0	27	
31 β	10	27	52.9	+36	42	26	4.21	1.8	G8	31	
37	10	38	43.1	+31	58	34	4.71	−2.1	G2	230	
46	10	53	18.6	+34	12	53	3.83	1.7	K0	23	Præcipua

VARIABLE STARS

	R.A. h	m	Dec. deg	min	Range	Type	Period, d.	Spec.
R	09	45.6	+34	31	6.3–13.2	Mira	371.9	M
S	09	53.7	+34	55	7.9–14.3	Mira	233.8	M
RW	10	16.1	+30	34	6.9–10.1	Mira	?	N

DOUBLE STAR

	R.A. h	m	Dec. deg	min	P.A. deg	Sep. sec	Mags.
β	10	27.9	+36	42	250	0.2	4.4, 6.1 Binary, 37.2 y

LEO MINOR (continued)

GALAXIES

M	C	NGC	R.A.		Dec.		Mag.	Dimensions	Type
			h	m	deg	min		min	
		3003	09	48.6	+33	25	11.7	5.9 × 1.7	SBc
		3245	10	27.3	+28	30	10.8	3.2 × 1.9	E5
		3254	10	29.3	+29	30	11.5	5.1 × 1.9	Sb
		3294	10	36.3	+37	20	11.7	3.3 × 1.8	Sc
		3344	10	43.5	+24	55	9.9	6.9 × 6.5	Sc
		3414	10	51.3	+27	59	10.7	3.6 × 2.7	SBa
		3430	10	52.2	+32	57	11.5	3.9 × 2.3	Sc
		3432	10	52.5	+36	37	11.2	6.2 × 1.5	SB
		3486	11	00.4	+28	58	10.3	6.9 × 5.4	Sc

LEPUS

(Abbreviation: Lep).

An original constellation. In mythology, it was said that Orion was particularly fond of hunting hares – and so a hare was placed beside him in the sky.

Lepus is fairly distinctive.

The very red variable R Leporis is known as 'the Crimson Star'.

There are eight stars above the fourth magnitude.

	Mag.	Luminosity Sun = 1	Dist lt-yrs
α	2.58	7500	945
β	2.84	600	316
ε	3.19	110	160
μ	3.31	180	215
ζ	3.55	17	78
γ	3.60	2	26
η	3.71	17	65
δ	3.81	58	156

BRIGHTEST STARS

Star	R.A.			Dec.			Mag.	Abs. mag.	Spec.	Dist pc	
	h	m	s	deg	min	sec					
2 ε	05	05	27.6	−22	22	16	3.19	−0.3	K5	50	
3 ι	05	12	17.8	−11	52	09	4.45	−0.2	B8	85	
5 μ	05	12	55.8	−16	12	20	3.31	−0.8	B9	66	
4 κ	05	13	13.8	−12	56	30	4.36	−1.2	B8	130	
6 λ	05	19	34.4	−13	10	36	4.29	−4.5	B0.5	520	(Measure uncertain)
HD 34968	05	20	26.8	−21	14	23	4.71	0.6	A0	66	
9 β	05	28	14.7	−20	45	35	2.84	−2.1	G2	97	Nihal
11 α	05	32	43.7	−17	49	20	2.58	−4.7	F0	290	Arneb
13 γ	05	44	27.7	−22	26	55	3.60	4.1	F6	8.1	
14 ζ	05	46	57.2	−14	49	20	3.55	1.7	A3	24	
15 δ	05	51	19.2	−20	52	45	3.81	0.3	G8	48	
16 η	05	56	24.2	−14	10	04	3.71	1.7	F0	20	
18 θ	06	06	09.3	−14	56	07	4.67	0.8	A0	18	

Also above mag. 5:

	Mag.	Abs. mag.	Spec.	Dist
17	4.93		A2	

VARIABLE STARS

	R.A.		Dec.		Range	Type	Period, d.	Spec.
	h	m	deg	min				
R	04	59.6	−14	48	5.5–11.7	Mira	432.1	N
T	05	04.8	−21	54	7.4–13.5	Mira	368.1	M
RX	05	11.4	−11	51	5.0–7.0	Irreg.	—	M
S	06	05.8	−24	12	7.1–8.9	Semi-reg.	90	M

LEPUS (continued)

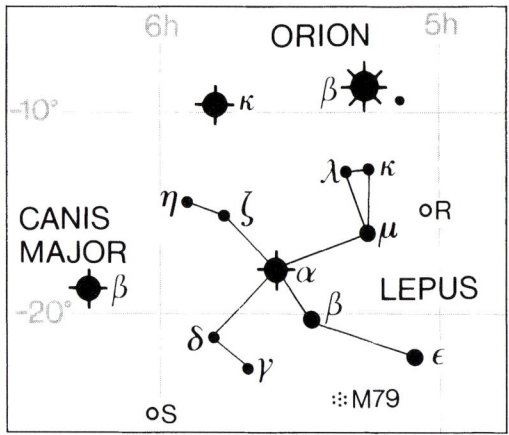

DOUBLE STARS

	R.A.		Dec.		P.A.	Sep.	Mags.
	h	m	deg	min	deg	sec	
ι	05	12.3	−11	52	337	12.7	4.5, 10.8
κ	05	13.2	−12	56	357	2.2	4.5, 7.4
β	05	28.2	−20	46	AB 330	2.5	2.8, 7.3
					AC 145	64.3	11.8
					AD 075	206.4	10.3
					AE 058	241.5	10.3
γ	05	44.5	−22	27	350	96.3	3.7, 6.3

GLOBULAR CLUSTER

M	NGC	R.A.		Dec.		Diameter	Mag.
		h	m	deg	min	min	
79	1904	05	24.5	−24	33	8.7	8.0

PLANETARY NEBULA

M	NGC	R.A.		Dec.		Diameter	Mag.	Mag. of
		h	m	deg	min	sec		central star
	IC 418	05	27.5	−12	42	12	10.7	10.7

GALAXIES

M	NGC	R.A.		Dec.		Mag.	Dimensions	Type
		h	m	deg	min		min	
	1744	05	00.0	−26	01	11.2	6.8 × 4.1	SBc
	1964	05	33.4	−21	57	10.8	6.2 × 2.5	Sb

LIBRA

(Abbreviation: Lib).

One of the Zodiacal constellations. It is, however, decidedly obscure. It was originally known as Chelæ Scorpionis (the Scorpion's Claws). Some Greek legends associate it, though rather vaguely, with Mochis, the inventor of weights and measures.

There are six stars above the fourth magnitude.

	Mag.	Luminosity	Dist
		Sun = 1	lt-yrs
β	2.61	105	121
α²	2.75	28	72
σ	3.29	120	166
υ	3.58	110	127
τ	3.66	310	326
γ	3.91	16	75

σ was formerly included in Scorpius, as γ Scorpii. τ and υ were also included in Scorpius. δ is one of the few Algol-type eclipsing binaries visible with the naked eye.

β is said to be the only bright single star which is green in colour, though most observers will certainly regard it as white.

LIBRA (continued)

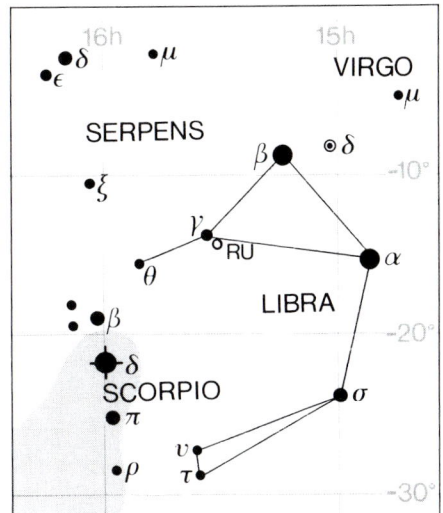

BRIGHTEST STARS

Star	R.A.			Dec.			Mag.	Abs. mag.	Spec.	Dist pc	
	h	m	s	deg	min	sec					
9 α^2	14	50	52.6	−16	02	30	2.75	1.2	A3	22	Zubenelgenubi
16	14	57	10.9	−04	20	47	4.49	1.7	F0	31	
20 σ	15	04	04.1	−25	16	55	3.29	−0.5	M4	51	Zubenelgubi
24 ι	15	12	13.2	−19	47	30	4.54	−0.3	B9	93	
27 β	15	17	00.3	−09	22	58	2.61	−0.2	B8	37	Zubenelchemale
38 γ	15	35	31.5	−14	47	23	3.91	1.8	G8	23	Zubenelhakrabi
39 υ	15	37	01.4	−28	08	06	3.58	−0.3	K5	39	
40 τ	15	38	39.3	−29	46	40	3.66	−1.4	B4	100	
43 κ	15	41	56.7	−19	40	44	4.74	−0.3	K5	91	
46 θ	15	53	49.4	−16	43	46	4.15	1.8	G8	25	

Also above mag. 5:

	Mag.	Abs. mag.	Spec.	Dist
δ	4.8 (max)	0.6	A2	73
υ	4.83	−0.3	A0	120
31 ε	4.94	3.4	F5	23
42	4.96	−0.3	K4	110
48	4.88	−0.8	B9	120

VARIABLE STARS

	R.A.		Dec.		Range	Type	Period, d.	Spec.
	h	m	deg	min				
δ	15	01.1	−08	31	4.9–5.9	Algol	2.33	B
Y	15	11.7	−06	10	7.6–14.7	Mira	275.0	M
S	15	21.4	−20	23	7.5–13.0	Mira	192.4	M
RS	15	24.3	−22	55	7.0–13.0	Mira	217.7	M
RU	15	33.3	−15	20	7.2–14.2	Mira	316.6	M
RR	15	56.4	−18	18	7.8–15.0	Mira	277.0	M

DOUBLE STARS

	R.A.		Dec.		P.A.	Sep.	Mags.
	h	m	deg	min	deg	sec	
μ	14	49.3	−14	09	355	1.8	5.8, 6.7
α	14	50.9	−16	02	314	231.0	2.8, 5.2
ι	15	12.2	−19	47	111	57.8	5.1, 9.4
κ	15	41.9	−19	41	279	172.0	4.7, 9.7

GLOBULAR CLUSTER

M	C	NGC	R.A.		Dec.		Diameter	Mag.
			h	m	deg	min	min	
		5897	15	17.4	−21	01	12.6	8.6 H.IV. 19

LUPUS

(Abbreviation: Lup).

An original constellation, though no definite legends seem to be attached to it. There are 11 stars above the fourth magnitude.

	Mag.	Luminosity Sun = 1	Dist lt-yrs
α	2.30	5000	6800
β	2.68	830	360
γ	2.78	450	258
δ	3.22	1320	587
ε	3.37	700	456
ζ	3.41	58	137
η	3.41	830	490
φ¹	3.56	110	183
κ	3.72	60	127
π	3.89	400	424
χ	3.95	110	220

BRIGHTEST STARS

Star	R.A. h	m	s	Dec. deg	min	sec	Mag.	Abs. mag.	Spec.	Dist pc	
ι	14	19	24.1	−46	03	28	3.55	−1.7	B3	110	
τ¹	14	26	08.1	−45	13	17	4.56	−3.0	B2	310	
τ²	14	26	10.7	−45	22	45	4.35	4.0	dF8	28	
σ	14	32	36.8	−50	27	25	4.42	−2.5	B2	130	
ρ	14	37	53.1	−49	25	32	4.05	−1.1	B5	110	
α	14	41	55.7	−47	23	17	2.30	−4.4	BI	210	Men
o	14	51	38.3	−43	34	31	4.32	−1.9	B6	180	
β	14	58	31.8	−43	08	02	2.68	−2.5	B2	110	KeKouan
π	15	05	07.1	−47	03	04	3.89	−1.6	B5	130	
λ	15	08	50.5	−45	16	47	4.05	−2.3	B3	190	
κ	15	11	56.0	−48	44	16	3.72	0.2	B9	39	
ζ	15	12	17.0	−52	05	57	3.41	0.3	G8	42	
2 f	15	17	49.7	−30	08	55	4.34	0.1	gK0	55	
μ	15	18	31.9	−47	51	30	4.27	−0.2	B8	77	
δ	15	21	22.2	−40	38	51	3.22	−3.0	B2	180	
φ¹	15	21	48.3	−36	15	41	3.56	−0.3	K5	56	
ε	15	22	40.7	−44	41	21	3.37	−2.3	B3	140	
φ²	15	23	09.2	−36	51	30	4.54	−2.3	B3	220	
κ	15	25	20.1	−38	44	01	4.60	0.0	A0	80	
γ	15	35	08.4	−41	10	00	2.78	−1.7	B3	79	
ω	15	38	03.1	−42	34	02	4.33	−0.4	M0	69	
3 ψ¹	15	39	45.9	−34	24	42	4.67	0.3	gG5	63	
g	15	41	11.2	−44	39	40	4.64	2.8	F5	22	
4 ψ²	15	42	40.9	−34	42	37	4.75	−0.9	B6	140	
5 χ	15	50	57.4	−33	37	38	3.95	−0.3	B9	68	
η	16	00	07.1	−38	23	48	3.41	−2.5	B2	150	
θ	16	06	35.4	−36	48	08	4.23	−2.3	B3	200	

Also above mag. 5:

	Mag.	Abs. mag.	Spec.	Dist
I	4.91	−6.6	F0	1800
ν¹	5.00	3.4	F3	20
ξ	4.6 (5.1 + 5.6)		A + A	

VARIABLE STARS

	R.A. h	m	Dec. deg	min	Range	Type	Period, d.	Spec.
S	14	53.4	−46	37	7.8–13.5	Mira	342.7	S
GG	15	18.9	−40	47	5.4–6.0	Beta Lyrae	2.16	B + A

LUPUS (continued)

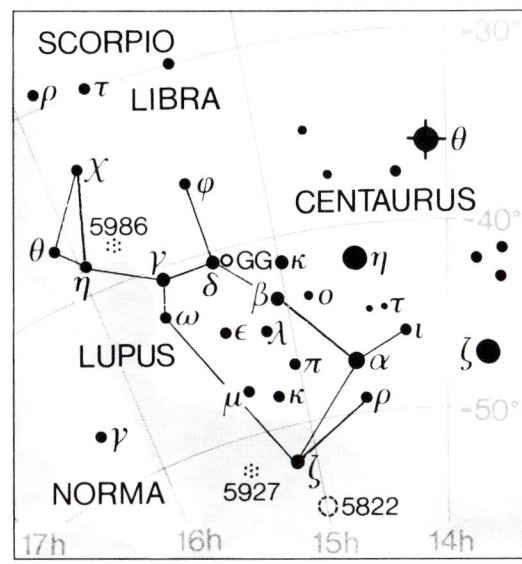

DOUBLE STARS

	R.A. h	m	Dec. deg	min	P.A. deg	Sep. sec	Mags.
τ^1	14	26.1	−45	13	204	148.2	4.6, 9.3
π	15	05.1	−47	03	067	1.7	4.6, 4.7
κ	15	11.9	−48	44	144	26.8	3.9, 5.8
μ	15	18.5	−47	53	AB 142	1.2	4.6, 4.7
					AC 130	23.7	7.2
ε	15	22.7	−44	41	247	0.6	3.7, 7.2
υ	15	24.7	−39	43	038	1.4	5.4, 10.9
ξ	15	56.9	−33	58	049	10.4	5.3, 5.8
η	16	00.1	−38	24	020	15.0	3.6, 7.8

OPEN CLUSTERS

M	C	NGC	R.A. h	m	Dec. deg	min	Diameter min	Mag.	No. of stars
		5749	14	48.9	−54	31	8	8.8	30
		5822	15	05.2	−54	21	40	6.5	150

GLOBULAR CLUSTERS

M	C	NGC	R.A. h	m	Dec. deg	min	Diameter min	Mag.
		5824	15	04.0	−33	04	6.2	9.0
		5927	15	28.0	−50	40	12.0	8.3
		5986	15	46.1	−37	47	9.8	7.1

PLANETARY NEBULÆ

M	C	NGC	R.A. h	m	Dec. deg	min	Diameter sec	Mag.	Mag. of central star
		IC 4406	14	22.4	−44	09	28	10.6	14.7
		5882	15	16.8	−45	39	7	10.5	12.0

GALAXIES

M	NGC	R.A. h	m	Dec. deg	min	Mag.	Dimensions min	Type
	5643	14	32.7	−44	10	10.7	4.6 × 4.1	SB0

LYNX

(Abbreviation: Lyn).

A very ill-defined and obscure northern constellation. It was added to the sky by Hevelius, and has no mythological associations. There are two stars above the fourth magnitude.

	Mag.	Luminosity Sun = 1	Dist lt-yrs
α	3.13	115	166
38	3.92	17	88

See chart for Ursa Major.

BRIGHTEST STARS

Star	R.A. h	m	s	Dec. deg	min	sec	Mag.	Abs. mag.	Spec.	Dist pc	
2	06	19	37.3	+59	00	39	4.48	1.4	A2	35	
15	06	57	16.5	+58	25	21	4.35	1.8	G5	32	
21	07	26	42.8	+49	12	42	4.64	0.3	A1	74	
31	08	22	50.1	+43	11	17	4.25	−0.3	K5	76	Alsciaukat
HD 77912	09	06	31.7	+38	27	08	4.56	−3.3	G8	370	
38	09	18	50.6	+36	48	09	3.92	1.7	A3	27	
40 α	09	21	03.2	+34	23	33	3.13	−0.4	M0	51	

Also above mag. 5:

	Mag.	Abs. mag.	Spec.	Dist
12	4.87	1.2	A2	
16	4.90	1.4	A2	50
24	4.99	0.0	A3	100
27	4.84	1.4	A2	60

VARIABLE STARS

	R.A. h	m	Dec. deg	min	Range	Type	Period, d.	Spec.
RR	06	26.4	+56	17	5.6–6.0	Algol	9.95	A
R	07	01.3	+55	20	7.2–14.5	Mira	378.7	S
Y	07	28.2	+45	59	7.8–10.3	Semi-reg.	110	M

DOUBLE STARS

	R.A. h	m	Dec. deg	min	P.A. deg	Sep. sec	Mags.
4	06	22.1	+59	22	124	0.8	6.2, 7.7
12	06	45.2	+59	27	AB 070	1.7	5.4, 6.0
					AC 308	8.7	7.3
					AD 256	170.0	10.6
19	07	22.9	+55	17	AB 315	14.8	5.6, 6.5
					AD 003	214.9	8.9
					BC 287	74.2	10.9
38	09	18.8	+36	48	AB 229	2.7	3.9, 6.6
					BC 212	87.7	10.8
					BD 256	177.9	10.7

GLOBULAR CLUSTER

M	C	NGC	R.A. h	m	Dec. deg	min	Diameter min	Mag.	
	25	2419	7	38.1	+38	53	4.1	10.4	'Intergalactic Tramp'

GALAXIES

M	C	NGC	R.A. h	m	Dec. deg	min	Mag.	Dimensions min	Type	
		2537	08	13.2	+46	00	12.3	1.7 × 1.5	S	Bear Paw Galaxy
		2541	08	14.7	+49	04	11.7	6.6 × 3.5	S	
		2683	08	52.7	+33	25	9.7	9.3 × 2.5	Sb	
		2776	09	12.2	+44	57	11.6	2.9 × 2.7	Sc	

LYRA

(Abbrevation: Lyr).

A small constellation, but a very interesting one; it is graced by the presence of the brilliant blue Vega, as well as the prototype eclipsing binary β Lyræ, and the quadruple ε Lyræ, and the 'Ring Nebula' M.57. Mythologically it represents the harp which Apollo gave to the great musician Orpheus.

There are several stars above the fourth magnitude. Of these, one (β) is the famous variable; the combined magnitude of the quadruple ε is about 3.9, though keen-sighted people can see the two main components as separated. The brightest stars are:

	Mag.	Luminosity Sun = 1	Dist lt-yrs
α	0.03	52	26
γ	3.24	180	192
β	3.3 (max)	130	300
ε	3.9 (combined)	17+11	124

Vega has been described as 'steely blue', and is associated with cool material which may be planet-forming. Its diameter has been given as 3 700 000 km.

ε is the famous quadruple star. The two pairs are certainly associated, but their orbital period must be very long indeed, amounting to hundreds of thousands of years.

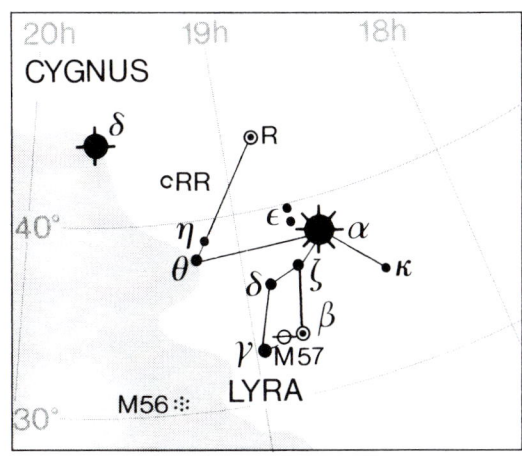

BRIGHTEST STARS

Star	R.A. h	m	s	Dec. deg	min	sec	Mag.	Abs. mag.	Spec.	Dist pc	
1 κ	18	19	51.5	+36	03	52	4.33	−0.1	K2	77	
3 α	18	36	56.2	+38	47	01	0.03	0.6	A0	8.1	Vega
4 ε¹	18	44	20.1	+39	40	15	4.67	1.7	A3	38	
5 ε²	18	44	22.7	+39	36	46	5.1	2.1	A5	38	
6 ζ¹	18	44	46.2	+37	36	18	4.36	1.2	A3	64	
7 ζ²	18	44	48.0	+37	35	40	5.73	2.8	F0	64	
10 β	18	50	04.6	+33	21	46	var.	−0.6	B7	92	Sheliak
12 δ²	18	54	30.0	+36	53	56	4.30v	−2.4	M4	220	
13 R	18	55	19.9	+43	56	46	var.	var.	M5	40	
14 γ	18	58	56.4	+32	41	22	3.24	−0.8	B9	59	Sulaphat
20 η	19	13	45.3	+39	08	46	4.39	−3.0	B2	270	Aladfar
21 θ	19	16	21.9	+38	08	01	4.36	−2.1	K0	170	

Also above mag. 5:

	Mag.	Abs. mag.	Spec.	Dist
15 λ	4.93	−2.3	K3	260

LYRA (continued)

VARIABLE STARS

	R.A. h	m	Dec. deg	min	Range	Type	Period, d.	Spec.
W	18	14.9	+36	40	7.3–13.0	Mira	196.5	M
T	18	32.3	+37	00	7.8–9.6	Irreg.	—	R
β	18	50.1	+33	22	3.3–4.3	Beta Lyræ	12.94	B + A
R	18	55.3	+43	57	3.9–5.0	Semi-reg.	46	M
RR	19	25.5	+42	47	7.1–8.1	RR Lyræ	0.57	A–F

DOUBLE STARS

	R.A. h	m	Dec. deg	min	P.A. deg	Sep. sec	Mags.	
ε	18	44.3	+39	40	AB + CD 173	207.7	4.7, 5.1	
					ε^1 = AB 357	2.6	5.0, 5.1	Binary, 1165 y
					ε^2 = CD 094	2.3	5.2, 5.5	Binary, 585 y
ζ	18	44.8	+37	36	150	43.7	4.3, 5.9	
β	18	50.1	+33	22	149	45.7	var., 8.6	
δ^1	18	53.7	+36	58	020	174.6	5.6, 9.3	
δ^2	18	54.5	+36	54	349	86.2	4.5, 11.2	
η	19	13.8	+39	09	082	28.1	4.4, 9.1	

OPEN CLUSTER

M	C	NGC	R.A. h	m	Dec. deg	min	Diameter min	Mag.	No. of stars
		6791	19	20.7	+37	51	16	9.5	300

GLOBULAR CLUSTER

M	C	NGC	R.A. h	m	Dec. deg	min	Diameter min	Mag.
56		6779	19	16.6	+30	11	7.1	8.2

PLANETARY NEBULA

M	C	NGC	R.A. h	m	Dec. deg	min	Dimensions sec	Mag.	Mag. of central star	
57		6720	18	53.6	+33	02	70 × 150	9.7	14.8	Ring Nebula

MENSA

(Abbreviation: Men).

A very dim constellation, introduced by Lacaille in 1752 under the name of Mons Mensæ (the Table Mountain). A small part of the Large Magellanic Cloud extends into it. There are no stars brighter than the fifth magnitude, and no objects to be listed. For the record, the brightest star is α (5.09). Next comes γ: R.A. 5 h 31 m 53 s.1, dec. −76 deg 20 min 28 sec, mag. 5.19, absolute mag. −0.3. Spectrum K4. Distance 130 pc. It has an optical companion of mag. 11, at P.A. 107 deg, separation 38″.2. See chart for Musca.

VARIABLE STARS

	R.A. h	m	Dec. deg	min	Range	Type	Period, d.	Spec.
U	04	09.6	−81	51	8.0–10.9	Mira	407	M
TY	05	26.9	−81	35	7.7–8.2	W UMa	0.46	A
TZ	05	30.2	−84	47	6.2–6.9	Algol	8.57	B

MICROSCOPIUM

(Abbreviation: Mic).

A small southern constellation. γ (4.67) is the brightest star.
See chart for Grus.

BRIGHTEST STARS

Star	R.A. h	m	s	Dec. deg	min	sec	Mag.	Abs. mag.	Spec.	Dist pc
γ	21	01	17.3	−32	15	28	4.67	0.3	G4	70
ε	21	17	56.1	−32	10	21	4.71	2.1	A2p	

γ was formerly known as 1 PsA and ε as 4 PsA.

MICROSCOPIUM (continued)

Also above mag. 5:

	Mag.	Abs. mag.	Spec.	Dist
α	4.90	0.3	G6	73
θ¹	4.82	−0.6	A2p	

VARIABLE STARS

	R.A.		Dec.		Range	Type	Period, d.	Spec.
	h	m	deg	min				
T	20	27.9	−28	16	7.7–9.6	Semi-reg.	344	M
U	20	29.2	−40	25	7.0–14.4	Mira	334.2	M
S	21	26.7	−29	51	7.8–14.3	Mira	208.9	M

DOUBLE STARS

	R.A.		Dec.		P.A.	Sep.	Mags.
	h	m	deg	min	deg	sec	
α	20	50.0	−33	47	166	20.5	5.0, 10.0
θ²	21	24.4	−41	00	AB 267	0.5	6.4, 7.0
					AC 066	78.4	10.5

GALAXIES

M	C	NGC	R.A.		Dec.		Mag.	Dimensions	Type
			h	m	deg	min		min	
		6923	20	31.7	−30	50	12.1	2.5 × 1.4	Sb
		6925	20	34.3	−31	59	11.3	4.1 × 1.6	Sb

MONOCEROS

(Abbreviation: Mon).

Not an ancient constellation, and though it represents the fabled unicorn there are no definite legends attached to it. It is crossed by the Milky Way, and the general area is decidedly rich. There are four stars above the fourth magnitude; of these, β is a double, and the magnitude is combined. The stars are:

	Mag.	Luminosity Sun = 1	Dist lt-yrs
β	3.7 (combined)	900+450	720
30	3.90	50	150
α	3.93	60	176
γ	3.98	105	215

See Chart for Orion.

BRIGHTEST STARS

Star	R.A.			Dec.			Mag.	Abs. mag.	Spec.	Dist pc	
	h	m	s	deg	min	sec					
5 γ	06	14	51.3	−06	16	29	3.98	−0.2	K3	66	
8 ε	06	23	46.0	+04	35	34	4.33	0.3	A5	54	
11 β	06	28	48.9	−07	01	58	3.7	{−2.6 / −1.7}	B2 / B3	220	
13	06	32	54.2	+07	19	58	4.50	−5.2	A0	860	
15 S	06	40	58.6	+09	53	45	var.	−5.5	O7	920	
18	06	47	51.6	+02	24	44	4.47	0.2	K0	59	
22 δ	07	11	51.8	−00	29	34	4.15	0.0	A0	64	
26 α	07	41	14.8	−09	33	04	3.93	0.2	K0	54	
28	08	01	13.2	−01	23	33	4.68	−0.3	K4	88	
29 ζ	08	08	35.6	−02	59	02	4.34	−4.5	G2	560	
30	08	25	39.5	−03	54	23	3.90	0.6	A0	46	
31	08	43	40.4	−07	14	01	4.62	−4.5	G2	670	= F Hydræ

Also above mag. 5:

	Mag.	Abs. mag.	Spec.	Dist
17	4.77	−0.3	K4	100
19	4.99	−3.5	B1	500
27	4.93	−0.1	K2	97

MONOCEROS (continued)

VARIABLE STARS

	R.A. h	m	Dec. deg	min	Range	Type	Period, d.	Spec.
V	06	22.7	−02	12	6.0–13.7	Mira	333.8	M
T	06	25.2	+07	05	6.0–6.6	Cepheid	27.02	F–K
S	06	41.0	+09	54	4–5?	Irreg.	—	07
X	06	57.2	−09	04	6.9–10.0	Semi-reg.	156	M
RY	07	06.9	−07	33	7.7–9.2	Semi-reg.	466	N
U	07	30.8	−09	47	6.1–8.1	RV Tauri	92.3	F–K

DOUBLE STARS

	R.A. h	m	Dec. deg	min	P.A. deg	Sep. sec	Mags.
ε	06	23.8	+04	36	027	13.4	4.5, 6.5
β	06	28.8	−07	02	AB 132	7.3	4.7, 5.2
					AC 124	10.0	6.1
					AD 056	25.9	12.2
S (15)	06	41.0	+09	54	AB 213	2.8	4.7v, 7.5
					AC 013	16.6	9.8
					AD 308	41.3	9.6
					AE 139	73.9	9.9
					AF 222	156.0	7.7
					AK 056	105.6	8.1

OPEN CLUSTERS

M	C	NGC	R.A. h	m	Dec. deg	min	Diam. min	Mag.	No. of stars	
		2215	06	21.0	−07	17	11	8.4	40	
	50	2244	06	32.4	+04	52	24	4.8	100	In Rosette Neb.
		2251	06	34.7	+08	22	10	7.3	30	
		2251	06	34.7	+0.8	22	10	7.3	30	
		2286	06	47.6	−03	10	15	7.5	50	
		2301	06	51.8	+00	28	12	6.0	80	
50		2323	07	03.2	−08	20	16	5.9	80	
		2335	07	06.6	−10	05	12	7.2	35	
		2343	07	08.3	−10	39	7	6.7	20	
		2353	07	14.6	−10	18	20	7.1	30	
	54	2506	08	00.2	−10	47	7	7.6	150	

NEBULÆ

M	C	NGC	R.A. h	m	Dec. deg	min	Dimensions min	Mag. of illuminating star	
		2149	06	03.5	−09	44	3 × 2	9	
	49	2237–9	06	32.3	+05	03	80 × 60		Rosette Nebula
	46	2261	06	39.2	+08	44	2 × 1	10v	R Monocerotis
		2264	06	40.9	+09	54	60 × 30	4v	S Monocerotis. Cone Nebula

MUSCA AUSTRALIS

(Abbreviation: Mus).

This small but fairly distinctive constellation, not far from Crux, is generally known simply as 'Musca'. There are five stars above the fourth magnitude.

	Mag.	Luminosity Sun = 1	Dist lt-yrs
α	2.69v	700	326
β	3.05	450	290
δ	3.62	96	176
λ	3.64	11	52
γ	3.87	230	192

α is actually variable over a very small range (2.66–2.73).

MUSCA AUSTRALIS (continued)

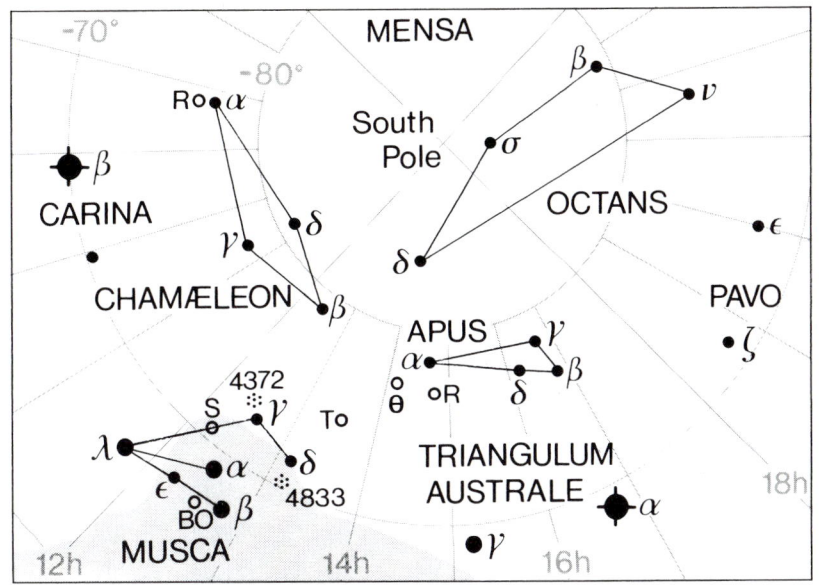

BRIGHTEST STARS

Star	R.A.			Dec.			Mag.	Abs.	Spec.	Dist
	h	m	s	deg	min	sec		mag.		pc
λ	11	45	36.4	−66	43	43	3.64	2.1	A5	16
μ	11	48	14.4	−66	48	53	4.72	−0.5	gM2	110
ε	12	17	34.2	−67	57	38	4.11	2.2	gM6	12
γ	12	32	28.1	−72	07	58	3.87	−1.1	B5	59
α	12	37	11.0	−69	08	07	2.69v	−2.3	B3	100
β	12	46	16.9	−68	06	29	3.05	−1.7	B3	89
δ	13	02	16.3	−71	32	56	3.62	−0.1	K2	54

Also above mag. 5:

	Mag.	Abs. mag.	Spec.	Dist
η	4.80	−0.2	B8	100

VARIABLE STARS

	R.A.		Dec.		Range	Type	Period, d.	Spec.
	h	m	deg	min				
S	12	12.8	−70	09	5.9–6.4	Cepheid	9.66	F
B0	12	34.9	−67	45	6.0–6.7	Irreg.	—	M
R	12	42.1	−69	24	5.9–6.7	Cepheid	7.48	F
T	13	21.2	−74	27	7.1–9.0	Semi-reg.	93	N

DOUBLE STARS

	R.A.		Dec.		P.A.	Sep.	Mags.
	h	m	deg	min	deg	sec	
ζ²	12	22.1	−67	31	130	32.4	5.2, 10.6
α	12	37.2	−69	08	316	29.6	2.7, 12.8
β	12	46.3	−68	06	039	1.2	3.7, 4.0
θ	13	08.1	−65	18	187	5.3	5.7, 7.3

OPEN CLUSTER

M	C	NGC	R.A.		Dec.		Diam.	Mag.	No. of stars
			h	m	deg	min	min		
		4463	12	30.0	−64	48	5	7.2	30

MUSCA AUSTRALIS (continued)

GLOBULAR CLUSTERS

M	C	NGC	R.A.		Dec.		Diameter	Mag.
			h	m	deg	min	min	
		4372	12	25.8	−72	40	18.6	7.8
		4833	12	59.6	−70	53	13.5	7.3

PLANETARY NEBULÆ

M	C	NGC	R.A.		Dec.		Diameter	Mag.	Mag. of central star	
			h	m	deg	min	sec			
		IC 4191	13	08.8	−67	39	5	12.0	—	
		5189	13	33.5	−65	59	153	10	14	Gum 47

NORMA

(Abbreviation: Nor).

An obscure constellation, once known as Quadra Euclidis (Euclid's Quadrant); it was added to the sky by Lacaille. It contains no star above the fourth magnitude.

See charts for Ara and Lupus.

BRIGHTEST STARS

Star	R.A.			Dec.			Mag.	Abs. mag.	Spec.	Dist pc
	h	m	s	deg	min	sec				
η	16	03	12.6	−49	13	47	4.65	0.3	gG4	64
ι^1	16	03	31.9	−57	46	31	4.63	2.1	A5	30
δ	16	06	29.3	−45	10	24	4.72	0.6	A0	18
γ^2	16	19	50.3	−50	09	20	4.02	0.3	G8	40
ε	16	27	10.9	−47	33	18	4.47	−1.7	B3	150

Also above mag. 5:

	Mag.	Abs. mag.	Spec.	Dist
κ	4.94	0.3	G4	61
γ^1	4.99	−6.3	F8	1700

VARIABLE STARS

	R.A.		Dec.		Range	Type	Period, d.	Spec.
	h	m	deg	min				
R	15	36.0	−49	30	6.5–13.9	Mira	492.7	M
T	15	44.1	−54	59	6.2–13.6	Mira	242.6	M
S	16	18.9	−57	54	6.1–6.8	Cepheid	9.75	F–G

DOUBLE STARS

	C	R.A.		Dec.		P.A.	Sep.	Mags.	
		h	m	deg	min	deg	sec		
ι^1		16	03.5	−57	47	100	0.2	5.3, 5.5	Binary, 26.9 y
ε		16	27.2	−47	33	335	22.8	4.8, 7.5	

OPEN CLUSTERS

M	C	NGC	R.A.		Dec.		Diam.	Mag.	No. of stars	
			h	m	deg	min	min			
		5925	15	27.7	−54	31	15	8.4	120	
		5999	15	52.2	−56	28	5	9.0	40	
		6031	16	07.6	−54	04	2	8.5	20	
	89	6067	16	13.2	−54	13	13	5.6	100	
		6087	16	18.9	−57	54	12	5.4	40	S Normæ cluster
		H.10	16	19.9	−54	59	30	—	30	
		6134	16	27.7	−49	09	7	7.2	—	
		6152	16	32.7	−52	37	30	8.1	70	
		6167	16	34.4	−49	36	8	6.7	—	

PLANETARY NEBULA

M	NGC	R.A.		Dec.		Diameter	Mag.	Mag. of central star
		h	m	deg	min	sec		
	Sp-1	15	51.7	−51	31	76	13.6	13.8

OCTANS

(Abbrevations: Oct).

The south polar constellation. It is very obscure, and contains only one star above the fourth magnitude: ν.

	Mag.	Luminosity Sun = 1	Dist lt-yrs
ν	3.76	60	104

The star lettered α Octantis is only of magnitude 5.2.

See chart for Musca.

BRIGHTEST STARS

Star	R.A. h	m	s	Dec. deg	min	sec	Mag.	Abs. mag.	Spec.	Dist pc
δ	14	26	55.0	−83	40	04	4.32	−0.1	gK2	60
ν	21	41	28.6	−77	23	24	3.76	0.2	K0	32
β	22	46	03.1	−81	22	54	4.15	2.6	dF0	20

The South Pole Star is σ Octantis; R.A. 21 h 08 m 44 s .9, dec. −88 deg 57 min 24 sec (epoch 2000), magnitude 5.47, absolute magnitude 2.7, spectrum A7, distance 37 parsecs (121 light-years). The polar distance was 45 sec in 1900, but has now increased to over one degree.

Also above mag. 5:

	Mag.	Abs. mag.	Spec.	Dist
0	4.78	−0.1	K2	78

VARIABLE STARS

	R.A. h	m	Dec. deg	min	Range	Type	Period, d.	Spec.
R	05	26.1	−86	23	6.4–13.2	Mira	405.6	M
U	13	24.5	−84	13	7.1–14.1	Mira	302.6	M
S	18	08.7	−86	48	7.3–14.0	Mira	258.9	M
ε	22	20.0	−80	26	4.9–5.4	Semi-reg.	55	M

DOUBLE STARS

	R.A. h	m	Dec. deg	min	P.A. deg	Sep. sec	Mags.
ι	12	55.0	−85	07	230	0.6	6.0, 6.5
μ^2	20	41.7	−75	21	017	17.4	7.1, 7.6
λ	21	50.9	−82	43	070	3.1	5.4, 7.7

OPHIUCHUS

(Abbreviations: Oph).

This constellation is also sometimes known as Serpentarius. It commemorates Æsculapius, son of Apollo and Coronis, who became so skilled in medicine that he was even able to restore the dead to life. To avoid depopulation of the Underworld, Jupiter reluctantly disposed of Æsculapius with a thunderbolt, but relented sufficiently to place him in the sky.

Ophiuchus is a very large constellation, with 13 stars above the fourth magnitude.

	Mag.	Luminosity Sun = 1	Dist lt-yrs
α	2.08	58	62
η	2.43	26	59
ζ	2.56	5000	550
δ	2.74	120	140
β	2.77	96	121
κ	3.20	96	117
ε	3.24	58	104
θ	3.27	1320	590
ν	3.34	60	137
72	3.73	14	91
γ	3.75	50	114
λ	3.82	28	108
67	3.97	16 000	2400

OPHIUCHUS (continued)

Barnard's Star (Munich 15040) lies in Ophinchus near 67 and 70; R.A. 17 h 55 m .4, dec +04 deg 33 min. The apparent magnitude is 9.54, spectral type M5. The annual proper motion is 10″.29, and the star is moving due north; it covers one degree in 351 years.

The region of the celebrated binary 70 Ophiuchi was once separated out into the constellation Taurus Poniatowski, which has, however, been deleted from modern maps.

BRIGHTEST STARS

Star	R.A. h	m	s	Dec. deg	min	sec	Mag.	Abs. mag.	Spec.	Dist pc	
1 δ	16	14	20.6	−03	41	39	2.74	−0.5	M1	43	Yed Prior
2 ε	16	18	19.1	−04	41	33	3.24	0.3	G8	32	Yed Post
4 ψ	16	24	06.0	−20	02	15	4.50	0.2	K0	72	
5 ρ	16	25	34.9	−23	26	46	4.59	−2.5	B2	230	
7 χ	16	27	01.3	−18	27	23	4v	−2.5v	B2p	150	
3 υ	16	27	48.1	−08	22	18	4.63	1.4	A2	21	
10 λ	16	30	54.7	+01	59	02	3.82	1.2	A1	33	Marfik
8 φ	16	31	08.2	−16	36	46	4.28	0.3	G8	63	
9 ω	16	32	08.0	−21	27	59	4.45	1.7	A7	19	
13 ζ	16	37	09.4	−10	34	02	2.56	−4.4	O9.5	170	Han
20	16	49	49.9	−10	46	59	4.65	0.7	F6	56	
25 ι	16	54	00.4	+10	09	55	4.38	−0.6	B8	99	
27 κ	16	57	40.0	+09	22	30	3.20	−0.1	K2	36	
35 η	17	10	22.5	−15	43	30	2.43	1.4	A2	18	Sabik
36	17	15	20.7	−26	36	04	4.31	6.4	K0	5.4	
41	17	16	36.5	−00	26	43	4.73	−0.1	K2	93	
40 ξ	17	21	00.0	−21	06	46	4.39	3.0	F2	19	
42 θ	17	22	00.4	−24	59	58	3.27	−3.0	B2	180	
44 b	17	26	22.1	−24	10	31	4.17	2.5	A9	25	
49 σ	17	26	30.7	+04	08	25	4.34	−2.3	K3	190	
47	17	26	37.7	−05	05	12	4.54	3.1	F3	21	
45 d	17	27	21.1	−29	52	01	4.29	2.1	F5	27	
55 α	17	34	55.9	+12	33	36	2.08	0.3	A5	19	Rasalhague
57 μ	17	37	50.5	−08	07	08	4.62	−0.2	B8	73	
60 β	17	43	28.2	+04	34	02	2.77	−0.1	K2	37	Cheleb
62 γ	17	47	53.4	+02	42	26	3.75	0.6	A0	35	
64 ν	17	59	01.4	−09	46	25	3.34	0.2	K0	42	
66	18	00	15.5	+04	22	07	4.64	−2.5	B2	250	
67	18	00	38.5	+02	55	53	3.97	−5.7	B5	740	
68	18	01	45.0	+01	18	19	4.45	1.2	A1	45	
70 p	18	05	27.2	+02	29	58	4.03	5.7	K0	5.1	
71	18	07	18.2	+08	44	02	4.64	1.8	G8	34	
72	18	07	20.8	+09	33	50	3.73	1.9	A4	28	

Also above mag. 5:

	Mag.	Abs. mag.	Spec.	Dist
30	4.82	−0.3	K4	93
58	4.87	3.4	F5	20
69 τ	4.79	2.6	F0	22
74	4.86	0.3	G8	82

VARIABLE STARS

	R.A. h	m	Dec. deg	min	Range	Type	Period, d.	Spec.
V	16	26.7	−12	26	7.3–11.6	Mira	298.0	N
χ	16	27.0	−18	27	4.2–5.0	Irreg. (γc)	—	B
SS	16	57.9	−02	46	7.8–4.5	Mira	80.0	M
R	17	07.8	−16	06	7.0–13.8	Mira	302.6	
U	17	16.5	+01	13	5.9–6.6	Algol	1.68	B + B
Z	17	19.5	+01	31	7.6–14.0	Mira	348.7	K–M
RS	17	50.2	−06	43	5.3–12.3	Recurrent nova	—	O + M (Outbursts 1933, 1958, 1967)
Y	17	52.6	−06	09	5.9–6.4	Cepheid	17.12	F–G
RY	18	16.6	+03	42	7.5–13.8	Mira	150.5	M
X	18	38.3	+08	50	5.9–9.2	Mira	334.4	M + K

OPHIUCHUS (continued)

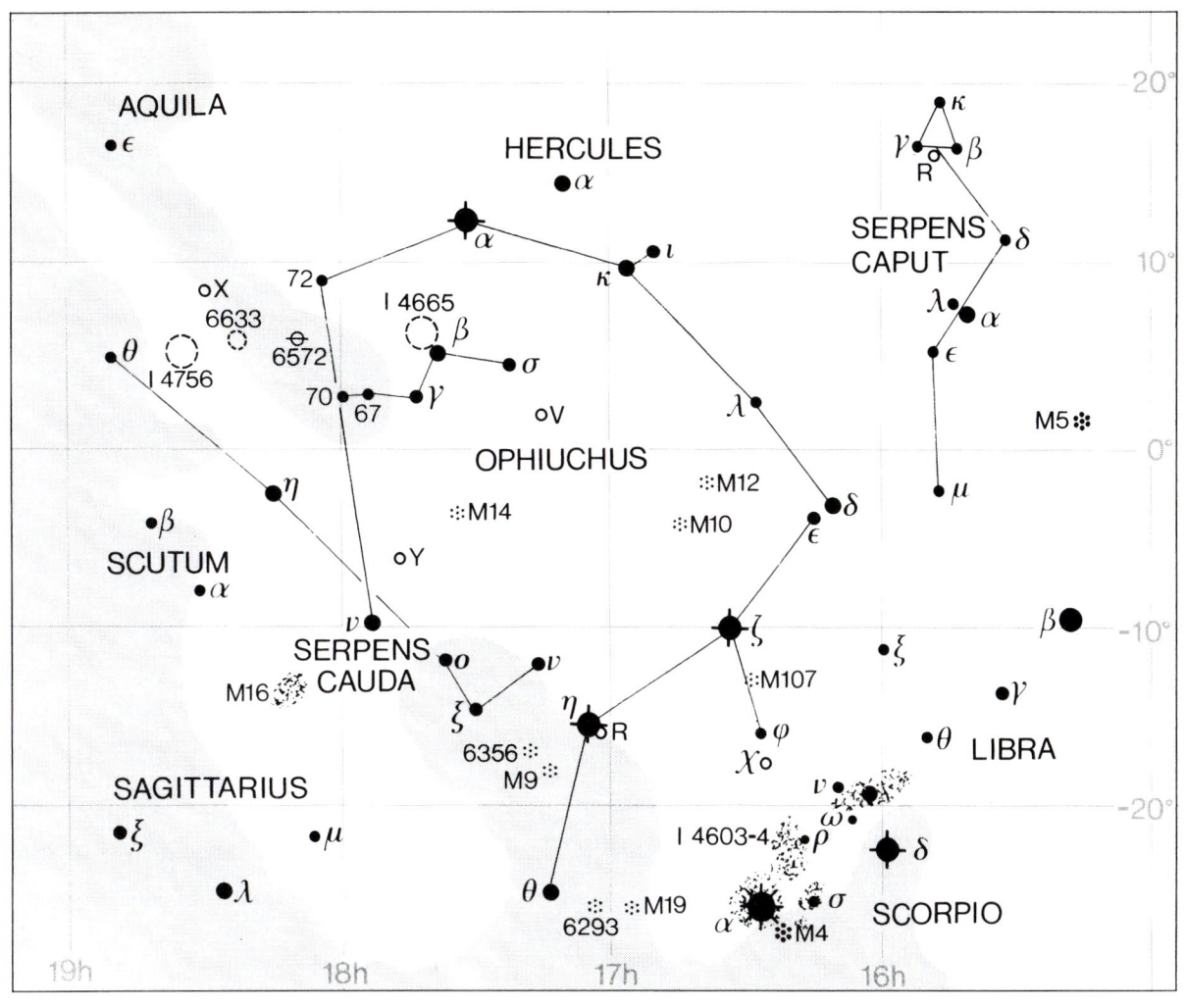

DOUBLE STARS

	R.A. h	m	Dec. deg	min	P.A. deg	Sep. sec	Mags.	
ρ	16	25.6	−23	27	344	3.1	5.3, 6.0	
υ	16	27.8	−08	22	095	1.0	4.6, 7.8	
λ	16	30.9	+01	59	AB 022	1.5	4.2, 5.2	Binary, 129.9 y
					AB+C 170	119.2	11.1	
					AD 246	313.8	9.9	
φ	16	31.1	−16	37	037	34.4	4.3, 12.8	
19	16	47.2	+02	04	089	23.4	6.1, 9.4	
η	17	10.4	−15	43	247	0.5	3.0, 3.5	Binary, 84.3 y
36	17	15.3	−26	36	150	4.7	5.1, 5.1	Binary, 549 y
41	17	16.6	−00	27	346	1.0	4.8, 7.8	
53	17	34.6	+09	35	191	41.2	5.8, 8.5	
τ	18	03.1	−08	11	AB 280	1.8	5.2, 5.9	Binary, 280 y
					AC 127	100.3	9.3	
70	18	05.5	+02	30	152	3.4	4.2, 6.0	Binary, 88.1 y
73	18	09.6	+04	00	300	0.4	6.1, 7.0	Binary, 270 y
X	18	38.3	+08	50	150	0.4	var., 8.6	Binary, 485 y

OPHIUCHUS (continued)

OPEN CLUSTERS

M	C	NGC	R.A.		Dec.		Diam.	Mag.	No. of stars
			h	m	deg	min	min		
		IC 4665	17	46.3	+05	43	41	4.2	30
		6633	18	27.7	+06	34	27	4.6	30

GLOBULAR CLUSTERS

M	C	NGC	R.A.		Dec.		Diameter	Mag.
			h	m	deg	min	min	
107		6171	16	32.5	−13	03	10.0	8.1
12		6218	16	47.2	−01	57	14.5	6.6
10		6254	16	57.1	−04	06	15.1	6.6
62		6266	17	01.2	−30	07	14.1	6.6
19		6273	17	02.6	−26	16	13.5	7.1
		6304	17	14.5	−29	28	6.8	8.4
		6316	17	16.6	−28	08	4.9	9.0
9		6333	17	19.2	−18	31	9.3	7.9
		6356	17	23.6	−17	49	7.2	8.4
		6355	17	24.0	−26	21	5.0	9.6
14		6402	17	37.6	−03	15	11.7	7.6
		6401	17	38.6	−23	55	5.6	9.5

PLANETARY NEBULÆ

M	C	NGC	R.A.		Dec.		Dimensions	Mag.	Mag. of
			h	m	deg	min	sec		central star
		6309	17	14.1	−12	55	14 × 66	10.8	14.4
		6572	18	12.1	+06	51	8 × 8	9.0	13.6

GALAXY

M	C	NGC	R.A.		Dec.		Mag.	Dimensions	Type
			h	m	deg	min		min	
		6384	17	32.4	+07	04	10.6	6.0 × 4.3	Sb

ORION

(Abbreviation: Ori).

One of the most magnificent constellations in the sky; it represents the mythological hunter who boasted that he could kill any creature on earth, but who was fatally stung by a scorpion. The two leading stars are Rigel, which is actually variable over a very small range (0.08 to 0.20) and the red variable Betelgeux – a name which may also be spelled Betelgeuse or Betelgeuze. The gaseous nebula M.42, in the Sword, is the most famous example of its type, and is easily visible with the naked eye. Altogether there are 15 stars above the fourth magnitude:

	Mag.	Luminosity Sun = 1	Dist lt-yrs
β	0.12v	60 000	910
α	0.5v	15 000v	310
γ	1.64	2200	360
ε	1.70	23 000	1200
ζ	1.77	19 000	1100
κ	2.06	49 000	2100
δ	2.23v	22 000	2350
ι	2.76	20 900	1860
π^3	3.19	37	270
η	3.36v	2100	750
λ	3.39	9000	1800
τ	3.60	650	420
π^4	3.69	2200	910
π^5	3.72v	2200	940
σ	3.73	5000	1800

The distance of Betelgeux given here is according to the Cambridge catalogue. Some other catalogues increase this to 510 light-years. Observations with the VLA (Very Large Array) in New Mexico show surface features; huge convective plumes rise into the star's atmosphere, and the atmosphere itself extends to many times the diameter of the star – which was given as 600 times that of the Sun, much larger than the orbit of Mars.

ORION (continued)

Rigel is very luminous and massive; it may be 25 times as massive as the Sun. It and its companion share common motion in space, and are presumably associated. The companion is a spectroscopic binary with a period of 9.9 days. The brighter component is 100 times as luminous as the Sun.

BRIGHTEST STARS

Star	R.A.			Dec.			Mag.	Abs. mag.	Spec.	Dist pc	
	h	m	s	deg	min	sec					
1 π^3	04	49	50.3	+06	57	41	3.19	3.8	F6	7.7	
2 π^2	04	50	36.6	+08	54	01	4. 6	0.6	A0	55	
3 π^4	04	51	12.3	+05	36	18	3.69	−3.6	B2	280	
4 O^1	04	52	31.9	+14	15	02	4.74		M3	32	
8 π^5	04	54	15.0	+02	26	26	3.72v	−3.6	B2	290	
7 π^1	04	54	53.7	+10	09	03	4.65	0.9	A0p	83	
9 o^2	04	56	22.2	+13	30	52	4.07	−0.1	K2	68	
10 π^6	04	58	32.8	+01	42	51	4.47	−2.2	K2	190	
11	05	04	34.1	+15	24	14	4.68	0.0	A0p	26	
17 ρ	05	13	17.4	+02	51	40	4.46	−0.2	K3	86	
19 β	05	14	32.2	−08	12	06	0.12v	−7.1	B8	280	Rigel
20 τ	05	17	36.3	−06	50	40	3.60	−2.2	B5	130	
22 o	05	21	45.7	−00	22	57	4.73	−3.0	B2	340	
29 e	05	23	56.7	−07	48	29	4.14	0.3	G8	58	
28 η	05	24	28.6	−02	23	50	3.36	−3.5	B1	230	Algjebbah
24 γ	05	25	07.8	+06	20	59	1.64	−3.6	B2	110	Bellatrix
30 ψ	05	26	50.2	+03	05	44	4.59	−3.0	B2	330	
31 Cl	05	29	43.9	−01	05	32	4.71	−0.3	K5	93	
32 A	05	30	47.0	+05	56	53	4.20	−1.6	B5	150	
36 υ	05	31	55.8	−07	18	05	4.62	−4.1	B0	560	Thabit
34 δ	05	32	00.3	−00	17	57	2.23v	−6.1	09.5	720	Mintaka
37 ϕ^1	05	34	49.2	+09	29	22	4.41	−4.6	B0	570	
39 λ	05	35	08.2	+09	56	02	3.39	−51	O8	550	Heka
42 c	05	35	23.1	−04	50	18	4.59	−3.6	B2	135	(near θ)
44 ι	05	35	25.9	−05	54	36	2.76	−6.0	09	570	Hatysa
46 ε	05	36	12.7	−01	12	07	1.70	−6.2	B0	370	Alnilam
40 ϕ^2	05	36	54.3	+09	17	27	4.09	0.2	K0	60	
48 σ	05	38	44.7	−02	36	00	3.73	−4.4	09.5	550	
47 ω	05	39	11.0	+04	07	17	4.57	−2.9	B3	310	
50 ζ	05	40	45.5	−01	56	34	1.77	−5.9	09.5	340	Alnitak
53 κ	05	47	45.3	−09	40	11	2.06	−6.9	B0.5	650	Saiph
54 χ^1	05	54	22.9	+20	16	34	4.41	4.4	G0	9.9	
58 α	05	55	10.2	+07	24	26	var.	−5.6v	M2	95	Betelgeux
HD											
40657	06	00	03.3	−03	04	27	4.53	−0.1	K2	80	
61 μ	06	02	22.9	+09	38	51	4.12	1.2	A0	37	
62 χ^2	06	03	55.1	+20	08	18	4.63	−6.8	B2	1000	
67 υ	06	07	34.2	+14	46	06	4.42	−1.7	B3	170	
70 ξ	06	11	56.3	+14	12	31	4.48	−1.7	B3	170	

δ has long been classed as a variable, but is in fact an eclipsing star with a range of only 2.20 to 2.35 (period 5.73 days). η is also an eclipsing star with a range of only 0 m.2. π^5 is an ellipsoidal variable with a range of 0.1 mag.

Also above mag. 5:

	Mag.	Abs. mag.	Spec.	Dist
15	4.82	1.9	F2	41
23 m	5.00	−3.5	B1	470
25	4.95	−3.5	B1	490
49 d	4.80	1.0	A4	57
56	4.78	−2.2	K2	240
51	4.91	0.0	K1	87
69	4.95	−1.1	B5	160

VARIABLE STARS

	R.A.		Dec.		Range	Type	Period, d.	Spec.
	h	m	deg	min				
W	05	05.4	+01	11	5.9–7.7	Semi-reg.	212	N
S	05	29.0	−04	42	7·5–13.5	Mira	419.2	M
CK	05	30.3	+04	12	5.9–7.1	Semi-reg.	120	K
α	05	55.2	+07	24	0.1–0.9	Semi-reg.	2110	M
U	05	55.8	+20	10	4.8–12.6	Mira	372.4	M

ORION (continued)

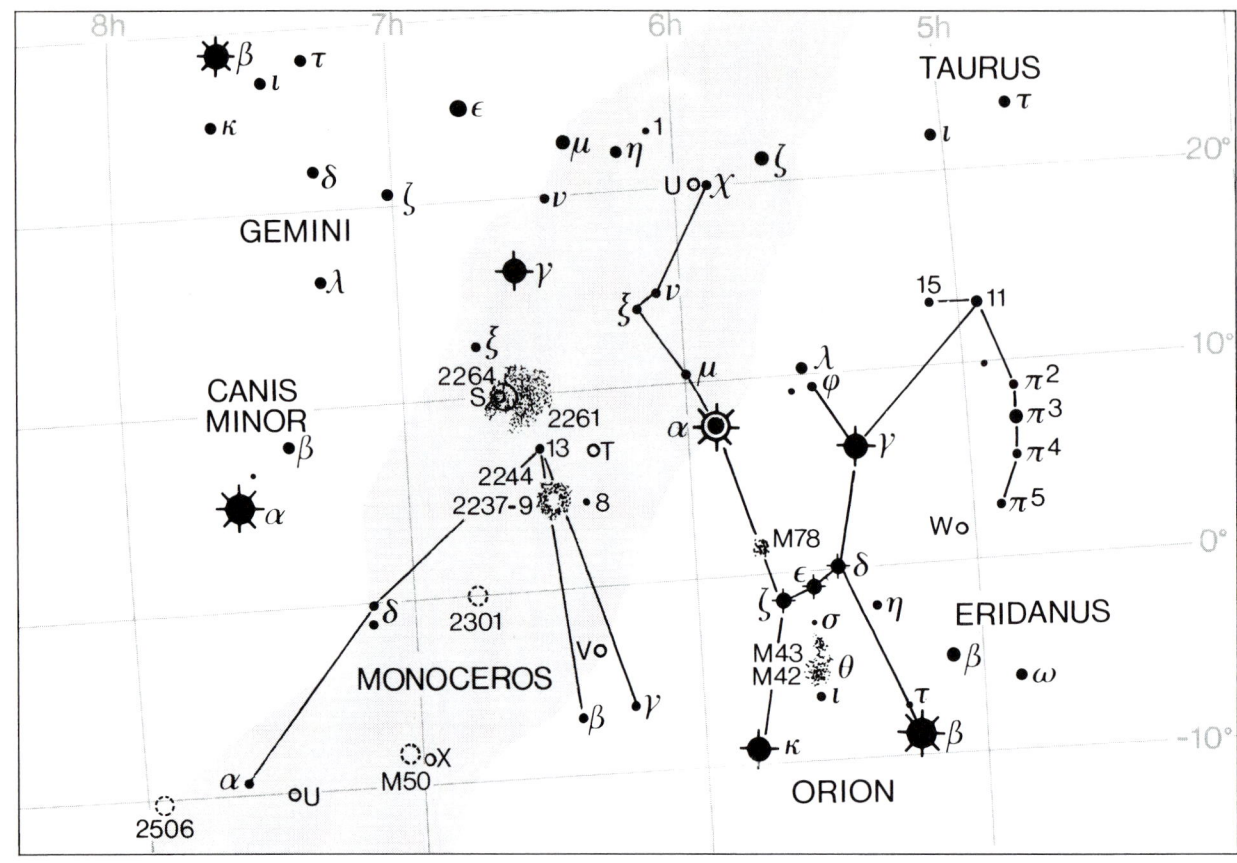

DOUBLE STARS

	R.A. h	m	Dec. deg	min	P.A. deg	Sep. sec	Mags.	
π^3	04	49.8	+06	58	138	94.6	3.2, 8.7	
β	05	14.5	−08	12	202	9.5	0.1, 6.8	
ρ	05	15.3	+02	54	064	7.0	4.5, 8.3	
η	05	24.5	−02	24	AB 078	1.6	3.8, 4.8	
					AC 051	115.1	9.4	
δ	05	32.0	−00	18	359	52.6	2.2v, 6.3	
λ	05	35.1	+09	56	043	4.4	3.6, 5.5	
σ	05	38.7	−02	36	AB 137	0.2	4.0, 6.0	Binary, 170 y
					AB + C 238	11.4	10.3	
					AB + D 084	12.9	7.5	
					AB + E 061	42.6	6.5	
θ	05	35.3	−05	23	AB 031	8.8	6.7, 7.9	
					AC 132	12.8	5.1	
					AD 096	21.5	6.7	
ι	05	35.4	−05	55	141	11.3	2.8, 6.9	
ζ	05	40.8	−01	57	AB 162	2.4	1.9, 4.0	Binary, 1509 y
					AC 010	57.6	9.9	
μ	06	02.4	+09	39	023	0.4	4.4, 6.0	

ORION (continued)

OPEN CLUSTERS

M	C	NGC	R.A. h	m	Dec. deg	min	Diam. min	Mag.	No. of stars
		1981	05	35.2	−04	26	25	4.6	20
		2112	05	53.9	+00	24	11	9.1	5
		2175	06	09.8	+20	19	18	6.8	60
		2186	06	12.2	+05	27	4	8.7	30

NEBULÆ

M	C	NGC	R.A. h	m	Dec. deg	min	Dimensions min	Mag. of illuminating star	
42		1976	05	35.4	−05	27	66 × 60	5	Great Nebula
43		1982	05	35.6	−05	16	20 × 15	7	Extension of M.42
78		2068	05	46.7	+00	03	8 × 6	10	Nebula is mag. 8
		IC 434	05	41.0	−02	24	60 × 10	2	(ζ). Behind Horse's Head dark nebula. Barnard 33

PAVO

(Abbreviation: Pav).

One of the 'Southern Birds'. The brightest star, α, is somewhat isolated from the main pattern; the most celebrated object is κ, often called a Cepheid even though it is, strictly speaking, a W Virginis type variable.

Pavo includes six stars above the fourth magnitude.

	Mag.	Luminosity Sun = 1	Dist lt-yrs
α	1.94	700	230
β	3.42	28	100
δ	3.56	1.0	18.6
η	3.62	82	162
κ	3.9 (max)	3.8v	75
ε	3.96	50	134

κ is the only naked-eye W Virginis variable.

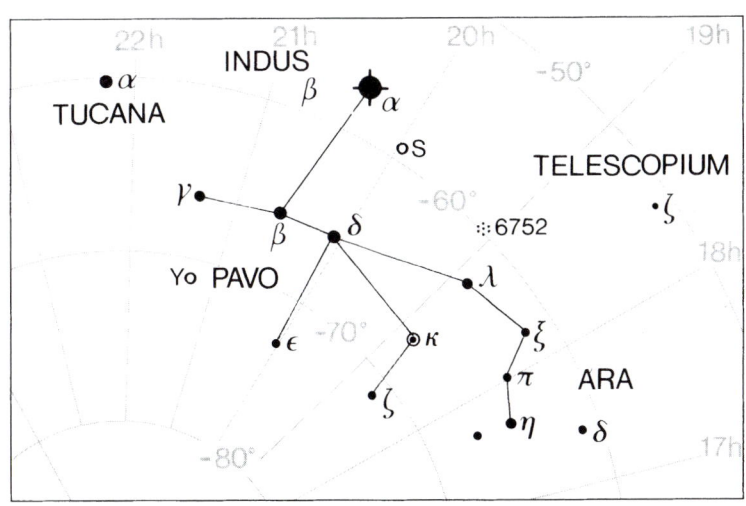

PAVO (continued)

BRIGHTEST STARS

Star	R.A. h	m	s	Dec. deg	min	sec	Mag.	Abs. mag.	Spec.	Dist pc
η	17	45	43.8	−64	43	25	3.62	0.0	K1	45
π	18	08	34.6	−63	40	06	4.35	1.6	A	17
ξ	18	23	13.3	−61	29	38	4.36	−0.5	M1	94
ν	18	31	22.2	−62	16	42	4.64	−1.2	B8	150
ζ	18	43	02.0	−71	25	42	4.01	−0.1	K2	66
λ	18	52	12.8	−62	11	16	4.22v	−3.5	B1	350
κ	18	56	56.9	−67	14	01	var.	3.4v	F5v	23
ε	20	00	35.4	−72	54	38	3.96	0.6	A0	41
δ	20	08	43.3	−66	10	56	3.56	4.8	G5	5.7
α	20	25	38.7	−56	44	06	1.94	−2.3	B3	71
β	20	44	57.4	−66	12	12	3.42	1.2	A5	28
γ	21	26	26.6	−65	21	59	4.22	4.5	F6	86

Also above mag. 5:

	Mag.	Abs. mag.	Spec.	Dist
ϕ^1	4.76		F0	16
ρ	4.88	3.4	F5	24

VARIABLE STARS

	R.A. h	m	Dec. deg	min	Range	Type	Period, d.	Spec.
R	18	12.9	−63	37	7.5–13-8	Mira	229.8	M
λ	18	52.2	−62	11	3.4–4.3	Irreg. (γc)	—	B
κ	18	56.9	−67	14	3.9–4.7	W Virginis	9.09	F
T	19	50.7	−71	46	7.0–14.0	Mira	244.0	M
S	19	55.2	−59	12	6.6–10.4	Semi-reg.	386	M
Y	21	24.3	−69	44	5.7–8.5	Semi-reg.	233	N
SX	21	28.7	−69	30	5.4–6.0	Semi-reg.	50	M

DOUBLE STAR

	R.A. h	m	Dec. deg	min	P.A. deg	Sep. sec	Mags.
ξ	18	23.2	−61	30	154	3.3	4.4, 8.6

GLOBULAR CLUSTER

M	C	NGC	R.A. h	m	Dec. deg	min	Diameter min	Mag.
	93	6752	19	10.9	−59	59	20.4	5.4

GALAXIES

M	C	NGC	R.A. h	m	Dec. deg	min	Mag.	Dimensions min	Type
		IC 4662	17	47.1	−64	38	11.4	2.2 × 1.4	Irreg.
		6684	18	49.0	−65	11	10.4	3.7 × 2.7	SB0
	101	6744	19	09.8	−63	51	9.0	15.5 × 10.2	SBb
		6753	19	11.4	−57	03	11.9	2.5 × 2.2	Sb

PEGASUS

(Abbreviation: Peg).

One of the most distinctive of the northern constellations. It commemorates the flying horse which the hero Bellerophon rode during an expedition to slay the fire-breathing Chimæra. The main stars of Pegasus make up a square; three of these are α, β and γ. The fourth is Alpheratz, which used to be included in Pegasus as δ Pegasi, but has been officially – and, frankly, illogically – transferred to Andromeda, as α Andromedæ. In Pegasus, excluding Alpheratz, there are nine stars above the fourth magnitude.

	Mag.	Luminosity Sun = 1	Dist lt-yrs
ε	2.38	5000	520
β	2.4 (max)	310v	176
α	2.49	60	100
γ	2.83v	1320	490
η	2.94	200	173
ζ	3.40	82	156
μ	3.48	60	147
θ	3.53	26	82
ι	3.76	3.8	39

PEGASUS (continued)

The celebrated group of galaxies known as Stephan's Quintet lies in Pegasus; NGC 7317, 7318 A and B, 7319 and 7320. All are faint, but are of special interest.

NGC	R.A.		Dec.		Dimensions	Mag.	Type
	h	m	deg	min	min		
7317	22	35.9	+33	57	1.0 × 0.8	14.6	E2
7318A	22	35.9	+33	58	1.0 × 1.0	14.3	E2p
7318B	22	36.0	+33	58	1.9 × 1.3	14.0	SBb+p
7319	22	36.1	+33	49	1.7 × 1.3	14.0	SBb+p
7320	22	36.1	+33	57	2.2 × 1.2	13.3	Sd

Though the galaxies seem to be connected, their red shifts are not the same; NGC 7320 has a red shift much less than the remaining members. Neither are the radial velocities the same.

The first star shown to be attended by a planet was 51 Pegasi. R.A. 22 h 57 m 273.0 s, dec. +20° 45′ 08″. Mag. 5.49, absolute mag. +5.0, type G4, distance 13 parsecs (42 light-years). The planet is said to have a mass of 0.47 that of Jupiter, and to have a distance of 0.08 a.u. from the star.

BRIGHTEST STARS

Star	R.A.			Dec.			Mag.	Abs.	Spec.	Dist	
	h	m	s	deg	min	sec		mag.		pc	
1	21	22	05.0	+19	48	16	4.08	0.0	K1	63	
2	21	29	56.8	+23	38	20	4.57	−0.5	M1	97	
8 ε	21	44	11.0	+09	52	30	2.38	−4.4	K2	160	Enif
9	21	44	30.5	+17	21	00	4.34	−4.5	G5	490	
10 κ	21	44	38.5	+25	38	42	4.13	2.1	F5	27	
24 ι	22	07	00.5	+25	20	42	3.76	3.4	F5	13	
29 π	22	09	59.1	+33	10	42	4.29	−0.6	F5	96	
26 θ	22	10	11.8	+06	11	52	3.53	1.4	A2	25	Biham
42 ζ	22	41	27.6	+10	49	53	3.40	0.0	B8.5	48	Homan
44 η	22	43	00.0	+30	13	17	2.94	−0.9	G2	53	Matar
47 λ	22	46	31.7	+23	33	56	3.95	−0.9	G8	33	
46 ξ	22	46	41.4	+12	10	22	4.19	3.8	F7	14	Al Suud al Nujam
48 μ	22	50	00.0	+24	36	06	3.48	0.2	K0	45	Sadalbari
53 β	23	03	46.3	+28	04	58	var.	−1.4v	M2	54	Scheat
54 α	23	04	45.5	+15	12	19	2.49	0.2	B9	31	Markab
55	23	07	00.1	+09	24	34	4.52	−0.5	M2	100	
62 τ	23	20	38.1	+23	44	25	4.60	1.2	A5	47	Kerb
68 υ	23	25	22.7	+23	24	15	4.40	2.4	F8	22	
70 q	23	29	09.1	+12	45	38	4.55	0.3	G8	70	
84 ψ	23	57	45.4	+25	08	29	4.66	−0.5	M3	110	
88 γ	00	13	14.1	+15	11	01	2.83v	−3.0	B2	150	Algenib

Also above mag. 5:

	Mag.	Abs. mag.	Spec.	Dist
22 ν	4.84	−0.3	K4	99
32	4.81	−0.2	B8	90
35	4.79	0.2	K0	77
43 o	4.79	1.2	A1	52
50 ρ	4.90	1.2	A1	55
56	4.76	−2.1	K0	240
72	4.98	−0.3	K4	110
78	4.93	0.2	K0	88
89 χ	4.80	−0.5	M2	120

VARIABLE STARS

	R.A.		Dec.		Range	Type	Period, d.	Spec.
	h	m	deg	min				
AG	21	51.0	+12	38	6.0–9.4	Z Andromedæ	830	WN+M
AW	21	52.3	+24	01	7.8–9.2	Algol	10.62	A+F
V	22	01.0	+06	07	7.0–15.0	Mira	302.3	M
TW	22	04.0	+28	21	7.0–9.2	Semi-reg.	956	M
RZ	22	05.9	+33	30	7.6–13.6	Mira	439.4	M
β	23	03.8	+28	05	2.3–2.8	Semi-reg.	38	M
R	23	06.6	+10	33	6.9–13.8	Mira	378.0	M
W	23	19.8	+26	17	7.9–13.0	Mira	344.9	M
S	23	20.6	+08	55	7.1–13.8	Mira	319.2	M
Z	00	00.1	+25	53	7.7–13.6	Mira	325.5	M

PEGASUS (continued)

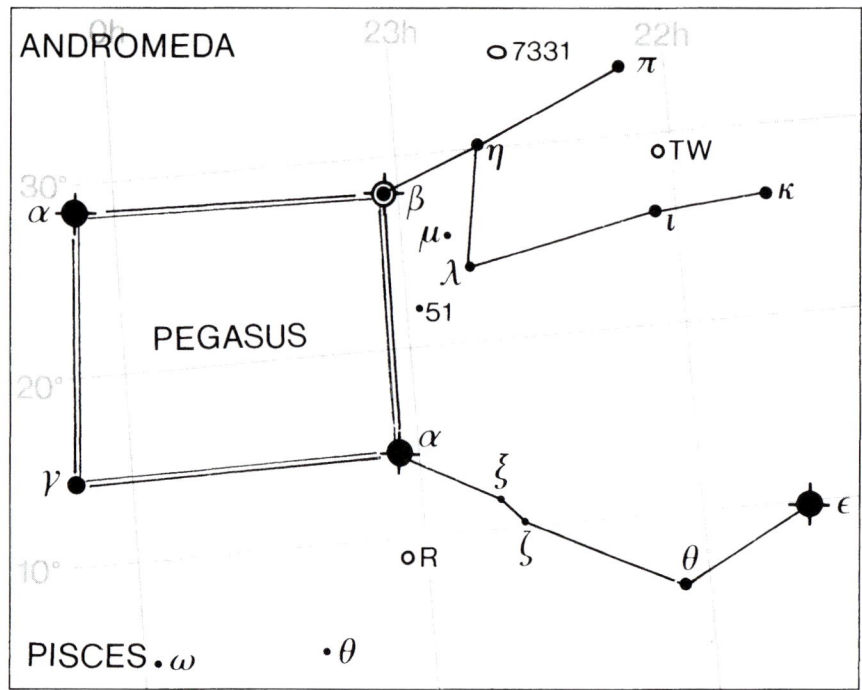

DOUBLE STARS

	R.A.		Dec.		P.A.	Sep.	Mags.	
	h	m	deg	min	deg	sec		
ε	21	44.2	+09	52	AB 325	81.8	2.4, 11.2	
					AC 320	142.5	9.4	
κ	21	44.6	+25	39	095	0.3	4.7, 5.0	Binary, 11.6 y
35	22	27.9	+04	42	AB 210	98.3	4.8, 9.8	
					AC 241	181.5	9.7	
37	22	30.0	+04	26	118	0.9	5.8, 7.1	Binary, 140 y
η	22	43.0	+30	13	339	90.4	2.9, 9.9	B is a close dble
β	23	03.8	+28	05	AB 211	108.5	2v, 11.6	
					AC 098	253.1	9.4	

GLOBULAR CLUSTER

M	C	NGC	R.A.		Dec.		Diameter	Mag.
			h	m	deg	min	min	
15		7078	21	30.0	+12	10	12.3	6.3

GALAXIES

M	C	NGC	R.A.		Dec.		Mag.	Dimensions	Type
			h	m	deg	min		min	
	30	7331	22	37.1	+34	25	9.5	10.7 × 4.0	Sb
		7332	22	37.4	+23	48	11.8	4.2 × 1.3	E7
	44	7479	23	04.9	+12	19	11.0	4.1 × 3.2	SBb
	43	7814	00	03.3	+16	09	10.5	6.3 × 2.6	Sb

PERSEUS

(Abbreviation: Per).

A prominent constellation, containing the prototype eclipsing star Algol as well as the superb Sword-Handle cluster (H.VI.33–4, C.14). Mythologically, Perseus was the hero of one of the most famous of all legends; he killed the Gorgon, Medusa, and married Andromeda, daughter of Cepheus and Cassiopeia. The Gorgon's Head is marked by the winking 'Demon Star', Algol. Perseus includes 12 stars above the fourth magnitude.

	Mag.	Luminosity Sun = 1	Dist lt-yrs
α	1.80	6000	620
β	2.12 (max)	105	95
ζ	2.85	16 000	1100
ε	2.89	2600	680
γ	2.93	58	110
δ	3.01	650	326
ρ	3.2 (max)	120v	196
η	3.76	5000	820
ν	3.77	520	460
κ	3.80	60	173
o	3.83	5000	1010
τ	3.95	58	176

β (Algol) is the prototype eclipsing binary. The main component (A) is of type B, around 4 000 000 km in diameter, and just over 100 times as luminous as the Sun. The secondary (B) is of type G, over 5 000 000 km in diameter and 3 times as luminous as the Sun, though its mass is less than that of A. Eclipses are not total. The secondary minimum has an amplitude of no more than 0.1 magnitude.

BRIGHTEST STARS

Star	R.A. h	m	s	Dec. deg	min	sec	Mag.	Abs. mag.	Spec.	Dist pc	
φ	01	43	39.6	+50	41	20	4.07	−3.9	B1	350	
13 θ	02	44	11.9	+49	13	43	4.12	3.8	F7	13	
16	02	50	34.9	+38	19	07	4.23	0.6	F2	50	
15 η	02	50	41.8	+55	53	44	3.76	−4.4	K3	250	Miram
17	02	51	30.8	+35	03	35	4.53	−0.3	K5	150	
18 τ	02	54	15.4	+52	45	45	3.95	0.3	G4	54	Kerb
22 π	02	58	45.6	+39	39	46	4.70	1.4v	A2	44	Gorgonea Secunda
23 γ	03	04	47.7	+53	30	23	2.93	0.3	G8	34	
25 ρ	03	05	10.5	+38	50	25	var.	−0.5v	M4	60	Gorgonea Terti
26 β	03	08	10.1	+40	57	21	2.12v	−0.2	B8	29	Algol
ι	03	09	03.9	+49	36	49	4.05	3.7	G0	12	
27 κ	03	09	29.7	+44	51	27	3.80	0.2	K0	53	Misam
28 ω	03	11	17.3	+39	36	42	4.63	0.0	K0	180	Gorgonea Quarta
33 α	03	24	19.3	+49	51	40	1.80	−4.6	F5	190	Mirphak
34	03	29	22.0	+49	30	32	4.67	−2.3	B3	210	
35 σ	03	30	34.4	+47	59	43	4.35	−0.2	K3	71	
37 ψ	03	36	29.3	+48	11	34	4.23	−1.2	B5e	126	
39 δ	03	42	55.4	+47	47	15	3.01	−2.2	B5	100	
38 o	03	44	19.1	+32	17	18	3.83	−4.4	B1	310	Ati
41 ν	03	45	11.6	+42	34	43	3.77	−2.0	F5	140	
44 ζ	03	54	07.8	+31	53	01	2.85	−5.7	B1	340	Atik
45 ε	03	57	51.1	+40	00	37	2.89	−3.7	B0.5	208	
46 ξ	03	58	57.8	+35	47	28	4.04	−5.4	O7	210	Menkib
47 λ	04	06	35.0	+50	21	05	4.29	0.2	B9	62	
48 υ	04	08	39.6	+47	42	45	4.04	−1.7	B3	140	Nembus
52 f	04	14	53.3	+40	29	01	4.71	−4.5	G5 + A5	560	
51 μ	04	14	53.8	+48	24	33	4.14	−4.5	G0	460	
b	04	18	14.6	+50	17	44	4.62v	var.	A2	55	

Also above mag. 5:

	Mag.	Abs. mag.	Spec.	Dist
12	4.91	4.2	F9	15
24	4.93	−0.1	K2	92
32	4.95	1.4	A2	51
40	4.97	0.5	B5	
54	4.93	0.3	G8	84
53	4.85	−1.7	B3	170

PERSEUS (continued)

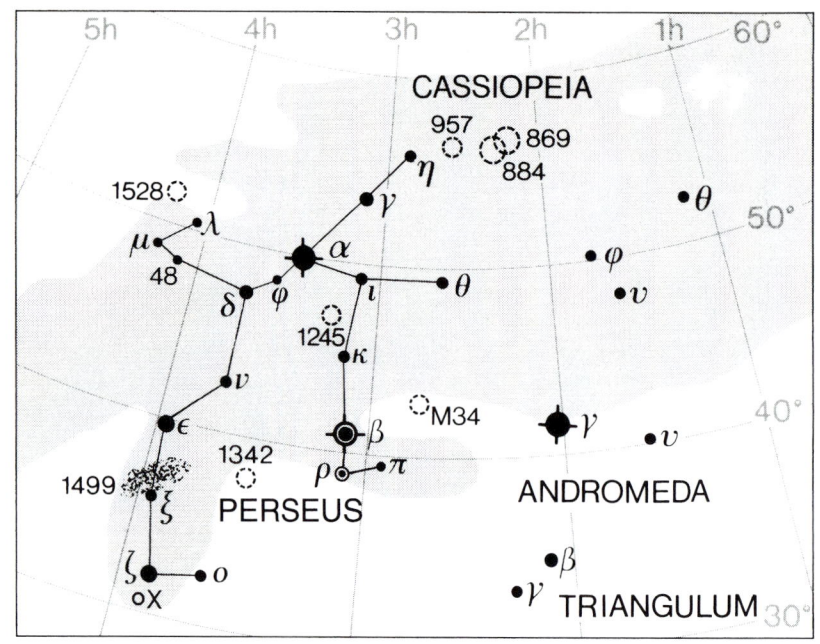

VARIABLE STARS

	R.A.		Dec.		Range	Type	Period, d.	Spec.	
	h	m	deg	min					
U	01	59.6	+54	49	7.4–12.3	Mira	321.0	M	
S	02	22.9	+58	35	7.9–11.5	Semi-reg.	Long	M	
ρ	03	05.2	+38	50	3–4	Semi-reg.	33 to 55	M	
β	03	08.2	+40	57	2.1–3.4	Algol	2.87	B + G	
R	03	30.1	+35	40	8.1–14.8	Mira	210.0	M	
X	03	55.4	+31	03	6.0–7.0	Irreg.	—	09.5m	X-ray source
AW	04	47.8	+36	43	7.1–7.8	Cepheid	6.46	F–G	

DOUBLE STARS

	R.A.		Dec.		P.A.	Sep.	Mags.	
	h	m	deg	min	deg	sec		
η	02	50.7	+55	54	300	28.3	3.3, 8.5	
θ	02	44.2	+49	14	215	19.8	4.1, 9.9	Binary, 2720 y
γ	03	04.8	+53	30	326	57.0	2.9, 10.6	
ζ	03	54.1	+31	53	AB 208	12.9	2.9, 9.5	
					AC 286	32.8	11.3	
					AD 195	94.2	9.5	
					AE 185	120.3	10.2	
ε	03	57.9	+40	01	010	8.8	2.9, 8.1	

OPEN CLUSTERS

M	C	NGC	R.A.		Dec.		Diam.	Mag.	No. of stars	
			h	m	deg	min	min			
		744	01	58.4	+55	29	11	7.9	20	
	14	869	02	19.0	+57	09	30	4.3	200 ⎤	Sword
		884	02	22.4	+57	07	30	4.4	150 ⎦	Handle
		957	02	33.6	+57	32	11	7.6	30	
34		1039	02	42.0	+42	47	35	5.2	60	
		1245	03	14.7	+47	15	10	8.4	200	
		1444	03	49.4	+52	40	4	6.6	—	
		1513	04	10.0	+49	31	9	8.4	50	
		1528	04	15.4	+51	14	24	6.4	40	
		1545	04	20.9	+50	15	8	6.2	20	

PERSEUS (continued)

PLANETARY NEBULA

M	C	NGC	R.A.		Dec.		Diameter	Mag.	Mag. of	
			h	m	deg	min	sec		central star	
76		650–1	01	42.4	+51	34	65 × 290	12.2	17	Little Dumbbell

NEBULÆ

M	C	NGC	R.A.		Dec.		Diameter	Mag. of illuminating star	
			h	m	deg	min	min		
		1333	03	29.3	+31	25	9 × 7	9.5	(Near dark nebula B.205)
		1499	04	00.7	+36	37	145 × 40	4	California Nebula

GALAXIES

M	C	NGC	R.A.		Dec.		Mag.	Dimensions	Type	
			h	m	deg	min		min		
		1003	02	39.3	+40	52	11.5	5.4 × 2.1	Sc	
		1023	02	40.4	+39	04	9.5	8.7 × 3.3	E7p	
	24	1275	03	19.8	+41	31	11.6	2.6 × 1.9	Pec.	Perseus A

PHŒNIX

(Abbreviation: Phe).

One of the 'Southern birds'. It is not very distinctive, and Ankaa is the only bright star; there are however seven stars above the fourth magnitude.

	Mag.	Luminosity Sun = 1	Dist lt-yrs
α	2.39	60	78
β	3.31	58	130
γ	3.41	5000	910
ζ	3.6 (max)	105	220
ε	3.88	60	75
κ	3.94	17	62
δ	3.95	17	91

ζ is the brightest Algol-type eclipsing binary apart from Algol itself and λ Tauri.

BRIGHTEST STARS

Star	R.A.			Dec.			Mag.	Abs. mag.	Spec.	Dist pc	
	h	m	s	deg	min	sec					
ι	23	35	04.4	−42	36	54	4.71	?	A	29	
ε	00	09	24.6	−45	44	51	3.88	0.2	K0	23	
κ	00	26	12.1	−43	40	48	3.94	1.7	A3	19	
α	00	26	17.0	−42	18	22	2.39	0.2	K0	24	Ankaa
μ	00	41	19.5	−46	05	06	4.59	0.3	G8	70	
η	00	43	21.2	−57	27	48	4.36	0.0	A0	71	
β	01	06	05.0	−46	43	07	3.31	0.3	G8	40	
ζ	01	08	23.0	−55	14	45	var.	−0.2	B8	67	
γ	01	28	21.9	−43	19	06	3.41	−4.4	K5	280	
δ	01	31	15.0	−49	04	22	3.95	1.7	K0	28	
ψ	01	53	38.7	−46	18	09	4.41	−0.5	M4	96	

Also above mag. 5:

	Mag.	Abs. mag.	Spec.	Dist
λ¹	4.77	0.6	A0	130
ν	4.96	4.0	F8	15
HD 12055	4.83	2.6	G5	90

VARIABLE STARS

	R.A.		Dec.		Range	Type	Period, d.	Spec.
	h	m	deg	min				
SX	23	46.5	−41	35	6.8–7.5	Delta Scuti	0.055	A
R	23	56.5	−49	47	7.5–14.4	Mira	267.9	M
S	23	53.1	−56	35	7.4–8.2	Semi-reg.	141	M
ζ	01	08.4	−55	15	3.6–4.1	Algol	1.67	B + B

PHŒNIX (continued)

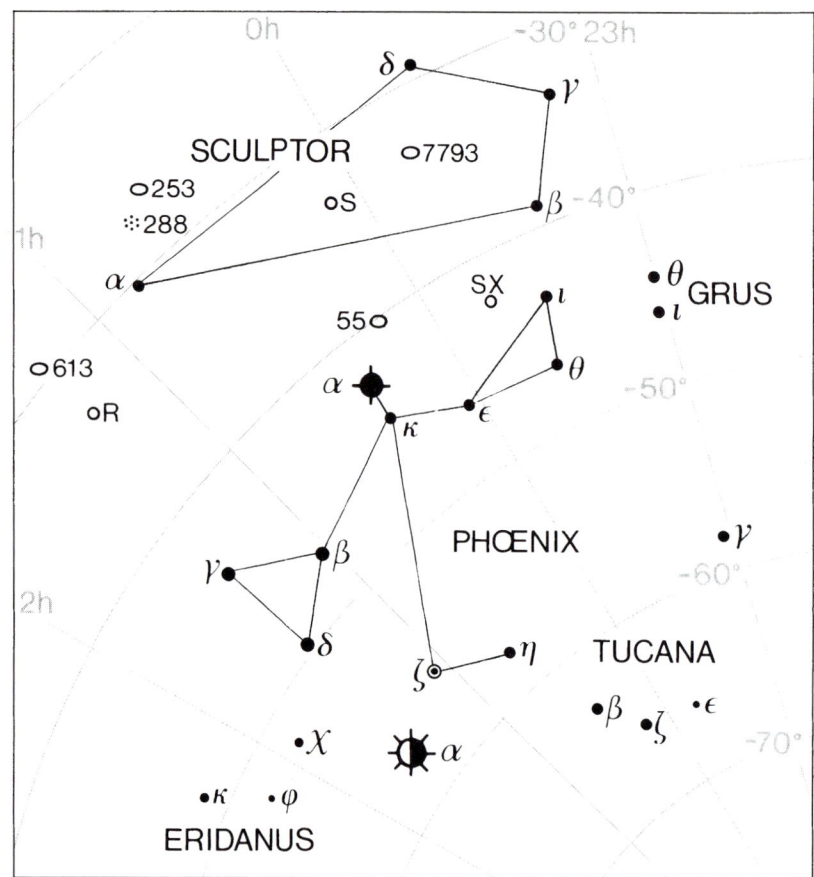

DOUBLE STARS

	R.A. h	m	Dec. deg	min	P.A. deg	Sep. sec	Mags.
ξ	00	41.8	−56	30	253	13.2	5.8, 10.2
η	00	43.4	−57	28	217	19.8	4.4, 11.4
β	01	06.1	−46	43	346	1.4	4.0, 4.2

PICTOR

(Abbreviation: Pic).

An unremarkable constellation near Canopus, known originally under the cumbersome name of Equuleus Pictoris (the Painter's Easel). There are two stars above the fourth magnitude.

	Mag.	Luminosity Sun = 1	Dist lt-yrs
α	3.27	11	52
β	3.85	58	78

β Pictoris is surrounded by a cloud of cool, possibly planet-forming material, recorded in infra-red and also visually. See chart for Carina.

PICTOR (continued)

BRIGHTEST STARS

Star	R.A.			Dec.			Mag.	Abs.	Spec.	Dist
	h	m	s	deg	min	sec		mag.		pc
β	05	47	17.1	−51	03	59	3.85	0.3	A5	24
γ	05	49	49.6	−56	10	00	4.51	0.0	K1	80
δ	06	10	17.9	−54	58	07	4.7v		B1	
α	06	48	11.4	−61	56	29	3.27	2.1	A5	16

VARIABLE STARS

	R.A.		Dec.		Range	Type	Period, d.	Spec.
	h	m	deg	min				
R	04	46.2	−49	15	6.7–10.0	Semi-reg.	164	M
S	05	11.0	−48	30	6.5–14.0	Mira	426.6	M
T	05	15.1	−46	55	7.9–14.4	Mira	200.6	M
δ	06	10.3	−54	58	4.7–4.9	Beta Lyræ	1.67	B

DOUBLE STARS

	R.A.		Dec.		P.A.	Sep.	Mags.
	h	m	deg	min	deg	sec	
ι	04	50.9	−53	28	058	12.3	5.6, 6.4
θ	05	24.8	−52	19	AB 152	0.2	6.9, 7.2
					AB + C 287	38.2	6.8
μ	06	32.0	−58	45	231	2.4	5.8, 9.0

PISCES

(Abbreviation: Psc).

A large but faint Zodiacal constellation; it now contains the Vernal Equinox. Its mythological associations are rather vague, but it may represent the fishes into which Venus and Cupid once changed themselves in order to escape from the monster Typhon. There are three stars above the fourth magnitude.

	Mag.	Luminosity Sun = 1	Dist lt-yrs
η	3 62	58	144
γ	3.69	58	156
α	3.79	26	98

BRIGHTEST STARS

Star	R.A.			Dec.			Mag.	Abs.	Spec.	Dist	
	h	m	s	deg	min	sec		mag.		pc	
4 β	23	03	52.5	+03	49	12	4.53	−0.4	B5p	101	
6 γ	23	17	09.7	+03	16	56	3.69	0.3	G8	48	
10 θ	23	27	57.9	+06	22	44	4.28	0.0	K1	72	
17 ι	23	39	56.9	+05	37	35	4.13	3.8	F7	13	
18 λ	23	42	02.6	+01	46	48	4.50	2.4	A7	26	
19 TX	23	46	23.3	+03	29	13	4.3–5.1	−2.0v	N	330	
28 ω	23	59	18.5	+06	51	48	4.01	0.8	F4	26	
30	00	01	57.5	−06	00	51	4.41	0.4	M3	30	
33	00	05	20.1	−05	42	27	4.61	0.0	K1	84	
63 δ	00	48	40.9	+07	35	06	4.43	−0.3	K5	88	
71 ε	01	02	56.5	+07	53	24	4.28	0.2	K0	66	
74 ψ	01	05	40.9	+21	28	24	4.7	−0.1, 0.2	B9.5, B9	120	
84 χ	01	11	27.1	+21	02	05	4.66	0.2	K0	75	
83 τ	01	11	39.5	+30	05	23	4.51	1.7	K0	29	
85 φ	01	13	44.8	+24	35	01	4.65	0.2	K0	74	
99 η	01	31	28.9	+15	20	45	3.62	0.3	G8	44	Alpherg
106 ν	01	41	25.8	+05	29	15	4.44	−0.2	K3	42	
110 o	01	45	23.5	+09	09	28	4.26	0.2	K0	65	Torcular
111 ξ	01	53	33.3	+03	11	15	4.62	0.2	K0	77	
113 α	02	02	02.7	+02	45	49	3.79	1.4	A2	30	Al Rischa

PISCES (continued)

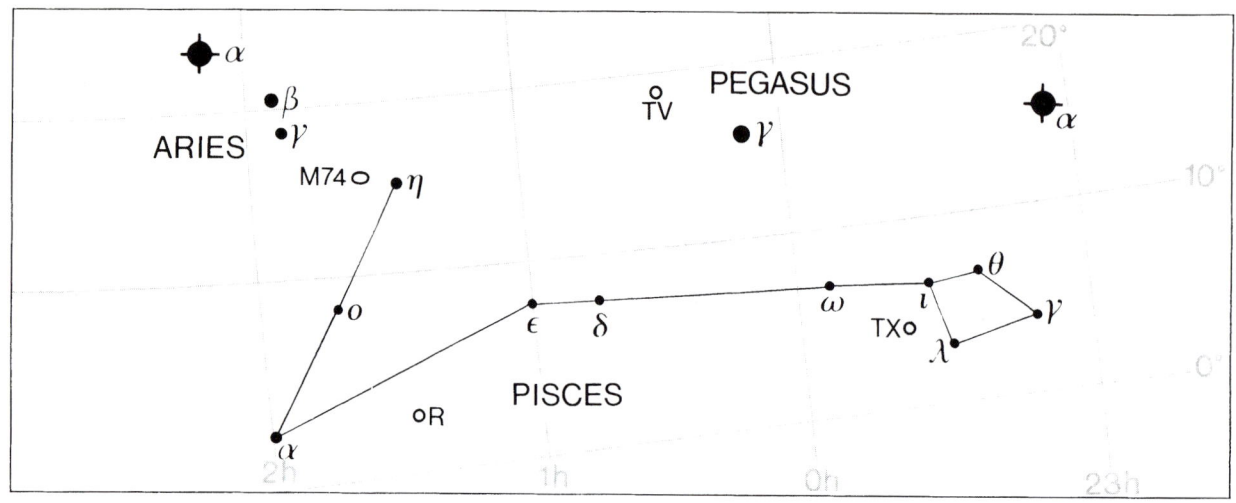

Also above mag. 5:

	Mag.	Abs. mag.	Spec.	Dist
κ	4.94	2.2	A2	
27	4.86	0.2	G9	86
86 ζ	4.86	2.1	F6	33
90 υ	4.76	1.4	A2	47
98 μ	4.84	−0.3	K4	67
47 TV	4.8v	−0.5	M3	140

VARIABLE STARS

	R.A. h	m	Dec. deg	min	Range	Type	Period, d.	Spec.
TX	23	46.4	+03	29	6.9–7.7	Irreg.	—	N
TV	00	28.0	+17	24	4.6–5.4	Semi-reg.	70	M
Z	01	16.1	+25	46	7.0–7.9	Semi-reg.	144	N
R	01	30.6	+02	53	7.1–14.8	Mira	344.0	M

DOUBLE STARS

	R.A. h	m	Dec. deg	min	P.A. deg	Sep. sec	Mags.	
ζ	01	13.7	+07	35	063	23.0	5.6, 6.5	
ψ¹	01	05.6	+21	28	AB 159	30.0	5.6, 5.8	
					AC 123	92.6	11.2	
α	02	02.0	+02	46	279	1.9	4.2, 5.1	Binary, 933 y

GALAXIES

M	NGC	R.A. h	m	Dec. deg	min	Mag.	Dimensions min	Type
	470	01	19.7	+03	25	11.9	3.0 × 2.0	Sc
	474	01	20.1	+03	25	11.1	7.9 × 7.2	S0
	488	01	21.8	+05	15	10.3	5.2 × 4.1	Sb
	524	01	24.8	+09	32	10.6	3.2 × 3.2	E1
74	628	01	36.7	+15	47	9.2	10.2 × 9.5	Sc

Piscis Australis

(Abbreviation: PsA).

Also known as Piscis Austrinus. No specific mythological legends have been associated with it. It contains Fomalhaut, which is, incidentally, the most southerly of the first-magnitude stars to be visible from England, but there are no other stars above the fourth magnitude.

	Mag.	Luminosity Sun = 1	Dist lt-yrs
α	1.16	13	22

Fomalhaut is one of the stars known to be associated with cool, possibly planet-forming material.

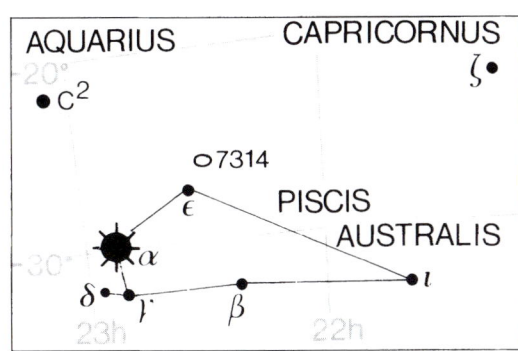

BRIGHTEST STARS

Star	R.A. h	m	s	Dec. deg	min	sec	Mag.	Abs. mag.	Spec.	Dist pc	
9 ι	21	44	56.7	−33	01	33	4.34	0.6	A0	42	
14 μ	22	08	22.8	−32	59	19	4.50	1.4	A2	42	
17 β	22	31	30.1	−32	20	46	4.29	0.6	A0	53	Fum el Samakah
18 ε	22	40	39.2	−27	02	37	4.17	−0.2	B8	75	
22 γ	22	52	31.4	−32	52	32	4.46	0.6	A0	59	
23 δ	22	55	56.8	−32	32	23	4.21	0.3	gG4	49	
24 α	22	57	38.9	−29	37	20	1.16	2.0	A3	6.7	Fomalhaut

Also above mag. 5:

	Mag.	Abs. mag.	Spec.	Dist
υ	4.99	−0.3	K5	110
15 τ	4.92	3.4	F5	11

VARIABLE STARS

	R.A. h	m	Dec. deg	min	Range	Type	Period, d.	Spec.
S	22	03.8	−28	03	8.0–14.5	Mira	271.7	M
V	22	55.3	−29	37	8.0–9.0	Semi-reg.	148	M

DOUBLE STARS

	R.A. h	m	Dec. deg	min	P.A. deg	Sep. sec	Mags.	
η	22	00.8	−28	27	115	1.7	5.8, 6.8	
β	22	31.5	−32	21	172	30.3	4.4, 7.9	(optical)
γ	22	52.5	−32	53	262	4.2	4.5, 8.0	
δ	22	55.9	−32	32	244	5.0	4.2, 9.2	

GALAXIES

M	C	NGC	R.A. h	m	Dec. deg	min	Mag.	Dimensions min	Type	
		7172	22	02.0	−31	52	11.9	2.2 × 1.3	S	
		7174	22	02.1	−31	59	12.6	1.3 × 0.7	S	
		7314	22	35.8	−26	03	10.9	4.6 × 2.3	Sc	Arp 14

PUPPIS

(Abbreviation: Pup).

The poop of the dismembered ship, Argo Navis. There are 12 stars above the fourth magnitude.

	Mag.	Luminosity Sun = 1	Dist lt-yrs
ζ	2.25	60 000	2400
π	2.70	110	130
ρ	2.81v	525	300
τ	2.93	60	82
υ	3.17	250	245
σ	3.25	110	166
ξ	3.34	5200	750
L²	3.4 (max)	1400v	75
c	3.59	8200	1730
a	3.73	58	125
k	3.82	150 + 130	360
3	3.96	70 000	5540

ρ is a variable with very small range (2.72–2.87).

L² is one of the brightest of the red semi-regular variables, and is a naked-eye object throughout its cycle. It has a 9.5 mag. companion at a distance of about 1′, but the separation is increasing, and the two probably do not make up a binary system.

See chart for Carina.

BRIGHTEST STARS

Star	R.A. h	m	s	Dec. deg	min	sec	Mag.	Abs. mag.	Spec.	Dist pc	
υ	06	37	45.6	−43	11	45	3.17	−1.2	B8	75	
τ	06	49	56.1	−50	36	53	2.93	0.2	K0	25	
l	07	12	33.6	−46	45	34	4.49	−2.6	F0	24	
L²	07	13	13.3	−45	10	59	var.	−3.1v	M5	23	
π	07	17	08.5	−37	05	51	2.70	−0.3	K5	40	
υ¹	07	18	18.4	−36	44	03	4.66	−1.7	B3	180	
σ	07	29	13.8	−43	18	05	3.25	−0.3	K5	51	
HD 60532	07	34	03.1	−22	17	46	4.45	2.3	F7	26	
p	07	35	22.8	−28	22	10	4.64	0.1	B8	77	
f	07	37	22.0	−34	58	07	4.53	−0.2	B8	88	
m	07	38	17.9	−25	21	53	4.67	−0.1	B8	92	
k	07	38	49.7	−26	48	13	3.82	−0.7, −0.6	B8	110	
1	07	43	32.3	−28	24	40	4.59	−0.3	gK5	79	
3	07	43	48.4	−28	57	18	3.96	−7.5	A2	1700	
c	07	45	15.2	−37	58	07	3.59	−5.0	cK	530	
o	07	48	05.1	−25	56	14	4.50	−3.5	B1	390	
Q	07	48	20.2	−47	04	39	4.71	0.2	K0	72	
P	07	49	14.3	−46	22	24	4.11	−5.6	B0	820	
7 ξ	07	49	17.6	−24	51	35	3.34	−6	G3	230	Asmidiske
a	07	52	13.0	−40	34	33	3.73	0.3	G5	38	
J	07	53	03.7	−49	36	47	4.63	−3.6	B2	440	
11	07	56	51.5	−22	52	48	4.20	−2.0	F8	150	
V	07	58	14.3	−49	14	42	var.	−4.0v	B1 + B3	470	
232G	07	59	52.0	−18	23	58	4.61	1.7	A3	38	
ζ	08	03	35.0	−40	00	12	2.25	−7.1	O5.8	740	Suhail Hadar
ρ	08	07	32.6	−24	18	15	2.81v	−2.0	F6	92	Turais
16	08	09	01.5	−19	14	42	4.40	−1.1	B5	130	
19	08	11	16.2	−12	55	37	4.72	0.2	K0	45	
h¹	08	11	21.5	−39	37	07	4.45	−5.9	cK	850	
h²	08	14	02.8	−40	20	52	4.44	−0.1	gK2	81	
q	08	18	33.2	−36	39	34	4.45	0.5	A7	62	

Also above mag. 5:

	Mag.	Abs. mag.	Spec.
Y	5.00	−0.2	G7
A	4.85		M3
d¹	4.91	−0.9	B3
r	4.77	−1.3	B3
W	4.94	1.8	M0

PUPPIS (continued)

VARIABLE STARS

	R.A. h	m	Dec. deg	min	Range	Type	Period, d.	Spec.
L²	07	13.5	−44	39	2.6–6.2	Semi-reg.	140	M
Z	07	32.6	−20	40	7.2–14.6	Mira	499.7	M
VX	07	32.6	−21	56	7.7–8.5	Cepheid	3.01	F
X	07	32.8	−20	55	7.8–9.2	Cepheid	25.96	F–G
W	07	46.0	−41	12	7.3–13.6	Mira	120.1	M
AP	07	57.8	−40	07	7.1–7.8	Cepheid	5.08	F
V	07	58.2	−49	15	4.7–5.2	Beta Lyræ	1.45	B + B
AT	08	12.4	−36	57	7.5–8.4	Cepheid	6.66	F–G
RS	09	13.1	−34	35	6.5–7.6	Cepheid	41.39	F–G

DOUBLE STAR

	R.A. h	m	Dec. deg	min	P.A. deg	Sep. sec	Mags.
σ	07	29.2	−43	18	074	22.3	3.3, 9.4
κ	07	38.8	−24	48	318	9.8	4.5, 4.7

OPEN CLUSTERS

M	C	NGC	R.A. h	m	Dec. deg	min	Diam. min	Mag.	No. of stars	
		2383	07	24.8	−20	56	6	8.4	40	
		2421	07	36.3	−20	37	10	8.3	70	
47		2422	07	36.6	−14	30	30	4.4	30	
		Mel 71	07	37.5	−12	04	9	7.1	80	
		Mel 72	07	38.4	−10	41	9	10.1	40	
		2432	07	40.9	−19	05	8	10.2	50	
		2439	07	40.8	−31	39	10	6.9	80	R Puppis. Asterism
46		2437	07	41.8	−14	49	27	6.1	100	
93		2447	07	44.6	−23	52	22	6.2	80	
		2451	07	45.4	−37	58	45	2.8	40	
	71	2477	07	52.3	−38	33	27	5.8	160	
		2479	07	55.1	−17	43	7	9.6	45	
		2489	07	56.2	−30	04	8	7.9	45	
		2509	08	00.7	−19	04	8	9.3	70	
		2527	08	05.3	−28	10	22	6.5	40	
		2533	08	07.0	−29	54	3.5	7.6	60	
		2539	08	10.7	−12	50	22	6.5	50	
		2546	08	12.4	−37	38	41	6.3	40	
		2567	08	18.6	−30	38	10	7.4	40	
		2571	08	18.9	−29	44	13	7.0	30	
		2580	08	21.6	−30	19	8	9.7	50	
		2587	08	23.5	−29	30	9	9.2	40	

PLANETARY NEBULÆ

M	C	NGC	R.A. h	m	Dec. deg	min	Dimensions sec	Mag.	Mag. of central star	
		2438	07	41.8	−14	44	66	10.1	17.7	In cluster NGC 2437
		2440	07	41.9	−18	13	14 × 32	10.8	14.3	Protoplanetary?

NEBULA

M	C	NGC	R.A. h	m	Dec. deg	min	Dimensions min	Mag. of illuminating star	
		2467	07	52.5	−26	24	8 × 7	9.2	Gum 9

PYXIS

(Abbreviation: Pyx).

Also originally part of Argo. The only stars above the fourth magnitude are α and β.

	Mag.	Luminosity Sun = 1	Dist lt-yrs
α	3.68	5000	1340
β	3.97	58	150

See chart for Carina.

BRIGHTEST STARS

Star	R.A. h	m	s	Dec. deg	min	sec	Mag.	Abs. mag.	Spec.	Dist pc
β	08	40	06.1	−35	18	30	3.97	0.3	G4	46
α	08	43	35.5	−33	11	11	3.68	−4.4	B2	410
γ	08	50	31.9	−27	42	36	4.01	−0.3	K4	73
κ	09	08	02.8	−25	51	30	4.58	−0.4	gM0	97
θ	09	21	29.5	−25	57	55	4.72	−0.5	M1	100
λ	09	23	12.1	−28	50	02	4.69	0.3	gG7	76

Also above mag. 5:

	Mag.	Abs. mag.	Spec.	Dist
ζ	4.89	0.3	G4	75
δ	4.89	1.7	A3	42

VARIABLE STARS

	R.A. h	m	Dec. deg	min	Range	Type	Period, d.	Spec.	
TY	08	59.7	−27	49	6.9–7.5	Eclipsing	3.20	G+G	
T	09	04.7	−32	23	6.3–14.0	Recurrent nova	—	Q	Outbursts 1890, 1902, 1920, 1944, 1966
S	09	05.1	−23	05	8.0–14.2	Mira	206.4	M	

DOUBLE STARS

	R.A. h	m	Dec. deg	min	P.A. deg	Sep. sec	Mags.
ζ	08	39.7	−29	34	061	52.4	4.9, 9.1
δ	08	55.5	−27	41	AB 268	23.8	4.9, 14.0
					CD 017	2.5	11.0, 11.0
ε	09	09.9	−30	22	A+BC 147	17.8	5.6, 10.5
					BC 088	0.3	10.5, 10.8
					AD 340	35.4	5.6, 13.5
κ	09	08.0	−25	52	263	2.1	4.6, 9.8

OPEN CLUSTERS

M	C	NGC	R.A. h	m	Dec. deg	min	Diam. min	Mag.	No. of stars
		2627	08	37.3	−29	57	11	8.4	60
		2658	08	43.4	−32	39	12	9.2	80

PLANETARY NEBULA

M	C	NGC	R.A. h	m	Dec. deg	min	Diameter sec	Mag.	Mag. of central star
		2818	09	16.0	−36	28	38	13.0	13.0

GALAXY

M	C	NGC	R.A. h	m	Dec. deg	min	Mag.	Dimensions min	Type
		2613	08	33.4	−22	58	10.4	7.2 × 2.1	Sb

RETICULUM

(Abbreviation: Ret).

Originally Reticulum Rhomboidalis (the Rhomboidal Net). A small but quite distinctive constellation of the far south. There are two stars above the fourth magnitude.

	Mag.	Luminosity Sun = 1	Dist lt-yrs
α	3.35	600	390
β	3.85	4.5	55

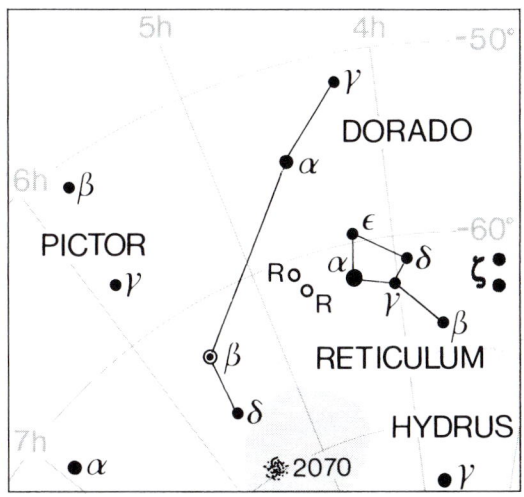

BRIGHTEST STARS

Star	R.A. h	m	s	Dec. deg	min	sec	Mag.	Abs. mag.	Spec.	Dist pc
κ	03	29	22.7	−62	56	15	4.72	3.4	F5	20
β	03	44	12.0	−64	48	26	3.85	3.2	K0	17
δ	03	58	44.7	−61	24	01	4.56	−0.5	M2	97
γ	04	00	53.8	−62	09	34	4.51	?	Mb	?
α	04	14	25.5	−62	28	26	3.35	−2.1	G6	120
ε	04	16	28.9	−59	18	07	4.44	−0.3	gK5	21

ζ is a wide, easy double; both components are of type G. They have common motion in space, and so are presumably associated.

Also above mag. 5:

	Mag.	Abs. mag.	Spec.	Dist
ι	4.97	−0.4	M0	120

VARIABLE STAR

	R.A. h	m	Dec. deg	min	Range	Type	Period, d.	Spec.
R	04	33.5	−63	02	6.5–14.0	Mira	278.3	M

DOUBLE STARS

	R.A. h	m	Dec. deg	min	P.A. deg	Sep. sec	Mags.	
β	03	44.2	−64	48	following	1440	3.9, 13.1	
ζ	03	18.2	−62	30	218	310.0	4.7, 5.2	Common proper motion

GALAXIES

M	NGC	R.A. h	m	Dec. deg	min	Mag.	Dimensions min	Type
	1313	03	18.3	−66	30	9.4	8.5 × 6.6	SBd
	1559	04	17.6	−62	47	10.4	3.3 × 2.1	SBc

SAGITTA

(Abbreviation: Sge).

An original constellation; small though it is, it is quite distinctive. It has been identified with Cupid's bow, and also with the arrow used by Apollo against the one-eyed Cyclops. There are two stars above the fourth magnitude.

	Mag.	Luminosity Sun = 1	Dist lt-yrs
γ	3.47	110	166
δ	3.82	800	550

See chart for Aquila.

BRIGHTEST STARS

Star	R.A. h	m	s	Dec. deg	min	sec	Mag.	Abs. mag.	Spec.	Dist pc	
5 α	19	40	05.6	+18	00	50	4.37	−2.0	G0	190	Alsahm
6 β	19	41	02.8	+17	28	33	4.37	−2.1	G8	200	
7 δ	19	47	23.0	+18	32	03	3.82	−2.4	M2	170	
12 γ	19	58	45.3	+19	29	32	3.47	−0.3	K5	51	

Also above mag. 5:

	Mag.	Abs. mag.	Spec.	Dist
8 ζ	5.00	1.7	A3	46

VARIABLE STARS

	R.A. h	m	Dec. deg	min	Range	Type	Period, d.	Spec.	
U	19	18.8	+19	37	6.6–9.2	Algol	3.38	B–K	
S	19	56.0	+16	38	5.3–6.0	Cepheid	8.38	F–G	
X	20	05.1	+20	39	7.9–8.4	Semi-reg.	196	N	
WZ	20	07.6	+17	42	7.0–15.5	Recurrent nova	—	Q	Outbursts 1913, 1946, 1978

DOUBLE STAR

	R.A. h	m	Dec. deg	min	P.A. deg	Sep. sec	Mags.	
ζ	19	49.0	+19	09	AB + C 311	8.6	5.5, 8.7	
					AB 163	0.3	5.5, 6.2	Binary, 22.8 y

OPEN CLUSTER

M	C	NGC	R.A. h	m	Dec. deg	min	Diam. min	Mag.	No. of stars
		H.20	19	53.1	+18	20	7	7.7	15

GLOBULAR CLUSTER

M	C	NGC	R.A. h	m	Dec. deg	min	Diameter min	Mag.
71		6838	19	53.8	+18	47	7.2	8.3

PLANETARY NEBULÆ

M	C	NGC	R.A. h	m	Dec. deg	min	Diameter sec	Mag.	Mag. of central star
		6879	20	10.5	+16	55	5	13.0	15
		IC 4997	20	20.2	+16	45	2	11.6	13 (v?)

SAGITTARIUS

(Abbreviation: Sgr)

The southernmost of the Zodiacal constellations, and not wholly visible from England. Mythologically it has been associated with Chiron, the wise centaur who was tutor to Jason and many others; but it would certainly be more logical to associate Chiron with Centaurus, and another version states that Chiron merely invented the constellation Sagittarius to help in guiding the Argonauts in their quest of the Golden Fleece. The centre of the Milky Way lies behind the star-clouds here, and the whole area is exceptionally rich; it abounds in Messier objects. It is worth commenting that the stars lettered α and β are relatively faint. There are 16 stars above the fourth magnitude.

	Mag.	Luminosity Sun = 1	Dist lt-yrs
ε	1.85	110	85
σ	2.02	525	209
ζ	2.59	50	78
δ	2.70	96	81
λ	2.81	96	98
π	2.89	525	310
γ	2.99	60	117
η	3.11	800	420
ϕ	3.17	250	244
τ	3.32	82	130
ξ^2	3.51	82	144
o	3.77	58	140
μ	3.86v	60 000	3900
ρ^1	3.93	17	91
β^1	3.93	105	218
α	3.97	105	117

μ is an Algol binary with a very small range (3.8 to 3.9).

The Milky Way is very rich in Sagittarius, and the 'star-clouds' mask our view of the centre of the Galaxy.

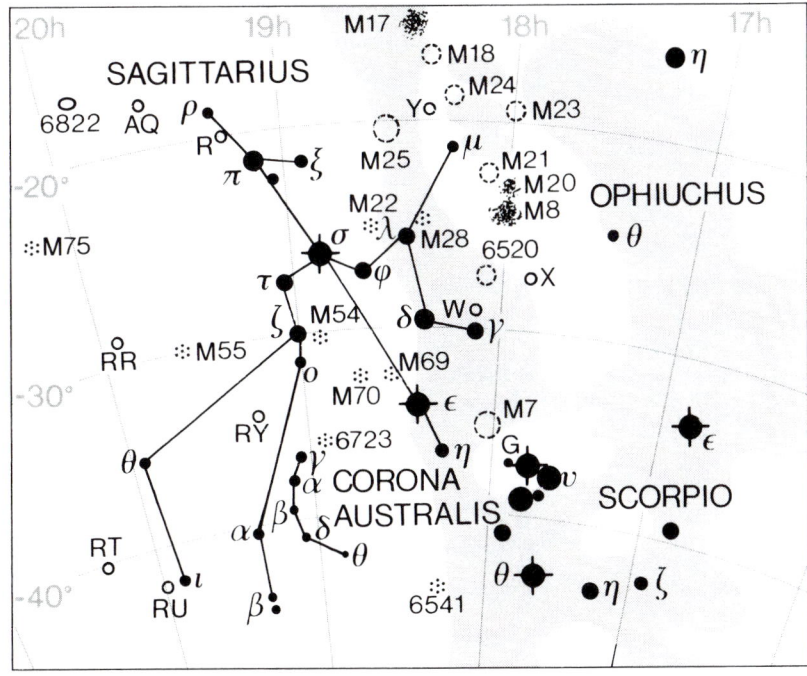

SAGITTARIUS (continued)

BRIGHTEST STARS

Star	R.A. h	m	s	Dec. deg	min	sec	Mag.	Abs. mag.	Spec.	Dist pc	
3 X	17	47	33.4	−27	49	51	var.	−2.0	F7	200	
W	18	05	01.1	−29	34	48	var.	var.	F8	400	
10 γ	18	05	48.3	−30	25	26	2.99	0.2	K0	36	Alnasr
HD 165634	18	08	04.8	−28	27	25	4.57	1.0	Gp	52	
13 μ	18	13	45.6	−21	03	32	3.86v	−7.1	B8	1200	Polis
η	18	17	37.5	−36	45	42	3.11	−2.4	M3	130	
HD 167818	18	18	03.0	−27	02	33	4.65	−0.3	gK5	84	
19 δ	18	20	59.5	−29	49	42	2.70	−0.1	K2	25	Kaus Meridionalis
20 ε	18	24	10.2	−34	23	05	1.85	−0.3	B9	26	Kaus Australis
22 γ	18	27	58.1	−25	25	18	2.81	−0.1	K2	30	Kaus Borealis
27 φ	18	45	39.2	−26	59	27	3.17	−1.2	B8	75	
34 σ	18	55	15.7	−26	17	48	2.02	−2.0	B3	64	Nunki
37 ξ²	18	57	43.6	−21	06	24	3.51	0.0	K1	44	
38 ζ	19	02	36.5	−29	52	49	2.59	0.6	A2	24	Ascella
39 o	19	04	40.8	−21	44	30	3.77	0.3	gG8	43	
40 τ	19	06	56.2	−27	40	13	3.32	0.0	K1	40	
41 π	19	09	45.6	−21	01	25	2.89	−2.0	F2	95	Albaldah
44 ρ¹	19	21	40.2	−17	50	50	3.93	1.7	F0	28	
46 υ	19	21	43.5	−15	57	18	4.61v	3.0	F2	21	
β¹	19	22	38.1	−44	27	32	3.93	−0.2	B8	67	Arkab
β²	19	23	12.9	−44	47	59	4.29	0.6	F0	52	
α	19	23	53.0	−40	36	58	3.97	−0.2	B8	36	Rukbat
52 h¹	19	36	42.3	−24	53	01	4.60	0.9	B9	55	
ι	19	55	15.5	−41	52	06	4.13	0.2	K0	55	
58 ω	19	55	50.2	−26	17	58	4.70	5.2	dG5	11	Terebellum
59	19	56	56.6	−27	10	12	4.52	−0.2	gK3	69	
θ¹	19	59	44.0	−35	16	35	4.37	−2.3	B3	210	
62 c	20	02	39.4	−27	42	35	4.58	−0.5	M4	100	

Also above mag. 5:

	Mag.	Abs. mag.	Spec.	Dist
4	4.76	0.6	A0	21
1	4.98	0.2	K0	83
21	4.81	0.2	K0	57
32 ν¹	4.83	−6.0	K2	1400
36 ξ¹	5.00	0.6	A0	76
42 ψ	4.85	3.4	F5	21
43	4.90	−2.1	G8	210
60	4.83	0.3	G5	80

VARIABLE STARS

	R.A. h	m	Dec. deg	min	Range	Type	Period, d.	Spec.
X	17	47.6	−27	50	4.2–4.8	Cepheid	7.01	F
W	18	05.0	−29	35	4.3–5.1	Cepheid	7.59	F–G
VX	18	08.1	−22	13	6.5–12.5	Semi-reg.	732	M
RS	18	17.6	−34	06	6.0–6.9	Algol	2.41	B + A
Y	18	21.4	−18	52	5.4–6.1	Cepheid	5.77	F
RV	18	27.9	−33	19	7.2–14.8	Mira	317.5	M
U	18	31.9	−19	07	6.3–7.1	Cepheid	6.74	F–G
YZ	18	49.5	−16	43	7.0–7.7	Cepheid	9.55	F–G
UX	18	54.9	−16	31	7.6–8.4	Semi-reg.	100	M
ST	19	01.5	−12	46	7.6–16.0	Mira	395.1	S
T	19	16.3	−16	59	7.6–12.9	Mira	392.3	S
RY	19	16.5	−33	31	6.0–15	R Coronæ	—	Gp
R	19	16.7	−19	18	6.7–12.8	Mira	268.8	M
AQ	19	34.3	−16	22	6.6–7.7	Semi-reg.	200	N
RR	19	55.9	−29	11	5.6–14.0	Mira	334.6	M
RU	19	58.7	−41	51	6.0–13.8	Mira	240.3	M
RT	20	17.7	−39	07	6.0–14.1	Mira	305.3	M

SAGITTARIUS (continued)

DOUBLE STARS

	R.A. h	m	Dec. deg	min	P.A. deg	Sep. sec	Mags.	
21	18	25.3	−20	32	289	1.8	4.9, 7.4	
ζ	19	02.6	−29	53	320	0.3	3.3, 3.4	Binary, 21.2 y
η	18	17.6	−36	46	105	3.6	3.2, 7.8	
π	19	09.8	−21	01	AB 150	0.1	3.7, 3.7	
					AB + C 122	0.4	5.9	
β¹	19	22.6	−44	28	077	28.3	3.9, 8.0	Wide naked-eye pair with β²
κ²	20	23.9	−42	25	234	0.8	6.9, 6.9	

OPEN CLUSTERS

M	C	NGC	R.A. h	m	Dec. deg	min	Diam. min	Mag.	No. of stars	
		6469	17	52.9	−22	21	12	8.2	50	
23		6494	17	56.8	−19	01	27	5.5	150	
		6520	18	03.4	−27	54	6	7.6	60	In M.20
21		6531	18	04.6	−22	30	13	5.9	70	
		6530	18	04.8	−24	20	15	4.6	—	In M.20
		6546	18	07.2	−23	20	13	5.9	70	
		6568	18	12.8	−21	36	13	8.6	50	
24		—	18	16.9	−18	29	90	4.5	—	Star-cloud; not a true cluster
18		6613	18	19.9	−17	08	9	6.9	20	
25		IC 4725	18	31.6	−19	15	32	4.6	30	υ Sagittarii cluster
		6645	18	32.6	−16	54	10	8.5	40	
		6716	18	54.6	−19	53	7	6.9	20	

GLOBULAR CLUSTERS

M	C	NGC	R.A. h	m	Dec. deg	min	Diameter min	Mag.
		6522	18	03.6	−30	02	5.6	8.6
		6544	18	07.3	−25	00	8.9	8.2
		6553	18	09.3	−25	54	8.1	8.2
		6558	18	10.3	−31	46	3.7	—
		6569	18	13.6	−31	50	5.8	8.7
		6624	18	23.7	−30	22	5.9	8.3 H.I.50
28		6626	18	24.5	−24	52	11.2	6.9
69		6637	18	31.4	−32	21	7.1	7.7
		6638	18	30.9	−25	30	5.0	9.2 H.I.5 1
		6652	18	35.8	−32	59	3.5	8.9
22		6656	18	36.4	−23	54	24.0	5.1
54		6715	18	55.1	−30	29	9.1	7.7
70		6681	18	43.2	−32	18	7.8	8.1
55		6809	19	40.0	−30	58	19.0	6.9
75		6864	20	06.1	−21	55	6.0	8.6

PLANETARY NEBULÆ

M	C	NGC	R.A. h	m	Dec. deg	min	Diameter sec	Mag.	Mag. of central star
		6567	18	13.7	−19	05	8	11.1	15.0
		6629	18	25.7	−23	12	15	11.6	12.8
		6644	18	32.6	−25	08	3	12.2	15.9
		6818	19	44.0	−14	09	17	9.9	13.0

NEBULÆ

M	C	NGC	R.A. h	m	Dec. deg	min	Dimensions min	Mag. of illuminating star	
20		6514	18	02.6	−23	02	29 × 27	7.6	Trifid Nebula
8		6523	18	03.8	−24	23	90 × 40	6.0	Lagoon Nebula
17		6618	18	20.8	−16	11	46 × 37	7.0	Omega Nebula

GALAXY

M	C	NGC	R.A. h	m	Dec. deg	min	Mag.	Dimensions min	Type	
	57	6822	19	44.9	−14	48	9.3	10.2 × 9.5	Irreg.	Barnard's Galaxy

SCORPIUS

(Abbreviation: Sco).

Alternatively, and less correctly, known as Scorpio. Mythologically it is usually associated with the scorpion which Juno caused to attack and kill the great hunter Orion. Note that Orion and Scorpius are now on opposite sides of the sky – placed there, it is said, so that the creature could do Orion no further damage!

Scorpius is one of the most magnificent of all constellations, and one of the few which gives at least a vague impression of the creature it is meant to represent. It is dominated by Antares, but the whole area is exceptionally rich. The 'sting', which includes Shaula – only just below the first magnitude – is to all intents and purposes invisible from England.

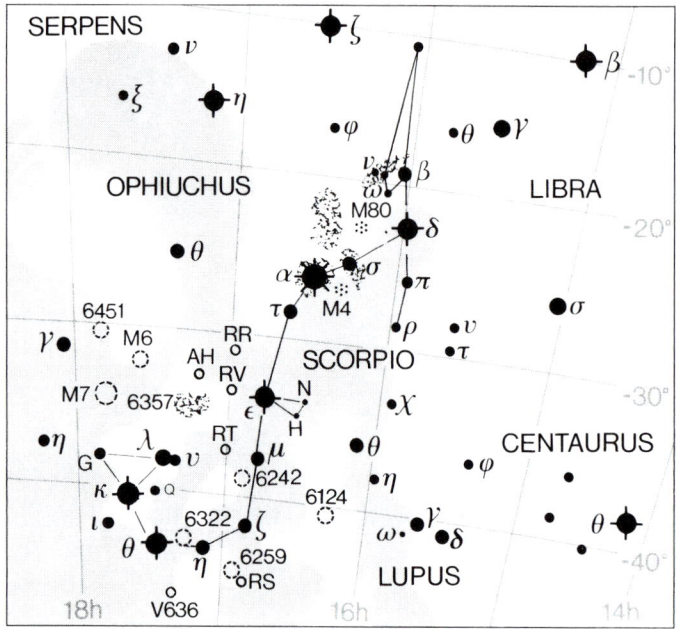

There are 20 stars above the fourth magnitude. Antares itself is very slightly variable (range 0.86–1.02). The leaders are:

	Mag.	Luminosity Sun = 1	Dist lt-yrs
α	0.96v	7500	330
λ	1.63	1300	275
θ	1.87	14 000	900
ε	2.29	96	65
δ	2.32	3800	550
κ	2.41	1300	390
β	2.64	2600	815
υ	2.69	16 000	1560
τ	2.82	3800	780
σ	2.89v	5000	590
π	2.89	2100	620
ι^1	3.03	200 000	5500
μ^1	3.04	1300	520
G	3.21	96	150
η	3.33	50	68
μ^2	3.57	1300	680
ζ^2	3.62	110	160
ρ	3.88	830	620
ω^1	3.96	2100	810
υ	4.00	1000 + 100	550

σ is very slightly variable (2.82–2.90).

SCORPIUS (continued)

Antares is a red supergiant, diameter of the order of 300 000 000 km, but its mass is probably no more than 10 times that of the Sun. It is very slightly variable (mag. 0.86–1.06). The companion appears green, no doubt mainly by contrast; the separation is decreasing. The Antares system is a radio source.

ζ is a wide double, but not a binary system. The fainter component, ζ^1, is exceptionally luminous and remote.

γ has been transferred to Libra, as σ Libræ.

BRIGHTEST STARS

Star	R.A. h	m	s	Dec. deg	min	sec	Mag.	Abs. mag.	Spec.	Dist pc	
1 b	15	50	58.6	−25	45	05	4.64	−1.7	B3	170	
2 A	15	53	36.6	−25	19	38	4.59	−2.1	B2.5	220	
5 ρ	15	56	53.0	−29	12	50	3.88	−2.5	B2	190	
6 π	15	58	51.0	−26	06	50	2.89	−3.5	B1	190	
7 δ	16	00	19.9	−22	37	18	2.32	−4.1	B0	170	Dschubba
ξ	16	04	22.0	−11	22	24	4.16	2.2	F6	26	
8 β	16	05	26.1	−19	48	19	2.64	−3.7	B0.5 + B2	250	Graffias
9 ω^1	16	06	48.3	−20	40	09	3.96	−3.5	B1	250	Jabhat al Akrab
10 ω^2	16	07	24.2	−20	52	07	4.32	0.4	gG2	53	
14 ν	16	11	59.6	−19	27	38	4.00	0.0, −2.8	A0 + B2	170	Jabbah
13 c^2	16	12	18.1	−27	55	35	4.58	−2.1	B2.5	220	
19 O	16	20	38.0	−24	10	10	4.55	−2.1	A5	92	
20 σ	16	21	11.2	−25	35	34	2.89v	−4.4	B1	180	Alniyat
21 α	16	29	24.3	−26	25	55	0.96v	−4.7	M1	100	Antares
N	16	31	22.8	−34	42	15	4.23	−2.5	B2	220	
23 τ	16	35	52.8	−28	12	58	2.82	−4.1	B0	240	
H	16	36	22.4	−35	15	21	4.16	0.6	M0	55	
26 ε	16	50	09.7	−34	17	36	2.29	−0.1	K2	20	Wei
μ^1	16	51	52.1	−38	02	51	3.04	−3.0	B1.5	160	
μ^2	16	52	20.0	−38	01	03	3.57	−3.0	B2	210	
ζ^1	16	53	59.6	−42	21	44	4.73	−8.7	B1.5	5000	
ζ^2	16	54	34.9	−42	21	41	3.62	−0.3	K5	50	
η	17	12	09.0	−43	14	21	3.33	0.6	F2	21	
34 υ	17	30	45.6	−37	17	45	2.69	−5.7	B3	480	Lesath
35 λ	17	33	36.4	−37	06	14	1.63	−3.0	B2	84	Shaula
Q	17	36	32.6	−38	38	07	4.29	0.2	gK0	58	
θ	17	37	19.0	−42	59	52	1.87	−5.6	F0	280	Sargas
κ	17	42	29.0	−39	01	48	2.41	−3.0	B2	120	Girtab
ι^1	17	47	34.9	−40	07	37	3.03	−8.4	F2	1700	
G	17	49	51.3	−37	02	36	3.21	−0.1	K2	46	

Also above mag. 5:

	Mag.	Abs. mag.	Spec.	Dist
22	4.79	−2.5	B2	260
ι^2	4.81	−0.6	A2	
d	4.78	1.4	A2	44
k	4.87	−4.8	B1	
15 ψ	4.94	1.1	A0	40

VARIABLE STARS

	R.A. h	m	Dec. deg	min	Range	Type	Period, d.	Spec.
RT	17	03.5	−36	55	7.0–16.0	Mira	449.0	M
FV	17	13.7	−32	51	7.9–8.6	Algol	5.72	B
RY	17	50.9	−33	42	7.5–8.4	Cepheid	20.31	F–G
RR	16	55.6	−30	35	5.0–12.4	Mira	279.4	M
RS	16	56.6	−45	06	6.2–13.0	Mira	320.0	M
RV	16	58.3	−33	37	6.6–7.5	Cepheid	6.06	F–G
BM	17	41.0	−32	13	6.8–8.7	Semi-reg.	850	K
RU	17	42.4	−43	45	7.8–13.7	Mira	369.2	M

SCORPIUS (continued)

DOUBLE STARS

	R.A. h	m	Dec. deg	min	P.A. deg	Sep. sec	Mags.	
2	15	53.6	−25	20	274	2.5	4.7, 7.4	
π	15	58.9	−26	07	132	50.4	2.9, 12.1	
ξ	16	04.4	−11	22	AB 040	0.8	4.8, 5.1	
					AC 051	17.6	7.3	
β	16	05.4	−19	48	AC 021	3.6	2.6, 4.9	A is a close dble
11	16	07.6	−12	45	257	3.3	5.6, 9.9	
ν	16	12.0	−19	28	AB 003	0.9	4.3, 6.8	Binary, 45.7 y
					AC 337	41.1	6.4	
12	16	12.3	−28	25	073	4.0	5.9, 7.9	
σ	16	21.2	−25	36	273	20.0	2.9, 8.5	
α	16	29.4	−26	26	274	2.7	1.2, 5.4	Binary, 878 y

OPEN CLUSTERS

M	C	NGC	R.A. h	m	Dec. deg	min	Diam. min	Mag.	No. of stars	
	75	6124	16	25.6	−40	40	29	5.8	100	
		6178	16	35.7	−45	38	4	7.2	12	
		6192	16	40.3	−43	22	8	8.5	60	
	76	6231	16	54.0	−41	48	15	2.6	—	
		6242	16	55.6	−39	30	9	6.4	—	
		6259	17	00.7	−44	40	10	8.0	120	
		6268	17	02.4	−39	44	6	9.5	—	
		6281	17	04.8	−37	54	8	5.4	—	
		6383	17	34.8	−32	34	5	5.5	40	(Nebulosity)
		6400	17	40.8	−36	57	8	8.8	60	
6		6405	17	40.1	−32	13	15	4.2	50	Butterfly Cluster
		6416	17	44.4	−32	21	18	5.7	40	
		6451	17	50.7	−30	13	8	8.2	80	
7		6475	17	53.9	−34	49	80	3.3	80	
		6322	17	18.5	−42	57	10	6.0	30	

GLOBULAR CLUSTERS

M	C	NGC	R.A. h	m	Dec. deg	min	Diameter min	Mag.
80		6093	16	17.0	−22	59	8.9	7.2
4		6121	16	23.6	−26	32	26.3	5.9
		6388	17	36.3	−44	44	8.7	6.8

PLANETARY NEBULÆ

M	C	NGC	R.A. h	m	Dec. deg	min	Diameter sec	Mag.	Mag. of central star	
		6153	16	31.5	−40	15	25	11.5	—	
	69	6302	17	13.7	−37	06	50	12.8	—	Bug Nebula
		6337	17	22.3	−38	29	48	—	14.7	

SCULPTOR

(Abbreviation: Scl).

Originally Apparatus Sculptoris. There is no star above the fourth magnitude.

See chart for Phœnix.

The south galactic pole lies in Sculptor, not far from α.

BRIGHTEST STARS

Star	R.A. h	m	s	Dec. deg	min	sec	Mag.	Abs. mag.	Spec.	Dist pc
γ	23	28	49.3	−32	31	55	4.41	0.3	G8	47
β	23	32	58.0	−37	49	07	4.37	0.0	B9	77
δ	23	48	55.4	−28	07	49	4.51	0.6	A0	9.1
α	00	58	36.3	−29	21	27	4.31	−1.2	B8	130

Also above mag. 5:

	Mag.	Abs. mag.	Spec.	Dist
η	4.81		M5	

SCULPTOR (continued)

VARIABLE STARS

	R.A.		Dec.		Range	Type	Period, d.	Spec.
	h	m	deg	min				
Y	23	09.1	−30	08	7.5–9.0	Semi-reg.	300	M
S	00	15.4	−32	03	5.5–13.6	Mira	365.3	M
R	01	27.0	−32	33	5.8–7.7	Semi-reg.	370	N

DOUBLE STARS

	R.A.		Dec.		P.A.	Sep.	Mags.	
	h	m	deg	min	deg	sec		
δ	23	48.9	−28	08	AB 243	3.9	4.5, 11.5	
					AC 297	74.3	9.3	
ζ	00	02.3	−29	43	320	3.0	5.0, 13.0	
κ¹	00	09.3	−27	59	265	1.4	6.1, 6.2	
λ¹	00	42.7	−38	28	003	0.7	6.7, 7.0	
ε	01	45.6	−25	03	028	4.7	5.4, 8.6	Binary, 1192 y

GLOBULAR CLUSTER

M	C	NGC	R.A.		Dec.		Diameter	Mag.
			h	m	deg	min	min	
		288	00	52.8	−26	35	13.8	8.1

GALAXIES

M	C	NGC	R.A.		Dec.		Mag.	Dimensions	Type
			h	m	deg	min		min	
		IC 5332	23	34.5	−36	06	10.6	6.6 × 5.1	Sd
		7713	23	36.5	−37	56	11.6	4.3 × 2.0	SBd
		7755	23	47.9	−30	31	11.8	3.7 × 3.0	SBd
		7793	23	57.8	−32	35	9.1	9.1 × 6.6	Sd
		24	00	09.9	−24	58	11.5	5.5 × 1.6	Sb
	72	55	00	14.9	−39	11	8.2	32.4 × 6.5	SB
		134	00	30.4	−33	15	10.1	8.1 × 2.6	SBb
	65	253	00	47.6	−25	17	7.1	25.1 × 7.4	Scp
	70	300	00	54.9	−37	41	8.7	20.0 × 14.8	Sd
		613	01	34.3	−29	25	10.0	5.8 × 4.6	SBb

SCUTUM

(Abbreviation: Sct).

Originally Scutum Sobieskii or Clypeus Sobieskii (Sobieski's Shield). It has only one star brighter than the fourth magnitude, but is a rich area bordering Aquila, and contains the glorious 'Wild Duck' cluster, M.11.

	Mag.	Luminosity Sun = 1	Dist lt-yrs
α	3.85	105	180

See chart for Aquila.

R Scuti is much the brightest of the RV Tauri variables.

BRIGHTEST STARS

Star	R.A.			Dec.			Mag.	Abs. mag.	Spec.	Dist pc
	h	m	s	deg	min	sec				
ζ	18	23	39.3	−08	56	03	4.68	0.2	K0	79
γ	18	29	11.7	−14	33	57	4.70	1.4	A2	45
α	18	35	12.1	−08	14	39	3.85	−0.2	K3	55
δ	18	42	16.2	−09	03	09	4.71v	1.3	F3	49
β	18	47	10.3	−04	44	52	4.22	−2.1	G5	150

Also above mag. 5:

	Mag.	Abs. mag.	Spec.	Dist
ε	4.90	−2.1	G8	250

SCUTUM (continued)

VARIABLE STARS

	R.A. h	m	Dec. deg	min	Range	Type	Period, d.	Spec.
RZ	18	26.6	−09	12	7.3–8.8	Algol	15.19	B
R	18	47.5	−05	42	4.4–8.2	RV Tauri	140	G–K
S	18	50.3	−07	54	7.0–8.0	Semi-reg.	148	N

OPEN CLUSTERS

M	C	NGC	R.A. h	m	Dec. deg	min	Diam. min	Mag.	No. of stars	
		6664	18	36.7	−08	13	16	7.8	50	EV Scuti cluster
26		6694	18	45.2	−09	24	15	8.0	30	
		6704	18	50.9	−05	12	6	9.2	30	
11		6705	18	51.1	−06	16	14	5.8	500	Wild Duck cluster

GLOBULAR CLUSTER

M	C	NGC	R.A. h	m	Dec. deg	min	Diameter min	Mag.
		6288	00	52.8	−26	35	13.8	8.1
		6712	18	53.1	−08	42	7.2	8.2

NEBULA

M	C	NGC	R.A. h	m	Dec. deg	min	Dimensions min	Mag. of illuminating star
		IC 1287	18	31.3	−10	50	4.4 × 3.4	5.5

SERPENS

(Abbreviation: Ser).

A curious constellation inasmuch as it is divided into two parts: Caput (the Head) and Cauda (the Body). It evidently represents the serpent with which Ophiuchus is struggling, since it has been pulled in half! Caput contains six stars above the fourth magnitude, and Cauda three. θ is a very wide, easy double, with equal components.

	Mag.	Luminosity Sun = 1	Dist lt-yrs
Caput			
α	2.65	96	85
μ	3.54	50	144
β	3.67	50	121
ε	3.71	19	107
δ	3.8 (combined)	17	88
γ	3.85	3.2	39
Cauda			
η	3.26	17	52
θ	3.4 (combined)	12 + 12	102
ξ	3.54	17	75

See chart for Ophiuchus.

SERPENS (continued)

BRIGHTEST STARS

Star	R.A. h	m	s	Dec. deg	min	sec	Mag.	Abs. mag.	Spec.	Dist pc	
CAPUT											
13 δ	15	34	48.0	+10	32	21	3.80	1.7	F0	27	Tsin
21 ι	15	41	33.0	+19	40	13	4.52	1.2	A1	36	
24 α	15	44	16.0	+06	25	32	2.65	−0.1	K2	26	Unukalhai
28 β	15	46	11.2	+15	25	18	3.67	0.6	A2	37	
27 λ	15	46	26.5	+07	21	12	4.43	4.4	G0	11	
35 κ	15	48	44.3	+18	08	29	4.09	−0.5	M1	78	
32 μ	15	49	37.1	−03	25	49	3.54	0.6	A0	44	
37 ε	15	50	48.9	+04	28	40	3.71	1.6	A2	33	
41 γ	15	56	27.1	+15	39	42	3.85	3.7	F6	12	
CAUDA											
53 ν	17	20	49.4	−12	50	48	4.33	1.2	A1	42	
55 ξ	17	37	35.0	−15	23	55	3.54	1.7	F0	23	
56 o	17	41	24.7	−12	52	31	4.26	1.4	A2	36	
57 ζ	18	00	28.7	−03	41	25	4.62	3.1	F3	22	
58 η	18	21	18.4	−02	53	56	3.26	1.7	K0	16	Alava
63 θ	18	56	13.0	+04	12	13	4.5	2.1	A5	31	Combined mag. 3.4, Alya
	18	56	14.5	+04	12	07	4.5	2.1	A5		

Also above mag. 5:

	Mag.	Abs. mag.	Spec.	Dist
38 ρ	4.76	−0.3	K5	98
44 π	4.83	1.7	A3	42

VARIABLE STARS

	R.A. h	m	Dec. deg	min	Range	Type	Period, d.	Spec.
S	15	21.7	+14	19	7.0–14.1	Mira	368.6	M
τ⁴	15	36.5	+15	06	7.5–8.9	Irreg. (Lb)	—	M
R	15	50.7	+15	08	5.1–14.4	Mira	356.4	M
U	16	07.3	+09	56	7.8–14.7	Mira	237.9	M
d	18	27.2	+00	12	4.9–5.9	?	?	G + A

DOUBLE STARS

	R.A. h	m	Dec. deg	min	P.A. deg	Sep. sec	Mags.	
δ	15	34.8	+10	32	175	4.0	4.1, 5.2	Binary, 3168 y
β	15	46.2	+15	25	265	30.6	3.7, 9.9	
ν	17	20.8	−12	51	028	46.3	4.3, 8.3	
d	18	27.2	+00	12	318	3.8	5.3v, 7.6	
θ	18	56.2	+04	12	104	22.4	4.5, 4.5	Common proper motion

OPEN CLUSTERS

M	C	NGC	R.A. h	m	Dec. deg	min	Diam. min	Mag.	No. of stars
		6611	18	18.8	−13	47	7	6.0	In M.16
		6604	18	18.1	−12	14	2	7.0	30

GLOBULAR CLUSTER

M	C	NGC	R.A. h	m	Dec. deg	min	Diameter min	Mag.
5		5904	15	18.6	+02	05	17.4	5.8

NEBULA

M	C	NGC	R.A. h	m	Dec. deg	min	Dimensions min	Mag. star
16		6611	18	18.8	−13	47	35 × 28 Eagle Nebula	6.4

GALAXIES

M	C	NGC	R.A. h	m	Dec. deg	min	Mag.	Dimensions min	Type
		6118	16	21.8	−02	17	12.3	4.7 × 2.3	Sb

SEXTANS

(Abbreviations: Sxt).

A very obscure constellation, with no star as bright as the fourth magnitude.
See chart for Hydra.

BRIGHTEST STAR

Star	R.A.			Dec.			Mag.	Abs.	Spec.	Dist
	h	m	s	deg	min	sec		mag.		pc
15 α	10	07	56.2	−00	22	18	4.49	−1.1	B5	100

VARIABLE STAR

	R.A.		Dec.		Range	Type	Period, d.	Spec.
	h	m	deg	min				
S	10	34.9	−00	20	8.2–13.5	Mira	261.0	M

DOUBLE STAR

	R.A.		Dec.		P.A.	Sep.	Mags.	
	h	m	deg	min	deg	sec		
γ	09	52.5	−08	06	AB 067	0.6	5.5, 6.1	Binary, 75.6 y
					AC 325	35.8	12.0	

GALAXIES

M	C	NGC	R.A.		Dec.		Mag.	Dimensions	Type	
			h	m	deg	min		min		
		2967	09	42.1	+00	20	11.6	3.0 × 2.9	Sc	
	53	3115	10	05.2	−07	43	9.1	8.3 × 3.2	E6	Spindle Galaxy
		3166	10	13.8	+03	26	10.6	5.2 × 2.7	SBa	
		3169	10	14.2	+03	28	10.4	4.8 × 3.2	Sb	

TAURUS

(Abbreviation: Tau).

One of the brightest of the Zodiacal constellations. Mythologically it has been said to represent the bull into which Jupiter transformed himself when he wished to carry off Europa, daughter of the King of Crete. Taurus includes the reddish first-magnitude star Aldebaran, and also the two most famous open clusters in the sky: the Pleiades and the Hyades. Altogether there are 16 stars above the fourth magnitude.

	Mag.	Luminosity Sun = 1	Dist lt-yrs
α	0.85v	100	68
β	1.65	400	130
η	2.87	400	238
ζ	3.00	1300	490
λ	3.3 (max)	450	330
θ²	3.42	52	125
ε	3.54	60	145
o	3.60	58	150
27	3.63	250	290
γ	3.63	60	166
17	3.70	500	390
ξ	3.74	96	98
δ¹	3.76	60	166
θ¹	3.85	60	176
20	3.88	400	390
ν	3.91	28	111

Aldebaran has a very small range (0.78–0.93). β was formerly included in Auriga, as γ Aurigæ.

The Hyades cluster is rather overpowered by the bright glare of Aldebaran, which is not a member of the cluster and simply lies in the same line of sight. At a distance of 151 light-years, as measured by the astrometric satellite Hipparcos, the Hyades form the nearest open cluster to us apart from the Ursa Major, which is spread widely over the sky. All member stars are moving to a point E of Betelgeux, at R.A. 6 h.08, dec. +9°.1 their radial velocity is +43 km/s (+ indicates recession). The central group is about 10 light-years in diameter, while outlying members spread out to 80 light-years. The age is about 660 million years.

Aldebaran is a red giant, diameter around 56 000 000 km. It has two faint companions, at 31.4 and 121 arcsec respectively, but these do not seem to be genuinely related to Aldebaran.

λ Tauri is the brightest of the eclipsing Algol binaries, apart from Algol itself. The Crab Nebula (M1) lies close to ζ Tauri.

TAURUS (continued)

BRIGHTEST STARS

Star	R.A. h	m	s	Dec. deg	min	sec	Mag.	Abs. mag.	Spec.	Dist pc	
1 o	03	24	48.7	+09	01	44	3.60	0.3	G8	46	
2 ξ	03	27	10.1	+09	43	58	3.74	−0.1	B8p	30	
5	03	30	52.3	+12	56	12	4.11	−0.9	K0	97	
10	03	36	52.3	+00	24	06	4.28	4.0	F8	13	
17	03	44	52.5	+24	06	48	3.70	−1.9	B6	120	Electra (Pleiades)
19	03	45	12.4	+24	28	02	4.30	−0.9	B6	110	Taygete (Pleiades)
20	03	45	49.5	+24	22	04	3.88	−1.6	B7	120	Maia (Pleiades).
23	03	46	19.5	+23	56	54	4.18	−1.3	B6	120	Merope (Pleiades)
25 η	03	47	29.0	+24	06	18	2.87	−1.6	B7	73	Alcyone (Pleiades)
27	03	49	09.7	+24	03	12	3.63	−1.2	B8	90	Atlas (Pleiades)
28 BU	03	49	11.1	+24	08	12	var.	var.	B8p	29	Pleione (Pleiades)
35 λ	04	00	40.7	+12	29	25	var.	−1.7	B3	100	
38 ν	04	03	09.3	+05	59	21	3.91	1.2	A1	34	
37 A¹	04	04	41.7	+22	04	55	4.36	0.2	K0	60	
49 μ	04	15	32.0	+08	53	32	4.29	−1.7	B3	140	
54 γ	04	19	47.5	+15	37	39	3.63	0.2	K0	51	Hyadum Primus (Hyades)
61 δ¹	04	22	56.0	+17	32	33	3.76	0.2	K0	51	(Hyades)
65 κ	04	25	22.1	+22	17	38	4.22	2.4	A7	23	
68 δ³	04	25	29.3	+17	55	41	4.30	1.4	A2	38	(Hyades)
69 υ	04	26	18.4	+22	48	49	4.29	1.1	F0	42	(Hyades)
71	04	26	20.7	+15	37	06	4.49	2.6	F0	24	
73 π	04	26	36.4	+14	42	49	4.69	0.3	G8	72	
77 θ¹	04	28	34.4	+15	57	44	3.85	0.2	K0	54	(Hyades)
74 ε	04	28	36.9	+19	10	49	3.54	0.2	K0	45	Ain. (Hyades)
78 θ²	04	28	39.6	+15	52	15	3.42	0.5	A7	38	(Hyades)
86 ρ	04	33	50.8	+14	50	40	4.65	2.6	F0	26	
88 d	04	35	39.2	+10	09	39	4.25	2.4	A3	25	
87 α	04	35	55.2	+16	30	33	0.85	−0.3	K5	21	Aldebaran
90 c¹	04	38	09.4	+12	30	39	4.27	2.1	A5	27	
92 σ²	04	39	16.4	+15	55	05	4.68	2.1	A5	33	
94 τ	04	42	14.6	+22	57	25	4.28	−1.7	B3	150	
102 ι	05	03	05.7	+21	35	24	4.64	2.4	A7	28	
112 β	05	26	17.5	+28	36	27	1.65	−1.6	B7	40	Al Nath. = γ Aurigæ
119 CE	05	32	12.7	+18	35	39	4.3v	−4.8v	M2	290	
123 ζ	05	37	38.6	+21	08	33	3.00	−3.0	B2	150	Alheka
136	05	53	19.6	+27	36	44	4.58	0.4	B9.5	64	

Also above mag. 5:

	Mag.	Abs. mag.	Spec.	Dist
47	4.84	0.3	G5	81
50 ω	4.94	2.3	A5	
52 φ	4.95	0.0	K1	90
64 δ²	4.80	2.4	A7	30
75	4.97	−0.1	K2	100
104	4.92	5.0	G4	13
109	4.94	0.3	G8	85
111	4.98	4.0	F8	16
114	4.88	−1.7	B3	210
126	4.86	−2.3	B3	250
132	4.86	0.3	G8	73
134	4.91	0.3	B9	110
139	4.82	−5.7	B1	1100

VARIABLE STARS

	R.A. h	m	Dec. deg	min	Range	Type	Period, d.	Spec.	
BU	03	49.2	+24	08	4.8–5.5	Irreg.	—	Bp	Pleione
λ	04	00.7	+12	29	3.3–3.8	Algol	3.95	B+A	
T	04	22.0	+19	32	8.4–13.5	T Tauri	Irreg.	G–K	
R	04	28.3	+10	10	7.6–14.7	Mira	323.7	M	
HU	04	38.3	+20	41	5.9–6.7	Algol	2.06	A	
ST	05	45.1	+13	35	7.8–8.6	W Virginis	4.03	F–G	
TU	05	45.2	+24	25	5.9–8.6	Semi-reg.	190	N	
SU	05	49.1	+19	04	9.1–16.0	R Coronæ	—	G0p	

TAURUS (continued)

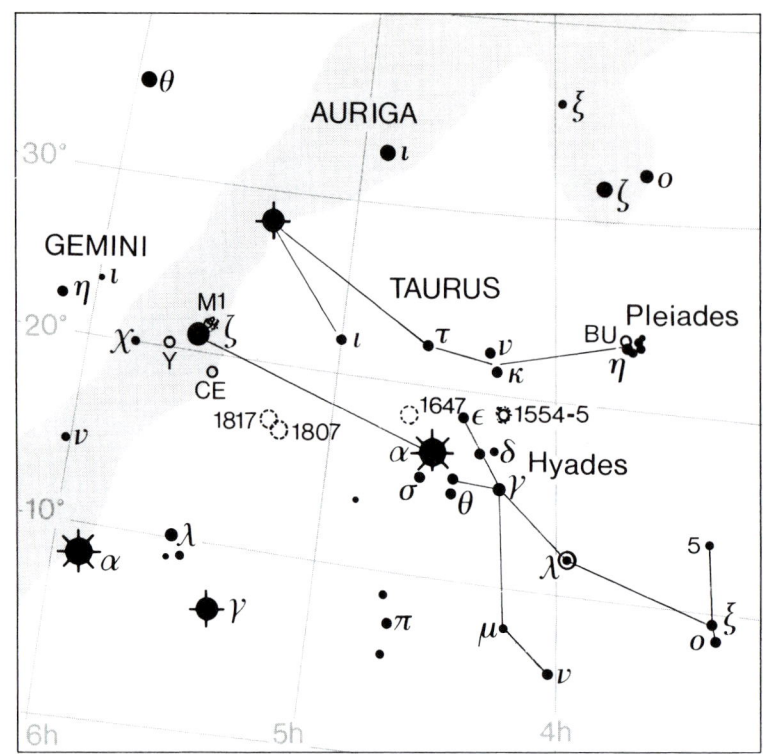

DOUBLE STARS

	R.A.		Dec.		P.A.	Sep.	Mags.	
	h	m	deg	min	deg	sec		
ϕ	04	20.4	+27	21	250	52.1	5.0, 8.4	
χ	04	22.6	+25	38	024	19.4	5.5, 7.6	
66	04	23.9	+09	28	265	0.1	5.8, 5.9	Binary, 51.6 y
κ +67	04	25.4	+22	18	173	339	4.2, 5.3	
θ	04	28.7	+15	32	346	337.4	3.4, 3.8	
σ	04	39.3	+15	55	193	431.2	4.7, 5.1	
126	05	41.3	+16	32	238	0.3	5.3, 5.9	

OPEN CLUSTERS

M	C	NGC	R.A.		Dec.		Diam.	Mag.	No. of stars
			h	m	deg	min	min		
45		1432/5	03	47.0	+24	07	110	1.2	300 + Pleiades
	41	—	04	27	+16		330	1	200 + Hyades
		1647	04	46.0	+19	04	45	6.4	200
		1746	05	03.6	+23	49	42	6.1	20
		1807	05	10.7	+16	32	17	7.0	20 Asterism?
		1817	05	12.1	+16	42	16	7.7	60

PLANETARY NEBULA

M	C	NGC	R.A.		Dec.		Diameter	Mag.	Mag. of central star
			h	m	deg	min	sec		
		1514	04	09.2	+30	47	114	10	9.4

NEBULÆ

M	C	NGC	R.A.		Dec.		Dimensions	Mag. of illuminating star	
			h	m	deg	min	min		
		1554–5	04	21.8	+19	32	var.	9v	Hind's Variable Nebula (T Tauri)
1		1952	05	34.5	+22	01	6 × 4	16	Crab Nebula: SNR

TELESCOPIUM

(Abbreviation: Tel).

A small constellation with only one star above the fourth magnitude: α.

	Mag.	Luminosity Sun = 1	Dist lt-yrs
α	3.51	1200	587

See chart for Ara.

BRIGHTEST STARS

Star	R.A. h	m	s	Dec. deg	min	sec	Mag.	Abs. mag.	Spec.	Dist pc
ε	18	11	13.6	−45	57	15	4.53	0.3	G5	57
α	18	26	58.2	−45	58	06	3.51	−2.9	B3	180
ζ	18	28	49.7	−49	04	15	4.13	0.3	gG8	51

Also above mag. 5:

	Mag.	Abs. mag.	Spec.	Dist
λ	5.00	−0.8	B9	140
L	4.90	0.2	G9	74
ξ	4.94	−0.5	M2	120

VARIABLE STARS

	R.A. h	m	Dec. deg	min	Range	Type	Period, d.	Spec.
BL	19	06.6	−51	25	7.7–9.8	Eclipsing	778.1	F + M
RR	20	04.2	−55	43	6.5–16.5	Z Andromedæ	—	F5p
R	20	14.7	−46	58	7.6–14.8	Mira	461.9	M

PLANETARY NEBULA

M	C	NGC	R.A. h	m	Dec. deg	min	Diameter sec	Mag.
		IC 4699	18	18.5	−45	59	10	11.9

TRIANGULUM

(Abbreviation: Tri).

A small but original constellation – and its main stars really do form a triangle! There are two stars above the fourth magnitude.

	Mag.	Luminosity Sun = 1	Dist lt-yrs
β	3.00	58	114
α	3.41	10	59

M.33 is on the fringe of naked-eye visibility, and is much the brightest of the spirals apart from M.31 (disregarding the Large Cloud of Magellan, which is not a well-marked spiral).

See chart for Andromeda.

BRIGHTEST STARS

Star	R.A. h	m	s	Dec. deg	min	sec	Mag.	Abs. mag.	Spec.	Dist pc	
2 α	01	53	04.8	+29	34	44	3.41	2.2	F6	18	Rasalmothallah
4 β	02	09	32.5	+34	59	14	3.00	0.3	A5	35	
9 γ	02	17	18.8	+33	50	50	4.01	0.6	A0	46	

Also above mag. 5:

	Mag.	Abs. mag.	Spec.	Dist
6	4.94	0.3	G5	85
8 δ	4.87	4.4	G0	10

VARIABLE STAR

	R.A. h	m	Dec. deg	min	Range	Type	Period, d.	Spec.
R	02	37.0	+34	16	5.4–12.6	Mira	266.5	M

TRIANGULUM (continued)

DOUBLE STARS

	R.A.		Dec.		P.A.	Sep.	Mags.
	h	m	deg	min	deg	sec	
6	02	12.4	+30	18	071	3.9	5.3, 6.9
ι	02	15.9	+33	21	240	2.3	5.4, 7.0

GALAXIES

M	C	NGC	R.A.		Dec.		Mag.	Dimensions	Type	
			h	m	deg	min		min		
33		598	01	33.9	+30	39	5.7	62 × 39	Sc	Pinwheel Galaxy
		925	02	27.3	+33	35	10.0	9.8 × 6.0	SBc	

TRIANGULUM AUSTRALE

(Abbreviation: TrA).

This 'triangle' also merits its name. There are four stars above the fourth magnitude.

	Mag.	Luminosity Sun = 1	Dist lt-yrs
α	1.92	96	55
β	2.85	5.2	33
γ	2.89	50	91
δ	3.85	600	360

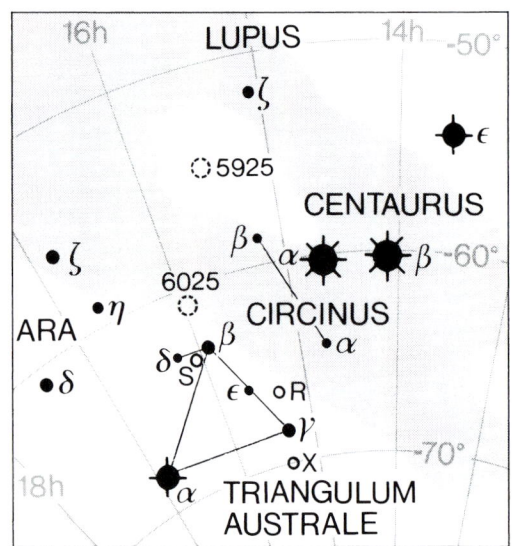

BRIGHTEST STARS

Star	R.A.			Dec.			Mag.	Abs. mag.	Spec.	Dist pc	
	h	m	s	deg	min	sec					
γ	15	18	54.5	−68	40	46	2.89	0.6	A0	28	
ε	15	36	43.1	−66	19	02	4.11	0.2	K0	44	
β	15	55	08.4	−63	25	50	2.85	3.0	F5	10	
δ	16	15	26.2	−63	41	08	3.85	−2.1	G2	110	
α	16	48	39.8	−69	01	39	1.92	−0.1	K2	17	Atria

Also above mag. 5:

	Mag.	Abs. mag.	Spec.	Dist
ζ	4.91	4.4	G0	12

TRIANGULUM AUSTRALE (continued)

VARIABLE STARS

	R.A. h	m	Dec. deg	min	Range	Type	Period, d.	Spec.
X	15	14.3	−70	05	8.1–9.1	Irreg.	–	N
R	15	19.8	−66	30	6.4–6.9	Cepheid	3.39	F–G
S	16	01.2	−63	47	6.1–6.8	Cepheid	6.32	F
U	16	07.3	−62	55	7.5–8.3	Cepheid	2.57	F

DOUBLE STARS

	R.A. h	m	Dec. deg	min	P.A. deg	Sep. sec	Mags.
ε	15	36.7	−66	19	218	83.2	4.1, 9.5
ι	16	28.0	−64	03	016	29.6	5.3, 10.3

OPEN CLUSTER

M	C	NGC	R.A. h	m	Dec. deg	min	Diam. min	Mag.	No. of stars
	95	6025	16	03.7	−60	30	12	5.1	60

TUCANA

(Abbreviation: Tuc).

The dimmest of the 'Southern Birds', but graced by the presence of the glorious globular cluster 47 Tucanæ – inferior only to ω Centauri. The only star above the fourth magnitude is α (2.86), but the combined magnitude of β^1 and β^2 is 3.7. β^1 and β^2 have common proper motion, and are presumably associated.

	Mag.	Luminosity Sun = 1	Dist lt-yrs
α	2.86	105	114
$\beta^1 + \beta^2$	3.7	105	108

47 Tucanæ, at a distance of 15 000 light-years, contains a number of 'blue stragglers', formed either by collisions of unrelated stars or, more probably, by the slow merger of the two components of a binary system.

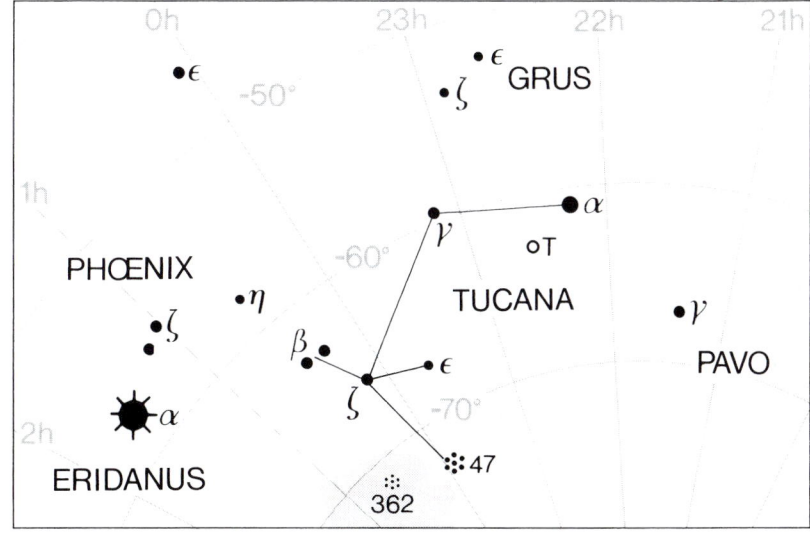

TUCANA (continued)

BRIGHTEST STARS

Star	R.A.			Dec.			Mag.	Abs.	Spec.	Dist	
	h	m	s	deg	min	sec		mag.		pc	
α	22	18	30.1	−60	15	35	2.86	−0.2	K3	35	
δ	22	27	19.9	−64	58	00	4.48	−0.2	B8	76	
γ	23	17	25.6	−58	14	08	3.99	0.6	F0	45	
ε	23	59	54.9	−65	34	38	4.50	0.0	B8.5	67	
ζ	00	20	04.4	−64	52	30	4.23	5.0	G0	7.1	
β	{ 00	32	32.7	−62	57	30	4.37	−0.2	B8 }	33	Combined mag. 3.7
	{ 00	32	33.6	−62	57	57	4.53		A2 }		

Also above mag. 5:

	Mag.	Abs. mag.	Spec.	Dist
η	5.00		A2	

VARIABLE STARS

	R.A.		Dec.		Range	Type	Period, d.	Spec.
	h	m	deg	min				
T	22	40.6	−61	33	7.7–13.8	Mira	250.8	M
S	00	23.1	−61	40	8.2–15.0	Mira	240.7	M
U	00	57.2	−75	00	8.0–14.8	Mira	259.5	M

DOUBLE STARS

	R.A.		Dec.		P.A.	Sep.	Mags.	
	h	m	deg	min	deg	sec		
δ	22	27.3	−64	58	282	6.9	4.5, 9.8	
β¹	00	32.7	−62	58	169	27.1	4.4, 4.8	
β²	00	33.6	−62	58	295	0.6	4.8, 6.0	Binary, 44.4 y
κ	01	15.8	−68	53	336	5.4	5.1, 7.3	

GLOBULAR CLUSTERS

M	C	NGC	R.A.		Dec.		Diameter	Mag.	
			h	m	deg	min	min		
	106	104	00	24.1	−72	05	30.9	4.0	47 Tucanæ
	104	362	01	03.2	−70	51	12.9	6.6	

GALAXY

	R.A.		Dec.		Mag.	Dimensions
	h	m	deg	min		min
Small Cloud of Magellan	00	53	−72	50	2.3	280 × 160

URSA MAJOR

(Abbreviation: UMa).

The most famous of all northern constellations; circumpolar in England and the northern United States. Mythologically it represents Callisto, daughter of King Lycaon of Arcadia. Her beauty surpassed that of Juno, which so infuriated the goddess that she ill-naturedly changed Callisto into a bear. Years later Arcas, Callisto's son, found the bear while out hunting, and was about to shoot it when Jupiter intervened, swinging both Callisto and Arcas – also transformed into a bear – up to the sky: Callisto as Ursa Major, Arcas as Ursa Minor. The sudden jolt explains why both bears have tails stretched out to decidedly un-ursine length!

The seven main stars of Ursa Major are often called the Plough; sometimes King Charles' Wain, and, in America, the Big Dipper. One of the Plough stars is Mizar, the most celebrated naked-eye double in the sky since it makes a pair with Alcor; Mizar is itself a compound system, and the two main components are easily separable with a small telescope.

Ursa Major contains 19 stars above the fourth magnitude.

URSA MAJOR (continued)

	Mag.	Luminosity Sun = 1	Dist lt-yrs
ε	1.77	60	62
α	1.79	60	75
η	1.86	450	108
ζ	2.09	56 + 11	59
β	2.37	28	62
γ	2.44	50	75
ψ	3.01	82	121
μ	3.05	115	156
ι	3.14	8	49
θ	3.17	10	46
δ	3.31	17	65
o	3.36	200	230
λ	3.45	50	121
ν	3.48	105	150
κ	3.60	130	91
h	3.67	17	81
χ	3.71	60	121
ξ	3.79	0.9	25
υ	3.80v	14	85

Several of the Plough stars have alternative proper names; thus η may also be called Benetnasch, while γ may be Phekda or Phecda. However, Mizar is the only star whose proper name is generally used.

47 Ursæ Majoris (R.A. 10 h 59 m 27 s.9, dec. +40° 25′ 49″, mag. 5.05, absolute mag. +4.4, type G0, distance 13 parsecs or 42 light-years) is believed to be attended by a planet with 2.4 times the mass of Jupiter, orbiting the star at a distance of 0.01 astronomical units.

ξ Ursæ Majoris was the first binary to have its orbit accurately computed (by F. Savary, in 1830).

Five of the stars in the 'Plough' pattern make up a moving cluster; the two exceptions are α and η.

β and α are known as the Pointers, because they show the way to the Pole Star.

BRIGHTEST STARS

Star	R.A. h	m	s	Dec. deg	min	sec	Mag.	Abs. mag.	Spec.	Dist pc	
0 o	08	30	15.8	+60	43	05	3.36	−0.9	G4	71	Muscida
4 π²	08	40	12.9	+64	19	40	4.60	−0.1	K2	87	Ta Tsun
9 ι	08	59	12.4	+48	02	29	3.14	2.4	A7	15	Talita
10	09	00	38.3	+41	46	57	3.97	3.4	F5	14	
12 κ	09	03	37.5	+47	09	23	3.60	−0.6	A0	28	Al Kaprah
15 f	09	08	52.2	+51	36	16	4.48	+2.1	A0	29	
14 τ	09	10	55.0	+63	30	49	4.67	0.5	F6+A5	70	
23 h	09	31	31.7	+63	03	42	3.67	1.7	F0	25	
25 θ	09	32	51.3	+51	40	38	3.17	2.2	F6	14	
24 d	09	34	28.8	+69	49	49	4.56	3.0	G2	21	
26	09	34	49.4	+52	03	05	4.50	1.4	A2	42	
29 υ	09	50	59.3	+59	02	19	3.80v	1.9	F2	26	
30 φ	09	52	06.3	+54	03	41	4.59	0.9	A3	13	
33 λ	10	17	05.7	+42	54	52	3.45	0.6	A2	37	Tania Borealis
34 μ	10	22	19.7	+41	29	58	3.05	−0.4	M0	48	Tania Australis
48 β	11	01	50.4	+56	22	56	2.37	1.2	A1	19	Merak
50 α	11	03	43.6	+61	45	03	1.79	0.2	K0	23	Dubhe
52 ψ	11	09	39.7	+44	29	54	3.01	0.0	K1	37	
53 ξ	11	18	10.9	+31	31	45	3.79	4.9	G0	7.7	Alula Australis
54 ν	11	18	28.7	+33	05	39	3.48	−0.2	K3	46	Alula Borealis
63 χ	11	46	03.0	+47	46	45	3.71	0.2	K0	37	Alkafzah
64 γ	11	53	49.7	+53	41	41	2.44	0.6	A0	23	Phad
69 δ	12	15	25.5	+57	01	57	3.31	1.7	A3	20	Megrez
77 ε	12	54	01.7	+55	57	35	1.77v	0.2	A0	19	Alioth
79 ζ	13	23	55.5	+54	55	31	2.09	0.4, 2.1	A2 + A6	18	Mizar
80 g	13	25	13.4	+54	59	17	4.01	2.1	A5	25	Alcor
83	13	40	44.1	+54	40	54	4.66	−0.5	M2	100	
85 η	13	47	32.3	+49	18	48	1.86	−1.7	B3	33	Alkaid

URSA MAJOR (continued)

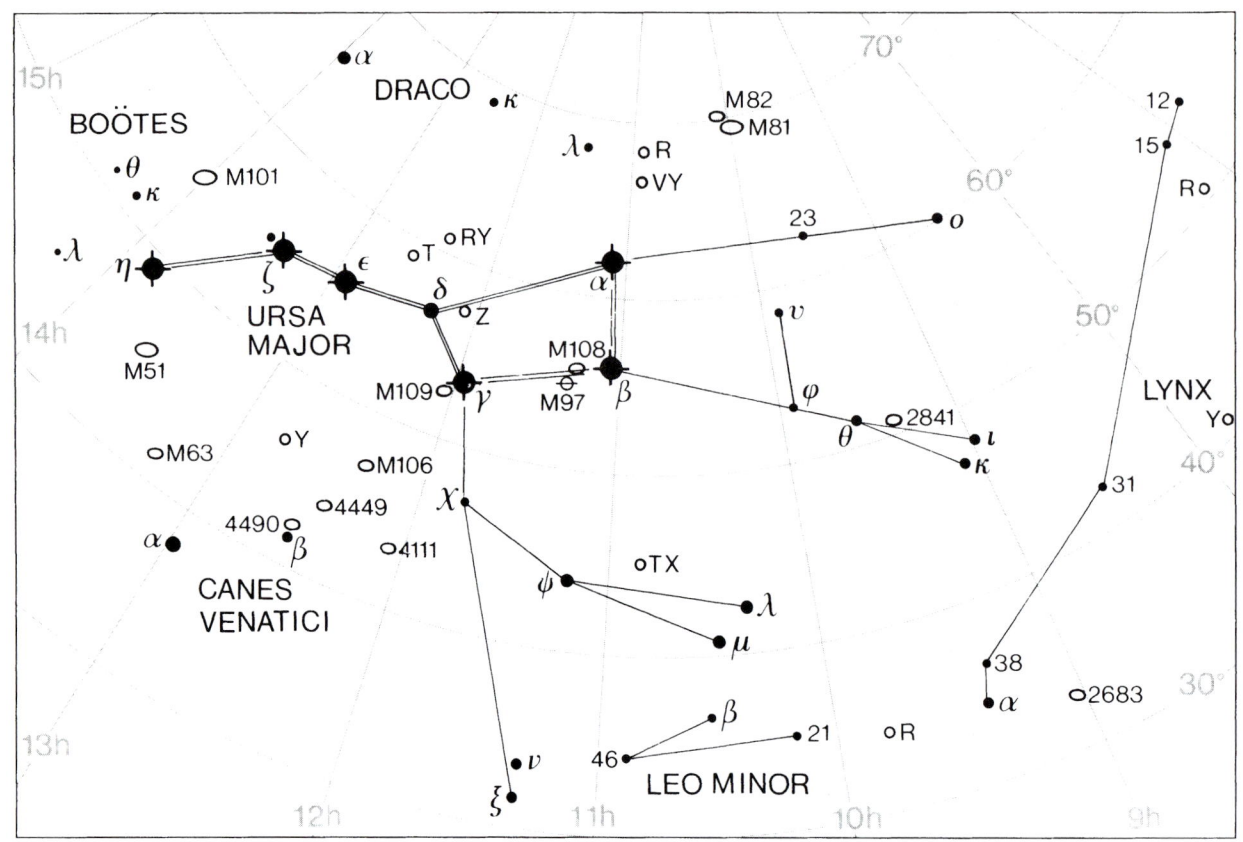

Also above mag. 5:

	Mag.	Abs. mag.	Spec.	Dist
8 ρ	4.76		M0	
13 σ^2	4.80	3.1	F7	22
18	4.83	2.1	A5	34
55	4.78	1.4	A2	42
56	4.99	−2.1	G8	260
78	4.93	3.0	F2	24

VARIABLE STARS

	R.A. h	m	Dec. deg	min	Range	Type	Period, d.	Spec.
X	08	40.8	+50	08	8.0–14.8	Mira	248.8	M
W	09	43.8	+55	57	7.9–8.6	W UMa	0.33	F + F
R	10	44.6	+68	47	6.7–13.4	Mira	301.7	M
TX	10	45.3	+45	34	7.1–8.8	Algol	3.06	B + F
VY	10	45.7	+67	25	5.9–6.5	Irreg.	—	N
VW	10	59.0	+69	59	6.8–7.7	Semi-reg.	125	M
ST	11	27.8	+45	11	7.7–9.5	Semi-reg.	81	M
CF	11	53.0	+37	43	8.5–12	Flare	—	B
Z	11	56.5	+57	52	6.8–9.1	Semi-reg.	196	M
RY	12	20.5	+61	19	6.7–8.5	Semi-reg.	311	M
T	12	36.4	+59	29	6.6–13.4	Mira	256.5	M
S	12	43.9	+61	06	7.0–12.4	Mira	225.0	S

URSA MAJOR (continued)

GALAXIES

M	C	NGC	R.A. h	m	Dec. deg	min	Mag.	Dimensions min	Type	
		2681	08	53.5	+51	19	10.3	3.8 × 3.5	Sa	
		2685	08	55.6	+58	44	11.0	5.2 × 3.0	Sbp	
		2768	09	11.6	+60	02	10.0	6.3 × 2.8	E5	
		2787	09	19.3	+69	12	10.8	3.4 × 2.3	Sap	
		2841	09	22.0	+50	58	9.3	8.1 × 3.8	Sb	
		2976	09	47.3	+67	55	10.1	4.9 × 2.5	Scp	
		2985	09	50.4	+72	17	10.5	4.5 × 3.4	Sb	
81		3031	09	55.6	+69	04	6.9	25.7 × 14.1	Sb	Bode's Nebula
82		3034	09	55.8	+69	41	8.4	11.2 × 4.6	Pec.	
		3077	10	03.3	+68	44	9.8	4.6 × 3.6	E2p	
		3079	10	02.0	+55	41	10.6	7.6 × 1.7	Sb	
		3184	10	18.3	+41	25	9.7	6.9 × 6.8	Sc	
		3198	10	19.9	+45	33	10.4	8.3 × 3.7	Sc	
		3310	10	38.7	+53	30	10.9	3.6 × 3.0	SBc	
108		3556	11	08.7	+55	57	10.0	8 × 1	Sc	
		3359	10	46.6	+63	13	10.4	6.8 × 4.3	SBc	
		3610	11	18.4	+58	47	10.7	3.2 × 2.5	E2p	
		3631	11	21.0	+53	10	10.4	4.6 × 4.1	Sc	
		3675	11	26.1	+43	35	10.9	5.9 × 3.2	Sb	
		3687	11	28.0	+29	31	12.6	2.0 × 2.0	Sb	
		3718	11	32.6	+53	04	10.5	8.7 × 4.5	SBap	
		3726	11	33.3	+47	02	10.4	6.0 × 4.5	Sc	
		3877	11	46.1	+47	30	11.6	5.4 × 1.5	Sb	
		3898	11	49.2	+56	05	10.8	4.4 × 2.6	Sb	
		3945	11	53.2	+60	41	10.6	5.5 × 3.6	SBa	
		3949	11	53.7	+47	52	11.0	3.0 × 1.8	Sb	
		3953	11	53.8	+52	20	10.1	6.6 × 3.6	Sb	
109		3992	11	55.0	+53	39	95	7 × 0.4	Sb	
		3998	11	57.9	+55	27	10.6	3.1 × 2.5	E2p	
		4026	11	59.4	+50	58	11.7	5.1 × 1.4	S0	
		4036	12	01.4	+61	54	10.6	4.5 × 2.0	E6	
		4041	12	02.2	+62	08	11.1	2.8 × 2.7	Sc	
		4051	12	03.2	+44	32	10.3	5.0 × 4.0	Sc	
		4062	12	04.1	+31	54	11.2	4.3 × 2.0	Sb	
		4088	12	05.6	+50	33	10.5	5.8 × 2.5	Sc	
		4096	12	06.0	+47	29	10.6	6.5 × 2.0	Sc	
		4100	12	06.2	+49	35	11.5	5.2 × 1.9	Sb	
		4605	12	40.0	+61	37	11.0	5.5 × 2.3	SBcp	
		5308	13	47.0	+60	58	11.3	3.5 × 0.8	S0	
		5322	13	49.3	+60	12	10.0	5.5 × 3.9	E2	
101		5457	14	03.2	+54	21	7.7	26.9 × 26.3	Sc	Pinwheel
		5475	14	05.2	+55	45	12.4	2.2 × 0.6	Sa	

PLANETARY NEBULA

M	NGC	R.A. h	m	Dec. deg	min	Diameter sec	Mag.	Mag. of central star	
97	3587	11	14.8	+55	01	194	12.0	15.9	Owl Nebula

DOUBLE STARS

	R.A. h	m	Dec. deg	min	P.A. deg	Sep. sec	Mags.	
ι	08	59.2	+48	02	100	1.8	3.1, 10.2	Binary, 818 y
κ	09	03.6	+47	09	258	0.1	4.2, 4.4	Binary, 70 y
σ²	09	10.4	+67	08	000	3.4	4.8, 8.2	Binary, 1067 y
φ	09	52.1	+54	04	188	0.2	5.3, 5.4	Binary, 105.5 y
α	11	03.7	+61	45	283	0.7	1.9, 4.8	Binary, 44.7 y
ξ	11	18.2	+31	32	060	1.6	4.3, 4.8	Binary, 59.8 y
ν	11	18.5	+33	06	147	7.2	3.5, 9.9	
78	13	00.7	+56	22	057	1.5	5.0, 7.4	Binary, 116 y
ζ	13	23.9	+54	56	AB 152	14.4	2.3, 4.0	
					AC 071	708.7	2.1, 4.0	

URSA MINOR

(Abbreviation: UMi).

The north polar constellation. Mythologically it represents Arcas, son of Callisto (see Ursa Major). There are three stars above the fourth magnitude.

	Mag.	Luminosity Sun = 1	Dist lt-yrs
α	1.99v	6000	680
β	2.08	110	95
γ	3.05	230	225

The closest approach of Polaris to the celestial pole will be on 24 March 2100. Its declination will then be +89 deg 32 min 51 sec. Some catalogues give the distance of Polaris as 480 light-years. It is a Cepheid with very small amplitude. It and its 9th-magnitude companion have common proper motion, and are presumably associated, but the orbital period must amount to thousands of years.

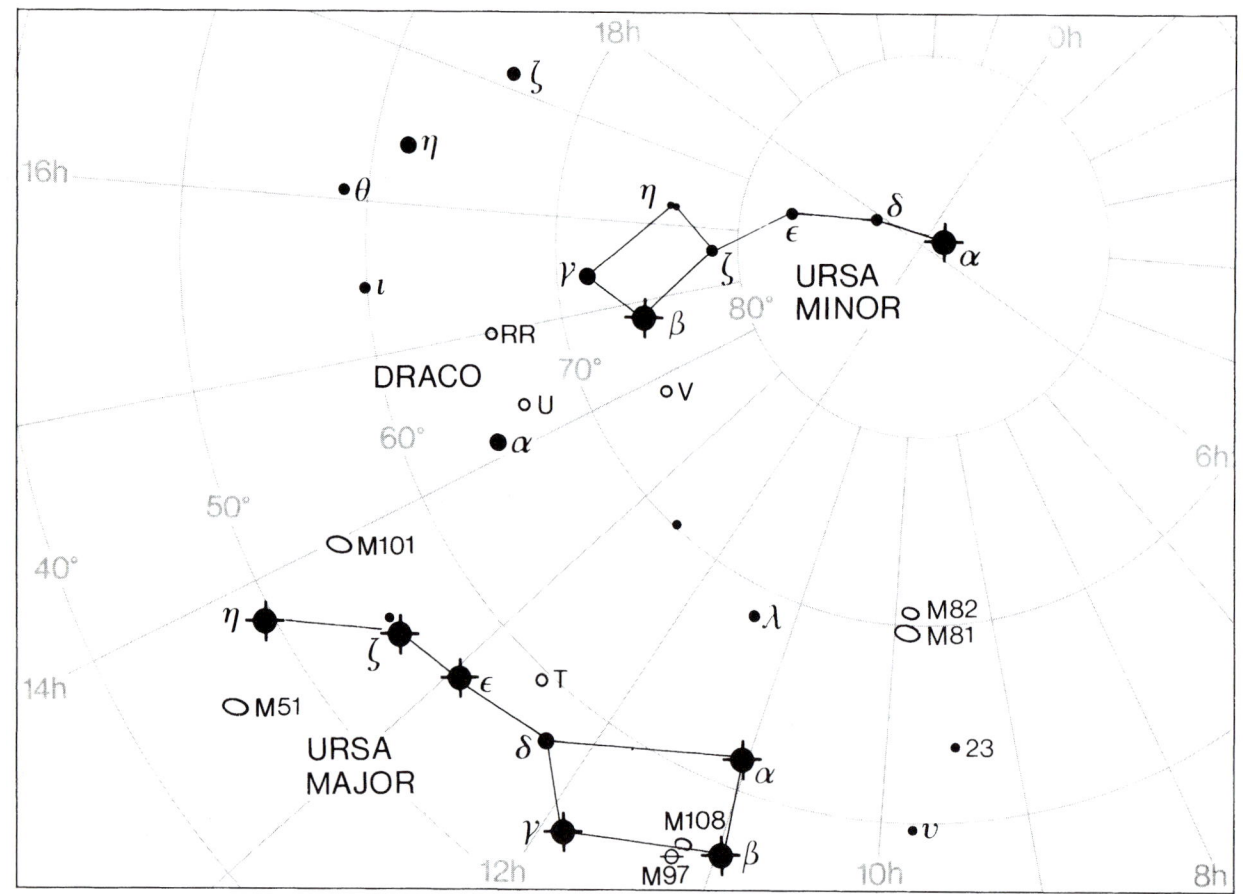

URSA MINOR (continued)

BRIGHTEST STARS

Star	R.A. h	m	s	Dec. deg	min	sec	Mag.	Abs. mag.	Spec.	Dist pc	
2	01	01	31	+85	59	24	4.52	0.0	K0	77	
1 α	02	31	50.4	+89	15	51	1.99v	−4.6	F8	208	Polaris
4 π²	08	40	12.9	+64	19	40	4.60	−0.1	K2	87	
5	14	27	31.4	+75	41	45	4.25	−0.3	K4	79	
7 β	14	50	42.2	+74	09	19	2.08	−0.3	K4	29	Kocab
13 γ	15	20	43.6	+71	50	02	3.05	−1.1	A3	69	Pherkad Major
16 ζ	15	44	03.3	+77	47	40	4.32	1.7	A3	33	Alifa
22 ε	16	45	57.8	+82	02	14	4.23	0.3	G5	61	
23 δ	17	32	12.7	+86	35	11	4.36	1.2	A1	44	Yildun

Also above mag. 5:

	Mag.	Abs. mag.	Spec.	Dist	
4	4.82	−0.2	K3	90	
21 η	4.95	2.6	F0	28	Alasco

VARIABLE STARS

	R.A. h	m	Dec. deg	min	Range	Type	Period, d.	Spec.
T	13	34.7	+73	26	8.1–15.0	Mira	313.9	M
V	13	38.7	+74	19	7.4–8.8	Semi-reg.	72	M
U	14	17.3	+66	48	7.4–12.7	Mira	326.5	M
RR	14	57.6	+65	56	6.0–6.5	Semi-reg.?	40?	M
S	15	29.6	+78	38	7.7–12.9	Mira	326.2	N
R	16	30.0	+72	17	8.8–11.0	Semi-reg.	324	M

DOUBLE STAR

	R.A. h	m	Dec. deg	min	P.A. deg	Sep. sec	Mags.
α	02	31.8	+89	16	218	18.4	2.0, 9.0

VELA

(Abbreviation: Vel).

The Sails of the dismembered ship Argo. There are 14 stars above the fourth magnitude.

	Mag.	Luminosity Sun = 1	Dist lt-yrs
γ	1.78	3800	520
δ	1.96	50	68
λ	2.21v	5000	490
κ	2.50	1320	390
μ	2.69	58	98
N	3.13	110	147
φ	3.54	20 900	2510
ψ	3.60	14	62
o	3.62v	450	390
c	3.75	96	186
p	3.84	2	75
b	3.84	200 000	6200
q	3.85	26	102
a	3.91	130	245

δ and κ make up the 'False Cross' with ε and ι Carinae.

Vela contains the Gum Nebula; R.A. 8h 30 m, dec. −45 deg. It measures 1200 × 720 minutes of arc. This is the remnant of the Vela supernova and contains the pulsar PSR 0833–45.

γ (Regor) is the brightest of the Wolf-Rayet stars. It is a splendid double; the 4.2-mag. companion is probably associated with the bright star, but the fainter companions seem to be optical.

See chart for Carina.

VELA (continued)

BRIGHTEST STARS

Star	R.A. h	m	s	Dec. deg	min	sec	Mag.	Abs. mag.	Spec.	Dist pc	
γ	08	09	31.9	−47	20	12	1.78	−4.1	WC7	160	Regor
e	08	37	38.6	−42	59	21	4.14	−5.8	A9	950	
o	08	40	17.6	−52	55	19	3.62v	−1.7	B3	120	
δ	08	44	42.2	−54	42	30	1.96	0.6	A0	21	Koo She
b	08	40	37.6	−46	38	55	3.84	−8.4	F2	1900	
a	08	46	01.7	−46	02	30	3.91	−0.6	A0	75	
W	09	00	05.4	−41	15	14	4.45	2.4	F8	20	
c	09	04	09.2	−47	05	02	3.75	−0.1	K2	57	
λ	09	07	59.7	−43	25	57	2.21v	−4.4	K5	150	Al Suhail al Wazn
k²	09	15	45.1	−37	24	47	4.62	2.5	F3	22	
κ	09	22	06.8	−55	00	38	2.50	−3.0	B2	120	Markeb
ψ	09	30	41.9	−40	28	00	3.60	1.9	F2	19	
N	09	31	13.3	−57	02	04	3.13	−0.3	K5	45	
M	09	36	49.7	−49	21	18	4.35	2.1	dA5	28	
m	09	51	40.7	−46	32	52	4.58	0.3	gG6	46	
ϕ	09	56	51.7	−54	34	03	3.54	−6.0	B5	770	
q	10	14	44.1	−42	07	19	3.85	1.4	A2	31	
HD 89682	10	19	36.8	−55	01	46	4.57	−5.9	cK	910	
HD 89890	10	20	55.4	−56	02	36	4.50	−2.3	B3	210	
p	10	37	18.0	−48	13	32	3.84	2.0	F4	23	
x	10	39	18.3	−55	36	12	4.28	−2.1	G2	150	
μ	10	46	46.1	−49	25	12	2.69	0.3	G5	30	
i	11	00	09.2	−42	13	33	4.39	0.6	A2	52	

Also above mag. 5:

	Mag.	Abs. mag.	Spec.	Dist
HD 74272	4.77	−2.3	A3	260
HD 92036	4.89	−0.3	K5	93

VARIABLE STARS

	R.A. h	m	Dec. deg	min	Range	Type	Period, d.	Spec.
AH	08	12.0	−46	39	5.5–5.9	Cepheid	4.23	F
AI	08	14.1	−44	34	6.4–7.1	Delta Scuti	0.11	A–F
RZ	08	37.0	−44	07	6.4–7.6	Cepheid	20.40	G
T	08	37.7	−47	22	7.7–8.3	Cepheid	4.64	F
SW	08	43.6	−47	24	7.4–9.0	Cepheid	23.47	K
SX	08	44.9	−46	21	8.0–8.6	Cepheid	9.55	G
CV	09	00.6	−51	33	6.5–7.3	Algol	6.89	B + B
SY	09	12.4	−43	47	7.6–8.1	Semi-reg.	63	M
RW	09	20.3	−49	31	7.8–12.0	Mira	451.7	M
V	09	22.3	−55	58	7.2–7.9	Cepheid	4.37	F
S	09	33.2	−45	13	7.7–9.5	Algol	5.93	A + K
U	09	33.2	−45	31	7.9–8.2	Semi-reg.	37	M
Z	09	52.9	−54	11	7.8–14.8	Mira	421.6	M
SV	10	44.9	−56	17	7.9–9.1	Cepheid	14.10	F–G

DOUBLE STARS

	R.A. h	m	Dec. deg	min	P.A. deg	Sep. sec	Mags.	
γ	08	09.5	−47	20	AB 220	41.2	1.9, 4.2	
					AC 151	62.3	8.2	
					AD 141	93.5	9.1	
					DE 146	1.8	12.5	
b	08	40.6	−46	39	058	37.5	3.8, 10.2	
δ	08	44.7	−54	43	AB 153	2.6	2.1, 5.1	
					AC 061	69.2	11.0	
					CD 102	6.2	13.5	
μ	10	46.8	−49	25	057	2.6	2.7, 6.4	Binary, 116 y

VELA (continued)

OPEN CLUSTERS

M	C	NGC	R.A.		Dec.		Diam.	Mag.	No. of stars	
			h	m	deg	min	min			
		2547	08	10.7	−49	16	20	4.7	80	
	85	IC 2391	08	40.2	−53	04	50	2.5	30	o Velorum cluster
		IC 2395	08	41.1	−48	12	8	4.6	40	
		2669	08	44.9	−52	58	12	6.1	40	
		2670	08	45.5	−48	47	9	7.8	30	
		IC 2488	09	27.6	−56	59	15	7.4	70	
		2910	09	30.4	−52	54	5	7.2	30	
		2925	09	33.7	−53	26	12	8.3	40	
		2972	09	40.3	−50	20	4	9.9	25	
		3033	09	48.8	−56	25	5	8.8	50	
		3228	10	21.8	−51	43	18	6.0	15	

GLOBULAR CLUSTER

M	C	NGC	R.A.		Dec.		Diameter	Mag.
			h	m	deg	min	min	
		3201	10	17.6	−46	25	18.2	6.7

PLANETARY NEBULA

M	C	NGC	R.A.		Dec.		Diameter	Mag.	Mag. of central star
			h	m	deg	min	sec		
	74	3132	10	07.7	−40	26	47	8.2	10.1

VIRGO

(Abbreviation: Vir).

A very large constellation, representing Astræa – the goddess of justice, daughter of Jupiter and Themis. The 'bowl' of Virgo abounds in faint galaxies. Of the nine stars down to the fourth magnitude, Spica is an eclipsing variable over a small range (0.91–1.01), and γ (Arich) used to be one of the most spectacular binary pairs in the sky, though it is closing up and will have become very difficult to separate by the end of the 20th century. The leading stars are:

	Mag.	Luminosity Sun = 1	Dist lt-yrs
α	0.98v	2100	260
γ	2.75	7	36
ε	2.83	60	104
ζ	3.37	17	75
δ	3.38	120	147
β	3.61	3.4	33
109	3.72	50	124
μ	3.88	14	85
η	3.89	26	104

70 Virginis (R.A. 13 h 28 m 25 s.7, dec. +13° 46′ 43″, mag. 5.01, absolute mag. +5.2, type G5, distance 10 parsecs or 33 light-years, is believed to be attended by a planet or brown dwarf 6.6 times as massive as Jupiter, orbiting the star at a distance of 0.43 astronomical units.

In 1889 H. Vogel found Spica to be a spectroscopic binary; visually it is very difficult to resolve. Both components are of type B; the primary seems to be about 15 300 000 km in diameter and 11 times as massive as the Sun, while the secondary has 6 times the diameter of the Sun and 4 times the solar mass. The real centre-to-centre separation is 18 000 000 km; period 40.14 days. The mutual eclipses cause variation of very low amplitude (mag. 0.97–1.04).

VIRGO (continued)

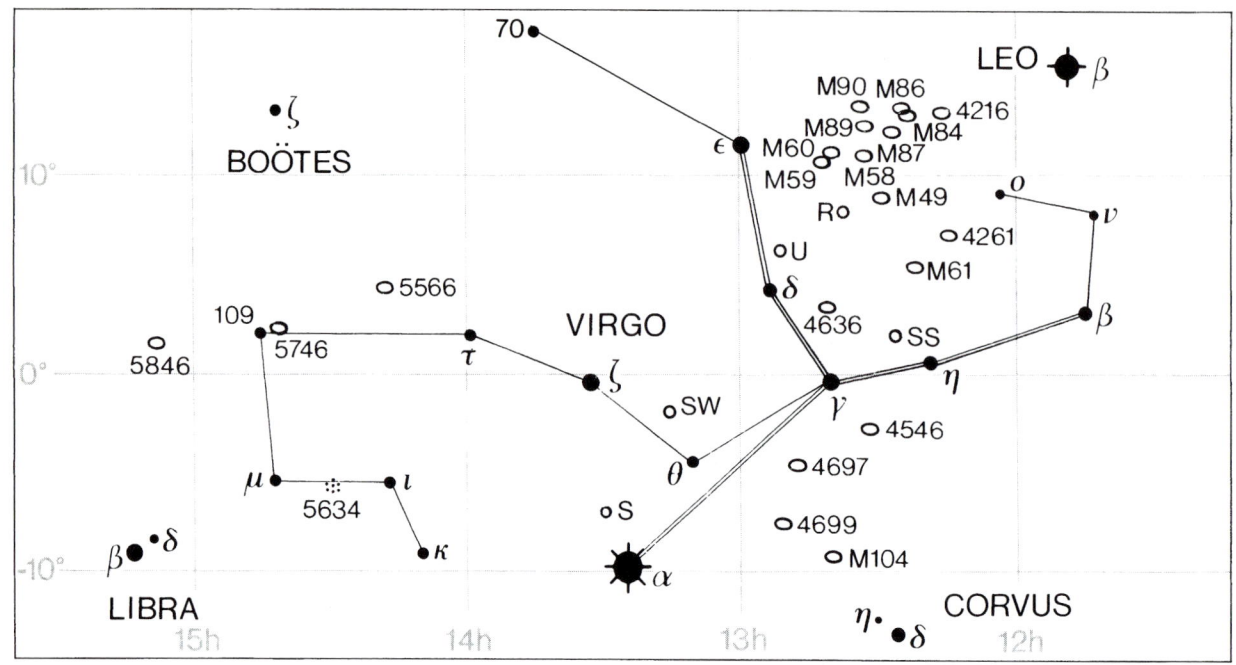

BRIGHTEST STARS

Star	R.A. h	m	s	Dec. deg	min	sec	Mag.	Abs. mag.	Spec.	Dist pc	
3 ν	11	45	51.5	+06	31	45	4.03	−0.5	M1	51	
5 β	11	50	41.6	+01	45	53	3.61	3.6	F8	10	Zavijava
8 π	12	00	52.3	+06	36	51	4.66	1.7	A3	37	
9 o	12	05	12.5	+08	43	59	4.12	0.3	G8	34	
15 η	12	19	54.3	−00	40	00	3.89	1.4	A2	32	Zaniah
26 χ	12	39	14.7	−07	59	45	4.66	−0.1	K2	78	
29 γ	12	41	39.5	−01	26	57	2.75	2.6	F0+F0	11	Arich
43 δ	12	55	36.1	+03	23	51	3.38	−0.5	M3	45	Minelauva
47 ε	13	02	10.5	+10	57	33	2.83	0.2	G9	32	Vindemiatrix
51 θ	13	09	56.9	−05	32	20	4.38	1.2	A1	43	Apami-Atsa
61	13	18	24.2	−18	18	21	4.74	5.1	G6	8.4	
67 α	13	25	11.5	−11	09	41	0.98v	−3.5	B1	79	Spica
74	13	31	57.8	−06	15	21	4.69	−0.5	gM3	110	
79 ζ	13	34	41.5	−00	35	46	3.37	1.7	A3	23	Heze
93 τ	14	01	38.7	+01	32	40	4.26	1.7	A3	32	
98 κ	14	12	53.6	−10	16	25	4.19	−0.2	K3	71	
99 ι	14	16	00.8	−06	00	02	4.08	0.7	F6	22	Syrma
100 λ	14	19	06.5	−13	22	16	4.52	2.3	A0	20	Khambalia
107 μ	14	43	03.5	−05	39	30	3.88	1.9	F3	26	Rijl al Awwa
109	14	46	14.9	+01	53	34	3.72	0.6	A0	38	
110	15	02	53.9	+02	05	28	4.40	0.2	K0	65	

Also above mag. 5:

	Mag.	Abs. mag.	Spec.	Dist
2 ξ	4.85		A3	
16	4.96	0.2	K0	67
30 ρ	4.88	0.6	A0	63
40 ψ	4.80	−0.5	M3	120
60 σ	4.80	−0.5	M2	100
69	4.76	0.0	K1	32
78	4.94	0.9	A2	26

VIRGO (continued)

VARIABLE STARS

	R.A.		Dec.		Range	Type	Period, d.	Spec.
	h	m	deg	min				
X	12	01.9	+09	04	7.3–11.2	?	—	F
SS	12	25.3	+00	48	6.0–9.6	Mira	354.7	N
R	12	38.5	+06	59	6.0–12.1	Mira	145.6	M
U	12	51.1	+05	33	7.5–13.5	Mira	206.8	M
S	13	33.0	−07	12	6.3–13.2	Mira	377.4	M
RS	14	27.3	+04	41	7.0–14.4	Mira	352.8	M

DOUBLE STARS

	R.A.		Dec.		P.A.	Sep.	Mags.	
	h	m	deg	min	deg	sec		
17	12	22.5	+05	18	337	20.0	6.6, 9.4	
γ	12	41.7	−01	27	287	3.0	3.5, 3.5	Binary, 171.4 y
θ	13	09.9	−05	32	343	7.1	4.4, 9.4	
73	13	32.0	−18	44	183	0.1	6.7, 6.9	
84	13	43.1	+03	32	229	2.9	5.5, 7.9	
τ	14	01.6	+01	33	290	80.0	4.3, 9.6	
φ	14	28.2	−02	14	110	4.8	4.8, 9.3	

GALAXIES

M	NGC	R.A.		Dec.		Mag.	Dimensions	Type	
		h	m	deg	min		min		
	4216	12	15.9	+13	09	10.0	8.3 × 2.2	Sb	
	4261	12	19.4	+05	49	10.3	3.9 × 3.2	E2	
61	4303	12	21.9	+04	28	9.7	6.0 × 5.5	Sc	
85	4382	12	22.9	+08	28	9.3	3.0 × 3.0	S0	
84	4374	12	25.1	+12	53	9.3	5.0 × 4.4	E1	
	4429	12	27.4	+11	07	10.2	5.5 × 2.6	S0	
	4438	12	27.8	+13	01	10.1	9.3 × 3.9	Sap	
	4442	12	28.1	+09	48	10.5	4.6 × 1.9	E5p	
86	4406	12	26.2	+12	57	9.2	7.4 × 5.5	E3	
49	4472	12	29.8	+08	00	8.4	8.9 × 7.4	E4	
87	4486	12	30.8	+12	24	8.6	7.2 × 6.8	E1	Virgo A
	4699	12	49.0	−08	40	9.6	3.5 × 2.7	Sa	
	4535	12	34.3	+08	12	9.8	6.8 × 5.0	SBc	
	4546	12	35.5	−03	48	10.3	3.5 × 1.7	E6	
	4527	12	34.1	+02	39	11.3	6.3 × 2.3	Sb	
	4536	12	34.5	+02	11	11.0	7.4 × 3.5	Sc	
	4546	12	35.5	−03	48	11.3	3.5 × 1.7	E6	
89	4552	12	35.7	+12	33	9.8	4.2 × 4.2	E0	
90	4569	12	36.8	+13	10	9.5	9.5 × 4.7	Sb	
58	4579	12	37.7	+11	49	9.8	5.4 × 4.4	Sb	
104	4594	12	40.0	−11	37	8.3	8.9 × 4.1	Sb	Sombrero Hat
	4596	12	39.9	+10	11	10.5	3.9 × 2.8	SBa	
59	4621	12	42.0	+11	39	9.8	5.1 × 3.4	E3	
60	4649	12	43.7	+11	33	8.8	7.2 × 6.2	E1	
	4654	12	44.0	+13	08	10.5	4.7 × 3.0	Sc	
	4636	12	42.8	+02	41	9.6	6.2 × 5.0	E1	
	4660	12	44.5	+11	11	11.9	2.8 × 1.9	E5	
	4697	12	48.6	−05	48	9.3	6.0 × 3.8	E4	
	4699	12	49.0	−08	40	9.6	3.5 × 2.7	Sa	
	4753	12	52.4	−01	12	9.9	5.4 × 2.9	Pec.	
	4762	12	52.9	+11	14	10.2	8.7 × 1.6	SB0	
	4856	12	59.3	−15	02	10.4	4.6 × 1.6	SBa	
	5247	13	38.1	−17	53	10.5	5.4 × 4.7	Sb	
	5363	13	56.1	+05	15	10.2	4.2 × 2.7	Ep	
	5364	13	56.2	+05	01	10.4	7.1 × 5.0	SB+p	
	5068	13	18.9	−21	02	10.8	6.9 × 6.3	SBc	
	5850	15	07.1	+01	33	11.7	4.3 × 3.9	SBb	

VOLANS

(Abbreviation: Vol).

Originally Piscis Volans. A small constellation, intruding into Carina. There are four stars above the fourth magnitude:

	Mag.	Luminosity Sun = 1	Dist lt-yrs
γ	3.6 (combined)	37 + 6	75
β	3.77	96	192
ζ	3.95	60	176
δ	3.98	12 000	2380

See chart for Carina.

BRIGHTEST STARS

Star	R.A. h	m	s	Dec. deg	min	sec	Mag.	Abs. mag.	Spec.	Dist pc	
γ	07	08	42.3	−70	29	50	5.7	2.8	dF4	23	Combined mag. 3.6
	07	08	45.0	−70	29	57	3.8	0.9	G8		
δ	07	16	49.8	−67	57	27	3.98	−5.4	F8	730	
ζ	07	41	49.3	−72	36	22	3.95	0.2	K0	54	
ε	08	07	55.9	−68	37	02	4.35	−1.1	B5	120	
β	08	25	44.3	−66	08	13	3.77	−0.1	K2	59	
α	09	02	26.9	−66	23	46	4.00	2.1	A5	24	

VARIABLE STAR

	R.A. h	m	Dec. deg	min	Range	Type	Period, d.	Spec.
S	07	29.8	−73	23	7.7–13.9	Mira	395.8	M

DOUBLE STARS

	R.A. h	m	Dec. deg	min	P.A. deg	Sep. sec	Mags.
γ²	07	08.8	−70	30	300	13.6	4.0, 5.9
ζ	07	41.8	−72	36	116	16.7	4.0, 9.8
ε	08	07.9	−68	37	024	6.1	4.4, 8.0
κ	08	19.8	−71	31	AB 057	65.0	5.4, 5.7
					BC 030	37.7	8.5
θ	08	39.1	−70	23	108	45.0	5.3, 10.3

GALAXY

M	NGC	R.A. h	m	Dec. deg	min	Mag.	Dimensions min	Type
	2442	07	36.4	−69	32	11.2	6.0 × 5.5	SBb

VULPECULA

(Abbreviation: Vul).

Originally Vulpecula et Anser, the Fox and Goose – nowadays the goose has disappeared (possibly the fox has eaten it). Vulpecula contains no star above magnitude 4.4, and is notable only because of the presence of the Dumbbell Nebula and the fact that several novæ have appeared within the boundaries of the constellation.

See chart for Cygnus.

BRIGHTEST STARS

Star	R.A. h	m	s	Dec. deg	min	sec	Mag.	Abs. mag.	Spec.	Dist pc
6 α	19	28	42.2	+24	39	54	4.44	0.0	M0	26
13	19	53	27.5	+24	04	47	4.58	−0.6	A0	110
15	20	01	05.9	+27	45	13	4.64	1.7	A0	18
23	20	15	49.1	+27	48	51	4.52	−0.2	K3	88

Also above mag. 5:

	Mag.	Abs. mag.	Spec.	Dist
12	4.95	−1.7	B3	210

VULPECULA (continued)

VARIABLE STARS

	R.A. h	m	Dec. deg	min	Range	Type	Period, d.	Spec.
RS	19	17.7	+12	26	6.9–7.6	Algol	4.46	B + A
Z	19	21.7	+25	34	7.4–9.2	Algol	2.45	B + A
U	19	36.6	+20	20	6.8–7.5	Cepheid	7.99	F–G
T	20	51.5	+28	15	5.4–6.1	Cepheid	4.44	F–G
SV	19	51.5	+27	28	6.7–7.7	Cepheid	45.03	F–K
R	21	04.4	+23	49	7.0–14.3	Mira	136.4	M

DOUBLE STARS

	R.A. h	m	Dec. deg	min	P.A. deg	Sep. sec	Mags.
2 (ES)	19	17.7	+23	02	127	1.8	5.4, 9.2
α–8	19	28.7	+24	40	028	413.7	4.4, 5.8

OPEN CLUSTERS

M	C	NGC	R.A. h	m	Dec. deg	min	Diam. min	Mag.	No. of stars	
		Cr 399	19	25.4	+20	11	60	3.6	40	Coat-hanger (Brocchi's Cluster)
		6823	19	43.1	+23	18	12	7.1	30	
		6830	19	51.0	+23	04	12	7.9	20	
	37	6885	20	12.0	+26	29	7	5.7	30	
		6940	20	34,6	+28	18	31	6.3	60	

PLANETARY NEBULA

M	NGC	R.A. h	m	Dec. deg	min	Dimensions sec	Mag.	Mag. of central star	
27	6853	19	59.6	+22	43	350 × 910	7.6	13.9	Dumbbell Nebula

30 TELESCOPES AND OBSERVATORIES

OBSERVATORIES

Strictly speaking, an observatory is any place from which astronomical studies are carried out. It is even possible to claim that Stonehenge was an observatory, because there is little doubt that it is astronomically aligned. The oldest observatory building now standing seems to be that at Chomsong-dae, in Kyingju, South Korea; it dates from 632 AD. Later, elaborate measuring instruments were built by the Arabs and the Indians; some of these still exist, such as the great observatory at Delhi. In 1576 Tycho Brahe erected his elaborate observatory at Hven, in the Baltic, and used the equipment to draw up an amazingly accurate star catalogue. In the modern sense, observatories are of course associated with telescopes of some kind or another. A list of some great modern observatories is given in Table 30.2.

National observatories date back for centuries; the oldest seems to be that of Leiden in Holland (1632). The oldest truly national observatory was that at Copenhagen in Denmark, although unfortunately the original buildings were destroyed by fire.

The national British observatory, at Greenwich, was founded in 1675 by order of King Charles II, mainly so that a new star catalogue could be drawn up for the use of British seamen. The original buildings were designed by Wren, and are now known as Flamsteed House (after the first Astronomer Royal). Greenwich became the 'timekeeping centre' of the world, and the zero for longitude passes through it. Until 1971 the Director of the Royal Greenwich Observatory also held the post of Astronomer Royal; a list of Astronomers Royal is given in Table 30.1. When Greenwich Park became unsuitable as an observing site, in the mid-20th century, the main equipment was transferred to Herstmonceux in Sussex, and the largest telescope was later taken to La Palma in the Canary Islands, where observing conditions are far better than in England. In 1992 Herstmonceux was closed, and the RGO transferred to an office block in Cambridge. Finally, in October 1998, the Government closed the RGO completely, ostensibly as an economy measure – probably the twentieth century's worst act of scientific vandalism.

Most modern observatories are sited at high altitude, where seeing conditions are good; the summit of Mauna Kea, for example, lies at over 4000 m. The summit of Los Muchachos, at La Palma in the Canary Islands, is also excellent, as are sites in the Atacama Desert of Northern Chile. The United States national observatory is at Kitt Peak in Arizona.

Aircraft have been used; the Kuiper Airborne Observatory, brought into use in 1975, was a Lockheed C141 Starliner jet transport aircraft, in which was mounted an 0.915 m Cassegrain telescope. It was an outstanding success, and functioned until 1995, when it was abandoned for the usual reasons of economy. It could fly at 12 km, above 85% of the Earth's atmosphere and 99% of the water vapour. In 1990 the Hubble Space Telescope was launched into Earth orbit. In contrast, the Homestake Mine Observatory, in South Dakota, is located in a gold-mine, and is 1.5 km below ground level. The 'telescope' is a 100 000 gallon tank of cleaning fluid, to trap solar neutrinos. The experiment was started in 1965, and is still operating.

Many new observatories are being planned; the emphasis is on high-altitude installations, with a preference for the southern hemisphere, since many of the most important objects (such as the Magellanic Clouds) are inaccessible from Europe or the United States. It is by no means unlikely that in the reasonably near future it will be possible to set up an observatory on the surface of the Moon.

TELESCOPES

We cannot be sure when the first telescope was built. According to research by the British scientific historian C. A. Ronan, a telescope may have been constructed in England by Leonard Digges, some time between 1545 and 1599; this may have used a combination of mirrors and lenses. However, we have no definite proof, and there is no information as to whether it was ever turned to the sky.

The first refracting telescope whose existence can definitely be proved was made by H. Lippershey, in Holland, in 1608. Before long, telescopes were used for astronomical work; for example, a telescopic map of the Moon was drawn by Thomas Harriot in 1609, but the first great telescopic observer was of course Galileo, who made a telescope for himself and began his work in January 1610. With it, he made a series of spectacular discoveries: the craters of the Moon, the phases of Venus, the strange appearance of Saturn, the four principal satellites of Jupiter and the 'myriad stars' of the Milky Way.

The first reflecting telescope was made by Isaac Newton, and was presented to the Royal Society in 1671, so that it was probably made in 1668 or 1669. Telescopes of both kinds were steadily improved, and really large instruments became possible; in 1789 William Herschel produced a reflector with a 49 inch (124.5 cm) mirror. This was surpassed in 1845 by the 182.9 cm (72 inch) reflector made by the third Earl of Rosse, and set up at Birr Castle in Ireland. It was a strange instrument, mounted between two massive stone walls and capable of swinging for only a limited distance to either side of the meridian, but with it Lord Rosse discovered the spiral forms of the objects we know to be galaxies. The 'Leviathan', as it was nicknamed, was in fact the largest telescope in the world until 1917. It is now (2000) in full operation again.

Early reflectors, including those of Herschel and Rosse, had metal mirrors, but glass mirrors coated with some reflective substance such as silver or aluminium were far better. Large refractors were built during the latter part of the 19th century; the largest of all, the Yerkes Observatory 40 inch (101 cm) was brought into use in 1895. However, a lens has to be supported around its edges, and if too large will distort under its own weight; probably the Yerkes refractor will never be surpassed. A 124.5 cm object-glass was made and shown at the Paris Exposition of 1901, but was never used for serious research. A 41 inch object glass destined for the Pulkovo Observatory in Russia was never even mounted. Clearly the future belonged to the reflector. Master-minded by George Ellery Hale, of the United States, a 60 inch reflector was set up

Table 30.1. Astronomers Royal.

John Flamsteed	1675–1719
Edmond Halley	1720–1742
James Bradley	1742–1762
Nathaniel Bliss	1762–1764
Nevil Maskelyne	1765–1811
John Pond	1811–1835
Sir George Airy	1835–1881
Sir William Christie	1881–1910
Sir Frank Dyson	1910–1933
Sir Harold Spencer Jones	1933–1955
Sir Richard Woolley	1956–1971
Sir Martin Ryle	1972–1982
Sir Francis Graham-Smith	1982–1990
Sir Arnold Wolfendale	1991–1995
Sir Martin Rees	1995–

The first Astronomer Royal for Scotland was Thomas Henderson (1834–1844). Then followed Charles Piazzi Smyth, Ralph Copeland, Frank Dyson, Ralph Sampson, William Greaves, Hermann Brück, Vincent Reddish, Malcolm Longair, and the present (2000) holder of the office, John Campbell Brown.

in 1908 at Mount Wilson, in California; in 1917 came the Hooker 100 inch (254 cm) reflector, also at Mount Wilson. This remained in a class of its own until 1948, with the completion of the Hale 200 inch (508 cm) on Mount Palomar, also in California.

Today there are reflecting telescopes of many kinds. Some have segmented mirrors; for example the first Keck telescope, on Mauna Kea in Hawaii, has a mirror 9.8 m in diameter, made of 36 segments fitted together to form the correct optical curve. Keck II, by its side, is identical; when working together, as an interferometer they could in theory distinguish the headlights of a car, separately, over a range of 25 000 km. The first multiple-mirror telescope (MMT) was set up at the Whipple Observatory at Mount Hopkins, Arizona; it used six 183 cm mirrors in conjunction, so that the total light-grasp was equal to a single 442 cm mirror. It performed well, but the six mirrors have now been replaced by a single 650 cm mirror constructed by Roger Angel, using his new spin-casting technique. The Gemini project is an international partnership to build two 8 m telescopes, one on Mauna Kea and the other at Cerro Pachon in Chile; these will work together. The Mauna Kea telescope, Gemini North, is now operative, and Gemini South is nearing completion. There are telescopes with mirrors made from liquid mercury, spinning so as to produce a perfect parabola. The Hobby-Eberly Telescope has a spherical main mirror whose optical axis is tilted to the zenith at an angle of 35°; the mirror and telescope are mounted on a frame which turns 360° in the azimuthal direction, so that during an observation the telescope is fixed in the azimuth and objects are tracked by moving a spherical aberration corrector to follow the reflected light. The main mirror is composed of 91 segments, each hexagonal in shape and 1 m across the flats.

All these are dwarfed by the VLT or Very Large Telescope at Cerro Parañal, Chile. When the four 8 m mirrors are working together, they will provide an instrument far more powerful than anything previously built. Three of the mirrors (Antu, Kueyen and Melipal) were operating by 2000, and were producing spectacular results. Meanwhile, the European Southern Observatory is planning OWL – the Overwhelmingly Large Telescope – with an aperture of 100 metres. This would indeed be a telescope suited to the 21st century!

Until the 1970s, all large telescopes were mounted equatorially. The first large altazimuth telescope was the Russian 6 m reflector. Optically it has never been a real success, but its mounting was a new departure, and all the major telescopes are now altazimuths; modern computers can deal with guidance with no trouble at all. All major telescopes now make use of active optics (rapid adjustments of the shape of the mirror to allow for changes in elevation) and adaptive optics (to counteract rapid variations in the atmosphere). Electronic devices, such as CCDs (charge-coupled devices) have superseded photography; indeed, the last photographic plate taken with the Hale reflector was exposed as long ago as September 1989. Fibre optics are also widely used. There are, of course, specialist telescopes, such as solar telescopes; the largest of these is at Kitt Peak in Arizona. It uses a 203 cm heliostat, and has an inclined tunnel 146 m long. It was completed in 1962, and although designed solely for solar research it proved capable of being used for other types of observation as well – an unexpected bonus.

The Hubble Space Telescope was launched from the Space Shuttle on 25 April 1990. It has a 240 cm mirror and orbits at an altitude of 600 km in a period of 95 min; the orbital inclination is 28.5°. The total weight is 11 360 kg and the length 13.3 m; the diameter is 12 m with the solar arrays extended. Initially the telescope had a faulty mirror, but a servicing mission by astronauts introduced corrective optics and the telescope has now performed even better than had originally been hoped. With perfect seeing all the time, it can out-perform any Earth-based telescope in many ways. Three servicing missions to the telescope have now been carried out – the last in December 1999 (launch 19 December, return 27 December).

PLANETARIA

A planetarium is purely an educational device; an artificial sky is projected onto the inside of a large dome, by means of a very complex projector. The planetarium is really a development of the orrery, a device to show the movements of the planets. The first orrery was made in the 18th century by George Graham, for Prince Eugene of Savoy; the first American orrery was made by David Rittenhouse in 1772.

The true ancestor of the modern planetarium was the Gottorp Globe, made by H. Busch in Denmark about 1654–6. It was 4 m in diameter. The audience sat inside it; the stars were painted on the inside of the globe. However, the first modern-type planetarium, due to Dr. W. Bauersfeld of the Zeiss Optical Works, was opened at the Deutsches Museum in Bonn in 1923. Today most major cities have planetaria; the largest is that of Tokyo, at the Miyazumi Science Center, which has a 27 m dome, and was opened in 1987. In Britain, there are major planetaria at Armagh in Northern Ireland, and at Madame Tussaud's in London; the South Downs Planetarium at Chichester, in Sussex, is scheduled to open in 2000.

The most northerly planetarium is the Northern Lights Planetarium at Tromsø, in Norway – latitude 69°48′.

Table 30.2. Some of the world's great observatories.

Location	Name	Latitude °	Latitude ′	Latitude ″	Longitude °	Longitude ′	Longitude ″	Altitude (m)
Aarhus, Denmark	Ole Rømer Obs.	56N	07	40.0	10E	11	48	50
Alma Ata, Kazakhstan	Mountain Obs. of Academy of Sciences	43N	11	16.9	76E	57	24	1450
Ann Arbor, Michigan, USA	Univ. of Michigan	42N	16	48.7	83W	43	48	282
Arcetri (Florence), Italy	Astrophysical Obs.	43N	45	44.7	11E	15	18	184
Armagh, N Ireland	Armagh Obs.	54N	21	11.1	06W	38	54	64
Athens, Greece	National Obs.	37N	58	19.7	23E	43	0	110
Auckland, New Zealand	Auckland Public Obs.	36S	54	28.0	174E	46	36	80
Barcelona, Spain	Fabra Obs.	41N	24	59.3	02E	07	36	415
Beijing, China	Beijing University Obs.	39N	57	24	116E	21	36	76
Beirut, Lebanon	American University Obs.	33N	54	22.0	35E	28	12	38
Belgrade, Yugoslavia	Obs. of Academy of Science	44N	48	13.2	20E	30	48	253
Belo Horizonte, Brazil	Piedade Obs.	19S	49	18	43W	30	42	1746
Berlin, Germany	Wilhelm Förster Obs.	52N	28	30.0	13E	25	30	40
Berlin-Treptow, Germany	Archenhold Obs.	52N	29	07.0	13E	28	36	38
Berne, Switzerland	Astr. Inst. of Univ.	46N	57	12.7	07E	25	42	563
Big Bear Lake, California, USA	Big Bear Solar Obs.	34N	15	12	116W	54	54	2067
Bloemfontein, S Africa	Boyden Obs.	29S	02	18.0	26E	24	18	1387
Bochum, Germany	Astronomical Station	51N	27	54.8	07E	13	24	132
Bogotá, Colombia	National Astr. Obs.	04N	35	54	74W	04	54	2640
Bologna, Italy	San Vittore Obs.	44N	28	06	11E	20	30	280
Bombay, India	Government Obs.	18N	53	36.2	72E	48	54	14
Bonn, Germany	University Obs.	50N	43	45.0	07E	05	48	62
Bordeaux, France	Obs. of Univ. of Bordeaux	44N	50	07	00E	31	36	73
Bosque Alegro, Argentina	Cordoba Obs. Station	31S	35	54	64W	32	48	1250
Boulder, Colorado, USA	Sommers-Bausch Obs.	40N	00	13.0	105W	15	42	1648
Brno, Czech Republic	Nicolaus Copernicus Obs.	49N	12	18	16E	35	18	310
Bro, Sweden	Kvistaberg Obs.	59N	30	06	17E	36	24	0
Brussels, Belgium	Astronomy and Astrophys. Inst.	50N	48	48	04E	23	00	147
Bucharest, Romania	National Obs.	44N	24	49.4	26E	05	48	83
Budapest, Hungary	Konkoly Obs.	47N	29	58.6	18E	57	54	474
Buenos Aires, Argentina	Naval Obs.	34S	37	18.3	58W	21	18	6
Calar Alto, Spain	German-Spanish Ast. Centre	37N	13	48	02E	32	12	2168
Cambridge, England	University Obs.	52N	12	51.6	00E	05	42	28
Cambridge, Massachusetts, USA	Harvard College Obs.	42N	22	47.6	71W	07	48	24
Cape, South Africa	S African Astr. Obs.	33S	56	02.5	18E	28	36	10
Caracas, Venezuela	Cagigal Obs.	10N	30	24.3	66W	55	42	1042
Castleknock, Ireland	Dunsink Obs.	53N	23	18	06W	20	12	85
Catania, Sicily	Astrophysical Obs.	37N	31	42.0	15E	04	42	193
Cerro Las Campanas, Chile	Las Campanas Obs.	29S	00	30	70W	42	00	2282
Cerro Le Silia, Chile	European Southern Obs.	29S	15	24	70W	43	48	2347
Cerro Parañal, Chile	VLT	24S	38	00	70W	24	00	2635
Cerro Tololo, Chile	Cerro Tololo Inter-American Obs.	30S	09	54	70W	48	54	2215
Chung-li, Taiwan	National Central Univ. Obs.	24N	58	12	121E	11	12	152
Cincinnati, Ohio, USA	Cincinnati Obs.	39N	08	19.8	84W	25	18	247
Cocoa, Florida, USA	Brevard Community Astr. Obs.	28N	33	06	80W	45	42	17
Copenhagen, Denmark	Urania Obs.	55N	41	19.2	12E	32	18	10
Corboda, Argentina	National Obs.	31S	25	16.4	64W	11	48	434
Crimea, Ukraine	Crimean Astro. Obs.	44N	43	42.0	34E	01	00	550
Edinburgh, Scotland	Royal Obs.	55N	55	30.0	03W	11	00	146

Table 30.2. (Continued)

Location	Name	Latitude °	Latitude ′	Latitude ″	Longitude °	Longitude ′	Longitude ″	Altitude (m)
Flagstaff, Arizona, USA	US Naval Obs.	35N	11	00.0	111W	44	24	2310
Flagstaff, Arizona, USA	Lowell Obs.	35N	12	06.0	111W	39	48	2210
Göttingen, Germany	University Obs.	51N	31	48.2	09E	56	36	161
Greenbelt, Maryland, USA	Goddard Research Obs.	39N	01	11.5	76W	49	36	49
Groningen, Holland	Kapteyn Astron. Lab.	53N	13	13.8	06E	33	48	4
Haleakala, Hawaii, USA	Univ. of Hawaii	20N	42	22.0	156W	15	24	3054
Hamburg, Germany	Bergedorf Obs.	53N	28	46.9	10E	14	24	41
Hartebeespoort, S Africa	Republic Obs. Annexe	25S	46	22.4	27E	52	36	1220
Helsinki, Finland	University Obs.	60N	09	42.3	24E	57	18	33
Helwan, Egypt	Helwan Obs.	29N	51	31.1	31E	20	30	115
Hyderabad, India	Nizamiah Obs.	17N	25	53.0	78E	27	12	293
Jena, Germany	Karl Schwarzschild Obs.	50N	58	51.0	11E	42	48	331
Johannesburg, S Africa	Republic Obs.	26S	10	55.3	28E	04	30	1806
Juvisy, France	Flammarion Obs.	48N	41	37.0	02E	22	18	92
Kharkov, Ukraine	Kharkov Univ. Obs.	50N	00	02	36E	13	54	138
Kiev, Ukraine	Kiev Univ. Obs.	50N	27	12	30E	29	54	184
Kiso, Japan	Kiso Obs.	35N	47	36	137E	37	42	1130
Kitt Peak, Arizona, USA	Kitt Peak National Obs.	31N	57	48	111W	36	00	2120
Kitt Peak, Arizona, USA	Steward Obs. Station	31N	57	48	111W	36	00	2071
Kodaikanal, India	Astrophysical Obs.	10N	13	50.0	77E	28	06	2343
Kracow, Poland	University Obs.	50N	03	52.0	19E	57	36	221
Kyoto, Japan	Kwasan Obs.	34N	59	40.8	135E	47	36	234
La Palma, Canary Islands	Roque de los Muchachos Obs.	28N	45	36	17W	52	54	2326
La Plata, Argentina	La Plata Obs.	34S	54	30	57W	55	54	17
Las Cruces, New Mexico, USA	Corralitos Obs.	32N	22	48	107W	02	36	1453
Leyden, Holland	University Obs.	52N	09	19.8	04E	29	00	6
Lisbon, Portugal	Lisbon Astr. Obs.	38N	42	42	09N	11	12	111
Los Angeles, California, USA	Griffith Obs.	34N	06	46.8	118W	18	06	357
Lund, Sweden	Royal University Obs.	55N	41	51.6	13E	11	12	34
Madison, Wisconsin, USA	Washburn Obs.	43N	04	36	89W	24	30	292
Madrid, Spain	Astronomical Obs.	40N	24	30.0	03W	41	18	655
Mauna Kea, Hawaii, USA	Mauna Kea Obs.	19N	46	36	155W	28	18	4220
Meudon, France	Obs. of Physical Astr.	48N	48	18.0	02E	13	54	162
Milan, Italy	Brera Obs.	45N	27	59.2	09E	11	30	120
Mill Hill, London, England	Univ. of London Obs.	51N	36	46.3	00W	14	24	82
Minneapolis, Minnesota, USA	Univ. of Minnesota Obs.	44N	58	40.0	93W	14	18	260
Mitaka, Japan	National Astr. Obs.	35N	40	18	139E	32	30	58
Montevideo, Uruguay	National Obs.	34S	54	33.0	56W	12	42	24
Montreal, Quebec, Canada	McGill Univ. Obs.	45N	30	20.0	73W	34	42	57
Moscow, Russia	Sternberg Inst. Obs.	55N	45	19.8	37E	34	12	166
Mt Aragatz, Armenia	Byurakan Astrophys. Obs.	40N	20	06	44E	17	30	1500
Mt Bigelow, Arizona, USA	Steward Obs., Catalina Stn	32N	25	00	110W	43	54	2510
Mt Fowlkes, Texas, USA	Hobby-Eberly Obs.	30N	40	00	104W	01	00	2072
Mt Hamilton, California, USA	Lick Obs.	37N	20	25.3	121W	38	42	1283
Mt Hopkins, Arizona, USA	Fred L. Whipple Obs.	31N	41	18	110W	53	06	2608
Mt John, New Zealand	Mt John Univ. Obs.	43S	59	12	170E	27	54	1027
Mt Lemmon, Arizona, USA	Steward Obs., Catalina Stn	32N	26	30	110W	47	30	2776
Mt Locke, Texas, USA	McDonald Obs.	30N	40	18	104W	01	06	2075

Table 30.2. (Continued)

Location	Name	Latitude			Longitude			Altitude (m)
		°	′	″	°	′	″	
Mt Semirodriki, Caucasus, Russia	Zelenchukskaya	43N	49	32.0	41E	35	24	973
Mt Stromlo, Canberra, Australia	Mount Stromlo Obs.	35S	19	12	149E	00	30	767
Mt Wilson, California, USA	Mt Wilson Obs.	34N	12	59.5	118W	03	36	1742
Nanking, China	Purple Mountain Obs.	32N	03	59.9	118E	49	18	367
Naples, Italy	Capodimonte Obs.	40N	51	45.7	14E	15	24	164
Nice, France	Nice Obs.	43N	43	17.0	07E	18	00	376
Ondřejov, Czech Republic	Astrophysical Obs.	49N	54	38.1	14E	47	00	533
Ottawa, Ontario, Canada	Dominion Obs.	45N	23	38.1	75W	43	00	87
Palermo, Sicily	University Astron. Obs.	38N	06	43.6	13E	21	30	72
Palomar, California, USA	Palomar Obs.	33N	21	22.4	116W	51	48	1706
Pic du Midi, France	Obs. of University of Toulouse	42N	56	12.0	00E	08	30	2862
Piikkiö, Finland	Turku Univ. Obs.	60N	25	00	22E	26	48	40
Pittsburgh, Pennsylvania, USA	Allegheny Obs.	40N	20	58.1	80W	01	18	370
Potsdam, Germany	Astrophysical Obs.	52N	22	56.0	13E	04	00	107
Pulkovo, Russia	Astron. Obs. of Academy of Sciences	59N	46	18.5	30E	19	36	75
Purple Mountain, China	Purple Mountain Obs.	32N	04	00	118E	39	18	367
Quezon City, Philippines	Manila Obs.	14N	38	12	121E	04	36	58
Quito, Ecuador	National Obs.	00S	14	00.0	78W	29	36	2908
Richmond Hill, Ontario, Canada	David Dunlap Obs.	43N	51	46.0	79W	25	18	244
Riga, Latvia	Latvian State Univ. Astr. Obs.	56N	57	06	24E	07	00	39
Rio de Janeiro, Brazil	National Obs.	22S	53	42.2	43W	13	24	33
Rome, Italy	Vatican Obs.	41N	44	48	12E	39	06	450
St Andrews, Scotland	University Obs.	56N	20	12.0	02W	48	54	30
St Michel, France	Obs. of Haute-Provence	43N	55	54	05E	42	48	665
St Petersburg, Russia	Pulkovo Obs.	59N	46	05.5	30E	19	24	70
São Paulo, Brazil	Astronomical and Geophysical Inst.	23S	39	06.9	46W	37	24	800
Siding Spring, NSW, Australia	Siding Spring Obs.	31S	16	37.3	149E	04	00	1165
Stockholm, Sweden	Saltsjöbaden Obs.	59N	16	18.0	18E	18	30	55
Sunspot, New Mexico, USA	Sacr. Peak Natl. Solar Obs.	32N	47	12	105W	49	12	2811
Sutherland, CP, S Africa	S.A. Astr. Obs.	32S	22	46.5	20E	48	36	1830
Sydney, Australia	Government Obs.	33S	51	41.1	151E	12	18	44
Tartu, Estonia	Wilhelm Struve Astr. Obs.	58N	16	00	26E	28	00	0
Tautenberg, Germany	Karl Schwarzschild Obs.	50N	58	54	11E	42	48	331
Tokyo, Japan	Tokyo Obs. at Mitaka	35N	40	21.4	139E	32	30	59
Tonantzintla, Mexico	National Astro. Obs.	19N	01	57.9	98W	18	48	2150
Toruń, Poland	Copernicus Univ. Obs.	53N	05	47.7	18E	33	18	90
Trieste, Italy	Trieste Astr. Obs.	45W	38	30	13E	52	30	400
Tucson, Arizona, USA	Steward Obs.	32N	14	00	110W	56	54	757
Uccle, Belgium	Royal Obs.	50N	47	55.0	04E	21	30	105
Uppsala, Sweden	Univ. Astron. Obs.	59N	51	29.4	17E	37	30	21
Utrecht, Holland	Sonnenborgh Obs.	52N	05	09.6	05E	07	48	14
Victoria, BC, Canada	Dominion Astro. Obs.	48N	31	15.7	123W	25	00	229
Vienna, Austria	University Obs.	48N	13	55.1	16E	20	18	240
Vilnius, Lithuania	Vilnius Astr. Obs.	54N	41	00	25E	17	12	122
Washington, District of Columbia, USA	US Naval Obs.	38N	55	14.0	77W	03	54	86
Wellington, New Zealand	Carter Obs.	41S	17	03.9	174E	45	54	129
Williams Bay, Wisconsin, USA	Yerkes Obs.	42N	34	13.4	88W	33	24	334
Zürich, Switzerland	Swiss Federal Obs.	47N	22	36	08E	33	06	469

Table 30.3. Large optical telescopes.

Aperture (m)	(in)	Name	Observatory	Latitude °	′	Longitude °	′	Elevation (m)	Completion
16.0	630	Very Large Telescope[a]	Cerro Parañal, Chile	24	38S	70	24W	2635	1998 (Antu), 1999 (Kueyen, Melipal)
10.0	387	Keck I	Mauna Kea, Hawaii	19	50N	155	28W	4123	1991 Mirror composed of 36 segments
10.0	387	Keck II	Mauna Kea, Hawaii	19	50N	155	28W	4123	1996 Twin of Keck I
9.2	362	Hobby-Eberly	Mount Fowlkes, Texas	30	40N	104	01W	2072	1986 Segmented mirror. Fixed elevation
8.3	327	Subaru	Mauna Kea, Hawaii	19	50N	155	28W	4100	1999 Japanese
8.0	315	Gemini North	Mauna Kea, Hawaii	19	50N	155	28W	4100	1999 Twin of Gemini South
6.5	256	Mono-Mirror Telescope	Mount Hopkins, Arizona	31	04N	110	53W	2608	1999 Succeeds 4.5 m Multiple Mirror Telescope
6.0	236	Bolshoi Teleskop Azimutalnyi	Nizhny Arkhyz, Russia	49	39N	41	26E	2070	1975 Large Altazimuth Telescope
5.08	200	Hale Telescope	Palomar, California	33	21N	116	52W	1706	1948 200 inch reflector
4.2	165	William Herschel Telescope	La Palma, Roque de los Muchachos	28	46N	17	53W	2332	1987 British
4.0	158	Victor Blanco Telescope	Cerro Tololo, Chile	30	10S	70	49W	2215	1976 Inter-American Observatory
3.89	153	Anglo-Australian Telescope	Siding Spring, NSW	31	17S	149	04E	1149	1975 Coonabarabran, Australia
3.81	150	Nicholas U Mayall Reflector	Kitt Peak, Arizona	31	57N	111	37W	2120	1973
3.80	150	UKIRT	Mauna Kea, Hawaii	19	50N	155	28W	4194	1978 United Kingdom Infra-Red Telescope
3.58	141	Canada–France–Hawaii Telescope	Mauna Kea, Hawaii	19	49N	155	28W	4200	1979 CFH Telescope
3.57	141	3.6 m telescope	La Silla, Chile	29	16S	70	44W	2387	1977 European Southern Observatory
3.50	141	3.5 m telescope	Calar Alto, Spain	37	13N	02	32W	2168	1984 Spanish
3.50	141	Astrophysical Research Consortium	Apache Point, New Mexico	32	47N	105	49W	2788	1993 Mainly remote controlled
3.50	141	WIYN Telescope	Kitt Peak, Arizona	31	57N	111	36W	2089	1994 Wisconsin–Indiana–Yale–NOAO Telescope
3.50	141	NTT	La Silla, Chile	29	16S	70	44W	2353	1989 New Technology Telescope
3.50	141	Starfire	Kirtland AFB, New Mexico		—		—	1900	— Military
3.50	141	Telescopio Nazionale Galileo	La Palma, Canary Islands	28	45N	17	53W	2370	1998 Italian

[a] Four 8.2 m mirrors, working together: Antu (the Sun), Kueyen (the Moon), Melipal (the Southern Cross) and Yepun (Sirius); the names come from the Mapuche language of the people of the area. Yepun is at present (2000) awaiting installation.

Table 30.3. (Continued)

Aperture		Name	Observatory	Latitude		Longitude		Elevation (m)	Completion
(m)	(in)			°	′	°	′		
3.05	120	C Donald Shane Telescope	Lick Observatory, Mt Hamilton, California	37	21N	121	38W	1290	1959
3.00	118	NASA Infra-Red Telescope Facility	Mauna Kea, Hawaii	19	50N	155	28W	4208	1979 IRTF
3.00	118	NODO	Laval University, Quebec	32	59N	105	44W	2758	1999 Liquid mirror telescope
2.72	107	Harlan Smith Telescope	McDonald Observatory, Mount Locke, Texas	30	40N	104	01W	2075	1969 107-inch telescope
2.70	106	UBC-Laval Telescope	Malcolm Knapp Research Forest, British Columbia	49	07N	122	35W	50	1992 Liquid mirror. Not steerable
2.64	104	Shajn Reflector	Crimean Astrophysical Observatory, Ukraine	44	44N	34	00E	550	1960
2.64	104	Byurakan Reflector	Byurakan, Armenia	40	20N	44	18E	1500	1976 Byurakan Astrophysical Observatory
2.56	101	Nordic Optical Telescope	La Palma, Canary Islands	28	45N	17	53W	2382	1989 Los Muchachos
2.56	101	Isaac Newton Telescope	La Palma, Canary Islands	28	46N	17	53W	2336	1984 Los Muchachos
2.54	100	Irénée du Pont Telescope	Las Campanas, Chile	29	00S	70S	42W	2282	1976
2.50	100	Sloan Digital Sky Survey	Apache Point, New Mexico	32	47N	105	49W	2788	1999 Wide-field detector
2.50	100	Hooker Telescope	Mount Wilson, California	34	13N	118	03W	1742	1917 100-inch reflector
2.40	94	Hubble Space Telescope	Orbital	—		—		—	1990 Average 600 km altitude
2.34	92	Hiltner Telescope	Michigan–Dartmouth–MIT Observatory, Kitt Peak, Arizona	31	57N	111	37W	1938	1986
2.30	91	Vainu Bappu Telescope	Kavalur, Tamil Nadu, India	12	35N	78	50E	725	1986
2.30	91	2.3 m Telescope	Mt Stromlo, Australia	31	16S	149	03E	1149	1984 Mount Stromlo and Siding Spring Observatory
2.30	91	Bok Telescope	Kitt Peak, Arizona	31	57N	111	37W	2100	Steward Observatory
2.30	91	Wyoming Infra-red Telescope	Jelm Mountain, Wyoming	41	06N	105	59W	2943	1977
2.20	87	University of Hawaii Telescope	Mauna Kea, Hawaii	19	50N	155	28W	4200	—
2.20	87	ESO Telescope	La Silla, Chile	29	15S	70	44W	2200	1977 – European Southern Observatory

Table 30.3. (Continued)

Aperture (m)	(in)	Name	Observatory	Latitude °	′	Longitude °	′	Elevation (m)	Completion
Refractors									
1.01	40	Yerkes 40 in Telescope	Yerkes Observatory, Williams Bay, Wisconsin, USA	42	34N	88	33W	334	1897
0.89	36	36 in Refractor	Lick Observatory, Mount Hamilton, California, USA	37	20N	121	39W	1290	1888
0.83	33	33 in Meudon Refractor	Paris Observatory, Meudon, France	48	48N	02	14E	162	1889
0.80	31	Potsdam Refractor	Potsdam Observatory, Germany	52	23N	13	04E	107	1899
0.76	30	Thaw Refractor	Allegheny Observatory, Pittsburgh, Pennsylvania, USA	40	29N	80	01W	380	1985
0.74	29	Lunette Bischoffschei	Nice Observatory, France	43	43N	07	18E	372	1886
0.68	27	Grosser Refraktor	Archenold Observatory, Treptow, Germany	52	29N	13	29E	41	1896
0.67	26	Grosser Refraktor	Vienna Observatory, Austria	48	14N	16	20E	241	1880
0.67	26	McCormick Refractor	Leander McCormick Observatory, Charlottesville, Virginia, USA	38	02N	78	31W	264	1883
0.66	26	26 in Equatorial	US Naval Observatory, Washington, DC, USA	38	55N	77	04W	92	1873
0.66	26	Thompson Refractor	Herstmonceux, England	51	29N	00	00	50	1897
0.66	26	Innes Telescope	Republic Observatory, Johannesburg, South Africa	26	10S	55	33E	1806	1926
Schmidt Telescopes									
1.34	53	2 m Telescope	Karl Schwarzschild Observatory, Tautenberg, Germany	50	59N	11	43E	331	1960
1.24	49	Oschin Telescope	Palomar Observatory, California, USA	33	21N	116	51W	1706	1948
1.24	49	United Kingdom Schmidt Telescope (UKS)	Siding Spring, Australia	31	16S	149	04E	1145	1973
1.05	41	Kiso Schmidt Telescope	Kiso Observatory, Kiso, Japan	35	48N	137	38E	1130	1975
1.00	39	3TA-10 Schmidt Telescope	Byurakan Astrophysics Observatory, Mt Aragatz, Armenia	40	20N	44	30E	1450	1961
1.00	39	Kvistaberg Schmidt Telescope	Uppsala University Observatory, Kvistaberg, Sweden	59	30N	17	36E	33	1963

Table 30.3. (Continued)

Aperture (m)	(in)	Name	Observatory	Latitude °	′	Longitude °	′	Elevation (m)	Completion
1.00	39	ESO 1 m Schmidt Telescope	European Southern Observatory, La Silla, Chile	29	15S	70	44W	2318	1972
1.00	39	Venezuela 1 m Schmidt Telescope	Centro F J Duarte, Merida, Venezuela	08	47N	70	52W	3610	1978
0.90	35	Télescope de Schmidt	Observatoire de Calern, Calern, France	43	45N	06	56W	1270	1981
0.84	33	Télescope Combiné de Schmidt	Royal Observatory, Brussels, Belgium	50	48N	04	21E	105	1958
0.80	31	Schmidt Telescope	Radiophysical Observatory, Riga, Latvia	56	47N	24	24E	75	1968
0.80	31	Calar-Alto-Schmidtspiegel	Calar Alto Observatory, Spain	37	13N	02	32W	2168	1980

31 NON-OPTICAL ASTRONOMY

NON-OPTICAL ASTRONOMY

Until comparatively modern times, astronomers were limited to studying radiation in the visible range of the electromagnetic spectrum. Then, in 1931, came the detection of radio waves from the Milky Way, and now it is possible to examine almost all wavelengths, from long radio waves down to the very short gamma-rays.

Wavelengths are measured in nanometres or Ångströms –one nanometre being one thousand-millionth of a metre, and 1 Ångström being one ten thousand millionth of a metre, so that 1 nm = 10 Å. The accepted regions are as follows:

Below 0.01 nm	Gamma-rays
0.01–10 nm	X-rays
	Hard: 0.01–0.1
	Soft: 0.1–10
10–400 nm	Ultra-violet
	EUV (Extreme Ultra-Vioiet)
	10–120 nm
400–700 nm	Visible light
	(=4000–7000 Å)
700 nm–5000 nm	Near infra-red
500 nm–0.3 mm	Mid–far infra-red
0.3 mm–1 mm	Sub-millimetre
1 mm–0.3 m	Microwaves
Over 0.3 m	Radio waves

Cosmic-ray astronomy

Cosmic rays were discovered in 1912 by the Austrian physicist Victor Hess, who was anxious to find out why electrometers at ground level always recorded a certain amount of background. He therefore flew electrometers in a balloon, rising to a height of 4800 m, and discovered evidence that cosmic rays are coming constantly from all directions in space.

They are not rays at all, but particles. Most cosmic-ray primaries are protons, although there is a tiny gamma-ray component. There are many α-particles (helium nuclei) and some heavier atomic nuclei up to uranium and perhaps beyond.

About 3% of the primaries are electrons. The primaries cannot pass directly through the Earth's atmosphere, but break up both themselves and the air particles, so that only the secondary cosmic rays reach ground level.

Cosmic rays are the only particles we can detect which have crossed the Galaxy, moving at almost the velocity of light. Unfortunately, the particles are affected by the magnetic fields in the Galaxy and this means that ordinarily it is not possible to determine the directions from which they come.

The origins of cosmic rays seem to be varied. Those of the highest energy may have come from quasars or AGN; those of lower energy originate within the Galaxy in supernova remnants, supernova outbursts and pulsars. Those of the lowest energy come from solar flares. However, the Sun is not a major source of cosmic rays. The flux we receive decreases at the time of solar maximum, because of the increase in the strength of the interplanetary magnetic field, which acts as a screen (the Forbush effect).

Modern cosmic-ray astronomy relies mainly upon space research methods, and many satellites and space probes carry cosmic-ray detectors.

Gamma-ray astronomy

Gamma rays represent the most energetic form of electromagnetic radiation, at wavelengths shorter than those of X-rays – that is to say, below 0.01 nm. They are absorbed in the Earth's upper atmosphere, and only the most energetic can reach ground level, so that virtually all gamma-ray astronomy depends upon equipment carried in satellites or space-craft (although balloons have also been used). Moreover, care must always be taken to distinguish gamma-rays from the much more plentiful cosmic rays.

The first detection of high-energy gamma-rays from space was achieved by the Explorer XI satellite, in 1961; it picked up fewer than 100 gamma-ray photons. They seemed to come from all directions, indicating a more or less uniform gamma-ray background; this has been fully confirmed since. Discrete gamma-ray sources were also identified; the first of these, in the constellation of Sagittarius, was found in 1969. Detectors on OSO 3

(Orbiting Solar Observatory 3) recorded gamma-rays coming from the direction of the centre of the Galaxy.

The first really important gamma-ray satellites were SAS 2, launched in 1972, and Cos-B, launched in 1975. SAS 2 failed after six months, but Cos-B operated until 1982. It was confirmed that the most intense gamma radiation came from the galactic plane, due to interactions between cosmic rays and interstellar gas, and new discrete sources were identified, including the Vela[1] pulsar, the Crab pulsar and a strange object in Gemini, which was not visible optically and which did not seem to radiate except at gamma-ray wavelengths. It became known as Geminga, partly as an abbreviation for 'Gemini gamma-ray source' and partly because in the Milanese dialect, Geminga means 'it is not there'. Only in 1991 was its true nature established. It is a neutron star, about 320 light-years away, thought to be the result of a supernova outburst about 300 000 years ago. It is not detectable as a pulsar, presumably because its beams of radiation do not sweep across the Earth. It is a powerful gamma-ray source, although the two strongest sources in the sky are the Vela pulsar and the Crab pulsar.

One of the most successful vehicles involved with gamma-ray research is the satellite originally called GRO (the Gamma-Ray Observatory), but now named after the great pioneer of this research, Arthur Holly Compton. It was launched from the Shuttle on 5 April 1991; it weighed nearly 17 tons, and filled two-thirds of the Shuttle bay. The instruments weighed six tons, all were capable of detecting gamma-ray photons, measuring their energies, and determining the directions from which they came. It was found that the Milky Way glows at gamma-ray wavelengths; in 1993 the Compton satellite completed the first all-sky map of gamma-ray sources, and it also pinpointed many discrete sources, including AGN (Active Galactic Nuclei) and quasars. It was deliberately crashed into the Pacific on 4 June 2000, as the equipment had started to fail.

Very high-energy gamma-rays may be detected indirectly from observations made at ground level. In passing through the atmosphere they generate electrons which travel faster than the speed of light in air, resulting in

blue Čerenkov radiation which may be detected with special equipment. This does not mean that Einstein's theory of relativity is wrong, because it is still the case that nothing can travel faster than light in a vacuum. The largest gamma-ray 'telescope' of this kind is at the Whipple Observatory on Mount Hopkins, Arizona; the 'dish' is 10 m in diameter. Intensive research is being carried out by scientists at the University of Durham, where the main equipment has been set up at Narrabri in Australia.

Of special interest are the gamma-ray bursters; the first of these was detected by the Vela satellite in 1967, but many hundreds have been recorded since. They are of immense power, and last only briefly –from a few seconds to a few minutes; they may appear in any direction at any moment. For a long time they remained a complete mystery, but a great advance was made in February 1997, when an Italian satellite, BeppoSAX, pinpointed a burst in the constellation of Orion. Telescopes were promptly trained on the area, and detected a rapidly fading star, no doubt the aftermath of a titanic explosion; close by was a dim blob believed to be a galaxy. Since then seven more optical identifications have been found, and one of these is associated with a very remote galaxy –redshift 3.4. It is therefore obvious that the gamma-ray bursters originate in the far reaches of the universe, and are unbelievably violent. Many have been recorded with BATSE, the Burst and Transient Source Experiment, carried in the Compton gamma-ray observatory.

We do not yet know precisely what causes gamma-ray bursters. One suggestion involves the collapse of a massive star into a black hole (hypernovæ?), but whether this would generate enough power is debatable. Collisions between neutron stars have also been proposed but despite the progress made during the last few years it is fair to say that gamma-ray astronomy is still in an early stage of development.

X-ray astronomy

X-rays were discovered in 1895 by the German scientist Wilhelm Röntgen (it must be admitted that the discovery was accidental). In the electromagnetic spectrum they lie between gamma-rays and ultra-violet. Cosmic X-ray sources range from the Sun to the hot outer atmospheres of normal stars, white dwarfs, active galaxies (including

[1] The Vela pulsar lies in the southern constellation of Vela, the sails of the dismembered Argo Navis. The name has nothing to do with the Vela series of artificial satellites.

quasars) and accretion disks around neutron stars and black holes.

For obvious reasons, no optical telescope can detect X-rays, because the X-ray photons would simply penetrate the telescope mirror. However, just as a bullet will ricochet off a wall when striking at a grazing angle, so X-rays will ricochet off a mirror when striking it at a narrow angle. An X-ray telescope is therefore quite unlike an optical telescope; the mirrors have to be aligned almost parallel to the incoming X-rays, so that the telescope is shaped like a barrel. Moreover, X-rays are absorbed by the Earth's upper atmosphere, so that observations from the ground level are virtually impossible.

The first observations of X-rays from the sky were made on 5 August 1948, when R. Burnright of the US Naval Research Laboratory detected solar X-rays from the darkening of a photographic emulsion carried to altitude on a V2 rocket. On 29 September 1948 H. Friedman and E. O. Hulbert, also using a V2, detected intense X-radiation from the Sun. In 1956 Friedman's team recorded results which could have been due to celestial X-rays, but they could not be sure, as they always had the Sun in their field of view. Then, in 1959, R. Giacconi and his colleagues published a paper in which they predicted X-rays from very hot stars and supernova remnants. They referred particularly to the Crab Nebula.

On 24 October 1961 a rocket was launched from White Sands to search for X-rays from the Moon. It was thought that these could be due to incident X-rays striking the lunar surface and causing X-ray fluorescence, together with X-rays due to the surface being struck by energetic electrons from the solar wind. The equipment failed, as the protective covers refused to open. A second rocket – an Aerobee – was launched on 18 June 1962. No lunar X-rays were found, but a discrete source was detected, later found to be Scorpius X-1. On 12 October 1962 a new launch recorded X-rays from the Crab Nebula, with an intensity of 15% of those from Scorpius X-1. By 1966 30 sources had been identified, including the first galaxy, M.87. The first imaging X-ray telescope was designed by Giacconi and his team at Cambridge, Massachusetts, and was flown on a small rocket. In 1965 it managed to pick up images of hot spots in the upper atmosphere of the Sun.

Many X-ray satellites have been launched; a selected list is given in Table 31.1. The first satellite designed purely for X-ray work was Uhuru, launched from Kenya on 12 December 1970 (the name means 'Freedom' in Swahili). Uhuru carried out a comprehensive all-sky survey and located 339 sources, including binary systems, supernova remnants, Seyfert galaxies and clusters of galaxies; it also discovered diffuse X-radiation from the material contained in clusters of galaxies.

In 1971 Uhuru showed that one source, Centaurus X-3, is a binary system; it was found that the intensity of the X-radiation varied rapidly in a period of 4.8 s, suggesting a neutron star, and a period of 2.087 days indicating that the neutron star was being eclipsed by a binary companion. The larger component was then optically identified, and a second binary, Hercules X-1, was identified in 1971.

X-ray novæ are temporary phenomena. The first, Centaurus X-4, was discovered in May 1969 by an X-ray detector carried in a Vela satellite. An X-ray nova seen in December 1974, near the radio galaxy Centaurus A (although certainly unconnected with it) reached its maximum on Christmas Day, and was inevitably nicknamed CenXmas; another, in May 1975 close to the Crab Nebula (again, certainly unconnected with it) was equally inevitably nicknamed Fresh Crab! X-ray novæ seem to be close binaries with low-mass optical companions.

One very successful X-ray satellite was HEAO 2, the Einstein Observatory, which was launched on 13 November 1978 and operated until 1981. It carried the first really large X-ray telescope, consisting of four nested paraboloids and four nested hyperboloids, with an outer diameter of 58 cm. The ambitious Japanese satellite Yohkoh ('sunlight'), sent up in 1991, was designed mainly for solar studies. The Röntgen satellite, Rosat, was launched in 1990 and operated until finally switched off in February 1999; it achieved more than 9000 observations of objects, including comets, quasars, black holes, clusters of galaxies, proto-stars and supernovæ, as well as carrying out the first high-resolution sky surveys at X-ray and extreme ultra-violet (EUV) wavelengths. Rosat was truly international; it used a German X-ray telescope and a British wide-field camera used for extreme ultra-violet work, while the launch was achieved by an American rocket. In 1990, almost 30 years

Table 31.1. A selected list of X-ray satellites.

Name	Re-named	Launch	End of mission	Nationality	Notes
Vela 58		23 May 1969	19 June 1969	US	X-ray burster detected.
SAS-1	Uhuru	12 Dec 1970	March 1973	US	Small Astronomy Satellite 1. First X-ray satellite.
OSO-7		19 Sept 1971	9 July 1974	US	Orbiting Solar Observatory. All-sky survey.
OAO-3	Copernicus	21 Aug 1972	Late 1980	US–UK	Orbiting Astronomical Observatory 3.
	Ariel-5	15 Oct 1974	14 Mar 1980	UK	X-ray transients. Spectra.
SAS-3		May 1975	1979	US	Bursters. Locations of sources. X-rays from Algol.
HEAO-1		12 Aug 1977	9 Jan 1979	US	High Energy Astronomy Observatory 1. All-sky survey.
HEAO-2	Einstein	12 Nov 1978	Apr 1981	US	First fully imaging X-ray telescope.
Corsa-B	Hakucho	21 Feb 1979	15 Apr 1985	Japan	Transient phenomena. Cygnus X-1 study.
Astro-B	Tenma	20 Feb 1983	Oct 1985	Japan	'Pegasus'. Four high-energy experiments.
	Exosat	26 May 1983	9 Apr 1986	ESA	AGN; X-ray binaries; SNR.
Astro-C	Ginga	5 Feb 1987	1 Nov 1991	Japan	X-ray transient phenomena.
	Granat	1 Dec 1989	27 Nov 1998	Russia	Deep imaging of galactic centre area.
Röntgen satellite	Rosat	1 June 1990	12 Feb 1999	US	Observations of all types.
	Yohkoh	31 Aug 1991	—	Japan	Solar X-rays and gamma-rays.
Astro-D	ASCA	20 Feb 1993	—	Japan	Advances Satellite for Cosmology and Astrophysics. First use of CCDs in X-ray astronomy.
RXTE	Rossi X-ray Timing Explorer	30 Dec 1995	—	US	Variability in X-ray sources.
	BeppoSAX	30 Apr 1996	—	Italy	Afterglow of gamma-ray bursts.
AXAF	Chandra	23 July 1999	—	US	Observations of all kinds.
XMM	Newton	10 Dec 1999	—	ESA	Detailed X-ray spectroscopy.

after the first search for lunar X-rays, the Moon was indeed imaged at X-ray wavelengths by Rosat. The X-rays were emitted by the Sun, and were scattered by atoms in the surface layers of the Moon. Rosat detected and catalogued about 100 000 separate X-ray sources, i.e. many more than the number of stars one can see with the naked eye.

In July 1999 the Chandra X-ray Observatory was launched from the Shuttle. (Its original name was AXAF, the Advanced X-ray Astrophysics Facility, but it was then re-named in honour of Subrahmanyan Chandrasekhar, the great Indian astrophysicist.) Chandra was put into an orbit which takes it from 9980 to 140 000 km above the Earth, so that for much of the time it is above the terrestrial radiation belts. It is a hundred times more sensitive than any other X-ray telescope, and quickly showed its potential; and early success was the detection of a brilliant ring inside the Crab Nebula. Chandra is designed to have an operational lifetime of five years.

The X-ray Multi-Mirror Mission (XMM), of ESA, was launched on 10 December 1999 from an Ariane rocket at Kourou in South America. It is designed to perform detailed spectroscopy of cosmic X-ray sources over a broad band of energies ranging from 0.1 keV to 10 keV. There are three highly-nested grazing incidence mirror modules of type Wolter 1, coupled to reflection grating spectrometers and X-ray CCD cameras. It is now known as the Newton Satellite, in honour of the scientist who first

showed that light was composed of many different colours (wavelengths).

Newton and Chandra are both powerful in their own way. Chandra is optimized to provide the sharpest X-ray images, whilst Newton is the most sensitive for X-ray spectra.

Ultra-violet astronomy

The existence of ultra-violet radiation from the Sun was demonstrated in 1801 by J. Ritter, by producing a spectrum with a prism and noting the darkening of paper soaked in sodium chloride held in the region beyond the violet. This was possible because ultra-violet radiation of between 300 and 400 nm can penetrate the Earth's atmosphere. Most research in ultra-violet astronomy has to be carried out from altitude – initially by rockets, and nowadays by satellites.

The first ultra-violet spectrograph was launched from White Sands on 28 June 1946, by a V2 rocket. It crashed, and the film was lost, but on 10 October 1946 a V2 soared to 88 km, and R. Tousey recorded the first recoverable solar ultra-violet spectra from above the atmosphere.

Ultra-violet observations were subsequently made from the American satellites of the OAO (Orbiting Astronomical Observatory) series. OAO-1 (8 April 1966) failed. OAO-2 (7 December 1968) operated until 13 February 1973; it moved in an orbit ranging between 770 and 780 km above the ground. Its 11 ultra-violet telescopes viewed 1930 objects. Discoveries included a hydrogen cloud round a comet (Tago-Sato-Kosaka), magnetic fields of stars and ultra-violet radiation from a supernova. OAO-3 (the Copernicus satellite) was launched on 21 August 1972, and discovered many new sources, including supernova remnants and pulsars. It operated until late 1980.

However, the most successful ultra-violet satellite to date has been the IUE (International Ultra-violet Explorer). It was launched on 26 January 1978, with an estimated lifetime of five years; in fact it continued to operate until deliberately switched off on 30 September 1996. It has provided material for more scientific papers than any other satellite.

IUE carried a 45 cm aperture telescope, which could feed two spectrographs (one in the 190–320 nm range, the other 115–200 nm). Although the IUE was not designed to produce ultra-violet pictures, it did provide a complete survey of ultra-violet spectra for virtually every kind of astronomical object – hot stars, cool stars, variable stars, nebulæ, the interstellar medium, extragalactic objects and members of the Solar System, including Halley's Comet. It moved in a synchronous orbit, at a distance of 42 164 km and an inclination of 34°. Its discoveries included many stars with magnetic fields and surface activity, measurements of stellar winds, mapping of low-density bubbles of gas around the Sun and nearby stars, and measurements of the composition of planetary nebulæ.

The work of the IUE is now continued using the Hubble Space Telescope, which in addition to its visible instruments, also carries onboard cameras and spectrographs which are sensitive to the ultra-violet.

Observations at extreme ultra-violet wavelengths (EUV) really began with the Rosat satellite in 1990. Previously it had – wrongly –been thought that no worthwhile observations in this part of the electromagnetic spectrum could be made, because of strong absorption by neutral hydrogen in the interstellar medium. The first object to be studied by Rosat was the binary system WFC1, made up of a very hot white dwarf, HZ43, which has a temperature of 200 000 °C and is one of the hottest stars known; it has a red dwarf companion. Rosat also obtained the first EUV record of the Moon.

The brightest star in the sky at EUV wavelengths is ϵ Canis Majoris (Adhara). It is almost 500 light-years away, and had the interstellar medium been as opaque to EUV as was originally thought, it would have meant that radiations from that range would have been blocked. However, it now seems that the Solar System lies in a relatively dense, cool part of the interstellar medium, and is surrounded by more rarefied material at a higher temperature – so that the atoms of hydrogen are ionized and do not absorb EUV.

The Extreme Ultra-Violet Explorer satellite (EUVE) was launched on 7 June 1992, into a 530 km Earth orbit; it was designed to study radiation at wavelengths from 70 and 760 Å. It carried four grazing-incidence telescopes and recorded over 600 sources, ranging from Solar System objects (notably Mars, Jupiter and Io) to cool stars with hot coronæ, white dwarfs with temperatures up to 100 000 °C, and the photospheres and stellar winds of very hot stars. The nuclei of some active galaxies were also recorded, showing that in our own Galaxy there are 'tunnels' through which EUV can pass. By 1993 EUVE had completed a full survey of the sky in this region of the electromagnetic spectrum.

The Far-Ultraviolet Spectroscopic Explorer (FUSE) was launched from Cape Canaveral on 24 June 1999. It records spectra of astronomical objects at the far-ultraviolet end of the electromagnetic spectrum (90 to 120 nanometres) – light which is nearly as energetic as X-rays. The FUSE equipment is about 10 000 times more sensitive than that of the Copernicus satellite, which operated from 1972 to 1980.

Infra-red astronomy

Infra-red astronomy has become a vitally important part of modern research. The original discovery of infra-red radiation from the Sun was made in 1801 by William Herschel, by placing a thermometer beyond the red end of the solar spectrum. The range extends from a micrometre (0.001 mm) to several hundreds of micrometres, beyond which comes the microwave region.

Some wavelengths of infra-red can penetrate through to the Earth's surface, particularly the 'near infra-red', towards the red end of the visible spectrum. Water vapour is the main enemy of the infra-red astronomer, so that the best sites are at high altitude, with low humidity – such as the top of Mauna Kea in Hawaii, and the lofty Cerro Parañal in the

Atacama Desert of Northern Chile. Aircraft and balloons have been used to carry infra-red equipment, but of course satellites are of great importance.

Infra-red radiations can penetrate dust clouds, and so provide us with information about star-forming regions and the central region of the Galaxy. For example, deep inside the Orion Nebula there are powerful stars which can be detected only in the infra-red; at optical wavelengths the nebulosity forms an impenetrable screen.

The first major survey of the infra-red sky was undertaken in the 1960s by G. Neugebauer and R. Leighton; they discovered 6000 discrete sources. Today there are large infra-red telescopes, such as UKIRT (the United Kingdom Infra-Red Telescope) on Mauna Kea, above most of the atmospheric water vapour. The aperture is 380 cm. The mirror is thin; theoretically a telescope designed for infra-red work need not be so accurate as an optical telescope, although in fact the UKIRT is so good that it can be used at optical wavelengths as well as in infra-red. Also on Mauna Kea is the NASA IRTF (Infra-Red Telescope Facility) with an aperture of 300 cm, and there are also other large telescopes designed to study this part of the electromagnetic spectrum. Of course, the Hubble Space Telescope, above the atmosphere, is very effective in the infra-red.

Of infra-red satellites, one of the most successful has been IRAS (the Infra-Red Astronomical Satellite), which was launched on 25 January 1983; the reflector had an aperture of 60 cm, enclosed in a cooling vessel containing 500 litres of liquid helium, holding the temperature not far above absolute zero ($-273\,°C$). IRAS operated for 300 days, and could have picked up the radiation from a speck of dust several kilometres away. There were eight staggered rows of detectors, and observations were made in four bands (12, 25, 60 and 100 μm).

IRAS continued to function until 21 November 1983, when it ran out of coolant. By then it had surveyed 97% of the sky, and had made many major discoveries. For example, it found 245 389 discrete infra-red sources, increasing the number of known sources by more than 100; it found the first dust ring in the Solar System, which takes the form of a torus and lies at a distance about the same as that of the main asteroid belt, although inclined at an angle of 10° to the main plane of the Solar System; it obtained

observations of comets, and detected the first cometary dust-tail in infra-red (Comet Tempel 2; the tail was 30 000 000 km long), and it made studies of infra-red 'cirrus', due to cool dust clouds in the Galaxy which, when displayed on computerized maps, shows up as wispy clouds resembling cirrus clouds – hence the name. They are made up of tiny dust grains, mainly graphite.

The most spectacular discovery made by IRAS was that of clouds of cool material around some stars, notably Vega, Fomalhaut and the southern β Pictoris. In fact the cloud around Vega, the first to be found, was detected fortuitously; the main investigators, H. Aumann and R. Gillett, discovered it while calibrating the on-board equipment.

IRAS also discovered many AGNs, and starburst galaxies (galaxies undergoing massive bouts of star formation) which are radiating strongly in the infra-red region of the spectrum.

Other major satellites included ISO (the Infra-red Space Observatory) and the ESA satellite which was launched on 17 November 1995 and continued to operate until May 1998. It observed the universe at wavelengths from 2.5 to 240 μm. Its detectors were about a thousand times more sensitive than those of IRAS, and achieved angular resolutions about a hundred times better (these figures relate to its observations at a wavelength of 12 μm). Among its many achievements, it discovered water in the atmospheres of planets and of Titan, and in star-forming regions such as the Orion Nebula; it detected newly-forming stars at very early stages of development. It investigated the properties of ultra-luminous infra-red galaxies and, much nearer home, examined Comet Hale–Bopp.

Infra-red studies can provide information which could not be obtained in any other way. It is fortunate that a considerable amount of work can be carried out from ground level, although for some areas of research it is essential to use satellites.

Microwave astronomy

This part of the electromagnetic spectrum extends from 0.3 mm to 30 cm, which is longer than infra-red but shorter than radio radiation: sub-millimetre from 0.3 mm, millimetre from 1 mm.

As noted earlier (page 319), the microwave background at 3° above absolute zero was detected in 1964 by A. Penzias and R. Wilson at the Bell Telephone Laboratories in Holmdel, New Jersey, fortuitously when they were calibrating a 7.35 cm wavelength receiver built for satellite communications. It is assumed to be the last manifestation of the Big Bang in which the universe was created; the overall temperature has now fallen to 2.7° above absolute zero, agreeing with theory. The first slight irregularities in the microwave background were detected in April 1992 with COBE, the Cosmic Background Explorer satellite. These were 'ripples', or rarefied wisps of material, said to be the largest and most ancient structures known in the universe. This discovery was hailed as being among the most important of modern times, and confirmed that it was indeed possible for the initial material to collect into galaxies of the kind we know today.

The largest sub-millimetre telescopes are the JCMT (James Clerk Maxwell Telescope) on Mauna Kea, the SEST (Swedish Submillimetre Telescope) at La Silla in Chile and the CSO (California Institute of Technology Sub-millimetre Telescope) on Mauna Kea. The first two have 15 m segmented mirrors, and the CSO has a 10.4 m segmented mirror.

Radio astronomy

Radio astronomy began in 1931, when the American radio engineer Karl Jansky detected radio waves from the Milky Way. The discovery was fortuitous; Jansky was investigating 'static' on behalf of the Bell Telephone Company, using a homemade aerial nicknamed the 'Merry go-Round' (part of the mounting was made from a dismantled Ford car). Jansky's first paper was published in 1932, but caused surprisingly little interest, and Jansky never followed up his discovery as he might have been expected to do. The first intentional radio telescope was made by the American amateur Grote Reber, whose first paper appeared in 1940; Reber's 'telescope' was a 9.5 m dish. At that time Reber was the only radio astronomer in the world.

Radio waves from the Sun were first detected in 1942 by a British team led by J. S. Hey; originally the effects were thought to be due to German jamming of British radar. However, in February 1946 it was established that the giant sunspot of that month was a strong radio source.

At Jodrell Bank, in Cheshire, work began in 1945, when radar was used to measure meteor trails; in the same year, radar echoes from the Moon were detected by an American team led by J. H. de Witt and, independently, by a Hungarian team led by Z. Bay. The first Jodrell Bank radio telescope was a fixed 66 m paraboloid, due to B. Lovell (now Sir Bernard Lovell).

In 1946 the first discrete radio source beyond the Solar System was detected, Cygnus A, and in 1948 M. Ryle and F. G. Smith (now Sir Francis Graham–Smith) detected the supernova remnant Cassiopeia A. In 1949 the first optical identifications of sources beyond the Solar System were made: Taurus A (the Crab Nebula), Virgo A (the galaxy M.87) and Centaurus A (the galaxy NGC 5128, C77). Radio waves from Jupiter were detected in 1955 by B. F. Burke and K. Franklin; the first quasar was identified in 1963 by its radio emissions, and the first pulsar in 1967, by Jocelyn Bell Burnell.

In the years after the war, large radio telescopes were built; that at Dwingeloo in Holland dates back to 1956 – it is a dish 25 m across. The Lovell Telescope at Jodrell Bank, with a diameter of 76 metres (250 feet), came into operation in 1957; originally it was known as the Mark I, and from 1970, after major modifications, as the Mark IA. A second large telescope, the Mark II, was completed at Jodrell Bank in 1964. It is designed to operate at the shorter centimetre wavelengths beyond the range of the Lovell Telescope itself. The bowl of the Mark II is elliptical, measuring 38.2 × 25.4 m; it can be used at wavelengths down to 3 cm.

The largest steerable 'dish' is at the Max Planck Institute at Effelsberg in Germany; it was completed in 1971, and has a diameter of 100 m. The largest non-steerable dish is at Arecibo, Puerto Rico; it is 304.8 m across, built in a natural hollow in the ground, and was completed in 1963. The largest array is the VLA or Very Large Array, 80 km west of Socorro in New Mexico; it is Y-shaped, each arm being 20.9 km long, with 27 movable antennæ, each 25 m across. It was completed in 1981.

Radio telescopes can be used in conjunction. In England there is the Multi-Element Radio-Linked

Table 31.2. Selected list of radio observatories.

Location	Name	Latitude °	′	″	Longitude °	′	″	Altitude (m)
Arcetri (Florence), Italy	Astrophysical Obs.	43N	45	14.4	11E	15	18	184
Arecibo, Puerto Rico	Arecibo Obs., Cornell Univ.	18N	20	36.6	66W	45	12	496
Big Pine, California, USA	Owens Valley Radio Obs.	37N	13	54	118W	16	54	1236
Bochum, Germany	Radio Telescope Stn.	51N	25	43.0	07E	11	30	160
Boulder, Colorado, USA	High Altitude Obs.	40N	04	42.0	105W	16	30	1692
Cambridge, England	Mullard Radio Astr. Obs	52N	09	45.0	00E	02	24	26
Cassel, California, USA	Hat Creek Radio Astr. Obs.	40N	49	06	121W	28	24	1043
Cebrerps, Spain	Deep Space Stn.	40N	27	18	04E	22	00	789
Columbia, South Carolina, USA	Univ. of S Carolina Radio Obs.	33N	59	48	81W	01	54	127
Crimea, Ukraine	Crimean Astrophys. Obs.	44N	43	42.0	34E	01	00	550
Culgoora, New South Wales, Australia	Aus. Tel. Nat. Facility	30S	18	54	149E	33	42	217
Delaware, Ohio, USA	Ohio State Obs.	40N	15	04.7	83W	02	54	282
Dwingeloo, Holland	Dwingeloo Radio Obs.	52N	48	48	06E	23	48	25
Effelsberg, Germany	Max Planck Inst. Radio Astr.	50N	31	36	06E	53	06	369
Eschweiler, Germany	Stockert (Bonn Univ.)	50N	34	14.0	06E	43	24	435
Fort Irwin, California, USA	Goldstone Complex	35N	23	24	116W	50	54	1036
Gauribidanur, India	Gauribidanur Radio Obs.	13N	36	12	77E	26	06	686
Green Bank, West Virginia, USA	Nat. Radio Astr. Obs.	38N	26	17.0	79W	50	12	823
Harestua, Norway	Univ. of Oslo Obs.	60N	12	30.0	10E	45	30	585
Hartebeeshoek, S Africa	Hartebeeshoek Radio Astr. Obs.	25S	53	24	27E	41	06	1391
Hoskinstown, New South Wales, Australia	Molongo Radio Obs.	35S	22	18	149E	25	24	732
Jodrell Bank, England	Nuffield Rad. Astr. Lab.	53N	14	11.0	02W	18	24	70
Kitt Peak, Arizona, USA	Nat. Radio Astr. Obs.	31N	57	11.0	111W	36	48	1920
Kunming, China	Yunnan Obs.	25N	01	30	102E	47	18	1940
Mauna Kea, Hawaii, USA	Caltech Submillimetre Obs.	19S	49	36	155W	28	18	4060
Mitaka, Japan	National Astr. Obs.	35N	40	18	139E	32	30	58
Nagoya, Japan	Nagoya Univ. Radio Astr. Lab.	35N	08	54	136E	58	24	75
Nançay, France	Paris Ons. Radio Astr. Stn	47N	22	48	02E	11	48	150
Nançay, France	Rad. Obs. of Nançay	47N	22	48.0	02E	11	48	150
Nederhorst den Berg, Holland	Rad. Astr. Obs.	52N	14	03.0	05E	04	36	0
Parkes, New South Wales, Australia	Australian Nat. Rad. Obs.	33S	00	00.4	148E	15	42	392
Pulkovo, Russia	Astron. Obs. Acad. Sciences	59N	46	05.5	30E	19	24	70
Richmond Hill, Ontario, Canada	David Dunlap Obs.	43N	51	44.0	79W	25	12	244
St Michel, France	Nat. Centre of Scientific Res.	43N	55	00.0	05E	42	30	614
Socorro, New Mexico, USA	National Radio Astr. Obs.	34N	04	42	107W	37	06	81
Sugar Grove, West Virginia, USA	Naval Research Lab. Radio Stn	38N	31	12	79W	16	24	705
Tidbinbilla, Australia	Deep Space Station	35S	24	06	148E	58	48	656
Tokyo, Japan	Tokyo Obs. at Mitaka	35N	40	18.2	139E	32	24	70
Tremsdorf, Germany	Tremsdorf Radio Astr. Obs.	52N	17	06	13E	08	12	35
Uchinoura, Japan	Kagoshima Space Centre	31N	13	42	131E	04	00	228
Washington, District of Columbia, USA	Radio Astr. Obs. National Lab.	38N	49	16.6	77W	01	36	30
Westerbork, Holland	Westerbork Radio Astr. Obs.	52N	55	00	06E	36	18	16
Westford, Massachusetts, USA	Haystack Obs.	42N	37	24	71W	29	18	146

Interferometer Network, MERLIN, with six observing stations which together form a powerful telescope with an effective aperture of over 216 km; MERLIN has a maximum resolution, at 6 cm wavelength, of 40 m arcsec, which is about 20 times better that can usually be achieved by ground-based telescopes and is comparable with the Hubble Space Telescope. Such a power is equivalent to measuring the diameter of a £1 coin from a distance of 100 km.

The base telescope is at Jodrell Bank, either the Lovell Telescope or the Mark II. Some way away are 25 m dishes at Tabley and Darnhall. A third 25 m dish is at Knockin in Shropshire. More distant is the 25 m telescope at the Royal Signals and Radar Establishment at Defford, and finally there is the 32 m dish at Cambridge. Apart from Defford, all the MERLIN telescopes can work at wavelengths as short as 13 mm, achieving a resolution of 0.04 arcsec.

In Australia, CSIRO operates the Australia Telescope, which comprises eight radio-receiving antennæ. Six of these are at the Paul Wild Observatory near Narrabri, New South Wales. The other antennæ are the 64 m Parkes radio telescope and the 22 m antenna at Mopra, near Coonabarabran, again in New South Wales. Some of these can be networked with other radio telescopes in various parts of the world in order to make very detailed pictures of small areas of the sky. It has even been extended into space, with the Japanese orbiting radio telescope Halca, launched in 1997. This is known as Very Long Baseline Interferometry.

A selected list of radio astronomy observatories is given in Table 31.2.

To give every date of importance in the history of astronomy would be a mammoth undertaking. What I have therefore tried to do is to make a judicious selection, separating out purely space-research advances and discoveries.

It is impossible to say just when astronomy began, but even the earliest men capable of coherent thought must have paid attention to the various objects to be seen in the sky, so that it may be fair to say that astronomy is as old as *Homo sapiens*. Among the earliest peoples to make systematic studies of the stars were the Mesopotamians, the Egyptians and the Chinese, all of whom drew up constellation patterns. (There have also been suggestions that the constellations we use as a basis today were first worked out in Crete, but this is speculation only.) It seems that some constellation-systems date back to 3000 BC, probably earlier, but of course all dates in these very ancient times are uncertain.

The first essential among ancient civilizations was the compilation of a good calendar. Probably the first reasonably accurate value of the length of the year (365 days) was given by the Egyptians. (The first recorded monarch of all Egypt was Menes, who seems to have reigned around 3100 BC; he was eventually killed by a hippopotamus – possibly the only sovereign ever to have met with such a fate!) They paid great attention to the star Sirius (Sothis), because its 'heliacal rising', or date when it could first be seen in the dawn sky, gave a reliable clue to the time of the annual flooding of the Nile, upon which the Egyptian economy depended. The Pyramids are, of course, astronomically aligned, and arguments about the methods by which they were constructed still rage as fiercely as ever.

Obviously the Egyptians had no idea of the scale of the universe, and they believed the flat Earth to be all important. So too did the Chinese, who also made observations. It has been maintained that a conjunction of the five naked-eye planets recorded during the reign of the Emperor Chuan Hsü refers to either 2449 or 2446 BC. There is also the legend of the Court Astronomers, Hsi and Ho, who were executed in 2136 BC (or, according to some authorities, 2159 BC) for their failure to predict a total solar eclipse; since the Chinese believed eclipses to be due to attacks on the Sun by a hungry dragon, this was clearly a matter of extreme importance! However, this legend is discounted by modern scholars.

The earliest data collectors were the Assyrians; all students of ancient history know of the Library of Ashurbanipal (668–626 BC). This included the 'Venus Tablet', discovered by Sir Henry Layard and deciphered in 1911 by F. X. Kugler. It claims that when Venus appears, 'rains will be in the heavens'; when it returns after an absence of three months 'hostility will be in the land; the crops will prosper'. Early attempts at drawing up tables of the movements of the Moon and planets may well date from pre-Greek times, largely for astrological reasons; until relatively modern times astrology was regarded as a true science, and all the ancient astronomers (even Ptolemy) were also astrologers.

Babylonian astronomy continued well into Greek times, and some of the astronomers, such as Naburiannu (about 500 BC) and Kidinnu (about 380 BC) may have made great advances; but we know relatively little about them, and reliable dating begins only with the rise of Greek science.

Little progress was made in the following centuries, though there were some interesting Indian writings (Aryabhāta, 5th century AD), and in 570 AD Isidorus, Bishop of Seville, was the first to draw a definite distinction between astronomy and astrology. The revival of astronomy was due to the Arabs. In 813 Al-Ma'mūn founded the Baghdad school of astronomy, and various star catalogues were drawn up, the most notable being that of Al-Sûfī (born about 903). During this period two supernovæ were observed by Chinese astronomers; the star of 1006 (in Lupus) and 1054 (in Taurus, the remnant of which is today seen as the Crab Nebula).

The improved 'Alphonsine Tables' of planetary motions were published in 1270 by order of Alphonso X of Castile. In 1433 Ulūgh Beigh set up an elaborate observatory at Samarkand, but unfortunately he was a firm believer in astrology, and was told that his eldest son Abdallatif was destined to kill him. He therefore banished his son, who duly returned at the head of an army and had Ulūgh Beigh murdered. This marked the end of the Arab school of astronomy, and subsequent developments were mainly European. Some of the important dates in the history of astronomy are as follows:

1543 Publication of Copernicus' book *De Revolutionibus Orbium Cælestium*. This sparked off the 'Copernican revolution' which was not really complete until the publication of Newton's *Principia* in 1687.

1545–59 First telescope built by Leonard Digges?

1572 Tycho Brahe observed the supernova in Cassiopeia.

1576–96 Tycho worked at Hven, drawing up the best star catalogue of pre-telescopic times.

1600 Giordano Bruno burned at the stake in Rome, partly because of his defence of the theory that the Earth revolves round the Sun.

1603 Publication of Johann Bayer's star catalogue, *Uranometria*.

1604 Appearance of the last supernova to be observed in our Galaxy (Kepler's Star, in Ophiuchus).

1608 Telescope built by H. Lippershey, in Holland.

1609 First telescopic lunar map, drawn by Thomas Harriot. Serious telescopic work begun by Galileo, who made a series of spectacular discoveries in 1609–10 (phases of Venus, satellites of Jupiter, stellar nature of the Milky Way). Publication of Kepler's first two Laws of Planetary Motion.

1618 Publication of Kepler's third Law of Planetary Motion.

1627 Publication by Kepler of improved planetary tables (the Rudolphine Tables).

1631 First transit of Mercury observed by Gassendi (following Kepler's prediction of it).

1632 Publication of Galileo's *Dialogue*, which amounted to a defence of the Copernican system. In 1633 he was condemned by the Inquisition in Rome, and was forced into a completely hollow recantation. Founding of the first official observatory (the tower observatory at Leiden, Holland).

1637 Founding of the first national observatory (Copenhagen, Denmark).

1638 Identification of the first variable star (Mira Ceti, by Phocylides Holwarda in Holland).

1639 First transit of Venus observed (by Horrocks and Crabtree, in England, following Horrocks' prediction of it).

1647 Publication of Hevelius' map of the Moon.

1651 Publication of Riccioli's map of the Moon, introducing the modern-type lunar nomenclature.

1655 Discovery of Saturn's main satellite, Titan, by C. Huygens, who announced the correct explanation of Saturn's ring system in the same year.

1656 Founding of the second Copenhagen Observatory.

1659 Markings on Mars seen for the first time (by Huygens).

1663 First description of the principle of the reflecting telescope, by the Scottish mathematician James Gregory.

1665 Newton's pioneering experiments on light and gravitation, carried out at Woolsthorpe in Lincolnshire while Cambridge University was temporarily closed because of the Plague.

1666 First observation of the Martian polar caps, by G. D. Cassini.

1667 Founding of the Paris Observatory, with Cassini as Director. (It was virtually in action by 1671).

1668 First reflector made, by Newton. (This is the probable date. It was presented to the Royal Society in 1671, and still exists).

1675 Founding of the Royal Greenwich Observatory. Velocity of light measured, by O. Rømer (Denmark).

1676 First serious attempt at cataloguing the southern stars, by Edmond Halley from St Helena.

1685 First astronomical observations made from South Africa (Father Guy Tachard, at the Cape).

1689 Publication of Newton's *Principia*, finally proving the truth of the theory that the Sun is the centre of the Solar System.

1704 Publication of Newton's other major work, *Opticks*.

1705 Prediction of the return of a comet, by Halley (for 1758).

1723 Construction of the first really good reflecting telescope (a 6 in, by Hadley).

1725 Publication of the final version of the star catalogue by Flamsteed, drawn up at Greenwich. (Publication was posthumous).

1728 Discovery of the aberration of light, by James Bradley.

1750 First extensive catalogue of the southern stars, by Lacaille at the Cape. (His observations extended from 1750 to 1752). Wright's theory of the origin of the Solar System.

1758 First observation of a comet at a predicted return (Halley's Comet, discovered on 25 December by Palitzsch. Perihelion occurred in 1759). Principle of the achromatic refractor discovered by Dollond. (Previously described, by Chester Moor Hall in 1729, but his basic theory was erroneous, and his discovery had been forgotten.)

1761 Discovery of the atmosphere of Venus, during the transit of that year, by M. V. Lomonosov in Russia.

1762 Completion of a new star catalogue by James Bradley; it contained the measured positions of 60 000 stars.

1767 Founding of the *Nautical Almanac*, by Nevil Maskelyne.

1769 Observations of the transit of Venus made from many stations all over the world, including Tahiti (the expedition commanded by James Cook).

1774 First recorded astronomical observation by William Herschel.

1779 Founding of Johann Schröter's private observatory at Lilienthal, near Bremen.

1781 Publication of Charles Messier's catalogue of clusters and nebulæ. Discovery of the planet Uranus, by William Herschel.

1783 First explanation of the variations of Algol, by Goodricke. (Algol's variability had been discovered by Montanari in 1669.)

1784 First Cepheid variable discovered; δ Cephei itself, by Goodricke.

1786 First reasonably correct description of the shape of the Galaxy given, by William Herschel.

1789 Completion of Herschel's great reflector, with a mirror 49 in (124.5 cm) in diameter and a focal length of 40 ft (12.2 m).

1796 Publication of Laplace's 'Nebular Hypothesis' of the origin of the Solar System.

1799 Great Leonid meteor shower, observed by W. Humboldt.

1800 Infra-red radiation from the Sun detected by W. Herschel.

1801 First asteroid discovered (Ceres, by Piazzi at Palermo).

1802 Second asteroid discovered (Pallas, by Olbers). Existence of binary star systems established by W. Herschel.

Dark lines in the solar spectrum observed by W. H. Wollaston.

1804 Third asteroid discovered (Juno, by Harding).

1807 Fourth asteroid discovered (Vesta, by Olbers).

1814–18 Founding of the Calton Hill Observatory, Edinburgh.

1815 Fraunhofer's first detailed map of the solar spectrum (324 lines).

1820 Foundation of the Royal Astronomical Society.

1821 Arrival of F. Fallows at the Cape, as Director of the first observatory in South Africa. Founding of the Paramatta Observatory by Sir Thomas Brisbane, Governor of New South Wales. (This was the first Australian observatory. It was dismantled in 1847).

1822 First calculated return of a short-period comet (Encke's, recovered by Rümker at Paramatta).

1824 First telescope to be mounted equatorially, with clock drive (the Dorpat refractor, made by Fraunhofer).

1827 First calculation of the orbit of a binary star (ξ Ursæ Majoris, by Savary).

1829 Completion of the Royal Observatory at the Cape.

1834–8 First really exhaustive survey of the southern stars, carried out by John Herschel at Feldhausen (Cape).

1835 Second predicted return of Halley's Comet.

1837 Publication of the famous lunar map by Beer and Mädler. Publication of the first good catalogue of double stars (W. Struve's *Mensuræ Micrometricæ*).

1838 First announcement of the distance of a star (61 Cygni, by F. W. Bessel).

1839 Pulkovo Observatory completed.

1840 First attempt to photograph the Moon (by J. W. Draper).

1842 Important total solar eclipse, from which it was inferred that the corona and prominences are solar rather than lunar. First attempt to photograph totality (by Majocci), though he recorded only the partial phase.

1843 First daguerreotype of the solar spectrum obtained (by Draper).

1844 Founding of the Harvard College Observatory (first official observatory in the United States). The 15 in refractor was installed in 1847.

1845 Completion of Lord Rosse's 72 in reflector at Birr Castle, and the discovery with it of the spiral forms of galaxies ('spiral nebulæ'). Daguerreotype of the Sun taken by Fizeau and Foucault, in France. Discovery of the fifth asteroid (Astræa, by Hencke).

1846 Discovery of Neptune, by Galle and D'Arrest at Berlin, from the prediction by Le Verrier. The large satellite of Neptune (Triton) was discovered by W. Lassell in the same year.

1850 First photograph of a star (Vega, from Harvard College Observatory). Castor was also photographed, and the image was extended, though the two components were not shown separately. Discovery of Saturn's Crêpe Ring (Bond, at Harvard).

1851 First photograph of a total solar eclipse (by Berkowski). Schwabe's discovery of the solar cycle established by W. Humboldt.

1857 Clerk Maxwell proved that Saturn's rings must be composed of discrete particles. Founding of the Sydney Observatory. First good photograph of a double star (Mizar, with Alcor, by Bond, Whipple and Black).

1858 First photograph of a comet (Donati's, photographed by Usherwood).

1859 Explanation of the absorption lines in the solar spectrum given by Kirchhoff and Bunsen. Discovery of the Sun's differential rotation (by Carrington).

1860 Total solar eclipse. Final demonstration that the corona and prominences are solar rather than lunar.

1861–2 Publication of Kirchhoff's map of the solar spectrum.

1862 Construction of the first great refractors, including the Newall 25 in made by Cooke. (It was for many years at Cambridge, and is now in Athens.) Discovery of the Companion of Sirius (by Clark, at Washington). Completion of the *Bonner Durchmusterung*.

1863 Secchi's classification of stellar spectra published.

1864 Huggins' first results in his studies of stellar spectra. First spectroscopic examination of a comet (Tempel's, by Donati). First spectroscopic proof that 'nebulæ' are gaseous (by Huggins). Founding of the Melbourne Observatory. (The 'Great Melbourne Reflector' completed 1869).

1866 Association between comets and meteors established (by G. V. Schiaparelli). Great Leonid meteor shower. Announcement by J. Schmidt of an alteration in the lunar crater Linné. (Though the reality of change is now discounted, regular lunar observation dates from this time.)

1867 Studies of 'Wolf–Rayet' stars by Wolf and Rayet, at Paris.

1868 First description of the method of observing the solar prominences at times of non-eclipse (independently by Janssen and Lockyer). Publication of a detailed map of the solar spectrum, by A. Ångström.

1870 First photograph of a solar prominence (by C. Young).

1872 First photograph of the spectrum of a star (Vega, by H. Draper).

1874 Transit of Venus; solar parallax redetermined. (Another transit occurred in 1882, but the overall results were disappointing). Founding of observatories at Meudon (France) and Adelaide (Australia).

1876 First use of dry gelatine plates in stellar photography; spectrum of Vega photographed by Huggins.

1877 Discovery of the two satellites of Mars (by Hall, at Washington). Observations of the 'canals' of Mars (by Schiaparelli, at Milan).

1878 Publication of the elaborate lunar map by J. Schmidt (from Athens). Completion of the Potsdam Astrophysical Observatory.

1879 Founding of the Brisbane Observatory.

1880 First good photograph of a gaseous nebula (M.42, by Draper).

1882 Gill's classic photograph of the Great Comet of 1882, showing so many stars that the idea of stellar cataloguing by photography was born.

1885 Founding of the Tokyo Observatory. (An earlier naval observatory in Tokyo had been established in 1874.) Supernova in M.31, the Andromeda Galaxy (S Andromedæ). This was the only recorded extragalactic supernova to reach the fringe of naked-eye visibility until 1987.

1886 Photograph of M.31 (the Andromeda Galaxy) by Roberts, showing spiral structure. (A better photograph was obtained by him in 1888.)

1887 Completion of the Lick 36 in refractor.

1888 Publication of J. L. E. Dreyer's *New General Catalogue* of clusters and nebulæ. Vogel's first spectrographic measurements of the radial velocities of stars.

1889 Spectrum of M.31 photographed by J. Scheiner, from Potsdam. Discovery at Harvard of the first spectroscopic binaries (ζ Ursæ Majoris and β Aurigæ). First photographs of the Milky Way taken (by E. E. Barnard).

1890 Foundation of the British Astronomical Association. Unsuccessful attempts to detect radio waves from the Sun, by Edison. (Sir Oliver Lodge was equally unsuccessful in 1896.) Publication of the Draper Catalogue of stellar spectra.

1891 Completion of the Arequipa southern station of Harvard College Observatory. Spectroheliograph invented by G. E. Hale. First photographic discovery of an asteroid (by Max Wolf, from Heidelberg).

1892 First photographic discovery of a comet (by E. E. Barnard).

1893 Completion of the 28 in Greenwich refractor.

1894 Founding of the Lowell Observatory at Flagstaff, in Arizona.

1896 Publication of the first lunar photographic atlas (Lick), Founding of the Perth Observatory. Completion of the Meudon 33 in (83 cm) refractor. Completion of the new Royal Observatory at Blackford Hill, Edinburgh. Discovery of the predicted Companion to Procyon (by Schaeberle).

1897 Completion of the Yerkes Observatory.

1898 Discovery of the first asteroid to come well within the orbit of Mars (433 Eros, discovered by Witt at Berlin).

1899 Spectrum of the Andromeda Galaxy (M.31) photographed by Schelner.

1900 Publication of Burnham's catalogue of 1290 double stars. Horizontal refractor, of 49 in aperture, focal length 197 ft (60 m), shown at the Paris Exhibition. (It was never used for astronomical research.)

1905 Founding of the Mount Wilson Observatory (California).

1908 Giant and dwarf stellar divisions described by E. Hertzsprung (Denmark). Completion of the Mount Wilson 60 in reflector. Fall of the Siberian meteorite.

1912 Studies of short-period variables in the Small Magellanic Cloud, by Miss H. Leavitt, leading on to the period-luminosity law of Cepheids.

1913 Founding of the Dominion Astrophysical Observatory, Victoria (British Columbia). H. N. Russell's theory of stellar evolution announced.

1915 W. S. Adams' studies of Sirius B, leading to the identification of White Dwarf stars.

1917 Completion of the 100 in Hooker reflector at Mount Wilson (the largest until 1948).

1918 Studies by H. Shapley leading him to the first accurate estimate of the size of the Galaxy.

1919 Publication of Barnard's catalogue of dark nebulæ.

1920 The Red Shifts in the spectra of galaxies announced by V. M. Slipher.

1923 Proof given (by E. Hubble) that the galaxies are true independent systems rather than parts of our Milky Way system. Invention of the spectrohelioscope, by Hale.

1925 Establishment of the Yale Observatory at Johannesburg. (It was finally dismantled in 1952, its work done.)

1927 Completion of the Boyden Observatory at Bloemfontein, South Africa.

1930 Discovery of Pluto, by Clyde Tombaugh at Flagstaff. Invention of the Schmidt camera, by Bernhard Schmidt (Estonia).

1931 First experiments by K. Jansky at Holmdel, New Jersey, with an improvised aerial, leading on to the founding of radio astronomy. Jansky published his first results in 1932, and in 1933 found that the radio emission definitely came from the Milky Way.

1932 Discovery of carbon dioxide in the atmosphere of Venus (by T. Dunham).

1933–5 Completion of the David Dunlap Observatory near Toronto (Canada).

1937 First intentional radio telescope built (by Grote Reber); it was a 'dish' 31 ft (9.4 m) in diameter.

1938 New (and correct) theory of stellar energy proposed by H. Bethe and, independently, by C. von Weizsäcker.

1942 Solar radio emission detected by M. H. Hey and his colleagues (27–28 February). The emission had previously been attributed to intentional jamming by the Germans!

1944 Suggestion, by H. C. van de Hulst, that interstellar hydrogen must emit radio waves at a wavelength of 21.2 cm.

1945 Thermal radiation from the Moon detected at radio wavelengths (by R. H. Dicke).

1945–6 First radar contact with the Moon, by Z. Bay (Hungary) and independently by the US Army Signal Corps Laboratory.

1946 Work at Jodrell Bank begun (radar reflections from the Giacobinid meteor trails, 10 October). Beginning of radio astronomy in Australia (solar work by a team led by

E. G. Bowen). Identification of the radio source Cygnus A by Hey, Parsons and Phillips.

1947–8 Photoelectric observations of variable stars in the infra-red carried out by Lenouvel, using a Lallemand electronic telescope.

1948 Completion of the 200 in Hale reflector at Palomar (USA). Identification of the radio source Cassiopeia A, by M. Ryle and F. G. Smith.

1949 Identification of further radio sources; Taurus A (the Crab Nebula), Virgo A (M.87), and Centaurus A (NGC 5128). These were the first radio sources beyond the Solar System to be identified with optical objects.

1950 M.31 (the Andromeda Galaxy) detected at radio wavelengths by M. Ryle, F. G. Smith and B. Elsmore. Funds for the building of the great Jodrell Bank radio telescope obtained by Sir Bernard Lovell.

1951 Discovery by H. Ewen and E. Purcell of the 21 cm emission from interstellar hydrogen thus confirming van de Hulst's prediction. Optical identification of Cygnus A and Cassiopeia A (by Baade and Minkowski, using the Palomar reflector, from the positions given by Smith).

1952 W. Baade's announcement of an error in the Cepheid luminosity scale, showing that the galaxies are about twice as remote as had been previously thought. Electronic images of Saturn and θ Orionis obtained by Lallemand and Duchesne (Paris). Tycho's supernova of 1572 identified at radio wavelengths by Hanbury Brown and Hazard.

1953 I. Shklovskii explains the radio emission from the Crab Nebula as being due to synchroton radiation.

1955 Completion of the 250 ft radio 'dish' at Jodrell Bank. First detection of radio emissions from Jupiter (by Burke and Franklin). Construction of a radio interferometer by M. Ryle, and also the completion of the 2nd Cambridge catalogue of radio sources. (The 3rd Cambridge catalogue was completed in 1959).

1958 Observations of a red event in the lunar crater Alphonsus, by N. A. Kozyrev (Crimean Astrophysical Observatory, USSR). Venus detected at radio wavelengths (by Mayer).

1959 Radar contact with the Sun (Eshleman, at the Stanford Research Institute, USA).

1960 Aperture synthesis method developed by M. Ryle and A. Hewish.

1961 Completion of the Parkes radio telescope, 330 km west of Sydney.

1962 Thermal radio emission detected from Mercury, by Howard, Barrett and Haddock, using the 85 ft radio telescope at Michigan. First radar contact with Mercury (Kotelnikov, USSR). First X-ray source detected (in Scorpius). Sugar Grove fiasco; the US attempt to build a 600 ft fully steerable radio 'dish'. Work had begun in 1959, and when discontinued had cost $ 96 000 000.

1963 Announcement by P. van de Kamp, of a planet attending Barnard's Star (later found to be spurious). Identification of quasars (M. Schmidt, Palomar).

1965–6 Identification of the 3 °K microwave radiation, as a result of theoretical work by Dicke and experiments by Penzias and Wilson.

1967 Completion of the 98 in Isaac Newton reflector at Herstmonceux. Identification of the first pulsar, CP 1919, by Jocelyn Bell at Cambridge.

1968 Identification of the Vela pulsar (Large, Vaughan and Mills).

1969 First optical identification of a pulsar; the pulsar in the Crab Nebula by Cocke, Taylor and Disney at the Steward Observatory, USA.

1970 Completion of the 100 m radio 'dish' at Bonn (Germany). Completion of the large reflectors for Kitt Peak (Arizona) and Cerro Tololo (Chile); each 158 in (401 cm) aperture. First large reflector to be erected on Mauna Kea, Hawaii; an 88 in (224 cm).

1973 Opening of the Sutherland station of the South African Astronomical Observatories.

1974 Completion of the 153 in (389 cm) reflector at the Siding Spring Observatory, Australia.

1976 Completion of the 236 in (600 cm) reflector at Mount Semirodriki (USSR).

1977 Optical identification of the Vela pulsar (at Siding Spring). Discovery of Chiron (by C. Kowal, USA). Discovery of the rings of Uranus.

1978 Completion of the new Russian underground neutrino telescope. Discovery of Charon, the satellite of Pluto (J. Christy, USA). Discovery of the first satellite of an asteroid (Herculina). Rings of Uranus recorded from Earth (Matthews, Neugebauer, Nicholson). Discovery of X-rays from SS Cygni (HEAO 1).

1979 Official opening of the observatory at La Palma. Pluto and Charon recorded separately (Bonneau and Foy, Mauna Kea, thereby confirming Charon's independent existence). First comet observed to hit the Sun.

1980 Discovery of the first scintar (SS 433).

1981 Five asteroids contacted by radar from Arecibo, including two Apollos (Apollo itself, and Quetzalcoatl).

1982 Discovery of the remote quasar PKS 2000-330 (Wright and Launcey, Parkes). Recovery of Halley's Comet.

1983 Discovery of the fastest-vibrating pulsar, PKS 1937+215 in Vulpecula: period 1.557 806 449 022 milliseconds – twenty times shorter than the Crab pulsar. It spins 642 times per second.

1984 Isaac Newton Telescope installed on La Palma.

1986 Return of Halley's Comet.

1987 Completion of the William Herschel telescope at La Palma. Completion of the James Clerk Maxwell telescope on Mauna Kea. Supernova seen in the Large Cloud of Magellan.

1988 Completion of the Australia Telescope (radio astronomy network). Collapse of the Green Bank radio telescope.

1989 NTT (New Technology Telescope) brought into action at the European Southern Observatory, La Silla. Identification of the 'Great Wall' of galaxies.

1990 First brown dwarf identified by M. Hawkins. First surface details on a star (Betelgeux) detected from La Palma Observatory. First light on the Keck Telescope (Mauna Kea). White spot discovered on Saturn (24 September). End of the Pluto-Charon mutual phenomena (24 September). Sir Francis Graham-Smith retires as Astronomer Royal.

1991 Professor Arnold Wolfendale appointed Astronomer Royal. First really reliable measurement made of the distance of the Large Magellanic Cloud. Outburst of Halley's Comet (12 February). Fall of the Glatton Meteorite (5 May). First space image obtained of an asteroid, Gaspra (13 November).

1992 Completion of the Keck I telescope on Mauna Kea, Hawaii. Discovery of the first Kuiper Belt object (Jewitt and Luu, 31 August).

1993 Discovery of the first asteroidal satellite, Dactyl (25 August). Start of SETI, the Search for Extra-terrestrial Intelligence (12 October). Galileo exonerated of heresy by the Pope (30 October). (Galileo had been condemned for heresy on 22 June 1633!)

1994 Hooker telescope on Mt Wilson reopened. SETI cancelled by US Congress (14 March). Impact of Comet Shoemaker–Levy 9 on Jupiter (17–22 July). Discovery of the nearest galaxy, 80 000 light-years away (Ibata, 4 August).

1995 Professor Sir Martin Rees succeeds Sir Arnold Wolfendale as Astronomer Royal (January). First light on the VATT (Mount Graham) (July). Mayor and Queloz announce the discovery of a planet round 51 Pegasi (October). First maps of Vesta (HST) (November). Galileo probe impacts Jupiter (December 7).

1996 Hipparcos catalogue completed (February). HST sends back images of star-forming regions in M.16 (Eagle Nebula) (February). First surface details recorded on Pluto (HST) (March). Dedication of Keck II Telescope (May 8).

1997 First images of Mars from Pathfinder (July 4). Dedication of Hobby–Eberly Telescope (October 8).

1998 First light on Antu (first mirror of VLT) (May). Closure of the Royal Greenwich Observatory (October 31).

1999 First images released from Suburu telescope (January). First light on Kueyen (second mirror of VLT) (March). New mirror installed at the Rosse telescope, Birr (June 22). Inauguration of the Gemini North telescope on Mauna Kea (June 25–27). Total solar eclipse seen from Cornwall and Devon (August 11). Brilliant Leonid meteor shower (November 18). December 1999 reopening of the Rosse Telescope at Birr. First visual confirmation of an extra-solar planet (orbiting τ Boötis) (December).

2000 First light on Melipal (3rd mirror of VLT) (January 26). Asteroid Eros mapped from close range by the Shoemaker probe (February). First detection of an isolated black hole.

HISTORY OF SPACE RESEARCH

It is no longer possible to separate what may be called 'pure astronomy' from space research. The Space Age began on 4 October 1957 with the launch of Russia's first artificial satellite, Sputnik 1. The following list has been restricted to the more 'astronomical' events. Full lists of all lunar and planetary probes are not given here, as they will be found elsewhere in this book.

Pre-1957

c 150 Lucian of Samosata's *True History* about a journey to the Moon – possibly the first of all science-fiction stories.

1232 Military rockets used in a battle between the Chinese and the Mongols.

1865 Publication of Jules Verne's novel *From the Earth to the Moon*.

1881 Early rocket design by N. I. Kibaltchitch. (Unwisely, he made the bomb used to kill the Czar of Russia, and was predictably executed.)

1891 Public lecture about space-flight by the eccentric German inventor Hermann Ganswindt.

1895 First scientific papers about space-flight by K. E. Tsiolkovskii. He published important papers in 1903 and subsequent years. The Russians refer to him as 'the father of space-flight'.

1919 Monograph, *A Method of Reaching Extreme Altitudes*, published by R. H. Goddard in America. This included a suggestion of sending a small vehicle to the Moon, and adverse Press comments made Goddard disinclined to expose himself to further ridicule.

1924 Publication of *The Rocket into Interplanetary Space*, by H. Oberth. This was the first truly scientific account of space-research techniques.

1926 First liquid-propelled rocket launched, by Goddard.

1927 Formation of the German rocket group, *Verein für Raumschiffahrt*.

1931 First European firing of a liquid-propelled rocket (Winkler, in Germany).

1937 First rocket tests at the Baltic research station at Peenemünde. One of the leaders of the team was Wernher von Braun.

1942 First firing of the A4 rocket (better known as the V2) from Peenemünde. 1944–5; many V2s fell upon Southern England.

1945 White Sands proving ground established in New Mexico. Idea of synchronous artificial satellites for communications purposes proposed by Arthur C. Clarke.

1949 First step-rocket fired from White Sands; it reached an altitude of almost 400 km. Rocket testing ground established at Cape Canaveral, Florida.

1955 Announcement of the US 'Vanguard' project for launching artificial satellites.

The Space Age

1957 4 October; launching of the first artificial satellite, *Sputnik 1* (USSR).

1958 First successful US artificial satellite (*Explorer 1*). Instruments carried in it were responsible for the detection of the Van Allen radiation zones surrounding the Earth.

1959 First lunar probes; *Lunas 1, 2* and *3* (all USSR). *Luna 1* by-passed the Moon, *Luna 2* crash-landed there, and *Luna 3* went on a round trip, sending back pictures of the Moon's far side.

1960 First television weather satellite (*Tiros 1*, USA).

1961 First attempted Venus probe (USSR); contact with it lost. First manned space-flight (Yuri Gagarin, USSR). First manned US space-flight (A. Shepard; sub-orbital).

1962 First American to orbit the Earth (J. Glenn). First British-built satellite (*Ariel 1*, launched from Cape Canaveral). First transatlantic television pictures relayed by satellite (*Telstar*). First attempted Mars probe (USSR; contact lost). First successful planetary probe: *Mariner 2* to Venus.

1963 First occasion when two manned space-craft were in orbit simultaneously (Nikolayev and Popovich, USSR). First space-woman: Valentina Tereshkova-Nikolayeva (USSR).

1964 First good close-range photographs of the Moon (*Ranger 7*, USA).

1965 First 'space-walk' (A. Leonov, USSR). First successful Mars probe (*Mariner 4*, USA).

1966 First soft landing on the Moon by an automatic probe (*Luna 9*, USSR). First landing of a probe on Venus (*Venera 3*, USSR), though contact with it was lost. First soft landing of an American probe on the Moon (*Surveyor 1*). First circum-lunar probe (*Luna 10*, USSR). First really good close-range lunar pictures (*Orbiter 1*, USA).

1967 First soft landing of an unmanned probe on Venus (*Venera 4*, USSR).

1968 First recovery of a circum-lunar probe (*Zond 5*, USSR). First manned Apollo orbital flight (*Apollo 7*; Schirra, Cunningham, Eisele, USA). First manned flight round the Moon; *Apollo 8* (Borman, Lovell, Anders, USA).

1969 First testing of the lunar module in orbit round the Moon (*Apollo 10*; Stafford, Cernan, Young, USA). 21 July, first lunar landing (*Apollo 11*; N. Armstrong, E. Aldrin, USA).

1970 First Chinese and Japanese artificial satellites.

1971 First capsule landed on Mars, from the USSR probe *Mars 2*.

1971–2 First detailed pictures of Martian volcanoes, obtained from the probe *Mariner 9*, which entered orbit round the planet (USA).

1972 End of the Apollo programme, with *Apollo 17* (Cernan, Schmitt, Evans, USA).

1973–4 Operational 'life' of the US space-station Skylab, manned by three successive three-man crews, and from which much pioneer astronomical work was carried out.

1973 First close-range information from Jupiter (including pictures) obtained from the fly-by probe *Pioneer 10*. (*Pioneer 11* repeated the experiments in 1974). *Pioneer 10* was also the first probe to escape from the Solar System, while *Pioneer 11* was destined to be the first probe to by-pass Saturn (in 1979) (USA).

1974 First pictures of the cloud-tops of Venus from close range, from the two-planet probe *Mariner 10*; the probe then encountered Mercury, and sent back the first pictures of the cratered surface.

1975 First pictures received from the surface of Venus, from the Russian probes *Venera 9* and *Venera 10*.

1976 First successful soft landings on Mars (*Vikings 1* and *2*) sending back direct pictures and information from the surface of the planet (USA).

1977 Launching of *Voyagers 1* and *2* to the outer planets. (USA). Death of Wernher von Braun.

1978 Launch of two Pioneer probes to Venus – the first American attempts to put a vehicle into orbit round the planet and to land capsules there. Launching of the X-ray 'Einstein Observatory', which operated successfully for over two years.

1979 Fly-by of Jupiter by *Voyagers 1* and *2* (USA). Decay of *Skylab* in the Earth's atmosphere (11 July).

1980 *Voyager 1* fly-by of Saturn, obtaining data from Titan and other satellites during its pass.

1981 Successful *Voyager 2* pass of Saturn (USA).

1982 Landings of *Veneras 13* and *14* on Venus, obtaining improved pictures and data (USSR). Longest space-mission undertaken by the Russians (2 months in orbit; A. Berezevoy and V. Lebedev in *Salyut 7*).

1983 Launch of IRAS (Infra-Red Astronomical Satellite).

1986 *Voyager 2* fly-by of Uranus. Shuttle disaster, with the destruction of the *Challenger*. Five probes to Halley's Comet; two Japanese, two Russian, and one European (Giotto). 20 February, launch of Mir space station.

1987 Longest space mission (326 days) completed by Yuri Romanenko, on the space-station Mir.

1988 Longest space mission (366 days on the Mir station) completed by V. Titov and M. Manorov. Failure of the *Phobos 1* probe (contact lost, 29 August).

1989 Failure of the *Phobos 2* probe (contact lost, 29 March). *Magellan* probe to Venus launched from the Shuttle *Atlantis* (5 May). *Hipparcos* astrometric satellite launched (8 August). *Voyager 2* passed Neptune (24 August). *Galileo* probe launched to Jupiter (18 October).

1990 First Japanese launch to the Moon (Muses-A, 24 January). Hubble Space Telescope launched from the Shuttle *Discovery* (25 April). Launch of Ulysses (6 October).

1991 Gamma-Ray Observatory (Arthur Holly Compton Observatory) launched, 5 April. First survey of the sky in Extreme Ultra-Violet (Rosat satellite). Spectacular images of *Venus* sent back by the Magellan probe. Launch of Yohkoh (30 August).

1992 Ulysses (solar polar probe) flew past Jupiter (8 February). Giotto encountered comet Grigg-Skjellerup (10 July). Discovery in April of slight variations in the microwave background radiation (COBE). First space-walk by three astronauts simultaneously (13 May repair of Intelsat-6 satellite). Mars Observer launched (25 September). Pioneer Venus ceased transmitting (9 October).

1993 First gravitational wave test using satellites (21 March–12 April). Japanese probe Hiten impacted on the Moon (10 April), Hipparcos completed an all-sky survey at extreme ultra-violet wavelengths (18 November). Repair to the Hubble Space Telescope by astronauts (2–13 December). COBE switched off (23 December).

1994 Launch of lunar probe Clementine (24 January); it entered lunar orbit in February, but failed to complete its programme by going on to a rendezvous with the asteroid Geographos. Mars Observer lost (2 August). Decay of Magellan probe in Venus' atmosphere (10 October).

1995 Kuiper Airborne Observatory taken out of service (September 29). Launch of ISO (November 1). Launch of SOHO (December 2).

1996 Loss of Cluster mission (June). IUE shut down after 18 years (September 30). Launch of Mars Global Surveyor (November 7). Launch of Pathfinder to Mars (December 4).

1997 HST relaunched after second servicing mission (February 19). *Pioneers 6, 7* and *8* cease to be funded and tracked (March 31). Collision between Mir and Progress rocket (June 25). NEAR pass of asteroid Mathilde (June 27). Pathfinder lands on Mars (July 4). Mars Global Surveyor arrives in Martian orbit (September 11). Last full transmission from Pathfinder (September 27). Launch of Cassini mission to Saturn (October 15).

1998 Launch of Lunar Prospector (January 6). NEAR passes Earth at 148 000 km (January 23). End of ISO mission (April 8). Cassini passes Venus at 284 km (April 26). Launch of Nozomi, Japanese Mars probe (July 3). Launch of Mars Climate Orbiter (December 11). NEAR (Shoemaker probe) passes Eros, and sends back images (December 23).

1999 Launch of Mars Polar Lander (January 3). Launch of Stardust probe to Comet P/Wild 2 (January 15). Cassini passes Venus at 600 km (June 24). Launch of Chandra X-ray probe (July 23). Crash of Prospector on the Moon (August 30). Loss of Mars Climate Orbiter (September 3). Leonid meteor storm (November 18). Launch of Shenzhou, Chinese space-craft in preparation for a manned vehicle (November 21). Mars Polar Lander lands on Mars (December 3) but no contact re-established. Launch of Newton X-ray satellite (December 10). Third servicing mission to the Hubble Space Telescope (December 19–27).

2000 Shoemaker probe in orbit round Eros (February 14). Galileo probe flies past Io at a range of 200 km (February 22). Compton satellite brought down in the Pacific (June 4).

Selecting a limited number of astronomers for short biographical notes may be some what invidious. However, the list given here includes most of the great pioneers and researchers. No astronomers still living at the time of writing are included. All dates are AD unless otherwise stated.

Abul Wafa, Mohammed. 959–88. Last of the famous Baghdad school of astronomers. He wrote a book called *Almagest*, a summary of Ptolemy's great work also called the *Almagest*, in Arabic.

Adams, John Couch. 1819–92. English astronomer, *b.* Lidcot, Cornwall. He graduated brilliantly from Cambridge in 1843, but had already formulated a plan to search for a new planet by studying the perturbations of Uranus. By 1845 his results were ready, but no quick search was made, and the actual discovery was due to calculations by U. Le Verrier. Later he became Director of the Cambridge Observatory, and worked upon lunar acceleration, the orbit of the Leonid meteor shower, and upon various other investigations.

Airy, George Biddell. 1801–92. English astronomer. Born in Northumberland, he graduated from Cambridge 1823; and was Professor of Astronomy there 1826–35. On becoming Astronomer Royal (1835–81) he totally reorganized Greenwich Observatory and raised it to its present eminence. He re-equipped the Observatory and ensured that the best use was made of its instruments; it is ironical that he is probably best remembered for his failure to instigate a prompt search for Neptune when receiving Adams' calculations.

Aitken, Robert Grant. 1864–1951. American astronomer, *b.* Jackson, Cal. In 1895 he joined the staff of Lick Observatory, and specialized in double star work; he discovered 31000 new pairs, and wrote a standard book on the subject. From 1930 until his retirement in 1935 he was Director of the Lick Observatory.

Albategnius. *c.* 850–929. This is the Latinized form of the name of the Arab prince Al Battani. *b.* Batan, Mesopotamia; he drew up improved tables of the Sun and Moon, and found a more accurate value for the precession of the equinoxes. His *Movements of the Stars* enabled Hevelius, in the 17th century, to discover the secular variation in the Moon's motion. Albategnius was also a pioneer mathematician; in trigonometry, he introduced the use of sines.

Alphonso X. 1223–84. King of Castile. At Toledo he assembled many of the leading astronomers of the world, and drew up the famous Alphonsine Tables, which remained the standard for three centuries.

Alfvén, Hannes. 1908–95. Swedish physicist (Nobri Laureate. 1970) and the founder of the science of magnetohydrodynamics. One of his achievements was to prove the existence of an overall galactic magnetic field.

Alhazen (Abu Ali al Hassan). 987–1038. Arab mathematician, *b.* Basra. He went to Cairo, where he made his observations and also wrote the first important book on optics since the time of Ptolemy.

Allen, Clabon. 1905–88. Australian astronomer who specialized in studies of the Sun. For 20 years he lived in London, and was Director of the University of London Observatory. His book *Astronomical Quantities* (1955) is a classic.

Allen, David. 1946–94. Cambridge-born, he graduated from the University there, and concentrated upon infrared astronomy, in which he made many major contributions. He went to Australia, and worked with the Anglo–Australian Telescope at Coonabarabran. He was an outstanding writer of popular books as well as technical works. Sadly, he died of cancer when still in his forties.

Al-Ma'mūn, Abdalla. ?–833. Often referred to as Almanon. He was Caliph of Baghdad, son of Harun al Raschid; he collected and translated many Greek and Persian works, and built a major observatory in 829.

Al-Sūfī. 903–86. A Persian nobleman, who compiled an invaluable catalogue of 1018 stars, giving their approximate positions, magnitudes and colours.

Ambartsunian, Viktor Amazaspovich. 1909–95. Outstanding Russian astronomer; he set up the astrophysics programme at Leningrad University before Stalin's purges, and managed to survive, though many of his colleagues were executed. After the war he became Director of the Byurakan Observatory in Armenia. He made many original contributions; he introduced the concept of stellar associations, and was among the first to realize that T Tauri stars are very young, and are in the pre-Main Sequence stage.

Anaxagoras. 500–428 BC. *b.* Clazomenæ, Ionia. In Athens he became a friend of Pericles, and it was because of this friendship that he was merely banished, rather than being condemned to death, for teaching that the Moon contains plains, valleys and mountains, while the Sun is a blazing stone larger than the Peloponnesus (the peninsula upon which Athens stands).

Anaximander. *c.* 611–547 BC. Greek philosopher, *b.* Miletus. He believed the Earth to be a cylinder, suspended freely in the centre of a spherical universe. He attempted to draw up a map of the world, and introduced the gnomon into Greece.

Anaximenes. *c.* 585–525 BC. Greek philosopher, *b.* Miletus. He believed the Sun to be hot because of its quick motion round the Earth, that the stars were too remote to send us detectable heat, and the stars were fastened on to a crystal sphere.

Ångström, Anders. 1814–74. Swedish physicist, who graduated from Uppsala. He mapped the solar spectrum, and was the first to examine the spectra of auroræ. The Ångström unit (100-millionth part of a centimetre) is named in his honour.

Antoniadi, Eugenios. 1870–1944. Greek astronomer, who spent most of his life in France and became a naturalized Frenchman. He worked mainly at the Juvisy Observatory (with Camille Flammarion) and at Meudon, near Paris, where he used the 83 cm, refractor to make classic observations of the planets. Before the Space Age, his maps of Mars and Mercury were regarded as the standard works. He died in Occupied France during World War II.

Apian, Peter Bienewitz. 1495–1552. *b.* Leisnig, Saxony. Became professor of mathematics at Ingolstädt. He observed five comets, and was the first to note that their tails always point away from the Sun. The 1531 comet is known to be Halley's, and his observations of it enabled Edmond Halley to identify it with the comets of 1607 and 1682.

Apollonius. *c.* 250–200 BC. *b.* Perga, Asia Minor, but lived in Alexandria. An expert mathematician, he was one of the first to develop the theory of epicycles to represent the movement of the Sun, Moon and planets.

Arago, François Jean Dominique. 1786–1853. Director of the Paris Observatory from 1830. He made many important contributions, including a recognition of the importance of photography in astronomy. He made an exhaustive study of the great total solar eclipse of 1842, and maintained (correctly!) that the Sun is wholly gaseous.

Argelander, Friedrich Wilhelm August. 1799–1875. German astronomer, who became Director of the Bonn Observatory 1836. Here he drew up his atlas of the northern heavens (the *Bonn Durchmusterung*), containing the positions of 324 198 stars down to the ninth magnitude. This standard work was published in 1863.

Aristarchus. *c.* 310–250 BC. Greek astronomer, *b.* Samos. One of the first (quite possibly the very first) to maintain that the Earth moves round the Sun, and he also tried to measure the relative distances of the Sun and Moon by a method which was sound in theory, though inaccurate in practice.

Aristotle. (384–322 BC) believed in a finite, spherical universe. He further developed the theory of concentric spheres, and gave the first practical proofs that the Earth cannot be flat.

Baade, Wilhelm Heinrich Walter. 1893–1959. German astronomer. In 1920, while assistant at Hamburg Observatory, he discovered the unique asteroid 944 Hidalgo. In 1931 he went to America, and joined the staff of Mount Wilson. In 1952 his work upon the two classes of 'Cepheid' short period variables enabled him to show that the galaxies are approximately twice as remote as had previously been thought.

Bailey, Solon Irving. 1854–1931. American astronomer, *b.* New Hampshire. He joined the Harvard staff in 1879, and was for many years in charge of the Harvard southern station at Arequipa, Peru. His studies of globular clusters led to the discovery of 'cluster variables', now known as RR Lyrae stars.

Bappu, Manali Kallat Vainu. 1927–82. One of the most distinguished of Indian astronomers; he was Director of the Kodaikanal Observatory, and modernized it. He was a specialist in stellar spectroscopy.

Barnard, Edward Emerson. 1857–1923. American astronomer, b. Nashville, Tennessee. He was self taught, but joined the staff at Lick Observatory in 1888, moving to Yerkes in 1897. He was a renowned comet-hunter; he discovered the fifth satellite of Jupiter (Amalthea) and the swift-moving star in Ophiuchus now called Barnard's Star. He also specialized in studies of dark nebulæ.

Barrow, Isaac. 1630–77. English mathematician. He made various important contributions to science, but is perhaps best known because in 1669 he resigned his post as Lucasian Professor at Cambridge so that his pupil Isaac Newton could succeed him.

Bayer, Johann. 1572–1625. German astronomer; a lawyer by profession and an amateur in science. He is remembered for his 1603 star catalogue, in which he introduced the system of allotting Greek letters to the stars in each constellation – the system still in use today.

Beer, Wilhelm. 1797–1850. A Berlin banker, who set up a private observatory and collaborated with Mädler in the great map of the Moon published in 1837–8. This map remained the standard for many years. Beer was the brother of Meyerbeer, the famous composer.

Belopolsky, Aristarch. 1854–1934. Russian astronomer, who went from Moscow to the Pulkova Observatory in 1888. He became Director of the Observatory in 1916, but resigned in 1918. He specialized in spectroscopic astronomy and in studies of variable stars.

Bessel, Friedrich Wilhelm. 1784–1846. German astronomer. He went to Lilienthal as assistant to Schröter, but in 1810 became Director of the Königsberg Observatory, retaining the post until his death. He determined the position of 75 000 stars by reducing Bradley's observations; was the first to obtain a parallax value for a star (61 Cygni, in 1838), and predicted the positions of the then unknown companions of Sirius and Procyon.

Biela, Wilhelm von. 1782–1856. Austrian army officer, and amateur astronomer, who is remembered for his discovery (in 1826) of the now-defunct periodical comet which bears his name.

Bode, Johann Elert. 1747–1826. German astronomer. b. Hamburg. Appointed director of Berlin Observatory 1772. In the same year he drew attention to the 'law' of planetary distances which had been discovered by Titius of Wittenberg; rather unfairly, perhaps, this is known as Bode's Law. He published a star catalogue, did much to popularize astronomy, and for 50 years edited the *Berlin Astronomisches Jahrbuch.*

Bolton, John. 1922–1993. Bolton was a Yorkshireman, born in Sheffield, but spent most of his career in Australia. He was a pioneer of radio astronomy, and was the first to identify the Crab Nebula as a radio source. He spent some years as Director of the National Radio Astronomy Observatory in Australia, retiring in 1971.

Bond, George Phillips. 1825–65. Son of W. C. Bond. b. Massachusetts. In 1859 succeeded his father as Director of the Harvard Observatory. He was a pioneer of planetary and cometary photography, and was the first to assert upon truly scientific principles that Saturn's rings could not be solid.

Bond, William Cranch. 1789–1859. American astronomer, b. Maine. He began his career as a watchmaker, but his fame as an amateur astronomer led to his appointment as Director of the newly-founded Harvard Observatory. In 1848 he discovered Saturn's satellite Hyperion, and in 1850 he discovered Saturn's Crêpe Ring. He was also a pioneer of astronomical photography.

Bouvard, Alexis. 1767–1843. A shepherd boy, born in a hut at Chamonix. He went to Paris, taught himself mathematics, and was appointed assistant to Laplace. He made contributions to lunar theory and drew up tables of the motions of the outer planets, as well as discovering several comets.

Bradley, James. 1692–1762. English astronomer (Astronomer Royal, 1742–62). He was educated in Gloucestershire, and entered the Ministry, becoming Vicar of Bridstow in 1719; in 1721 he went to Oxford as Professor of Astronomy, and remained there until his appointment to Greenwich, mainly on the recommendation of his close friend Halley. He discovered the aberration of light and the nutation of the Earth's axis, but his greatest work was his catalogue of the positions of 60 000 stars.

Brorsen, Theodor. 1819–95. Danish astronomer, who discovered several comets, and in 1854 made the first scientific observations of the Gegenschein.

Brown, Ernest William. 1866–1938. English astronomer, who graduated from Cambridge and then went to USA. His chief work was on lunar theory, and his tables of the Moon's motion are still recognized as the standard.

Burnham, Sherburne Wesley. 1838–1921. American astronomer, who began as an amateur and then went successively to Lick (1888) and Yerkes (1897). He specialized in double star work, and discovered over 1300 new pairs. His *General Catalogue of Double Stars* remains a standard reference work.

Campbell, William Wallace, 1862–1938. American astronomer, *b.* Ohio. He joined the staff at Lick Observatory in 1891, and was Director from 1900 until his retirement in 1930. His main work was in spectroscopy; he discovered 339 spectroscopic binaries (among them Capella), and determined the radial velocities of stars and of 125 nebulæ as well as carrying out spectroscopic observations of the planets.

Cannon, Annie Jump. 1863–1941. Outstanding American woman astronomer, *b.* Delaware. In 1896 she joined the staff at Harvard College Observatory, where she worked unceasingly on the classification of stellar spectra; the present system is due largely to her. She also discovered five novæ and over 300 variable stars. From 1938 she was William Cranch Bond Astronomer.

Carrington, Richard Christopher. 1826–75. English amateur astronomer, who had his Observatory at Redhill, Surrey. He concentrated upon the Sun, and made many contributions, including the first observation of a solar flare and the independent discovery of Spörer's Law concerning the distribution of sunspots throughout a cycle.

Cassini, Giovanni Domenico. 1625–1712. Italian astronomer; Professor of Astronomy, Bologna 1650–69, when he went to Paris as the first Director of the Observatory there. He discovered four of Saturn's satellites as well as the main division in the rings; he drew up new tables of Jupiter's satellites, made pioneer observations of Mars, and made the first reasonably good measurement of the distance of the Sun.

Cassini, Jacques J. 1677–1756. Son of G. D. Cassini; *b.* Paris. He succeeded his father as Director of the Paris Observatory. He confirmed Halley's discovery of the proper motions of certain stars, and played an important part in measuring an arc of meridian from Dunkirk to the Pyrenees in order to determine the figure of the Earth.

Challis, James. 1803–62. English astronomer; Professor of Astronomy at Cambridge from 1836. He accomplished much useful work, but, unfortunately, is remembered as the man who failed to discover Neptune before the success by Galle and D'Arrest at Berlin.

Chandrasekhar, Subrahmanyan. 1910–95. Indian astrophysicist, born at Lahore; he graduated from Cambridge in 1933. In 1937 he emigrated to Chicago, and remained in the United States, becoming a US citizen. He made major contributions to astrophysics, and showed that there is a maximum possible mass for a white dwarf star – the Chandrasekhar Limit. In 1983 he was awarded the Nobel Prize for Physics.

Charlier, Carl Vilhelm Ludwig. 1862–1934. Swedish cosmologist; Professor of Astronomy at Lund from 1897. He accomplished outstanding work with regard to the distribution of stars in our Galaxy.

Christie, William Henry Mahoney. 1845–1922. English astronomer (Astronomer Royal 1881–1910). He modernized Greenwich Observatory, and fully maintained its great reputation, achieved under Airy.

Clairaut, Alexis Claude. 1713–65. French mathematical genius, who published his first important paper at the age of 12. He studied the motion of the Moon, and worked out the perihelion passage of Halley's Comet in 1759 to within a month of the actual date.

Clavius, Christopher Klau. 1537–1612. German Jesuit mathematical teacher, who laid down the calendar reform of 1582 at the request of Pope Gregory.

Copernicus, Nicolaus. 1473–1543. The Latinized name of Mikołaj Kopernik, *b.* Toruń, Poland. He entered the Church, and became Canon of Frombork. He had a varied career, including medicine and also the defence of his country against the Teutonic Knights, but is remembered for his great book *De Revolutionbus Orbium Cælestium*, finally published during the last days of his life (he had previously withheld it because he was well aware of Church opposition). It was this book which revived the heliocentric theory according to which the Earth moves round the Sun, and sparked off the 'Copernican revolution' which came to its end with the work of Newton more than a century later.

Curtis, Heber Doust. 1872–1942. American astronomer, who worked at Lick, Allegheny and Michigan Observatories. He was an outstanding spectroscopist, and in 1920 took part in the 'Great Debate' with Shapley about the size of the Galaxy and the status of the resolvable nebulæ; Curtis was wrong about the size of the Galaxy, but correct in maintaining that the spiral nebulæ were independent galaxies. He also played a major rôle in the establishment of the McMath-Hulbert Observatory, renowned for its solar research.

D'Arrest, Heinrich Ludwig. 1822–75. German astronomer, *b.* Berlin. While Assistant at the Berlin Observatory, he joined Galle in the successful search for Neptune. From 1857 he worked at Copenhagen Observatory. He specialized in comet and asteroid work, and also published improved positions for about 2000 nebulæ.

Darwin, George Howard. 1845–1912. Son of Charles Darwin. From 1883 he was Professor of Astronomy at Cambridge, and drew up his famous though now rejected tidal theory of the origin of the planets. He was knighted in 1906.

Dawes, William Rutter. 1799–1868. English clergyman, and a keen-eyed amateur observer who specialized in observations of the Sun, planets and double stars. He discovered Saturn's Crêpe Ring independently of Bond.

Delambre, Jean-Baptiste Joseph. 1749–1822. French astronomer, best remembered for his work on the history of the science but also a skilled computer of planetary tables.

De la Rue, Warren. 1815–89. English astronomer, born in Guernsey. He was a pioneer of astronomical photography; in 1852 he obtained the first good photographs of the Moon, and in 1857 of the Sun. His photographs of the total solar eclipse of 1860 finally proved that the prominences are solar rather than lunar.

De Vaucouleurs, Gerard. 1918–95. Outstanding French planetary observer, who specialized in research concerning Mars; he was also active in cosmological research – although the value which he derived for the Hubble Constant is now known to be much too high. He was the author of many popular books as well as technical works.

Delaunay, Charles. 1816–72. French astronomer, who specialized in studies of the Moon's motion. He became Director of the Paris Observatory in 1870, but was drowned in a boating accident two years later.

Democritus. *c.* 460–360 BC. Greek philosopher, *b.* Abdera, Thrace. He adopted Leucippus' atomic theory, and was the first to claim that the Milky Way is made up of stars.

Descartes, René. 1596–1650. French astronomer, author of the theory that matter originates as vortices in an all-pervading ether. He also made great improvements in optics. His books were published in Holland, but he died in Sweden.

De Sitter, Willem. 1872–1934. Dutch astronomer and cosmologist, *b.* Friesland; he went to the Cape, and from 1908 was Professor of Astronomy at Leiden. He studied the motions of Jupiter's satellites and also the rotation of the Sun, but is best remembered for his pioneer work in relativity theory. The 'De Sitter universe', finite but unbounded, was calculated to be 2000 million light-years in radius and to contain 80 000 million galaxies.

Deslandres, Henri Alexander. 1853–1948. French astronomer (originally an Army officer); from 1907 Director of the Meudon Observatory, and from 1927 Director of the Paris Observatory also. He was a pioneer spectroscopist, and developed the spectroheliograph independently of Hale.

Dollond, John. 1706–61. English optician, who reinvented the achromatic lens in 1758 and thus improved refractors beyond all recognition.

Donati, Giovanni Battista. 1826–73. Italian astronomer who discovered the great comet of 1858, and was the first to obtain the spectrum of a comet (Tempel's of 1864). From 1859 Director of the observatory at Florence, in 1872 he was largely responsible for the creation of the now-celebrated observatory at Arcetri.

Dreyer, John Louis Emil. 1852–1926. Danish astronomer, *b.* Copenhagen, who went to Ireland as astronomer to Lord Rosse at Birr Castle and became Director of the Armagh Observatory in 1882. He was a great astronomical historian, but is best remembered for his *New General Catalogue* of Clusters and Nebulæ (the NGC) still regarded as a standard work. In 1916 he retired from Armagh and went to Oxford, where he lived for the rest of his life.

Dyson, Frank Watson. 1868–1939. English astronomer (Astronomer Royal, 1910–33). A great administrator as well as an energetic observer of eclipses; he also carried out important work in the field of astrophysics and stellar motions.

Eddington, Sir Arthur Stanley. 1882–1945. English astronomer, *b.* Kendal. After working at Cambridge and Greenwich, he was appointed Professor of Astronomy at

Cambridge, 1913. He was a pioneer of the theory of the evolution and constitution of the stars, and an outstanding relativist; in 1919 he confirmed Einstein's prediction of the displacement of star positions near the eclipsed Sun. He was knighted in 1930. In addition to his outstanding work, Eddington was one of the best of all writers of popular scientific books, and was a splendid broadcaster.

Einstein, Albert. 1879–1955. German Jew, whose name will be remembered as long as Newton's; in 1905 he laid down the Special Theory of Relativity, and from 1915–17 he developed the General Theory. In 1933 he left Germany, fearing persecution of the Jews, and settled in USA.

Elger, Thomas Gwyn. 1838–97. English amateur astronomer; he was first Director of the Lunar Section of the British Astronomical Association, and in 1895 published an excellent outline map of the Moon.

Empedocles of Agrigentum (*c.* 490–50 BC) believed the Sun to be a reflection of fire, but is credited with being the first to maintain that light has a finite velocity.

Encke, Johann Franz. 1791–1865. German astronomer; from 1825, Director of the Berlin Observatory. He was responsible for compiling the star maps which enabled Galle and D'Arrest to locate Neptune. In 1818 he computed the orbit of a faint comet, and successfully predicted its return; this was Encke's Comet, which has the shortest period of any known comet (3.3 years).

Eratosthenes. *c.* 276–196 BC. Greek philosopher, *b.* Cyrene; he became Librarian at Alexandria, and made a remarkably accurate measurement of the circumference of the Earth.

Eudoxus. *c.* 408–355 BC. Greek astronomer, *b.* Cnidus. He went to Athens, and attended lectures by Plato. Finally he settled in Sicily. He developed the theory of concentric spheres – the first truly scientific attempt to explain the movements of the celestial bodies.

Euler, Leonhard, 1707–83. Brilliant Swiss mathematician, *b.* Basle. He pioneered studies of the lunar theory, the movements of planets, comets, and the tides. He lost his sight in 1766, but this did not stop him from working; he undertook the complicated calculations mentally.

Fabricius, David. 1564–1617. Dutch minister and amateur astronomer, who observed Mira Ceti in 1596 (though without recognizing it as a variable) and made pioneer telescopic observations, notably of the Sun. In 1617 he

announced from the pulpit that he knew the identity of a member of his congregation who had stolen one of his geese – and he was presumably correct, since he was assassinated before he could divulge the name of the culprit!

Fabricius, Johann. 1587–1616. Son of David Fabricius, and also a pioneer observer of the Sun by telescopic means; he discovered sunspots independently of Galileo and Scheiner.

Fallows, Fearon. 1789–1831. English astronomer, *b.* Cumberland. He went to South Africa in 1821 as the first Director of the Cape Observatory. He established the observatory, working under almost incredible difficulties, but the primitive living conditions undermined his health. The reduction of his Cape observations was undertaken by Airy.

Fauth, Philipp Johann Heinrich. 1867–1943. German astronomer, who compiled a large map of the Moon. Unfortunately he believed in the absurd theory that the Moon is ice-covered, and this influenced all his work.

Ferguson, James. 1710–76. Scottish popularizer of astronomy, who began life as a shepherd-boy but whose books gained great influence. He was also one of the first to suggest an evolutionary origin of the Solar System.

Flammarion, Camille. 1842–1925. French astronomer. renowned both for his observations of Mars and for his popular books. He set up his own observatory at Juvisy, and founded the Société Astronomique de France.

Flamsteed, John. 1646–1720. English astronomer (Astronomer Royal, 1675–1720, though at first the title was 'unofficial'). His main work was the compilation of a new star catalogue, the final version of which was published posthumously. Flamsteed was also Rector of Burstow, Surrey.

Fleming, Wilhelmina. 1857–1911. Scottish woman astronomer, who emigrated to America and worked at Harvard College Observatory, where she was in charge of the famous Draper star catalogue. She discovered 10 novæ and 222 variable stars.

Fontana, Francisco. 1585–1656. Italian amateur (a lawyer by profession). He left sketches of Mars and Venus, though the 'markings' which he recorded were certainly illusory.

Fowler, Alfred. 1868–1940. English astronomer, whose spectroscopic work in connection with the Sun, stars and comets was of great importance.

Franklin-Adams, John. 1843–1912. English businessman who took up astronomy as a hobby at the age of 47, and compiled a photographic chart of the stars which is still regarded as a standard work.

Fraunhofer, Joseph von. 1787–1826. Outstanding German optical worker, orphaned in early childhood and rescued from poverty by the Elector of Bavaria. He joined the Physical and Optical Institute of Munich, and was Director from 1823. He invented the diffraction grating, constructed the best lenses in the world, and studied the dark lines in the solar spectrum (the 'Fraunhofer Lines'). He made the Dorpat refractor for Struve (the first telescope to be clock-driven) and also the Königsberg heliometer. His comparatively early death was a tragedy for science.

Galilei, Galileo. 1564–1642. The first great telescopic observer – and also the true founder of experimental mechanics. He worked successively at Pisa, Padua and Florence. The story of his remarkable telescopic discoveries (including the satellites of Jupiter, the phases of Venus and the gibbous aspect of Mars, the starry nature of the Milky Way and many more), and of how his defence of the Copernican theory brought him into conflict with the Church, is one of the most famous in scientific history. He was condemned by the Inquisition in 1633, and was kept a virtual prisoner in his villa at Arcetri; in his last years he also lost his sight.

Galle, Johann Gottfried. 1812–1910. German astronomer, best remembered as being the first (with D'Arrest) to locate Neptune in 1846. He discovered three comets, and in 1872, while director of the Breslau Observatory, was the first to use an asteroid for measuring solar parallax.

Gassendi, Pierre. 1592–1655. French mathematician and astronomer. In 1631 he made the first of all observations of a transit of Mercury.

Gauss, Karl Friedrich. 1777–1855. German mathematical genius. In 1801 he calculated the orbit of the first asteroid, Ceres, from a few observations, and enabled Olbers to recover it in the following year. He invented the 'method of least squares', known to every mathematician.

Gill, David. 1843–1914. Scottish astronomer. In 1877 he used observations of Mars to redetermine the solar parallax, and in 1879 went to South Africa as HM Astronomer at the Cape. It was his photograph of the comet of 1882 which showed him the importance of mapping the sky

photographically – since his plate showed many stars as well as the comet. He was also deeply involved in cataloguing the southern stars. He was knighted in 1900.

Goldschmidt, Hermann. 1802–66. German astronomer, who settled in Paris. Using small telescopes poked through his attic window, he discovered 14 asteroids between 1852–61.

Goodacre, Walter. 1856–1938. English amateur astronomer, who published an excellent map of the Moon in 1910.

Goodricke, John. 1764–86. Born of English parents in Holland. He was a deaf-mute, but with a brilliant brain. It was he who found that Algol is an eclipsing binary rather than true variable, and he also discovered the fluctuations of the intrinsic variable δ Cephei.

Gould, Benjamin Apthorp. 1824–96. American astronomer, who founded the *Astrophysical Journal*. From Cordoba Observatory, Argentina, he compiled the *Uranimetria Argentina*, the first major catalogue of the southern stars.

Green, Charles. 1735–71. English astronomer who went with Captain Cook to study the 1769 transit of Venus. He died on the return voyage.

Gregory, James. 1638–75. Scottish mathematician. In 1663 he described the principle of the reflecting telescope, but never actually made one.

Grimaldi, Francesco Maria. 1618–63. Italian Jesuit, who made observations of the Moon used in the lunar map compiled by his friend Riccioli. Grimaldi discovered the refraction of light.

Gruithuisen, Franz von Paula. 1744–1852. German astronomer; from 1826 Professor of Astronomy, Munich. He was an assiduous observer of the Moon and planets, but his vivid imagination tended to discredit his work; at one stage he even reported the discovery of artificial structures on the Moon. He also proposed the impact theory of lunar crater formation.

Gum, Colin. 1924–60. Australian astronomer, who carried out work of vital importance in the surveying of southern radio sources. The famous 'Gum Nebula' in Vela/Puppis is named after him. He was killed in a skiing accident at Zermatt in Switzerland.

Hadley, John. 1682–1743. English astronomer; friend of Bradley. He made the first really good reflecting telescope

(6 in aperture) in 1723, and 1731 constructed his 'reflecting quadrant', which replaced the astrolabe and the cross-staff in navigation.

Hale, George Ellery. 1868–1938. American astronomer. A pioneer solar observer, who invented the spectroheliograph and discovered the magnetic fields of sunspots. In 1897 he became Director of Yerkes Observatory, and transferred to Mount Wilson in 1905; he master-minded the building of the 60 in and 100 in reflectors, as well as the Yerkes refractor. He was mainly responsible for the building of the Palomar 200 in reflector, unfortunately not completed in his lifetime.

Hall, Asaph. 1829–1907. American astronomer, noted for his planetary work. At Washington, in 1877, he discovered the two satellites of Mars. From 1896 he was Professor of Astronomy at Harvard.

Halley, Edmond. 1656–1742. English astronomer (Astronomer Royal, 1720–42). Though best known for his prediction of the return of the great comet which now bears his name, Halley accomplished much other valuable work; he catalogued the southern stars from St Helena, studied star clusters and nebulæ, and discovered the proper motions of some of the bright stars. More importantly, he was responsible for the writing of Newton's *Principia* and personally financed its publication.

Harding, Karl Ludwig. 1765–1834. German astronomer, who was at first assistant to Schröter and then was appointed Professor of Astronomy at Göttingen. In 1804 he discovered the third asteroid, Juno.

Haro, Guillermo. (1900–90). Possibly Mexico's most famous astronomer, celebrated for his studies of flare stars. He noted bright nebulæ with spectra showing emission lines, independently of G. Herbig; these are now known as Herbig-Haro Objects. He was Director of the Mexican Institute of Astronomy.

Harriot, Thomas. 1560–1621. English scholar, once tutor to Sir Walter Raleigh. He compiled the first telescopic map of the Moon, and completed it some months before Galileo began his work.

Harrison, John. 1693–1776. English clockmaker, who invented the marine chronometer which revolutionized navigation. Several of his original chronometers are now on display in London.

Hartmann, Johannes Franz. 1865–1936. German astronomer. Director Göttingen Observatory 1909–21, when he went to Argentina to superintend the National Observatory there. His important work was connected with stellar and nebular radial velocities, in the course of which he discovered interstellar absorption lines in the spectrum of δ Orionis.

Hay, William Thompson. 1888–1949. 'Will Hay' was probably the only skilled amateur astronomer who was by profession a stage and screen comedian! In 1933 he discovered the famous white spot on Saturn – the most prominent ever seen on that planet.

Heis, Eduard. 1806–77. German astronomer; Professor at Münster from 1852. He was a leading authority on the Zodiacal Light, meteors and variable stars, and published a valuable star catalogue. He was renowned for his keen eyesight, and is said to have counted 19 naked-eye stars in the Pleiades.

Hencke, Karl Ludwig. 1793–1866. German amateur astronomer; postmaster at Driessen. In 1845, after 15 years' search, he discovered the fifth asteroid, Astræa.

Henderson, Thomas. 1798–1844. Scottish astronomer. 1832–3 HM Astronomer at the Cape. While there, he made the measurements which enabled him to measure the parallax of α Centauri. In 1834 he became the first Astronomer Royal for Scotland.

Heraclides of Pontus (c. 388–315 BC) declared that the apparent daily rotation of the sky is due to the real rotation of the Earth. He also discovered that Mercury and Venus revolve round the Sun, not round the Earth.

Heraclitus of Ephesus (born c. 544 BC) took fire to be the principal element, and maintained that the diameter of the Sun was about one foot.

Herschel, Friedrich Wilhelm (always known as William Herschel). 1738–1822. Probably the greatest observer of all time. He was born in Hanover, but spent most of his life in England. He was the best telescope maker of his day, and in 1781 became famous by his discovery of the planet Uranus. He made innumerable discoveries of double stars, clusters and nebulæ; he found that many doubles are physically-associated or binary systems, and he was the first to give a reasonable idea of the shape of the Galaxy. He was knighted in 1816, and received every honour that the scientific world could bestow. George III appointed him King's Astronomer (not Astronomer Royal).

Herschel, Caroline. 1750–1848. William Herschel's sister, and constant assistant in his astronomical work. She discovered eight comets.

Herschel, John Frederick William. 1792–1871. William Herschel's son. He graduated from Cambridge 1813 and from 1832–8 took a large telescope to the Cape to make the first really systematic observation of the southern heavens. He discovered 3347 double stars and 525 nebulæ, and may be said to have completed his father's pioneering work.

Hertzsprung, Ejnar. 1873–1967. Danish astronomer, who worked successively at Frederiksberg, Copenhagen, Göttingen, Mount Wilson and Leiden (Director Leiden Observatory from 1935). In 1905 he discovered the giant and dwarf subdivisions of late-type stars, and this led on to the compilation of H–R or Hertzsprung–Russell Diagrams, which are of fundamental importance in astronomy.

Hevelius. 1611–87. The Latinized name of Johannes Hewelcke of Danzig (now Gdańsk). From his private observatory he drew up a catalogue of 1500 stars, and observed planets, the Moon and comets, using the unwieldy long-focus, small aperture refractors of his day. His observatory was burned down in 1679, but he promptly constructed another. His original map of the Moon has been lost; tradition says that the copper engraving was melted down and made into a teapot after his death.

Hind, John Russell. 1823–95. English astronomer, who discovered 11 asteroids, the 1848 nova in Ophiuchus, and his 'variable nebula' round T Tauri. He also computed many cometary orbits, and from 1853 was superintendent of the *Nautical Almanac.*

Hipparchus. Fl. 140 BC. Great Greek astronomer, who lived in Rhodes. He drew up a star catalogue, later augmented by Ptolemy. Among his many discoveries was that of precession; he also constructed trigonometric tables. Unfortunately all his original works have been lost.

Hoffmeister, Cuno. 1892–1967. He was born in Sonneberg, Thuringia, and after working at two German observatories founded the Sonneberg Observatory, in 1925. He was a specialist in variable star work, and discovered almost 10 000 new variables. He was also an authority on meteoritic astronomy.

Hooke, Robert. 1653–1703. English scientific genius, contemporary with (though no friend of!) Newton. He built various astronomical instruments, and made some useful observations, including sketches of lunar craters.

Horrocks, Jeremiah. 1619–41. English astronomer who, with his friend Crabtree, was the first to observe a transit of Venus (1639). He also worked on lunar theory. His early death was a great tragedy for science.

Howse, Derek. 1919–98. Essentially a Naval officer (Lieutenant-Commander, who won the DSC during the war), Howse became Keeper of Astronomy and Navigation at the National Maritime Museum at Greenwich, retiring in 1982. He was the author of many books and papers on all aspects of navigational astronomy.

Hubble, Edwin Powell. 1889–1953. American astronomer, who served in the Army during World War I and was also a qualified lawyer. In 1923, using the Mount Wilson 100 in reflector, he discovered short-period variables in the Andromeda Spiral, and proved the Spiral to be an independent galaxy. He also established the velocity/distance relationship known as Hubble's Law.

Huggins, William. 1824–1910. Pioneer English spectroscopist, who had his private observatory at Tulse Hill, near London. Pioneer of stellar spectroscopy; he established that the irresolvable nebulæ are gaseous; he was the first to determine stellar radial motions by means of the Doppler shifts in their spectral lines, and carried out important solar and planetary work. He was knighted in 1897.

Humason, Milton La Salle. 1891–1972. *b.* Minnesota. He was mainly self-taught, but joined the staff of Mount Wilson Observatory in 1920, and from then on worked closely with Hubble, studying the forms, spectra, radial motions and nature of the galaxies; he also photographed the spectra of supernovæ in external systems. In 1919 he carried out a photographic search for a trans Neptunian planet at the request of W. H. Pickering, who had made independent calculations similar to Lowell's. Humason took several plates, but failed to locate the planet. When the plates were re-examined years later, after Pluto had been discovered at Flagstaff, it was found that Humason had recorded the planet twice – but once the image was masked by a star, and on the other occasion it fell on a flaw in the plate!

Huygens, Christiaan. 1629–95. Dutch astronomer; probably the best telescopic observer of his time. He discovered Saturn's brightest satellite (Titan) in 1655, and was the first to realize that the curious appearance of the planet was due to a system of rings. He was also the first to see markings on Mars. His activities extended into many fields of science; in particular, he invented the pendulum clock.

Innes, Robert Thorburn Ayton. 1861–1933. Scottish astronomer, who emigrated first to Australia (becoming a wine merchant) and then went to South Africa, as director of the Observatory at Johannesburg. He specialized in double star work, discovering more than 1500 new pairs; he also discovered Proxima Centauri, the nearest star beyond the Sun.

Janssen, Pierre Jules César. 1824–1907. French astronomer, who specialized in solar work (in 1870 he escaped from the besieged city of Paris by balloon to study a total eclipse). Independently of Lockyer, he discovered the means of observing the Sun's chromosphere and prominences without waiting for an eclipse. From 1876 he was Director of the Meudon Observatory, and in 1904 published an elaborate solar atlas, containing more than 8000 photographs. The square at the entrance to the Meudon Observatory is still called the Place Janssen, and his statue is to be seen there.

Jansky, Karl Guthe. 1905–49. American radio engineer, of Czech descent. He joined the Bell Telephone Laboratories, and was using an improvised aerial to investigate problems of static when he detected radio waves which he subsequently showed to come from the Milky Way. This was, in fact, the beginning of radio astronomy; but for various reasons Jansky paid little attention to it after 1937, and virtually abandoned the problem.

Jeans, Sir James Hopwood. 1877–1946. English astronomer. He elaborated the plausible but now rejected theory of the tidal origin of the planets, but his major work was in connection with stellar constitution, in which he made notable advances. He was also an expert writer of popular scientific books, and was famous as a lecturer and broadcaster.

Jeffreys, Sir Harold. 1892–1989. Though primarily a geophysicist, Jeffreys also made many important advances in astronomy, and it was he who first showed that the giant planets are not miniature suns. His great book, *The Earth; its Origin, History and Physical Constitution* (1924) was immensely influential.

Jones, Sir Harold Spencer. 1890–1960. English astronomer (Astronomer Royal 1933–55). A Cambridge graduate, who was HM Astronomer at the Cape from 1923 until his appointment to Greenwich. From the Cape he carried out much important work, mainly in connection with star catalogues and stellar radial velocities. While Astronomer Royal he redetermined the solar parallax by means of the world-wide observations of Eros, and published several excellent popular books as well as technical papers. He played a major rôle in the removal of the main equipment from Greenwich to the new site at Herstmonceux, in Sussex, and himself transferred to Herstmonceux in 1948, though it was not until 1958 that the move was completed. He was knighted in 1943.

Kant, Immanuel. 1724–1804. German philosopher, remembered astronomically for proposing a theory of the origin of the Solar System which had some points of resemblance to Laplace's later Nebular Hypothesis.

Kapteyn, Jacobus Corneleus. 1851–1922. Dutch astronomer and cosmologist. His most celebrated discovery was that of 'star-streaming'.

Kepler, Johannes. 1571–1630. German astronomer, *b.* Württemberg. He was the last assistant to Tycho Brahe, and after Tycho's death used the mass of observations to establish his three Laws of Planetary Motion. He observed the 1604 supernova, and also several comets, as well as making improvements to the refracting telescope, but his main achievements were theoretical. He ranks with Copernicus and Galileo as one of the main figures in the story of the 'Copernican revolution'.

Kirch, Gottfried. 1639–1710. German astronomer; Director of the Berlin Observatory from 1705. He was one of the earliest of systematic observers of comets, star-clusters and variable stars. In 1686 he discovered the variability of χ Cygni.

Kirchhoff, Gustav Robert. 1824–87. Professor of physics at Heidelberg. One of the greatest of German physicists, who explained the dark lines in the Sun's spectrum. His great map of the solar spectrum was published from Berlin in 1860.

Kirkwood, Daniel. 1814–95. American astronomer; an authority on asteroids and meteors. He drew attention to gaps in the asteroid belt, known today as the Kirkwood Gaps; they are due to the gravitational influence of Jupiter.

Kuiper, Gerard P. 1905–73. Dutch-American astronomer, who made notable advances in planetary and lunar work and was deeply involved with the programmes of sending probes beyond the Earth. The first crater to be identified on Mercury from *Mariner 10* was named in his honour.

Kulik, Leonid. 1883–1942. Russian scientist, trained as a forester, who achieved fame because of his work in meteorite research. In particular, he led several expeditions to study the Tunguska object of 1908. He died in a German prison camp in 1942.

Lacaille, Nicolas Louis de. 1713–62. French astronomer, who went to the Cape to draw up the first good southern-star catalogue.

Lagrange, Joseph Louis de. 1736–1813. French mathematical genius, and author of the classic *Mécanique Analytique.* He wrote numerous astronomical papers, dealing, among other topics, with the Moon's libration and the stability of the Solar System.

Laplace, Pierre Simon. 1749–1827. French mathematician who made great advances in dynamical astronomy. In 1796 he wrote *Systéme du Monde*, in which he outlined his Nebular Hypothesis of the origin of the planets. Discarded in its original form, but modern theories have many points of resemblance to it.

Lassell, William. 1799–1880. English astronomer. He discovered Triton, Neptune's larger satellite, and (independently of Bond) Hyperion, the 7th satellite of Saturn, as well as two satellites of Uranus (Ariel and Umbriel). He set up a 24 in reflector in Malta, and with it discovered 600 nebulæ.

Leavitt, Henrietta Swan. 1868–1921. American woman astronomer, best remembered for her observations of Cepheids in the Small Magellanic Cloud (1912), based on photographs taken in S America; these led on to the discovery of the vital period-luminosity law for Cepheids. She also discovered four novæ, several asteroids, and over 2400 variable stars.

Lemaître, Georges. 1894–1966. Belgian priest, who was a leading mathematician; from 1927, Professor at Louvain University. His most important paper, leading to what is now called the 'Big Bang' theory of the universe, appeared in 1927, but did not become well-known until publicized by Eddington three years later. During World War I Lemaître served in the Belgian Army, and won the Croix du Guerre.

Le Monnier, Pierre Charles. 1715–99. French astronomer who was concerned in star cataloguing. He observed the planet Uranus several times, but did not check his observations, and missed the chance of a classic discovery. It was said that he never failed to quarrel with anyone whom he met!

Le Verrier, Urbain Jean Joseph. 1811–77. French astronomer, whose calculations led in 1846 to the discovery of Neptune. He was an authority on meteors, and in 1867 computed the orbit of the Leonids. He also developed solar and planetary theory, and believed in the existence of a planet (Vulcan) closer to the Sun than Mercury – now known to be a myth. He was forced to resign the Directorship of the Paris Observatory in 1870 because of his irritability, but was reinstated on the death by drowning of his successor, Delaunay.

Levin, Boris Yuljevich. 1912–89. Born in Moscow. At first he was concerned with meteors, but then worked with O. Schmidt in developing his theories of the origin of the Solar System. He also made valuable contributions to cometary astronomy.

Lexell, Anders John. 1740–84. Finnish astronomer, *b.* Abö. He became Professor of Mathematics at St Petersburg. He discovered the periodical comet of 1770 (now lost), and was one of the first to prove that the object discovered by Herschel in 1781 was a planet rather than a comet.

Lindsay, Eric Mervyn. 1907–74. Irish astronomer, Director of the Armagh Observatory from 1936 until his death. His main work was in connection with the Magellanic Clouds and with quasars. He had close connections with the Boyden Observatory in South Africa (where he had previously been assistant astronomer) and forged close links between it, Harvard, Dunsink (Dublin) and Armagh. He was an excellent lecturer on popular astronomy, and founded the Armagh Planetarium in 1966.

Lockyer, Sir Joseph Norman. 1836–1920. English astronomer, and an independent discoverer of the method of studying the solar chromosphere and prominences at time of non-eclipse. He was knighted in 1897. He founded the

Norman Lockyer Observatory at Sidmouth in Devon, which still exists and is open to the public by arrangement, and was also founder of the periodical *Nature*.

Lohrmann, Wilhelm Gotthelf. 1796–1840. German land surveyor, who began an elaborate lunar map but was unable to complete it owing to ill health. The map was completed 40 years later by Julius Schmidt.

Lomonosov, Mikhail. 1711–65. Russian astronomer; he was also termed 'the founder of Russian literature'. His father was a fisherman. In 1735 he went to the University of St Petersburg, and then to Marburg in Germany to study chemistry. On his return to Russia in 1741 he insulted some of his colleagues at the St Petersburg Academy and was imprisoned for several months, during which time he wrote two of his most famous poems. How ever, he later became Professor of Chemistry at St Petersburg, and in 1746 became a Secretary of State. He drew up the first accurate map of the Russian Empire, described a 'solar furnace', and investigated,electrical phenomena. He also studied auroræ. In 1761 he observed the transit of Venus, and rightly concluded that Venus has a considerable atmosphere. His most important contribution was his championship of the Copernican theory and of Newton's theories, neither of which had really taken root in Russia before Lomonosov's work.

Lowell, Percival. 1855–1916. American astronomer, who founded the Lowell Observatory at Flagstaff, Arizona, in 1894. He paid great attention to Mars, and believed the 'canals' to be artificial waterways. His calculations led to the discovery of the planet Pluto, though the planet was not actually found until 1930 – by Clyde Tombaugh, at the Lowell Observatory. Lowell himself was a great astronomer who did much for science, and it is regrettable that he is today remembered mainly because of his erroneous theories about the Martian canals.

Lyot, Bernard. 1897–1953. Great French astronomer; Director of the Meudon Observatory. He made many advances in instrumental techniques, and invented the coronagraph, which enables the inner corona to be studied at times of non-eclipse. He died suddenly while taking part in an eclipse expedition to Africa.

Maclear, Sir Thomas. 1794–1879. Irish astronomer, who in 1833 succeeded Henderson as HM Astronomer at the Cape. He made an accurate measurement of an arc of meridian as well as verifying Henderson's parallax of α Centauri; he also studied comets and nebulæ. He was knighted in 1860.

Mädler, Johann Heinrich von. 1794–1874. German astronomer, who was the main observer in the great lunar map by himself and Beer, published in 1837–8 – a map which remained the standard for several decades. In 1840 he left his Berlin home to become Director of the Dorpat Observatory in Estonia. He erroneously believed that η Tauri (Alcyone) was the star lying at the centre of the Galaxy. He retired in 1865, and spent his last years in Hanover.

Maraldi, Giacomo Filippo. 1665–1729. Italian astronomer; nephew of G. D. Cassini. He was renowned for his observations of the planets, particularly Mars, and assisted his uncle at the Paris Observatory.

Maskelyne, Nevil. 1732–1811. English astronomer (Astronomer Royal 1765–1811). Educated at Cambridge; he then went to St Helena, at the suggestion of Bradley, to observe the transit of Venus, and decided to make a serious study of navigation. During his régime as Astronomer Royal he founded the *Nautical Almanac*.

Masursky, Harold. 1923–90. A leading planetary geologist, who was closely involved with the NASA missions, playing a key rôle in their planning. He spent his entire career with the US Geological Survey.

McCrea, Sir William. 1904–99. Much of McCrea's career was spent in mathematical departments, and his first astronomical appointment – at the University of Sussex – came in 1966. However, his interest in astronomy dated back much further. In 1929 he was able to confirm that hydrogen is dominant in the atmosphere of the Sun, and he made many contributions to astrophysics following his graduation from Cambridge in 1923. He was also active in the field of theoretical cosmology.

McVittie, George. 1904–88. British astronomer, born in Smyrna (his mother was Greek). He graduated from Edinburgh, and worked in England and America. He was a pioneer in studies of relativity, and made many important contributions to cosmology.

Méchain, Pierre François Andre. 1744–1805. French astronomer, who discovered eight comets 1781–99.

Menzel, Donald H. 1901–76. American astronomer, celebrated for his research into problems of the Sun and planets as well as in stellar studies. He was also an excellent lecturer, and a skilled writer of popular books.

Messier, Charles. 1730–1817. French astronomer, interested mainly in comets. Though he discovered 13 comets, he is remembered chiefly because of his catalogue of star-clusters and nebulæ, published in 1781.

Michell, John. 1725–93. English clergyman, and an amateur astronomer who made the first suggestion that many double stars may be physically associated or binary systems.

Milne, Edward Arthur. 1896–1950. English astronomer, who graduated from Cambridge and then went successively to Manchester and Oxford. He made important contributions to astrophysics, and developed his theory of 'kinematic relativity', which was for a time regarded as an alternative to general Einsteinian relativity.

Minkowski, Rudolf. 1895–1976. German astronomer, who went to Mount Wilson in 1935 and remained there. He was one of the leading authorities on novæ and planetary nebulæ, and after the war became a pioneer in the new science of radio astronomy. His studies of rapidly-moving gases in radio galaxies led to the rejection of the 'colliding galaxies' theory.

Mitchell, Maria. 1812–88. America's first woman astronomer. She was the first woman to be elected to the American Academy of Arts and Sciences; she also discovered a comet.

Montanari, Geminiario. 1633–87. Italian astronomer, who worked at Bologna and then at Padua. In 1669 he discovered the variability of Algol.

Nevill, Edmund Neison. 1851–1940. English astronomer, who published an important book and map concerning the Moon in 1876; he wrote under the name of Neison. He was Director of the Natal Observatory at Durban in South Africa 1882–1910, returning to England when the Observatory was closed.

Newcomb, Simon. 1835–1909. American astronomer, for some years head of the American Nautical Almanac office. His chief work was in mathematical astronomy, to which he made valuable contributions. He is also remembered as the man who proved to his own satisfaction that no heavier-than-air machine could ever fly!

Newton, Sir Isaac. 1643–1727. Probably the greatest of all astronomers. To list all his contributions here would be pointless; suffice to say that his *Principia*, published in 1687, has been described as the greatest mental effort ever made by one man. In addition to his scientific work, he sat briefly in Parliament, and served as Master of the Mint. He was knighted in 1705, and on his death was buried in Westminster Abbey.

Olbers, Heinrich Wilhelm Matthias. 1758–1840. German doctor; also a skilled amateur astronomer; established his private observatory in Bremen. He discovered two of the first four asteroids (Pallas and Vesta) and rediscovered the first (Ceres); he carried out important work in connection with cometary orbits, and discovered a periodical comet which has a period of 69.5 years, and last returned in 1956. Olbers also wrote about his celebrated paradox: 'Why is it dark at night?'

Oort, Jan Hendrick. 1900–92. Outstanding Dutch astronomer, who interpreted Kapteyn's two streams of stars as evidence that the Galaxy is rotating. With C. A. Muller, he confirmed the detection by Ewen and Purcell of the 21 cm background radiation. He proposed that comets come from a huge spherical shell surrounding the Solar System, now known as the Oort Cloud.

Parmenides of Elea (second half of the 6th century BC) believed the stars to be of compressed fire, but agreed that the Earth was spherical, and in equilibrium because it was equidistant from all points on the sphere representing the universe.

Penston, Michael Victor. 1943–90. He was born in London, but spent much of his career at Cambridge. He made many contributions to astrophysics, and in 1983 was able to 'weigh' a black hole in the centre of the galaxy NGC 4151. He described himself as one of the LAGS – Lovers of Active Galaxies! Sadly, he died in 1990 after a long and brave fight against cancer.

Perrine, Charles Dillon. 1867–1951. American astronomer, who discovered two of Jupiter's satellites as well as nine comets. He worked at the Lick Observatory until 1909, when he became Director of the Cordoba Observatory in Argentina, where he constructed a 30 in reflector and made many observations of southern galaxies. He also planned a major star catalogue, but was politically unpopular, and after a narrow escape from assassination he retired (1936).

Peters, Christian Heinrich Friedrich. 1813–90. Danish astronomer, who emigrated to America in 1848. He discovered 48 asteroids.

Piazzi, Giuseppe. 1746–1826. Italian astronomer, who became Director of the Palermo Observatory in Sicily. During the compilation of a star catalogue he discovered the first asteroid, Ceres (on 1 January 1801, the 1st day of the new century).

Pickering, Edward Charles. 1846–1919. American astronomer; for 43 years, from 1876, Director of the Harvard College Observatory. He concentrated upon photometry, variable stars, and above all stellar spectra; in the famous Draper Catalogue, the stars were classified according to their spectra. During his régime the Harvard Observatory was modernized, and a southern outstation was setup at Arequipa in Peru.

Pickering, William Henry. 1858–1938. Brother of E. C. Pickering, who worked with him at Harvard and also served for a while as astronomer in charge of the Arequipa out-station. In 1898 he discovered Saturn's 9th satellite, Phœbe. He made extensive studies of the Moon and Mars, mainly from the Harvard station in Jamaica which was set up in 1900. Independently of Lowell, he calculated the position of the planet Pluto.

Plutarch. *c.* 46–120. Greek biographer, mentioned here because of his authorship of *De Facie in Orbe Lunæ?* – On the Face in the Orb of the Moon – in which he claimed that the Moon is a world of mountains and valleys.

Pond, John. 1767–1836. English astronomer (Astronomer Royal, 1811–35). Though an excellent and painstaking astronomer, Pond was handicapped by ill-health during the latter part of his régime at Greenwich, and was eventually asked to resign. He tried unsuccessfully to obtain star distances by the parallax method.

Pons, Jean Louis. 1761–1831. French astronomer, whose first post at an observatory (Marseilles) was that of caretaker! He was self-taught and concentrated on hunting for comets; he found 36 in all, and ended his career as Director of the Museum Observatory in Florence.

Proctor, Richard Anthony. 1837–88. English astronomer, who was an excellent cosmologist but is best known for his many popular books. In 1881 he emigrated to America, and remained there for the rest of his life. Proctor paid considerable attention to the planets, and constructed a map of Mars.

Ptolemy (Claudius Ptolemaeus). *c.* 120–180. The 'Prince of Astronomers', who lived and worked in Alexandria. Nothing is known about his life, but his great work has come down to us through its Arab translation (the *Almagest*). Ptolemy's star catalogue was based on that of Hipparchus but with many contributions of his own; he also brought the geocentric system to its highest state of perfection, so that it is always known as the Ptolemaic theory. He constructed a reasonable map of the Mediterranean world, and even showed Britain, though it is true that he joined Scotland on to England in a back-to-front position.

Purbach, Georg von. 1423–61. Austrian astronomer, who became a professor at Vienna in 1450. He founded a new school of astronomy, compiled tables, and began to write an *Epitome of Astronomy* based on Ptolemy's *Almagest*. After Purbach's premature death, the book was completed by his friend and pupil Regiomontanus.

Pythagoras. *c.* 572–500 BC. The great Greek geometer, mentioned here because he was one of the very first to maintain that the Earth is spherical rather than flat. He seems also to have studied the movements of the planets.

Rahe, Jurgen. 1940–97. German astronomer who emigrated to the United States and joined NASA, becoming Director of the Solar System Exploration Division; he also acted as a staff member of the California Institute of Technology. He was responsible for the overall general management, budget and strategic planning for many missions, including the Galileo probe to Jupiter. He was killed by a freak accident, when a tree fell on his car as he was driving near his home at Potomac in Maryland.

Ramsden, Jesse. 1735–1800. English maker of astronomical instruments. His meridian circles were the first to be lit through the hollow axis.

Rayet, Georges Antoine. 1839–1906. French astronomer; in 1867, with Wolf, drew attention to the Wolf-Rayet stars, which have bright lines in their spectra. He went from Paris to Bordeaux, and became Director of the Observatory there.

Redman, Richard Oliver. 1905–75. English astronomer, who graduated from Cambridge. He made extensive studies of the Sun, stellar velocities, galactic rotation and the photometry of galaxies. In 1937 he went to the Radcliffe Observatory, Pretoria, and designed the spectrograph for the 74 in reflector. In 1947 he returned to Cambridge as Professor of Astrophysics and Director of the Observatories. Many programmes were carried through, and Redman also devoted much time and energy in the planning and construction of the 153 in Anglo-Australian telescope at Siding Spring.

Regiomontanus. 1436–76. The Latinized name of Johann Müller, Purbach's pupil. He completed the *Epitome of Astronomy*, and at Nürnberg set up a printing press, publishing the first printed astronomical ephemerides. He died in Rome, where he had been invited to help in reforming the calendar.

Reinmuth, Karl. 1892–1979. He spent his entire career at the Königstuhl Observatory in Heidelberg, concentrating upon asteroids and comets. He discovered over 250 asteroids, among them the 1932 Earth-grazer Apollo.

Rhæticus, Georg Joachim. 1514–76. German astronomer, who became Professor of Astronomy, at Wittenberg in 1536. An early convert to the Copernican system, he visited Copernicus at Frombork, and persuaded him to send his great book for publication.

Riccioli, Giovanni Battista. 1598–1671. Italian Jesuit astronomer, who taught at Padua and Bologna. He was a pioneer telescopic observer, and drew up a lunar map, inaugurating the system of nomenclature which is still in use. Oddly enough, he never accepted the Copernican system.

Ridley, Harold Bytham. 1919–95. Harold Ridley, born in Outer London, was one of Britain's outstanding amateur astronomers; he specialized in meteor work, and obtained many meteor spectra. He was for many years Director of the Meteor Section of the British Astronomical Association (and was its President 1976–78).

Robinson, Romney. 1792–1882. Irish astronomer, who was Director of the Armagh Observatory from 1823 to his death. He published the Armagh catalogue of over 5000 stars, and made many other contributions; he also invented the cup anemometer. It is on record that when the railway company planned to build a line close to Armagh, Robinson managed to have it diverted, since he maintained that the trains would shake his telescopes!

Rømer, Ole. 1644–1710. Danish astronomer. In 1675 he used the eclipse times of Jupiter's satellites to make an accurate measurement of the velocity of light. In 1681 he became Director of the Copenhagen Observatory. Among his numerous inventions are the transit instrument and the meridian circle.

Ronan, Colin Alastair. 1920–95. Colin Ronan was essentially an astronomical historian. During the war he served with the Army, and rose to the rank of Major; he made a very important contribution to the war effort, inventing a new method of blooming lenses to increase light transmission. He wrote many books, and was an outstanding lecturer; he was for many years Director of the Historical Section of the British Astronomical Association (President 1989–91). He was an active researcher, and produced plausible evidence that the telescope was invented in England over half a century before Galileo's time.

Rosse, 3rd Earl of. 1800–67. Irish amateur astronomer, who in 1845 completed the building of a 72 in reflector and erected it at his home at Birr Castle. The 72 in with its metal mirror, was much the largest telescope ever built up to that time, and has now (2000) been brought back into use. His greatest discovery was that many of the galaxies are spiral in form.

Rosse, 4th Earl of. 1840–1908. Continued his father's work, and was also the first man to measure the tiny quantity of heat coming from the Moon.

Rowland, Henry Augustus. 1848–1901. American scientist; Professor of Physics in Baltimore from 1876. His great map of the solar spectrum was published in 1895–7; it showed 20 000 absorption lines.

Runcorn, S. Keith. 1922–98. Born in Southport, and graduated in engineering from Cambridge in 1942. After a period working at radar research, he joined Manchester University, and carried out major researches into palæomagnetism and all other aspects of planetary magnetic phenomena.

Russell, Henry Norris. 1877–1957. American astronomer; Director, Princeton Observatory from 1908. He devoted much of his energy to studies of stellar constitution and evolution, and independently of Hertzsprung he discovered the giant and dwarf sub-divisions of stars of late spectral type. This led on to the H–R or Hertzsprung–Russell Diagram, in which luminosity (or the equivalent) is plotted against spectral type.

Rutherfurd, Lewis Morris. 1816–92. American barrister, who gave up his profession to devote himself to astronomy.

A pioneer in astronomical photography, his pictures of the Moon were outstanding; his ruled solar gratings for solar spectra were the best of their time.

Ryle, Sir Martin. 1918–84. British pioneer of radar and radio astronomy. He graduated from Oxford, and then went to Cambridge as Professor of Radio Astronomy. In 1972 he succeeded Woolley as Astronomer Royal. He developed the technique of aperture synthesis, and made many very important contributions.

Sagan, Carl. 1934–96. Carl Sagan was born in Brookyln, and spent all his career in the United States. He made major contributions to astrophysics, and was also a leading member of the Planetary Society, but is perhaps best remembered for his popular works, and in particular the best-selling television series 'Cosmos'. Sadly, he died of cancer when still in his early fifties. The Pathfinder station on the surface of Mars was named in his honour.

Scheiner, Christoph. 1575–1650. German Jesuit, who was for some time a professor of mathematics in Rome. He discovered sunspots independently of his contemporaries, and wrote a book, *Rosa Ursina*, which contains solar drawings and observations for the years 1611–25. He was unfriendly towards Galileo, and played a rather discreditable part in the events leading up to Galileo's trial and condemnation.

Schiaparelli, Giovanni Virginio. 1835–4910. Italian astronomer, who graduated from Turin and became Director of the Brera Observatory in Milan. He discovered the connection between meteors and comets, but his most famous work was in connection with the planets. It was he who first drew attention to the 'canal network' on Mars, in 1877.

Schlesinger, Frank. 1871–1943. American astronomer, *b.* New York. His main work was in connection with stellar parallaxes. He was Director successively of the Yale and Allegheny Observatories, and was responsible for the Yale 'southern station' in Johannesburg. His major works, *General Catalogue of Parallaxes* and its supplement, dealt with more than 2000 stars. He also pioneered the determination of star positions by using wide-angle cameras.

Schmidt, Julius (actually Johann Friedrich Julius). 1825–84. German astronomer, who became Director of the Athens Observatory in 1858 and spent most of his life in Greece. He concentrated upon lunar work, producing an elaborate map (based on Lohrmann's early work) and making great improvements in selenography. He drew attention to the alleged change in the lunar crater Linné, in 1866. He also discovered the outburst of the recurrent nova T Coronæ, in 1866.

Schönfeld, Eduard. 1828–91. German astronomer, who collaborated with Argelander in preparing the *Bonn Durchmusterung* and later extended it to the southern hemisphere.

Schröter, Johann Hieronymus. 1745–1816. Chief magistrate of Lilienthal, near Bremen. He set up a private observatory, and made outstanding observations of the Moon and planets. Unfortunately many of his notebooks were lost in 1813, with the destruction of his observatory by the invading French troops.

Schwabe, Heinrich. 1789–1875. German apothecary, who became a noted amateur astronomer concentrating on the Sun. His great discovery was that of the 11-year sunspot cycle.

Schwarzschild, Karl. 1873–1916. German astronomer, who worked successively at Vienna, Göttingen and (as Observatory Director) Potsdam. His early work dealt with photometry, but he was also a pioneer of theoretical astrophysics. Military service during World War I broke his health and led to his premature death.

Schwarzschild, Martin. 1912–97. German astronomer, who spent much of his career at Princeton University in the United States. He carried out much original research with regard to stellar structure and evolution.

Secchi, Angelo. 1818–78. Italian Jesuit astronomer; one of the great pioneers of stellar spectroscopy, classifying the stars into four types (a system superseded later by that of Harvard). He was also an authority in solar work, and his planetary observations were equally outstanding.

Seyfert, Carl. 1911–60. American astronomer, who concentrated upon studies of galaxies. In 1942 he drew attention to those galaxies with very condensed nuclei, now always termed Seyfert galaxies.

Shajn, Grigorij Abramovich. 1892–1956. After serving in the Russian Army during World War I he then joined the staff of the Pulkovo Observatory, working upon meteoric astronomy. In 1924 he became Director of the Simeis Observatory, and began his work on stellar spectroscopy.

During most of his latter years he was concerned mainly with the varied distribution of the faint galactic nebulæ.

Shapley, Harlow. 1885–1972. Great American astronomer, who began his main work at Princeton under H N Russell. In 1914 he advanced the pulsation theory of Cepheid variables, and was soon able to use the variables in globular clusters to give the first accurate picture of the shape and size of the Galaxy. In 1921 he became Director of the Harvard College Observatory. In later years he concentrated upon studies of galaxies and upon the international aspect of astronomy. He was also an excellent lecturer, and a skilled writer of popular books.

Shklovskii, Iosif. 1916–85. He graduated from Moscow University, and became professor there, later heading the radio astronomy department at the Sternberg Institute. He specialized in extraterrestrial radio sources, and showed that the emission from the Crab Nebula is synchrotron radiation. With Carl Sagan, he wrote about 'intelligent life in the universe'.

Shoemaker, Eugene. 1928–97. American astronomer/geologist, who became the leading expert on meteoritics and impact craters; he was deeply involved in all the earlier planetary missions. He was also a devoted hunter of comets and near-Earth asteroids. He was killed in a car accident in Australia; subsequently his ashes were scattered on the surface of the Moon.

Smyth, William Henry. 1788–1865. English naval officer, rising to the rank of Admiral, who in 1830 established a private observatory at Bedford and made numerous astronomical observations. He is best remembered for his famous book *Cycle of Celestial Objects*.

Smyth, Piazzi (actually Charles Piazzi). 1819–1900. The son of Admiral Smyth. Astronomer Royal for Scotland from 1844 until his death. He was a skilled astronomer who carried out much valuable work, including spectroscopic examinations of the Zodiacal Light, but he was also an eccentric who wrote a large and totally valueless volume about the significance of the Great Pyramid.

Sosigenes. Greek astronomer, who flourished about 46 BC. He was entrusted by Julius Cæsar with the reform of the calendar. Nothing is known about his life.

South, James. 1785–1867. English amateur astronomer, who founded a private observatory in Southwark and collaborated with John Herschel in studies of double stars.

In 1822 he observed an occultation of a star by Mars, and the virtually instantaneous disappearance convinced him that the Martian atmosphere must be extremely tenuous.

Spitzer, Lyman. 1914–97. American astrophysicist, who influenced several generations of researchers through his writing and lecturing as well as his own personal contributions. He was founder and first Director of the Princeton Plasma Physics Laboratory, the leader of the group developing the Copernicus satellite.

Spörer, Friedrich Wilhelm Gustav. 1822–95. German astronomer, who joined the staff at the Potsdam Observatory. He concentrated mainly upon the Sun, and discovered the variation in latitude of spot zones over the course of a solar cycle (Sporer's Law).

Steavenson, William Herbert. 1894–1975. Steavenson was never a professional astronomer; he was a medical doctor whose practice was in Outer London. He was an expert observer, and one of the few amateurs to serve as President of the Royal Astronomical Society (1957–9). His knowledge of telescopes and optics was encyclopædic, and his advice was often sought by professional astronomers.

Strømgren, Bengt. 1908–87. Born in Sweden, but raised in Denmark. He became Director of the Royal Copenhagen Observatory in succession to his father Elias. He went to America to direct the Yerkes and McDonald Observatories during the 1950s, returning to Denmark in 1967. He was particularly known for his studies of H.II (ionized hydrogen) regions round hot stars often called Strømgren spheres.

Struve, Friedrich Georg Wilhelm. 1793–1864. German astronomer, *b.* Altona. He went to Dorpat in Estonia and in 1818 became Director of the Observatory. Using the 9 in Fraunhofer refractor – the first telescope to be clock-driven – he began his classic work on double stars. In 1839 he went to Pulkova, to become director of the new observatory set up by Tsar Nicholas. Here he continued his double-star work, and his *Mensuræ Micrometricæ* gives details of over 3000 pairs. Strove also measured the parallax of Vega; his value was announced in 1840.

Struve, Otto (Wilhelm). 1819–1905. Son of F. G. W. Strove; *b.* Dorpat. He became assistant to his father, accompanying him to Pulkova. He continued his father's work, and became a leading authority on double stars. He succeeded to the directorship of Pulkova Observatory in 1861, retiring in 1889 and returning to Germany.

Struve, Karl Hermann. 1854–1920. Son of Otto Strove; *b.* Pulkova, later becoming assistant to his father. His main work was in connection with planetary satellites. In 1895 he went to Königsberg, and in 1904 became Director of the Berlin Observatory, which was reorganized by him and transferred to Babelsberg during his term of office.

Struve, Gustav Wilhelm Ludwig. 1858–1920. Son of Otto Strove, and brother of Karl. He too was born at Pulkova and acted as assistant to his father. He went to Dorpat in 1886, and from 1894 was Director of the Kharkov Observatory. He was concerned mainly with statistical astronomy and with the motion of the Sun.

Struve, Otto. 1897–1963. Son of Gustav; often known as Otto Strove II. He was born in Kharkov, and fought during World War I; joined the White Army under Wrangel and Derrikin, and after their defeat reached Constantinople, where he worked as a labourer. Finally he was offered a post at the Yerkes Observatory, where he arrived in 1921. He spent the rest of his life in America; in 1932 he became Director at Yerkes, after which he founded the McDonald Observatory in Texas and was its director from 1939–47, when he became Chairman of the Department of Astronomy at Chicago. In 1959 he began a new career as the first Director of the National Radio Astronomy Observatory, but ill-health forced his resignation in 1962. He was a brilliant astrophysicist, dealing mainly with spectroscopic binaries, stellar rotation and interstellar matter; he was also an author of popular books. He is (so far!) the last of the famous Strove astronomers. It is notable that all four were in succession awarded the Gold Medal of the Royal Astronomical Society – a sequence unique in astronomical history.

Swift, Lewis. 1820–1913. American astronomer who specialized in hunting for comets and nebulæ. He found 13 comets (including the Great Comet of 1862) and 900 nebulæ.

Tempel, Ernest Wilhelm. 1821–89. German astronomer, who became Director of the Arcetri Observatory. In 1859 he discovered the nebula in the Pleiades; he also discovered six asteroids and several comets, including the comet of 1865–6 which is associated with the Leonid meteors.

Thales. *c.* 624–547 BC. The first of the great Greek philosophers. He believed the Earth to be flat, and floating in an ocean, but he was a pioneer mathematician and observer,

and successfully predicted the eclipse of 585 BC which stopped the war between the Lydians and the Medes.

Timocharis. Fl. *c.* 280 BC. Greek astronomer, who made some accurate measurements of star positions; one of these (of Spica) enabled Hipparchus, 150 years later, to demonstrate the precession of the equinoxes.

Tombaugh, Clyde. 1906–97. While still a young amateur, Tombaugh was called to the Lowell Observatory to search for 'Planet X', and in 1930 he discovered Pluto. He remained at the Lowell Observatory for many years, and searched for further planets and minor Earth satellites, although without success. During the war he worked at White Sands, developing telescopic methods of tracking ballistic missiles. He then went to Las Cruces university, where he remained for the rest of his career, latterly as Professor Emeritus; he was a tireless and inspiring teacher.

Turner, Herbert Hall. 1861–1930. English astronomer who played an important rôle in the organization and preparation of the International Astrographic Chart. In 1903 he discovered Nova Geminorum.

Tycho Brahe. 1546–1601. The great Danish observer – probably the best of pre-telescopic times. He studied the supernova of 1572 and from 1576 to 1596 worked at his observatory at Hven, an island in the Baltic, making amazingly accurate measurements of star positions and the movements of the planets, particularly Mars. His observatory – Uraniborg – became a scientific centre, but Tycho was haughty and tactless (during his student days he had part of his nose sliced off in a duel, and made himself a replacement out of gold, silver and wax!), and after quarrels with the Danish Court he left Hven and went to Prague as Imperial Mathematician to the Holy Roman Emperor, Rudolph II. Here he was joined by Kepler, who acted as his assistant. When Tycho died, Kepler came into possession of the Hven observations, and used them to prove that the Earth moves round the Sun – something which Tycho himself could never accept.

Ulugh Beigh (more properly Ulugbek). 1394–1449. Grandson of the Oriental conqueror Tamerlane. About 1420 he established an observatory at Samarkand, which became an astronomical centre; he compiled a star catalogue, and compiled tables of the Moon and planets. He was assassinated in 1449.

Van Maanen, Adriaan. 1884–1947. Dutch astronomer, who emigrated to America and joined the Mount Wilson staff in 1912. He specialized in stellar parallaxes and proper motions, and accomplished much valuable work, though his alleged detection of movements in the spiral arms of galaxies later proved to be erroneous. He also discovered the white dwarf still known as Van Maanen's Star.

Vehrenberg, Hans. 1910–91. German amateur astronomer, whose superb stellar photographs, contained in his Atlas Stellarum, are widely used.

Vogel, Hermann Carl. 1842–1907. German astronomer, born and educated in Leipzig. He went to Potsdam in 1874, and concentrated upon stellar spectroscopy, pioneering research into spectroscopic binaries. In 1883 he published the first catalogue of stellar spectra.

Walther, Bernard. 1430–1504. (Often spelled 'Walter'.) German amateur astronomer, who lived in Namberg; he financed Regiomontanus' equipment, and carried on the work when Regiomontanus died. He was a very accurate observer, whose measurements of star and planetary positions were of great value to later astronomers.

Wargentin, Pehr Vilhelm. 1717–83. Swedish astronomer, and Director of the Stockholm Observatory. His best work was in the preparation of accurate tables of Jupiter's satellites.

Webb, Thomas William. 1806–85. Vicar of Hardwicke in Herefordshire. He was an excellent observer, but is best remembered for his book *Celestial Objects for Common Telescopes*.

Wilkins, Hugh Percy. 1896–1960. Welsh amateur astronomer (by profession a Civil Servant) who concentrated upon lunar observation, and produced a 300 in map of the Moon. He was for many years Director of the Lunar Section of the British Astronomical Association.

Wolf, Maximilian Franz Joseph Cornelius. 1863–1932. (Better known as Max Wolf.) German astronomer, who was born and lived in Heidelberg. He studied comets, and discovered his periodical comet in 1884; he was the first to hunt for asteroids photographically. He also carried out research into dark nebulæ, and discovered well over 1000.

Woolley, Sir Richard van der Riet. 1906–. Pioneer astrophysicist, who graduated from Cambridge; he succeeded Spencer Jones as Astronomer Royal, and upon retirement returned to his native South Africa to become Director of the South African National Observatories.

Wright, Thomas. 1711–85. *b.* nr Durham. Trained as a clockmaker, though he afterwards taught mathematics. He is remembered for his book published in 1750, in which he suggested that the Galaxy is disk-shaped. He also believed Saturn's rings to be composed of small particles.

Xenophanes. *c.* 570–478 BC. Greek philosopher, *b.* Colophon. His astronomical theories sound strange today; an infinitely thick flat Earth, a new Sun each day, and celestial bodies which – apart from the Moon – were made of fire!

Zach, Franz Xavier von. 1754–1832. Hungarian baron, who became renowned as an amateur astronomer. He published tables of the Sun and Moon, and was one of the chief organizers of the 'Celestial Police' who banded together to hunt for the supposed planet between Mars and Jupiter. He became Director of the Seeberg Observatory at Gotha, and did much for international co-operation among astronomers.

Zwicky, Fritz. 1898–1974. Swiss astronomer; *b.* Bulgaria, but remained a Swiss citizen throughout his life. He graduated from Zürich, and in 1925 went to the California Institute of Technology, where he remained permanently, becoming Professor of Astrophysics from 1942 until his retirement in 1968. He became famous for his studies of galaxies and intergalactic matter; he predicted the existence of neutron stars (1934) and even black holes. He discovered many supernovæ in external galaxies, and masterminded a catalogue of compact galaxies. He was also active in the development of astronomical instrumentation, and was a pioneer worker with Schmidt telescopes. He received the Gold Medal of the Royal Astronomical Society in 1973.

34 GLOSSARY

Aberration of starlight. As light does not move infinitely fast, but at a rate of practically 300 000 km/s, and as the Earth is moving round the Sun at an average velocity of 25 km/s, the stars appear to be shifted slightly from their true positions. The best analogy is to picture a man walking along in a rainstorm, holding an umbrella. If he wants to keep himself dry, he will have to slant the umbrella forward; similarly, starlight seems to reach us 'from an angle'. Aberration may affect a star's position by up to 20.5 seconds of arc.

Ablation. The erosion of a surface by friction or vaporization.

Absolute magnitude. The apparent magnitude that a star would have if it could be observed from a standard distance of 10 parsecs (32.6 light years).

Absolute zero. The coldest possible temperature: −273.16 °C.

Accretion disk. A disk structure which forms round a spinning object when material falls on to it from beyond.

Achromatic object-glass. An object-glass which has been corrected so as to eliminate chromatic aberration or false colour as much as possible.

Aerolite. A meteorite whose main composition is stony.

Airglow. The light produced and emitted by the Earth's atmosphere (excluding meteor trains, thermal radiation, lightning and auroræ).

Albedo. The reflecting power of a planet or other non-luminous body. The Moon is a poor reflector; its albedo is a mere 7% on average.

Alpha particle. The nucleus of a helium atom, made up of two protons and two neutrons.

Altazimuth mounting for a telescope. A mounting on which the telescope may swing freely in any direction.

Altitude. The angular distance of a celestial body above the horizon.

Analemma. The figure-of-eight shape resulting if the Sun's position in the sky is recorded at the same time of day throughout the year.

Ångström unit. One hundred-millionth part of a centimetre.

Anorthosite. An igneous rock composed largely of anorthite, a calcium-rich plagioclase feldepar. It is rich in aluminium and calcium, and is the main constituent of the ancient highland crust of the Moon.

Apastron. The orbital positions of the two members of a binary star system when at their greatest separation.

Aphelion. The furthest distance of a planet or other body from the Sun in its orbit.

Apogee. The furthest point of the Moon from the Earth in its orbit.

Apparent magnitude. The apparent brightness of a celestial body. The lower the magnitude, the brighter the object: thus the Sun is approximately −27, the Pole Star +2, and the faintest stars detect able by modern techniques around +30.

Appulse. The apparent close approach of one celestial object to another. If one object covers the other, the appulse becomes an occultation.

Apsides. The two points of an elliptical orbit lying closest to and furthest from the centre of mass of the system.

Areography. The official name for 'the geography of Mars'.

Armillary sphere. A celestial globe in which the celestial sphere is represented by a skeletal framework of intersecting circles, with the Earth in the central position.

Array. An arrangement of a number of linked radio antennæ.

Ascending node. The point where an orbiting body crosses the plane of reference for its orbit, moving from south to north (the opposite point is known as the descending node).

Asterism. A pattern of stars which does not rank as a separate constellation.

Asteroids. One of the names for the minor planet swarm.

Astrobleme. A very old, very eroded crater, frequently of impact origin.

Astrograph. An astronomical telescope designed specially for astronomical photography.

Astrolabe. An ancient instrument used to measure the altitudes of celestial bodies.

Astronomical unit. The mean distance between the Earth and the Sun. It is equal to 149 598 500 km.

Aurora. Auroræ are 'polar lights'; Aurora Borealis (northern) and Aurora Australis (southern). They occur in the Earth's upper atmosphere, and are caused by charged particles emitted by the Sun.

Azimuth. The bearing of an object in the sky, measured from north ($0°$) through east, south and west.

Bailly's beads. Brilliant points seen along the edge of the Moon just before and just after a total solar eclipse. They are caused by the sunlight shining through valleys at the Moon's limb.

Barycentre. The centre of gravity of the Earth Moon system. Because the Earth is 81 times as massive as the Moon, the barycentre lies well inside the Earth's globe.

Basalt. A dark grey fine-grained volcanic rock, with a silica content of from 44 and 50%; basalt is the most widespread volcanic rock found on the surfaces of the terrestrial planets.

Billion. (American) One thousand million. (British) One million million. The American version is now generally used.

Binary star. A stellar system made up of two stars, genuinely associated, and moving round their common centre of gravity. The revolution periods range from millions of years for very widely separated visual pairs down to less than half an hour for pairs in which the components are almost in contact with each other. With very close pairs, the components cannot be seen separately, but may be detected by spectroscopic methods.

Black body. A body which absorbs all the radiation falling on it.

Black hole. A region round a very small, very massive collapsed star from which not even light can escape.

Blazar. Objects such as quasars, which show violent variations in light output.

BL Lacertæ objects. Variable objects which are powerful emitters of infra-red radiation, and are very luminous and remote. They are of the same nature as quasars.

Bode's law. A mathematical relationship linking the distances of the planets from the Sun. It may or may not be genuinely significant. Strictly speaking it should be called Titius' Law, since it was discovered by J. D. Titius some years before J. E. Bode popularized it in 1772.

Bolide. A brilliant exploding meteor.

Bolometer. An instrument for measuring the total amount of energy received from a source of electromagnetic radiation.

Bow shock. The edge of the magnetosphere of a planetary body, where the solar wind is deflected.

Caldera (pl calderæ). A large depression, usually found at the summit of a shield volcano, due to the withdrawal of magma from below.

Carbon stars. Red stars of spectral types R and N with unusually carbon-rich atmospheres.

Carbonaceous chondrites. Primitive stony meteorites (aerclites), containing carbonaceous compounds and hydrated silicates.

Cassegrain reflector. A reflecting telescope in which the secondary mirror is convex; the light is passed back through a hole in the main mirror. Its main advantage is that it is more compact than the Newtonian reflector.

Celestial sphere. An imaginary sphere surrounding the Earth, whose centre is the same as that of the Earth's globe.

Cepheid. A short-period variable star, very regular in behaviour; the name comes from the prototype star, Delta Cephei. Cepheids are astronomically important because there is a definite law linking their variation periods with their real luminosities, so that their distances may be obtained by sheer observation.

Čerenkov radiation. Electromagnetic radiation produced when electrically charged particles move through a medium at a velocity greater than the velocity of light in that medium.

Chandrasekhar limit. The maximum possible mass limit for a white dwarf star: 1.4 times the mass of the Sun.

Chondrite. Stony asteroite containing chondrules. Chondrites make up over 90% of stony meteorites (aerolites).

Chondrules. Spherical incursions found in chondrites. They are composed mainly of pyroxene and olivine, with some glass.

Chromatic aberration. A defect in all lenses, due to the fact that light is a mixture of all wavelengths – and these wavelengths are refracted unequally, so that false colour is produced round a bright object such as a star. The fault may be reduced by making the lens a compound arrangement, using different kinds of glasses.

Chromosphere. That part of the Sun's atmosphere which lies above the bright surface or photosphere.

Circumpolar star. A star which never sets. For instance, Ursa Major (the Great Bear) is circumpolar as seen from England; Crux Australis (the Southern Cross) is circumpolar as seen from New Zealand.

Cluster variables. An obsolete name for the stars now known as RR Lyrae variables.

Cœlostat. An optical instrument making use of two mirrors, one of which is fixed, while the other is movable and is mounted parallel to the Earth's axis; as the Earth rotates, the light from the star (or other object being observed) is caught by the rotatable mirror and is reflected in a fixed direction on to the second mirror. The result is that the eyepiece of the instrument need not move at all.

Collapsar. The end product of a very massive star, which has collapsed and has surrounded itself with a black hole.

Colour index. The difference between a star's visual magnitude and its photographic magnitude. The redder the star, the greater the positive value of the colour index; bluish stars have negative colour indices. For stars of type A0, colour index is zero.

Colures. Great circles on the celestial sphere.

Commensurability. A property of two orbits in which the period of one orbit is equal to, or a simple fraction of, the period of the other.

Conduction. A method of heat transfer in which the heat is transferred through solids by molecular impact.

Conjunction. (1) A planet is said to be in conjunction with a star, or with another planet, when the two bodies are apparently close together in the sky. (2) For the inferior planets, Mercury and Venus, inferior conjunction occurs when the planet is approximately between the Earth and the Sun; superior conjunction, when the planet is on the far side of the Sun and the three bodies are again lined up. Planets beyond the Earth's orbit can never come to inferior conjunction, for obvious reasons.

Convection. Heat transfer within a flowing material, in which hot material from lower levels rises because it is less dense. Cooler material at higher levels then sinks. The overall motion thus generated is known as a convection cell.

Corona. The outermost part of the Sun's atmosphere, made up of very tenuous gas. It is visible with the naked eye only during a total solar eclipse.

Coronagraph. A device used for studying the inner corona at times of non-eclipse.

Cosmic rays. High-velocity particles reaching the Earth from outer space. The heavier cosmic-ray particles are broken up when they enter the upper atmosphere.

Cosmic year. The time taken for the Sun to complete one revolution round the centre of the Galaxy: about 225,000,000 years.

Cosmogony. The study of the origin and evolution of the universe.

Cosmology. The study of the universe considered as a whole.

Counterglow. The English name for the sky-glow more generally called by its German name of the Gegenschein.

Culmination. The maximum altitude of a celestial body above the horizon.

Cusp. The pointed extremity of a crescent shape.

Cytherean. Relating to the planet Venus (an alternative to *Venusian*).

Dall–Kirkham telescope. A form of Cassegrain telescope using an ellipsoidal primary mirror and a spherical secondary mirror.

Dawes limit. The practical limit for the resolving power of a telescope; it is $4.56/d$ arcsec where d is the aperture of the telescope in inches, and $11.6/d$ where d is the aperture of the telescope in centimetres.

Day, sidereal. The interval between successive meridian passages, or culminations, of the same star: $23^h 56^m 4^s.091$.

Day, solar. The mean interval between successive meridian passages of the Sun: 24 h 3 m 56 s.555 of mean sidereal time. It is longer than the sidereal day because the Sun seems to move eastward against the stars at an average rate of approximately one degree per day.

Decametric radiation. Low-frequency radio waves, with wavelengths of tens or hundreds of metres.

Declination. The angular distance of a celestial body north or south of the celestial equator. It corresponds to latitude on the Earth.

Dewcap. An open tube fitted to the upper end of a refracting telescope. Its rôle is to prevent condensation upon the object-glass.

Dichotomy. The exact half-phase of the Moon or an inferior planet.

Diffraction grating. A device used for splitting up light; it consists of a polished metallic surface upon which thousands of parallel lines are ruled. It may be regarded as an alternative to the prism.

Direct motion. Movement of revolution or rotation in the same sense as that of the Earth.

Doppler effect. The apparent change in wavelength of the light from a luminous body which is in motion relative to the observer. With an approaching object, the wavelength is apparently shortened, and the spectral lines are shifted to the blue end of the spectral band; with a receding body there is a red shift, since the wavelength is apparently lengthened.

Double star. A star made up of two components – either genuinely associated (binary systems) or merely lined up by chance (optical pairs).

Driving clock. A mechanism for driving a telescope round at a rate which compensates for the axial rotation of the Earth, so that the object under observation remains fixed in the field of view.

Dune. An elongated mound of sand produced by wind activity. Dunes are found, for example, on Mars as well as on the Earth.

Dwarf novæ. A term sometimes applied to the U Geminorum (or SS Cygni) variable stars.

Earthshine. The faint luminosity on the night side of the Moon, frequently seen when the Moon is in its crescent phase. It is due to light reflected on to the Moon from the Earth.

Eclipse, lunar. The passage of the Moon through the shadow cast by the Earth. Lunar eclipses may be either total or partial. At some eclipses, totality may last for approximately $1\frac{3}{4}$ hours, though most are shorter.

Eclipse, solar. The blotting-out of the Sun by the Moon, so that the Moon is then directly between the Earth and the Sun. Total eclipses can last for over 7 minutes under exceptionally favourable circumstances. In a partial eclipse, the Sun is in completely covered. In an annular eclipse, exact alignment occurs when the Moon is in the far part of its orbit, and so appears smaller than the Sun; a ring of sunlight is left showing round the dark body of the Moon. Strictly speaking, a solar 'eclipse' is the occultation of the Sun by the Moon.

Eclipsing variable (or Eclipsing Binary). A binary star in which one component is regularly occulted by the other, so that the total light which we receive from the system is reduced. The prototype eclipsing variable is Algol (Beta Persei).

Ecliptic. The apparent yearly path of the Sun among the stars. It is more accurately defined as the projection of the Earth's orbit on to the celestial sphere.

Electron. Part of an atom; a fundamental particle carrying a negative electric charge.

Electron density. The number of free electrons per unit volume of space. In interstellar space the value is around 30 000 per cubic metre.

Elongation. The angular distance of a planet from the Sun, or of a satellite from its primary planet.

Ephemeris. A table showing the predicted positions of a celestial body such as a comet, asteroid or planet.

Epoch. A date chosen for reference purposes in quoting astronomical data.

Equator, celestial. The projection of the Earth's equator on to the celestial sphere.

Equatorial mounting for a telescope. A mounting in which the telescope is set up on an axis which is parallel with the axis of the Earth. This means that one movement only (east to west) will suffice to keep an object in the field of view.

Equinox. The equinoxes are the two points at which the ecliptic cuts the celestial equator. The vernal equinox or First Point of Aries now lies in the constellation of Pisces; the Sun crosses it about 21 March each year. The autumnal equinox is known as the First Point of Libra; the Sun reaches it about 22 September yearly.

Escape velocity. The minimum velocity which an object must have in order to escape from the surface of a planet, or other celestial body, without being given any extra impetus.

Evection. An inequality in the Moon's motion, due to slight changes in the shape of the lunar orbit.

Event horizon. The 'boundary' of a black hole. No light can escape from inside the event horizon.

Exosphere. The outermost part of the Earth's atmosphere.

Extinction. The apparent reduction in brightness of a star or planet when low down in the sky, so that more of its light is absorbed by the Earth's atmosphere. With a star 1° above the horizon, extinction amounts to 3 magnitudes.

Eyepiece (or Ocular). The lens, or combination of lenses, at the eye-end of a telescope. It is responsible for magnifying the image of the object under study. With a positive eyepiece (for instance, a Ramsden, Orthoscopic or Monocentric) the image plane lies between the eyepiece and the object glass (or main mirror); with a negative eyepiece (such as a Huyghenian or Tolles) the image plane lies inside the eyepiece. A Barlow lens is concave, and is mounted in a short tube which may be placed between the eyepiece and the object-glass (or mirror). It increases the effective focal length of the telescope, thereby providing increased magnification.

Faculæ. Bright, temporary patches on the surface of the sun.

Field star. A star which is seen close to a stellar cluster, but is not a cluster member. It may be much closer or much more remote.

Filar micrometer. A device used for measuring very small angular distances as seen in the eyepiece of a telescope.

Finder. A small, wide-field telescope attached to a larger one, used for sighting purposes.

Fireball. A brilliant meteor. There is no set definition, but a meteor with a magnitude of brighter than −5 will be classed as a fireball.

Flares, solar. Brilliant eruptions in the outer part of the Sun's atmosphere. Normally they can be detected only by spectroscopic means (or the equivalent), though a few have been seen in integrated light. They are made up of hydrogen, and emit charged particles which may later reach the Earth, producing magnetic storms and displays of auroræ. Flares are generally, though not always, associated with sunspot groups.

Flare stars. Faint Red Dwarf stars which show sudden, short-lived increases in brilliancy, due to intense flares above their surfaces.

Flash spectrum. The sudden change-over from dark to bright lines in the Sun's spectrum, just before the onset of totality in a solar eclipse. The phenomenon is due to the fact that at this time the Moon has covered up the bright surface of the Sun, so that the chromosphere is shining 'on its own'.

Flocculi. Patches of the Sun's surface, observable with spectroscopic equipment. They are of two main kinds: bright (calcium) and dark (hydrogen).

Forbidden lines. Lines in the spectrum of a celestial body which do not appear under normal conditions, but may be seen in bodies where conditions are exceptional.

Fraunhofer lines. The dark absorption lines in the spectrum of the Sun or any other star.

Galaxies. Systems made up of stars, nebulæ, and interstellar matter. Many, though by no means all, are spiral in form.

Galaxy, the. The system of which our Sun is a member. It contains approximately 100 000 million stars, and is a rather loose spiral.

Gamma-rays. Radiation of extremely short wavelength.

Gauss. Unit of measurement of a magnetic field. (The Earth's field, at the surface, is on average about 0.3 to 0.6 gauss.)

Gegenschein. A faint sky-glow, opposite to the Sun and very difficult to observe. It is due to thinly-spread interplanetary material.

Geocorona. The outermost part of the Earth's atmosphere, made up of a halo of hydrogen gas extending to around 15 Earth radii.

Geodesy. The study of the shape, size, mass and other characteristics of the Earth.

Gibbous phase. The phase of the Moon or planet when between half and full.

Geosynchronous orbit. An orbit round the Earth at an altitude of 35 900 km, where the period will be the same as the Earth's sidereal rotation period – 23h 56m 4.1s – assuming that the orbit is circular and lies in the plane of the Earth's equator.

Glitch. A sudden change in the rotation period of a pulsar, due probably to starquakes.

Globules. Small dark patches inside gaseous nebulæ. They are probably embryo stars.

Gnomon. In a sundial, it is a pointer whose function is to cast the Sun's shadow on to the dial. The gnomon always points to the celestial pole.

Graben. A downfaulted block of crust on a planetary surface bounded by a pair of normal faults.

Great circle. A circle on the surface of a sphere whose plane passes through the centre of that sphere.

Green Flash. Sudden, brief green light seen as the last segment of the Sun disappears below the horizon. It is purely an effect of the Earth's atmosphere. Venus has also been known to show a Green Flash.

Gregorian reflector. A telescope in which the secondary mirror is concave, and placed beyond the focus of the main mirror. The image is erect. Few Gregorian telescopes are in use nowadays.

H.I and H.II regions. Clouds of hydrogen in the Galaxy. In H.I regions the hydrogen is neutral; in H.II regions the hydrogen is ionized, and the presence of hot stars will make the cloud shine as a nebula.

Halo, galactic. The spherical-shaped cloud of stars round the main part of the Galaxy.

Hayashi track. The evolutionary track of a protostar on the HR diagram – before it joins the Main Sequence.

Heliacal rising. The rising of a star or planet at the same time as the Sun, though the term is generally used to denote the time when the object is first detectable in the dawn sky.

Heliosphere. The area round the Sun extending to between 50 and 100 a.u. where the Sun's influence is dominant. The boundary, where the solar wind merges with the interstellar medium, is called the heliopause.

Herbig–Haro object. A nebulous object associated with a newly-forming star.

Herschelian reflector. An obsolete type of telescope in which the main mirror is tilted, thus removing the need for a secondary mirror.

Hertzsprung gap. A region in the HR diagram, between the giant branch and the Main Sequence, containing comparatively few stars, because this stage in the evolution of a star is short.

Hertzsprung–Russell diagram (usually known as the H–R Diagram). A diagram in which stars are plotted according to the spectral types and their absolute magnitudes.

High-velocity star. A star travelling at more than around 65 km s^{-1} in relation to the Sun. These are old stars, which do not share the Sun's motion round the galactic centre, but travel in more elliptical orbits.

Horizon. The great circle on the celestial sphere which is everywhere 90 degrees from the observer's zenith.

Hour angle (of a celestial object). The time which has elapsed since the object crossed the meridian. If RA = right ascension of the object and LST = the local sidereal time, then Hour Angle = LST − RA.

Hour circle. A great circle on the celestial sphere, passing through both celestial poles. The zero hour circle coincides with the observer's meridian.

Hubble Constant. A measure of the rate at which a galaxy is receding. The current value is about 70 km s^{-1} megaparsec^{-1}.

Igneous rock. A rock formed by the crystallization of a magma.

Inferior planets. Mercury and Venus, whose distances from the Sun are less than that of the Earth.

Infra-red radiation. Radiation with wavelength longer than that of visible light (approximately 7500 Ångströms).

Interferometer, stellar. An instrument for measuring star diameters. The principle is based upon light-interference.

Ion. An atom which has lost or gained one or more of its planetary electrons, and so has respectively a positive or negative charge.

Ionosphere. The region of the Earth's atmosphere lying above the stratosphere.

Irradiation. The effect which makes very brilliant bodies appear larger than they really are.

Julian day. A count of the days, starting from 12 noon on 1 January 4713 BC. Thus 1 January 1977 was Julian Day 2 443 145. (The name 'Julian' has nothing to do with Julius Cæsar. The system was invented in 1582 by the mathematician Scaliger, who named it in honour of his father, Julius Scaliger.)

Kelvin scale. A scale of temperature. 1 K is equal to $1\,°C$, but the Kelvin scale starts at absolute zero ($-273.16\,°C$).

Kepler's laws of planetary motion. These were laid down by Johannes Kepler, from 1609 to 1618. They are: (1) The planets move in elliptical orbits, with the Sun occupying one focus. (2) The radius vector, or imaginary line joining the centre of the planet to the centre of the Sun, sweeps out equal areas in equal times. (3) With a planet, the square of the sidereal period is proportional to the cube of the mean distance from the Sun.

Kiloparsec. One thousand parsecs (3260 light-years).

Kirkwood gaps. Gaps in the main asteroid belt, where the periods would be commensurate with that of Jupiter – so that Jupiter keeps these areas 'swept clear'.

KREEP. Basaltic rocks found on the Moon, rich in potassium (K), rare earth elements (REE) and phosphorus (P).

Kreutz comet. A sun-grazing comet.

Kuiper belt. A belt of asteroid sized objects beyond the orbit of Neptune. It is believed that short-period comets also come from the Kuiper belt.

Latitude, celestial. The angular distance of a celestial body from the nearest point on the ecliptic.

Libration. The apparent 'tilting' of the Moon as seen from Earth. There are three librations: latitudinal, longitudinal and diurnal. The overall effect is that at various times an observer on Earth can see a total of 59%, of the total surface of the Moon, though, naturally, no more than 50%. at any one moment.

Light-year. The distance travelled by light, in vacuum, in one year: 9.4607 million million km, 63 240 a.u. or 0.306.6 parsecs.

Lithosphere. The rigid outermost layer of a planetary body.

Local group. A group of more than two dozen galaxies, one member of which is our own Galaxy. The largest member of the Local Group is the Andromeda Spiral, M.31.

Local Supercluster. A supercluster of galaxies, centred on the Virgo cluster and including our Local Group. It was first defined by G. de Vaucouleurs in 1956.

Long-period comet. A comet with an orbital period of over 200 years.

Longitude, celestial. The angular distance of a celestial body from the vernal equinox, measured in degrees eastward along the ecliptic.

Lunation. The interval between successive new moons: 29 d 12 h 44 m. (Also known as the Synodic Month.)

Magma. Molten rock or mineral, formed in the lower crust or mantle of the Earth or other planetary bodies. When it solidifies, it is known as igneous rock; when it emerges through the crust it is known as lava.

Magnetosphere. The region of the magnetic field of a planet or other body. In the Solar System, only the Earth, Jupiter, Mercury, Saturn, Ganymede, Uranus and Neptune are known to have detectable magnetospheres.

Main Sequence. A band along an H–R Diagram, including most normal stars except for the giants.

Maksutov telescope. An astronomical telescope involving both mirrors and lenses.

Mantle. The layer within a planetary body separating the crust from the core.

Mass. The quantity of matter that a body contains. It is not the same as 'weight'.

Mean Sun. An imaginary sun travelling eastward along the celestial equator, at a speed equal to the average rate of the real Sun along the ecliptic.

Megaparsec. One million parsecs.

Meridian, celestial. The great circle on the celestial sphere which passes through the zenith and both celestial poles.

Meteor. A small particle, friable in nature and usually smaller than a sand grain, moving round the Sun, and visible only when it enters the upper atmosphere and is destroyed by friction. Meteors may be regarded as cometary débris.

Meteorite. A larger object, which may fall to the ground without being destroyed in the upper atmosphere. A meteorite is fundamentally different from a meteor. Meteorites are not associated with comets, but are closely related to asteroids.

Micrometeorite. A very small particle of interplanetary material, too small to cause a luminous effect when it enters the Earth's upper atmosphere.

Micrometer. A measuring device, used together with a telescope to measure very small angular distances – such as the separations between the components of double stars.

Micron. One-thousandth of a millimetre. The usual symbol is μ.

Molecular cloud. A cloud of interstellar matter in which the gas is mainly in molecular form.

Month. (1) Anomalistic: the interval between successive perigee passages of the Moon (27.55 days). (2) Sidereal: the revolution period of the Moon with reference to the stars (27.32 days). (3) Synodical: the interval between successive new moons (29.53 days). (4) Nodical or Draconitic: the interval between successive passages of the Moon through one of its nodes (27.21 days). (5) Tropical: the time taken for the Moon to return to the same celestial longitude (about 7 seconds shorter than the sidereal month).

Nadir. The point on the celestial sphere directly below the observer.

Nebula. A cloud of gas and dust in space. Galaxies were once known as 'spiral nebulæ' or 'extragalactic nebulæ'.

Neutrino. A fundamental particle with little or no mass and no electric charge.

Neutron. A fundamental particle with no electric charge, but a mass practically equal to that of a proton.

Neutron star. The remnant of a massive star which has exploded as a supernova. Neutron stars send out rapidly-varying radio emissions, and are therefore called 'pulsars'. Only two (the Crab and Vela pulsars) have as yet been identified with optical objects.

Newtonian reflector. A reflecting telescope in which the light is collected by a main mirror, reflected on to a smaller flat mirror set at an angle of 45°, and thence to the side of the tube.

Nocturnal. An instrument for telling the time at night, by using the positions of the Pointers in Ursa Major (Dubhe and Merak) relative to the Pole Star.

Nodes. The points at which the orbit of the Moon, a planet or a comet cuts the plane of the ecliptic; south to north (Ascending Node) or north to south (Descending Node).

Nova. A star which suddenly flares up to many times its normal brilliancy, remaining bright for a relatively short time before fading back to obscurity.

Nutation. A slow, slight 'nodding' of the Earth's axis, due to the gravitational pull of the Moon on the Earth's equatorial bulge.

Object-glass (or Objective). The main lens of a refracting telescope.

Objective prism. A small prism placed in front of the object-glass of a telescope. It produces small-scale spectra of the stars in the field of view.

Obliquity of the ecliptic. The angle between the ecliptic and the plane of the Earth's orbit. Its present value is $23°26'45''$, but it can range between $21°55'$ and $24°18'$.

Occultation. The covering-up of one celestial body by another.

Ocular. Alternative name for a telescope eyepiece.

Olivine. A silicate of magnesium and iron: $(Mg,Fe)SiO_4$.

Oort cloud. An assumed spherical shell of comets surrounding the Solar system, at a range of around one light-year.

Opposition. The position of a planet when exactly opposite to the Sun in the sky; the Sun, the Earth and the planet are then approximately lined up.

Orbit. The path of a celestial object.

Orrery. A model showing the Sun and the planets, capable of being moved mechanically so that the planets move round the Sun at their correct relative speeds.

Outgassing. The process by which volatiles within a planet gradually escape through the surface.

Palæomagnetism. The study of the fossil or remenant magnetization of rocks of all ages.

Palimpsests. High-albedo circular features on the surfaces of some planetary bodies, notably Callisto and Ganymede. Low-walled, bright-floored circular structures on the Moon (mainly on the hemisphere turned away from Earth) have also been referred to as palimpsests.

Parallax, trigonometrical. The apparent shift of an object when observed from two different directions.

Parsec. The distance at which a star would have a parallax of one second of arc: 3.26 light-years, 206 265 astronomical units, or 30.857 million million kilometres.

Penumbra. (1) The area of partial shadow to either side of the main cone of shadow cast by the Earth. (2) The lighter part of a sunspot.

Periastron. In a binary star system, the point of closest approach of the two components.

Perigee. The position of the Moon in its orbit when closest to the Earth.

Perihelion. The position in orbit of a planet or other body when closest to the Sun.

Perturbations. The disturbances in the orbit of a celestial body produced by the gravitational effects of other bodies.

Phases. The apparent changes in shape of the Moon and the inferior planets from new to full. Mars may show a gibbous phase, but with the other planets there are no appreciable phases as seen from Earth.

Photoelectric cell. An electronic device; light falling on the cell produces an electric current, the strength of which depends upon the intensity of the light.

Photoelectric photometer. A photoelectric cell used together with a telescope for measuring the magnitudes of celestial bodies.

Photometer. An instrument used to measure the intensity of light from any particular source.

Photometry. The measurement of the intensity of light.

Photon. The smallest 'unit' of light.

Photosphere. The bright surface of the Sun.

Planetary nebula. A small, dense, hot star surrounded by a shell of gas. The name is ill-chosen, since planetary nebulæ are neither planets nor nebulæ!

Planetoid. Obsolete name for an asteroid (minor planet).

Plasma. An ionized gas: a mixture of electrons and atomic nuclei.

Plerion. A supernova remnant with no shell structure. One such object is the Crab Nebula.

Poles, celestial. The north and south points of the celestial sphere.

Populations, stellar. Two main types of star regions: I (in which the brightest stars are hot and bluish), and II (in which the brightest stars are old Red Giants.

Position angle. The apparent direction of one object with reference to another, measured from the north point of the main object through east, south and west.

Poynting–Robertson effect. The effect of solar radiation upon small particles orbiting the Sun, causing them to spiral slowly inward.

Precession. The apparent slow movement of the celestial poles. This also means a shift of the celestial equator, and hence of the equinoxes; the vernal equinox moves by 50 sec of arc yearly, and has moved out of Aries into Pisces. Precession is due to the pull of the Moon and Sun on the Earth's equatorial bulge.

Prime Meridian. The meridian on the Earth's surface which passes through the Airy Transit Circle at Greenwich Observatory. It is taken as longitude 0°.

Prominences. Masses of glowing gas rising from the surface of the Sun. They are made up chiefly of hydrogen.

Proper motion, stellar. The individual movement of a star on the celestial sphere.

Proton. A fundamental particle with a positive electric charge. The nucleus of the hydrogen atom is made up of a single proton.

Protoplanet. A body forming by the accretion of material, which will ultimately develop into a planet.

Protostar. The earliest stage in the formation of a star.

Pulsar. A rotating neutron star, often a strong radio source. Not all pulsars can be detected by radio, since the radiation is emitted in beams, and it depends upon whether these beams sweep over the Earth.

Purkinje effect. This is due to the change of the colour sensitivity of the human eye. If two sources are equal, and are then dimmed, the redder of the two will appear the fainter.

Quadrant. An ancient astronomical instrument used for measuring the apparent positions of celestial bodies.

Quadrature. The position of the Moon or a planet when at right-angles to the Sun as seen from the Earth.

Quantum. The amount of energy possessed by one photon of light.

Quasar. The core of a very powerful, remote active galaxy. The term QSO (quasi-stellar object) is also used.

Radial velocity. The movement of a celestial body toward or away from the observer; positive if receding, negative if approaching.

Radiant. The point in the sky from which the meteors of any particular shower seem to radiate.

Regolith. A layer of loose rock and mineral grains on the surface of a planetary body. With the addition of organic material it becomes a soil.

Regression of the nodes. The nodes of the Moon's orbit move slowly westward, making one complete revolution in 18.6 years. This regression is caused by the gravitational pull of the Sun.

Retardation. The difference in the time of moonrise between one night and the next.

Retrograde motion. Orbital or rotational movement in the sense opposite to that of the Earth's motion.

Reversing layer. The gaseous layer above the Sun's photosphere.

Right ascension. The angular distance of a celestial body from the vernal equinox, measured eastward. It is usually given in hours, minutes and seconds of time, so that the right ascension is the time-difference between the culmination of the vernal equinox and the culmination of the body.

Roche limit. The distance from the centre of a planet within which a second body would be broken up by the planet's gravitational pull. Note, however, that this would be the case only for a body which had no appreciable gravitational cohesion.

Saros. The period after which the Earth, Moon and Sun return to almost the same relative positions: 18 years 11.3 days. The Saros may be used in eclipse prediction, since it is usual for an eclipse to be followed by a similar eclipse exactly one Saros later.

Schmidt camera (or Schmidt telescope). An instrument which collects its light by means of a spherical mirror; a correcting plate is placed at the top of the tube. It is a purely photographic instrument.

Schwarzschild radius. The radius that a body must have if its escape velocity is to be equal to the velocity of light.

Scintillation. Twinkling of a star; it is due to the Earth's atmosphere. Planets may also show scintillation when low in the sky.

Secular acceleration of the Moon. The apparent speeding-up of the Moon in its orbit as measured over a long period of time, caused by the gradual slowing of the Earth's rotation (by 0.000 000 02 second per day).

Selenography. The study of the Moon's surface.

Selenology. The lunar equivalent of geology.

Sextant. An instrument used for measuring the altitude of a celestial object.

Seyfert galaxies. Galaxies with relatively small, bright nuclei and weak spiral arms. Some of them are strong radio emitters.

Sidereal period. The revolution period of a planet round the Sun, or of a satellite round its primary planet.

Sidereal time. The local time reckoned according to the apparent rotation of the celestial sphere. When the vernal equinox crosses the observer's meridian, the sidereal time is 0 hours.

Solar nebula. The cloud of interstellar gas and dust from which the Solar System was formed – around 5000 million years ago.

Solar wind. A flow of atomic particles streaming out constantly from the Sun in all directions.

Solstices. The times when the Sun is at its maximum declination of approximately $23\frac{1}{2}$ degrees; around 22 June (summer solstice, with the Sun in the northern hemisphere of the sky) and 22 December (winter solstice, Sun in the southern hemisphere).

Specific gravity. The density of any substance, taking that of water as 1. For instance, the Earth's specific gravity is 5.5, so that the Earth 'weighs' 5.5 times as much as an equal volume of water would do.

Speckle interferometry. A technique designed to reduce the blurring of star images due to turbulence in the Earth's atmosphere.

Spectroheliograph. An instrument used for photographing the Sun in the light of one particular wavelength only.

The visual equivalent of the spectroheliograph is the spectrohelioscope.

Spectroscopic binary. A binary system whose components are too close together to be seen individually, but which can be studied by means of spectroscopic analysis.

Speculum. The main mirror of a reflecting telescope.

Spherical aberration. Blurring of a telescope image; it is due to the fact that the lens (or mirror) does not bring the light-rays falling on its edge and on its centre to exactly the same focal point.

Starburst galaxy. A galaxy in which there is an exceptionally high rate of star formation.

Stellar wind. A continuous outflow of particles from a star, resulting in loss of mass.

Superior planets. All the planets lying beyond the orbit of the Earth in the Solar System (that is to say, all the principal planets apart from Mercury and Venus).

Superluminal motion. The apparent movement of material at a velocity greater than that of light. It is purely a geometrical effect.

Supernova. A colossal stellar outburst, involving (1) the total destruction of the white dwarf member of a binary system, or (2) the collapse of a very massive star.

Synchronous rotation. If the rotation period of a planetary body is equal to its orbital period, the rotation is said to be synchronous (or captured). The Moon and most planetary satellites have synchronous rotation.

Synchrotron radiation. Radiation emitted by a charged particle travelling at almost the velocity of light, moving in a strong magnetic field.

Synodic period. The interval between successive oppositions of a superior planet.

Syzygy. The position of the Moon in its orbit when new or full.

Tektites. Small, glassy objects found in a few localized areas of the Earth. They are not now believed to be meteoritic.

Terminator. The boundary between the day- and night-hemispheres of the Moon or a planet.

Thermocouple. An instrument used for measuring very small amounts of heat.

Torus. A three-dimensional ring shape, such as that of a quoit (or a doughnut).

Transit. (1) The passage of a celestial body across the observer's meridian. (2) The projection of Mercury or Venus against the face of the Sun.

Transit instrument. A telescope mounted so that it can move only in declination; it is kept pointing to the meridian, and is used for timing the passages of stars across the meridian. Transit instruments were once the basis of all practical timekeeping. The Airy transit instrument at Greenwich is accepted as the zero for all longitudes on the Earth.

Troposphere. The lowest part of the Earth's atmosphere; its top lies at an average height of about 11 km. Above it lies the stratosphere; and above the stratosphere come the ionosphere and the exosphere.

Twilight. The state of illumination when the sun is below the horizon by less than 18 degrees.

Umbra. (1) The main cone of shadow cast by the Earth. (2) The darkest part of a sunspot.

Van Allen zones. Zones of charged particles around the Earth. There are two main zones; the outer (made up chiefly of electrons) and the inner (made up chiefly of protons).

Variable stars. Stars which change in brilliancy over short periods. They are of various types.

Variation. An inequality in the Moon's motion, due to the fact that the pull of the Sun on the Moon is not constant for all positions in the lunar orbit.

White dwarf. A very small, very dense star which has used up its nuclear energy, and is in a very late stage of its evolution.

Widmanstätten patterns. If an iron meteorite is cut, polished and then etched with acid, characteristic figures of the iron crystals appear; these are the Widmanstätten patterns.

Wilson effect. Perspective effect of a sunspot near the solar limb; the limbward penumbra is broadened relative to the penumbra on the opposite side, indicating that the 'spot' is a depression. Not all sunspots show the effect, however.

WIMP. An electrically neutral, weakly interacting massive particle. As yet WIMPs are theoretical concepts only.

Wolf-Rayet stars. Very hot, greenish-white stars which are surrounded by expanding gaseous envelopes. Their spectra show bright (emission) lines.

Wormhole. A hypothetical tunnel-like structure in the fabric of spacetime.

Year. (1) Sidereal: the period taken for the Earth to complete one journey round the Sun (365.26 days). (2) Tropical: the interval between successive passages of the Sun across the vernal equinox (365.24 days). (3) Anomalistic: the interval between successive perihelion passages of the Earth (365.26 days; slightly less than 5 minutes longer than the sidereal year, because the position of the perihelion point moves along the Earth's orbit by about 11 seconds of arc every year). (4) Calendar: the mean length of the year according to the Gregorian calendar (365.24 days, or 365 d 5 h 49 m 12 s).

Zenith. The observer's overhead point (altitude 90°).

Zenith distance. The angular distance of a celestial object from the Zenith.

Zenithal hourly rate. The number of naked-eye shower meteors which would be seen by an observer under ideal conditions, with the meteor radiant at the zenith. In practice, these conditions are never attained.

Zodiac. A belt stretching round the sky, 8° to either side of the ecliptic, in which the Sun, Moon and principal planets are to be found at any time. (Pluto is the only planet which can leave the Zodiac, though many asteroids do so.)

Zodiacal Light. A cone of light rising from the horizon and stretching along the ecliptic; visible only when the Sun is a little way below the horizon. It is due to thinly-spread interplanetary material near the main plane of the Solar System.

Zone of avoidance. The sky region near the plane of the Milky Way in which few galaxies can be seen, because of obscuration by interstellar dust in our Galaxy.

Zürich number (or Wolf number). A measure of solar activity. It is expressed as $R = k(10g + f)$, where R is the Zürich number, g is the number of spot-groups and f is the number of individual spots; k is a constant, usually with a value of around 1, depending upon the observer and the instrument used.

INDEX